CHILTON'S MECHANICS' HANDBOOK

VOLUME 1: EMISSION DIAGNOSIS TUNE-UP AND ALIGNMENT

Managing Editor	Kerry A. Freeman, S.A.E.
Senior Editor	Richard J. Rivele

OFFICERS

President	William A. Barbour
Executive Vice President	James A. Miades
Vice President & General Manager	John P. Kushnerick

CHILTON BOOK COMPANY
Chilton Way, Radnor, Pa. 19089

Manufactured in USA
© 1980 Chilton Book Company
ISBN 0-8019-6995-6 (softcover)
ISBN 0-8019-7009-1 (hardcover)
Library of Congress Catalog
  Card No. 80-964

567890   987654

Index

Repair & Tune Up Notes ... 4

Trouble Diagnosis 420

American Motors:
- Tune Up Specifications 10
- Distributor Specifications 11
- Carburetor Specifications 12
- Ser. No. & Engine Identification . 13
- Emission Equipment 14
- Idle Speed & Mixture 15
- Initial Timing 17
- Vacuum Advance 17
- Spark Plugs 17
- Emission Control Systems 18

Vacuum Circuits 26

Buick:
- Tune Up Specifications 31
- Distributor Specifications 33
- Carburetor Specifications 36
- Ser. No. & Engine Identification . 39
- Emission Equipment 39
- Idle Speed & Mixture 40
- Initial Timing 42
- Vacuum Advance 43
- Spark Plugs 43
- Emission Control Systems 44

Vacuum Circuits 51

Cadillac:
- Tune Up Specifications 66
- Distributor Specifications 66
- Carburetor Specifications 67
- Ser. No. & Engine Identification . 68
- Emission Equipment 68
- Idle Speed & Mixture 69
- Initial Timing 70
- Vacuum Advance 70
- Spark Plugs 70
- Emission Control Systems 71

Vacuum Circuits 80

Chevrolet 6 & V-8, Monza V-8
- Tune Up Specifications 88
- Distributor Specifications 90
- Carburetor Specifications 94
- Ser. No. & Engine Identification . 99
- Emission Equipment 100
- Idle Speed & Mixture 101
- Initial Timing 103
- Vacuum Advance 104
- Spark Plugs 103
- Emission Control Systems ... 104

Vacuum Circuits 114

Chevrolet Vega & Monza 4-cyl.:
- Tune Up Specifications 123
- Distributor Specifications 124
- Carburetor Specifications 126
- Ser. No. & Engine Identification . 128
- Emission Equipment 128
- Idle Speed & Mixture 129
- Ignition Timing 130
- Vacuum Advance 130
- Spark Plugs 130
- Emission Control Systems ... 131

Vacuum Circuits 141

Chevette:
- Tune Up Specifications 144
- Distributor Specifications 144
- Carburetor Specifications 145
- Ser. No. & Engine Identification . 146
- Emission Equipment 146
- Idle Speed & Mixture 147
- Initial Timing 148
- Vacuum Advance 148
- Spark Plugs 148
- Emission Control Systems ... 148

Vacuum Circuits 160

Oldsmobile:
- Tune Up Specifications 165
- Distributor Specifications 167
- Carburetor Specifications 171
- Ser. No. & Engine Identification . 175
- Emission Equipment 176
- Idle Speed & Mixture 177
- Initial Timing 180
- Vacuum Advance 181
- Spark Plugs 181
- Emission Control Systems ... 182

Vacuum Circuits 197

Pontiac:
- Tune Up Specifications 205
- Distributor Specifications 206
- Carburetor Specifications 209
- Ser. No. & Engine Identification . 213
- Emission Equipment 214
- Idle Speed & Mixture 215
- Initial Timing 217
- Vacuum Advance 218
- Spark Plugs 218
- Emission Control Systems ... 219

Vacuum Circuits 227

Astre & Sunbird 4-cyl.:
- Tune Up Specifications 123
- Distributor Specifications 124
- Carburetor Specifications 126
- Ser. No. & Engine Identification . 128
- Emission Equipment 128
- Idle Speed & Mixture 129
- Initial Timing 130
- Vacuum Advance 130
- Spark Plugs 130
- Emission Control Systems ... 131

Vacuum Circuits 141

Chrysler Corporation:
- Tune Up Specifications 244
- Distributor Specifications 245
- Carburetor Specifications 246
- Ser. No. & Engine Identification . 249
- Emission Equipment 250
- Idle Speed & Mixture 251
- Initial Timing 253
- Vacuum Advance 254
- Spark Plugs 254
- Emission Control Systems ... 254

Vacuum Circuits 267

Horizon & Omni:
- Tune Up Specifications 288
- Distributor Specifications 288
- Carburetor Specifications 288
- Ser. No. & Engine Identification . 289
- Emission Equipment 289
- Idle Speed & Mixture 289
- Initial Timing 290
- Vacuum Advance 290
- Spark Plugs 290
- Emission Control Systems ... 290

Vacuum Circuits 297

Ford Motor Company:
- Tune Up Specifications 298
- Distributor Specifications 300
- Carburetor Specifications 304
- Ser. No. & Engine Identification . 308
- Emission Equipment 308
- Idle Speed & Mixture 309
- Initial Timing 313
- Vacuum Advance 315
- Spark Plugs 314
- Emission Control Systems ... 315

Vacuum Circuits 348

Index

Pinto, Mustang II, Bobcat 4-cyl. & V-6:
- Tune Up Specifications 361
- Distributor Specifications 361
- Carburetor Specifications 304
- Ser. No. & Engine Identification . 363
- Emission Equipment 363
- Idle Speed & Mixture 363
- Initial Timing 366
- Vacuum Advance 366
- Spark Plugs 366
- Emission Control Systems 367

Vacuum Circuits **379**

Datsun:
- Tune Up Specifications 383
- Ser. No. & Engine Identification . 385
- Emission Control Systems 385

Honda:
- Tune Up Specifications 392
- Ser. No. & Engine Identification . 393
- Emission Control Systems 394

Subaru:
- Tune Up Specifications 396
- Ser. No. & Engine Identification . 396
- Emission Control Systems 397

Toyota:
- Tune Up Specifications 401
- Ser. No. & Engine Identification . 402
- Emission Control Systems 402

Volkswagen:
- Tune Up Specifications 414
- Ser. No. & Engine Identification . 415
- Emission Control Systems 417

Brakes and Wheel Alignment
- Drum Brake Service 457
- Disc Brake Service 461
- Drum Brake Performance Diagnosis 462
- Disc Brake Performance Diagnosis 463
- Wheel Alignment 464
- Specifications 466
    - Domestic Cars 466
    - Import Cars and Trucks 490

SAFETY NOTICE

Proper service and repair procedures are vital to the safe, reliable operation of all motor vehicles, as well as the personal safety of those performing repairs. This manual outlines procedures for servicing and repairing vehicles using safe effective methods. The procedures contain many NOTES, CAUTIONS and WARNINGS which should be followed along with standard safety procedures to eliminate the possibility of personal injury or improper service which could damage the vehicle or compromise its safety.

It is important to note that repair procedures and techniques, tools and parts for servicing motor vehicles, as well as the skill and experience of the individual performing the work vary widely. It is not possible to anticipate all of the conceivable ways or conditions under which vehicles may be serviced, or to provide cautions as to all of the possible hazards that may result. Standard and accepted safety precautions and equipment should be used when handling toxic or flammable fluids, and safety goggles or other protection should be used during cutting, grinding, chiseling, prying, or any other process that can cause material removal or projectiles.

Some procedures require the use of tools specially designed for a specific purpose. Before substituting another tool or procedure, you must be completely satisfied that neither your personal safety, nor the performance of the vehicle will be endangered.

Although information in this manual is based on industry sources and is as complete as possible at the time of publication, the possibility exists that some car manufacturers made later changes which could not be included here. While striving for total accuracy, Chilton Book Company cannot assume responsibility for any errors, changes, or omissions that may occur in the compilation of this data.

Repair & Tune Up Notes

CAUSES OF EMISSION INSPECTION FAILURES

When a customer drives into your shop and hands you a piece of paper that shows his car has failed an emission inspection, it doesn't mean you are in for a lot of trouble. If your infra-red exhaust emission tester is calibrated frequently, and checked often against a known gas, then you have nothing to fear. The first thing to do is hook up your infra-red and do a normal idle CO and HC check to see if the car has high emissions. Don't forget to disconnect the air pump and plug the opening in the exhaust manifold to keep air from being drawn in.

If your infra-red does not show high emissions, or in other words does not agree with the inspection report the customer has brought in, the best thing you can do is send the car back to the inspection lane for a retest. If you attempt to reduce the emissions from the car, when your equipment already shows there is nothing wrong, then you are asking for trouble.

But in most cases (unless somebody has worked on the car before you see it) a car that has failed an emission inspection will also fail, or show high emissions on your shop equipment. If it does, here are some of the reasons for the high emission readings.

High CO

Imagine that the engine is divided into two areas or sections. One area is the intake system, from the fresh air tube, through the air cleaner, the carburetor, and ending at the carburetor flange. This area is what is responsible for high carbon monoxide (CO). There is only one way that an engine can put out too much CO. It has to burn too much fuel for the amount of air. This is the same as saying that the mixture is too rich. Anything that makes the mixture too rich will cause high CO. Here is a list of some specific causes in this area.
1. Restricted air cleaner or intake tube.
2. Dirty air filter.
3. Rich choke setting.
4. Choke stuck closed.
5. High fuel level in the carburetor bowl.
6. Dirty air bleeds in the carburetor.
7. Drilled out jets or improperly adjusted metering rods (tampering).
8. Excessive blowby which feeds a rich mixture into the engine through the PCV valve.
9. Idle mixture screws adjusted for performance instead of emissions.

As you can see, anything that richens the mixture will cause high CO. The biggest cause of high CO is incorrect adjustment of the idle mixture screws. Chrysler Corporation cars must be adjusted with an infra-red analyzer. Other cars are adjusted with an analyzer, or by the speed drop method. Almost any engine will have high CO and fail a state inspection if you adjust by the sound of the engine.

High HC

The second area that we divide the engine into consists of everything from the base of the carburetor (which overlaps the first area) into the engine, including the intake manifold, combustion chamber, valves, rings, and camshaft. Hydrocarbons come only from unburned fuel, so when we say that an engine has high HC, it means simply that there is unburned fuel coming out the exhaust, usually in the form of fuel vapor. It's a little harder to understand the reasons for high HC, because some of them involve engine design. It may be easier if you remember that an engine was designed to burn up all its fuel in the combustion chamber. If anything in the engine wears out, it can easily result in some of the fuel passing out of the combustion chamber without being burned. Here are some specific causes.

1. Leaking exhaust valves. Unburned fuel is forced out through the leaking valve into the exhaust.
2. Idle mixture screws adjusted lean. This causes lean misfire at idle.
3. Misfiring plugs. Fuel passes into the exhaust without being burned.
4. Incorrect spark timing. Too much spark advance lets the exhaust system cool off, and the HC does not burn up in the exhaust. This can be caused not only by incorrect initial timing, but also by faulty advance mechanisms.
5. Lean carburetion. If the carburetor is too lean at part throttle, it will cause lean misfire.
6. Worn camshaft lobes. If an intake cam lobe is worn, it does not allow as much charge into the cylinder, which reduces the pressure in the cylinder and the fuel doesn't burn up. The worn lobe won't necessarily show up in the way the engine runs, but can be easily spotted by removing the rocker cover and checking the lift at the pushrod.
7. Vacuum leaks. Too much air leans the mixture and causes lean misfire.
8. Worn piston rings. If the rings are worn enough, the pressure in the cylinder will be low, and the fuel doesn't burn up.

High NOx

Very few states are now testing for nitrogen oxides (NOx), but as more states begin inspection programs, they will probably include NOx sniffing in their inspection

NOx is controlled mainly by the exhaust gas recirculation system. Too much spark advance can increase NOx, but usually the cause of an NOx failure is that the EGR system is not working at all. Normal testing and checking of the EGR system will usually show that an EGR passageway is blocked, or that the EGR valve is not getting vacuum.

NOx sniffing is not possible in the field, because there are no exhaust analyzers that will detect it. Until such analyzers are made, most of the field testing will have to be with an infra-red, concentrating on HC and CO.

With a quality infra-red analyzer, frequently calibrated and checked often against a known gas, there is no reason why you can't tackle emission inspection failures with the same confidence you go after any other automotive problem.

HOW TO IDENTIFY A CALIFORNIA CAR

California cars are usually made with different emission control equipment, and adjusted to different specifications, such as idle speed and timing. There is only one positive way of identifying a California car. If the engine decal definitely states that the car conforms to California standards or regulations, then it is a "California car." If the decal says nothing about California, or states that the car conforms, except for California, then it is a "49-State car."

Some people think that the location where a car was assembled indicates whether it is a California or 49-State car. This is not true in all cases. It is also not true that a car first sold outside of California has to be a 49-State car. Cars built to California emission standards can be sold anywhere in the United States, and frequently are. The California emission hardware is usually listed on dealer order forms as an option group. All the dealer has to do is tell the factory, and they will build the car to California emission standard's, no matter where the dealer happens to be. Sometimes a dealer has no control over the car he gets. It may be easier for the factory to send an out-of-California dealer California cars, even though he didn't order them that way. Also, dealers frequently send cars to other dealers,

Repair & Tune Up Notes

and there is nothing to prevent a California dealer from sending surplus cars to another state. It does not work the other way, however. 49-State cars can not be sold in California, although they are welcome as used cars, after they have been first sold in another state.

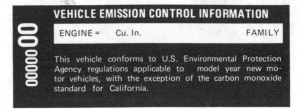

Chrysler Corporation label on 49-State car

The complete Chrysler Corporation label has tune-up information

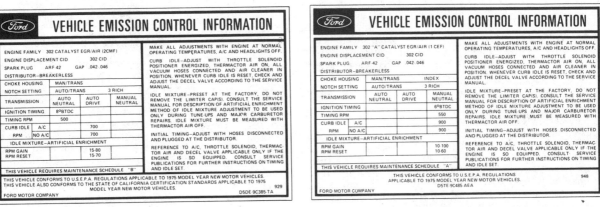

Ford Motor Co. California car label   Ford Motor Co. 49-State label

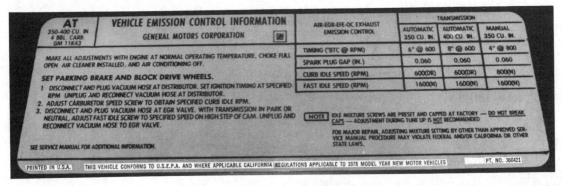

General Motors 1975 label for California cars

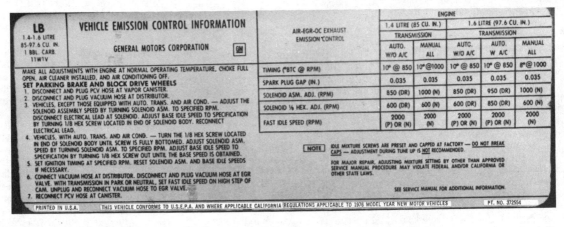

General Motors 1976 label for California cars

Repair & Tune Up Notes

THE "GRAMS PER MILE" LABEL

Every new car is required to have a label on the outside of the car (usually on a side window) giving emission test information about the car. Most of these labels list the grams per mile of hydrocarbons or carbon monoxide that were emitted in the federal test. This grams per mile figure has nothing to do with testing in the field using an infra-red unit. There is no way that this grams per mile figure can be translated into percentage of CO or parts per million of HC.

If the car maker recommends measuring the HC or CO in the field (some don't) the figures you need will be on the underhood label. The label can usually be found on the radiator brace, the underside of the hood, or on the rocker cover.

HC and CO READINGS

Some carmakers want you to measure HC and CO at the tailpipe with the air pump disconnected. Whenever you disconnect the pump prior to taking an HC-CO reading, you must plug the opening to the exhaust so that air will not be sucked in. Exhaust pulsations can draw in enough air through the check valve to change the CO readings if a hose connection is left open. When plugging the hose or connection, be sure you are plugging the opening that leads to the exhaust, not the hose that goes back to the pump.

CO READINGS 1977 ENGINES

If you take a reading on a 1977 car with the air pump disconnected (opening plugged) you may be surprised to find that the tailpipe CO reading is high. Hook the pump back up and the reading will drop to normal. This does not indicate that anything is wrong with the converter. It needs the extra air from the pump to do its job. Unless the reading is extremely high, there is probably nothing wrong with the engine, either.

The catalytic converter does such a good job of cleaning up the exhaust that it is possible to design engines for more power. This makes the exhaust dirtier, but the converter cleans it up. When the converters were first used in 1975, the engineers were reluctant to depend too much on the converter until they had more experience with it. After two years, the converter proved itself so well that some 1977 models now come from the factory with considerably more horsepower than before. The fuel mixtures have been richened and the spark has been advanced, which makes the engine much more responsive and easier to drive. On 49-State cars, the increased emissions from the engine can be cleaned up by the converter alone, but California engines usually need an air pump.

Because the converter is doing more to clean up the exhaust, an infra-red tester has a better chance of detecting a failed converter. Before the '77 models came out, you could be sure that high tailpipe readings were caused by the engine, because an engine in good shape was clean. Now, a bad converter can cause high readings.

The right way to correct high emissions is to check the engine first. If you are sure the engine is okay, then the problem is in the converter.

BACK PRESSURE EGR VALVE 1977

Many 1977 cars use a new EGR valve with a back pressure transducer built in. The transducer is really nothing more than a bleed in the diaphragm, controlled by exhaust pressure. When the exhaust pressure is low or non-existent, the bleed is open, which lets air into the top side of the diaphragm and bleeds off the vacuum so the valve will not open. When the exhaust pressure is high enough, the bleed closes. Then the ported vacuum from the carburetor can build up and open the valve so the exhaust gas will recirculate.

Back pressure EGR valves cannot be tested by applying vacuum with the engine off. Unless there is exhaust pressure, the bleed will be open, and the application of vacuum will not open the valve. It is possible to test the valve by removing it, blowing a gentle stream of air into the exhaust opening, and applying vacuum from a continuous source, such as a running engine. If the valve is not stuck or leaking, the diaphragm will move up. A stuck valve can be made to work by scraping the deposits away.

It has been common to test EGR valves for a leaking diaphragm by using a hand vacuum pump. Back pressure EGR valves cannot be tested this way. They all "leak" unless there is exhaust pressure to close the bleed.

GENERAL MOTORS DIAGNOSTIC CONNECTOR

Full size General Motors cars now have the same underhood diagnostic connectors that originally appeared on the Chevette. One of the connectors is for the air conditioning, and the other is for the engine. The connectors are the same on every car, so that once you learn which terminal is which, you will be able to make a quick diagnosis regardless of which G.M. division produced the car.

Many of the equipment companies now have adapter plugs that make it easy to hook up shop equipment to the diagnostic connector. There is even one very simple plug that AC-Delco sells for less than $10.00. It plugs in and provides labeled connectors so that you can attach alligator clips or make up your own harness for testing.

PELLET TYPE CATALYTIC CONVERTERS

The General Motors type of catalytic converter, which is also used on American Motors cars, has always had a screw plug in the bottom for changing the pellets. If you have ever tried to get one of the plugs out, you know that it is very difficult. Now, the screw plug has been eliminated, and a soft plug is pressed into the hole. To remove the plug, you just spear it with a punch and pry it out.

After the pellets are changed, you must leave the vacuum machine attached to the tailpipe to keep the pellets from falling out. Then you insert a special plug with a bridge to hold it in. Finally, the vacuum machine is removed, and the job is done.

Once the soft plug is removed, it cannot be replaced, so you must have the new bridge-type plug on hand before attempting to change the pellets.

ALTITUDE CARS FOR 1977

For many years, cars have been made one way for sale in the 49-States, and a second way for sale in California. Now, there is a third type of car. Counties that are mostly above 4000 feet must use cars that are made to run best at that altitude. Dealers in those high altitude counties must find out where the customer lives, or where he is going to use the car, to determine if an "altitude" car is required. California is exempt from the altitude requirement, so any 1977 car has to be either a California car, a 49-State car, or an Altitude car.

The only way you can tell what car you have is to look on the underhood label. If it says that the car meets the requirements for California or Altitude, you know what it is. If the label says only that it meets federal standards, then it has to be a 49-State car.

Repair & Tune Up Notes

GENERAL MOTORS ENGINES 1977

Buick, Oldsmobile, and Pontiac divisions of General Motors are now using Chevrolet engines, as well as swapping engines among themselves. The only division that continues to use its own engines exclusively is Chevrolet.

If you are not able to identify the various engines by looking at them, you must refer to the vehicle identification number visible through the lower left corner of the windshield. The fifth character in the vehicle identification number is a letter identifying the engine. We give the letters at the beginning of each Buick, Oldsmobile, and Pontiac chapter in this supplement. The same letters are given in the specifications whenever it is necessary, usually on the 350 V-8's, because there are so many of them.

For convenience, we repeat here the entire list of code letters for the General Motors passenger car divisions.

VIN Code	Engine Size	Source
A	231 V-6 (even firing)	Buick
B	140 4-Cyl.	Chevrolet
C	231 V-6	Buick
D	250 6-Cyl.	Chevrolet
E	1.6 Litre 4-Cyl.	Chevrolet
F	260 V-8	Oldsmobile
G	350 2-bbl. V-8	Chevrolet
H	350 2-bbl. V-8	Buick
I	1.4 litre 4-Cyl.	Chevrolet
J	350 4-bbl. V-8	Buick
K	403 4-bbl. V-8	Oldsmobile
L	350 4-bbl. V-8	Chevrolet
P	350 4-bbl. V-8	Pontiac
R	350 4-bbl. V-8	Oldsmobile
R	350 Seville V-8	Oldsmobile
S	425 4-bbl. V-8	Cadillac
T	425 Fuel Inj. V-8	Cadillac
U	305 2-bbl. V-8	Chevrolet
V	151 4-Cyl.	Pontiac
X	350 H.P. Corvette V-8	Chevrolet
Y	301 2-bbl. V-8	Pontiac
Z	400 4-bbl. V-8	Pontiac

COOL CATS FOR 1978

On 1978 cars, the catalytic converters will run about 300° less outer skin temperature than they did three years ago in 1975. The cats have to be hot, to do their job of oxydizing pollutants, but not as hot as was originally thought. The reason for the reduction in temperature is the increased spark advance that most manufacturers are now using. When the cats first came out, car engineers kept the spark retarded to heat up the cat and clean the exhaust. Now they have found out that the cat does a good job, even with the spark advanced for better performance. And we get the benefits by not having to work around such a hot exhaust system.

1978 HEATED AIR CLEANERS

Cars with heated air cleaners have always used a bimetal temperature sensitive valve inside the air cleaner. The valve controls the vacuum to the motor that operates the hot-cold air door in the snout of the air cleaner. Some 1978 General Motors cars will use a bimetal valve with an internal check valve. The check valve traps vacuum and holds the air door in the heat-on position. This prevents a stumble if the engine is accelerated cold. After the engine warms up and the bimetal valve comes into action, the check valve is bypassed.

The bimetal valve with the internal check valve can be identified by the small vacuum hole in one hose connection. To work properly, the small hole must be connected to manifold vacuum.

1978 M2MC210 ROCHESTER CARBURETOR

The M2MC Rochester 2-bbl. has been around for a few years. It is the 2-bbl. that was made out of a 4-bbl. casting. Up to now, it has always looked like a 4-bbl. with the extra two holes missing. In 1978, Rochester decided that the M2MC was here to stay and they stopped making it out of a 4-bbl. casting. The new M2MC 210 is basically the same carburetor, but with a lot of excess weight removed.

ALTITUDE ENGINES FOR 1978

Production of 49-State, Altitude, and California engine was well under way when the 95 Congress amended the Clean Air Act and eliminated the altitude engine. This meant that high altitude areas such as Denver, Colorado were allowed to use the same engines used in all other areas, except California. As soon as the law was amended, some manufacturers stopped making altitude engines. But others kept on making them because they ran so much better at altitude. The important thing to remember is that 1978 engines must be identified as "49-State," "Altitude", or "California," and adjusted to the correct tune-up specs. In 1979, the Altitude engine will probably be dropped by all manufacturers.

GENERAL MOTORS 1978 PROPANE ENRICHMENT

Ford and Chrysler were the first to use propane enrichment for setting idle mixture. Now, General Motors is also using it, but for a slightly different reason. The idle mixture passages on all the 1978 General Motors (Rochester) carburetors, are so small that backing the needles out will not richen the mixture enough to make the normal speed drop setting. Ordinarily, the mixture screws are backed out until the engine is at its highest rpm, and then the screws are turned in until the rpm drops by a specific amount. It's a simple procedure, but you can't do it on the 1978 Buick, Cadillac, Chevrolet, Oldsmobile, and Pontiac models because the fuel passages are not big enough. Backing the needles all the way out will not increase the engine speed to the highest rpm. To do it, you have to add propane. After shutting off the propane, the engine speed will drop. Then you can adjust the mixture needles until the total drop is precisely what is called for in the specifications.

The reason for the limited adjustment on the mixture needles is to lessen the effect of tampering. If the needles will not make the engine over rich, then those people who are always twisting the needles will do less polluting than they used to.

NON-ADJUSTABLE 1978 CARBURETORS

Some 1978 cars are actually on the street with hidden idle mixture needles. The needles are recessed, and covered by a pressed-in cap. To remove the cap, the carburetor must be removed from the engine, because of limited space to work in. In 1979, more of these non-adjustable carburetors will appear. In a few years, all carburetors will have only one adjustment, idle speed.

GENERAL MOTORS VAPOR CONTROL SYSTEM 1978

A new 5-hose canister is used on all G.M. carburetor engines. Hoses connect to the canister from the fuel tank dome, the PCV system, which supplies man-

Repair & Tune Up Notes

ifold vacuum, the EGR port, and the carburetor bowl. Vapor comes to the canister through the fuel tank hose, which is a direct connection to the canister, open at all times. Vapor also comes to the canister through the hose from the carburetor bowl, which connects to a vapor vent valve built into the top of the canister. The vapor vent valve is open when the engine is shut down, so that vapor can leave the carburetor bowl and accumulate in the canister. The top of the vent valve is connected to engine manifold vacuum through the PCV system. When the engine starts, the vacuum closes the vent valve.

A second valve built into the top of the canister is the old familiar purge valve. The top of the valve is connected to the EGR port on the carburetor. The bottom of the purge valve is connected to manifold vacuum through the PCV system. At idle the purge valve is closed. Above idle the EGR port is exposed to vacuum and the purge valve opens. This allows the manifold vacuum to draw air through the canister and purge it of fuel vapors. The advantage of the purge valve is that purging is done above idle when the engine is running fast enough to absorb the vapor without ill effects. If the purging were done continually at idle, the engine would run rich sometimes and lean sometimes, depending on how much vapor there was in the canister. Canister hose connections are as follows:

Canister Marking	Connects to
Control Vac	EGR port
Fuel Tank	Fuel tank dome
PCV	PCV Manifold vacuum
Carb Bowl	Carburetor bowl vent
Man. Vac.	"T" in PCV hose

ELECTRONIC FUEL CONTROL SYSTEM (EFC) 1978

The EFC system is new for 1978, and is designed to control the exhaust emissions by regulating the air/fuel mixture with the use of the following components;
1. Exhaust gas oxygen sensor
2. Electric control unit
3. Vacuum modulator
4. Controlled air-fuel ratio carburetor.
5. Phase II catalytic converter

The exhaust gas sensor is placed in the exhaust gas stream and generates a voltage which varies with the oxygen content of the exhaust gases.

The voltage signal is sent to the electronic control unit, where the voltage signal is monitored. The electronic control unit transmits a signal to the vacuum modulator.

To assist in maintaining proper electrical signal regulation during periods of unstable engine operation, a bi-metal switch is located in the cylinder head and provides an open or ground circuit for the electrical signal, when the temperature is above or below the calibration level value.

If below the calibration point, the electronic control unit uses this signal to restrict the amount of leanness that the carburetor can go. A temperature sensitive diode with-in the ECU, assists in the determination of the strength of the electrical signal. When the engine warms up, this feature drops out, indicating the engine is warm and the driveability will not be impaired while the carburetor is in the maximum lean range.

This system also uses the Phase II catalytic converter, which is selective to the reduction of oxides of nitrogen while improving the characteristics of inducing oxidation of hydrocarbons and carbon monoxide.

ENGINES 1979

A number of engines have been dropped by the various automobile manufacturers for the model year 1979.

The engines can be identified by referring to the vehicle identification number, visible through the left side of the windshield.

The fifth character of the vehicle identification number represents the engine that is being used in that particular vehicle for General Motors, Chrysler Corporation, and Ford Motor Co. The seventh character in the vehicle identification number represents the engine that is being used in vehicles that are being made by the American Motors Corporation.

For complete information and description about the particular vehicle identification number codes used by each individual automobile manufacturer, refer to the beginning of the various car sections in this book.

CATALYTIC CONVERTER (PHASE II) 1979

The Phase II catalytic converter helps to reduce all three exhaust emission pollutants, Hydrocarbons, Carbon Monoxide, and Oxides of Nitrogen, but is particularly selective in the reduction of oxides of nitrogen. This converter is designed to maintain high efficiency, thereby making it necessary to closely control the air fual ratio. The Phase II catalytic converter beads are coated with Platinum and Rhodium rather than Platinum and Palladium, which are used in the other types of catalytic converters.

TURBO CHARGER 1979

Buick Motor Division uses a turbo charged V6 engine in some of it's 1979 models. The turbo charger is used to increase power on a demand basis, thus allowing a smaller engine to do the work of a larger and more powerful engine. Buick's turbo V6 engine is more complicated than it's non-turbo V6 engine, but it does incorporate many of the design features utilized by the non-turbo engine.

New for Ford Motor Co. this year is the turbo charged 2.3 Liter 4 cyl. engine. This engine is available in the Ford Mustang and Mercury Capri.

When the turbo charger is in operation a green light will glow on the instrument panel, If, during turbo operation, a problem should occur, a red warning light and buzzer will warn the driver of a malfunction within the system. Also, should the engine oil temperature become critically high, the red light will flash on the instrument panel.

FORD MOTOR COMPANY ELECTRONIC ENGINE CONTROL SYSTEM 1979

The electronic engine control system, called EEC-II, is used on all Mercury models equipped with the 5.8 Liter V8 engine. The system will also be used on all 5.8 Liter V8 engines for the Ford models produced for sale in the state of California. The new system controls spark, exhaust gas recirculation, and air/fuel ratio, in incorpating an electronically controlled "feedback" carburetor and a three way catalyst system.

NON-ADJUSTABLE CARBURETORS 1979

General Motors Corporation for 1979, has made it impossible to adjust the air/fuel mixtures on it's carburetors. The only way adjustment can be accomplished is at the time of major carbureton overhaul. On American Motors Corporation cars,

Repair & Tune Up Notes

no adjustments can be made, with regards to the air/fuel mixture on the 4 cyl. engine, but the 6 and 8 cyl. engines are adjustable, within a narrow range.

GENERAL MOTORS FRONT WHEEL DRIVE (X BODY CARS) 1979

General Motors Corporation is introducing a mid year production automobile in it's Buick, Pontiac, Oldsmobile, and Chevrolet Divisions. This new vehicle will be equipped with front wheel drive and powered by a 60°, V6 engine, which is a performance option, or the standard 151 4 cyl. engine, which is now being made by Pontiac Motor Division. The terminology that is being used to identify these new cars is "X-body". The name is derived from the construction design of the automobile's chassis.

OMNI AND HORIZON HIGHLIGHTS

These two new subcompact cars are the first US-built subcompacts ever made by Chrysler Corporation. They are also the first front-wheel drive subcompacts ever built in the U.S. The engine is mounted across the car, with the front pulley on the passenger side, and the flywheel on the driver's side. The differential and drive axles are underneath and slightly to the rear of the engine. Power is transmitted from the transmission to the axles through gears, with no chains used.

Both cars are mechanically identical, with only trim and nameplate differences. The Omni, is sold by Dodge dealers. The Horizon, is sold by Plymouth dealers.

Emission controls on the Omni and Horizon have the same basic design as the controls used on larger cars. See the illustrations for details on the controls.

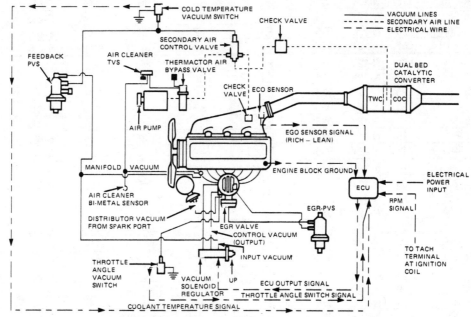

Typical System Connections—2.3L Electronic Engine Control System

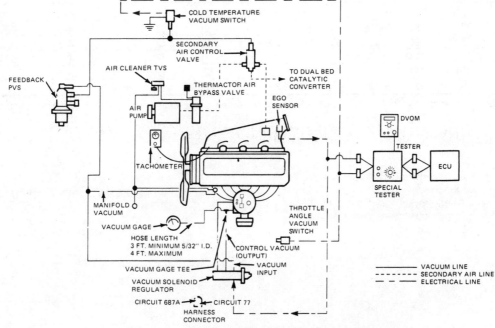

Test Connections—2.3L I-4 Electronic Engine Control System

9

American Motors

Ambassador • AMX/Javelin • Concord • Gremlin • Hornet • Matador • Pacer • Spirit

TUNE UP SPECIFICATIONS

Engine Code 1973, 5th character of the VIN number
Engine Code starting 1974, 7th character of the VIN number

Year	Eng. V.I.N. Code	ENGINE No. Cyl. Disp. (cu in)	H.P.	SPARK PLUGS Orig. Type	Gap (in)	DIST. Point Dwell (deg)	Point Gap (in)	IGNITION TIMING (deg) Man Trans •	Auto Trans	VALVES Intake Opens (deg)B.	FUEL PUMP Pres. (psi)	IDLE SPEED • (rpm) Man Trans	Auto Trans In Drive
75	E	6-232	100	N-12Y	.035	E.I.		5B	5B	12	4-5	600	550(700)
	A	6-258	110	N-12Y	.035	E.I.		3B	3B	12	4-5	600	550(700)
	H	8-304	150	N-12Y	.035	E.I.		5B	5B	14¾	5-6½	750	700
	N	8-360	175	N-12Y	.035	E.I.		5B	5B	14¾	5-6½	750	700
	P	8-360	195	N-12Y	.035	E.I.		5B	5B	14¾	5-6½	750	700
	P	8-360	220	N-12Y	.035	E.I.		5B	5B	14¾	5-6½	750	700
	Z	8-401	255	N-12Y	.035	E.I.		5B	5B	25½	5-6½	750	700
76	E	6-232	90	N-12Y	.035	E.I.		8B	8B	12	4-5	850	550(700)
	A	6-258	95	N-12Y	.035	E.I.		6B	8B	12	4-5	850①	550(700)
	H	6-258	120	N-12Y	.035	E.I.		6B	8B	12	4-5	850	550(700)
	N	8-304	120	N-12Y	.035	E.I.		5B	10B(5B)	14¾	5-6½	750	700
	P	8-360	140	N-12Y	.035	E.I.		—	10B(5B)	14¾	5-6½	—	700
	P	8-360	180	N-12Y	.035	E.I.		—	10B(5B)	14¾	5-6½	—	700
	Z	8-401	215	N-12Y	.035	E.I.		—	10B(5B)	25½	5-6½	—	700
77	G	4-121(1)	80	N-8L	.035	47	.018	12B	12B(8B)	41¾	4-6	900	800
	E	6-232	88	N-12Y	.035	E.I.		8B(10B)	10B	12	4-5	600(850)	550(700)
	A	6-258	98	N-12Y	.035	E.I.		6B	8B	12	4-5	600	550(700)
	C	6-258	114	N-12Y	.035	E.I.		6B	8B	12	4-5	600	550(700)
	H	8-304	121	N-12Y	.035	E.I.		—	10B(5B)	14¾	5-6½	—	600(700)
	N	8-360	129	N-12Y	.035	E.I		—	10B(5B)	14¾	5-6½	—	600(700)
78	G	4-121(1)	80	N-8L	.035	47	.018	12B	12B(8B)	41¾	4-6	900	800
	E	6-232	88	N-13L	.035	E.I.		8B	10B	12	4-5	600	550
	A	6-258	98	N-13L	.035	E.I.		10B(6B)	10B(8B)	12	4-5	600(850)	550(700)
	C	6-258	114	N-13L	.035	E.I.		6B	8B	14½	4-5	600	550
	H	8-304	121	N-12Y	.035	E.I		—	10B(5B)	14¾	5-6½	—	600(700)(4)
	N	8-360	129	N-12Y	.035	E.I		—	10B	14¾	5-6½	—	600(650)(5)
79	G	4-121(1)	80	N-8L	.035	47	.018	(2)	(2)	41¾	4-6	900	800
	E	6-232	88	N-13L	.035	E.I.		(2)	(2)	12	4-5	600	550
	A	6-258	98	N-13L	.035	E.I.		(2)	(2)	12	4-5	600(850)	550(700)
	C	6-258	114	N-13L	.035	E.I.		(2)	(2)	14½	4-5	600	550
	H	8-304	121	N-12Y	.035	E.I.		(2)	(2)	14¾	5-6½	—	600(700)(4)
	N	8-360	129	N-12Y	.035	E.I.		(2)	(2)	14¾	5-6½	—	600(650)(5)

NOTE: Should the information provided in this manual deviate from the specifications on the underhood tune-up label, the label specifications should be used, as they may reflect production changes.

- • Figure in parentheses indicates California engine
- — Not applicable
- E.I. —ELECTRONIC IGN.
- (1) VALUE CLEARANCE "HOT," INTAKE .006–.009— EXHAUST .016–.019.
- (2) SEE DECAL UNDER HOOD
- (3) HIGH ALTITUDE AND CALIFORNIA ONLY (N.A. 49 STATES)
- (4) HIGH ALTITUDE AND CALIFORNIA—700
- (5) 700 IN HIGH ALTITUDE AREAS
- (6) AT SENDING UNIT

American Motors

Ambassador • AMX/Javelin • Concord • Gremlin • Hornet • Matador • Pacer • Spirit

DISTRIBUTOR SPECIFICATIONS — AMERICAN MOTORS

Year	Distributor Identification	Centrifugal Advance Start Dist. Deg. @ Dist. RPM	Centrifugal Advance Maximum Dist. Deg. @ Dist. RPM	Vacuum Advance Start @ In. Hg.	Vacuum Advance Maximum Dist. Deg. @ In. Hg.
1975	3224746	0–1.5 @ 400	14–16 @ 2300	4–6	6.5–8.5 @ 12.7
	3224965	0–1 @ 400	15–17 @ 2200	5–7	6.5–8.5 @ 12.7
	3224966	0–2.5 @ 450	15–17 @ 2200	4–6	6.5–8.5 @ 12.7
	3224968	0–2 @ 600	13–15 @ 2250	5–7	7–9 @ 13.1
	3224969	0–2 @ 500	12–14 @ 2250	5–7	7–9 @ 13.1
1976	3227331	0–0.4 @ 500	5.25–7.25 @ 2200	5–7	7–9 @ 12.6
	3228263	0–2.8 @ 500	11–13 @ 2200	4–6	6.5–8.5 @ 11.7–12.8
	3228264	0–2.4 @ 500	13.5–15.5 @ 2200	5–7	6.5–8.5 @ 11.5–12.5
	3228265	0–2.7 @ 500	7.5–9.5 @ 2200	4–6	6.5–8.5 @ 11.7–12.8
	3228266	0–3.75 @ 500	10–12 @ 2200	4–6	6.5–8.5 @ 11.7–12.8
	3229719	0–1.75 @ 500	8–11 @ 2200	4–7	7–9 @ 12.7–13
1977	3227331	0–0.4 @ 500	5.25–7.25 @ 2200	5–7	7–9 @ 12.6
	3228263	0–2.8 @ 500	11–13 @ 2200	4–6	6.5–8.5 @ 11.7–12.8
	3228264	0–2.4 @ 500	13.5–15.5 @ 2200	5–7	6.5–8.5 @ 11.5–12.5
	3228265	0–2.7 @ 500	7.5–9.5 @ 2200	4–6	6.5–8.5 @ 11.7–12.8
	3228266	0–3.75 @ 500	10–12 @ 2200	4–6	6.5–8.5 @ 11.7–12.8
	3229719	0–1.75 @ 500	8–11 @ 2200	4–7	7–9 @ 12.7–13
'78	3230443	8–1.8 @ 475	10.8–13.2 @ 2200	1.5–5.5	6.8–9.5 @ 13.2
	3231340	1–1.8 @ 475	13.5–16 @ 2200	2–6.5	6.8–9.3 @ 12.8
	3231341	9–1.9 @ 500	7–9.5 @ 2200	1.5–5.5	6.8–9.5 @ 13.2
	3231915	1–1 @ 500	8–10.8 @ 2200	1.5–3.5	10.8–13.2 @ 11.5
	3232434	1–1 @ 500	8–10.5 @ 2200	2.2–6.5	6.8–9.3 @ 13.5
	3233173	1–2 @ 450	10.8–13.2 @ 2200	1.5–4	10.8–13.2 @ 13.0
	3233174	9–1.5 @ 475	7–9.5 @ 2200	1.5–4	10.8–13.2 @ 13.0
	3250163	0–3 @ 675	15–1.7 @ 2250	2–4	7–9 @ 11.0

VEHICLE IDENTIFICATION NUMBER (VIN)

Engine Code 1973, 5th character of the VIN number
Engine Code starting 1974, 7th character of the VIN number

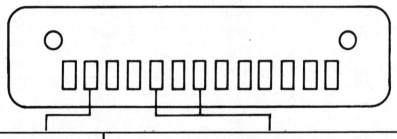

MODEL YEAR CODE		ENGINE CODE			
CODE	YEAR	CODE	DISP. (cu. in.)	CONFIG.	CARB.
		G	121	L-4	2V
3	1973	E	232	L-6	1V
4	1974	A	258	L-6	1V
5	1975	C	258	L-6	2V
6	1976	H	304	V-8	2V
7	1977	N	360	V-8	2V
8	1978	P	360	V-8	4V
9	1979	Z	401	V-8	4V

American Motors

CARTER BBD SPECIFICATIONS
AMERICAN MOTORS

Year	Model ④	Float Level (in.)	Accelerator Pump Travel (in.)	Bowl Vent (in.)	Choke Unloader (in.)	Choke Vacuum Kick (in.) ①	Fast Idle Cam Position ①	Fast Idle Speed (rpm)	Automatic Choke Adjustment
1976	8067	¼	0.500	—	0.250	0.128	0.095	1700	2 Rich
	8073	¼	0.500	—	0.250	0.128	0.095	1700	1 Rich
1977	8103	¼	0.496	—	0.280	0.150	0.120	1600	1 Rich
	8104	¼	0.520	—	0.280	0.128	0.095	1500	1 Rich
	8117	¼	0.480	—	0.280	0.152	0.112	1600	1 Rich
1978	8128	¼	0.496	—	0.280	0.150	0.110	1600	Index
	8129	¼	0.520	—	0.280	0.128	0.095	1500	1 Rich

① Indicates the drill bit number.
② Throttle closed
③ At idle
④ Model numbers located on the tag or casting

CARTER YF, YFA SPECIFICATIONS
AMERICAN MOTORS

Year	Model ①	Float Level (in.)	Fast Idle Cam (in.)	Unloader (in.)	Choke
1975	All	0.476	0.190	0.275	1 Rich
1976	7083, 7085, 7112	0.476	0.185	0.275	1 Rich
	7084, 7086	0.476	0.185	0.275	2 Rich
1977	7151	0.476	0.195	0.275	1 Rich
	7152	0.476	0.195	0.275	1 Rich
	7153	0.476	0.195	0.275	Index
	7195	0.476	0.195	0.275	1 Rich
	7223	0.476	0.195	0.275	Index
	7111	0.476	0.201	0.275	2 Rich
	7189	0.476	0.201	0.275	1 Rich
1978	7201	0.476	0.195	0.275	Index
	7228	0.476	0.195	0.275	1 Rich
	7229	0.476	0.195	0.275	1 Rich
	7235	0.476	0.195	0.275	Index
	7267	0.476	0.195	0.275	1 Rich
	7232	0.476	0.201	0.275	2 Rich
	7233	0.476	0.201	0.275	1 Rich

FORD, AUTOLITE, MOTORCRAFT MODELS 4300, 4350 SPECIFICATIONS
American Motors

Year	(9510)* Carburetor Identification ①	Dry Float Level (in.)	Pump Hole Setting	Choke Plate Pulldown (in.)	Fast Idle Cam Linkage (in.)	Fast Idle (rpm)	Dechoke (in.)	Choke Setting	Dashpot (in.)
1975	5TA4	0.90	Lower	0.140	0.160	1600	0.325	2 Rich	—
1976	6TA4	0.090	Lower	0.130	0.135	1600	0.325	2 Rich	—

American Motors

Model 5210-C
AMC OHC 4 Cylinder

Year	Carb. Part No. ① ②	Float Level (Dry) (in.)	Float Drop (in.)	Pump Position	Fast Idle Cam (in.)	Choke Plate Pulldown* (in.)	Secondary Vacuum Break (in.)	Fast Idle Setting (rpm)	Choke Unloader (in.)	Choke Setting
1977	7711	0.420	—	—	0.140	0.246	—	1600	0.300	1 Rich
	7712	0.420	—	—	0.140	0.246	—	1600	0.300	1 Rich
	7799	0.420	—	—	0.135	0.215	—	1600	0.300	Index
	7846	0.420	—	—	0.101	0.204	—	1600	0.300	1 Rich
1978	8163	.420	—	—	.193	.191	—	1800	.300	1 NR
	8164	.420	—	—	.204	.202	—	1800	.300	1 NR
	8165	.420	—	—	.177	.180	—	1800	.300	Index

FORD, AUTOLITE, MOTORCRAFT MODELS 2100, 2150 SPECIFICATIONS
American Motors

Year	(9510)* Carburetor Identification	Dry Float Level (in.)	Wet Float Level (in.)	Pump Setting Hole # ①	Choke Plate Pulldown (in.)	Fast Idle Cam Linkage Clearance (in.)	Fast Idle (rpm)	Dechoke (in)	Choke Setting	Dashpot (in.)
1975	5DA2	13/32	3/4	3	0.140	0.130	1600	0.250	1 Rich	—
	5DMS	13/32	3/4	3	0.130	0.130	1600	0.250	2 Rich	3/32
	5RAS	13/32	3/4	3	0.140	0.130	1600	0.250	1 Rich	—
1976	6DA2	13/32	3/4	3	0.140	0.130	1600	0.250	1 Rich	—
	6DM2	35/64	15/16	3	0.130	0.120	1600	0.250	2 Rich	—
	6RA2	13/32	3/4	3	0.140	0.130	1600	0.250	1 Rich	—
1977	7RA2	5/16	0.780	3	0.136	0.126	1600	0.250	1 Rich	—
	7RA2C	5/16	0.780	3	0.130	0.120	1800⑥	0.250	1 Rich	—
	7DA2	5/16	0.780	3	0.136	0.126	1600	0.250	Index	—
	7RA2A	5/16	0.780	3	0.104	0.089	1800	0.250	1 Rich	—
1978	8DA2	0.555	0.780	3	0.136	0.126	1600	0.250	Index	—
	8RA2	0.555	0.780	3	0.136	0.126	1600	0.250	1 Rich	—
	8RA2C	0.555	0.780	3	0.136	0.120	1800	0.250	1 Rich	—
	8RA2A	0.555	0.780	3	0.089	0.078	1800	0.170	2 Rich	—
	8DA2A	0.555	0.930	3	0.089	0.078	1600	0.170	2 Rich	—

CAR SERIAL NUMBER AND ENGINE IDENTIFICATION

1976

Mounted behind the windshield on the driver's side is a plate with the vehicle identification number. The second character is the model year, with 6 for 1976. The seventh character is the engine code, as follows:

A .... 258 1-bbl. 6-cyl.
C .... 258 2-bbl. 6-cyl.
E .... 232 1-bbl. 6-cyl.
H 304 2-bbl. V-8
N 360 2-bbl. V-8
P 360 4-bbl. V-8
Z 401 4-bbl. V-8

Engines can be identified by a build date code attached to the right bank rocker cover on V-8, or to the right side of the block between numbers 2 and 3 cylinder on the 6-cylinder. The fourth character in the code is a letter identifying the engine. The letters are the same as in the vehicle identification number, listed above.

1977

The vehicle identification number is mounted on a plate behind the windshield on the driver's side. The second character is the model year, with 7 for 1977. The seventh character is the engine code, as follows:

A .... 258 1-bbl. 6-cyl.
C .... 258 2-bbl. 6-cyl.
E .... 232 1-bbl. 6-cyl.
H 304 2-bbl. V-8
N 360 2-bbl. V-8
G .... 121 2-bbl. 4-cyl.

Engines can be identified by a build date code attached to the right bank rocker cover on V-8, or to the right side of the block between numbers 2 and 3 cylinder on the 6-cylinder. The fourth character in the code is a letter identifying the engine. The letters are the same as in the vehicle identification number, listed above.

1978

Mounted behind the windshield on the driver's side is a plate with the vehicle identification number. The second character is the model year, with 8 for 1978. The seventh character is the engine code, as follows:

American Motors

CAR SERIAL NUMBER AND ENGINE IDENTIFICATION

A .... 258 1-bbl. 6-cyl.
C .... 258 2-bbl. 6-cyl.
E .... 232 1-bbl. 6-cyl.
G .... 2-liter 2-bbl. 4-cyl.
H 304 2-bbl. V-8
N 360 2-bbl. V-8

NOTE: *The 2-liter 4-cylinder engine is actually 1984 cubic centimeters, or 121 cubic inches.*

Engines can be identified by a build date code attached to the right bank rocker cover on the V-8, or to the right side of the block between numbers 2 and 3 cylinder on the 6-cylinder. The 4-cylinder engine has the code on the left rear of the block near the dipstick. The fourth character in the code is a letter identifying the engine. The letters are the same as in the vehicle identification number listed above.

1979

Mounted behind the windshield on the driver's side is a plate with the vehicle identification number. The second character is the model year, with 9 for 1979. The seventh character is the engine code, as follows:

A .... 258 1-bbl. 6-cyl.
C .... 258 2-bbl. 6-cyl.
E .... 232 1-bbl. 6-cyl.
G .... 2-liter 2-bbl. 4-cyl.
H 304 2-bbl. V-8

NOTE: *The 2-liter 4-cylinder engine is actually 1984 cubic centimeters, or 121 cubic inches.*

Engines can be identified by a build date code attached to the right bank rocker cover on the V-8, or to the right side of the block between numbers 2 and 3 cylinder on the 6-cylinder. The 4-cylinder engine has the code on the left rear of the block near the dipstick. The fourth character in the code is a letter identifying the engine. The letters are the same as in the vehicle identification number listed above.

EMISSION EQUIPMENT

1976

All Models
Air pump
  49-States
    Used on all V-8 engines
    Pacer and Hornet 6-cyl. Man. Trans. only.
    Matador, all 6-cyl.
  Calif.
    All models
Closed positive crankcase ventilation
Emission calibrated carburetor
Emission calibrated distributor
Single diaphragm vacuum advance
Exhaust gas recirculation
Vapor control, canister storage
Heated air cleaner
Transmission controlled spark
  49-States
    Not used
  Calif.
    All models
Catalytic converter, single
  49-States
    Matador 258 1-bbl. Man. Trans. only
    All 2-bbl. V-8
  Calif.
    All 6-cylinder
Catalytic converter, dual
  49-States
    All 4-bbl. V-8
  Calif.
    All V-8
Electric choke
  49-States
    Hornet 6-cyl. man. trans. only
    Pacer 1-bbl. 6-cyl. man. trans. only
    Matador 6-cyl. man. trans. only
    All 4-bbl. V-8
  Calif.
    Hornet 6-cyl. auto. trans. only
    Gremlin 6-cyl. auto. trans. only
    Pacer 1-bbl. 6-cyl. auto. trans. only
    All 4-bbl. V-8

1977

All Models
Air pump
Closed positive crankcase ventilation
Emission calibrated carburetor
Emission calibrated distributor
Single diaphragm vacuum advance
Exhaust gas recirculation
Vapor control, canister storage
Heated air cleaner
Transmission controlled spark
  49-States
    6-cylinder Gremlin, Hornet, Pacer without catalytic converter only
    All 6-cylinder Matador
  Calif.
    All models, except
    Not used on 4-cyl.
  Altitude
    Auto. trans. only
    Not used on 4-cyl.
Catalytic converter, warm-up
  49-State
    Not used
  Calif.
    All 6-cyl. & V-8 models (with pellet type.)
    Not used on 4-cyl.
  Altitude
    Not used
Catalytic converter, pellet type
  49-States
    All models except,
    Not used on 6-cylinder Gremlin, Hornet, Pacer with TCS
  Calif.
    All 6-cyl. & V-8 models (with warm-up type.)
    4-cyl. uses pellet type only
  Altitude
    All models
Electric choke
  49-States
    All 4-cyl.
    Not used on 6-cyl. & V-8 models
  Calif.
    All 4-cyl. models
    Matador, V-8 only
  Altitude
    All 4-cyl. models
    Matador, V-8 only
Altitude compensation carburetor
  49-State
    Not used
  Calif.
    Not used
  Altitude
    All 6-cyl. & V-8 models
    Not used on 4-cyl.

1978

All Models
Air pump
Closed positive crankcase ventilation
Emission calibrated carburetor
Emission calibrated distributor
Single diaphragm vacuum advance
Exhaust gas recirculation
Vapor control, canister storage
Heated air cleaner
Transmission controlled spark
  49-States
    Not used
  Calif.
    All models, except
    Not used on 4-cyl.
  Altitude
    Not used
Catalytic converter, warm-up
  49-States
    Not used
  Calif.
    All 6-cyl. & V-8 models (with pellet type.)
    Not used on 4-cyl.
  Altitude
    Not used
Catalytic converter, pellet type
  49-States
    All models
  Calif.
    All 6-cyl. & V-8 models (with warm-up type.)

American Motors

EMISSION EQUIPMENT

1978

All 4-cyl.
  Altitude
    All models
Electric choke
  49-States
    All 4-cyl.
    Not used on 6-cyl. & V-8
  Calif.
    All 4-cyl. models
    360 2-bbl. V-8
    Not usd on 6-cyl.
  Altitude
    All 4-cyl. models
    All V-8 models
    Not used on 6-cyl.
Altitude compensation carburetor
  49-States
    Not used
  Calif.
    Not used
  Altitude
    All 6-cyl. & V-8 models
    Not used on 4-cyl.

1979

All Models
Air pump
Closed positive crankcase ventilation
Emission calibrated carburetor
Emission calibrated distributor
Single diaphragm vacuum advance
Exhaust gas recirculation
Vapor control, canister storage
Heated air cleaner
Thermostatically Controlled Air Cleaner (Vacuum Controlled)

Transmission controlled spark
  49-States
    Not used
  Calif.
    All models,
Catalytic converter, warm-up
  49-States
    Not used
  Calif.
    All 6-cyl. & V-8 models (with pellet type.)

Not used on 4-cyl.
Catalytic converter, pellet type
  49-States
    All models
  Calif.
    All 6-cyl. & V-8 models (with warm-up type.)
    All 4-cyl.

Electric choke
  49-States
    All 4-cyl.
    Not used on 6-cyl. & V-8
  Calif.
    Not used

Throttle Solenoid
  49-States
    4-cyl. models
    6-cyl. 258 CID w/Auto
    V-8 All
  Calif.
    All models

IDLE SPEED AND MIXTURE ADJUSTMENTS

1976

Air cleaner .... On
Air Cond. .... Off
Auto. Trans. .... Drive
Mix. adj. ...... See rpm drop below
Idle CO (cars without cat. converter)
6-Cyl.
  No air pump ...... 1.0% Max.
  With air pump
    Auto. trans. .... 0.8% Max.
    Man. trans. ..... 0.5% Max.

NOTE: *When checking CO on cars with an air pump, the pump should be connected on automatic transmission cars. Manual transmission cars should have the pump hose disconnected and plugged.*

1-Bbl. 6-Cylinder
  49-States Auto. Trans. 550
    Mixture adj. .... 25 rpm drop
      (575-550)
  Calif. Auto. Trans.
    Solenoid connected 700
    Solenoid disconnected 500
    Mixture adj. .... 25 rpm drop
      (725-700)
  All Man. Trans.
    Matador 600
      Mixture adj. .. 50 rpm drop
      (650-600)
    Hornet, Gremlin, Pacer
      Solenoid connected 850
      Solenoid disconnected .. 500
      Mixture adj. .. 50 rpm drop
      (900-850)

2-Bbl. 6-Cylinder
  49-States Auto. Trans. 700
    Mixture adj. .... 25 rpm drop
      (725-700)
  Calif. Auto. Trans.
    Solenoid connected 700
    Solenoid disconnected 500
    Mixture adj. .... 25 rpm drop
      (725-700)
  All Man. Trans.
    Solenoid connected 600
    Solenoid disconnected 500
    Mixture adj. .... 50 rpm drop
      (650-600)
  V-8 Auto. Trans.
    Solenoid connected 700
    Solenoid disconnected 500
    Mixture adj. .... 20 rpm drop
      (720-700)
  V-8 Man. Trans. 750
    Mixture adj. ... 100 rpm drop
      (850-750)

1977

Air cleaner ....On
Air cond. ....Off
Auto. Trans. ....Drive
Mix adj. ....See rpm drop below
Idle CO (cars without cat. converter)
6-Cyl. ....1.0% Max.

NOTE: *When checking idle CO, leave air pump connected.*

2-Bbl. 4-Cyl.
  All auto. trans.
    Solenoid connected 800
    Solenoid disconnected 500
    Mixture adj. .....120 rpm drop
      (920-800)
  49-States man. trans.
    Solenoid connected 900
    Solenoid disconnected 500
    Mixture adj. .....120 rpm drop
      (1020-900)

Altitude man. trans.
  Solenoid connected 900
  Solenoid disconnected 500
  Mixture adj. ......75 rpm drop
    (975-900)

1-Bbl. 6-Cylinder
  49-States with cat. conv.
    Auto. trans. 550
    Mixture adj.
      Gremlin, Hornet, Pacer .....175 rpm drop
      (725-550)
      Matador ......25 rpm drop
      (575-550)
    Man. trans. 600
      Mixture adj. ....50 rpm drop
      (650-600)
  49-States no cat. conv.
    Auto. trans. 550
    Mixture adj. ..1.0% CO Max.
  Calif Auto. Trans.
    Solenoid connected 700
    Solenoid disconnected 500
    Mixture adj. ......25 rpm drop
      (725-700)
  Calif. Man. Trans.
    Solenoid connected 850
    Solenoid disconnected 500
    Mixture adj. .....100 rpm drop
      (950-850)
  Altitude Auto. Trans. 550
    Mixture adj. ......25 rpm drop
      (575-550)
  Altitude Man. Trans. 600
    Mixture adj. ......50 rpm drop
      (650-600)

2-Bbl. 6-Cylinder
  49-States with cat. conv.
    Auto. trans.
      Solenoid connected 600
      Solenoid disconnected 500
      Mixture adj. ....50 rpm drop
      (650-600)

American Motors

IDLE SPEED MIXTURE ADJUSTMENTS

1977

Man. trans. 600
  Mixture adj. .... 50 rpm drop
    (650-600)
49-States no cat. conv.
  Auto. trans.
    Solenoid connected 600
    Solenoid disconnected 500
    Mixture adj. ...1.0% CO Max.
  Calif. Auto. Trans.
    Solenoid connected 700
    Solenoid disconnected 500
    Mixture adj. ...... 25 rpm drop
    (725-700)
V-8 Auto. Trans.
  49-States
    Solenoid connected 600
    Solenoid disconnected 500
    Mixture adj. ...... 20 rpm drop
    (620-600)
  Calif.
    Solenoid connected 700
    Solenoid disconnected 500
    Mixture adj. ...... 20 rpm drop
    (720-700)
  Altitude
    Solenoid connected 700
    Solenoid disconnected 500
    Mixture adj. ...... 20 rpm drop
    (720-700)

1978

Air cleaner .... In place
Air cond. .... Off
Auto. trans. .... Drive
Mix. adj. .... See rpm drop below
Idle CO .... Not used

Idle Speec
121 2-bbl. 4-Cyl.
  All auto. trans.
    Solenoid connected 800
    Solenoid disconnected 500
    Mixture adj. .... 120 rpm drop
    (920-800)
  49-States man. trans.
    Solenoid connected 900
    Solenoid disconnected 500
    Mixture adj. .... 120 rpm drop
    (1020-900)
  Calif. man. trans.
    Solenoid connected..............
    Solenoid disconnected ....
    Mixture adj. .... 120 rpm drop
    (-)
  Altitude man. trans.
    Solenoid connected 900
    Solenoid disconnected 500
    Mixture adj. .... 75 rpm drop
    (975-900)
232 1-bbl. 6-Cyl.
  49-States auto. trans. 550
    Mixture adj. .... 25 rpm drop
    (575-550)
  49-States man. trans. 600
    Mixture adj. .... 50 rpm drop
    (650-600)
258 1-bbl. 6-Cyl.
  Calif. auto. trans.
    Solenoid connected 700

    Solenoid disconnected 500
    Mixture adj. .... 25 rpm drop
    (725-700)
  Calif. man. trans.
    Solenoid connected 850
    Solenoid disconnected 500
    Mixture adj. .... 50 rpm drop
    (850-800)
  Altitude auto. trans. 550
    Mixture adj. .... 25 rpm drop
    (575-550)
  Altitude man. trans. 600
    Mixture adj. .... 50 rpm drop
    (650-600)
258 2-bbl. 6-Cyl.
  49-States auto. trans.
    Solenoid connected 600
    Solenoid disconnected 500
    Mixture adj. .... 25 rpm drop
    (625-600)
  49-States man. trans. 600
    Mixture adj. .... 50 rpm drop
    (650-600)
258 1-bbl. 6-Cyl.
  Calif. auto. trans.
    Solenoid connected 700
    Solenoid disconnected 500
    Mixture adj. .... 25 rpm drop
    (725-700)
  Calif. man. trans.
    Solenoid connected 850
    Solenoid disconnected 500
    Mixture adj. .... 50 rpm drop
    (850-800)
  Altitude auto. trans. 550
    Mixture adj. .... 25 rpm drop
    (575-550)
  Altitude man. trans. 600
    Mixture adj. .... 50 rpm drop
    (650-600)
258 2-bbl. 6-Cyl.
  49-States auto. trans.
    Solenoid connected 600
    Solenoid disconnected 500
    Mixture adj. .... 25 rpm drop
    (625-600)
  49-States man. trans. 600
    Mixture adj. .... 50 rpm drop
    (650-600)
304 V-8
  49-States auto. trans.
    Solenoid connected 600
    Solenoid disconnected 500
    Mixture adj. .... 20 rpm drop
    (620-600)
  Calif. auto. trans.
    Solenoid connected 700
    Solenoid disconnected 500
    Mixture adj. .... 20 rpm drop
    (720-700)
  Altitude auto. trans.
    Solenoid connected 700
    Solenoid disconnected 500
    Mixture adj. .... 20 rpm drop
    (720-700)
360 V-8
  49-States auto. trans.
    Solenoid connected 600
    Solenoid disconnected 500
    Mixture adj. .... 20 rpm drop
    (620-600)

  Calif. auto. trans.
    Solenoid connected 650
    Solenoid disconnected 500
    Mixture adj. .... 20 rpm drop
    (670-650)
304 V-8
  49-States auto. trans.
    Solenoid connected 600
    Solenoid disconnected 500
    Mixture adj. .... 20 rpm drop
    (620-600)
  Calif. auto. trans.
    Solenoid connected 700
    Solenoid disconnected 500
    Mixture adj. .... 20 rpm drop
    (720-700)
  Altitude auto. trans.
    Solenoid connected 700
    Solenoid disconnected 500
    Mixture adj. .... 20 rpm drop
    (720-700)
360 V-8
  49-States auto. trans.
    Solenoid connected 600
    Solenoid disconnected 500
    Mixture adj. .... 20 rpm drop
    (620-600)
  Calif. auto. trans.
    Solenoid connected 650
    Solenoid disconnected 500
    Mixture adj. .... 20 rpm drop
    (670-650)
  Altitude auto. trans.
    Solenoid connected 700
    Solenoid disconnected 500
    Mixture adj. .... 20 rpm drop
    (720-700)

1979

Air cleaner .... In place
Air cond. .... Off
Auto. trans. .... Drive
Mix. adj. ...... See rpm drop below
Idle CO .... Not used
Idle Spec
121 2-bbl. 4-Cyl.
  All auto. trans.
    Solenoid connected 800
    Solenoid disconnected 500
    Mixture adj. ...... 45 rpm drop
    (845-800)
  49-States man. trans.
    Solenoid connected 900
    Solenoid disconnected 500
    Mixture adj. .... 120 rpm drop
    (1020-900)
  Code "EH" only ①
    Solenoid connected 1000
    Solenoid disconnected 500
    Mixture adj. .... 120 rpm drop
    (1120-1000)
  Calif. man. trans.
    Solenoid connected 1000
    Solenoid disconnected 500
    Mixture adj. .... 120 rpm drop
    (1120-1000)
232 1-bbl. 6-Cyl.
  49-States auto. trans. 550
    Mixture adj. ...... 25 rpm drop
    (575-550)

American Motors

IDLE SPEED AND MIXTURE ADJUSTMENTS

1979
49-States man. trans. 600
    Mixture adj. ......50 rpm drop
                          (650-600)

258 1-bbl. 6-Cyl.
  Calif. auto. trans.
    Solenoid connected 700
    Solenoid disconnected 500
    Mixture adj. ......25 rpm drop
                           (725-700)

258 2-bbl. 6-Cyl.
  49-States auto. trans.
    Solenoid connected 600
    Solenoid disconnected 500
    Mixture adj. ......25 rpm drop
                           (625-600)
  49-States man. trans. 600
    Mixture adj. ......50 rpm drop
                           (650-600)

304 V-8
  49-States all trans.
    Solenoid connected 600
    Solenoid disconnected 500
    Mixture adj. ......40 rpm drop
                           (640-600)

①—"EH" noted on upper right corner of vehicle emission control information label

1976
NOTE: *Distributor vacuum hose must be disconnected and plugged. Set timing at 500 rpm. Variation allowed: Plus or minus 2°.*

232 6-Cyl. .... 8° BTDC
258 6-Cyl.
  Auto. Trans. .... 8° BTDC
  Man. Trans. .... 6° BTDC
304 V-8
  49-States Auto. Trans. 10° BTDC
  Calif. Auto. Trans. .... 5° BTDC
  All Man. Trans. ...... 5° BTDC
360 V-8
  49-States Auto. Trans. .10° BTDC
  Calif. Auto. Trans. .... 5° BTDC
401 V-8
  49-States Auto. Trans. 10° BTDC
  Calif. Auto. Trans. .... 5° BTDC

1977
NOTE: *Distributor vacuum hose must be disconnected and plugged. Set timing at 500 rpm. Variation allowed: Plus or minus 2°.*

4-Cyl.
  49-States ....12° BTDC
  Calif. .... 8° BTDC
  Altitude ....12° BTDC
232 6-Cyl.
  Auto. trans. ....10° BTDC
  Man. trans.
    49-States .... 8° BTDC
    Calif. ....10° BTDC
    Altitude ....10° BTDC
258 1-Bbl. 6-Cyl.
  Auto. trans.
    49-States
      Gremlin, Hornet,
        Pacer .... 8° BTDC
      Matador .... 6° BTDC
    Calif. .... 8° BTDC

INITIAL TIMING

    Altitude ....10° BTDC
      Below 4000 feet .. 8° BTDC
  Man. trans.
    49-States .... 6° BTDC
    Altitude ....10° BTDC
      Below 4000 feet .. 6° BTDC
258 2-Bbl. 6-Cyl.
  Auto. Trans. .... 8° BTDC
  Man. trans. .... 6° BTDC
304 2-Bbl. V-8
  49-States ....10° BTDC
  Calif. .... 5° BTDC
360 2-Bbl. V-8
  49-States ....10° BTDC
  Calif. .... 5° BTDC
  Altitude ....10° BTDC

1978
NOTE: *Distributor vacuum hose must be disconnected and plugged. Set timing at idle speed. Variation allowed: Plus or minus 2°. On auto. trans. cars, idle speed in Neutral may be fast enough to bring in mechanical advance. To prevent this, make timing setting with transmission in Drive, and wheels safely blocked.*

4-Cyl.
  49-States................ 12° BTDC
  Calif. .... 8° BTDC
  Altitude ....12° BTDC
232 6-Cyl.
  Auto. trans. ....10° BTDC
  Man. trans. .... 8° BTDC
258 1-Bbl. 6-Cyl.
  Auto. trans.
    Calif. .... 8° BTDC
    Altitude ....10° BTDC
  Man. trans.
    Calif. .... 6° BTDC
    Altitude ....10° BTDC
258 2-Bbl. 6-Cyl.
  Auto. trans. ....8° BTDC
  Man. trans. .... 6° BTDC

304 2-bbl. V-8
  49-States................ 10° BTDC
  Calif. .... 5° BTDC
  Altitude ....10° BTDC
360 2-bbl. V-8
  49-States................ 10° BTDC
  Calif. .... 10° BTDC
  Altitude
    Above 4,000 ft. ....10° BTDC
    Below 4,000 ft. .... 5° BTDC

1979
NOTE: *Distributor vacuum hose must be disconnected and plugged. Set timing at idle speed. Variation allowed: Plus or minus 2°. On auto. trans. cars, idle speed in Neutral may be fast enough to bring in mechanical advance. To prevent this, make timing setting with transmission in Drive, and wheels safely blocked.*

4-Cyl.
  49-States ....12° BTDC
  Code "EH" only ① ..16° BTDC
  Calif. .... 8° BTDC
232 6-Cyl.
  49-States
    Auto. trans. ....10° BTDC
    Man. trans. .... 8° BTDC
    Code "EH" only ① ..12 BTDC
258 1-bbl. 6-Cyl.
  Calif.
    Auto. trans. .... 8° BTDC
258 2-bbl. 6-Cyl.
  49-States
    Auto. trans. .... 8° BTDC
    Man. trans. .... 4° BTDC
304 2-bbl. V-8
  49-States
    Auto. trans. .... 8° BTDC
    Man. trans. .... 5° BTDC

①—"EH" noted on upper right corner of vehicle emission control information label

VACUUM ADVANCE

All engines .... Ported

SPARK PLUGS

All 4-Cyl. .... CH-N8L .... .035
All 6-Cyl. .... CH-N12Y ... .035
All V-8 .... CH-RN12Y . .035

NOTE: *Either resistor or non-resistor plugs may be used.*

American Motors

New for 1976 is a 2-bbl. carburetor available as an option on the 6-cylinder engine in Pacer only. The new carburetor is a Carter BBD, similar to the 1¼-inch BBD used on Chrysler products, except that the choke coil is mounted on the lower flange.

CAUTION: *The curb idle and fast idle screws are side by side, and it is easy to get on the wrong one when setting idle on cars without a throttle stop solenoid. The screw for idle speed is the longest of the two. On cars with a solenoid, curb idle is set with the solenoid adjustment, and the throttle screw is used only for the solenoid-disconnected speed.*

Some YF carburetors (see Emission Equipment list) now use the same electric choke as 4-bbl. V-8 engines. It operates the same as on the V-8, getting current from a 7-volt connection on the alternator.

ROLLOVER FUEL CONTROL 1976

This is not an emission control, but a system designed to limit fuel leakage after a collision and roll-over. The system varies with different models, but usually consists of check valves to prevent fuel from running out if the car is upside down.

On Pacer and Matador models, the fuel tank filler cap is the usual pressure-vacuum release type, but with a rollover check valve added. If the car gets upside down, a stainless steel ball in the cap falls onto the vent hole and prevents the fuel from running out. If the car is righted, the ball falls away from the hole and the pressure-vacuum release works normally.

A rollover check valve is also used on all cars in the fuel tank vent line near the rear axle. It keeps fuel from running forward and spilling out of the carbon canister while the car is upside down. Engines with 4-bbl. carburetors also have a check valve in the fuel return line next to the fuel filter.

Some publications show a rollover check valve in the fuel return line near the rear axle. That is incorrect. The rollover valve is in the vent line.

TRANSMISSION CONTROLLED SPARK 1977

On some cars, a check valve has

EMISSION CONTROL SYSTEMS 1976

FUEL RETURN SYSTEM 1976

This is not an emission control, but a system for eliminating vapor lock in the fuel line. An inline fuel filter is used between the fuel pump and the carburetor. In addition to the filter inlet and outlet hoses, a third hose connects an outlet at the top of the filter to a return connection on the fuel tank sending unit. Pressure from the fuel pump forces a continuous stream of fuel back to the tank whenever the engine is running. The outlet on the filter is restricted so that only a small amount of fuel goes back to the tank.

Because the return outlet is at the top of the filter, any vapors that are caused by heat on the fuel lines are passed back to the tank, instead of going on to the carburetor where they might upset the mixture.

During idle or slow speed traffic conditions, the heat under the car can be very high, and can boil the fuel in the supply line to the pump. The fuel can evaporate so fast that the fuel pump gets nothing but vapor. The fuel pump then becomes vapor locked, and then the engine will die. The fuel return system prevents this by allowing a continuous flow of fuel through the pump, which cools the lines and prevents vapor lock.

Returning fuel to the tank reduces the amount of fuel that the pump can deliver to the carburetor, but the amount of fuel returned is so small that fuel delivery is still adequate. Fuel pump testing is done with the return line plugged.

CAUTION: *The inline filter must be installed with the return hose connection at the top, so the vapors will rise to the return hose.*

4-barrel carburetors have a check valve in the return line at the filter, for rollover fuel control.

Some publications show a rollover check valve in the return line near the rear axle. That is incorrect. The rollover check valve is in the vent line.

Carter BBD 2-bbl. carburetor, optional on 1976 Pacer

EMISSION CONTROL SYSTEMS 1977

been added in the vacuum hose a few inches from its connection at the intake manifold. The check valve traps the vacuum in the hose and keeps the distributor in the maximum advance position while the engine is cold and the distributor advance is operating on manifold vacuum. When the temperature at the Coolant Temperature Override Switch (CTO) is 160° F. or more, the CTO switches the advance to ported vacuum, and the check

American Motors

valve no longer has any effect. The purpose of the check valve is to keep the spark advanced for better driveability when the engine is cold.

ALTITUDE COMPENSATION 1-BBL CARBURETOR 1977

Six-cylinder engines with a 1-bbl. carburetor use the Carter YF. In cars sold for use over 4000 feet elevation, the YF comes with a built-in altitude compensation. A compensation plug on the side of the bowl cover near the fuel inlet is used to open a passageway that lets in additional air to lean out the mixture for high altitude use. The same plug is also used to shut off the additional air for low altitude use.

The YF altitude compensation system supplies extra air to both the idle and main metering systems, so that the mixture at high altitude will be correct for all operating conditions. Whenever the altitude compensation setting is changed, the spark initial timing must be reset according to the following chart.

It takes about 2-1/2 turns from the inner seat to the outer seat. The plug is not removable. If you attempt to unscrew it all the way out, you will damage the carburetor.

ALTITUDE COMPENSATION 2-BBL CARBURETOR 1977

Eight-cylinder engines with a 2-bbl. carburetor use the Motorcraft (Ford) 2100. In cars sold for use over 4000 feet elevation, the 2100 comes with built-in altitude compensation, and is called the 2150. The system is controlled by an aneroid (bellows), which opens an air passage at high altitude, and closes the passage at low altitude. The aneroid is completely automatic in operation, and is factory adjusted. Adjustment in the field is not permitted, therefore the factory does not give any method of adjustment. Since the aneroid moves in relation to atmospheric pressure, it may open at low altitude if the atmospheric pressure is low enough, such as during bad weather.

The extra air that leans the mixture enters the venturi between the main discharge nozzle and the throttle plate. The vacuum in this area of the venturi is very low, so the amount of extra air that comes in is not very much.

But all the extra air that comes in is pure air, with no fuel mixed in, so it mixes with the air that has picked up fuel from the main nozzle, and leans the mixture. Because the extra air enters the carburetor throat above the throttle plate, it has no effect on the idle system. It only leans out the main metering mixture, which is in use when the throttle is open above idle.

The factory warns against making any adjustment in the field, but there is an adjusting screw in the end of the aneroid housing. The locknut must be loosened first, and then the screw can be turned. Turning the screw in will push the aneroid against the valve and open it. Turning the screw out will allow the aneroid to move away from the valve and close it.

CATALYTIC CONVERTER 1977

Catalytic converters are used on all

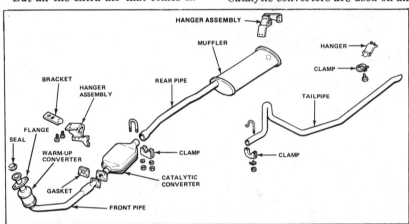

Six cylinder with single converter and warm-up converter
(© American Motors)

Transmission	Altitude Setting	Plug Position	Initial Timing
All	Above 4000	Outer seat	10° BTDC
Manual	Below 4000	Inner seat	6° BTDC
Auto.	Below 4000	Inner seat	8° BTDC

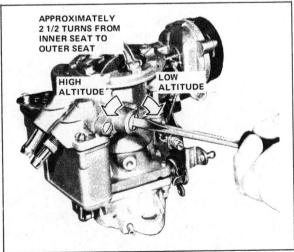

Compensator adjustment (© American Motors)

cars, except those made for Canada, and some Gremlin, Hornet, and Pacer 49-State cars. The Gremlin, Hornet, and Pacer 49-State cars with automatic transmission are made both with and without catalytic converter. When the catalytic converter is used, they do not use the transmission controlled spark system. Without catalytic converter, the cars come with the transmission controlled spark. Both "with" and "without" cars are certified by EPA, and the production line may switch back and forth, turning out either, for sale anywhere in the 49-States. The only way to tell if a Gremlin, Hornet, or Pacer 49-State car has a converter or not is to look at the exhaust system, or look for the TCS system in the engine compartment.

CAUTION: *All AMC 1977 cars, with or without catalytic converter, must use no-lead fuel. The no-lead fuel is required on non-converter cars because that is the way they were certified.*

American Motors

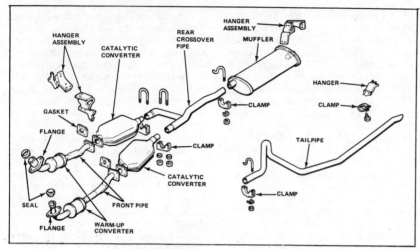

V-8 with dual converters and warm-up converters
(© American Motors)

California cars (except 4-cylinder) all use a new warm-up converter, in addition to the normal converter. V-8's use two normal converters and two warm-up converters. The warm-up converter is a cellular (honeycomb) type and is welded into the exhaust pipe close to the exhaust manifold. The normal converter is a pellet type, and is mounted under the floor of the passenger compartment, the same as before.

The honeycomb inside the warmup converter cannot be removed or serviced. If a warmup converter needs to be changed, the entire exhaust pipe must be changed. The pellet type underfloor converter continues to be serviceable as before, if you have the pellet changing equipment. The front of the pellet converter has a flange that bolts to another flange on the exhaust pipe. A gasket must be used between the flanges to prevent exhaust leaks.

2-LITRE 4-CYLINDER ENGINE, 1977

The AMC 2-litre engine was originally a European design. Much of the engine has been Americanized, but not all. If you are used to working on imported cars, you will see many things that are familiar. The design and operation of the emission controls is the same as on the 6-cylinder and V-8 engines, with the exception of the positive crankcase ventilation (PCV) system.

A vacuum solenoid is used with the 4-cylinder PCV system. When the ignition switch is ON, the vacuum passage through the solenoid is open, and the PCV valve is connected to intake manifold vacuum for normal operation. When the ignition switch is OFF, the solenoid closes the vacuum passage to the PCV valve, preventing any air from entering the intake manifold through the PCV hose. This setup is necessary because the engine can get enough air and fuel vapor through the PCV valve to keep running after the key is off. An anti-dieseling solenoid is used on the carburetor, also. It is the usual design that allows the throttle to close when the ignition key is OFF.

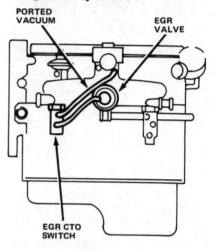

Four cylinder EGR system
(© American Motors)

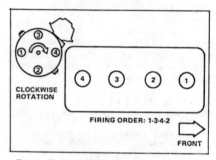

Four cylinder—distributor wiring sequence and firing order (© American Motors)

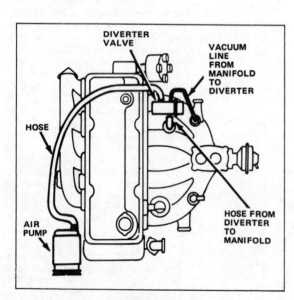

Four cylinder air pump
(© American Motors)

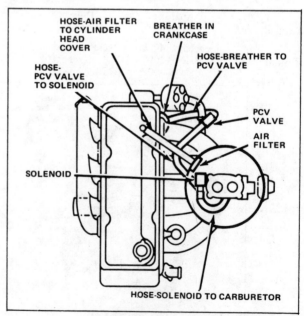

Four cylinder PCU system
(© American Motors)

American Motors

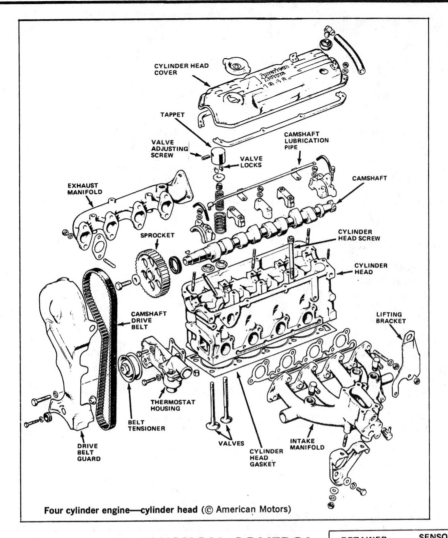

Four cylinder engine—cylinder head (© American Motors)

EMISSION CONTROL SYSTEMS 1978

SOLID STATE IGNITION 1978

All 6-Cylinder and 8-Cylinder 1978 AMC cars use the Motorcraft Solid State Ignition, manufactured by Ford Motor Co. This ignition uses a magnetic trigger wheel in place of the distributor cam, and a sensor coil to detect the position of the trigger wheel. There are no distributor points, and periodic service of the distributor is not required, except for inspecting the cap, rotor, and wires.

The control box and coil are part of the system purchased from Motorcraft, so they are of Ford design also. The primary connector at the coil is the typical Ford slip on design. Tachometers should be connected to the coil negative terminal, which has enough metal exposed so that an alligator clip can be connected without removing the coil connector.

The spark from this ignition can jump comparatively great distances. And like any ignition, the spark will jump to the closest metallic object. If the plug wires to cylinders No. 3 or 5 on a 6-cylinder, or

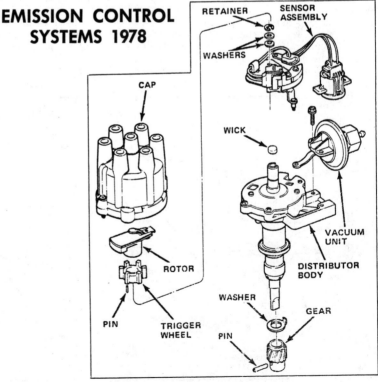

SSI distributor components—six cylinder shown (© American Motors)

American Motors

No. 3 or 4 on an 8-cylinder are removed with the engine running, the spark will probably jump inside the cap to the sensor bracket. This sudden surge of high voltage can damage the sensor. For this reason, those plug wires must never be removed while th engine is running, not even for an instant. When removing a plug wire to check available secondary voltage, do it with any plug wire other than those mentioned above.

VAPOR CONTROL SYSTEM 1978

Carbon canisters now have four hose connections, and a purge valve. The hose to the purge valve is connected to ported vacuum. When the engine is at idle, the port in the carburetor is above the throttle blade, and the purge valve is closed. Above idle, the port is exposed to vacuum, which acts on the purge valve in the canister and allows manifold vacuum to pull the fumes from the canister to the engine. As the throttle is opened farther, ported vacuum increases. When it gets to 12 in. Hg. or more, the purge valve opens into a second stage and allows a higher rate of purging.

Gasoline fumes are fed to the canister through two hoses, one from the tank and one from the carburetor. The carburetor hose connects to an external vent on the carburetor bowl. This vent is operated either by manifold vacuum or mechanical linkage, depending on the model of carburetor.

On the Model 5210 Holley-Weber carburetor, used on the 4-cylinder engine, a small diaphragm and valve are built into the bowl cover. The operating side of the diaphragm is connected to manifold vacuum, which closes the valve whenever the engine is running. If the engine is not running, a spring behind the diaphragm opens the valve and vents the bowl to the carbon canister.

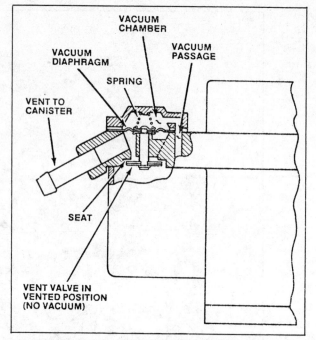

External bowl vent —Holley Weber 5210 (© American Motors)

Carter's YF 1-bbl. carburetor, used on the 6-cylinder engine, has an external bowl vent operated by mechanical linkage from the throttle. The vent opens only when the throttle is in the curb idle position. The vent is closed at all other throttle positions.

Because the YF bowl vent has no spring, high vapor pressure in the carburetor bowl can also open the vent. This will occur, when the throttle is above idle, if the pressure in the bowl goes above 0.14 in. of water.

Carter's 2-bbl. BBD model, used on the 6-cylinder, has two external bowl vents built into the bowl cover. The upper vent, with the plastic hose fitting, is pressure operated, and opens only if there is high pressure in the bowl. The lower vent, with the metal hose fitting, is operated by the throttle linkage, and opens only with the throttle in the curb idle position. Both of these vents are connected by short hoses to a "T", and then a single hose goes to the canister.

The Motorcraft 2100 2-bbl. used on the V-8 engines has a mechanical external bowl vent operated by the accelerating pump lever. The vent is a plunger type, with a small spring that holds it in the open position. Whenever the throttle is above idle, a lever with a stronger spring overrides the plunger spring and closes the vent. At curb idle, the lever lifts off the plunger, which opens the vent.

External vents not only help prevnt air pollution, but also keep gasoline vapors from entering the intake manifold through the internal vent that most carburetors have. This helps avoid hard starting after a car has been parked in the hot sun.

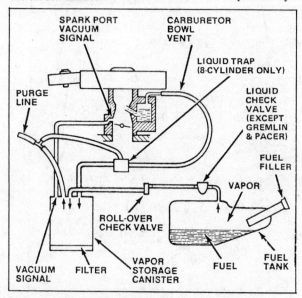

Fuel vapor control system—Typical (© American Motors)

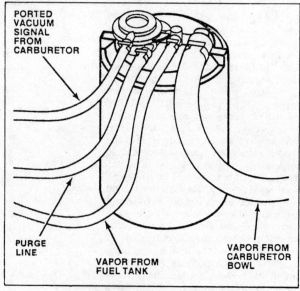

Charcoal canister and hoses (© American Motors)

American Motors

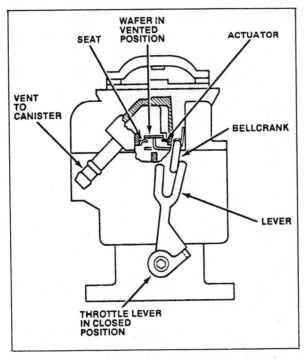

External fuel bowl vent—Carter YF (© American Motors)

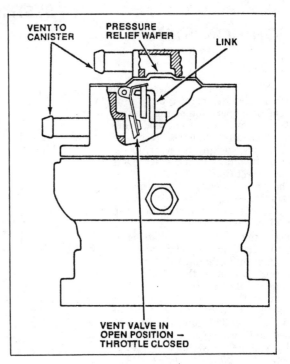

External bowl vent—Carter BBD (© American Motors)

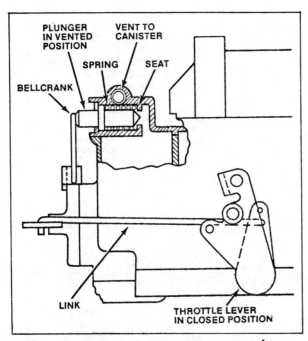

External bowl vent—Motorcraft 2100 (© American Motors)

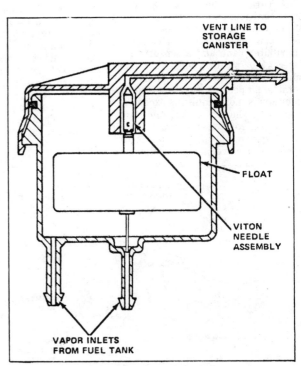

Liquid check valve—Typical (© American Motors)

American Motors

IGNITION SYSTEM 1979

4-cylinder

The four cylinder engines use the Bosch mechanical, point-type ignition system. The operation of the ignition system is in the conventional manner, which uses the three groups of components working together, to deliver high voltage to the spark plugs at the correct time.

The three groups consist of the distributor assembly, the ignition coil and the secondary distributor cap and wiring assembly.

The primary side of the distributor contains the ignition points and condensor, with mechanical and vacuum advances to control the opening and closing of the ignition points. The current to the ignition points is controlled by a resistance wire in the body harness. The coil is a self-contained unit which changes the primary voltage to secondary voltage through the induction process. The third group consists of the distributor rotor, distributor cap, secondary wiring and spark plugs.

6-Cylinder and V-8 Engines

All 6-Cylinder and 8-Cylinder 1979 AMC cars use the Motorcraft Solid State Ignition, manufactured by Ford Motor Co. This ignition uses a magnetic trigger wheel in place of the distributor cam, and a sensor coil to detect the position of the trigger wheel. There are no distributor points, and periodic service of the distributor is not required, except for inspecting the cap, rotor, and wires. The control box and coil are part of the system purchased from Motorcraft, so they are of Ford design also.

EMISSION CONTROL SYSTEMS 1979

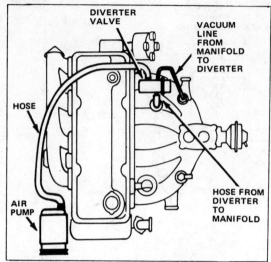

Air guard system—1979 4 cylinder engine

The primary connector at the coil is the typical Ford slip on design. Tachometers should be connected to the coil negative terminal, which has enough metal exposed so that an alligator clip can be connected without removing the coil connector.

The spark from this ignition can jump comparatively great distances. And like any ignition, the spark will jump to the closest metallic object. If the plug wires to cylinders No. 3 or 5 on a 6-cylinder, or No. 3 or 4 on an 8-cylinder are removed with the engine running, the spark will probably jump inside the cap to the sensor bracket. This sudden surge of high voltage can damage the sensor. For this reason, those plug wires must never be removed while the engine is running, not even for an instant. When removing a plug wire to check available secondary voltage, do it with any plug wire other than those mentioned above.

VAPOR CONTROL SYSTEM 1979

Carbon canisters have four hose connections, and a purge valve. The hose to the purge valve is connected to ported vacuum. When the engine is at idle, the port in the carburetor is above the throttle blade, and the purge valve is closed. Above idle, the port is exposed to vacuum, which acts on the purge valve in the canister and allows manifold vacuum to pull the fumes from the canister to the engine. As the throttle is opened farther, ported vacuum increases. When it gets to 12 in. Hg. or more, the purge valve opens into a second stage and allows a higher rate of purging.

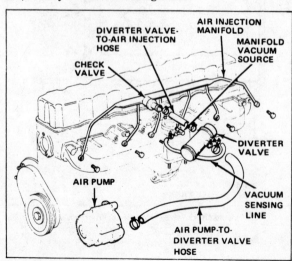

Air guard system—1979 6 cylinder engine

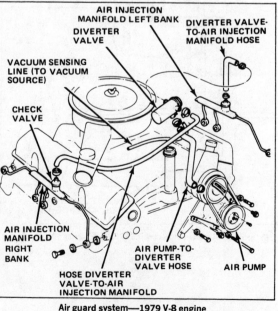

Air guard system—1979 V-8 engine

American Motors

Gasoline fumes are fed to the canister through two hoses, one from the tank and one from the carburetor. The carburetor hose connects to an external vent on the carburetor bowl. This vent is operated either by manifold vacuum or mechanical linkage, depending on the model of carburetor.

On the Model 5210 Holley-Weber carburetor, used on the 4-cylinder engine, a small diaphragm and valve are built into the bowl cover. The operating side of the diaphragm is connected to manifold vacuum, which closes the valve whenever the engine is running. If the engine is not running, a spring behind the diaphragm opens the valve and vents the bowl to the carbon canister.

Carter's YF 1-bbl. carburetor, used on the 6-cylinder engine, has an external bowl vent operated by mechanical linkage from the throttle. The vent opens only when the throttle is in the curb idle position. The vent is closed at all other throttle positions.

Because the YF bowl vent has no spring, high vapor pressure in the carburetor bowl can also open the vent. This will occur, when the throttle is above idle, if the pressure in the bowl goes above 0.14 in. of water.

Carter's 2-bbl. BBD model, used on the 6-cylinder, has two external bowl vents built into the bowl cover. The upper vent, with the plastic hose fitting, is pressure operated, and opens only if there is high pressure in the bowl. The lower vent, with the metal hose fitting, is operated by the throttle linkage, and opens only with the throttle in the curb idle position. Both of these vents are connected by short hoses to a "T", and then a single hose goes to the canister.

The Motorcraft 2100 2-bbl. used on the V-8 engines has a mechanical external bowl vent operated by the accelerating pump lever. The vent is a plunger type, with a small spring that holds it in the open position. Whenever the throttle is above idle, a lever with a stronger spring overrides the plunger spring and closes the vent. At curb idle, the lever lifts off the plunger, which opens the vent.

External vents not only help prevent air pollution, but also keep gasoline vapors from entering the intake manifold through the internal vent that most carburetors have. This helps avoid hard starting after a car has been parked in the hot sun.

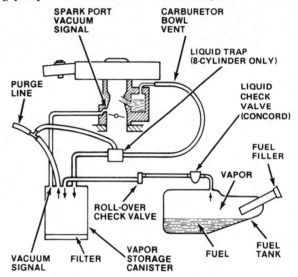

Fuel vapor control system—Typical (© American Motors)

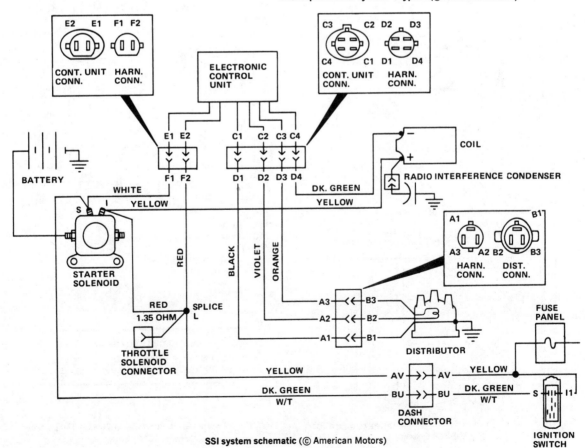

SSI system schematic (© American Motors)

25

American Motors

VACUUM CIRCUITS

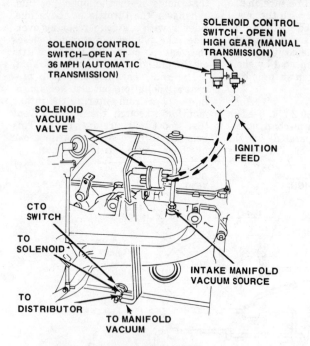

1976 Calif. 6-Cylinder System

1976 Calif. V-8 TCS System

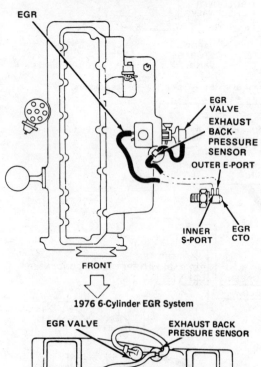

1976 6-Cylinder EGR System

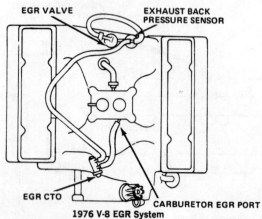

1976 V-8 EGR System

OPERATING CHARACTERISTICS
A. BELOW 160°F - MANIFOLD VACUUM
B. ABOVE 160°F - CARBURETOR PORTED

1976 49-State Distributor Vacuum Coolant Temperature Override (CTO) Switch

American Motors

VACUUM CIRCUITS

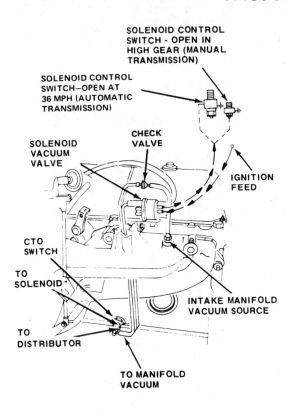

1977 Six cylinder vacuum spark control check valve—California
(© American Motors)

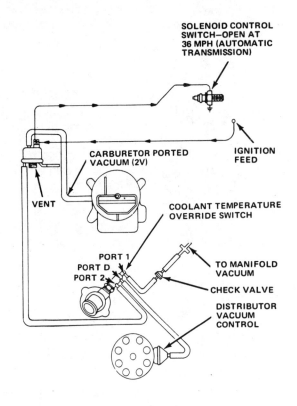

1977 V-8 vacuum spark control check valve—California
(© American Motors)

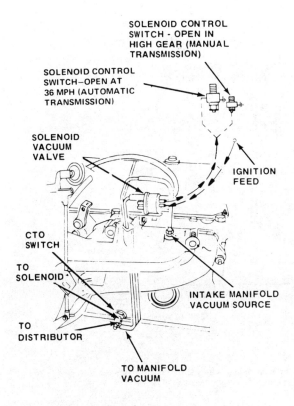

1977 Six cylinder TCS system (© American Motors)

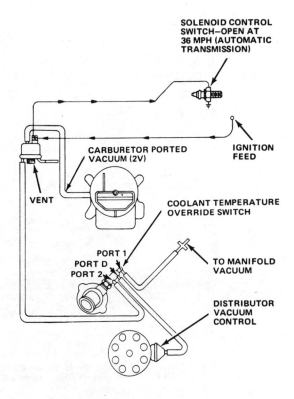

1977 V-8 engine TCS system (© American Motors)

American Motors

VACUUM CIRCUITS

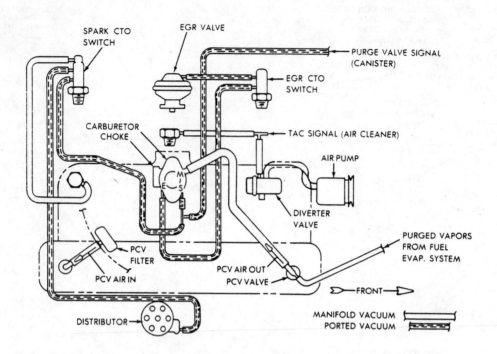

1978-79 Vacuum hose schematic—49 States—232 CID 6 cylinder engine with manual/automatic transmission

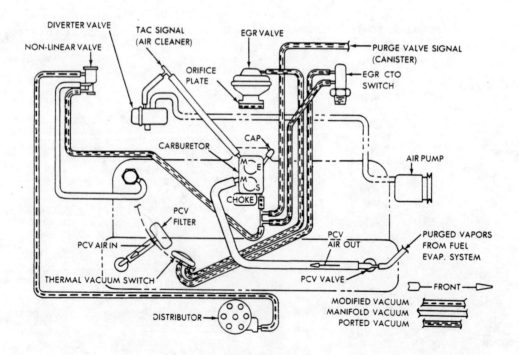

1978-79 Vacuum hose schematic—49 States—258 CID 6 cylinder engine with manual transmission (EGR valve orifice plate is deleted on the 258 CID engine with automatic transmission)

VACUUM CIRCUITS

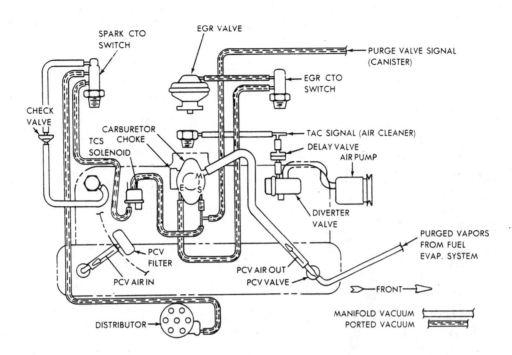

1978-79 Vacuum hose schematic—California 258 CID 6 cylinder engine with automatic transmission

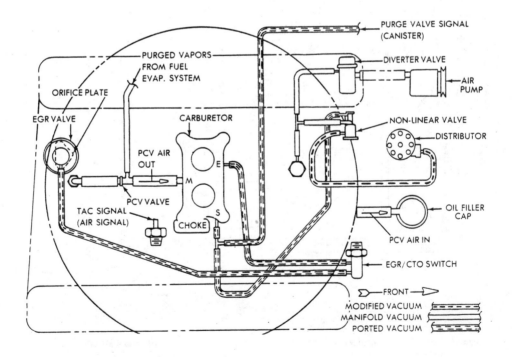

1978-79 Vacuum hose schematic—49 States—304 CID V-8 engine with automatic transmission (EGR valve orifice plate and the non-linear valve are deleted on the 304 CID V-8 engine with manual transmission)

American Motors

VACUUM CIRCUITS

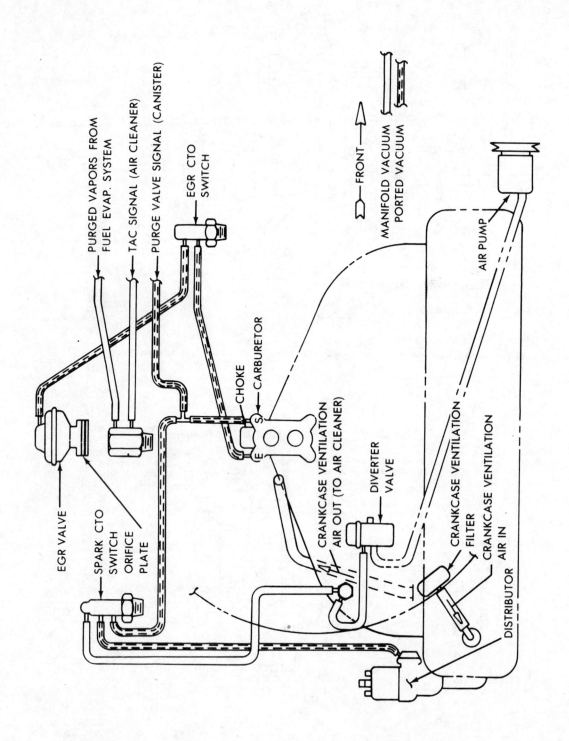

1978-79 Vacuum hose schematic—49 States—121 CID 4 cylinder engine with manual/automatic transmission

Buick

Full-Size • Intermediate • Compact

BUICK—ALL MODELS EXCEPT SKYHAWK

TUNE-UP SPECIFICATIONS
Engine Code, 5th character of the VIN number
Model Year Code, 6th character of the VIN number

Yr.	Eng. V.I.N. Code	Eng. No. Cyl. Disp. (cu in)	Eng.② Mfg.	Carb Bbl	H.P.	SPARK PLUGS Orig. Type	Gap (in)	DIST. Point Dwell (deg)	DIST. Point Gap (in)	IGNITION TIMING (deg BTDC) Man Trans	IGNITION TIMING (deg BTDC) Auto Trans	VALVES Intake Opens (deg BTDC)•■	FUEL PUMP Pres. (psi)	IDLE SPEED (rpm)① Man Trans	IDLE SPEED (rpm)① Auto Trans In Drive
'75	C	6-231	B	2	110	R44SX	.060	E.I.	—	12	12	17	3	800/600	700/500
	D	6-250	C	1	105	R46TX	.060	E.I.	—	10	10	14	4-5	850/425	550/425
	F	8-260	O	2	110	R46SX	.080	E.I.	—	18 (14)③	18 (14)③	22	5.5-6.5	—	550/600
	H	8-350	B	2	145	R45TSX	.060	E.I.	—	—	12	19	3	—	600
	J	8-350	B	4	165	R45TSX	.060	E.I.	—	—	12	19	3	—	600
	S	8-400	P	4	185	R45TSX	.060	E.I.	—	—	16	23	5.5-6.5	—	650
	T	8-455	B	4	205	R45TSX	.060	E.I.	—	—	12	10	4.5	—	600
'76	C	6-231	B	2	105	R445X	.060	E.I.	—	12	12	17	3	800/600	600
	F	8-260	O	2	110	R46SX	.080	E.I.	—	—	18 (14)③	14	5.5-6.5	—	550/600
	H	8-350	B	2	140	R45TSX	.060	E.I.	—	—	12	13.5	3	—	600
	J	8-350	B	4	155	R45TSX	.060	E.I.	—	—	12	13½	3	—	600
	T	8-455	B	4	205	R45TSX	.060	E.I.	—	—	12	10	4.5	—	600
'77	C	6-231	B	2	105	R46TSX	.060	E.I.	—	12	12	17	3	800	600
	Y	8-301	P	2	135	R46TSX	.060	E.I.	—	—	12	27	7-8.5	—	550
	U	8-305	C	2	145	R45TS	.045	E.I.	—	—	8	28	7.5-9	—	500
	H	8-350	B	2	140	R46TSX	.060	E.I.	—	—	12	13.5	3	—	600
	J	8-350	B	4	155	R46TSX	.060	E.I.	—	—	12	13.5	3	—	550
	L	8-350	C	4	170	R45TS	.045	E.I.	—	—	8	28	7.5-9	—	550④
	R	8-350	O	4	170	R46SZ	.060	E.I.	—	—	20⑤③	22	5.5-6.5	—	550④
	K	8-403	O	4	185	R46SZ	.060	E.I.	—	—	24⑥③	13.5	5.5-6.5	—	550④
'78	A	6-231	B	2	105	R46TSX	.060	E.I.	—	15	15	17	3	800/600	670/600
	C	6-196	B	2	95	R46TSX	.060	E.I.	—	15	15	18	3	600	600
	G	6-231	B	2	150	R46TSX	.060	E.I.	—	—	15	17	5	—	650
	3	6-231	B	4	165	R46TSX	.060	E.I.	—	—	15	17	5	—	650
	H	8-305	C	4	160	R45TS	.045	E.I.	—	—	4	28	7.5-9	—	600/500
	K	8-403	O	4	185	R46SZ	.060	E.I.	—	—	20③	16	5.5-6.5	—	650/550)
	L	8-350	C	4	170	R45TS	.045	E.I.	—	—	8	28	7.5-9	—	500
	X	8-350	B	4	155	R46TSX	.060	E.I.	—	—	15	13.5	3	—	550
	U	8-305	C	2	145	R45TS	.045	E.I.	—	—	4	28	7.5-9	—	500
	R	8-350	O	4	170	R46SZ	.060	E.I.	—	—	20③	16	5.5-6.5	—	550
	Y	8-301	P	2	165	R46TSX	.060	E.I.	—	—	12	27	7.0-8.5	—	650/550
'79	A	6-231	B	2	105*	R46TSX	.060	E.I.	—	⑦	⑦	17	3	800/600	670/600
	2	6-231	B	2	—	R46TSX	.060	E.I.	—	⑦	⑦	⑧	⑧	⑧	
	C	6-196	B	2	95*	R46TSX	.060	E.I.	—	⑦	⑦	18	3	600	600
	3	6-231	B	4	165*	R46TSX	.060	E.I.	—	⑦	⑦	17	5	—	650
	H	8-305	C	4	160*	R45TS	.045	E.I.	—	⑦	⑦	28	7.5-9	—	600/500
	K	8-403	O	4	185*	R46SZ	.060	E.I.	—	⑦	⑦	16	5.5-6.5	—	550 (500)
	L	8-350	C	4	170*	R45TS	.045	E.I.	—	⑦	⑦	28	7.5-9	—	500
	X	8-350	B	4	155*	R46TSX	.060	E.I.	—	⑦	⑦	13.5	3	—	550
	G	8-305	C	2	145*	R45TS	.045	E.I.	—	⑦	⑦	28	7.5-9	—	500
	R	8-350	O	4	170*	R46SZ	.060	E.I.	—	⑦	⑦	16	5.5-6.5	—	550
	Y	8-301	P	2	165*	R46TSX	.060	E.I.	—	⑦	⑦	27	7.0-8.5	—	650/550

Should the information in this manual deviate from the specifications on the underhood tune-up decal the decal specifications should be used as they may reflect production changes.

Buick

- Where more than one figure appears, the figure in parentheses is for California applications.
* Chilton Estimate
① LOWER FIGURE INDICATES IDLE SPEED WITH SOLENOID DISCONNECTED
② B-BUICK
  C-CHEVROLET
  O-OLDSMOBILE
  P-PONTIAC
③ AT 1100 RPM.
④ 600 ON HIGH ALTITUDE APPLICATIONS.
⑤ 18 ON SKYLARK AND LESABRE.
⑥ 20 ON CALIFORNIA AND HIGH ALTITUDE APPLICATIONS.
⑦ SEE UNDERHOOD TUNE-UP DECAL.

Skyhawk

TUNE-UP SPECIFICATIONS

Engine Code, 5th character of the VIN number
Model Year Code, 6th character of the VIN number

Yr.	Eng. V.I.N. Code	Engine No. Cyl. Disp. (cu in)	Eng.■ Mfg.	Carb Bbl	H.P.	SPARK PLUGS Orig. Type	Gap (in)	DIST. Point Dwell (deg)	Point Gap (in)	IGNITION TIMING (deg BTDC) Man	Auto	VALVES Intake Opens (deg BTDC)	FUEL Pump Pres. (psi)	IDLE SPEED (rpm) Man Trans	Auto Trans In Drive
75	A	4-140	C	1v	75	R43TSX	.035	Electronic		8 @ 700	10	22	3.0-4.5	1200	750
	B	4-140	C	2v	85	R43TSX	.060	Electronic		10 @ 700	12	28	3.0-4.5	700	750
	C	6-231	B	2v	110	R44SX	.060	Electronic		12	12	17	3②	800	700
	F	8-260	O	2v	110	R46SX	.080	Electronic		16 @ 1100*	⑦	14	6	750	550⑩
	G	8-262	C	2v	110	R44TX	.060	Electronic		8	8	26	7.0-8.0	800	600
76	O	4-122	C	FI	120	R43T8X	.035	Electronic		12	—	38	—	1600	—
	A	4-140	C	1v	75	R43TS	.035	Electronic		8	10	22	3.0-4.5	1200	750
	B	4-140	C	2v	85	R43TS	.035	Electronic		10	12	28	3.0-4.5	700	750
	C	6-231	B	2v	110	R44SX	.060	Electronic		12	12	17	3②	800	600
	F	8-260	O	2v	110	R46SX	.060	Electronic		16 @ 1100	⑦	14	5.5-6.5	750	550
	G	8-262	C	2v	110	R45TS	.045	Electronic		6	8	26	7.0-8.0	800	600
	Q	8-305	C	2v	140	R45TS	.045	Electronic		6	8	26	7.0-8.5	800	600
77	B	4-140	C	2v	84	R43TS	.035	Electronic		TDC @ 700⑫	2	34	3.0-4.5	1250③⑨	850③⑪
	V	4-151	P	2v	87	R44TSX	.060	Electronic		14 @ 1000*	14(12)	33	4.5-5.0	500	650
	C	6-231	B	2v	105	R46SX	.040	Electronic		12 @ 600	12	17	3②	600	550
	U	8-305	C	2v	145	R45TS	.045	Electronic		8	8	28	7.0-8.5	600	500
78	1	4-151	P	2v	85	R43TSX	.060	Electronic		14	14⑯	33	4-5.5	⑤	⑤
	V	4-151	P	2v	85	R43TSX	.060	Electronic		14	14	33	4-5.5	⑤	⑤
	C	6-196	B	2v	90	R46TSX	.060	Electronic		15	15	18	3-4.5⑬	⑤	⑤
	A	6-231	B	2v	105	R46TSX	.060	Electronic		15	15	17	3-4.5⑬	800	600
	U	8-305	C	2v	145	R45TS	.045	Electronic		4	6⑰	28	7.5-9.0	600	500⑮
79	1	4-151	P	2v	85	R43TSX	.060	Electronic		12	12	33	4-5.5	⑤	⑤
	V	4-151	P	2v	85	R43TSX	.060	Electronic		14	14	33	4-5.5	1000⑭	650⑭
	9	4-151	P	2v	85	R43TSX	.060	Electronic		14	14	33	4-5.5	⑤	⑤
	C	6-196	B	2v	90	R46TSX	.060	Electronic		⑤	⑤	18	3-4.5⑬	⑤	⑤
	A	6-231	B	2v	115	R46TSX	.060	Electronic		15	15	17	3-4.5⑬	800	600
	2	6-231	B	2v	115	R46TSX	.060	Electronic		15	15	17	3-4.5⑬	800	600
	G	8-305	C	2v	145	R45TS	.045	Electronic		4	4(4)	28	7.5-9.0	600	500⑮

Should the information provided in this manual deviate from the specifications on the underhood tune-up label, the label specifications should be used, as they may reflect production changes.

① THERE ARE TWO VERSIONS OF THE THIS ENGINE IN 1978, ONE FEATURES A SPECIAL CLOSED-LOOP EMISSION SYSTEM. THIS ENGINE CAN BE IDENTIFIED BY THE DEALER ORDER NUMBER UPC L36.
② 1975 SKYHAWK 3-4 PSI 1976-1977 SKYHAWK 3.-0-45 PSI.
• Figures in parenthesis are California specs.
③ ADJUST SPEEDUP SOLENOID TO OBTAIN SPECIFIED RPM, WITH AIR CONDITIONING UNIT TURNED ON AND COMPRESSOR CLUTCH WIRE DISCONNECTED.
④ AT 700 OR LESS RPM
⑤ SEE UNDERHOOD TUNE-UP DECAL.
⑥ –34
⑦ (14 @ 1100)
⑧ (1200)
⑨ (800)
⑩ (600)
⑪ (700)
⑫ 2ATDC @ 700
* Timing is with both automatic and manual transmissions.
⑬ AT 12.6 VOLTS
⑭ BASE IDLE 500 RPM.
⑮ HIGH ALT. BASE IDLE IS 600 RPM
⑯ WITH EGR VALVE: 12° WITHOUT EGR VALVE
⑰ HIGH ALT. IS 8°
■ B–Buick
  C–Chevrolet
  O–Oldsmobile
  P–Pontiac

Buick

Full-Size • Intermediate • Compact

BUICK—ALL MODELS EXCEPT SKYHAWK

DISTRIBUTOR SPECIFICATIONS

Year	Engine or Distributor Identification	Centrifugal Advance Start Dist. Deg. @ Dist. RPM	Centrifugal Advance Finish Dist. Deg. @ Dist. RPM	Vacuum Advance Start @ In. Hg.	Vacuum Advance Finish Dist. Deg. @ In. Hg.
'75	6-231	0 @ 500	8 @ 2,050	5-7	9 @ 10
	6-250	0 @ 500	8 @ 2,100	4	9 @ 12
	8-260	0 @ 325	14 @ 2,200	4	12 @ 15
	8-350	0 @ 375	5-7 @ 2,250	6.5-8.5	8 @ 11.5
	8-400	0 @ 600	8 @ 2,200	6-8	12.5 @ 13.5
	8-455	0 @ 375	7-9 @ 2,200	4-6	10 @ 11
'76	6-231	0 @ 770	9 @ 2,500	5.3	12.75 @ 12.8
	8-260	0 @ 325	14 @ 2,200	4.5	15 @ 11
	8-260①	0 @ 455	13 @ 2,230	6	10 @ 14.75
	8-350	0 @ 872	11 @ 2,500	6.9	11.75 @ 14.3
	8-455	0 @ 660	9 @ 2,500	4.5	12.75 @ 14.1
	8-455①	0 @ 660	9 @ 2,500	4.5	9.75 @ 12
'77	1103239	0 @ 600	10 @ 2,200		②
	1103244	0 @ 500	10 @ 1,900		③
	1103246	0 @ 600	11 @ 2,100	4	9 @ 12
	1103248	0 @ 600	10-12 @ 2,100	3-5	4-6 @ 7-9
	1103252	0 @ 500	10 @ 1,900	4	9 @ 12
	1103259	0 @ 500	9.5 @ 2,000	6	12 @ 13
	1103260	0 @ 500	6.5 @ 1,800	6	12 @ 13
	1103264	0 @ 500	6.5 @ 1,800	5	8 @ 11
	1103266	0 @ 500	9.5 @ 2,000	5	8 @ 11
	1103272	0 @ 415	10.7 @ 1,715	4	12.5 @ 12
	1103273	0 @ 500	9.5 @ 1,800	4	12.5 @ 12
	1110677	0 @ 700	10 @ 1,800	4	12 @ 11
	1110686	0 @ 700	10 @ 1,800	7	4 @ 9
	1110694	0 @ 760	8-11 @ 2,500	3-4	15 @ 20
'78	1110695	0 @ 840	6-9 @ 1,800	3-6	12 @ 10-13
	1110723	0 @ ④	4-6 @ 2,000	4-6	4 @ 5-7
	1110728	0 @ ④	7-9 @ 1,600	1-5	10 @ 4-8
	1110730	0 @ ④	9.5-11.5 @ 2,000	4-6	12 @ 7-10
	1110731	0 @ 840	6.5-8.5 @ 1,800	4-6	8 @ 7-9
	1110732	0 @ 840	6.5-8.5 @ 1,800	7-9	7 @ 11-13
	1110735	0 @ 445-620	9.5-11.5 @ 2,000	2-4	10 @ 11-13
	1110739	0 @ ④	6.5-8.5 @ 1,800	2-4	10 @ 11-13
	11103285	0 @ 600	10-12 @ 2,100	3-5	5 @ 7-9
	1103281	0 @ 500	9-11 @ 1,900	3-5	9 @ 11-13
	1103282	0 @ 500	9-11 @ 1,900	3-6	10 @ 9-12
	1103314	0 @ 413	9.5-11.5 @ 1,700	3-5	12.5 @ 11-13
	1103322	0 @ 300	13.5-15.5 @ 2,000	4-6	12 @ 12-14
	1103323	0 @ 500	8.5-10.5 @ 2,000	4-5	8 @ 10-12
	1103324	0 @ 300	10.5-12.5 @ 1,800	4-6	12 @ 12-14
	1103325	0 @ 500	6.5-7.5 @ 1,800	4-5	8 @ 10-12
	1103342	0 @ 950	7.5-9.5 @ 2,200	5-7	12 @ 11-13
	1103346	0 @ 500	8.5-10.5 @ 2,000	4-6	12 @ 12-14
	1103347	0 @ 500	6.5-7.5 @ 1,800	4-6	12 @ 12-14
	1103353	0 @ 1100	11-12 @ 2,250	3-6	10 @ 9-12

Buick

Full-Size • Intermediate • Compact

BUICK—ALL MODELS EXCEPT SKYHAWK

DISTRIBUTOR SPECIFICATIONS

Year	Engine or Distributor Identification	Centrifugal Advance		Vacuum Advance	
		Start Dist. Deg. @ Dist. RPM	Finish Dist. Deg. @ Dist. RPM	Start @ In. Hg.	Finish Dist. Deg. @ In. Hg.
'79	1103266	0 @ 500	10.5 @ 2,000	5	8 @ 11
	1103342	0 @ 950	8.5 @ 2,200	6.5	12 @ 14
	1110766	0 @ 840	7.5 @ 1,800	3	12 @ 18.35
	1110769	0 @ 500	7.5 @ 1,800	4	12 @ 11
	1110770	0 @ 840	7.5 @ 1,800	3	10 @ 9
	1110775	0 @ 500	7.5 @ 1,800	3	10 @ 9

① CALIFORNIA USAGE
② 0 @ 14
   7.5 @ 10
③ 0 @ 14
   10 @ 10
④ NOT AVAILABLE.

Skyhawk

DISTRIBUTOR SPECIFICATIONS — GM SMALL CARS (except Chevette)

Year	Distributor Identification	Centrifugal Advance		Vacuum Advance	
		Start Dist. Deg. @ Dist. RPM	Finish Dist. Deg. @ Dist. RPM	Start @ In. Hg.	Finish Dist. Deg. @ In. Hg.
75	1112862	0 @ 810	11 @ 2400	5	12 @ 12
	1112880	0 @ 600	11 @ 2100	4	9 @ 12
	1112862	0 @ 810	11 @ 2400	5	12 @ 12
	1112863	0 @ 550	8 @ 2100	3-5	3.25-9.75 @ 11-12.5
	1110650	0 @ 550	8 @ 2100	3-5	6.75-8.25 @ 11.5-12.
	1110651	0 @ 540	8 @ 2050	5-7	9 @ 10
	1112951	0 @ 325	14 @ 2200	4	12 @ 15
	1112956	0 @ 325	14 @ 2200	NONE	NONE
76	1112862	0 @ 810	11 @ 2400	5	12 @ 12
	1112983	0 @ 600	11 @ 2000	4	7.5 @ 10
	1112862	0 @ 450	11 @ 2400	5	10 @ 14
	1110668	0 @ 638	8 @ 1600	6	12 @ 12
	1110661	0 @ 525	8 @ 2050	6	9 @ 10
	1112863	0 @ 550	8 @ 2100	3-5	12 @ 14.5
	1112863	0 @ 550	8 @ 2100	4	9 @ 12
	1110666	0 @ 500	10 @ 2100	4	12 @ 15
	1110668	0 @ 638	8 @ 1588	6	12 @ 11.5
	1110661	0 @ 530	8 @ 2050	6	9 @ 10
	1112994	0 @ 325	14 @ 2200	4.5	12 @ 10.5
	1112995	0 @ 455	13 @ 2233	4	15 @ 11
	1112996	0 @ 550	14 @ 2200	—	—
77	1110538	0 @ 425	85 @ 1000	5	12 @ 10
	1110539	0 @ 425	8.5 @ 1000	5	12 @ 10
	1103239	0 @ 600	10 @ 2100	14	7.5 @ 10
	1103244	0 @ 500	10 @ 1900	4	10 @ 10
	1103252	0 @ 500	10 @ 1900	4	9 @ 12

Buick

Skyhawk

DISTRIBUTOR SPECIFICATIONS

Year	Distributor Identification	Centrifugal Advance		Vacuum Advance	
		Start Dist. Deg. @ Dist. RPM	Finish Dist. Deg. @ Dist. RPM	Start @ In. Hg.	Finish Dist. Deg. @ In. Hg.
'77	1103229	0 @ 600	10 @ 2200	3.5	10 @ 12
	1103263	0 @ 600	10 @ 2200	3.5	10 @ 9
	1103231	0 @ 600	10 @ 2200	3.5	10 @ 12
	1103230	0 @ 600	10 @ 2200	3.5	10 @ 9
	1103303	0 @ 600	10 @ 2200	9	10 @ 16
	1110677	0 @ 700	10 @ 1800	4	12 @ 11
	1110686	0 @ 700	10 @ 1800	7	4 @ 9
78	1103281	0 @ 500	10 @ 1900	4	9 @ 12
	1103282	0 @ 500	10 @ 1900	4	10 @ 10
	1103326	0 @ 850	10 @ 2325	3.4	10 @ 10.7
	1103328	0 @ 600	10 @ 2200	3.5	10 @ 9
	1103329	0 @ 600	10 @ 2200	3.5	10 @ 12
	1103365	0 @ 850	10 @ 2325	5.5	7 @ 9.5
	1110695	0-2 @ 1000	6-9 @ 1800	6	8 @ 9
	1110731	0-2 @ 1000	6-9 @ 1800	6	8 @ 9
	1110732	0-2 @ 1000	6-9 @ 1800	9	7 @ 13
79	1103229	0 @ 600	10 @ 2200	3.5	10 @ 12
	1103231	0 @ 600	10 @ 2200	3.5	10 @ 12
	1103239	0 @ 600	10 @ 2100	4	5 @ 8
	1103244	0 @ 500	10 @ 1900	4	10 @ 10
	1103282	0 @ 500	10 @ 1900	4	10 @ 10
	1103285	0 @ 600	11 @ 2100	4	5 @ 8
	1103365	0 @ 850	10 @ 2325	5	8 @ 11
	1110726	0 @ 500	9 @ 2000	4	10 @ 10
	1110757	0 @ 600	9 @ 2000	4	10 @ 10
	1110766	0 @ 810	7.5 @ 1800	3	10 @ 9
	1110767	0 @ 840	7.5 @ 1800	4	12 @ 11

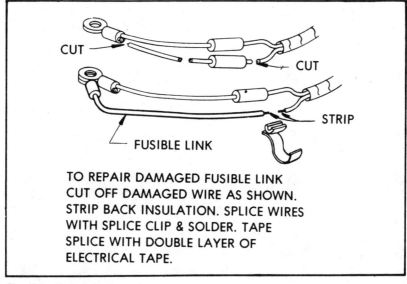

Fusible link repair

Buick

MV, 1MV CARBURETOR SPECIFICATIONS
BUICK

Year	Carburetor Identification①	Float Level (in.)	Metering Rod (in.)	Pump Rod	Idle Vent (in.)	Vacuum Break (in.)	Auxiliary Vacuum Break (in.)	Fast Idle Off Car (in.)	Choke Rod (in.)	Choke Unloader (in.)	Fast Idle Speed (rpm)
1975	7045012	11/32	0.080	—	—	0.200	0.215	—	0.160	0.275	1700②
	7045013	11/32	0.080	—	—	0.350	0.312	—	0.275	0.275	1800②
	7045314	11/32	0.080	—	—	0.275	0.312	—	0.230	0.275	1700②

① The Carburetor identification number is stamped on the float bowl, next to the fuel inlet nut.
② In Neutral or Park

2GC, 2GV, 2GE CARBURETOR SPECIFICATIONS
BUICK

Year	Carburetor Identification①	Float Level (in.)	Float Drop (in.)	Pump Rod (in.)	Idle Vent (in.)	Primary Vacuum Break (in.)	Secondary Vacuum Break (in.)	Automatic Choke (notches)	Choke Rod (in.)	Choke Unloader (in.)	Fast Idle Speed (rpm)
1975	7045145	15/32	1 9/32	1 15/32	—	0.120	0.120	Index	0.080	0.120	—
	7045146	15/32	1 9/32	1 15/32	—	0.120	0.120	—	0.080	0.120	—
	7045147	15/32	1 9/32	1 15/32	—	0.120	0.120	1 Lean	0.080	0.120	—
	7045148	15/32	1 9/32	1 15/32	—	0.120	0.120	1 Rich	0.080	0.120	—
	7045149	15/32	1 9/32	1 15/32	—	0.120	0.120	1 Rich	0.080	0.120	—
	7045446	15/32	1 9/32	1 15/32	—	0.120	0.120	—	0.080	0.120	—
	7045448	15/32	1 9/32	1 15/32	—	0.120	0.120	Index	0.080	0.120	—
	7045449	15/32	1 9/32	1 15/32	—	0.120	0.120	1 Lean	0.080	0.120	—
	7045143	15/32	1 9/32	1 15/32	—	0.140	0.120	1 Rich	0.080	0.140	—
	7045140	15/32	1 9/32	1 15/32	—	0.140	0.120	1 Rich	0.080	0.140	—
1976	17056447	7/16	1 9/32	1 19/32	—	0.130	0.100	1 Rich	0.080	0.140	—
	17056145	13/32	1 9/32	1 19/32②	—	0.110	0.100	1 Rich	0.080	0.140	—
	17056148	7/16	1 9/32	1 19/32	—	0.120	0.100	1 Rich	0.080	0.140	—
	17056149	7/16	1 9/32	1 19/32	—	0.120	0.100	1 Rich	0.800	0.140	—
	17056448	7/16	1 9/32	1 19/32	—	0.130	0.110	1 Rich	0.080	0.140	—
	17056449	7/16	1 9/32	1 19/32	—	0.130	0.110	1 Rich	0.080	0.140	—
	17056143	15/32	1 9/32	1 19/32	—	0.140	0.100	1 Rich	0.080	0.180	—
	17056140	15/32	1 9/32	1 19/32	—	0.140	0.100	1 Rich	0.080	0.180	—
1977	17057140	15/32	1 5/32	1 9/16	—	0.140	0.100	1 Rich	0.080	0.180	—
	17057141, 17057145, 17057147	7/16	1 5/32	1 1/2	—	0.110	0.040	1 Rich	0.080	0.140	—
	17057143, 17075144	7/16	1 5/32	1 17/32	—	0.130	0.100	1 Rich	0.080	0.140	—
	17057146, 17057148	7/16	1 5/32	1 17/32	—	0.110	0.040	1 Rich	0.080	0.140	—
	17057445	7/16	1 5/32	1 1/2	—	0.140	0.100	1 Rich	0.080	0.140	—
	17057446, 17057448	7/16	1 5/32	1 1/2	—	0.130	0.110	1 Rich	0.080	0.140	—
	17057447	7/16	1 5/32	1 1/2	—	0.130	0.100	1 Rich	0.080	0.140	—
1978	17058104	15/32	1 9/32	1 21/32	—	0.160	—	Index	0.260	0.325	—
	17058105	15/32	1 9/32	1 21/32	—	0.160	—	Index	0.260	0.325	—
	17058108	19/32	1 9/32	1 21/32	—	0.160	—	Index	0.260	0.325	—
	17058110	19/32	1 9/32	1 21/32	—	0.160	—	Index	0.260	0.325	—
	17058112	19/32	1 9/32	1 21/32	—	0.160	—	Index	0.260	0.325	—
	17058114	19/32	1 9/32	1 21/32	—	0.160	—	Index	0.260	0.325	—
	17058126	19/32	1 9/32	1 17/32	—	0.150	—	Index	0.260	0.325	—

Buick

2GC, 2GV, 2GE CARBURETOR SPECIFICATIONS
BUICK

Year	Carburetor Identification①	Float Level (in.)	Float Drop (in.)	Pump Rod (in.)	Idle Vent (in.)	Primary Vacuum Break (in.)	Secondary Vacuum Break (in.)	Automatic Choke (notches)	Choke Rod (in.)	Choke Unloader (in.)	Fast Idle Speed (rpm)
	17058128	19/32	1 9/32	1 17/32	—	0.150	—	Index	0.260	0.325	—
	17058404	1/2	1 9/32	1 21/32	—	0.160	—	1/2 Lean	0.260	0.325	—
	17058405	1/2	1 9/32	1 21/32	—	0.160	—	1/2 Lean	0.260	0.325	—
	17058408	21/32	1 9/32	1 21/32	—	0.160	—	1/2 Lean	0.260	0.325	—
	17058410	21/32	1 9/32	1 21/32	—	0.160	—	1/2 Lean	0.260	0.325	—
	17058412	21/32	1 9/32	1 21/32	—	0.160	—	1/2 Lean	0.260	0.325	—
	17058414	21/32	1 9/32	1 21/32	—	0.160	—	1/2 Lean	0.260	0.325	—
	17058140	7/16	1 5/32	1 19/32	—	0.070	0.110	1 Rich	0.080	0.140	—
	17058143	7/16	1 5/32	1 9/16	—	0.080	0.110	1 Rich	0.080	0.140	—
	17058144	7/16	1 5/32	1 5/8	—	0.060	0.110	1 Rich	0.080	0.140	—
	17058145	7/16	1 5/32	1 19/32	—	0.060	0.110	1 Rich	0.080	0.160	—
	17058148	7/16	1 5/32	1 19/32	—	0.080	0.110	1 Rich	0.080	0.150	—
	17058149	7/16	1 5/32	1 19/32	—	0.080	0.110	1 Rich	0.080	0.150	—
	17058141	7/16	1 5/32	1 19/32	—	0.100	0.140	1 Rich	0.080	0.140	—
	17058147	7/16	1 5/32	1 19/32	—	0.100	0.140	1 Rich	0.080	0.140	—
	17058182	7/16	1 5/32	1 19/32	—	0.080	0.110	1 Rich	0.080	0.140	—
	17058183	7/16	1 5/32	1 19/32	—	0.080	0.110	1 Rich	0.080	0.140	—
	17058444	7/16	1 5/32	1 19/32	—	0.100	0.140	1 Rich	0.080	0.140	—
	17058446	7/16	1 5/32	1 19/32	—	0.110	0.130	1 Rich	0.080	0.140	—
	17058447	7/16	1 5/32	1 19/32	—	0.110	0.150	1 Rich	0.080	0.140	—
	17058448	7/16	1 5/32	1 9/16	—	0.100	0.140	1 Rich	0.080	0.140	—
	17058185	7/16	1 5/32	1 19/32	—	0.050	0.110	1 Rich	0.080	0.140	—
	17058187	7/16	1 5/32	1 19/32	—	0.050	0.110	1 Rich	0.080	0.140	—
	17058189	7/16	1 5/32	1 19/32	—	0.080	0.110	1 Rich	0.080	0.140	—
	17058188	7/16	1 5/32	1 5/8	—	0.050	0.120	1 Rich	0.080	0.140	—

① The carburetor identification number is stamped on the float bowl, next to the fuel inlet nut.
② 1¾ in. on Skyhawk.

2MC, M2MC, M2ME CARBURETOR SPECIFICATIONS
BUICK

Year	Carburetor Identification①	Float Level (in.)	Choke Rod (in.)	Choke Unloader (in.)	Vacuum Break Lean or Front (in.)	Vacuum Break Rich or Rear (in.)	Pump Rod (in.)	Choke Coil Lever (in.)	Automatic Choke (notches)
1975	7045156	5/32	0.130	0.285	0.235	0.150	9/32②	0.120	1 Rich
	7045248	5/32	0.130	0.285	0.235	0.150	9/32②	0.120	1 Rich
	7045358	3/16	0.130	0.285	0.300	0.150	5/16③	0.120	1 Rich
	7045354	3/16	0.130	0.285	0.300	0.150	5/16③	0.120	1 Rich
1976	17056156	1/8	0.105	0.210	0.175	0.110	9/32②	0.120	1 Rich
	17056158	1/8	0.105	0.210	0.175	0.110	9/32②	0.120	1 Rich
	17056458	1/8	0.105	0.210	0.175	0.110	3/16③	0.120	1 Rich
	17056454	1/8	0.105	0.210	0.175	0.110	3/16③	0.120	1 Rich
1977	17057172	11/32	0.075	0.240	0.135	0.240	3/8③	0.120	2 Rich
	17057173	11/32	0.075	0.240	0.165	0.240	3/8③	0.120	2 Rich
1978	17058160	11/32	0.133	0.220	0.149	0.227	1/4③	0.120	2 Lean
	17058192	1/4	0.074	0.350	0.117	0.103	9/32②	0.120	1 Rich
	17058496	1/4	0.077	0.243	0.136	0.211	3/8③	0.120	1 Rich

37

Buick

QUADRAJET CARBURETOR SPECIFICATIONS
BUICK

Year	Carburetor Identification①	Float Level (in.)	Air Valve Spring (turn)	Pump Rod (in.)	Primary Vacuum Break (in.)	Secondary Vacuum Break (in.)	Secondary Opening (in.)	Choke Rod (in.)	Choke Unloader (in.)	Fast Idle Speed ④ (rpm)
1975	7045240	7/16	7/16	9/32	0.135	0.120	②	0.095	0.240	1800
	7045548	7/16	7/16	9/32	0.135	0.120	②	0.095	0.240	1800
	7045244	5/16	3/4	15/32	0.130	0.115	②	0.095	0.240	1800
	7045246	5/16	3/4	15/32	0.130	0.115	②	0.095	0.240	1800
	7045544	5/16	3/4	15/32	0.145	0.130	②	0.095	0.240	1800
	7045546	5/16	3/4	15/32	0.145	0.130	②	0.095	0.240	1800
1976	17056240	15/32	7/16	3/8	0.135	0.120	②	0.095	0.250	1800
	17056540	15/32	7/16	3/8	0.135	0.120	②	0.095	0.250	1800
	17056244	5/16	3/4	3/8	0.130	0.120	②	0.095	0.250	1800
	17056246	5/16	3/4	3/8	0.130	0.120	②	0.095	0.250	1800
	17056544	5/16	3/4	3/8	0.130	0.130	②	0.095	0.250	1800
	17056546	5/16	3/4	3/8	0.130	0.130	②	0.095	0.250	1800
1977	17057241	5/16	3/4	3/8	0.120	0.105	②	0.095	0.240	⑤
	17057250, 17057253, 17057255, 17057256	13/32	1/2	9/32	0.120	0.170	②	0.095	0.205	⑤
	17057258	13/32	1/2	9/32	0.125	0.215	②	0.095	0.205	⑤
	17057550, 17057553	13/32	1/2	9/32	0.125	0.215	②	0.095	0.200	⑤
1978	17058240	7/32	3/4	9/32	0.117	0.117	②	0.074	0.243	⑤
	17058241	5/16	3/4	3/8	0.120	0.103	②	0.096	0.243	⑤
	17058250	13/32	1/2	9/32	0.129	0.183	②	0.096	0.220	⑤
	17058253	13/32	1/2	9/32	0.129	0.183	②	0.096	0.220	⑤
	17058254	15/32	1/2	9/32	0.136	—	②	0.103	0.220	⑤
	17058257	13/32	1/2	9/32	0.136	0.231	②	0.103	0.220	⑤
	17058258	13/32	1/2	9/32	0.136	0.231	②	0.103	0.220	⑤
	17058259	13/32	1/2	9/32	0.136	0.231	②	0.103	0.220	⑤
	17058582	15/32	7/8	9/32	0.179	—	②	0.314	0.277	⑤
	17058584	15/32	7/8	9/32	0.179	—	②	0.314	0.277	⑤
	17058282	15/32	7/8	9/32	0.157	—	②	0.314	0.277	⑤
	17058284	15/32	7/8	9/32	0.157	—	②	0.314	0.277	⑤
	17058228	15/32	1	9/32	0.179	—	②	0.314	0.277	⑤
	17058502	15/32	7/8	9/32	0.164	—	②	0.314	0.277	⑤
	17058504	15/32	7/8	9/32	0.164	—	②	0.314	0.277	⑤
	17058202	15/32	7/8	9/32	0.157	—	②	0.314	0.277	⑤
	17058204	15/32	7/8	9/32	0.157	—	②	0.314	0.277	⑤
	17058540	7/32	3/4	9/32	0.117	0.117	②	0.074	0.243	⑤
	17058550	13/32	1/2	9/32	0.136	0.231	②	0.103	0.220	⑤
	17058553	15/32	1/2	9/32	0.129	0.231	②	0.096	0.220	⑤
	17058559	15/32	1/2	9/32	0.136	0.231	②	0.096	0.231	⑤

① The carburetor identification number is stamped on the float bowl, near the secondary throttle lever.
② No measurement necessary on two point linkage;
③ Manual/Automatic
④ On low step of cam, automatic in Drive through 1974; on high step of cam, automatic in Park starting 1975.
⑤ 3 turns after contacting lever for preliminary setting

Buick

CAR SERIAL NUMBER AND ENGINE IDENTIFICATION

1976

Mounted behind the windshield on the driver's side is a plate with the vehicle identification number. The sixth character is the model year, with 6 for 1976. The fifth character is the engine code, as follows.

C 231 2-bbl. V-6
F 260 2-bbl. V-8
H 350 2-bbl. V-8
J 350 4-bbl, V-8
T 455 4-bbl. V-8

The V-6, 350 V-8 and 455 V-8 are Buick engines, with the distributor in the front. The 260 V-8 is an Oldsmobile engine, which has the distributor in the rear. The inline 6-cylinder engine is not used in 1976 Buicks.

Engines can be identified by a two-letter code. The V-6 code is on the top surface of the block, near the first spark plug on the right side. On the 260 V-8, the code is on the oil filler tube. On the 350 V-8, the code is near No. 1 spark plug, on the top surface of the block. The 455 V-8 has the code on the top surface of the block near No. 5 plug.

231 2-bbl. V-6

1977

The vehicle identification number is mounted on a plate behind the windshield on the driver's side. The sixth character is the model year, with 7 for 1977. The fifth character is the engine code, as follows.

A ...... 231 2-bbl. V-6 LD-5 Buick
C ...... 231 2-bbl. V-6 LD-7 Buick
H ...... 350 2-bbl. V-8 L-32 Buick
J ...... 350 4-bbl. V-8 L-77 Buick
K ...... 403 4-bbl. V-8 L-80 Olds.
L ...... 350 4-bbl. V-8 LM-1 Chev.
R ...... 350 4-bbl. V-8 L-34 Olds.
U ...... 305 2-bbl. V-8 LG-3 Chev.
Y ...... 301 2-bbl. V-8 L-27 Pont.

1978

Behind the windshield on the driver's side is a plate with the vehicle identification number. The sixth character is the model year, with 8 for 1978. The fifth character is the engine code, as follows:

2 .... 231 2-bbl. V-6 LC-6 Buick
3 .... 231 4-bbl. V-6 LC-8 Buick
A .... 231 2-bbl. V-6 LD-5 Buick
C .... 196 2-bbl. V-6 LC-9 Buick
G .... 231 2-bbl. V-6 LC-5 Buick
H .... 305 4-bbl. V-8 LG-4 Chev.
K .... 403 4-bbl. V-8 L-80 Olds.
L .... 350 4-bbl. V-8 LM-1 Chev.
R .... 350 4-bbl. V-8 L-34 Olds.
U .... 305 2-bbl. V-8 LG-3 Chev.
X .... 305 4-bbl. V-8 L-77 Buick
Y .... 301 2-bbl. V-8 L-27 Pont.

NOTE: *Codes "3" and "G" are the turbocharged engines. Code "A" is the standard 2 bbl/V-6 engine. Code "2" is the 2 bbl/V-6 engine equipped with the electronic fuel control system.*

1979

The vehicle identification number is stamped on a plate which is attached to the top left side of the instrument panel. It is visible through the lower left hand corner of the windshield. The sixth character designates the model year; 9 represents 1979. The fifth character designates the engine code, as follows:

A .... 3.8L (231 CID)
  2 bbl V-6 Buick
2 .... 3.8L (231 CID)
  2 bbl V-6 Buick
3 (turbocharged) .. 3.8L (231 CID)
  4 bbl V-6 Buick
C .... 3.2L (196 CID)
  2 bbl V-6 Buick
G (turbocharged) .. 5.0L (305 CID)
  2 bbl V-8 Buick
H .... 5.0L (305 CID)
  4 bbl V-8 Chevrolet
K .... 6.5L (403 CID)
  4 bbl V-8 Oldsmobile
L .... 5.7L (350 CID)
  4 bbl V-8 Chevrolet
R .... 5.7L (350 CID)
  4 bbl V-8 Oldsmobile
W .... 4.9L (301 CID)
  4 bbl V-8 Pontiac
X .... 5.7L (350 CID)
  4 bbl V-8 Buick
Y .... 4.9L (301 CID)
  2 bbl V-8 Pontiac

EMISSION EQUIPMENT

1976

All Models

Closed positive crankcase ventilation
Emission calibrated carburetor
Emission calibrated distributor
Heated air cleaner
Vapor control, canister storage
Exhaust gas recirculation
Catalytic converter
Single diaphragm vacuum advance
　49-States
　　All models
　Calif.
　　On all models except.
　　Not on 260 V-8
Air pump
　Not used
Cold air intake
　All 260 V-8
Early fuel evaporation
Spark delay valve
　Not used
Choke air modulator
　On all models except,
　Not used on 260 V-8
Spark advance vacuum modulator
　49-States
　　260 V-8 auto. trans. only
　Calif.
　　Not used

1977

All Models

Closed positive crankcase ventilation
Emission calibrated carburetor
Emission calibrated distributor
Heated air cleaner
Vapor control, canister storage
Exhaust gas recirculation
Catalytic converter
Early fuel evaporation
Single diaphragm vacuum advance
Air pump
　49-States
　　Not used
　Calif.
　　231 V-6
　　350 4-bbl. V-8 "L"
　　350 4-bbl. V-8 "R"
　　403 4-bbl. V-8 "K"
　Altitude
　　231 V-6
　　350 4-bbl. V-8 "R"
　　403 4-bbl. V-8 "K"
Choke air modulator
　49-States
　　350 2-bbl. V-8 "H"
　　350 4-bbl. V-8 "J"
　Calif.
　　Not used
　Altitude
　　Not used
Spark advance vacuum modulator
　49-States
　　403 4-bbl. V-8 "K" late
　　　production
　　231 V-6 Economy Special
　Calif.
　　231 V-6 Economy Special
　Altitude
　　231 V-6 Economy Special
Distributor spark vacuum modulator
　301 2-bbl. V-8 with auto. trans. and
　　air cond.
Spark delay valve, 4-nozzle
　301 2-bbl. V-8 with auto. trans. and
　　air cond.
Spark delay valve, 2-nozzle
　350 4-bbl. V-8 "R" Calif. only.
Temperature controlled choke vacuum break
　350 4-bbl. V-8 "R"
　301 2-bbl. V-8
　403 4-bbl. V-8
Vacuum reducer valve
　49-States
　　350 4-bbl. V-8 "R"

39

Buick

EMISSION EQUIPMENT

1977

403 4-bbl. V-8
  Calif.
    Not used
  Altitude
    Not used
Air cleaner thermal control valve
  301 2-bbl. V-8
Transmission controlled spark
  49-States
    Not used
  Calif.
    231 V-6 man. trans. only
  Altitude
    231 V-6 man. trans. only

1978

All Models
Closed positive crankcase ventilation
Emission calibrated carburetor
Emission calibrated distributor
Heated air cleaner
Vapor control, canister storage
Exhaust gas recirculation
Catalytic converter
Early fuel evaporation
Air pump
  49-States
    231 4-bbl. V-6 "3"
  Calif.
    231 2-bbl. V-6 "A"
    350 4-bbl. V-8 "L"
    350 4-bbl. V-8 "R"
    350 4-bbl. V-8 "K"
  Altitude
    231 2-bbl. V-6 "A"
    231 4-bbl. V-6 "3"
    350 4-bbl. V-8 "R"
    403 4-bbl. V-8 "K"
Spark advance vacuum modulator
  49-States
    403 4-bbl. V-8 "K" late
      production
Distributor spark vacuum modulator
  301 2-bbl. V-8 "Y" with auto trans.
    and air cond.
Spark delay valve, 4-nozzle
  301 2-bbl. V-8 with auto. trans. and
    air cond.
Spark delay valve, 2-nozzle
  305 2-bbl. V-8, Calif. only
Temperature controlled choke vacuum
  break
    301 2-bbl. V-8 "Y"
    350 4-bbl. V-8 "R"
    403 4-bbl. V-8 "K"
Vacuum reducer valve
  49-States
    350 4-bbl. V-8 "R"
    403 4-bbl. V-8 "K"
  Calif.
    Not used

Altitude
  Not used
Transmission controlled spark
  49-States

1979

The emission control systems for most engines consist of the following:
Calibrated carburetion
Calibrated spark distribution
Catalytic converter
Early fuel evaporation (EFE)
Exhaust gas recirculation (EGR)
Positive crankcase ventilation (PCV)
Calibrated carburetor choke
Thermostatic air cleaner (TAC)
Evaporative emission control system
  (EECS)
Included in some systems is an additional emission control device called an air injection reaction (AIR).
EFE thermal vacuum switch (EFE-
  TVS)
  305 CID (G)
  231 CID (3)
  301 CID (Y)
  305 CID (H)
  350 CID (L)
  301 CID (W)
EFE-EGR thermal vacuum switch
  (EFE-EGR-TVS)
  196 CID (C)
  231 CID (3)
  231 CID (A)
  231 CID (2)
  350 CID (X)
EFE check valve (EFE-CV)
  196 CID (C)
  305 CID (G)
  231 CID (3)
  231 CID (A)
  305 CID (H)
  350 CID (L)
EGR thermal control valve
  (EGR-TCV)
  350 CID (R)
  403 CID (K)
Canister purge thermal vacuum
  switch (CP-TVS)
  305 CID (G)
  305 CID (H)
  350 CID (X)
  350 CID (L)
Spark delay valve (SDV)
  305 CID (G)
  305 CID (H)
Early fuel evaporation distributor
  thermal vacuum switch
  (EFE-DTVS)
  305 CID (G) California only
Distributor vacuum regulating valve
  301 CID (Y)

Spark retard delay valve
  301 CID (Y)
Distributor thermal vacuum switch
  (DTVS)
  196 CID (C)
  305 CID (G)
  231 CID (A)
  301 CID (Y)
  305 CID (H)
  350 CID (R)
  403 CID
  301 CID (W)
EGR distributor thermal vacuum
  switch (EGR-DTVS)
  350 CID (R) California only
  403 CID (K)
Spark advance vacuum modulator
  system (SAVM)
  196 CID (C)
  231 CID (A)
Distributor spark vacuum modulator
  valve (DS-VMV)
  301 CID (W)
Distributor vacuum delay valve
  (DVDV)
  301 CID (W)
Choke thermal vacuum switch
  (CTVS)
  301 CID (Y)
  350 CID (X)
  350 CID (R)
  403 CID (K)
Secondary vacuum break thermal
  vacuum switch (SUB-TVS)
  231 CID (3) 49 states only
  301 CID (W)
Idle speed up solenoid
  196 CID (C) with automatic
    transmission and air
    conditioning
  231 CID (3) E series vehicles
  231 CID (A) with automatic
    transmission and air
    conditioning
Anti-dieseling solenoid
  196 CID (C) A, H, and X series
    vehicles with manual
    transmission
  231 CID (A) A, H, and X series
    vehicles with manual
    transmission
Air injection reaction system (AIR)
  305 CID (G) California only
  231 CID (3)
  231 CID (A) California only
  305 CID (H) California and
    high altitude only
  350 CID (L) California and
    high altitude only
  350 CID (R) California only
  403 CID (K) California only
Thermostatic air cleaner thermal
  check valve (TAC-TCV)
  231 CID (3) E series vehicles

IDLE SPEED AND MIXTURE ADJUSTMENTS

1976

Air cleaner
  231 V-6 .... On
  350 & 455 V-8 .... On

260 V-8 .... Off
Vapor hose .... Plugged
Auto. trans. .... Drive
Distributor vac. hose ..... Plugged
Air cond. .... Off

EGR vac. hose .... Plugged
Mixture adj. .. See rpm drop below

231 V-6
  Auto. trans. 600

Buick

IDLE SPEED AND MIXTURE ADJUSTMENTS

1976

Mixture adj. .... 80 rpm drop (880-800)
Man. trans.
  Solenoid connected 800
  Solenoid disconnected 600
  Mixture adj. ... 300 rpm drop (1100-800)
260 V-8 49-States
  Auto. trans. 550
  Mixture adj. ... 60 rpm drop (610-550)
  AC idle speedup 650
260 V-8 Calif.
  Auto. trans. 600
  Mixture adj. ... 100 rpm drop (700-600)
350 V-8 600
  Mixture adj. ...... 80 rpm drop (680-600)
455 V-8 600
  Mixture adj. ...... 80 rpm drop (680-600)

1977

Air cleaner .... Set aside with hoses connected
Auto level hose .... Plugged
Air cond. .... Off
Auto. trans. .... Drive
Other hoses .... See underhood label
Idle CO .... Not used
Mixture adj. .... See rpm drop below
231 V-6 49-States
  Auto. trans. 600
  Mixture adj. ..... 40 rpm drop (640-600)
  AC idle speedup 600
  Man. trans. 800
  Mixture adj. ..... 60 rpm drop (860-800)
  AC idle speedup 800
231 V-6 Calif.
  Auto. trans. 600
  Mixture adj. ...... 10 rpm drop (610-600)
  AC idle speedup 600
  Man. trans. 800
  Mixture adj. ...... 10 rpm drop (810-800)
  AC idle speedup 800
231 V-6 Altitude
  Auto. trans. 600
  Mixture adj. ...... 10 rpm drop (610-600)
  AC idle speedup 600
  Man. trans. 800
  Mixture adj. ...... 10 rpm drop (810-800)
  AC idle speedup 800
301 V-8 49-States
  Auto. trans. 550
  Mixture adj. ...... 40 rpm drop (590-550)
  AC idle speedup 650
305 V-8 49-States
  Auto. trans. 500
  Mixture adj. ...... 50 rpm drop (550-500)
  AC idle speedup 650
350 V-8 "H" 49-States
  Auto. trans. 600
  Mixture adj. ...... 60 rpm drop (660-600)
  AC idle speedup 600
350 V-8 "J" 49-States
  Auto. trans. 550
  Mixture adj. ...... 50 rpm drop (600-550)
350 V-8 "L" 49-States
  Auto. trans. 500
  Mixture adj. ...... 50 rpm drop (550-500)
  AC idle speedup 650
350 V-8 "L" Calif.
  Auto. trans. 500
  Mixture adj. ...... 50 rpm drop (550-500)
  AC idle speedup 650
350 V-8 "L" Altitude
  Auto. trans. 600
  Mixture adj. ...... 50 rpm drop (650-600)
  AC idle speedup 650
350 V-8 "R" 49-States
  Auto. trans. 550
  Mixture adj. ...... 30 rpm drop (580-550)
  AC idle speedup 650
350 V-8 "R" Calif.
  Auto. trans. 550
  Mixture adj. ...... 25 rpm drop (575-550)
  AC idle speedup 650
350 V-8 "R" Altitude
  Auto. trans. 600
  Mixture adj. ...... 25 rpm drop (625-600)
  AC idle speedup 650
403 V-8 49-States
  Auto. trans. 550
  Mixture adj. ...... 30 rpm drop (580-550)
  AC idle speedup 650
403 V-8 Calif.
  Auto. trans. 550
  Mixture adj. ...... 25 rpm drop (575-550)
  AC idle speedup 650
403 V-8 Altitude
  Auto. trans. 600
  Mixtue adj. ...... 25 rpm drop (625-600)
  AC idle speedup 650

1978

Air cleaner .... In place
Air cond. .... Off
Auto. trans. .... Drive
Hoses .... See underhood label
Idle CO .... Not used
Mixture adj. .. See propane rpm below
Idle Speed
196 V-6 49-States
  Auto. trans. 600
  Propane enriched 640
  AC idle speedup ....
Man. trans.
  Solenoid connected 800
  Solenoid disconnected 600
  Propane enriched 940
231 V-6 49-States
  Auto. trans. 600
  Propane enriched 650
  AC idle speedup 670
Man. trans.
  Solenoid connected 800
  Solenoid disconnected 600
  Propane enriched 940
231 V-6 Calif.
  Auto. trans. 600
  Propane enriched 615
  AC idle speedup 670
Man. trans.
  Solenoid connected 800
  Solenoid disconnected 600
  Propane enriched 880
231 V-6 Altitude
  Auto. trans. 600
  Propane enriched 615
  AC idle speedup 670
Man. trans.
  Solenoid connected 800
  Solenoid disconnected 600
  Propane enriched 880
231 2-bbl. Turbo. 49-States
  Auto. trans. 650
231 4-bbl. Turbo. 49-States
  Auto. trans. 650
231 4-bbl. Turbo. Altitude
  Auto. trans. 650
301 2-bbl. V-8 49-States
  Auto. trans. 550
  Propane enriched 580
  AC idle speedup 650
305 2-bbl. V-8 49-States
  Auto. trans. 500
  Propane enriched 520-540
  AC idle speedup 600
305 2-bbl. V-8 Calif.
  Auto. trans. 500
  Propane enriched 520-540
  AC idle speedup 650
305 2-bbl. V-8 Altitude
  Auto. trans. 600
  Propane enriched 620-640
  AC idle speedup 700
305 4-bbl. V-8 49-States
  Auto. trans. 500
  Propane enriched 530-570
  AC idle speedup 600
350 4-bbl. V-8 "L" Calif.
  Auto. trans. 500
  Propane enriched 530-570
  AC idle speedup 600
350 4-bbl. V-8 "L" Altitude
  Auto. trans. 600
  Propane enriched 630-670
  AC idle speedup 650
350 4-bbl. V-8 "R" Calif.
  Auto. trans. 550
  Propane enriched 575
  AC idle speedup 650
350 4-bbl. V-8 "R" Altitude
  Auto. trans. 600
  Propane enriched ....
  AC idle speedup 700
403 4-bbl. V-8 "K" 49-States
  Auto. trans. 550
  Propane enriched 635
  AC idle speedup 650

Buick

IDLE SPEED AND MIXTURE ADJUSTMENT

1978

403 4-bbl. V-8 "K" Calif.
 Auto. trans. 550
 Propane enriched 575
 AC idle speedup 650
403 4-bbl. V-8 "K" Altitude
 Auto. trans. 600
 Propane enriched ....
 AC idle speedup 700

1979

Air cleaner .... In place
A/C compressor lead .. Disconnected
Auto. trans. .... Drive
 NOTE: Refer to the emission label under the hood of the vehicle being serviced

RPM

196 V-6 (6) 49 states
 Auto. Trans.
  Propane enriched 575
  OFF idle 550
  Solenoid screw (Curb idle) . 670
 Man. Trans.
  Propane enriched 1000
  OFF idle 600
  Solenoid screw (Curb idle) . 800
  Fast idle 2200
231 V-6 (A) 49 states
 Auto. Trans.
  Propane enriched 575
  OFF idle 550
  Solenoid screw (Curb idle) . 670
  Fast idle (Park) 2200
 Man. Trans.
  Propane enriched 1000
  OFF idle 600
  Solenoid screw (Curb idle) . 800
  Fast idle 2200
231 V-6 (A) Calforna and high altitude
 Auto. Trans.
  Propane enriched 615
  OFF idle 600
  Solenoid screw (Curb idle) . NA
  Fast idle (Park) 2200
 Man. Trans.
  Propane enriched 840
  OFF idle 600
  Solenoid screw (Curb idle) 800
  Fast idle 2200

231 V-6 (2) California only
 Auto. Trans.
  OFF idle 580
  Solenoid screw (Curb idle) . 670
  Fast idle (Park) 2200
231 V-6 (3) 49 states
 Auto. Trans.
  OFF idle 650
  Fast idle (Park) 2500
231 V-6 (3) California and high altitude
 Auto. Trans.
  OFF idle 650
  Fast idle (Park) 2500
231 V-6 (3) 49 states and California "E" series only
 Auto. Trans.
  OFF idle 600
  Solenoid screw (Curb idle) . 650
  Fast idle (Park) 2500
301 V-8 (Y) 49 states
 Auto. Trans.
  Propane enriched 530
  OFF idle 500
  Solenoid screw (Curb idle) . 650
  Fast idle (Park) 2000
301 V-8 (W) 49 states
 Auto. Trans.
  Propane enriched 540
  OFF idle 500
  Solenoid screw (Curb idle) . 650
  Fast idle (Park) 2200
305 V-8 (G) 49 states
 Auto. Trans.
  Propane enriched 520-540
  OFF idle .... 550 A/C
   —500 without A/C
  Solenoid screw (Curb idle) . 600
  Fast idle (Park) 1600
305 V-8 (H) California only
 Auto. Trans.
  Propane enriched 520-560
  OFF idle 500
  Solenoid screw (Curb idle) . 600
  Fast idle (Park) 1600
305 V-8 (H) High altitude
 Auto. Trans.
  Propane enriched 630-670
  OFF idle 600
  Solenoid screw (Curb idle) . 650
  Fast idle (Park) 1750

350 V-8 (L) California only
 Auto. Trans.
  Propane enriched 520-560
  OFF idle 500
  Solenoid screw (Curb idle) . 600
  Fast idle (Park) 1600
350 V-8 (L) High altitude
 Auto. Trans.
  Propane enriched 630-670
  OFF idle 600
  Solenoid screw (Curb idle) . 650
  Fast idle (Park) 1750
350 V-8 (R) 49 states
 Auto. Trans.
  Propane enriched 625-640
  OFF idle 550
  Solenoid screw (Curb idle) . 650
  Fast idle (Park) 900
350 V-8 (R) California only
 Auto. Trans.
  Propane enriched 565-585
  OFF idle 500-550
  Solenoid screw (Curb idle) . 600
  Fast idle (Park) 1000
350 V-8 (R) High altitude
 Auto. Trans.
  Propane enriched 590
  OFF idle 550-600
  Solenoid screw Curb idle) . 700
  Fast idle (Park) 900
350 V-8 (X) 49 states
 Auto. Trans.
  Propane enriched 590
  OFF idle 550
  Fast idle (Park) 1500
403 V-8 (K) 49 states
 Auto. Trans.
  Propane enriched 625-645
  OFF idle 550
  Solenoid screw (Curb idle) . 650
  Fast idle (Park) 900
403 V-8 (K) California only
 Auto. Trans.
  Propane enriched 565-585
  OFF idle 500-550
  Solenoid screw (Curb idle) . 600
  Fast idle (Park) 1000
403 V-8 (K) High altitude
 Auto. Trans.
  Propane enriched 590
  OFF idle 600
  Solenoid screw (Curb idle) . 700
  Fast idle (Park) 1000

INITIAL TIMING

1976

NOTE: Distributor vacuum hose must be disconnected and plugged. Set timing at idle speed unless shown otherwise. Tolerance is plus or minus 2°:

231 V-6 .... 12° BTDC
260 V-8 (at 1100 rpm)
 49-States .... 18° BTDC
 Calif. .... 14° BTDC
350 V-8 .... 12° BTDC
455 V-8 .... 12° BTDC

1977

NOTE: Distributor vacuum hose must be disconnected and plugged. Set timing at idle speed unless shown otherwise.

231 V-6 .... 12° BTDC
301 V-8 .... 12° BTDC
305 V-8 .... 8° BTDC
350 V-8 "L" .... 8° BTDC
350 V-8 "H" .... 12° BTDC
350 V-8 "J" .... 12° BTDC

350 V-8 "R" (at 1100 rpm)
 49-States .... 20° BTDC
 Altitude .... 20° BTDC
 Calif.
  Skylark .... 18° BTDC
  LeSabre except
   wagon .... 18° BTDC
  All others .... 20° BTDC

403 V-8 (at 1100 rpm)
 49-States .... 24° BTDC
 Calif. .... 20° BTDC
 Altitude .... 20° BTDC

Buick

INITIAL TIMING

1978

NOTE: Distributor vacuum hose must be disconnected and plugged. Set timing at idle speed unless shown otherwise.

196 V-6	15° BTDC
231 V-6	15° BTDC
231 V-6 Turbo	15° BTDC
301 V-8	12° BTDC
305 2-bbl. V-8	
49-States	4° BTDC
Calif.	6° BTDC
Altitude	8° BTDC
305 4-bbl. V-8	4° BTDC
350 V-8 "X"	15° BTDC
350 V-8 "L"	8° BTDC
350 V-8 "R" (at 1100 rpm)	
All	20° BTDC
403 V-8 "K" (at 1100 rpm)	
All	20° BTDC

1979

NOTE: Distributor vacuum hose must be disconnected and plugged. Set timing at idle speed unless shown otherwise.

196 V-6	15° BTDC
231 V-6	15° BTDC
231 V-6 Turbo	15° BTDC
301 V-8	12° BTDC
305 2-bbl. V-8	
49-States	4° BTDC
Calif.	4° BTDC
Altitude	8° BTDC
305 4-bbl. V-8	4° BTDC
350 V-8 "X"	15° BTDC
350 V-8 "L"	8° BTDC
350 V-8 "R" (at 1100 rpm)	
All	20° BTDC
403 V-8 "K" (at 1100 rpm)	
All	20° BTDC

SPARK PLUGS

1976

231 V-6	AC-R44SX	.060
260 V-8	AC-R46SX	.080
350 V-8	AC-R45TSX	.060
455 V-8	AC-R45TSX	.060

1977

231 V-6	AC-R46TS	.040
231 V-6	AC-R46TSX	.060
301 V-8	AC-R46TSX	.060
305 V-8	AC-R45TS	.045
350 V-8 "L"	AC-R45TS	.045
350 V-8 "H"	AC-R46TS	.040
350 V-8 "H"	AC-R46TSX	.060
350 V-8 "J"	AC-R46TS	.040
350 V-8 "J"	AC-R46TSX	.060
350 V-8 "R"	AC-R46SZ	.060
403 V-8	AC-R46SZ	.060

NOTE: The Buick engines (V-6, V-8 "H" and "J") may come with either "TS" or "TSX" plugs (last letter) as shown above, but not mixed on the same engine. The "TS" plug must be gapped at .040", and the "TSX" plug at .060". When replacing worn plugs, the "TSX" plug is recommended on AC plug charts.

1978

196 V-6	AC-R46TSX	.060
231 V-6	AC-R46TSX	.060
231 V-6 Turbo	AC-R44TSX	.060
301 V-8	AC-R46TSX	.060
305 V-8	AC-R45TS	.045
350 V-8 "X"	AC-R46TSX	.060
350 V-8 "L"	AC-R45TS	.045
350 V-8 "R"	AC-R46SZ	.060
403 V-8	AC-R46SZ	.060

1979

196 V-6	AC-R46TSX—	.060
231 V-6	AC-R46TSX	.060
231 V-6 Turbo	AC-R44TSX	.060
301 V-8	AC-R46TSX	.060
305 V-8	AC-R45TS	.045
350 V-8 "X"	AC-R46TSX	.060
350 V-8 "L"	AC-R45TS	.045
350 V-8 "R"	AC-R46SZ	.060
403 V-8	AC-R46SZ	.060

VACUUM ADVANCE

1976

231 V-6	Ported
260 V-8 Auto. Trans.	
49-States	Modulated
Calif.	None
260 V-8 Man. Trans.	Ported
350 V-8	Ported
455 V-8	Ported

1977

Diaphragm type .... Single
Vacuum source

231 V-6	
49-States	Manifold
Economy Special	Modulated
Calif.	Ported
Economy Special	Modulated
Altitude	Manifold
Economy Special	Modulated
301 V-8	
Auto. trans. with A.C.	Manifold
Auto. trans. no A.C.	Modulated
Man. trans.	Ported
305 V-8	
49-States	Manifold
Calif.	Ported
Altitude	Manifold
350 V-8 "L"	Manifold
350 V-8 "H"	Manifold
350 V-8 "J"	Manifold
350 V-8 "R"	
49-States	Manifold
Calif.	Ported
Altitude	Ported
403 V-8	
49-States	Manifold
Calif.	Ported
Altitude	Ported

1978

Diaphragm type .... Single
Vacuum source

All V-6	Ported
301 V-8	
Auto. trans. no A.C.	Manifold
Auto. trans. A.C.	Modulated
305 V-8	
49-States	Manifold
Calif.	Ported
Altitude	Manifold
350 V-8 "X"	Ported
350 V-8 "L"	Manifold
350 V-8 "R"	
Calif.	Ported
Altitude	Ported
403 V-8	
49-States	Manifold
Calif.	Ported
Altitude	Ported

1979

Diaphragm type .... Single
Vacuum source

All V-6	Ported
301 V-8	
Auto. trans. no A.C.	Manifold
Auto. trans. A.C.	Modulated
305 V-8	
49-States	Manifold
Calif.	Ported
Altitude	Manifold
350 V-8 "X"	Ported
350 V-8 "L"	Manifold
350 V-8 "R"	
Calif.	Ported
Altitude	Ported
403 V-8	
49-States	Manifold
Calif.	Ported
Altitude	Ported

Buick

EMISSION CONTROL SYSTEMS 1976

SPARK ADVANCE VACUUM MODULATOR (SAVM) 1976

Description of Modulator

The SAVM has hoses connected to it from the distributor vacuum advance unit, manifold vacuum, and carburetor ported vacuum. The SAVM allows either ported vacuum, manifold vacuum, or 7" Hg. vacuum to operate the vacuum advance. At idle, when manifold vacuum in the engine is high, but ported vacuum is non-existent, the SAVM allows only 7" Hg. vacuum to go to the distributor. As the throttle is opened, and the ported vacuum increases, the distributor continues to receive 7" Hg. until the ported vacuum rises to 7" Hg. After the ported vacuum goes over 7" Hg. the SAVM switches the distributor over to ported vacuum only, whatever it may be. If both the ported vacuum and the manifold vacuum are below 7" Hg. as happens near wide open throttle, the distributor operates on manifold vacuum only.

In other words, at idle the distributor gets 7" Hg, at part throttle it gets ported vacuum, and near full throttle it gets manifold vacuum. This system gives more spark advance for better running, but not so much advance that emissions go up.

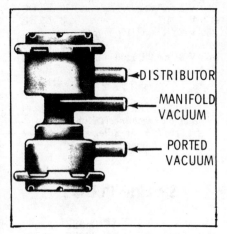

Spark advance vacuum modulator

Testing and Troubleshooting Modulator

Test 1: Connect a vacuum gauge to the distributor nozzle of the SAVM. Connect a vacuum pump to the intake manifold nozzle of the SAVM. Slowly pump up vacuum to a maximum of 20" Hg. The vacuum on the gauge should be equal to the pump gauge up to 7" Hg. As the pump gauge goes above 7" Hg. the vacuum gauge should stay at 7" Hg. If not, the SAVM is defective and must be replaced.

Test 2: Connect a vacuum gauge to the distributor nozzle of the SAVM. Connect a vacuum pump to the carburetor nozzle of the SAVM. Slowly pump up vacuum to a maximum of 20" Hg. The vacuum gauge should stay at zero until the pump gauge reaches 7" Hg. At 7" Hg. and above the vacuum gauge should read exactly the same as the pump gauge. If not, the SAVM is defective and must be replaced.

Test 3: Connect a vacuum gauge to the carburetor nozzle of the SAVM. Connect a vacuum pump to the distributor nozzle of the SAVM. Pump up several inches of vacuum. The vacuum gauge should stay at zero. If not, the SAVM is defective and must be replaced.

Repairing SAVM

Repairs are limited to replacement.
CAUTION: *There are several different valves that all look alike. The part number for the SAVM is 553952.*

TAILPIPE DIFFUSER, 1976

A few inches inside the end of the tailpipe is a small screen, on Century Wagons only. This screen is shown on some emission control charts, but its real purpose is to soften the noise of the exhaust so it will pass federal regulations. Attaching an HC-CO sniffer to the exhaust pipe may be difficult without special adapters, because the screen is so close to the end of the pipe. No maintenance or repairs are required on the screen.

EMISSION CONTROL SYSTEMS 1977

BUICK ENGINES 1977

Buick is still using its own 350 V-8 with both 2-bbl. and 4-bbl. carburetors, but some models will use V-8 engines from Chevrolet, Oldsmobile, and Pontiac. All of these engines can be identified from the fifth character in the engine code, under Car Serial Number And Engine Identification, in this section.

The V-8's can also be identified just by looking at them.

The Buick engines have the distributor in the front. All the other engines have the distributor in the rear. To recognize the Olds engines, look for the oil filler tube at the front of the intake manifold. The new Pontiac 301 is similar to other Pontiac V-8's but has a unique engine oil dipstick at the rear of the left bank. The Chevrolet engines are typical small block V-8's, with the PCV valve plugged into the left rocker cover. Buick will be using two V-6's in 1977. The "C" engine is the design with unevenly spaced firing intervals. The "A" engine is the new "Slick Six" with a redesigned crank, cam, and distributor which gives evenly spaced firing intervals.

Buick V-6—crankshaft—even firing design
(© GM Corporation)

Buick V-6—crankshaft—odd firing design
(© GM Corporation)

Buick

For a complete description of the new Pontiac 301 V-8, see the Pontiac section in this supplement.

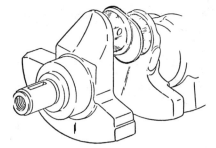

Buick V-6—crankshaft design—even firing
(© GM Corporation)

Buick V-6—Crankshaft design—1975-77
(© GM Corporation)

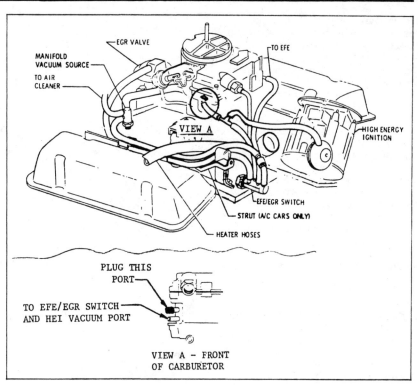

Buick V-6 49 state engine—late carburetor hookup change
(© GM Corporation)

ELECTRIC CHOKE 1977

V-6 engines now use the 2GE carburetor, which has an electric choke. This choke is similar to the choke used on other GM cars, such as Chevette, Cadillac, and the discontinued 454 Chevrolet engine.

The choke receives current from the oil pressure switch in all cars and is only being heated while the engine is running. A minor difference occurs when the V6 is used in the Skyhawk. The oil pressure switch supplies current to the electric fuel pump when the engine is either cranking or running. The choke is connected to the fuel pump circuit, so it also receives current when the engine is either cranking or running. The only other thing that changes between car models in some cases is the colors of the wires.

V-6 VACUUM ADVANCE 1978

49-State manual transmission V-6 engines (no turbocharger) have the distributor vacuum advance connected to the EFE-EGR Thermal Vacuum Switch. The vacuum advance hose connects, through a "T" to the EGR part of the switch. The switch is closed below approximately 120°F. engine coolant temperature, to eliminate exhaust gas recirculation when the engine is cold. Because the vacuum advance is connected to the EGR vacuum source, there is no vacuum advance until the engine warms up above 120°F. Above that temperature, the vacuum advance operates off of ported vacuum which is only available above idle.

With the vacuum advance and EGR diaphragms both connected to the same vacuum source, a leak in either diaphragm will affect the other. For example, a leaking EGR diaphragm might make the EGR valve stay closed, but it might also bleed off so much vacuum that the distributor vacuum unit does not advance the spark.

EMISSION CONTROL SYSTEMS 1978

TURBOCHARGED V-6 1978

A turbocharger is a compressor, driven by the engine exhaust. It compresses the air-fuel mixture and forces it into the combustion chamber for increased power and acceleration. Buick has started using turbochargers because their V-6 engines do not give enough power when installed in the larger cars. The turbocharged V-6 is supposed to give the power of a V-8, with the economy of a V-6

A turbocharger consists of two turbine wheels, each in its own separate housing, and connected by a shaft. One of the turbine wheels is driven by the pressure of the exhaust gas leaving the engine. Buick takes the exhaust gas from the rear of the left bank exhaust manifold and routes it into the rear of the right bank manifold. All the exhaust from the engine exits the front of the right exhaust manifold, and goes to the turbine housing.

The pressure and flow of the exhaust makes the turbine wheel turn, just like air blowing on a windmill. The turbine wheel is connected by a shaft to the compressor wheel. The compressor wheel accepts the air-fuel mixture from the carburetor at its inlet, and pushes this mixture into the engine through its outlet. The outlet of the compressor is bolted to the intake manifold to the engine cylinders the same as any intake manifold.

The two turbine wheels, connected by the shaft, are completely free to turn at any time. There is no clutch or brake of any kind that prevents them from turning. When the engine is idling, exhaust pressure is low, so the compressor wheel turns from the force of the incoming air-

Buick

fuel mixture going into the engine. The compressor contributes a pressure increase, known as "boost" only when the exhaust pressure is high. Usually, the only time boost occurs is during wide open throttle.

When the driver steps on the gas pedal to accelerate, he will get a normal amount of acceleration immediately. One or more seconds later, the exhaust pressure will build up enough so that the compressor starts supplying boost. Then he will get additional acceleration. This delay is always present, and does not indicate that anything is wrong. It takes time for the exhaust pressure to build up and start the boost.

The boost is greatest at the highest speeds. The system is designed to act as a kind of "passing gear" at highway speeds, so that V-6 owners can still have the highway passing ability they had with a V-8.

To keep the boost from getting too high, and possibly causing engine damaging detonation, a pressure-operated wastegate is used. A vacuum diaphram is connected to the pressure side of the compressor housing. When the pressure reaches 8 pounds, the diaphragm moves a rod attached to the wastegate. The wastegate then opens and lets some of the exhaust pressure bypass the exhaust on the turbine and the compressor wheel slows slightly, which lowers the pressure.

At 8 pounds of pressure, the wastegate actuating rod only moves about .008 in. This is such a small movement that you can barely see it move, but it is enough to open the wastegate slightly and lower the pressure on the turbine wheel. A threaded end on the rod allows adjustment.

The vacuum diaphragm actuator that moves the wastegate has two hoses attached to it. The hose on the end is from

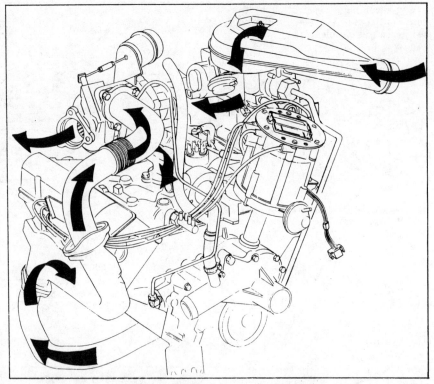

Air flow of 1979 Buick Turbocharged V-6 engine

the compressor housing and senses pressure. The hose on the side is connected to intake manifold vacuum. The amount of vacuum that acts on the diaphragm has very little effect. The main reason for the hose is to provide an escape path for gas vapors in case the diaphragm should rupture. If vapors should pass through the diaphragm, they will immediately be drawn into the engine.

The wastegate opening can be checked by applying 8 lbs. pressure with a hand pump to the actuator, or by temporarily installing a pressure gauge and driving the car. A full throttle acceleration from zero to 50 should produce 7-8 psi for the 2-bbl. engine, and 8-9 psi on the 4-bbl. engine. A maximum of 10 psi is allowed on the 4-bbl. engine.

Vacuum and pressure switches are mounted in the engine compartment and connected to the intake manifold with hoses. When the intake manifold has low vacuum, the vacuum switch illuminates a yellow light on the instrument panel, indicating moderate acceleration. When the intake manifold is under pressure, the pressure switch lights an orange light on the instrument panel, indicating full acceleration.

Because turbochargers spin at extremely high speeds, their shaft bearings need full pressure lubrication. An oil pressure line comes from the engine block to the turbocharger shaft. Oil

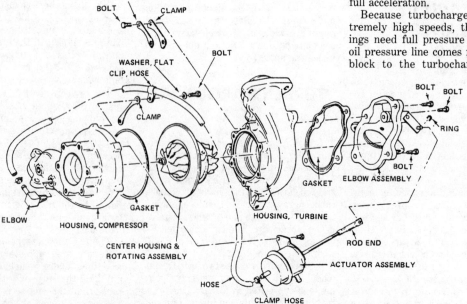

1978 Buick Turbo V-6 PCV valve and Westgate actuator tubing

Buick

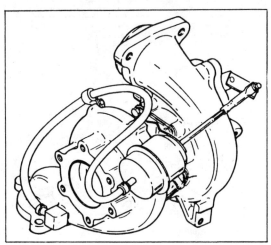

1978 Buick V-6 Turbocharger

ELECTRONIC SPARK CONTROL 1978

The boost pressure that a turbocharger provides is similar to increasing the compression ratio on an engine. Higher compression and added boost both require less spark advance to avoid detonation. Even though the spark advance on the turbocharged engine has been tailored to the engine, there is still a chance of detonation, mainly because of the low octane, non-leaded gas used.

To prevent detonation that might cause engine damage, Buick uses an Electronic Spark Control that is a retarder. Mounted on the intake manifold above the engine thermostat is a detonation sensor. If the engine detonates, the metal of the intake manifold will vibrate. The sensor picks up this vibration and sends a signal to an electronic black box on the fan shroud. The black box then retards the timing as much as 18-22°.

Testing the retarder is done by tapping on the intake manifold next to the sensor. Do not tap on the sensor itself. With the engine runing at 2000 rpm in Neutral, tapping on the manifold will make the spark retard. If you have a timing light hooked up you can actually see it retard while you are tapping. Within 20 seconds after the tapping stops, the timing will return to its normal setting.

runoff from the shaft bearings runs back into the engine through a rubber tube under the turbocharger shaft.

A glance at the illustrations will show that the turbocharger is downstream from the carburetor. With this design, the turbocharger not only pumps air-fuel mixture into the engine, but also creates a tremendous suction that pulls air through the carburetor. This suction is highest at wide open throttle, when the turbocharger is providing full boost. This suction creates high vacuum between the carburetor and the turbocharger. A high vacuum under the carburetor throttle blades at wide open throttle is exactly the opposite of what you get on an unblown engine, where the vacuum drops to zero at wide open throttle.

If the high vacuum at wide open throttle, is allowed to operate the distributor vacuum advance, the EGR valve, and the carburetor power piston, those units will act as if the engine is decelerating or cruising. The distributor would go to full vacuum advance, while the EGR valve opened wide, and the carburetor power piston would close the power valve. Of course, this is exactly the opposite of what is needed at wide open throttle. We need no vacuum advance, no EGR, and want the power valve wide open.

To get the action needed, an extra valve is added to the intake manifold. It senses intake manifold pressure, and shuts off the vacuum when the turbo is operating. On the 2-bbl. engine, the valve is called the Turbocharger Vacuum Bleed Valve (TVBV). When pressure in the intake manifold gets above 3 psi, the TVBV shuts off and bleeds down the vacuum to the distributor vacuum advance, EGR valve, and carburetor power piston.

On the 4-bbl. engine, the valve is called the Power Enrichment Control Valve. (PECV). Because the 4-bbl. carburetor has a larger area at wide open throttle, the vacuum does not go as high. It is only necessary to shut off vacuum to the carburetor power piston. Any time pressure in the intake manifold goes above 3 psi, the PECV shuts off the vacuum to the carburetor power piston.

The TVBV used on the 2-bbl. engine has seven hoses connected to it. Each pair of hoses brings vacuum to the valve, and then out to the EGR valve, distributor, or power piston. The seventh hose, in the middle of the valve, is the bleed. It is connected to the carburetor so that only clean air will enter the system.

The PECV used on the 4-bbl. engine has only three hoses connected to it. Two of them supply vacuum to the power valve, and the third, on the right side, is the bleed.

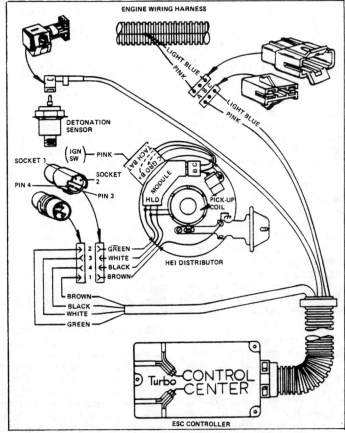

1978 Buick Turbo V-6 electronic spark control system

Buick

ELECTRONIC FUEL CONTROL SYSTEM (EFC) 1978

This system will not go into production until the summer of 1978. Only a few hundred cars will be produced, for sale in California. The system consists of an oxygen sensor in the exhaust, which signals an electronic controller whether the exhaust is rich or lean. The controller operates an electric vacuum valve which controls the vacuum to a special vacuum piston built into the carburetor. The vacuum piston is connected to a metering rod and an idle bleed needle. When the electric vacuum valve allows full manifold vacuum, the vacuum piston moves down. This pushes the metering rod into the jet and gives a lean mixture. At the same time the idle bleed needle is pulled out of the idle bleed orifice so that the idle system gets the full amount of bleed air, and the idle mixture goes lean.

When the electric vacuum valve reduces the vacuum, the spring under the vacuum piston moves the piston up. Both the metering rod and the idle bleed needle move in the rich direction.

Through precise regulation of the vacuum applied to the piston, both the idle mixture and the cruising mixture are controlled. Because the signal to the electronic controller comes from the oxygen sensor in the exhaust, and the carburetor controls the mixture that feeds the exhaust, the system forms a complete circle, or loop, and is known as a closed loop system. The system is similar to those used on other GM divisions, and Ford Motor Company. The outstanding difference in the Buick system is that it controls the idle mixture, while the others only control the cruising mixture. Because this description is being written before any cars have been produced, the final form of the system may vary.

Along with the electronic fuel control is a system of vacuum spark advance control which uses a double-acting vacuum unit on the distributor. A temperature switch and relay are used to turn on two vacuum solenoids. Below 150°F. engine temperature, the distributor advance is held in the full retard position. This helps heat up the catalytic converter so it works better. Above 150°F. the temperature switch and relay turn off the vacuum solenoids and the vacuum advance is operated on ported carburetor vacuum in the normal way.

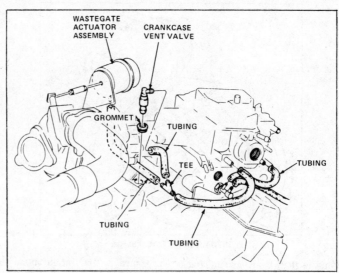
Electronic fuel control system

EMISSION CONTROL SYSTEMS 1979

V-6 VACUUM 1979

49-State manual transmission V-6 engines (no turbocharger) have the distributor vacuum advance connected to the EFE-EGR Thermal Vacuum Switch. The vacuum advance hose connects, through a "T" to the EGR part of the switch. The switch is closed below approximately 120°F. engine coolant temperature, to eliminate exhaust gas recirculation when the engine is cold. Because the vacuum advance is connected to the EGR vacuum source, there is no vacuum advance until the engine warms up above 120°F. Above that temperature, the vacuum advance operates off of ported vacuum which is only available above idle.

With the vacuum advance and EGR diaphragms both connected to the same vacuum source, a leak in either diaphragm will affect the other. For example, a leaking EGR diaphragm might make the EGR valve stay closed, but it might also bleed off so much vacuum that the distributor vacuum unit does not advance the spark.

TURBOCHARGER 1979

A turbocharger is a compressor, driven by the engine exhaust. It compresses the air-fuel mixture and forces it into the combustion chamber for increased power and acceleration.

A turbocharger consists of two turbine wheels, each in its own separate housing, and connected by a shaft. One of the turbine wheels is driven by the pressure of the exhaust gas leaving the engine. Buick takes the exhaust gas from the rear of the left bank exhaust manifold and routes it into the rear of the right bank manifold. All the exhaust from the engine exits the front of the right exhaust manifold, and goes to the turbine housing.

The pressure and flow of the exhaust makes the turbine wheel turn, just like air blowing on a windmill. The turbine wheel is connected by a shaft to the compressor wheel. The compressor wheel accepts the air-fuel mixture from the carburetor at its inlet, and pushes this mixture into the engine through its outlet. The outlet of the compressor is bolted to the intake manifold to the engine cylinders the same as any intake manifold.

The two turbine wheels, connected by the shaft, are completely free to turn at any time. There is no clutch or brake of any kind that prevents them from turning. When the engine is idling, exhaust pressure is low, so the compressor wheel turns from the force of the incoming air-fuel mixture going into the engine. The compressor contributes a pressure increase, known as "boost" only when the exhaust pressure is high. Usually, the only time boost occurs is during wide open throttle.

When the driver steps on the gas pedal to accelerate, he will get a normal amount of acceleration immediately. One or more seconds later, the exhaust presure will build up enough so that the compressor starts supplying boost. Then he will get additional acceleration. This delay is always present, and does not indicate that anything is wrong. It takes time for the exhaust pressure to build up and start the boost.

To keep the boost from getting too high, and possibly causing engine damaging detonation, a pressure-operated wastegate is used. A vacuum diaphram is connected to the pressure side of the compressor housing. When the pressure reaches 8 pounds, the diaphragm moves a rod attached to the wastegate. The wastegate then opens and lets some of the exhaust pressure bypass the exhaust on the turbine and the compressor wheel slows slightly, which lowers the pressure.

At 8 pounds of pressure, the wastegate actuating rod only moves about .008 in. This is such a small movement that you can barely see it move, but it is enough to open the wastegate slightly and lower the pressure on the turbine wheel. A threaded end on the rod allows adjustment.

The vacuum diaphragm actuator that moves the wastegate has two hoses attached to it. The hose on the end is from the compressor housing and senses pressure. The hose on the side is connected to intake manifold vacuum. The amount of vacuum that acts on the diaphragm has very little effect. The main reason for the hose is to provide the escape path for gas vapors in case the diaphragm should rupture. If vapors should pass through the diaphragm they will immediately be drawn into the engine.

The wastegate opening can be checked by applying 8 lbs. pressure with a hand pump to the acuator, or by temporarily installing a pressure gauge and driving the car. A full throttle acceleration from zero to 50 should produce 7-8 psi for the 2-bbl. engine, and 8-9 psi on the 4-bbl. engine. A maximum of 10 psi is allowed on the 4-bbl. engine.

Vacuum and pressure switches are mounted in the engine compartment and connected to the intake manifold with hoses. When the intake manifold has low vacuum, the vacuum switch illuminates a yellow light on the instrument panel, indicating moderate acceleration. When the intake manifold is under pressure, the pressure switch lights an orange light on the instrument panel, indicating full acceleration.

Because turbochargers spin at extremely high speeds, their shaft bearings need full pressure lubrication. An oil pressure line comes from the engine block to the turbocharger shaft. Oil runoff from the shaft bearings runs back into the engine through a rubber tube under the turbocharger shaft.

The turbocharger is downstream from the carburetor. With this design, the turbocharger not only pumps air-fuel mixture into the engine, but also creates a tremendous suction that pulls air through the carburetor. This suction is highest at wide open throttle, when the turbocharger is providing full boost. This suction creates high vacuum between the carburetor and the turbocharger. A high vacuum under the carburetor throttle blades at wide open throttle is exactly the opposite of what you get on an unblown engine, where the vacuum drops to zero at wide open throttle.

If the high vacuum at wide open throttle, is allowed to operate the distributor vacuum advance, the EGR valve, and the carburetor power piston, those units will act as if the engine is decelerating or cruising. The distributor would go to full vacuum advance, while the EGR valve opened wide, and the carburetor power piston would close the power valve. Of course, this is exactly the opposite of what is needed at wide open throttle. We need no vacuum advance, no EGR and want the power valve wide open.

To get the action needed, an extra valve is added to the intake manifold. It senses intake manifold pressure, and shuts off the vacuum when the turbo is operating. On the 2-bbl. engine, the valve is called the Turbocharger Vacuum Bleed Valve (TVBV). When pressure in the intake manifold gets above 3 psi. the TVBV shuts off and bleeds down the vacuum to the distributor vacuum advance, EGR valve, and carburetor power piston.

On the 4-bbl. engine, the valve is called the Power Enrichment Control Valve (PECV). Because the 4-bbl. carburetor has a large area at wide open throttle, the vacuum does not go as high. It is only necessary to shut off vacuum to the carburetor power piston. Any time pressure in the intake manifold goes above 3 psi, the PECV shuts off the vacuum to the carburetor power piston.

The TVBV used on the 2-bbl. engine has seven hoses connected to it. Each pair of hoses brings vacuum to the valve, and then out to the EGR valve, distributor, or power piston. The seventh hose, in the middle of the valve, is the bleed. It is connected to the carburetor so that only clean air will enter the system.

The PECV used on the 4-bbl. engine has only three hoses connected to it. Two of them supply vacuum to the power valve, and the third, on the right side, is the bleed.

ELECTRONIC SPARK CONTROL

The boost pressure that a turbocharger provides is similar to increasing the compression ratio on an engine. Higher compression and added boost both require less spark advance to avoid detonation. Even though the spark advance on the turbocharged engine has been tailored to the engine, there is stil a chance of detonation, mainly because of the low octane, non-

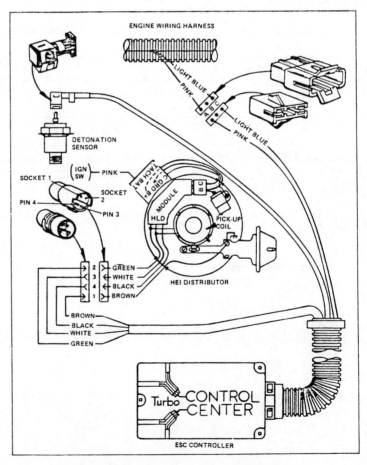

1979 Buick Turbo V-6 electronic spark control system

Buick

leaded gas used.

To prevent detonation that might cause engine damage, Buick uses an Electronic Spark Control that is a retarder. Mounted on the intake manifold above the engine thermostat is a detonation sensor. If the engine detonates, the metal of the intake manifold will vibrate. Then sensor picks up this vibration and sends a signal to an electronic black box on the fan shroud. The black box then retards the timing as much as 18-22°.

Testing the retarder is done by tapping on the intake manifold next to the sensor. Do not tap on the sensor itself. With the engine running at 2000 rpm in Neutral, tapping on the manifold will make the spark retard. If you have a timing light hooked up you can actually see it retard while you are tapping. Within 20 seconds after the tapping stops, the timing will return to its normal setting.

COMPUTER CONTROLLED CATALYTIC CONVERTER SYSTEM

C-4 System

This new, electronically regulated, emission control system is available on 1979 California "A" bodies with a 231 cu. in., V-6, VIN code "2" engine. This special exhaust system, also known as the closed loop system, is comprised of an exhaust gas oxygen sensor, an electronic control module (ECM), a controlled air-fuel ratio carburetor, a

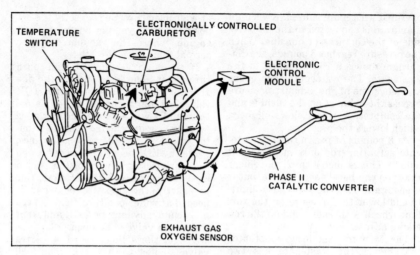

C-4 system 1979 V-6 vin code 2

Phase II catalytic converter, and an oxygen sensor maintenance reminder (located in the instrument panel) which signals when it is time to replace the oxygen sensor.

The oxygen sensor (zirconia sensor), located in the left side exhaust manifold, develops a voltage which varies correspondingly with the volume of oxygen present in the exhaust gas. A rise in the voltage indicates a decrease in the oxygen content, and a drop or decrease in voltage indicates an increase in the oxygn content.

Sensor voltage is monitored by the ECM, which, in turn, signals the mixture control solenoid in the carburetor. As input voltage to the ECM increases (indicating a rich mixture), the output

signal to the mixture control solenoid increases resulting in a leaner mixture at the carburetor. As input voltage to the ECM decreases (indicating a lean mixture), the output signal to the mixture control solenoid decreases resulting in a richer mixture at the carburetor.

Three conditions must be met before the ECM can begin regulating the carburetor air/fuel ratio: (1) a minimum of ten seconds must have elapsed since the engine was started before C-4 system operation can occur, (2) the coolant temperature must be higher than 90 degrees F., and (3) the ECM performs an oxygen sensor output-voltage check to determine when it has warmed up sufficiently to provide good information for C-4 system operation.

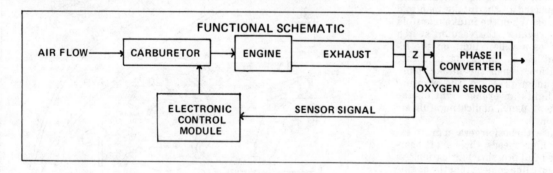

C-4 system schematic

Buick

VACUUM CIRCUITS

1976 231 V-6

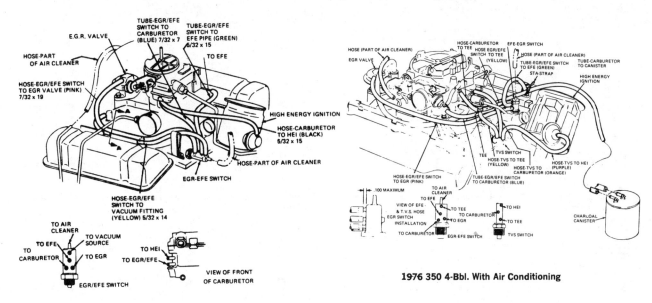

1976 350 4-Bbl. With Air Conditioning

1976 350 2-Bbl. With Air Conditioning

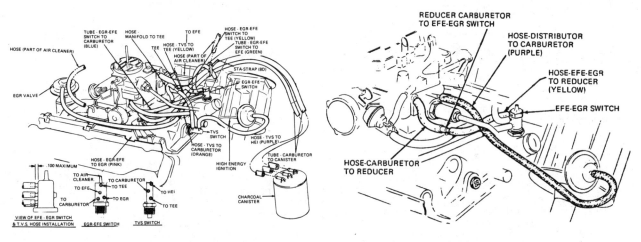

1976 350 4-Bbl. No Air Conditioning

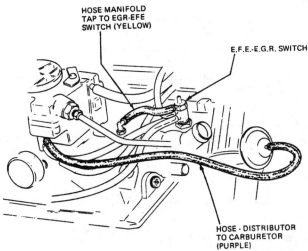

1976 350 2-Bbl. No Air Conditioning

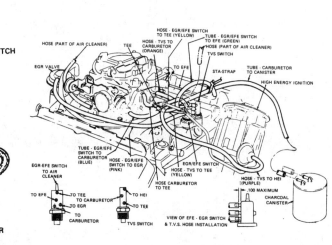

1976 455 4-Bbl.

Buick

VACUUM CIRCUITS

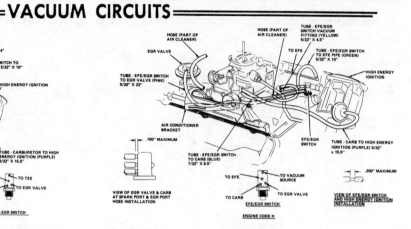

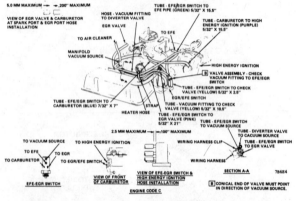

Buick 350 4bbl V-8 engine EFE-EGR switch and high energy ignition hose installation (© GM Corporation)

Buick 350 2bbl V-8 engine EGR valve and carb at spark port and EGR port hose installation (© GM Corporation)

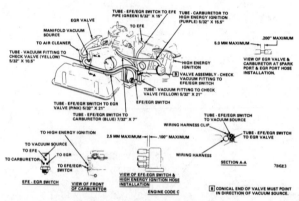

Buick V-6 California engine EFE-EGR switch and high energy ignition hose installation (© GM Corporation)

Buick V-6 49 state engine—early EFE-EGR switch and high energy ignition hose installation. (© GM Corporation)

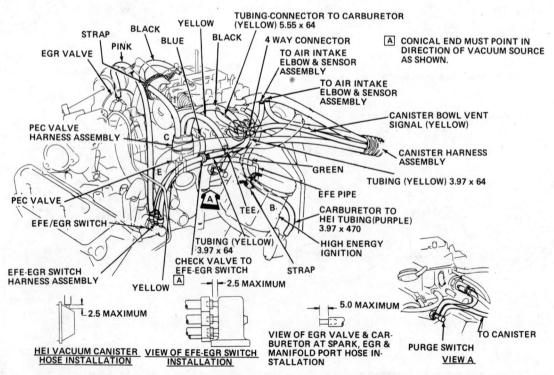

1978 Buick 231 4-bbl. Turbo V-6 vacuum hose routing

Buick

VACUUM CIRCUITS

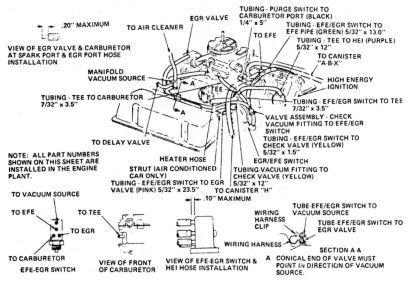

1978 California Buick 231 2-bbl. V-6 vacuum hose routing

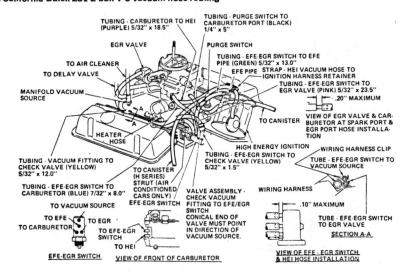

1978 Altitude Buick 231 2-bbl. V-6 vacuum hose routing

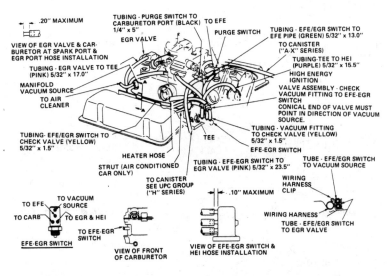

1978 49 States Buick manual trans. 196 and 231 2-bbl. V6 vacuum hose routing

Buick

VACUUM CIRCUITS

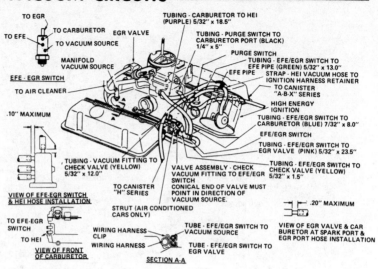

1978 49 States Buick automatic trans. 196 and 231 2-bbl. V-6 vacuum hose routing

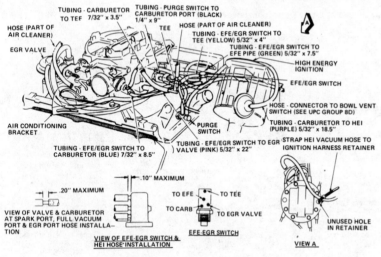

1978 Buick 350 4-bbl. V-8 vacuum hose routing

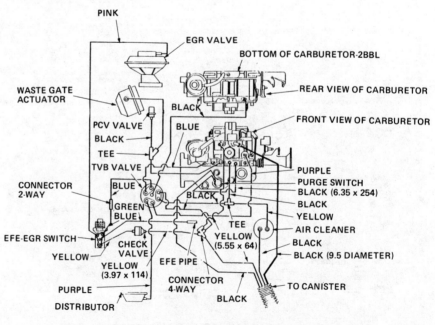

1978 Buick 231 2-bbl. Turbo V-6 vacuum hose routing

Buick

VACUUM CIRCUITS

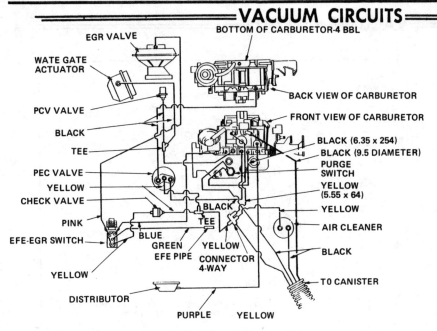

1978 Buick 231 2-bbl. Turbo V-6 vacuum hose routing

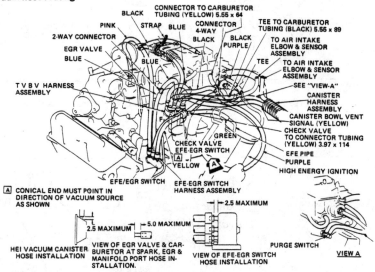

1978 Buick 231 4-bbl. Turbo V-6 vacuum hose routing

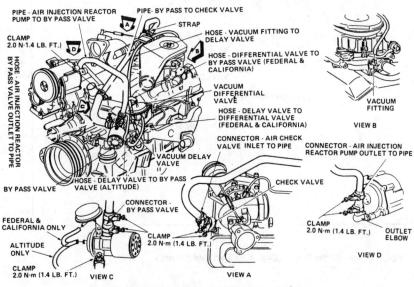

1978 Buick 231 4-bbl. Turbo V-6 air pump system

Buick

VACUUM CIRCUITS

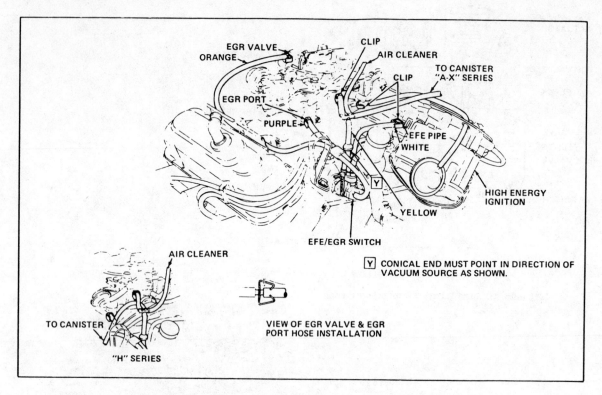

1979 Buick Vacuum hose routing—Code A, C—Series A, H, X (49 states/man. trans.)

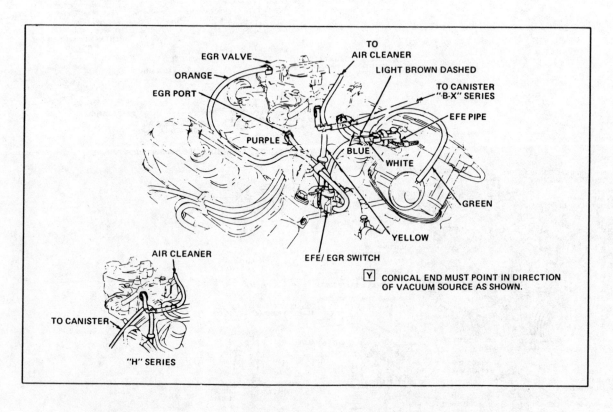

1979 Buick Vacuum hose routing—Code A—Series B, H, X (Calif. & high alt./auto. trans.)

Buick

VACUUM CIRCUITS

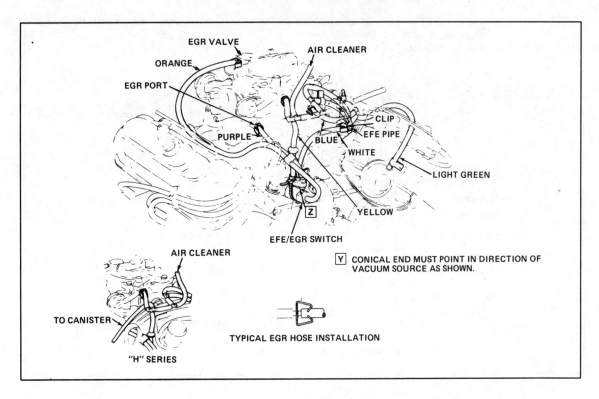

1979 Buick Vacuum hose routing—Code A, C—Series A, B, H, X (49 states/auto. trans.)

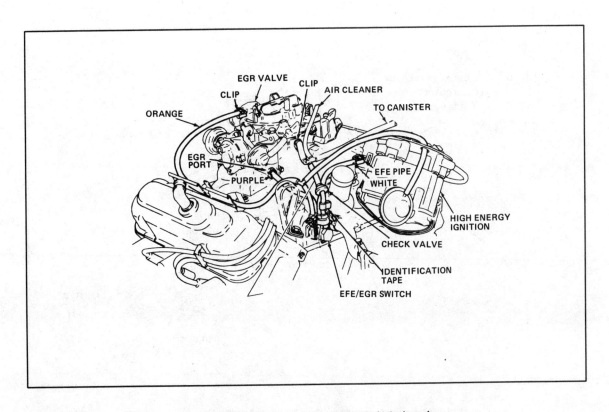

1979 Buick Vacuum hose routing—Code A—Series A (Calif./auto. trans.)

57

Buick

VACUUM CIRCUITS

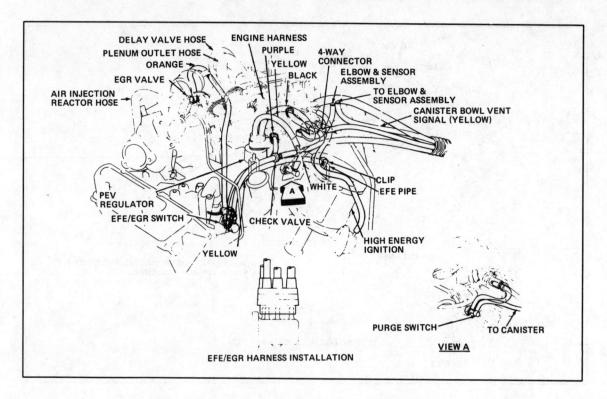

1979 Buick Vacuum hose routing—Code 3—Series A, B (49 states)

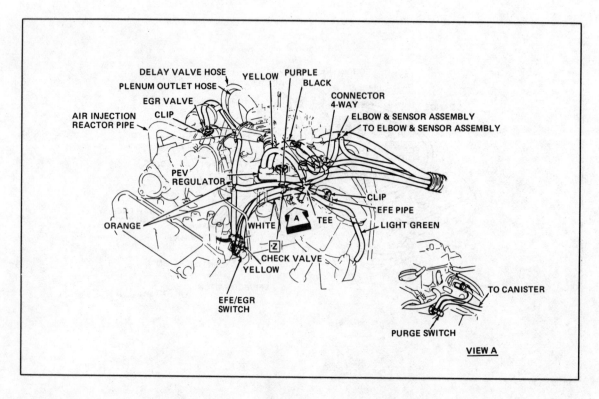

1979 Buick Vacuum hose routing—Code 3—Series A, B (Calif.)

Buick

VACUUM CIRCUITS

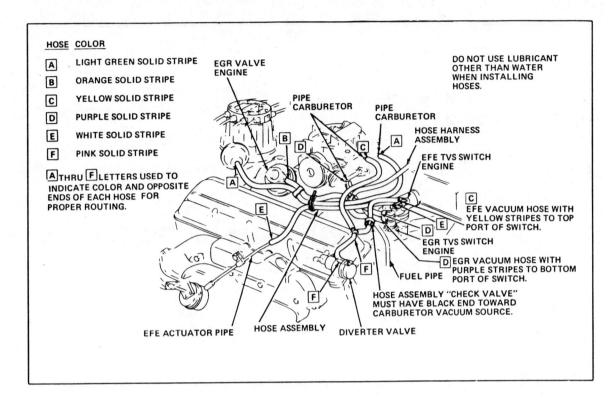

1979 Buick Vacuum hose routing—Code H, L—Series A

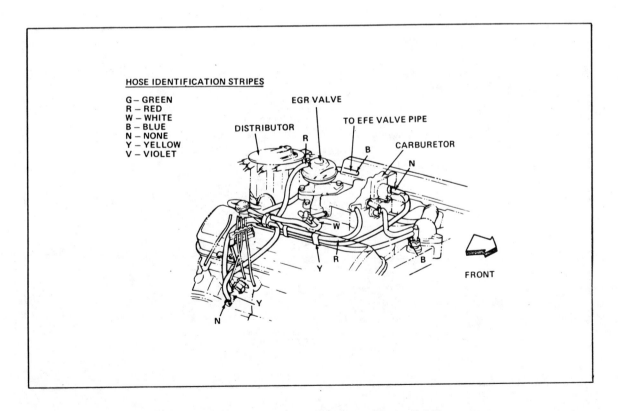

1979 Buick Vacuum hose routing—Code Y—Series A, B (49 states with A/C)

Buick

VACUUM CIRCUITS

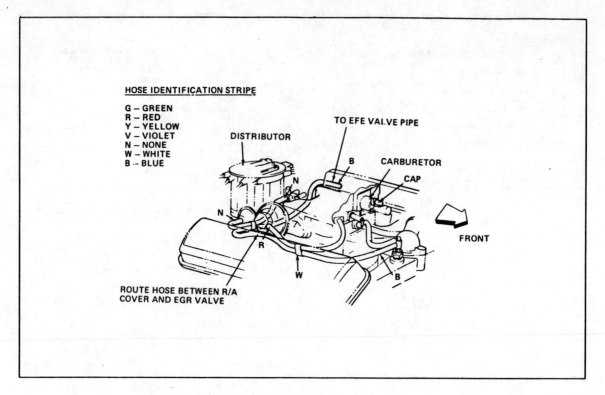

1979 Buick Vacuum hose routing—Code W—Series A (49 states)

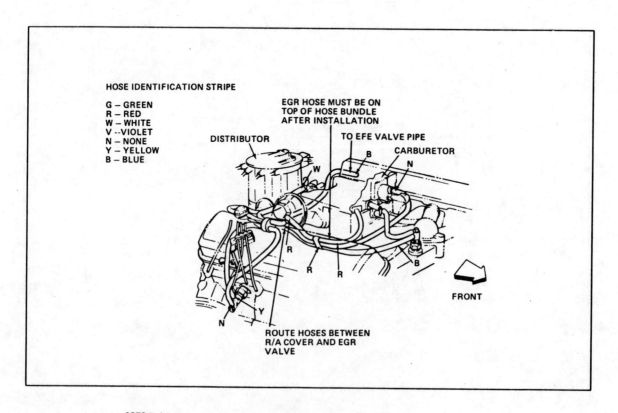

1979 Buick Vacuum hose routing—Code W—Series A (49 states with A/C)

Buick

VACUUM CIRCUITS

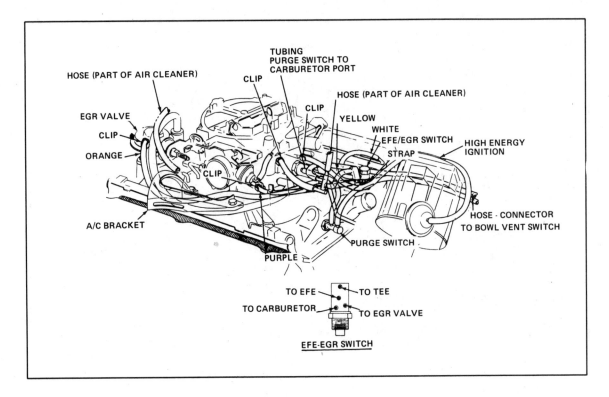

1979 Buick Vacuum hose routing—Code X—Series B, C

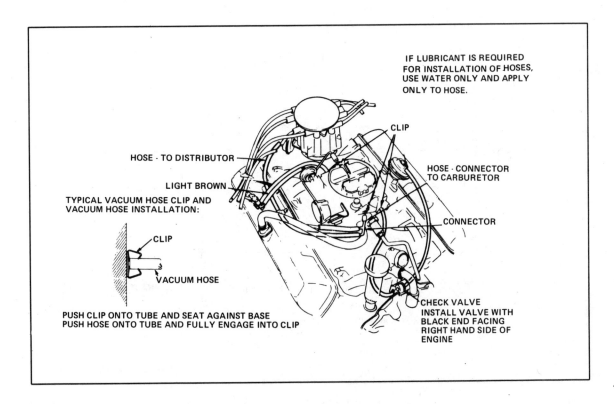

1979 Buick Vacuum hose routing—Code K—Series B, C (49 states)

Buick

VACUUM CIRCUITS

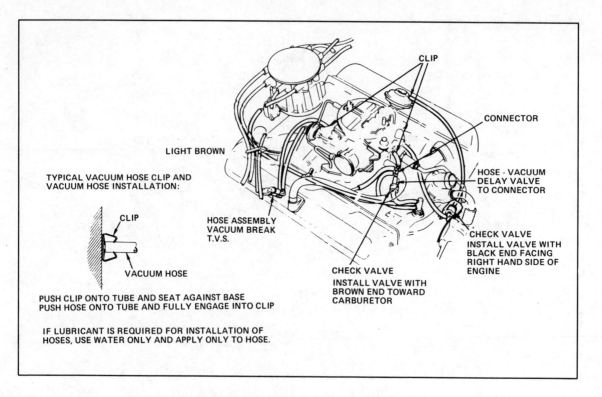

1979 Buick Vacuum hose routing—Code R, K—Series B, C (High alt.)

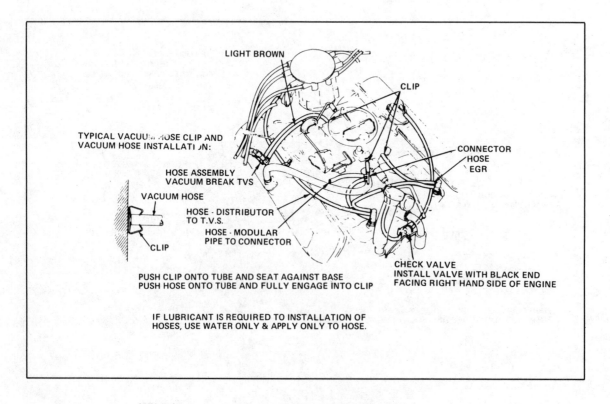

1979 Buick Vacuum hose routing—Code R, K—Series B, C (Calif.)

Buick

VACUUM CIRCUITS

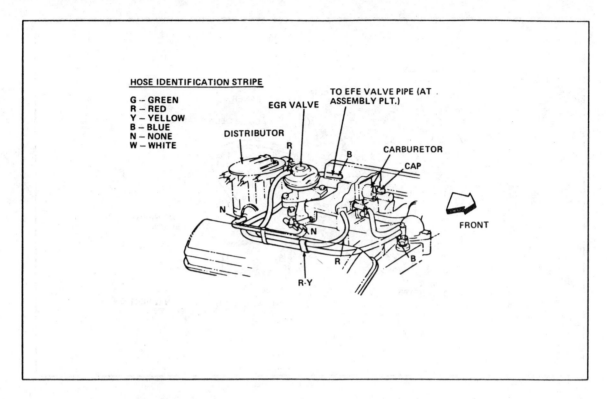

1979 Buick Vacuum hose routing—Code Y—Series A, B (49 states)

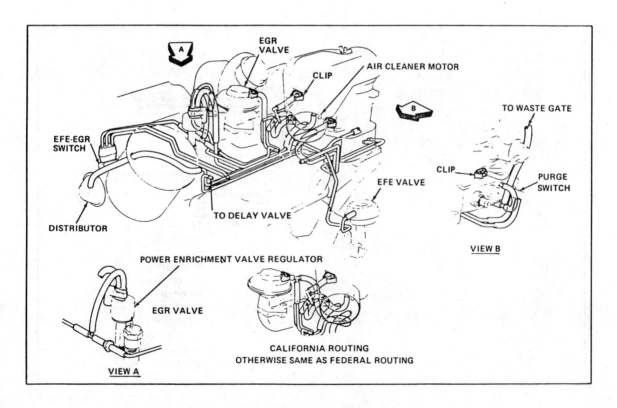

1979 Buick Vacuum hose routing—Code 3—Series E

Buick

VACUUM CIRCUITS

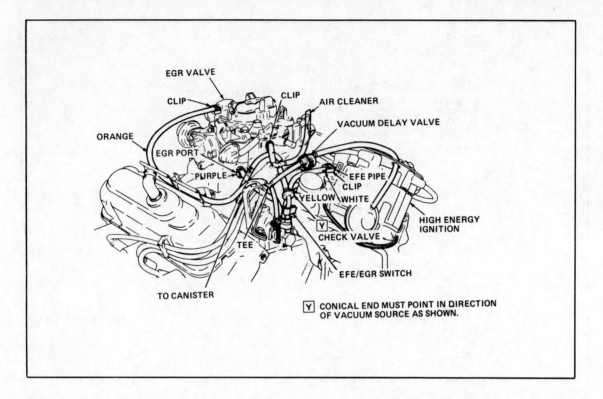

1979 Buick Vacuum hose routing—Code A—Series H (Calif./man. trans.)

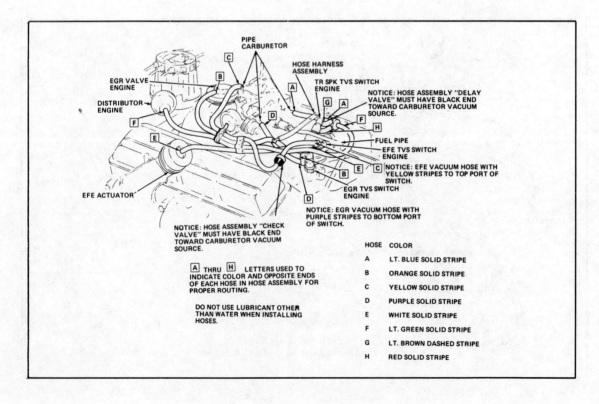

1979 Buick Vacuum hose routing—Code G—Series X

Buick

VACUUM CIRCUITS

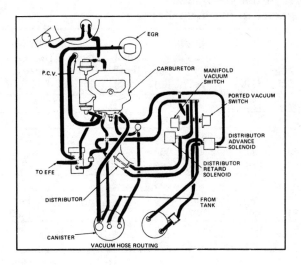

1979 Buick Vacuum hose schematic—C-4 System

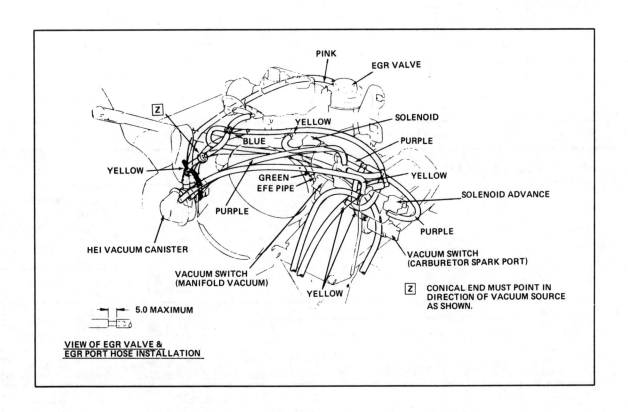

1979 Buick Vacuum hose routing—C-4 System

Cadillac

Calais • DeVille • Eldorado • Seville

TUNE-UP SPECIFICATIONS

Engine Code, 5th character of the VIN number
Model Year Code, 6th character of the VIN number

Year	Eng. V.I.N. Code	Engine No. Cyl. Disp. (cu in)	Eng. Mfg.	Carb Bbl	H.P.	SPARK PLUGS Orig. Type	Gap (In)	DIST. Point Dwell (deg)	DIST. Point Gap (In)	IGNITION TIMING (deg BTDC)	VALVES Intake Opens (deg BTDC)	FUEL PUMP Pres. (psi)	IDLE SPEED (rpm) Auto Trans In Drive
75	S	8-500	C	4v	190	R-45NSX	.060	E.I.		6	21	5.25-6.50	600/400
	S	8-500	C	E.F.I.	210	R-45NSX	.060	E.I.		6	21	5-6	600
76	R	8-350	O	E.F.I.	180	R-46SX	.080	E.I.		10 (6)	16	39±.78	600 (600)
	S	8-500	C	4v	190	R-45NSX	.060	E.I.		6	21	5.25-6.50	600
	S	8-500	C	E.F.I.	215	R-45NSX	.060	E.I.		12	21	39±.78	600
77	R	8-350	O	E.F.I.	180	R-47SX	.060	E.I.		10(8) @ 600	22	39±.78	650
	S	8-425	C	4v	180	R-45NSX	.060	E.I.		18 @ 1400	11	5.25-6.50	600
	T	8-425	C	E.F.I.	195	R-45NSX	.060	E.I.		18 @ 1400	11	39±.78	650
78	B	8-350	C	E.F.I.	170	R-47SX	.060	E.I.		10(8) @ 600	22	39±.78	600
	S	8-425	C	4v	180	R-45NSX	.060	E.I.		①	11	5.25-6.50	600
	T	8-425	C	E.F.I.	195	R-45NSX	.060	E.I.		18 @ 1400	11	39±.78	650
79	B	8-350	C	E.F.I.	170*	R-47SX	.060	E.I.		②	22	39±.78	600
	S	8-425	C	4v	180*	R-45NSX	.060	E.I.		②	11	5.25-6.50	600
	T	8-425	C	E.F.I.	195*	R-45NSX	.060	E.I.		②	11	39±.78	650

- FIGURES IN PARENTHESES INDICATES CALIFORNIA SPECIFICATION.
① ELDORADO—22° @ 1600 BROUGHAM AND DeVILLE W/4V—21°B @ 1600 RPM, FEDERAL AND CALIFORNIA FLEETWOOD W/4V—18°B @ 1600 RPM, FEDERAL AND CALIFORNIA HIGH ALTITUDE BROUGHAM, DeVILLE, AND FLEETWOOD W/4V 23°B @ 1600 RPM

② INFORMATION NOT AVAILABLE AT TIME OF PUBLICATION, SEE EMISSION CONTROL DECAL UNDER HOOD.
THE INFORMATION LISTED ON THE EMISSION CONTROL TUNE-UP DECAL SUPERCEDES ALL PUBLISHED INFORMATION AS IT MAY REFLECT PRODUCTION CHANGES.

Calais • DeVille • Eldorado • Seville

DISTRIBUTOR SPECIFICATIONS

Year	Engine	Eng. V.I.N. Code	Distributor Identification Number	Centrifugal Advance Start Dist. Deg. @ Dist. RPM	Centrifugal Advance Finish Dist. Deg. @ Dist. RPM	Vacuum Advance Start @ In. Hg.	Vacuum Advance Finish Dist. Deg. @ In. Hg.
75	500	S	1112891	0 @ 200-450	8-10 @ 2500	4.5-5.5	14.5 @ 16
	500(E.F.I.)	S	1112892	0 @ 200-525	6-8 @ 2500	—	—
	500	S	1112897	0 @ 200-450	8-10 @ 2500	5.5-6.5	10.5 @ 13
76	500	S	1112897	0 @ 200-450	8-10 @ 2500	5.5-6.5	10.5 @ 13
	500	S	1112924	0 @ 200-525	6-8 @ 2500	4.5-5.5	14.5 @ 15
	350	R	1112931	0 @ 450	21 @ 2500	5	14 @ 15.5
	500	S	1112954	0 @ 200-450	8-10 @ 2500	4.5-5.5	14.5 @ 15
	500	S	1113202	0 @ 200-525	8-16 @ 2500	5.5-6.5	10.5 @ 13
77	425	S	1103217	0 @ 350	8.5 @ 2000	5	14 @ 15.5
	425	S	1103219	0 @ 350	8.5 @ 2000	5	14 @ 15.5
	350	R	1103221	0 @ 450	21 @ 2500	7.5	12 @ 17.5
	350	R	1103222	0 @ 450	21 @ 2500	A-10 R-3	A-9 @ 18 R-4 @ 9
	425	S	1103297	0 @ 400	7 @ 2500	4	10 @ 12.6
	425	S	1103298	0 @ 400	7 @ 2500	5	10 @ 9.5

Cadillac

Calais • DeVille • Eldorado • Seville

DISTRIBUTOR SPECIFICATIONS

Year	Engine	Eng. V.I.N. Code	Distributor Identification Number	Centrifugal Advance Start Dist. Deg. @ Dist. RPM	Centrifugal Advance Finish Dist. Deg. @ Dist. RPM	Vacuum Advance Start @ In. Hg.	Vacuum Advance Finish Dist. Deg. @ In. Hg.
78	350	B	1103307(AB)	.5R-1.6 @ 450	20-22 @ 2500	9.5	8.5-9.5 @ 18.4
	425	S	1103331(AC)	.5-3.5 @ 450	6.4-9.5 @ 2500	3.5	13.5-14.5 @ 17.3
	425	S	1103332(AD)	.5-3.5 @ 450	6.4-9.5 @ 2500	3.5	9.5-10.5 @ 13.2
	425	T	1103334(AH)	.5-3.5 @ 450	7.0-9.5 @ 2500	4.4	13.5-14.5 @ 16.2
	425	T	1103335(AJ)	.5R-1.6 @ 450	6.0-8.0 @ 2500	3.5	9.5-10.5 @ 13.2
	425	S	1103345(AN)	.5-3.5 @ 450	6.4-9.5 @ 2500	3.5	7.5-8.5 @ 17.2
	350	B	1103348(AO)	.5R-1.6 @ 450	20-22 @ 2500	5.6	11.5-12.5 @ 12.5
	350	B	1103349(AP)	.5R-1.6 @ 450	20-22 @ 2500	5.6	13.5-14.5 @ 16.4
	425	S	1103352(AS)	.5-3.5 @ 450	6.4-9.5 @ 2500	5.6	4.5-5.5 @ 11.2
	425	S	1103389(AT)	.5-3.5 @ 450	6.4-9.5 @ 2500	3.5	27-29 @ 15.7

.5R = .5 Degrees Retarded
EFI = Electronic Fuel Injection

ROCHESTER QUADRAJET SPECIFICATIONS

CADILLAC

Year	Carburetor Identification①	Float Level (in.)	Air Valve Spring (turn)	Pump Rod (in.)	Primary Vacuum Break (in.)	Secondary Vacuum Break (in.)	Secondary Opening (in.)	Choke Rod (in.)	Choke Unloader (in.)	Fast Idle Speed② (rpm)
1973	7047331	1/4	1/2	11/32	0.200	—	③	0.090	0.015	—
	7047332	23/64	1/2	11/32	0.205	—	③	0.090	0.015	—
1974	7044230	1/4	3/8	1/4	0.185	—	③	0.110	0.312	1200-1500
	7044232	23/64	1/2	1/4	0.200	—	③	0.110	0.312	1200-1500
	7044530	1/4	3/8	1/4	0.185	—	③	0.110	0.312	1200-1500
	7044532	23/64	1/2	1/4	0.200	—	③	0.110	0.312	1200-1500
	7044234	1/4	7/16	11/32	0.185	—	③	0.110	0.312	1200-1500
	7044235	23/64	9/16	11/32	0.200	—	③	0.110	0.312	1200-1500
	7044233	19/64	3/8	11/32	0.185	—	③	0.110	0.312	1200-1500
1975	7045230	15/32	7/16	3/8	0.160	0.130	③	0.080	0.215	1200-1250
	7045530	15/32	1/2	3/8	0.230	0.230	③	0.080	0.215	1200-1250
1976	7056232	13/32	3/8	3/8	0.160	0.160	③	0.080	0.230	1400
	7056230	13/32	3/8	3/8	0.160	0.160	③	0.080	0.230	1400
	7056530	7/16	3/8	9/32	0.160	0.160	③	0.080	0.230	1400
1977	17057232 17057233	13/32	1/2	3/8	0.140	0.140	③	0.080	0.230	1400
	17057230	13/32	1/2	7/16	0.140	0.140	③	0.080	0.230	1400
	17057231	17/32	1/2	3/8	0.140	0.140	③	0.080	0.230	1400
	17057530	13/32	1/2	7/16	0.150	0.150	③	0.080	0.230	1500
1978	17058230	13/32	1/2	3/8	0.150	0.165	③	0.080	0.230	1500
	All others	13/32	1/2	3/8	0.140	0.250	③	0.080	0.230	1400

① The carburetor identification number is stamped on the float bowl, near the secondary throttle lever.
② On second step of cam.
③ No measurement necessary on two point linkage; see text.

Cadillac

CAR SERIAL NUMBER AND ENGINE IDENTIFICATION

1976

Mounted behind the windshield on the driver's side is a plate with the vehicle identification number. The sixth character is the model year, with 6 for 1976. The fifth character is the engine code, as follows.

S 500 V-8
R 350 Seville V-8

Part of the car serial number is stamped on the rear of the block behind the intake manifold on all except the Seville. The Seville has part of the serial number on a boss at the front left side of the block, below No. 1 spark plug. The second character is the model year.

1977

The vehicle identification number is mounted on a plate behind the windshield on the driver's side. The sixth character is the model year, with 7 for 1977. The fifth character is the engine code, as follows:

R .... 350 Seville F.I. V-8
S 425 4-bbl. V-8
T .... 425 F.I. V-8

Part of the car serial number is stamped on the rear of the block behind the intake manifold on all except the Seville. The Seville has part of the serial number on a boss at the front left side of the block, below No. 1 spark plug. The second character is the model year.

1978

The vehicle identification number is mounted on a plate behind the windshield on the driver's side. The sixth character is the model year, with 8 for 1978. The fifth character is the engine code, as follows:

B .... 350 Seville F.I. V-8
N .... 350 Seville diesel V-8
S 425 4-bbl. V-8
T .... 425 F.I. V-8

Part of the car serial number is stamped on the rear of the block behind the intake manifold on all except Seville. The Seville has part of the serial number on a boss at the front left side of the block, below No. 1 spark plug. The second character is the model year.

Cadillac 425-inch V-8's are made by Cadillac. The Seville 350 V-8 is based on a block assembly produced by Oldsmobile, which is similar to other Olds 350 V-8's. Cadillac adds the fuel injection to complete the Seville engine.

1979

The vehicle identification number is mounted on a plate behind the windshield on the driver's side. The sixth character is the model year, with 9 for 1979. The fifth character is the engine code, as follows:

B .... 350 F.I. V-8
N .... 350 diesel V-8
S 425 4-bbl. V-8
T .... 425 F.I. V-8

Part of the car serial number is stamped on the rear of the block behind the intake manifold on all except Seville. The Seville has part of the serial number on a boss at the front left side of the block, below No. 1 spark plug. The second character is the model year.

The 425-inch V-8's engines are made by Cadillac. The 350 V-8 engine is based on a block assembly produced by Oldsmobile, which is similar to other Olds 350 V-8's.

EMISSION EQUIPMENT

1976

Carburetor Models

Closed positive crankcase ventilation
Air pump
  49-States
    Commercial chassis with
      145-amp alternator only
  Calif.
    All models
Emission calibrated carburetor
Emission calibrated distributor
Heated air cleaner
Single diaphragm vacuum advance
Vapor control, canister storage
Exhaust gas recirculation
Catalytic converter
Electric choke
Early fuel evaporation

Fuel Injection Models

Closed positive crankcase ventilation
Air pump
Emission calibrated distributor
Single diaphragm vacuum advance
Vapor control, canister storage
Exhaust gas recirculation
Catalytic converter
  NOTE: *The electric choke, heated air cleaner, and early fuel evaporation systems are not used on fuel injection engines.*

1977

Carburetor Models

Closed positive crankcase ventilation
Air pump
  49-States
    Not used
  Calif.
    All models
  Altitude
    Not used
Emission calibrated carburetor
Emission calibrated distributor
Heated air cleaner
Vapor control, canister storage
Exhaust gas recirculation
Catalytic converter
Electric choke
Early fuel evaporation
Altitude compensation
  49-States
    Not used
  Calif.
    Not used
  Altitude
    All models

Fuel Injection Models

Closed positive crankcase ventilation
Air pump
Emission calibrated distributor
Vapor control, canister storage
Exhaust gas recirculation
Catalytic converter

1978

Carburetor Models

Closed positive crankcase ventilation
Emission calibrated carburetor
Emission calibrated distributor
Heated air cleaner
Vapor control, canister storage
Exhaust gas recirculation
Catalytic converter
Electric choke
Early fuel evaporation
Altitude compensation
  Not used
Air pump
  49-States
    Limousines and commercial chassis only
  Calif.
    All models
  Altitude
    All models
Temperature controlled choke vacuum break
  49-States
    All models
  Calif.
    Not used
  Altitude
    All models

Fuel Injection Models

Closed positive crankcase ventilation
Air pump
Emission calibrated distributor
Vapor control, canister storage
Exhaust gas recirculation
Catalytic converter

Diesel Models

Positive crankcase ventilation

Cadillac

EMISSION EQUIPMENT

1979

Carburetor Models
Closed positive crankcase ventilation
Emission calibrated carburetor

Sealed idle mixture screws
Emission calibrated distributor
Heated air cleaner

Vapor control, canister storage
Exhaust gas recirculation
Catalytic converter

Electric choke
Early fuel evaporation
Intake manifold with riser tubes

Air pump
  49-States
    Limousines and commercial chassis only
  Calif.
    All models
  Altitude
    All models
Temperature controlled choke vacuum break
  49-States
    All models
  Calif.
    Not used
  Altitude
    All models

Fuel Injection Models
Closed positive crankcase ventilation
Air pump, except Seville for Calif.

Emission calibrated distributor
Vapor control, canister storage
Exhaust gas recirculaion

Catalytic converter
"Closed Loop" Fuel Injection Sensor
  Calif.
    All Sevilles

Diesel Models
Positive crankcase ventilation

IDLE SPEED AND MIXTURE ADJUSTMENTS

1976

Air cleaner .... Removed
Auto trans. .... Drive
Air cond. .... Off
Dist. vac. adv. hose. ..... Connected
Parking brake vac. hose . Disconnect at cylinder and plug
Air leveling comp. hose . Disconnect at air cleaner and plug
All 600
Mixture adj.
  49-States .... 50 rpm drop (650-600)
  Calif. .... 20 rpm drop (620-600)

NOTE: *The air cleaner must be removed, but the hoses should be left connected, except for the air leveling compressor hose. The distributor hose must be connected, because the distributor uses manifold vacuum. The parking brake hose must be disconnected and plugged under the instrument panel, so that calibrated leakage in other parts of the system will be in operation during the adjustment. Mixture adjustments are made on carburetor engines only. Fuel injection engines have only one adjustment, for idle speed. See Fuel Injection, in this section, for idle adjustment procedures.*

1977

Air cleaner ....Set aside, with hoses connected
Auto. trans. ....Drive
Air cond. ....Off
Parking brake vac. hose ..Disconnect at cylinder and plug
Air leveling comp. hose ..Disconnect at air cleaner and plug
Other hoses ....See underhood label
Idle Speed
  All 600
Mixture adj.
  49-States ....70 rpm drop (670-600)
  Calif. ....30 rpm drop (630-600)
AC idle speedup 675

NOTE: *After adjusting the mixture so that the engine is at 600 rpm, check the underhood label. If the speed on the label is different, reset the idle speed with the throttle screw. The mixture adjustment is made as shown above, even though it may be necessary to correct the final idle speed to agree with the underhood label. The above idle speed is for all cars, including fuel injection. Mixture adjustments are made on carburetor engines only. AC idle speedup is made only on carburetor engines equipped with a solenoid.*

1978

Air cleaner .... In place
Auto. trans. ....Drive
Air cond. .... Off
Parking brake vac. hose .. Disconnect at cylinder and plug
Air leveling comp. hose .... Disconnect at air cleaner and plug
Other hoses ...... See underhood label
Idle speed
  Carburetor engines 600
  AC idle speedup 675
  Fuel inj. engines 600
  Diesel engines 575

1979

Air cleaner .... In place
Auto. trans. ....Drive
Air cond. ....Off
Parking brake vac. hose .Disconnect at cylinder and plug
Air leveling comp. hose ..Disconnect at air cleaner and plug
Other hoses ....See underhood label
Idle speed
  Carburetor engines 600
  AC idle speedup 675
  Fuel inj. engines 600
  Diesel engines 575

NOTE: *Idle Air/Fuel Adjustment should only be made at times of major carburetor overhaul, throttle base replacement or high idle CO as determined by state or local emission inspections.*

PROPANE ENRICHED MIXTURE ADJUSTMENTS 1978

Air cleaner................. In place
Auto. trans. .... Drive
Air cond. .... Off

Parking brake vac. hose ... Disconnect at cylinder and plug
Other hoses ...... See underhood label

Enriched RPM
Cadillac, except Eldorado
  49-States 630
  Calif. 640
  Altitude 625
Eldorado
  49-States 640
  Calif. 640

Altitude 625
Commercial and Limousine
  49-States 640
  Calif. 640
  Altitude 625

NOTE: *Mixture adjustments are made on carburetor engines only.*

Cadillac

INITIAL TIMING

1976

Carburetor engines ..... 6° BTDC
Fuel Injection
  Cadillac & Eldorado .. 12° BTDC
  Seville .... 10° BTDC

NOTE: *The distributor vacuum hose must be disconnected and plugged. Also disconnect and plug the auto level compressor hose at the air cleaner. The hole in the timing bracket on the engine front cover is for magnetic timing equipment only.*

1977

49-States
  Cadillac (1400 rpm) ..18° BTDC
  Seville (600 rpm) .....10° BTDC
Calif.
  Cadillac (1400 rpm) ...18° BTDC
  Seville (600 rpm) ..... 8° BTDC
Altitude
  Same as 49-States

NOTE: *The distributor vacuum hose must be disconnected and plugged. Also disconnect and plug the EGR vacuum hose on Cadillac engines, both 4-bbl. and F.I. The hole in the timing bracket on the engine front cover is for magnetic timing equipment only.*

1978

NOTE: *The distributor vacuum hose must be disconnected and plugged. Also disconnect and plug the EGR vacuum hose on Cadillac engines, both 4-bbl. and F.I. The hole in the timing bracket on the engine front cover is for magnetic timing equipment only. Set timing with engine running at rpm shown below.*

425 carburetor except Eldorado
  49-States (1600 rpm) ..... 21° BTDC
  Calif. (1600 rpm) .... 21° BTDC
  Altitude (1600 rpm) ...... 23° BTDC
425 fuel inj. engine
  All (1400 rpm) .... 18° BTDC
Eldorado
  49-States (1600 rpm) ..... 21° BTDC
  Calif. (1600 rpm) .... 18° BTDC
  Altitude (1600 rpm) ...... 23° BTDC
Com. chassis and limousine
  49-States (1600 rpm) ..... 18° BTDC
  Calif. (1600 rpm) .... 18° BTDC
  Altitude (1600 rpm) ...... 23° BTDC
Seville 350
  49-States (600 rpm) ...... 10° BTDC
  Calif. (600 rpm) .... 8° BTDC
  Altitude (600 rpm) .... 10° BTDC

1979

CAUTION: *Emission control adjustment changes are noted on the Vehicle Emission Information Label by the manufacturer. Refer to the label before any adjustments are made.*

NOTE: *The distributor vacuum hose must be disconnected and plugged. Also disconnect and plug the EGR vacuum hose on Cadillac engines, both 4-bbl. and F.I. The hole in the timing bracket on the engine front cover is for magnetic timing equipment only. Set timing with engine running at rpm shown below.*

425 carburetor except Eldorado
  49-States (1600 rpm) ..21° BTDC
  Calif. (1600 rpm) .....21° BTDC
  Altitude (1600 rpm) ..23° BTDC
425 fuel inj. engine
  All (1400) rpm ......18° BTDC
Eldorado
  49-States (1600 rpm) ..21° BTDC
  Calif. 1600 rpm) ......18° BTDC
  Altitude (1600 rpm) ..23° BTDC
Com. chassis and limousine
  49-States (1600 rpm) ..18° BTDC
  Calif. (1600 rpm) .....18° BTDC
  Altitude (1600 rpm) ..23° BTDC
Seville 350
  49-States (600 rpm) ..10° BTDC
  Calif. (600 rpm) ...... 8° BTDC
  Altitude (600 rpm) ....10° BTDC

SPARK PLUGS

1976

All models—AC-R45NSX ..... .060

1977

Cadillac ....AC-R45NSX .... .060
Seville ....AC-R47SX .... .060

1978

All 425 .... AC-R45NSX ...060
Seville .... AC-R47SX ...060

1979

All 425 ......AC-R45NSX ... .060
Seville and Eldorado w/EFI .. .060

VACUUM ADVANCE

1976

49-States
  Seville before 463462 ..... Ported
  Seville after 463462 .... Manifold
  Cadillac F.I. .... Ported
  Cadillac 4-bbl. .... Manifold
Calif.
  Seville .... Ported
  Cadillac F.I. .... Ported
  Cadillac 4-bbl. .... Manifold

NOTE: *Fuel injection engines with ported vacuum spark advance use a thermal vacuum switch (TVS) at the front of the block, connected to the distributor. Fuel injection engines with manifold vacuum spark advance do not have the TVS. Carburetor engines have a TVS that operates the EFE valve only.*

1977

Diaphragm type
  49-States ....Single
Calif.
  Cadillac ex. Eldo. ....Single
  Eldorado ....Dual
  Seville ....Dual
  Altitude ....Single
Vacuum source
  49-States ....Manifold
Calif.
  4-Bbl. carb. ....Manifold
  Fuel inj. ....Ported

1978

Diaphragm type .... Single
Vacuum source
  All with carburetor .... Manifold
  All fuel inj. .... Ported

1979

Diaphragm ....Single
Vacuum source
  All with carburetor ....Manifold
  All fuel inj. ....Ported

Cadillac

FUEL INJECTION, 1976

Cadillac's 1976 fuel injection, which is standard on the Seville, and optional on other Cadillac models, is the type known as port injection. Most of the injector system is built onto a special intake manifold, which has eight injectors mounted so that they squirt fuel into the intake manifold as close as possible to the cylinder head intake port. Air for the engine is admitted through a throttle body located in the same place as a conventional carburetor. The injectors open when they receive a signal from an electronic control unit, mounted behind the instrument panel above the glove box. The engine receives more or less fuel, depending on what it needs, by varying the length of time that the injectors are open. This variation in injector opening period is controlled by the electronic control unit, which receives information on manifold vacuum, inlet air temperature, and engine speed. Actually, what the electronic control unit does is meter the amount of fuel going into the engine according to how much air the engine takes in, which is exactly what a carburetor does.

Because the fuel is injected close to the intake valve, it has little or no chance to fall out of the airstream, so standard carburetor engine items such as intake manifold heat and finely atomized fuel-air mixtures are unnecessary. The exhaust heat valve is eliminated on fuel injection engines, and a blank spacer used in its place.

Injectors

An injector is a valve controlling the flow of fuel under high pressure. There are eight fuel injectors mounted on the intake manifold, each one close to the intake port for its

EMISSION CONTROL SYSTEMS 1976
FIRING ORDER, 1976

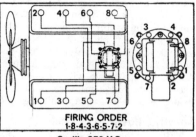

FIRING ORDER
1-8-4-3-6-5-7-2
Seville 350 V-8

cylinder. Each injector is connected to the fuel system so that it receives fuel under about 40 lbs of pressure whenever the fuel pump is running. When the electronic control unit sends electric current to the injectors, they open and the fuel squirts into the intake manifold.

The eight injectors are divided into two groups, with the outer cylinders (1,2,7,8) in the first group, and the inner cylinders (3,4,5,6) in the second group. The opening time for Cadillac group No. 1 is just before the intake valve for No. 2 cylinder opens. The second group opens just before the intake valve for cylinder No. 5. The opening time for Seville group No. 1 is just before the intake valve for No. 7 cylinder opens. The second group opens just before the intake valve for cylinder No. 4. All the injectors in each group open and close at the same time.

When a group of injectors opens, the intake valves on all cylinders in that group are closed. The fuel stays in the intake manifold until the valve opens. Fuel for the last cylinder in the group has waited in the intake manifold longer than the fuel for the other three, but this doesn't hurt anything. If you consider that all the injectors are doing is putting fuel into the intake manifold, the same as a carburetor, then the system is easier to understand.

Fuel Delivery

A booster pump inside the fuel tank pumps the fuel to a high pressure fuel pump mounted under the car on the frame rail. From the high pressure pump the fuel goes to a fuel filter mounted on the engine in the same position as the old engine driven fuel pump, which is no longer used. The fuel pipe then goes to the back of the engine to a "T" and runs down each side of the intake manifold where it connects to each injector. Another "T" at the front of the engine connects the fuel pipe to the fuel pressure regulator, and from the regulator a return line takes the fuel back to the tank.

Between the high pressure fuel pump and the regulator, the pressure is kept at a constant 39 lbs psi above the pressure in the intake manifold. The pressure regulator maintains this pressure by allowing only excess fuel to go back to the tank. The fuel pump has a capacity much higher than 39 lbs so that no matter how fast the engine uses fuel, the system is always kept under 39 lbs of pressure. The booster pump in the tank is a low pressure design whose only function is to keep the high pressure pump supplied with fuel.

Fuel Pressure Regulator

A spring-loaded diaphragm inside the pressure regulator is made so that it takes 39 lbs psi to open the orifice and allow fuel to go back to the tank. A vacuum chamber on the opposite side of the diaphragm is connected to the throttle body by a hose. As the pressure (vacuum) inside the intake manifold changes, it affects the strength of the spring and varies the pressure to maintain a constant 39 lbs psi above the pressure in the intake manifold.

When the engine is shut off, the fuel pumps stop, but there is a check valve in the high pressure pump that

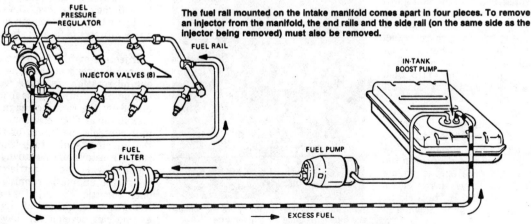

The fuel rail mounted on the intake manifold comes apart in four pieces. To remove an injector from the manifold, the end rails and the side rail (on the same side as the injector being removed) must also be removed.

Cadillac

Fuel pressure regulator (arrow) is mounted at the front near the distributor

holds the pressure between the pump and the regulator. In theory, this fuel will stay in the line under pressure until the next time the engine is started, no matter how long that might be.

Pressure Release

Because the system is designed to stay under pressure, even when the engine is off, you must take precautions to avoid being squirted with fuel when loosening any fuel fittings or changing the filter. Cover all fittings with a cloth to catch the spray of fuel while they are being loosened.

On later production cars a Schrader pressure release fitting is permanently installed in the fuel pipe at the rear of the intake manifold. This is the same type of valve used on air conditioning systems. To relieve the pressure, remove the valve cap, cover the valve with a cloth, and depress the core in the valve.

Fast Idle Valve

Air entering the engine during normal operation is controlled by the throttle body, which works the same as the throttle body on the bottom of an ordinary 2-bbl. carburetor. A choke is not used, because the system automatically feeds more fuel through the injectors when the engine is cold. A cold engine needs additional air to go with the extra fuel, and this air is provided by the fast idle valve.

On carburetors, the fast idle linkage provides additional air by holding the throttle open. The fuel injection fast idle valve provides additional air by opening a passage in the throttle body. As the engine warms up, the fast idle valve slowly closes off the additional air. It stays closed until the engine is shut down and cools off.

Inside the fast idle valve is an electric heater, a valve, and a temperature sensitive element. The heater is connected to the fuel pump circuit, so that whenever the fuel pump is operating, the heater is on. When the valve is cold, the element opens the fast idle passage. When the engine is running, the heater warms up the element, and the valve closes. The length of time it takes the valve to close depends on the outside air temperature. At 68° F. the valve will close in approximately 1½ minutes. At 20° below zero, it will take 5 minutes.

Throttle Position Switch

This switch is mounted on the throttle body and connected to the throttle shaft. It is electronically connected to the control unit, so that the control unit knows at all times the position of the throttle.

The mounting of the throttle position switch is adjustable. With the throttle held in the idle position, the switch should be rotated completely counterclockwise as far as it will go. Hold it in that position and tighten the mounting screws.

Temperature Sensors

Two temperature sensors are used, one for intake air temperature and the other for engine coolant. The intake air sensor is mounted on the intake manifold behind the throttle body. It threads into a hole in one of the intake air passages so that it senses the temperature of the air passing through the manifold.

The coolant sensor is mounted on the heater hose fitting at the rear of the right cylinder head. It senses the temperature of the coolant inside the hose. Both of the sensors are connected electrically to the electronic control unit. Both of the sensors are identical and interchangeable.

Speed Sensor

The High Energy Ignition used on fuel injection engines is the same electrically as other HEI ignitions.

Temperature sensors (arrows) are for coolant and intake manifold air temperature

But the distributor housing and shaft are different. Built into the housing underneath the bowl of the distributor is a speed sensor. The sensor is made up of two switches that are tripped by a magnet attached to the distributor shaft.

The sensor tells the electronic control unit how fast the engine is going, and synchronizes the injectors to the intake valve opening.

Electronic Control Unit

This unit, installed behind the instrument panel above the glove compartment, is the brains of the system. It receives the following signals from the various engine sensors.

Throttle position and change of position
Engine speed and firing position
Intake manifold pressure (vacuum)
Intake manifold air temperature
Engine coolant temperature

The intake manifold pressure sensor is built into the electronic control unit. A hose brings manifold vacuum from the throttle body to the electronic control unit. The unit is programmed to control the electric fuel pump, the fast idle valve, and the injectors according to the signals it gets. In addition the unit also controls the exhaust gas recirculation. A vacuum solenoid in the ported vacuum line to the EGR valve turns the valve on and off according to the commands from the electronic control unit. Vacuum to the EGR valve is shut off below 130° F. coolant temperature.

The control unit keeps the electric fuel pumps running all the time the engine is either cranking or running. If the engine is not turning over, the fuel pumps will operate for only one second after the ignition switch is turned on.

The fast idle valve is connected to the fuel pump circuit, so it will only operate for one second if the ignition switch is turned on without cranking the engine. This keeps the fast idle valve heater off and ready for a start even if the driver turns the switch on and gets busy with something else for a few minutes before he starts the engine.

Injection of fuel from the injectors always happens at the same point in engine rotation. In other words, the timing of the injectors is fixed, determined by the rotating magnet in the distributor. The electronic control unit controls the length of time that the injectors squirt fuel. This can be zero (injectors off) as during deceleration or wide open throttle cranking. When the engine is running at wide open throttle, the injectors are open the maximum amount of time, so the engine can develop full power.

Cadillac

Idle Speed Adjustment

The electronic control unit controls the injectors during engine idle so that the engine gets the right mixture. There are no idle mixture adjustments. Idle speed is controlled by a bypass screw on the throttle body. It is an Allen head screw with a locknut. The idle speed should be adjusted to 600 rpm in Drive, with the distributor vacuum hose disconnected at the distributor and plugged on ported vacuum cars, but left connected on manifold vacuum cars (late Sevilles.) The air leveling compressor hose should also be disconnected at the air cleaner and plugged.

CAUTION: *The parking brake disconnects automatically when the transmission is in Drive. Either disconnect and plug the hose at the vacuum release cylinder, or block the wheels securely.*

Testing and Troubleshooting

The troubleshooting chart, furnished by Cadillac, should be used only when the Kent-Moore Fuel Injection Analyzer is not available. The analyzer, No. J-25400, comes with a Diagnosis Guide. The same information is also in the 1976 Cadillac Shop Manual, but is not in the first edition of the 1976 Seville manual.

The usual tune-up items should be checked first, before assuming that something is wrong with the fuel injection system.

ENGINE CRANKS BUT WILL NOT START

NOTE: *The following problems assume that the rest of the car electrical system is functioning properly.*

1. Blown 10 amp in-line fuel pump fuse located below instrument panel near ECU connectors. To check the fuse, turn the ignition switch to the On position, without starting the engine. The fuel pump should run for one second. If not, check the fuse.
2. Open circuit in 12 purple wire between starter solenoid and ECU.
3. Open circuit in 18 dark green wire between generator "BAT" terminal and ECU (fusible link). An open circuit will cause the same trouble as a blown fuse. Check as described under No. 1, above.
4. Poor connection at ECU jumper harness (below instrument panel) or at ECU.
5. Poor connection at fuel pump jumper harness (below instrument panel near ECU), 14 dark green wire. A poor connection will cause the same trouble as a blown fuse. Check as described under No. 1, above.

Remove the air cleaner stud (arrow) and then the fast idle valve can be twisted and removed

The fast idle valve seat can be pulled out of the recess with your fingers

Except for the seat, this is what the fast idle valve looks like. Note the small pin. If it's missing, the valve won't work

Cadillac

Throttle position sensor is mounted on side of carburetor, and plugs into wiring harness

PCV valve (arrow) is in the same place on the 500 V-8, but now is hard to pull out because the wiring harness gets in the way

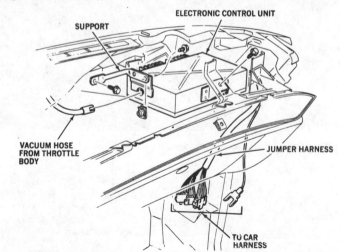

Acting as the brains of the Cadillac fuel injection system, the ECU (Electronic Control Unit) is found behind the instrument panel above the glove compartment

This screw on the front of the throttle body is used to adjust the engine idle speed. The locknut (arrow) must be loosened before moving the screw

6. Poor connection at engine coolant sensor or open circuit in sensor or wiring (cold engine only). To check the sensor for an open circuit, disconnect the wiring and use an ohmmeter. If the sensor resistance is greater than 1600 ohms, replace the sensor.
7. Poor connection at distributor trigger (speed sensor).
8. Distributor trigger (speed sensor) stuck closed.
9. Malfunction in chassis-mounted pump.
10. Malfunction in throttle position switch (W.O.T. section shorted). To check, disconnect switch—engine should start.
11. Fuel flow restriction.

HARD STARTING

1. Open engine coolant sensor (cold or partially warm engine only—starts ok hot). Disconnect the wiring and use an ohmmeter to test the sensor for an open circuit. If the sensor resistance is greater than 1600 ohms, replace the sensor.
2. Malfunction in throttle position switch (W.O.T. section shorted). To check, disconnect switch—engine should start normally.
3. Malfunction in chassis-mounted fuel pump. (Check valves leaking back.)
4. Malfunction in pressure regulator.

POOR FUEL ECONOMY

1. Disconnected or leaking MAP sensor hose.
2. Disconnected vacuum hose at fuel pressure regulator or at throttle body.
3. Malfunction of air or coolant sensor. Check the sensor for a short circuit. Disconnect the wiring and use an ohmmeter. If the sensor resistance is less than 700 ohms, replace the sensor.

ENGINE STALLS AFTER START

1. Open circuit in 12 black/yellow ignition signal wire between fuse

block and ECU or poor connection at connector (12 black/yellow wire) located below instrument panel near ECU.
2. Poor connection at engine coolant sensor or open circuit in sensor or wiring (cold or warm engine only). To check the sensor for an open circuit, disconnect the wiring and use an ohmmeter. If the sensor resistance is more than 1600 ohms, replace the sensor.

ROUGH IDLE

1. Disconnected, leaking or pinched MAP sensor hose. If plastic harness line requires replacement, replace entire EFI engine harness.
2. Poor connection at air or coolant sensor or wiring (cold engine only). To check the sensor for an open circuit, disconnect the wiring and use an ohmmeter. If the sensor resistance is more than 1600 ohms, replace the sensor.
3. Poor connection at injection valve(s).

4. Shorted engine coolant sensor. To check for a short in the sensor, disconnect the wiring and use an ohmmeter. If the sensor resistance is less than 700 ohms, replace the sensor.
5. Speed sensor harness is located close to secondary ignition wires.

PROLONGED FAST IDLE

1. Poor connection at fast idle valve or open circuit in heating element.
2. Throttle position switch misadjusted.
3. Vacuum leak.

NO FAST IDLE

1. Bent fast idle valve micro switch causing heater to malfunction and drive valve section down to locked closed position.

ENGINE HESITATES OR STUMBLES ON ACCELERATION

1. Disconnected, leaking on pinched MAP sensor hose. If plastic harness line requires replacement,

replace entire EFI engine harness.
2. Throttle position switch misadjusted.
3. Malfunction in throttle position switch.
4. Intermittent malfunction in distributor trigger (speed sensor).
5. Poor connection at 6 pin connector of ECU.
6. Poor connection at EGR solenoid or open solenoid (cold engine only).

LACK OF HIGH SPEED PERFORMANCE

1. Misadjusted throttle position switch (W.O.T. only).
2. Malfunction in throttle position switch.
3. Malfunction of chassis-mounted fuel pump.
4. Intermittent malfunction in distributor trigger (speed sensor).
5. Fuel filter blocked or restricted.
6. Open circuit in 12 purple wire between starter solenoid and ECU.

4-NOZZLE TVS 1977

49-State carburetor-equipped engines use a new 4-nozzle Thermal Vacuum Switch (TVS) in the usual position at the front left of the cylinder block. The two nozzles closest to the block control the vacuum to the EGR valve. Below 120° F. coolant temperature, the ported vacuum from the carburetor is cut off. Above 120° F. the TVS opens and allows the ported vacuum to pass through.

The two forward nozzles on the TVS control the vacuum to the Early Fuel Evaporation (EFE) valve. Below 120° F. coolant temperature, manifold vacuum passes through the TVS and operates the EFE valve, closing off the exhaust pipe and forcing the exhaust through the crossover in the intake manifold. Above 120° F. the TVS cuts off the vacuum, and vents the EFE actuator through a small foam-filled vent on the side of the TVS.

5-NOZZLE TVS 1977

California carburetor-equipped engines use a new 5-nozzle Thermal Vacuum Switch (TVS). The two nozzles closest to the block control the vacuum to the EGR valve, the same as described above under 4-Nozzle TVS. The two forward nozzles also

EMISSION CONTROL SYSTEMS 1977

operate the same way to control vacuum to the EFE valve.

The 5th nozzle, on the opposite side of the TVS, controls the manifold vacuum to the distributor vacuum advance. Above 120° F. coolant temperature the vacuum advance receives manifold vacuum for normal operation. Below 120° F. the manifold vacuum to the advance is completely shut off. This means that a cold engine, operating on fast idle, will have no vacuum advance. When adjusting fast idle on the car, the vacuum advance hose must be disconnected and plugged, to duplicate actual engine conditions. If the adjustment is made by mistake with the hose connected, then the fast idle speed will be too slow when the advance is shut off by the TVS, and this may cause engine dying.

DUAL DIAPHRAGM DISTRIBUTOR 1977

California Eldorados with fuel injection, and California Sevilles use a dual diaphragm distributor. The retard side of the diaphragm is connected to manifold vacuum through a normally open vacuum solenoid, which is hooked up to the fuel injection electronic control unit (ECU). The ECU knows at all times the temperature of the engine coolant. Below 115° F. on Eldorado, or 132° F. on Seville, the ECU de-energizes the solenoid to allow manifold vacuum to

act on the retard diaphragm. This retards the spark at idle and during deceleration. Above those temperatures the ECU energizes the solenoid, which closes, shutting off the vacuum and venting the retard diaphragm. With no vacuum on the retard diaphragm, the advance diaphragm goes to the neutral or no-advance position during idle and deceleration.

SPARK DELAY SOLENOID 1977

California Cadillacs with fuel injection (not Eldorado and not Seville) have a normally open vacuum solenoid in the hose that supplies ported vacuum to the distributor vacuum advance. The solenoid is connected to the fuel injection electronic control unit. (ECU). The ECU knows at all times the temperature of the engine coolant. Below a coolant temperature of 130° F. the ECU energizes the solenoid, which closes, preventing vacuum advance. Above 130° F. the solenoid is de-energized, and vacuum advance is normal.

EGR SYSTEM 1977

Cadillac has changed the recommended method for testing the EGR valve, on models with a separate back pressure transducer. Formerly, the test was to idle the engine on the second step of the fast idle cam and

Cadillac

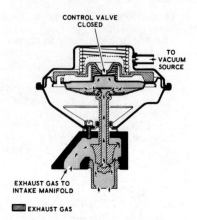

Back pressure EGR control valve—closed
(© GM Corporation)

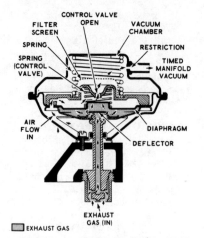

Back pressure EGR control valve—open
(© GM Corporation)

apply vacuum to the valve. The engine speed was supposed to drop off at least 250 rpm. Now, the recommended method is to run the engine at approximately 1500 rpm in NEUTRAL or PARK and disconnect the hose to the valve. When running at 1500 rpm, the valve should be open. Disconnecting the hose should cause the valve stem to move down toward the closed position. The engine speed should also increase, If not, the valve is leaking or stuck, or an exhaust passage is clogged.

Metering gaskets are used between the transducer flange and the mounting pad on the intake manifold for 1977. Following are the G.M. part numbers of the gaskets and the size of the metering hole, in inches.

Fuel Injection EGR Gaskets
49-States
   Cadillac, exc. Eldo. 1608508 . .600
   Eldorado 1608507 .... .640
   Seville 417 173 .... .390
Calif.
   Cadillac, exc. Eldo. 1608508 .600
   Eldorado 1608508 .... .600
   Seville 551 147 .... .422

EMISSION CONTROL SYSTEMS 1978

ELECTRONIC SPARK SELECTION 1978

All 1978 Seville engines have the Electronic Spark Selection (ESS) system. The system is added to a modified high energy ignition to provide advanced timing during cruising, normal timing in stop and go driving, and retarded timing when starting.

The system is actually an electronic retarder. It delays the shutoff of primary current to the coil, and thereby retards the spark. Basic timing is set with the retarder providing 6° retard. During cruising conditions, the retarder shuts off, and the spark advances 6°. During cranking, 8° retard is inserted for easier starting. On California cars, the retard during cranking stays in until the engine coolant gets up to 130°F.

The ESS system does not affect the vacuum or centrifugal spark curves that are built into the distributor. Actually, the distributor is a standard HEI unit, except that the connector has four wires instead of three. What the system does is add six engine degrees of advance. Whatever the total vacuum, centrifugal, and initial advance happen to be, the ESS system increases it by six degrees when it turns on.

The addition of the advance is controlled by an electronic decoder, mounted on the engine side of the dash, next to the wiper motor. It is connected to the distributor for rpm readings, and to a vacuum sensor for vacuum readings. The decoder knows at all times how fast the engine is turning, and the amount of engine vacuum. The decoder adds the six degrees of advance whenever the engine rpm is above 1450 ±150 rpm, and the vacuum is above 10 in. Hg. For 49 State Cavs. or 8 in Hg. for California and Altitude Cavs. Both vacuum and rpm must be above these values to get the six degrees additional advance. For example, if the car is cruising at 2000 rpm and 15 in. Hg. of vacuum, the spark would be advanced six degrees additional. But if the driver shoves his foot down, and the vacuum drops below 8 or 10 inches, the engine would lose the six degrees advance.

The vacuum sensor is the fuel economy switch, which is standard equipment on all Sevilles. This switch lights an amber light on the instrument panel below 8 or 12 in. Hg. vacuum, and turns on a green light above that vacuum. With this system, the amber light is much more important than in last year's Seville. It not only indicates that the driver has a heavy foot, but also that the engine is running six degrees retarded. When the green light is on, it means the spark is advanced, providing the engine speed is over 1450 ±150 rpm.

The decoder is also connected to the ignition switch. When the switch is in the cranking position, the decoder gets a signal and retards the spark eight degrees below normal timing. On California cars, the decoder is connected to the exhaust gas recirculation solenoid. The electronic control unit for the fuel injection keeps the solenoid closed below 130° F. engine coolant temperature. Whenever the solenoid is closed, the decoder gets a signal and holds the spark in the retarded position.

To test the advance mode of the system, run the engine in Park with the throttle blocked open at a steady 1800 rpm. Use a timing light to check the timing on the front pulley. While watching the pulley, pull the hose off of the fuel economy switch. The timing should instantly retard about 6°. Connect the hose and the timing should advance about 6° because the distributor vacuum and centrifugal advance changes slightly with changes in engine rpm. After you are familiar with the system, you can check it by holding the throttle steady at about 1800 rpm and listening for a speed change when you pull the hose off of the fuel economy switch.

To test the retard mode of the system, idle the engine in Park, with the distributor vacuum hose disconnected and plugged. The engine must be at operating temperature. Use a timing light to check the timing on the front pulley. Then disconnect the 3-pin connector (see illustration.) The timing should retard approximately 8°.

To bypass the system completely, disconnect the 4-wire plug between the distributor and the decoder, connect a jumper wire between pin No. 2 on the decoder plug, and pin No. 4 on the distributor plug. This will allow the engine to run without the ESS system, but the initial timing will be 6° higher than normal. If you are going to drive the car with the decoder bypassed, set the initial timing to the normal specification to prevent detonation. If the car will be driven for longer than a few minutes with the system bypassed, make a permanent connection that will not vibrate loose, and waterproof it with tape.

Problems with the system will usually be caused by wiring problems, usually poor connections. Kent-Moore sells an analyzer No. J-24642 that can be plugged into the wiring for testing the system. If you do not have the analyzer, and you

Cadillac

suspect a bad decoder, the only way to test it is by substituting a known good decoder.

EFE-EGR THERMAL VACUUM SWITCH 1978

Carburetor engines in 1977 used a four or five-nozzle thermal vacuum switch to control vacuum to the EGR and EFE systems. On California cars, the same switch also controlled vacuum to the distributor. In 1978, the switch is the same, and it still controls vacuum to the EGR system and to the distributor. But the EFE system is not controlled by a separate switch, mounted on the right side of the thermostat housing.

The new EFE switch applies vacuum to the EFE actuator below 165°F. instead of the 120°F. used in 1977. This lets the EFE system stay on longer for better cold driveability.

The EGR switch still keeps the EGR off below 120°F. It also blocks vacuum to the distributor so there is no vacuum advance until the engine coolant temperature goes over 120°F.

TEMPERATURE CONTROLLED CHOKE VACUUM BREAK SYSTEM

49-State cars use a thermal vacuum switch (TVS) mounted in the air cleaner to control vacuum to the secondary vacuum break at the rear of the carburetor. Below approximately 62°F. air cleaner temperature, the TVS is closed, preventing the secondary vacuum break from opening the choke. This delays the full opening of the choke for better cold operation.

The air cleaner thermal sensor also connects to the TVS. This connection is for convenience only. The vacuum path through the TVS to the thermal sensor is open at all times, and is not affected by the part of the TVS that shuts off vacuum to the vacuum break unit.

EGR VALVES 1978

49-State Cadillacs now use an EGR valve with a new design integral back pressure transducer. A bleed inside the valve regulates the opening.

To test the EGR system on carburetor engines, run the engine in NEUTRAL OR PARK at 200 rpm. Then disconnect the vacuum hose to the EGR valve. The engine should speed up, and the EGR valve stem should move down to the closed position. If not, the valve is leaking or stuck, or an exhaust passage is plugged.

SEVILLE DIESEL ENGINE 1978

This is the Oldsmobile diesel 350 V-8. For details on the engine, see the Oldsmobile chapter.

The idle speed on the diesel is 575 rpm, set with an adjustment on the side of the injection pump. Because the diesel has no electric ignition system, special equipment must be used to measure the idle rpm. A bracket for holding a magnetic tachometer pickup is on the front of the engine.

EMISSION CONTROL SYSTEMS 1979

ELECTRONIC SPARK SELECTION 1979

The ESS system is now standard on the Eldorado, Seville, Limousine, Commercial Chassis and carbureted California "C" cars.

The maximum advance RPM to put the ESS system into operation is as follows;
Eldorado 1,200
Seville 1,450
All other models 1,350

Because the EGR solenoids are deleted on the carbureted cars, a new three-way coolant temperature switch is used to send a signal to the ESS decoder, to control the spark retard during cold engine operation and to prevent over advance during hot engine operation.

EFE-EGR THERMAL VACUUM SWITCH 1979

Carburetor engines use a four or five-nozzle thermal vacuum switch to control vacuum to the EGR and EFE systems. On California cars, the same switch also controlled vacuum to the distributor. In 1979, the switch is the same, and it still controls vacuum to the EGR system and to the distributor. The EFE system is controlled by a separate switch, mounted on the right side of the thermostat housing.

The EFE switch applies vacuum to the EFE actuator below 165°F. This allows the EFE system stay on longer for better cold driveability.

The EGR switch keeps the EGR off below 120°F. which blocks vacuum to the distributor so there is no vacuum advance until the engine coolant temperature goes over 120°F.

TEMPERATURE CONTROLLED CHOKE VACUUM BREAK SYSTEM

All carbureted 49-State cars use a thermal vacuum switch (TVS) mounted in the air cleaner to control vacuum to the secondary vacuum break at the rear of the carburetor. Below approximately 62°F. air cleaner temperature, the TVS is closed, preventing the secondary vacuum break from opening the choke. This delays the full opening of the choke for better cold operation.

The air cleaner thermal sensor also connects to the TVS. This connection is for convenience only. The vacuum path through the TVS to the thermal sensor is open at all times, and is not affected by the part of the TVS that shuts off vacuum to the vacuum break unit.

EGR VALVES 1979

49-State Cadillacs use an EGR valve with an integral back pressure transducer. A bleed inside the valve regulates the opening.

To test the EGR system on carburetor engines, run the engine in NEUTRAL OR PARK at 200 rpm. Then disconnect the vacuum hose to the EGR valve. The engine should speed up, and the EGR valve stem should move down to the closed position. If not, the valve is leaking or stuck, or an exhaust passage is plugged.

EGR GASKET REVISION 1979

The EGR gaskets have been revised on the Electronic Fuel Injection engines, to meter the exhaust gases more evenly. The applications are as follows;

Application	Orifice Size	Part Number
All "C" cars (EFI)	0.600 in.	1608508
All Eldorados	0.500 in.	417174
Seville Federal	0.484 in.	1608505
Seville Calif.	0.375 in.	551516

Cadillac

CATALYTIC CONVERTER 1979

(Sevilles sold in California)

Rodium pellets are used in the converter, in addition to the platium and paladium pellets, to meet the stricter Oxides of Nitrogen standards in the state of California.

INTAKE MANIFOLDS 1979

(Carbureted "C" cars)

Riser tubes have been added to the floor of the intake manifold to more evenly distribute the exhaust gases, thereby making the EGR system more efficient.

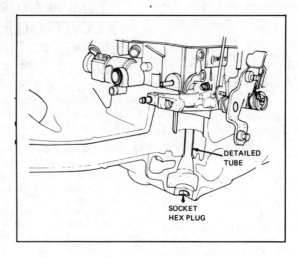

Riser tube location in the floor of the intake manifold

CLOSED LOOP EPI SYSTEM 1979

(All Sevilles sold in California)

The term "Closed Loop" is derived from the systems ability to sense the air/fuel ratio and correct it as necessary through the use of a feedback signal to the Electronic Control Unit (ECU). This process is accomplished by the use of an exhaust oxygen sensor, located in the top of the right exhaust manifold.

The sensor is an electro-chemical device, using Zirconium Dioxide to produce a variable voltage in relation to the oxygen content in the exhaust gases.

The sensor is exposed at one end to the exhaust gases as they leave the engine, while the sensor center is vented to the atmosphere, thereby comparing the oxygen content of both the exhaust gases and the atmosphere air. The variable voltage produced is dependent upon the richness or leanness of the exhaust gases of the oxygen content. The electrical signal is sent to the ECU, which adjusts the air/fuel mixture entering the engine to the ideal ratio of 14.7:1.

A "SENSOR" indicator dash light will operate at an interval of 15,000 miles. This indicates the Exhaust Oxygen sensor must be replaced. The dash "SENSOR" light is reset from behind the instrument cluster by pulling the sensor reset cable.

DIESEL ENGINE 1979

The idle speed on the diesel is 575 rpm, set with an adjustment on the side of the injection pump. Because the diesel has no electric ignition system, special equipment must be used to measure the idle rpm. A bracket for holding a magnetic tachometer pickup is on the front of the engine.

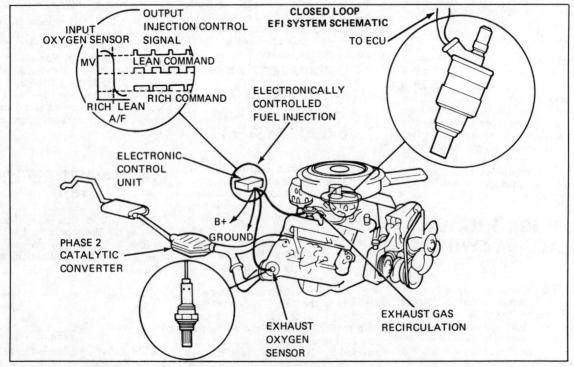

Closed loop EFI system

Cadillac

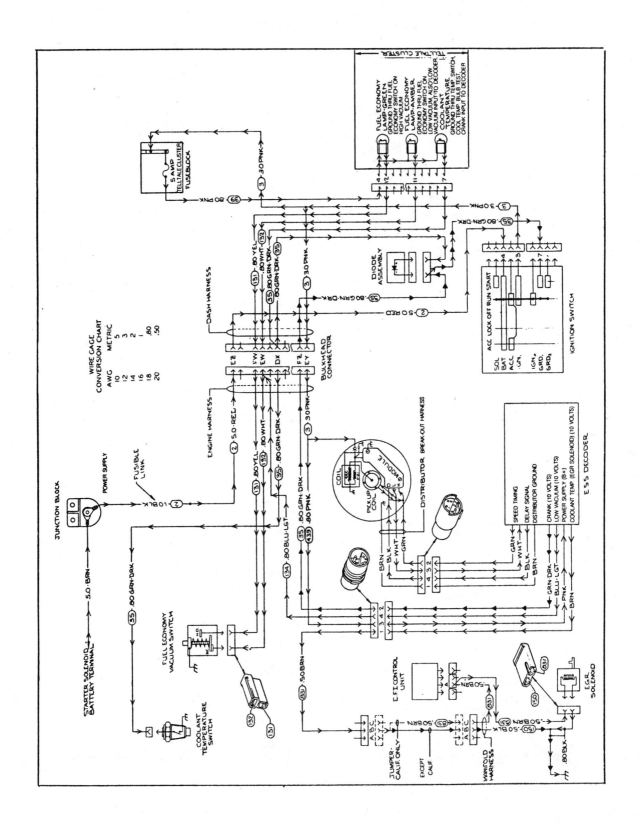

1978 Seville electronic spark selection circuit diagram

Cadillac

VACUUM CIRCUITS

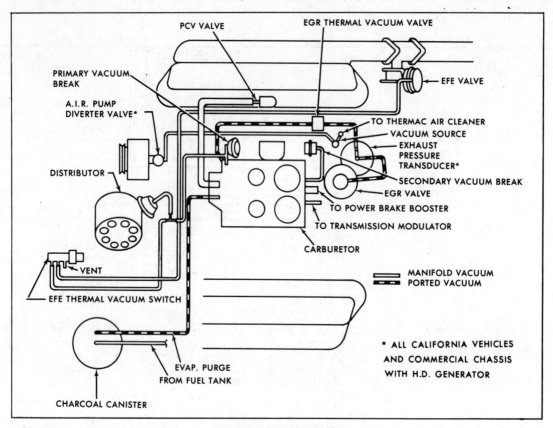

1976 Cadillac 500 4-Bbl. V-8

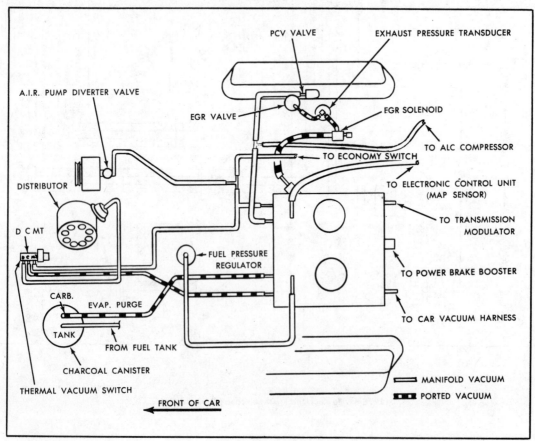

1976 Cadillac 500 Fuel Injection V-8

Cadillac

VACUUM CIRCUITS

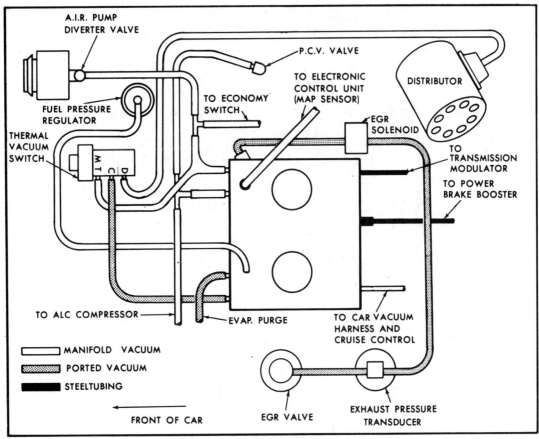

1976 Seville 350 V-8 with TVS

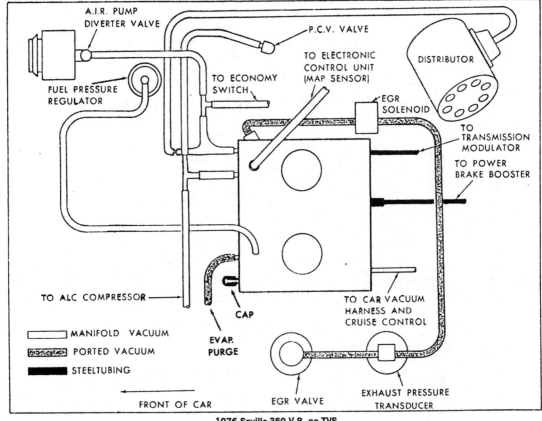

1976 Seville 350 V-8, no TVS

Cadillac

VACUUM CIRCUITS

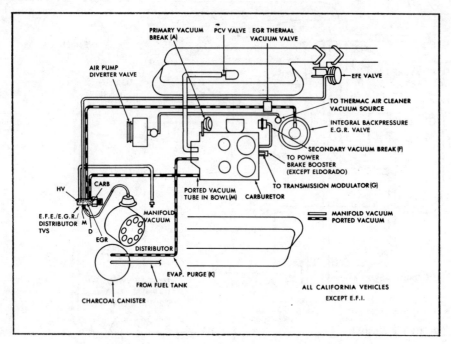

Vacuum hose routing 1977 Cadillac 425 V-8 4bbl California
(© GM Corporation)

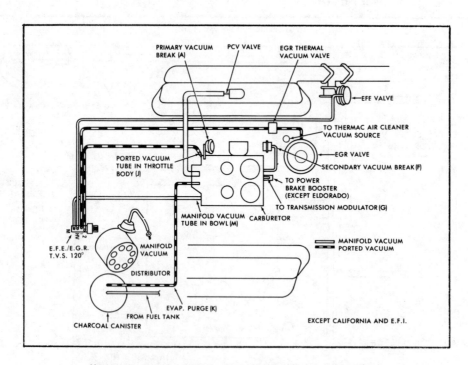

Vacuum hose routing 1977 Cadillac 425 V-8 4bbl 49 states engine
(© GM Corporation)

Cadillac

VACUUM CIRCUITS

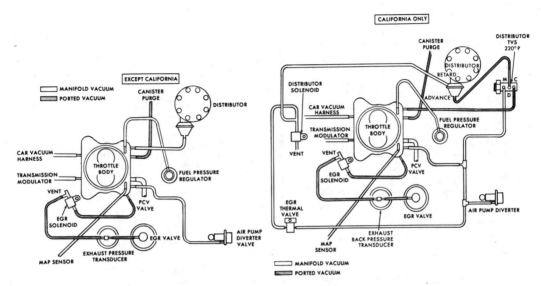

Vacuum hose routing 1977 Eldorado 425 V-8 F.I.
(© GM Corporation)

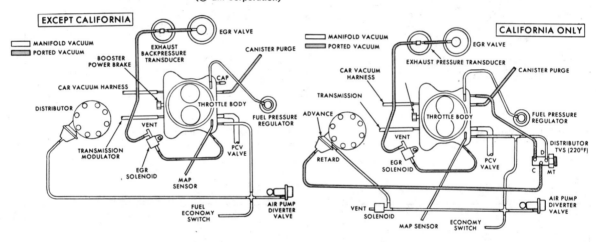

Vacuum hose routing 1977 Cadillac 425 V-8 F.I.
(© GM Corporation)

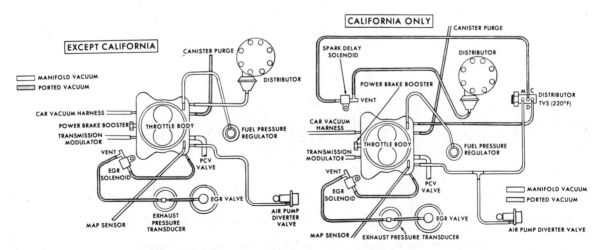

Vacuum hose routing 1977 Cadillac Seville 350 V-8 F.I.
(© GM Corporation)

Cadillac

VACUUM CIRCUITS

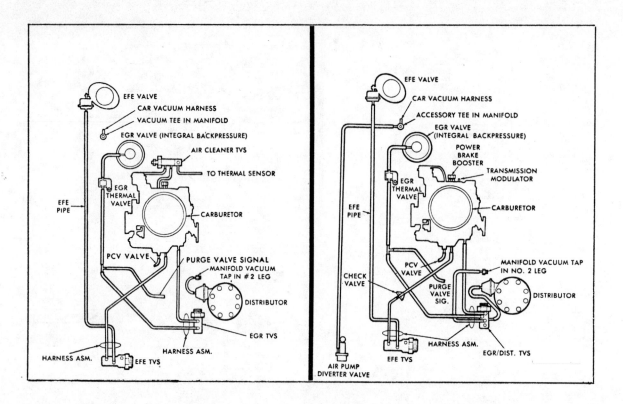

1978 49-States Cadillac exc. Eldorado 4-bbl. vacuum hose routing

1978 Altitude Cadillac and Eldorado 4-bbl. vacuum hose routing

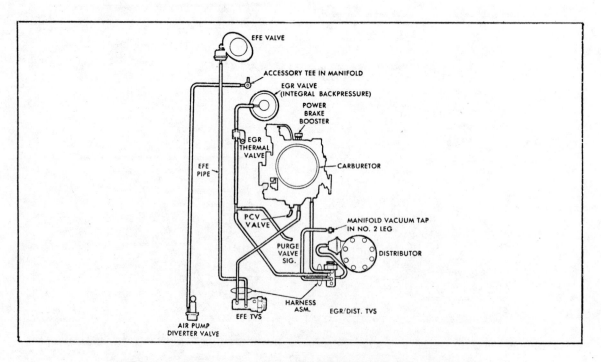

1978 49-States Eldorado, commercial chassis and limousine 4-bbl. vacuum hose routing
1978 California all 4-bbl. engines

Cadillac

VACUUM CIRCUITS

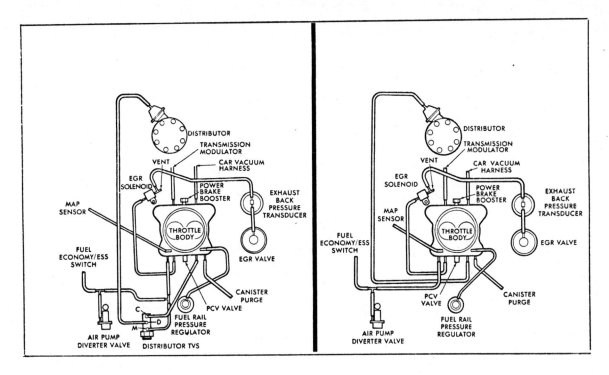

1978 California Seville vacuum hose routing

1978 49-States Seville vacuum hose routing

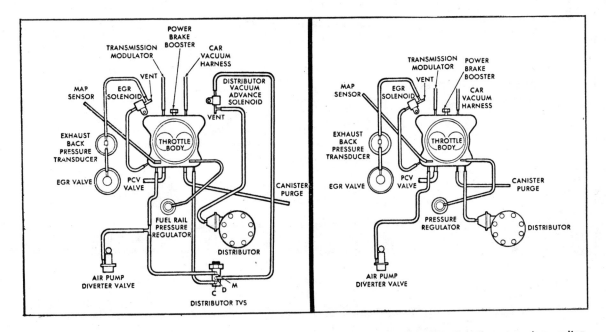

1978 California Cadillac 425 fuel injection vacuum hose routing

1978 49-States Cadillac 425 fuel injection vacuum hose routing

Cadillac

VACUUM CIRCUITS

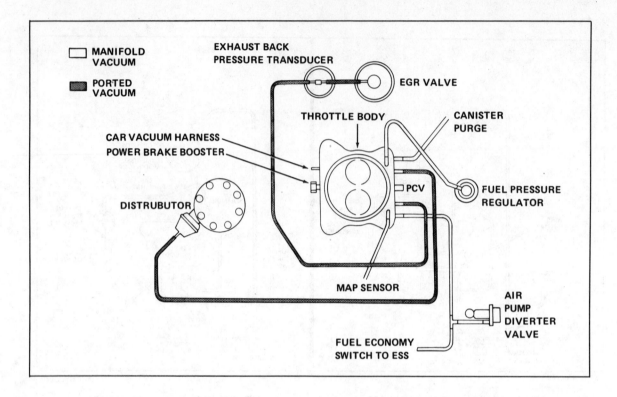

1979 Vacuum hose schematic—Eldorado for 49 States

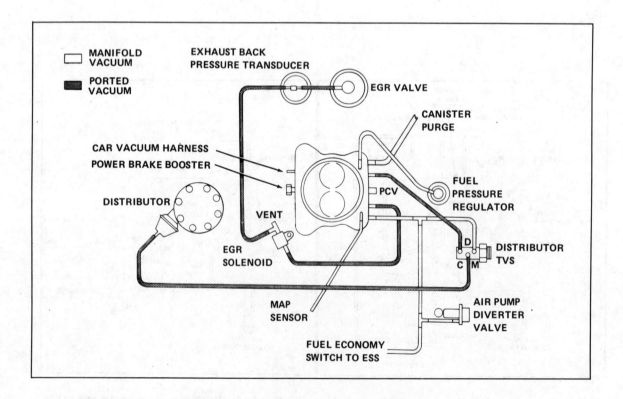

1979 Vacuum hose schematic—Eldorado for California

Cadillac

VACUUM CIRCUITS

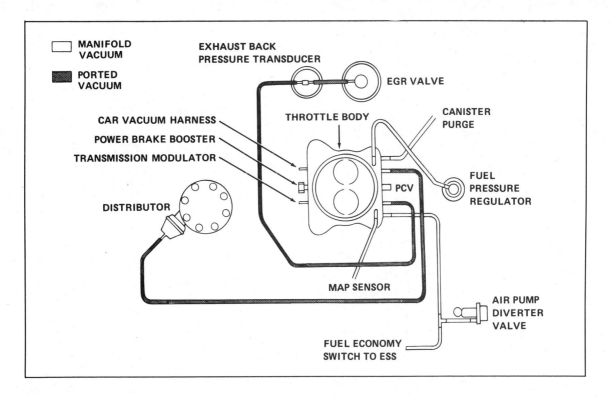

1979 Vacuum hose schematic—Seville for 49 States

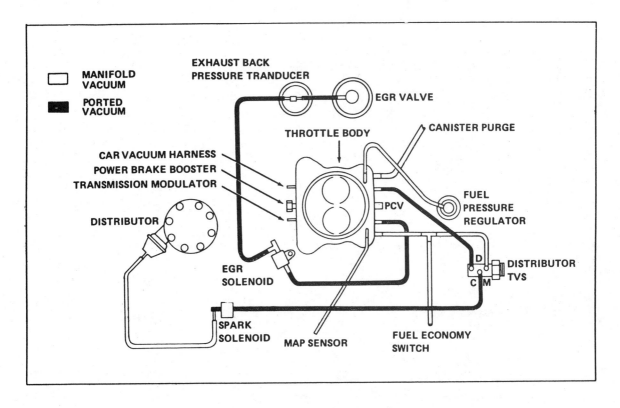

1979 Vacuum hose schematic—Seville for California

Chevrolet 6 Cyl. — V-8 • Monza V-8

Full-Size • Intermediate • Compact

TUNE-UP SPECIFICATIONS — CHEVROLET

Engine Code, 5th character of the VIN number
Model Year Code, 6th character of the VIN number

Yr.	Eng. V.I.N. Code	Eng. No. Cyl Disp. (cu in)	Eng. Mfg.	Carb Bbl	H.P.	SPARK PLUGS Orig. Type	Gap (in)	DIST. Point Dwell (deg)	DIST. Point Gap (in)	IGNITION TIMING (deg BTDC) Man Trans	IGNITION TIMING (deg BTDC) Auto Trans	VALVES Intake Opens (deg BTDC)	FUEL PUMP Pres. (psi)	IDLE SPEED (rpm) Man Trans	IDLE SPEED (rpm) Auto Trans In Drive
75	D	6-250	C	1	105	R46TX	.060	E.I.		10	10	16	3½-4½	850	550①
	D	6-250④	C	1	105	R46TX	.060	E.I.		—	8	16	3½-4½	—	600
	G	8-262	C	2	110	R44TX	.060	E.I.		8	8	26	7-8½	800	600
	H	8-350	C	2	145	R44TX	.060	E.I.		6 (4)	6	28	7-8½	800	600
	J	8-350	C	4	155	R44TX	.060	E.I.		6 (4)	8 (6)②	52	7-8½	800	600
	L	8-350	C	4	165	R44TX	.060	E.I.		6 (4)	8 (6)②	52	7-8½	800	600
	T	8-350	C	4	205	R44TX	.060	E.I.		6 (4)④	8 (6)②	52	7-8½	800	600
	U	8-400	C	4	175	R44TX	.060	E.I.		6 (4)	8 (8)②	28	7-8½	800	600
	Y	8-454	C	4	215	R44TX	.060	E.I.		—	16	55	7-8½	—	600
76	D	6-250	C	1	105	R46TS	.035	E.I.		6	6	16	3½-4½	850	550①
	D	6-250	C	1	105	R46TS	.035	E.I.		—	8	16	3½-4½	—	600
	Q	8-305	C	2	140	R45TS	.045	E.I.		6	8 (10)	28	7-8½	800	600
	V	8-350	C	2	145	R45TS	.045	E.I.		—	6	28	7-8½	—	600
	L	8-350	C	4	165	R45TS	.045	E.I.		8 (6)	8 (6)	28	7-8½	800	600
	L	8-350	C	4	180	R45TS	.045	E.I.		8 (6)	8 (6)	28	7-8½	1000	700
	X	8-350	C	4	205	R45TS	.045	E.I.		12	12	52	7-8½	1000	700
	U	8-400	C	4	175	R45TS	.045	E.I.		—	8	28	7-8½	—	600
	S	8-454	C	4	225	R45TS	.045	E.I.		—	12	55	7-8½	—	550
77	D	6-250	C	1	110	R46TS	.035	E.I.		6	8③	16	3½-4½	750	550
	U	8-305	C	2	145	R45TS	.045	E.I.		8	8 (6)	28	7-8½	600	500
	L	8-350	C	4	170	R45TS	.045	E.I.		8	8	28	7-8½	700	500
	X	8-350	C	4	180	R45TS	.045	E.I.		12	12	52	7-8½	800	700
78	M	V6-200	C	2	95	R45TS	.045	E.I.		8	8	34	4½-6	700	600
	A	V6-231	B	2	105	R46TSX	.060	E.I.		15	15	17	4¼-5¾	750	550
	D	6-250	C	1	110	R46TS	.035	E.I.		6	10(6)	16	3½-4½	600	500
	U	8-305	C	2	145	R45TS	.045	E.I.		4	4(6)	28	7-8½	700	500
	L	8-350	C	4	170	R45TS	.045	E.I.		6	6(8)	28	7-8½	800	700
79	M	V6-200	C	2	94	R45TS	.045	E.I.		8	14	34	4½-6	③	③
	A	V6-231	B	2	115	R46TSX	.060	E.I.		15	15	17	4¼-5¾	③	③
	D	6-250	C	1	115	R46TS	.035	E.I.		8	10(6)	16	3½-4½	③	③
	J	8-267	C	2	125	R45TS	.045	E.I.		4	10	34	7½-9	③	③
	G	8-305	C	2	130	R45TS	.045	E.I.		4	4	28	7½-9	③	③
	H	8-305	C	4	160	R43TS	.045	E.I.		4	4	28	7½-9	③	③
	L	8-350	C	4	170	R45TS	.045	E.I.		6	6	28	7½-9	③	③

Should the information provided in this manual deviate from the specifications on the underhood tune-up label, the label specifications should be used, as they may reflect production changes.

• Figures in parentheses indicate California engine
— Not applicable
① (600)
② 6 ON CORVETTE
③ SEE UNDERHOOD TUNE-UP DECAL

Chevrolet 6 Cyl. — V-8 • Monza V-8

Full-Size • Intermediate • Compact

TUNE-UP SPECIFICATIONS

CHEVROLET Corvette

Engine Code, 5th character of the VIN number
Model Year Code, 6th character of the VIN number

Year	Eng. V.I.N. Code	Eng No. Cyl. Disp. (cu in)	Eng. Mfg.⑤	Carb Bbl	Type	SPARK PLUGS Orig.	Gap (in)	DIST. Point Dwell (deg)	Point Gap (in)	IGNITION TIMING (deg BTDC) Man Trans	Auto (deg BTDC)	VALVES Intake Opens	FUEL PUMP Pres. (psi)	IDLE SPEED (rpm) Man Trans	Auto Trans In Drive
1975	J	8-350	C	4v	165	R44TX	.060	Electronic		6	6	28 (44)	7.0-8.5	800	600
	T	8-350	C	4v	205	R44TX	.060	Electronic		12	12	52	7.0-8.5	900	600
1976	L	8-350	C	4v	180	R45TS	.045	Electronic		8	8	28	7.0-8.5	800	600
	T	8-350	C	4v	210	R45TS	.045	Electronic		12	12	52	7.0-8.5	1000	700
1977	L	8-350	C	4v	180	R45TS	.045	Electronic		8	8	28	7.0-8.5	700	500①
	X	8-350	C	4v	210	R45TS	.045	Electronic		12	12	52	7.0-8.5	800	700
1978	L	8-350	C	4v	185	R45TS	.045	Electronic		6	6(8)	28	7.5-9.0	700	500③
	4	8-350	C	4v	220	R45TS	.045	Electronic		12	12	52	7.5-9.0	900	700④
1979	8	8-350	C	4v	185*	R45TS*	.045*	Electronic		②	②	28*	7.5-9.0*	700*	500③
	4	8-350	C	4v	220*	R45TS*	.045*	Electronic		②	②	52*	7.5-9.0*	900*	700④

Should the information provided in this manual deviate from the specifications on the underhood tune-up label, the label specifications should be used, as they may reflect production changes.

- ● Figures in parenthesis indicate California Spec.
- * Chilton Estimate
- ① AIR CONDITIONED MODELS EQUIPPED WITH IDLE SPEED-UP SOLENOID, TURN AC ON, DISCONNECT COMPRESSOR CLUTCH WIRE AND ADJUST SOLENOID TO OBTAIN 650 RPM.
- ② UNAVAILABLE AT TIME OF PUBLICATION. SEE UNDERHOOD TUNE-UP DECAL.
- ③ 49 STATE AND CALIFORNIA CURB IDLE SHOWN, SET BASE IDLE AT 600 RPM. HIGH ALTITUDE CURB, SET AT 600 RPM, BASE AT 650 RPM.
- ④ CURB IDLE SHOWN, SET BASE IDLE AT 800 RPM.
- ⑤ C-CHEVROLET

Monza

TUNE-UP SPECIFICATIONS

Engine Code, 5th character of the VIN number
Model Year Code, 6th character of the VIN number

Yr.	Eng. V.I.N. Code	Engine No. Cyl. Disp. (cu in)	Eng. Mfg.	Carb Bbl	H.P.	SPARK PLUGS Orig. Type	Gap (in)	DIST. Point Dwell (deg)	Point Gap (in)	IGNITION TIMING (deg BTDC) Man	Auto	VALVES Intake Opens (deg BTDC)	FUEL Pump Pres. (psi)	IDLE SPEED (rpm) Man Trans	Auto Trans In Drive
75	A	4-140	C	1v	75	R43TSX	.035	Electronic		8 @ 700	10	22	3.0-4.5	1200	750
	B	4-140	C	2v	85	R43TSX	.060	Electronic		10 @ 700	12	28	3.0-4.5	700	750
	C	6-231	B	2v	110	R44SX	.060	Electronic		12	12	17	3②		800	700
	F	8-260	O	2v	110	R46SX	.080	Electronic		16 @ 1100*	⑦	14	6	750	550⑩
	G	8-262	C	2v	110	R44TX	.060	Electronic		8	8	26	7.0-8.0	800	600
76	O	4-122	C	FI	120	R43T8X	.035	Electronic		12	—	38	—	1600	—
	A	4-140	C	1v	75	R43TS	.035	Electronic		8	10	22	3.0-4.5	1200	750
	B	4-140	C	2v	85	R43TS	.035	Electronic		10	12	28	3.0-4.5	700	750
	C	6-231	B	2v	110	R44SX	.060	Electronic		12	12	17	3②	800	600
	F	8-260	O	2v	110	R46SX	.060	Electronic		16 @ 1100	⑦	14	5.5-6.5	750	550
	G	8-262	C	2v	110	R45TS	.045	Electronic		6	8	26	7.0-8.0	800	600
	Q	8-305	C	2v	140	R45TS	.045	Electronic		6	8	26	7.0-8.5	800	600
77	B	4-140	C	2v	84	R43TS	.035	Electronic		TDC @ 700⑫	2	34	3.0-4.5	1250③⑨	850③①
	V	4-151	P	2v	87	R44TSX	.060	Electronic		14 @ 1000*	14(12)	33	4.5-5.0	500	650
	C	6-231	B	2v	105	R46TSX	.040	Electronic		12 @ 600	12	17	3②	600	550
	U	8-305	C	2v	145	R45TS	.045	Electronic		8	8	28	7.0-8.5	600	500

Chevrolet 6 Cyl. — V-8 • Monza V-8

Monza

TUNE-UP SPECIFICATIONS

Engine Code, 5th character of the VIN number
Model Year Code, 6th character of the VIN number

Yr.	Eng. V.I.N. Code	Eng. No. Cyl Disp. (cu in)	Eng. Mfg.	Carb Bbl	H.P.	SPARK PLUGS Orig. Type	SPARK PLUGS Gap (in)	DIST. Point Dwell (deg)	DIST. Point Gap (in)	IGNITION TIMING (deg BTDC) Man Trans	IGNITION TIMING (deg BTDC) Auto Trans	VALVES Intake Opens (deg BTDC)	FUEL PUMP Pres. (psi)	IDLE SPEED (rpm) Man Trans	IDLE SPEED (rpm) Auto Trans In Drive
78	1	4-151	P	2v	85	R43TSX	.060	Electronic		14	14⑯	33	4-5.5	⑤	⑤
	V	4-151	P	2v	85	R43TSX	.060	Electronic		14	14	33	4-5.5	⑤	⑤
	C	6-196	B	2v	90	R46TSX	.060	Electronic		15	15	18	3-4.5⑬	⑤	⑤
	A	6-231	B	2v	105	R46TSX	.060	Electronic		15	15	17	3-4.5⑬	800	600
	U	8-305	C	2v	145	R45TS	.045	Electronic		4	6⑰	28	7.5-9.0	600	500⑮
79	1	4-151	P	2v	85	R43TSX	.060	Electronic		12	12	33	4-5.5	⑤	⑤
	V	4-151	P	2v	85	R43TSX	.060	Electronic		14	14	33	4-5.5	1000⑭	650⑭
	9	4-151	P	2v	85	R43TSX	.060	Electronic		14	14	33	4-5.5	⑤	⑤
	C	6-196	B	2v	90	R46TSX	.060	Electronic		⑤	⑤	18	3-4.5⑬	⑤	⑤
	A	6-231	B	2v	115	R46TSX	.060	Electronic		15	15	17	3-4.5⑬	800	600
	2	6-231	B	2v	115	R46TSX	.060	Electronic		15	15	17	3-4.5⑬	800	600
	G	8-305	C	2v	145	R45TS	.045	Electronic		4	4(4)	28	7.5-9.0	600	500⑮

Should the information provided in this manual deviate from the specifications on the underhood tune-up label, the label specifications should be used, as they may reflect production changes.

① THERE ARE TWO VERSIONS OF THE THIS ENGINE IN 1978, ONE FEATURES A SPECIAL CLOSED-LOOP EMISSION SYSTEM. THIS ENGINE CAN BE IDENTIFIED BY THE DEALER ORDER NUMBER UPC L36.
② 1975 SKYHAWK 3-4 PSI 1976-1977 SKYHAWK 3.-0-45 PSI.
• Figures in parenthesis are California specs.
③ ADJUST SPEEDUP SOLENOID TO OBTAIN SPECIFIED RPM, WITH AIR CONDITIONING UNIT TURNED ON AND COMPRESSOR CLUTCH WIRE DISCONNECTED.
④ AT 700 OR LESS RPM
⑤ SEE UNDERHOOD TUNE-UP DECAL.
⑥ –34
⑦ (14 @ 1100)
⑧ (1200)
⑨ (800)
⑩ (600)
⑪ (700)
⑫ 2ATDC @ 700
* Timing is with both automatic and manual transmissions.
⑬ AT 12.6 VOLTS
⑭ BASE IDLE 500 RPM.
⑮ HIGH ALT. BASE IDLE IS 600 RPM
⑯ WITH EGR VALVE: 12° WITHOUT EGR VALVE
⑰ HIGH ALT. IS 8°
■ B–Buick
C–Chevrolet
O–Oldsmobile
P–Pontiac

DISTRIBUTOR SPECIFICATIONS — CHEVROLET Corvette

Year	Distributor Identification	Centrifugal Advance Start Dist. Deg. @ Dist. RPM	Centrifugal Advance Finish Dist. Deg. @ Dist. RPM	Vacuum Advance Start @ In. Hg.	Vacuum Advance Finish Dist. Deg. @ In. Hg.
1975	1112880	0 @ 600	11 @ 2,100	4	9 @ 12
	1112888	0 @ 550	8 @ 2,100	4	9 @ 12
	1112883	0 @ 550	11 @ 2,300	4	7.5 @ 10
1976	1103200	0 @ 1200	16 @ 2,000	4	10 @ 8
	1112905	0 @ 1200	22 @ 4,200	6	15 @ 12
1977	1103246	0 @ 1200	22 @ 4,200	4	18 @ 12
	1103248	0 @ 1200	22 @ 4,200	4	10 @ 8
	1103256	0 @ 1200	16 @ 2,000	4	10 @ 8
1978	1103285	0 @ 1200	22 @ 2400	4	10 @ 8
	1103291	0 @ 1200	16 @ 2000	4	10 @ 8
	1103337	0 @ 1100	16 @ 2400	4	24 @ 10
	1103353	0 @ 1100	16 @ 2400	4	20 @ 10

Chevrolet 6 Cyl. — V-8 • Monza V-8

DISTRIBUTOR SPECIFICATIONS — CHEVROLET

Year	Distributor Identification	Centrifugal Advance Start Dist. Deg. @ Dist. RPM	Centrifugal Advance Finish Dist. Deg. @ In. Hg.	Vacuum Advance Start @ In. Hg.	Vacuum Advance Finish Dist. Deg. @
75-76	1110652	0 @ 550	12 @ 2050	5	12 @ 15
	1110650	0 @ 550	8 @ 2100	4	9 @ 12
	1110662	0 @ 550	12 @ 2050	4	9 @ 12
	1103203	0 @ 500	7.5 @ 1400	4	9 @ 7
	1112882	0 @ 500	7.5 @ 1400	8	7.5 @ 15.5
	1112886	0 @ 900	6 @ 2100	4	9 @ 7
	1110666	0 @ 500	10 @ 2100	4	12 @ 15
	1112863	0 @ 550	8 @ 2100	4	9 @ 12
	1112977	0 @ 500	10 @ 1900	4	9 @ 12
	1112999	0 @ 540	10 @ 1900	4	5 @ 8
	1112880	0 @ 600	11 @ 2100	4	9 @ 12
	1112905	0 @ 600	11 @ 2100	4	6 @ 15
	1112888	0 @ 550	11 @ 2300	4	9 @ 12
	1112880	0 @ 600	11 @ 2100	4	9 @ 12
	1112880	0 @ 600	11 @ 2100	4	9 @ 12
	1112880	0 @ 550	10 @ 2100	4	12 @ 15
	1112905	0 @ 600	11 @ 2100	6	7.5 @ 12
	1112883	0 @ 550	11 @ 2300	4	7.5 @ 10
77	1110678	0 @ 500	10 @ 2050	4	12 @ 15
	1110681	0 @ 500	10 @ 2100	4	7.5 @ 12
	1110725	0 @ 500	10 @ 2100	5	12 @ 12
	1103239	0 @ 500	10 @ 2100	4	7.5 @ 10
	1103244	0 @ 500	10 @ 1900	4	10 @ 10
	1103246	0 @ 600	11 @ 2100	4	9 @ 12
	1103248	0 @ 600	11 @ 2100	4	5 @ 8
	1103256	0 @ 600	8 @ 1000	4	5 @ 8
78	1103281	0 @ 500	10 @ 1900	4	9 @ 12
	1103282	0 @ 500	10 @ 1900	4	10 @ 10
	1103285	0 @ 600	11 @ 2100	4	5 @ 8
	1103286	0 @ 550	11 @ 2300	4	9 @ 12
	1103337	0 @ 550	8 @ 1200	4	12 @ 10
	1103353	0 @ 550	8 @ 1200	4	10 @ 10
	1110695	0-6 @ 1000	6-9 @ 1800	7-9	12 @ 10-13
	1110696	0 @ 500	10 @ 1900	3	8 @ 6.5
	1110715	0 @ 500	10 @ 2100	4	12 @ 15
	1110716	0 @ 500	10 @ 2100	4	7.5 @ 12
	1110718	0 @ 500	10 @ 2100	4	9 @ 12
	1110731	0-4 @ 1000	6-9 @ 1800	4-6	8 @ 7-9
79	1103281	0 @ 500	10 @ 1900	4	9 @ 12
	1103282	0 @ 500	10 @ 1900	4	10 @ 10
	1103285	0 @ 600	22 @ 2100	4	5 @ 8
	1103337	0 @ 550	11 @ 2300	4	12 @ 10
	1103353	0 @ 550	11 @ 2300	4	10 @ 10
	1103370	0 @ 700	7 @ 1900	3	15 @ 9.5
	1103371	0 @ 500	10 @ 1900	3	15 @ 9.5
	1110695	0 @ 840	7.5 @ 1800	3.9	12 @ 10.9
	1110716	0 @ 500	10 @ 2100	4	7.5 @ 12
	1110731	0 @ 840	7.5 @ 1800	4.9	8 @ 8.3
	1110737	0 @ 500	10 @ 1900	3	15 @ 9.5
	1110748	0 @ 500	10 @ 2100	4	10 @ 10
	1110756	0 @ 700	7 @ 1900	3	15 @ 9.5

Chevrolet 6 Cyl. — V-8 • Monza V-8

DISTRIBUTOR SPECIFICATIONS — Monza

Year	Distributor Identification	Centrifugal Advance Start Dist. Deg. @ Dist. RPM	Centrifugal Advance Finish Dist. Deg. @ Dist. RPM	Vacuum Advance Start @ In. Hg.	Vacuum Advance Finish Dist. Deg. @ In. Hg.
75	1112862	0 @ 810	11 @ 2400	5	12 @ 12
	1112880	0 @ 600	11 @ 2100	4	9 @ 12
	1112862	0 @ 810	11 @ 2400	5	12 @ 12
	1112863	0 @ 550	8 @ 2100	3-5	3.25-9.75 @ 11-12.5
	1110650	0 @ 550	8 @ 2100	3-5	6.75-8.25 @ 11.5-12.
	1110651	0 @ 540	8 @ 2050	5-7	9 @ 10
	1112951	0 @ 325	14 @ 2200	4	12 @ 15
	1112956	0 @ 325	14 @ 2200	NONE	NONE
76	1112862	0 @ 810	11 @ 2400	5	12 @ 12
	1112983	0 @ 600	11 @ 2000	4	7.5 @ 10
	1112862	0 @ 450	11 @ 2400	5	10 @ 14
	1110668	0 @ 638	8 @ 1600	6	12 @ 12
	1110661	0 @ 525	8 @ 2050	6	9 @ 10
	1112863	0 @ 550	8 @ 2100	3-5	12 @ 14.5
	1112863	0 @ 550	8 @ 2100	4	9 @ 12
	1110666	0 @ 500	10 @ 2100	4	12 @ 15
	1110668	0 @ 638	8 @ 1588	6	12 @ 11.5
	1110661	0 @ 530	8 @ 2050	6	9 @ 10
	1112994	0 @ 325	14 @ 2200	4.5	12 @ 10.5
	1112995	0 @ 455	13 @ 2233	4	15 @ 11
	1112996	0 @ 550	14 @ 2200	—	—
77	1110538	0 @ 425	85 @ 1000	5	12 @ 10
	1110539	0 @ 425	8.5 @ 1000	5	12 @ 10
	1103239	0 @ 600	10 @ 2100	14	7.5 @ 10
	1103244	0 @ 500	10 @ 1900	4	10 @ 10
	1103252	0 @ 500	10 @ 1900	4	9 @ 12
	1103229	0 @ 600	10 @ 2200	3.5	10 @ 12
	1103263	0 @ 600	10 @ 2200	3.5	10 @ 9
	1103231	0 @ 600	10 @ 2200	3.5	10 @ 12
	1103230	0 @ 600	10 @ 2200	3.5	10 @ 9
	1103303	0 @ 600	10 @ 2200	9	10 @ 16
	1110677	0 @ 700	10 @ 1800	4	12 @ 11
	1110686	0 @ 700	10 @ 1800	7	4 @ 9
78	1103281	0 @ 500	10 @ 1900	4	9 @ 12
	1103282	0 @ 500	10 @ 1900	4	10 @ 10
	1103326	0 @ 850	10 @ 2325	3.4	10 @ 10.7
	1103328	0 @ 600	10 @ 2200	3.5	10 @ 9
	1103329	0 @ 600	10 @ 2200	3.5	10 @ 12
	1103365	0 @ 850	10 @ 2325	5.5	7 @ 9.5
	1110695	0-2 @ 1000	6-9 @ 1800	6	8 @ 9
	1110731	0-2 @ 1000	6-9 @ 1800	6	8 @ 9
	1110732	0-2 @ 1000	6-9 @ 1800	9	7 @ 13

Chevrolet 6 Cyl. — V-8 • Monza V-8

DISTRIBUTOR SPECIFICATIONS — Monza

Year	Distributor Identification	Centrifugal Advance Start Dist. Deg. @ Dist. RPM	Centrifugal Advance Finish Dist. Deg. @ Dist. RPM	Vacuum Advance Start @ In. Hg.	Vacuum Advance Finish Dist. Deg. @ In. Hg.
79	1103229	0 @ 600	10 @ 2200	3.5	10 @ 12
	1103231	0 @ 600	10 @ 2200	3.5	10 @ 12
	1103239	0 @ 600	10 @ 2100	4	5 @ 8
	1103244	0 @ 500	10 @ 1900	4	10 @ 10
	1103282	0 @ 500	10 @ 1900	4	10 @ 10
	1103285	0 @ 600	11 @ 2100	4	5 @ 8
	1103365	0 @ 850	10 @ 2325	5	8 @ 11
	1110726	0 @ 500	9 @ 2000	4	10 @ 10
	1110757	0 @ 600	9 @ 2000	4	10 @ 10
	1110766	0 @ 810	7.5 @ 1800	3	10 @ 9
	1110767	0 @ 840	7.5 @ 1800	4	12 @ 11

VEHICLE IDENTIFICATION NUMBER (VIN)

Typical Vehicle Identification Plate located top left side of dash panel visible through windshield.

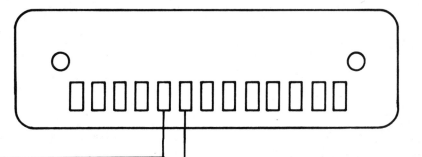

Code	Eng Disp. (cu in)	Eng. Config.	Carb	Eng. Mfgr	Model Year Code	Year
D	250	L6	1V	Chev	5	1975
G	262	V8	2V	Chev		
H	350	V8	2V	Chev		
J	350	V8	4V	Chev		
L	350	V8	2V	Chev		
T	350	V8	4V	Chev		
U	400	V8	4V	Chev		
Y-Z	454	V8	4V	Chev		
D	250	L6	1V	Chev	6	1976
Q	305	V8	2V	Chev		
V	350	V8	2V	Chev		
L	350	V8	4V	Chev		
X	350	V8	4V	Chev		
U	400	V8	4V	Chev		
S	454	V8	4V	Chev		
T	350	V8	4V	Chev		

Code	Eng Disp. (cu in)	Eng. Config.	Carb	Eng. Mfgr	Model Year Code	Year
D	250	L6	1V	Chev	7	1977
U	305	V8	2V	Chev		
L	350	V8	4V	Chev		
X	350	V8	4V	Chev		
M	200	V6	2V	Chev	8	1978
A	231	V6	2V	Buick		
D	250	L6	1V	Chev		
U	305	V8	2V	Chev		
L	350	V8	4V	Chev		
4	350	V8	4V	Chev		
C	196	V6	2V	Buick	9	1979
M	200	V6	2V	Chev		
A	231	V6	2V	Buick		
2	231	V6	2V	Buick		
D	250	L6	1V	Chev		
J	267	V8	2V	Chev		
G	305	V8	2V	Chev		
H	305	V8	4V	Chev		
L	350	V8	4V	Chev		
4	350	V8	4V	Chev		
8	350	V8	4V	Chev		

Chevrolet 6 Cyl. — V-8 • Monza V-8

Model 5210-C
CHEVROLET Monza,   Holley Carburetor

Year	Carb. Part No. ① ②	Float Level (Dry) (in.)	Float Drop (in.)	Pump Position	Fast Idle Cam (in.)	Choke Plate Pulldown* (in.)	Secondary Vacuum Break (in.)	Fast Idle Setting (rpm)	Choke Unloader (in.)	Choke Setting
1975	348659, 348663,	0.420	1	#2	0.110	0.325	—	1600⑥	—	3 Rich
	348661, 348665	0.420	1	#2	0.110	0.275	—	1600⑥	—	3 Rich
	348660, 348664	0.420	1	#2	0.110	0.300	—	1600⑥	—	4 Rich
	348662, 348666	0.420	1	#2	0.110	0.275	—	1600⑥	—	4 Rich
1976	366829, 366831	0.420	1	#3	0.320	0.313	—	2200	0.375	2 Rich
	366833, 366841	0.420	1	#3	0.320	0.268	—	2200	0.375	2 Rich
	366830, 366832	0.420	1	#2	0.320	0.288	—	2200	0.375	3 Rich
	366834, 366840	0.420	1	#2	0.320	0.268	—	2200	0.375	3 Rich
1977	458103, 458105	0.420	1	#2	0.120	0.250	—	2500	0.350	3 Rich
	458107, 458109	0.420	1	#2	0.120	0.275	—	2500	0.400	3 Rich
	458102, 458104	0.420	1	#1	0.085	0.250	—	2500	0.350	3 Rich
	458106, 458108	0.420	1	#1	0.120	0.275	—	2500	0.400	3 Rich
	458110, 458112	0.420	1	#1	0.120	0.300	—	2500	0.400	3 Rich
1978	see notes	.520	1	—	.150	⑦	.400	2500	.350	⑧

① Located on tag attached to the carburetor, or on the casting or choke plate
② Beginning 1974, GM identification numbers are used in place of the Holley numbers
③ 0.268 in California
④ #1 manual, #2 automatic
⑤ Not used
⑥ With no vacuum to the distributor
* Vacuum break initial choke valve clearance on AMC
⑦ Part #10001048, 10001050: .300
     #10001047, 10001049, 10001052, 10001054: .325
⑧ Part #10001047, 10001049: 1 Rich
     #10001048, 10001050, 10001052, 10001054: 2 Rich
⑨ Part #10001047, 10001049: .325
     #10004048, 10004049: .300
⑩ Part #10001047, 10001049: 2200
     #10004048, 10004049: 2400
⑪ Part #10001047, 10001049: 1 Rich
     #10004048, 10004049: 2 Rich
⑫ Part #10001047, 10001049: .325
     #10004048, 10004049: .300
⑬ Part #10001047, 10001049: 2200
     #10004048, 10004049: 2400
⑭ Part #10001047, 10001049: 1 Rich
     #10004048, 10004049: 2 Rich

MV, 1MV CARBURETOR SPECIFICATIONS
CHEVROLET   Rochester Carburetor

Year	Carburetor Identification①	Float Level (in.)	Metering Rod (in.)	Pump Rod	Idle Vent (in.)	Vacuum Break (in.)	Auxiliary Vacuum Break (in.)	Fast Idle Off Car (in.)	Choke Rod (in.)	Choke Unloader (in.)	Fast Idle Speed (rpm)
1975	7045013	11/32	0.080	—	—	0.200	0.215	—	0.160	0.215	1800④
	7045012	11/32	0.080	—	—	0.350	0.312	—	0.275	0.275	1800④
	7045314	11/32	0.080	—	—	0.275	0.312	—	0.230	0.275	1800④
1976	17056012	11/32	0.084	—	—	0.140	0.265	—	0.100	0.260	2200⑤
	17066013	11/32	0.082	—	—	0.140	0.325	—	0.140	0.260	2100
	17056016	11/32	0.080	—	—	0.140	WFO	—	0.115	0.260	2200⑤
	17056018	11/32	0.084	—	—	0.140	0.265	—	0.100	0.260	2200⑤
	17056314	11/32	0.083	—	—	0.150	0.325	—	0.135	0.260	1700

① The carburetor identification number is stamped on the float bowl, next to the fuel inlet nut.
② High step of cam.
③ Without vacuum advance.
④ 1700 rpm with automatic transmission in neutral.
⑤ 2100 rpm with integral intake manifold.

Chevrolet 6 Cyl. — V-8 • Monza V-8

2GC, 2GV, 2GE CARBURETOR SPECIFICATIONS
CHEVROLET MONZA
Rochester

Year	Carburetor Identification①	Float Level (in.)	Float Drop (in.)	Pump Rod (in.)	Idle Vent (in.)	Primary Vacuum Break (in.)	Secondary Vacuum Break (in.)	Automatic Choke (notches)	Choke Rod (in.)	Choke Unloader (in.)	Fast Idle Speed (rpm)
1975	7045101, 7045105	19/32	1 7/32	1 19/32	—	0.130	—	Index	0.375	0.350	—
	7045401, 7045405	21/32	1 7/32	1 19/32	—	0.130	—	Index	0.380	0.350	—
	7045102, 7045106	19/32	1 7/32	1 19/32	—	0.130	—	Index	0.375	0.350	—
	7045406	21/32	1 7/32	1 19/32	—	0.130	—	Index	0.380	0.350	—
1976	17056101	17/32	1 9/32	1 5/8	—	0.130	—	Index	0.260	0.325	—
	17056102	17/32	1 9/32	1 5/8	—	0.130	—	Index	0.260	0.325	—
	17056104	17/32	1 5/32	1 5/8	—	0.140	—	Index	0.260	0.325	—
	17056404	9/16	1 3/16	1 21/32	—	0.140	—	Index	0.260	0.325	—
1977	17057104, 17057105	1/2	1 9/32	1 21/32	—	0.150	—	Index	0.260	0.325	—
	17057107, 17057109	1/2	1 9/32	1 21/32	—	0.160	—	Index	0.260	0.325	—
	17057404, 17057405	1/2	1 9/32	1 21/32	—	0.160	—	1/2 Lean	0.260	0.325	—
1978	17058102	15/32	1 9/32	1 17/32	—	0.150	—	Index	0.260	0.325	—
	17058103	15/32	1 9/32	1 17/32	—	0.150	—	Index	0.260	0.325	—
	17058104	15/32	1 9/32	1 21/32	—	0.160	—	Index	0.260	0.325	—
	17058107	15/32	1 9/32	1 17/32	—	0.160	—	Index	0.260	0.325	—
	17058109	15/32	1 9/32	1 17/32	—	0.160	—	Index	0.260	0.325	—
	17058404	1/2	1 9/32	1 21/32	—	0.160	—	1/2 Lean	0.260	0.325	—
	17058405	1/2	1 9/32	1 21/32	—	0.160	—	Index	0.260	0.325	—
	17058447	7/16	1 5/32	1 5/8	—	0.110	0.150	1 Rich	0.080	0.140	—
	17058143	7/16	1 5/32	1 5/8	—	0.040	0.110	1 Rich	0.080	0.140	—
	17058147	7/16	1 5/32	1 5/8	—	0.100	0.140	1 Rich	0.080	0.140	—
	17058144	7/16	1 5/32	1 5/8	—	0.060	0.110	1 Rich	0.080	0.140	—

① The carburetor identification number is stamped on the float bowl, next to the fuel inlet nut.
② TCS disconnected for full vacuum advance.

MV, 1MV CARBURETOR SPECIFICATIONS
CHEVROLET MONZA
Rochester Carburetor

Year	Carburetor Identification①	Float Level (in.)	Metering Rod (in.)	Pump Rod (in.)	Vacuum Break (in.)	Auxiliary Vacuum Break (in.)	Fast Idle Off Car (in.)	Choke Rod (in.)	Choke Unloader (in.)	Fast Idle Speed (rpm)
1975	Manual	1/8	—	—	0.100	0.450	—	0.080	0.375	2000
	Automatic	1/8	—	—	0.100	0.450	—	0.080	0.375	2000
1976	Manual	1/8	—	—	0.060	0.450	—	0.045	0.215	1200
	Automatic	1/8	—	—	0.060	0.450	—	0.045	0.215	750

① The carburetor identification number is stamped on the float bowl, next to the fuel inlet nut.
② TCS disconnected for full vacuum advance.
③ No vacuum to distributor.

Chevrolet 6 Cyl. — V-8 • Monza V-8

2GC, 2GV, 2GE CARBURETOR SPECIFICATIONS
CHEVROLET
Rochester

Year	Carburetor Identification①	Float Level (in.)	Float Drop (in.)	Pump Rod (in.)	Idle Vent (in.)	Primary Vacuum Break (in.)	Secondary Vacuum Break (in.)	Automatic Choke (notches)	Choke Rod (in.)	Choke Unloader (in.)	Fast Idle Speed (rpm)
1975	7045105	19/32	1 7/32	1 19/32	—	0.130	—	—	0.375	0.350	—
	7045106	19/32	1 7/32	1 19/32	—	0.130	—	—	0.380	0.350	—
	7045111	21/32	31/32	1 5/8	—	0.130	—	—	0.400	0.350	—
	7045112	21/32	31/32	1 5/8	—	0.130	—	—	0.400	0.350	—
	7045114	21/32	31/32	1 5/8	—	0.130	—	—	0.400	0.350	—
	7045115	21/32	31/32	1 5/8	—	0.130	—	—	0.400	0.350	—
	7045123	21/32	31/32	1 5/8	—	0.130	—	—	0.400	0.350	—
	7045124	21/32	31/32	1 5/8	—	0.130	—	—	0.400	0.350	—
	7045405	21/32	1 7/32	1 19/32	—	0.130	—	—	0.380	0.350	—
	7045406	21/32	1 7/32	1 19/32	—	0.130	—	—	0.380	0.350	—
1976	17056108	9/16	1 9/32	1 21/32	—	0.140	—	Index	0.260	0.325	—
	17056110	9/16	1 9/32	1 21/32	—	0.140	—	Index	0.260	0.325	—
	17056111	9/16	1 9/32	1 21/32	—	0.140	—	Index	0.260	0.325	—
	17056112	9/16	1 9/32	1 21/32	—	0.140	—	Index	0.260	0.325	—
	17056113	9/16	1 9/32	1 21/32	—	0.140	—	Index	0.260	0.325	—
	17056114	21/32	31/32	1 11/16	—	0.130	—	1 Rich	0.260	0.325	—
	17056430	9/16	1 9/32	1 21/32	—	0.140	—	Index	0.260	0.325	—
	17056432	9/16	1 9/32	1 21/32	—	0.140	—	Index	0.260	0.325	—
1977	17057108, 17057110, 17057111, 17057112, 17057113, 17057114, 17057121, 17057123	19/32	1 9/32	1 21/32	—	0.160	—	Index	0.260	0.325	—
	17057408, 17057410, 17057412, 17057414	21/32	1 9/32	1 21/32	—	0.160	—	½ Lean	0.260	0.325	—
1978	17058102	15/32	1 9/32	1 17/32	—	0.150	—	Index	0.260	0.325	—
	17058103	15/32	1 9/32	1 17/32	—	0.150	—	Index	0.260	0.325	—
	17058104	15/32	1 9/32	1 21/32	—	0.160	—	Index	0.260	0.325	—
	17058107	15/32	1 9/32	1 17/32	—	0.160	—	Index	0.260	0.325	—
	17058109	15/32	1 9/32	1 17/32	—	0.160	—	Index	0.260	0.325	—
	17058404	½	1 9/32	1 21/32	—	0.160	—	½ Lean	0.260	0.325	—
	17058405	½	1 9/32	1 21/32	—	0.160	—	Index	0.260	0.325	—
	17058447	7/16	1 5/32	1 5/8	—	0.110	0.150	1 Rich	0.080	0.140	—
	17058143	7/16	1 5/32	1 5/8	—	0.040	0.110	1 Rich	0.080	0.140	—
	17058147	7/16	1 5/32	1 5/8	—	0.100	0.140	1 Rich	0.080	0.140	—
	17058144	7/16	1 5/32	1 5/8	—	0.060	0.110	1 Rich	0.080	0.140	—

① The carburetor identification number is stamped on the float bowl, next to the fuel inlet nut.
② This setting is with the low idle at 500 rpm with the clutch fan disengaged.

Chevrolet 6 Cyl. — V-8 • Monza V-8

1ME CARBURETOR SPECIFICATIONS
CHEVROLET PRODUCTS

Year	Caburetor Identification① Number	Float Level (in.)	Metering Rod (in.)	Fast Idle Speed (rpm)	Fast Idle Cam (in.)	Vacuum Break (in.)	Choke Unloader (in.)	Choke Setting (notches)
1976	17056036, 17056030, 17056031, 17056037	5/32	0.072	2000②	0.065	0.070	0.165	3 Rich
	17056032, 17056034, 17056033, 17056035	5/32	0.073	2000③	0.045	0.070	0.200	3 Rich
	17056330, 17056331	5/32	0.072	2000	0.065	0.070	0.165	3 Rich
	17056332, 17056333, 17056334	5/32	0.073	2000	0.045	0.070	0.200	3 Rich
	17056335	5/32	0.073	2000	0.045	0.120	0.200	3 Rich
1977	17057016	3/8	0.070	2000	0.095	0.125	0.325	1 Lean
	17057013, 17057015	3/8	0.070	2000	0.100	0.125	0.325	1 Rich
	17057018	3/8	0.070	2000	0.085	0.120	0.325	1 Rich
	17057014, 17057020	3/8	0.070	2000	0.085	0.120	0.120	2 Rich
	17057310, 17057312	3/8	0.070	1800	0.100	0.100	0.110	Index
	17057314, 17057318	3/8	0.070	1800	0.100	0.110	0.110	Index
	17057042, 17057044, 17047045, 17057332, 17057334, 17057335	5/32	0.080	2300	0.050	0.080	0.200	2 Rich
	17057030, 17057031, 17057032, 17057034, 17057035	5/32	0.080	2300	0.050	0.080	0.200	3 Rich
1978	17058013	3/8	0.080	2000	0.180	0.200	0.500	Index
	17058014	5/16	0.100	2100	0.180	0.200	0.500	Index
	17058020	5/16	0.100	2100	0.180	0.200	0.500	Index
	17058314	3/8	0.100	2000	0.190	0.245	0.400	Index
	17058031	5/32	0.080	2400	0.105	0.150	0.500	2 Rich
	17058032	5/32	0.080	2400	0.080	0.130	0.500	3 Rich
	17058033	5/32	0.080	2400	0.080	0.130	0.500	2 Rich
	17058034	5/32	0.080	2400	0.080	0.130	0.500	3 Rich
	17058035	5/32	0.080	2300	0.080	0.130	0.500	3 Rich
	17058036	5/32	0.080	2400	0.080	0.130	0.500	3 Rich
	17058037	5/32	0.080	2400	0.080	0.130	0.500	2 Rich
	17058038	5/32	0.080	2400	0.080	0.130	0.500	3 Rich
	17058042	5/32	0.080	2400	0.080	0.160	0.500	2 Rich
	17058044	5/32	0.080	2400	0.080	0.160	0.500	2 Rich
	17058045	5/32	0.080	2300	0.080	0.160	0.500	2 Rich
	17058332	5/32	0.080	2400	0.080	0.160	0.500	2 Rich
	17058334	5/32	0.080	2400	0.080	0.160	0.500	2 Rich
	17058335	5/32	0.080	2300	0.080	0.160	0.500	2 Rich

Chevrolet 6 Cyl. — V-8 • Monza V-8

2MC, M2MC, M2ME CARBURETOR SPECIFICATIONS

CHEVROLET — Rochester Carburetor

Year	Carburetor Identification①	Float Level (in.)	Choke Rod (in.)	Choke Unloader (in.)	Vacuum Break Lean or Front (in.)	Vacuum Break Rich or Rear (in.)	Pump Rod (in.)	Choke Coil Lever (in.)	Automatic Choke (notches)
1978	All	1/4	0.314	0.314	0.136	—	9/32②	0.120	Index

Model 6510-C

General Motors Corp.

Year	Part Number	Vacuum Break Adjustment (in.)	Fast Idle Cam Adjustment (in.)	Unloader Adjustment (in.)	Fast Idle Adjustment (rpm)	Float Level Adjustment (in.)	Choke Setting
1978	10001056, 10001058	.325	.150	.350	2400	.520	1 Rich

QUADRAJET CARBURETOR SPECIFICATIONS

CHEVROLET — Rochester Carburetor

Year	Carburetor Identification①	Float Level (in.)	Air Valve Spring (turn)	Pump Rod (in.)	Primary Vacuum Break (in.)	Secondary Vacuum Break (in.)	Secondary Opening (in.)	Choke Rod (in.)	Choke Unloader (in.)	Fast Idle Speed② (rpm)
1975	7045200	17/32	9/16	0.275	0.200	0.550	⑤	0.300	0.325	1000
	7045202	15/32	7/8	0.275	0.180	0.170	⑤	0.300	0.325	1600
	7045203	15/32	7/8	0.275	0.180	0.170	⑤	0.300	0.325	1600
	7045206	15/32	7/8	0.275	0.180	0.170	⑤	0.300	0.325	1600
	7045207	15/32	7/8	0.275	0.180	0.170	⑤	0.300	0.325	1600
	7045208	15/32	7/8	0.275	0.180	0.170	⑤	0.300	0.325	1600
	7045209	15/32	7/8	0.275	0.180	0.170	⑤	0.300	0.325	1600
	7045210	15/32	7/8	0.275	0.180	0.170	⑤	0.300	0.325	1600
	7045211	15/32	7/8	0.275	0.180	0.170	⑤	0.300	0.325	1600
	7045222	15/32	7/8	0.275	0.180	0.170	⑤	0.300	0.325	1600
	7045223	15/32	7/8	0.275	0.180	0.170	⑤	0.300	0.325	1600
	7045224	15/32	3/4	0.275	0.180	0.170	⑤	0.325	0.325	1600
	7045228	15/32	3/4	0.275	0.180	0.170	⑤	0.325	0.325	1600
	7045502	15/32	7/8	0.275	0.180	0.170	⑤	0.300	0.325	1600
	7045503	15/32	7/8	0.275	0.180	0.170	⑤	0.300	0.325	1600
	7045504	15/32	7/8	0.275	0.180	0.170	⑤	0.300	0.325	1600
	7045506	15/32	7/8	0.275	0.180	0.170	⑤	0.300	0.325	1600
	7044507	15/32	7/8	0.275	0.180	0.170	⑤	0.300	0.325	1600
1976	17056202	13/32	7/8	9/32	0.185	—	⑤	0.325	0.325	1600
	17056203	13/32	7/8	9/32	0.170	—	⑤	0.325	0.325	1600
	17056206	13/32	7/8	9/32	0.185	—	⑤	0.325	0.325	1600
	17056207	13/32	7/8	9/32	0.170	—	⑤	0.325	0.325	1600
	17056210	13/32	1.0	9/32	0.185	—	⑤	0.325	0.325	1600
	17056211	13/32	3/4	9/32	0.185	—	⑤	0.325	0.325	1600
	17056228	13/32	7/8	9/32	0.185	—	⑤	0.325	0.325	1600
	17056502	13/32	7/8	9/32	0.185	—	⑤	0.325	0.325	1600
	17056506	13/32	3/4	9/32	0.185	—	⑤	0.325	0.325	1600
	17056528	13/32	7/8	9/32	0.185	—	⑤	0.325	0.325	1600
	17056200	13/32	7/8	9/32	0.240	0.160	⑤	0.190	0.270	1600

Chevrolet 6 Cyl. — V-8 • Monza V-8

QUADRAJET CARBURETOR SPECIFICATIONS
CHEVROLET
Rochester Carburetor

Year	Carburetor Identification①	Float Level (in.)	Air Valve Spring (turn)	Pump Rod (in.)	Primary Vacuum Break (in.)	Secondary Vacuum Break (in.)	Secondary Opening (in.)	Choke Rod (in.)	Choke Unloader (in.)	Fast Idle Speed② (rpm)
1977	17057202, 17057204	15/32	7/8	15/32	0.180	—	⑤	0.325	0.280	1600
	17057203	15/32	7/8	15/32	0.180	—	⑤	0.325	0.280	1300
	17057502, 17057504	15/32	7/8	15/32	0.165	—	⑤	0.325	0.280	1600
	17057210, 17057510, 17057528	15/32	1	15/32	0.180	—	⑤	0.325	0.280	1600
	17057211	15/32	1	15/32	0.180	—	⑤	0.325	0.280	1300
	17057228	13/32	1	15/32	0.180	—	⑤	0.325	0.280	1600
	17057582, 17057584	15/32	7/8	13/32	0.180	—	⑤	0.325	0.280	1600
1978	17058202	15/32	7/8	9/32	0.179	—	②	0.314	0.277	⑤
	17058203	15/32	7/8	9/32	0.179	—	②	0.314	0.277	⑤
	17058204	15/32	7/8	9/32	0.179	—	②	0.314	0.277	⑤
	17058210	15/32	1/2	9/32	0.203	—	②	0.314	0.277	⑤
	17058211	15/32	1/2	9/32	0.203	—	②	0.314	0.277	⑤
	17058228	15/32	7/8	9/32	0.203	—	②	0.314	0.277	⑤
	17058502	15/32	7/8	9/32	0.187	—	②	0.314	0.277	⑤
	17058504	15/32	7/8	9/32	0.187	—	②	0.314	0.277	⑤
	17058582	15/32	7/8	9/32	0.203	—	②	0.314	0.277	⑤
	17058584	15/32	7/8	9/32	0.203	—	②	0.314	0.277	⑤

① The carburetor identification number is stamped on the float bowl, near the secondary throttle lever.
② Without vacuum advance.
③ With automatic transmission; vacuum advance connected and EGR disconnected and the throttle positioned on the high step of cam.
④ With manual transmission; without vacuum advance and the throttle positioned on the high step of cam.
⑤ No measurement necessary on two point linkage; see text.

CAR SERIAL NUMBER AND ENGINE IDENTIFICATION

1976

Mounted behind the windshield on the driver's side is a plate with the vehicle identification number. The sixth character is the model year, with 6 for 1976. The fifth character is the engine code, as follows.

D 250 1-bbl. 6-cyl. L-22
G 262 2-bbl. V-8 LV-1
L .... 350 4-bbl. V-8 LM-1 & L-48
Q 305 2-bbl. V-8 LG-3
S .... 454 4 bbl. V-8 LS-4
U 400 4-bbl. V-8 LT-4
V 350 2-bbl. V-8 L-65
X 350 4-bbl. V-8 L-82

NOTE: *The LM-1 and the L-48 are the same engine. When used as an optional engine in cars other than Corvette, it is known as an LM-1. When used as the standard engine in Corvette, it is known as the L-48. The L-82 is the high performance V-8, used only in the Corvette.*

Engines can be identified by a three-letter code. The 6-cylinder code is stamped on a pad at the rear of the distributor. The 262, 305, 350, and 400 V-8 codes are at the front of the right bank, below the cylinder head. The 454 V-8 code is at the front top of the block in front of the intake manifold.

CCB 250 6-Cyl. Auto. 49-States
CCC 250 6-Cyl. Auto. Calif.
CCD 250 6-Cyl. Man. All States; Auto. Calif.
CCF 250 6-Cyl. Auto. 49-States
CGL 262 V-8 Auto. Calif.
CHC 350 V-8 Man. L-82 49-States
CHS 350 V-8 Auto. LM-1 49-States
CHT 350 V-8 Man. LM-1 49-States
CHU 350 V-8 Auto. LM-1 49-States
CKC 350 V-8 Auto. L-82 49-States
CKK 350 V-8 Auto. LM-1 Pol/Taxi Calif.
CKU 350 V-8 Auto. LM-1 49-States
CKW 350 V-8 Man. L-48 49-States
CKX 350 V-8 Auto. L-48 49-States
CKY 350 V-8 Auto. LM-1 Pol/Taxi 49-States
CLF 350 V-8 Auto. L-65 Pol/Taxi 49-States
CLH 350 V-8 Auto. LM-1 Calif.
CLS 350 V-8 Auto. L-48 Calif.
CMH 350 V-8 Auto. LM-1 Calif.
CMJ 350 V-8 Auto. L-65 49-States
CML 350 V-8 Auto. LM-1 Calif.
CMM 350 V-8 Auto. LM-1 Calif.
CPA 305 V-8 Man. 49-States
CPB 305 V-8 Man. 49-States
CPJ 305 V-8 Auto. Calif.
CSA 400 V-8 Auto. Pol/Taxi Calif.
CSB 400 V-8 Auto. Calif.
CSF 400 V-8 Auto. Calif.
CSW 400 V-8 Auto. Pol/Taxi 49-States
CSX 400 V-8 Auto. 49-States
CTL 400 V-8 Auto. Calif.
CUF 350 V-8 Auto. LM-1 Pol/Taxi Calif.
CXX 454 V-8 Auto. 49-States
CXY 454 V-8 Auto. Pol/Taxi 49-States
CZL 262 V-8 Man. 49-States
CZM 262 V-8 Auto. 49-States
CZT 262 V-8 Man. 49-States
CZU 262 V-8 Auto. 49-States

Chevrolet 6 Cyl. — V-8 • Monza V-8

CAR SERIAL NUMBER AND ENGINE IDENTIFICATION

1977

The vehicle identification number is mounted on a plate behind the windshield on the driver's side. The sixth character is the model year, with 7 for 1977. The fifth character is the engine code, as follows:

D 250 1-bbl. 6-cyl. L-22
L ...... 350 4-bbl. V-8 LM-1 & L-48
U 305 2-bbl. V-8 LG-3
X 350 4-bbl. V-8 L-82

NOTE: *The LM-1 and the L-48 are the same engine. When used as an optional engine in cars other than Corvette, it is known as an LM-1. When used as the standard engine in Corvette, it is known as the L-48. The L-82 is the high performance V-8, used only in the Corvette.*

Engines can be identified by a three-letter code. The 6-cylinder code is stamped on a pad at the rear of the distributor. The V-8 codes are at the front of the right bank, below the cylinder head. Also stamped in the same area is part of the vehicle identification number. The second character is the year, with 7 for 1977.

CCC 250 6-Cyl. Auto. Calif.
CCD 250 6-Cyl. Man. 49-States
CCF 250 6-Cyl. Auto. 49-States
CCR 250 6-Cyl. Auto. 49-States & Altitude
CCS 250 6-Cyl. Auto. Calif.
CCY 250 6-Cyl. Auto. Pol/Taxi 49-States & Altitude
CHA 350 V-8 Auto. LM-1 Pol/Taxi Altitude
CHB 350 V-8 Auto. Pol/Taxi Calif.
CHX 350 V-8 Auto. LM-1 Pol/Taxi 49-States
CHY 350 V-8 Auto. LM-1 Pol/Taxi 49-States
CKA 350 V-8 Auto. LM-1 Altitude
CKB 350 V-8 Auto. LM-1 Calif.
CKH 350 V-8 Auto. LM-1 49-States
CKK 350 V-8 Auto. LM-1 Altitude
CKM 350 V-8 Auto. LM-1 Altitude
CKR 350 V-8 Auto. LM-1 Calif.
CKS 350 V-8 Man. LM-1 49-States
CKZ 350 V-8 Man. L-48 49-States
CLA 350 V-8 Auto. L-48 49-States
CLB 350 V-8 Auto. L-48 Altitude
CLC 350 V-8 Auto. L-48 Calif.
CLD 350 V-8 Man. L-48 49-States
CLF 350 V-8 Auto. L-82 49-States
CLK 350 V-8 Auto. LM-1 Pol/Taxi Calif.
CLL 350 V-8 Auto. LM-1 49-States
CLT 350 V-8 Auto. LM-1 Pol/Taxi Altitude
CMM 350 V-8 Auto. LM-1 Calif.
CPA 305 V-8 Man. 49-States
CPC 305 V-8 Auto. Calif.
CPK 305 V-8 Auto. 49-States
CPL 305 V-8 Auto. Calif.
CPM 305 V-8 Auto. Calif.
CPR 305 V-8 Auto. 49-States
CPS 305 V-8 Auto. Pol/Taxi 49-States
CPT 305 V-8 Auto. Pol/Taxi Calif.
CPU 305 V-8 Auto. Altitude
CPX 305 V-8 Man. 49-States
CPY 305 V-8 Auto. 49-States
CRA 305 V-8 Auto. 49-States
CRB 305 V-8 Auto. 49-States
CRF 305 V-8 Auto. Pol/Taxi 49-States
7SB 250 6-Cyl. Auto. Pol/Taxi 49-States & Altitude
CUB 350 V-8 Auto. LM-1 49-States
CUC 350 V-8 Auto. LM-1 Calif.

1978

The vehicle identification number is mounted on a plate behind the windshield on the driver's side. The sixth character is the model year, with 8 for 1978. The fifth character is the engine code, as follows:

A .... 231 2-bbl. V-6 LD-5 Buick
C .... 196 2-bbl. V-6 LC-9 Buick
D .... 250 1-bbl. 6-cyl. L-22 Chev.
4 .... 350 4-bbl. V-8 L-82 Chev.
L .. 350 4-bbl. V-8 LM-1 & L-48 Chev.
M .... 200 2-bbl. V-6 L-26 Chev.
U .... 305 2-bbl. V-8 LG-3 Chev.

NOTE: *The LM-1 and the L-48 are the same engine. When used as an optional engine in cars other than Corvette, it is known as an LM-1. When used as the standard engine in the Corvette, it is known as the L-48. The L-82 is the high performance V-8, used only in the Corvette.*

1979

The vehicle identification number is mounted on a plate behind the windshield on the driver's side. The sixth character is the model year, with 9 for 1979. The fifth character is the engine code, as follows:

A .... 231 2-bbl. V-6 LD-5 Buick
2 .... 231 2-bbl. V-6 LC-6 Buick
D ...... 250 1-bbl. 6-cyl. L-22 Chev.
H .... 305 4-bbl. V-8 LG-4 Chev.
L .. 350 4-bbl. V-8 LM-1 & L-48 Chev.
M .... 200 2-bbl. V-6 L-26 Chev.
G .... 305 2-bbl. V-8 LG-3 Chev.

NOTE: *The LM-1 and the L-48 are the same engine. When used as an optional engine in cars other than Corvette, it is known as an LM-1. When used as the standard engine in the Corvette, it is known as the L-48. The L-82 is the high performance V-8, used only in the Corvette.*

EMISSION EQUIPMENT

1976

All Models
Closed positive crankcase ventilation
Air pump
   49-States
     Corvette 350 4-bbl. V-8 Hp. only
     All 454 V-8
   Calif.
     All models except,
     Not on 350 4-bbl. V-8 165 Hp.
Emission calibrated carburetor
Emission calibrated distributor
Heated air cleaner
Single diaphragm vacuum advance
Vapor control, canister storage
Catalytic converter, single
Exhaust gas recirculation
Early fuel evaporation
   All models except,
     Not on 250 6-cyl. with 2-piece head and manifold.
Temp. controlled choke vacuum break
   6-Cyl. one-piece head and manifold only.
Electric choke
   All 454 V-8

1977

All Models
Closed positive crankcase ventilation
Emission calibrated carburetor
Emission calibrated distributor
Heated air cleaner
Vapor control, canister storage
Catalytic converter, underfloor
Catalytic converter, manifold
   6-cyl. Calif. only
Early fuel evaporation
Exhaust gas recirculation
Spark delay valve
   49-States
     Not used
   Calif.
     305 2-bbl. V-8
   Altitude
     Not used
Air pump
   49-States
     Corvette 350 4-bbl. V-8 "X"
   Calif.
     All models
   Altitude
     350 4-bbl V-8 "L"
     Monza 305 2-bbl. V-8 "U"
Electric choke
   All 6-cyl.

1978

All Models
Closed positive crankcase ventilation
Emission calibrated carburetor
Emission calibrated distributor
Heated air cleaner

Chevrolet 6 Cyl. — V-8 • Monza V-8

EMISSION EQUIPMENT

1978
Vapor control, canister storage
Exhaust gas recirculation
Catalytic converter, underfloor
Catalytic converter, manifold
  6-cyl. Calif. only
Early fuel evaporation
  All models, except
  Not used on Camaro 350 4-bbl. V-8

Vacuum delay valve, 4-nozzle
  49-States
    200 2-bbl. V-6
    250 6-cyl.
  Calif.
    250 6-cyl.
  Altitude
    Not used

Spark delay valve
  49-States
    Not used
  Calif.
    305 2-bbl. V-8
  Altitude
    Not used

Air pump
  49-States
    Corvette 350 4-bbl. V-8 4
  Calif.
    All models
  Altitude
    All models
Electric choke
  All 6-cyl.

1979

All Models
Closed positive crankcase ventilation
Emission calibrated carburetor
Emission calibrated distributor
Heated air cleaner
Vapor control, canister storage
Exhaust gas recirculation
Catalytic converter, underfloor
Catalytic converter, manifold
  6-cyl. Calif. only
Early fuel evaporation
  All models, except
  Not used on Camaro 350 4-bbl. V-8

Vacuum delay valve, 4-nozzle
  49-States
    200 2-bbl. V-6
    250 6-cyl.
  Calif.
    250 6-cyl.
  Altitude
    Not used
Spark delay valve
  49-States
    Not used
  Calif.
    305 2-bbl. V-8
  Altitude
    Not used
Air pump
  49-States
    Corvette 350 4-bbl. V-8 L-82
  Calif.
    All models
  Altitude
    All models
Pulse Air System
  All 250 6-cyl.
Electric choke
  All 6-cyl.

IDLE SPEED AND MIXTURE ADJUSTMENTS

1976 V-8 ENGINES

Air Cleaner .... On
Auto trans. .... Drive
Distributor vac. hose. ... Connected
Vapor can tank hose. . Disconnected
Air cond. .... Off
Mixture adj. . See rpm drop below
262 V-8 49-States
  Auto. Trans. 600
    Mix. Adj. .... 80 rpm drop
      (630-600)
  Man. Trans. 800
    Mix. Adj. ...... 100 rpm drop
      (900-800)
305 V-8
  Auto. Trans. 600
    Mix. Adj. .... 30 rpm drop
      (630-600)
  Man. Trans. 800
    Mix. Adj. ...... 100 rpm drop
      (900-800)
350 2-bbl. V-8 49-States
  Auto. Trans. 600
    Mix. Adj. .... 50 rpm drop
      (650-600)
350 4-bbl. V-8 LM-1 & L-48
  Auto. Trans. 600
    Mix. Adj. .... 50 rpm drop
      (650-600)
  Man. Trans. 800
    Mix. Adj. ...... 100 rpm drop
      (900-800)
350 4-bbl. V-8 L-82 49- States
  (High performance Corvette only)
  Auto. Trans. 700
    Idle speed up solenoid 700
    Mix. Adj. .... 50 rpm drop
      (750-700)
  Man. Trans. 1000
    Mix. Adj. ...... 100 rpm drop
      (1100-1000)

400 4-bbl. V-8
  Auto. Trans. 600
    Mix. Adj. .... 50 rpm drop
      (650-600)
454 4-bbl. V-8 49-States
  Auto Trans. 550
    Mix. Adj. .... 30 rpm drop
      (580-550)
  NOTE: *Corvettes with L-82 engine and auto. trans. with air cond. use an idle speed up solenoid that only operates when the air cond. is on. The solenoid speed is 700 rpm with the air cond. ON.*

1976 6-CYL. ENGINES

Air cleaner .... On
Auto trans. .... Drive
Air cond.
  For idle speed adj.
    Man. trans. .... Off
    Auto. trans. .... On
  For mixture adj. .... Off
Vapor canister hoses ...... Plugged
Distributor vac. hose
  One-piece manifold
    49-States .... Connected
    Calif. .... Plugged
  Two-piece manifold ..... Plugged
250 6-Cyl. 49-States
  One-piece manifold, Auto. Trans.
    Solenoid connected 550
    Solenoid disconnected 425
    Mix. adj. .... 25 rpm drop
      (575-550)
  One-piece manifold, Man. Trans.
    Solenoid connected 850
    Solenoid disconnected 425
    Mix. adj. ...... 350 rpm drop
      (1200-850)

Two-piece manifold, Auto. Trans.
  Solenoid connected 600
  Solenoid disconnected 425
  Mix. adj. .... 40 rpm drop
    (640-600)
250 6-Cyl. Calif.
  Auto. Trans.
    Solenoid connected 600
    Solenoid disconnected 425
    Mix. adj. .... 40 rpm drop
      (640-600)

1977

Air cleaner
  For speed adj. .... In place
  For mixture adj. ....Set aside with
    hoses connected
Other hoses ....See underhood label
Air cond. .... Off
Auto. trans. .... Drive
Idle CO .... Not used
Mixture adj. .... See rpm drop below
250 6-cyl. 49-States
  Auto. trans. no air cond.
    Solenoid connected 550
    Solenoid disconnected 425
    Mixture adj. ......25 rpm drop
      (575-550)
  Auto. trans. with air cond.
    Solenoid connected 600
    Solenoid disconnected 425
    Mixture adj. ......25 rpm drop
      (575-550, then
      reset to 600)
  Man. trans. no air cond.
    Solenoid connected 750
    Solenoid disconnected 425
    Mixture adj. .....200 rpm drop
      (950-750)
  Man. trans. with air cond.
    Solenoid connected 800

Chevrolet 6 Cyl. — V-8 • Monza V-8

IDLE SPEED AND MIXTURE ADJUSTMENTS

1977
Solenoid disconnected 425
Mixture adj. ..... 200 rpm drop
(950-750, then reset to 800)
250 6-cyl. Calif.
Auto. trans. no air cond.
  Solenoid connected 550
  Solenoid disconnected 425
  Mixture adj. ...... 40 rpm drop
  (640-600, then reset to 550)
Auto. trans. with air cond.
  Solenoid connected 600
  Solenoid disconnected 425
  Mixture adj. ...... 40 rpm drop
  (640-600)
250 6-cyl. Altitude
Auto. trans.
  Solenoid connected 600
  Solenoid disconnected 425
  Mixture adj. ...... 50 rpm drop
  (650-600)
305 2-bbl. V-8 except Monza
Auto. trans. 500
  Mixture adj. ..... 50 rpm drop
  (550-500)
  A.C. idle speedup 650
Man. trans. 600
  Mixture adj. ...... 50 rpm drop
  (650-600)
  A.C. idle speedup 700
305 2-bbl. V-8 Monza
49-States
  Auto. trans. 500
    Mixture adj. .... 50 rpm drop
    (550-500)
    A.C. idle speedup 700
  Man. trans. 600
    Mixture adj. .... 50 rpm drop
    (650-600)
    A.C. idle speedup 700
Calif.
  Auto. trans. 500
    Mixture adj. .... 50 rpm drop
    (550-500)
    A.C. idle speedup 700
Altitude
  Auto. trans. 600
    Mixture adj. .... 50 rpm drop
    (650-600)
    A.C. Idle speedup 700
350 4-bbl. V-8 "L" engine
49-States
  Auto. trans. 500
    Mixture adj. .... 50 rpm drop
    (550-500)
    A.C. idle speedup 650
  Man. trans. 700
    Mixture adj. ... 100 rpm drop
    (800-700)
Calif.
  Auto. trans. 500
    Mixture adj. .... 50 rpm drop
    (550-500)
    A.C. idle speedup 650
Altitude
  Auto. trans. 600
    Mixture adj. .... 50 rpm drop
    (650-600)
    A.C. idle speedup 650

350 4-bbl. V-8 "X" engine (Corvette)
49-States
  Auto. trans. 700
    Mixture adj. .... 50 rpm drop
    (750-700)
    A.C. idle speedup 800
  Man. trans. 800
    Mixture adj. ... 100 rpm drop
    (900-800)

1978
Air cleaner .... In place
Air cond. .... Off
Auto. trans. .... Drive
Hoses .... See underhood label
Idle CO .... Not used
Mixture adj. .. See propane rpm below
196 V-6 49-States
Auto. trans. 600
  Propane enriched ....
Man. trans.
  Solenoid connected 800
  Solenoid disconnected 600
  Propane enriched 940
200 V-6 49-States
Auto. trans. 600
  Propane enriched 620-640
  AC idle speedup 700
Man. trans. 700
  Propane enriched 800-840
  AC idle speedup 800
231 V-6 49-States
Auto. trans. 600
  Propane enriched ....
Man. trans.
  Solenoid connected 800
  Solenoid disconnected 600
  Propane enriched ....
231 V-6 Calif.
Auto. trans. 600
  Propane enriched ....
Man. trans.
  Solenoid connected 800
  Solenoid disconnected 600
  Propane enriched ....
231 V-6 Altitude
Auto. trans. 600
  Propane enriched ....
Man. trans.
  Solenoid connected 800
  Solenoid disconnected 600
  Propane enriched ....
250 6-cyl. 49-States
Auto. trans. 550
  Propane enriched 600-630
  AC idle speedup 600
Man. trans.
  Solenoid connected 800
  Solenoid disconnected 425
  Propane enriched 800-1000
250 6-cyl. Calif.
Auto. trans.
  Solenoid connected 600
  Solenoid disconnected 400
  Propane enriched 600-605
305 2-bbl. V-8 49-States
Auto. trans. 500
  Propane enriched 520-540
  AC idle speedup 600
Man. trans. 600
  Propane enriched 700-740

AC idle speedup 700
305 2-bbl. V-8 Calif.
Auto. trans. 500
  Propane enriched 520-540
  AC idle speedup 650
305 2-bbl. V-8 Altitude
Auto. trans. 600
  Propane enriched 620-640
  AC idle speedup 700
350 4-bbl. V-8 "L" 49-States
Auto. trans. 500
  Propane enriched 530-570
  AC idle speedup 600
Man. trans. 700
  Propane enriched 850-900
350 4-bbl. V-8 "L" Calif.
Auto. trans. 500
  Propane enriched 530-570
  AC idle speedup 600
350 4-bbl. V-8 "L" Altitude
Auto. trans. 600
  Propane enriched 630-670
  AC idle speedup 650
350 4-bbl. V-8 "H" 49-States
Auto. trans. 700
  Propane enriched 760-800
  AC idle speedup 750
Man. trans. 900
  Propane enriched 1050-1100
  Mixture adj. .... 50 rpm drop
  (550-500)
  A.C. idle speedup 650
Altitude
  Auto. trans. 600
  Mixture adj. .... 50 rpm drop
  (650-600)
  A.C. idle speedup 650

350 4-bbl. V-8 "X" engine (Corvette)
49-States
  Auto. trans. 700
  Mixture adj. .... 50 rpm drop
  (750-700)
  A.C. idle speedup 800
  Man. trans. 800
  Mixture adj. ... 100 rpm drop
  (900-800)

1979
CAUTION: *Emission control adjustment changes are noted on the Vehicle Emission Information Label by the manufacturer. Refer to the label before any adjustments are made.*
Air cleaner .... In place
Air cond. .... Off
Auto. trans. .... Drive
Hoses .... See underhood label
Idle CO .... Not used
Mixture adj. .. See propane rpm below
200 V-6 49-States
Auto. trans. 600
  Propane enriched 620-640
  AC idle speedup 700
Man. trans. 700
  Propane enriched 800-840
  AC idle speedup 800
231 V-6 49-States
Auto. trans. 600
  Propane enriched ....

Chevrolet 6 Cyl. — V-8 • Monza V-8

IDLE SPEED AND MIXTURE ADJUSTMENTS

1979
Man. trans.
  Solenoid connected 800
  Solenoid disconnected 600
  Propane enriched ....
231 V-6 Calif.
  Auto. trans. 600
    Propane enriched ....
  Man. trans.
    Solenoid connected 800
    Solenoid disconnected 600
    Propane enriched ....
231 V-6 Altitude
  Auto. trans. 600
    Propane enriched ....
  Man. trans.
    Solenoid connected 800
    Solenoid disconnected 600
    Propane enriched ....
250 6-cyl. 49-States
  Auto. trans. 550
    Propane enriched 600-630
    AC idle speedup 600

Man. trans.
  Solenoid connected 800
  Solenoid disconnected 425
  Propane enriched 800-1000
250 6-cyl. Calif.
  Auto. trans.
    Solenoid connected 600
    Solenoid disconnected 400
    Propane enriched 600-605
305 2-bbl. V-8 49-States
  Auto. trans. 500
    Propane enriched 520-540
    AC idle speedup 600
  Man. trans. 600
    Propane enriched 700-740
    AC idle speedup 700
305 2-bbl. V-8 Calif.
  Auto. trans. 500
    Propane enriched 520-540
    AC idle speedup 650
305 2-bbl. V-8 Altitude
  Auto. trans. 600
    Propane enriched 620-640
    AC idle speedup 700

350 4-bbl. V-8 "L" 49-States
  Auto. trans. 500
    Propane enriched 530-570
    AC idle speedup 600
  Man. trans. 700
    Propane enriched 850-900
350 4-bbl. V-8 "L" Calif.
  Auto. trans. 500
    Propane enriched 530-570
    AC idle speedup 600
350 4-bbl. V-8 "L" Altitude
  Auto. trans. 600
    Propane enriched 630-670
    AC idle speedup 650
350 4-bbl. V-8 L82 Engine
49-States
  Auto. trans. 700
    Propane enriched 760-800
    AC idle speedup 750
  Man. trans. 900
    Propane enriched 1050-1100

INITIAL TIMING

1976
250 6-Cyl. 1-piece
  head/manifold .... 6° BTDC
250 6-Cyl. 2-piece
  head/manifold .... 8° BTDC
262 V-8 .... 8° BTDC
305 V-8 49-States
  Auto. Trans. .... 8° BTDC
  Man. Trans. .... 6° BTDC
305 V-8 Calif. .... 0° TDC
350 2-bbl. V-8 .... 6° BTDC
350 4-bbl. V-8 LM-1 & L-48
  49-States .... 8° BTDC
  Calif. .... 6° BTDC
350 4-bbl. V-8 L-82 (Corvette)
  49-States .... 12° BTDC
400 4-bbl. V-8 .... 8° BTDC
454 4-bbl. V-8 .... 12° BTDC

1977
250 6-cyl.
  49-States
    Auto. trans. .... 8° BTDC
    Man. trans. .... 6° BTDC
  Calif.
    Auto. trans. .... 6° or 8° BTDC
    (See label)

Altitude .... 10° BTDC
305 2-bbl. V-8
  49-States .... 8° BTDC
  Calif. .... 6° BTDC
  Altitude .... 6° or 8° BTDC
    (See label)
350 4-bbl. V-8 "L" engine
  All .... 8° BTDC
350 4-bbl. V-8 "X" engine (Corvette)
  All .... 12° BTDC

1978
196 2-bbl. V-6 .... 15° BTDC
200 2-bbl. V-6 .... 8° BTDC
231 2-bbl. V-6 .... 15° BTDC
250 6-cyl.
  49-States
    Auto. trans. ...... 10° or 8° BTDC
    (See label)
  Calif.
    Auto. trans. .... 6° BTDC
305 2-bbl. V-8
  49-States .... 4° BTDC
  Calif. .... 6° BTDC
  Altitude .... 8° BTDC
350 4-bbl. V-8 "L" engine
  49-States .... 6° BTDC

Calif. .... 8° BTDC
Altitude .... 6° or 8° BTDC
  (See label)
350 4-bbl. V-8 "4" engine (Corvette)
  All .... 12° BTDC

1979
200 2-bbl. V-6 .... 8° BTDC
231 2-bbl. V-6 .... 15° BTDC
250 6-cyl.
  49-States
    Auto. trans. ... 10° or 8° BTDC
    (See label)
Calif.
  Auto trans. .... 6° BTDC
305 4-bbl. V-8 .... 4° BTDC
305 2-bbl. V-8
  49-States .... 4° BTDC
  Calif. .... 6° BTDC
  Altitude .... 8° BTDC
350 4-bbl. V-8 "L" engine
  49-States .... 6° BTDC
  Calif. .... 8° BTDC
  Altitude .... 6° or 8° BTDC
    (See label)
350 4-bbl. V-8 "H" engine (Corvette)
  All .... 12° BTDC
267 2-bbl. V-8 .... 10° BTDC

SPARK PLUGS

1976
250 6-Cyl.—AC-R46TS .... .035
262 V-8—AC-R45TS .... .045
305 V-8—AC-R45TS .... .045
350 V-8—AC-R45TS .... .045
400 V8—AC-R45TS .... .045
454 V-8—AC-R45TS .... .045

1977
250 6-cyl. .... AC-R46TS .... .035
305 V-8 .... AC-R45TS .... .045
350 V-8 .... AC-R45TS .... .045

1978
196 V-6 .... AC-R46TSX ... .060
200 V-6 .... AC-R45TS ... .045
231 V-6 .... AC-R46TSX ... .060
250 6-cyl. .... AC-R46TS ... .035
305 V-8 .... AC-R45TS ... .045
350 V-8 .... AC-R45TS ... .045

Chevrolet 6 Cyl. — V-8 • Monza V-8

SPARK PLUGS

1979

200 V-6	AC-R45TS	.045
231 V-6	AC-R46TSX	.060
250 6-cyl.	AC-R46TS	.035
305 V-8	AC-R45TS	.045
(4-bbl.)	AC-R43TS	.045
350 V-8	AC-R45TS	.045
267 V-8	AC-B45TS	.045

VACUUM ADVANCE

1977

Diaphragm type ....Single
Vacuum source
250 6-cyl.
  49-States ....Manifold
  Calif. ....Ported
  Altitude ....Manifold

305 V-8
  49-States ....Manifold
  Calif. ....Ported
  Altitude ....Manifold
350 V-8 ....Manifold

1978

Diaphragm type ....Single
Vacuum source
196 V-6 ....Manifold
200 V-6 ....Manifold
231 V-6
  49-States ....Manifold
  Calif. ....Ported
  Altitude ....Ported
250 6-cyl. ....Manifold
305 V-8
  49-States ....Manifold
  Calif. ....Ported
  Altitude ....Manifold
350 V-8 ....Manifold

1979

Diaphragm type ....Single
Vacuum source
200 V-6 ....Manifold
231 V-6
  49-States ....Manifold
  Calif. ....Ported
  Altitude ....Ported
250 6-cyl. ....Manifold
305 V-8
  49-States ....Manifold
  Calif. ....Ported
  Altitude ....Manifold
350 V-8 ....Manifold

EMISSION CONTROL SYSTEMS 1976

EARLY FUEL EVAPORATION SYSTEM (EFE) 1976

Description of System

6-Cylinder with two-piece head and manifold: These old-style engines do not use the EFE system.

6-Cylinder with one-piece head and manifold: The vacuum solenoid used in 1975 has been eliminated. Vacuum to the EFE valve actuator is now controlled by a thermal vacuum switch (TVS) screwed into the oil gallery on the right side of the engine. Tubes and hoses take the vacuum from the rear of the intake manifold to the TVS, and then to the EFE valve actuator. The TVS is open below 150° engine oil temperature, allowing vacuum to operate the EFE valve. Above 150° the TVS closes, and shuts off the vacuum to the EFE actuator. A vent on the side of the TVS is open at all times. It allows outside air to enter the vacuum line when the TVS is open, which purges the line of fuel vapors. When the TVS is closed, the vent allows outside air to enter the actuator line so the actuator is not trapped in the closed position. A foam filter and cover help keep dirt out of the vent. The vacuum line from manifold vacuum connects to the side nozzle on the TVS. The end nozzle connects to the EFE actuator.

The EFE thermal vacuum switch (arrow) is on the right side of the block, screwed into an oil passage

The thermal vacuum switch is a new design, with a hooded vent containing a foam filter

Chevrolet 6 Cyl. — V-8 • Monza V-8

262, 305, 350, 400 V-8: The system is the same as in 1975. A TVS on the thermostat elbow is open below 180° engine coolant temperature, and closed above that temperature. On the 262 V-8, the EFE thermal vacuum switch is the one on the forward side of the elbow. The one on the rear is for the EGR system. On all of these small V-8's, the upper nozzle on the thermal vacuum switch connects to manifold vacuum. The lower nozzle connects to the EFE actuator.

454 V-8: The vacuum solenoid used in 1975 is discontinued. Vacuum to the EFE valve actuator is now controlled by a thermal vacuum switch (TVS) screwed into the thermostat elbow. The TVS has three nozzles. The middle nozzle connects to manifold vacuum at the carburetor. The top nozzle connects to the EFE actuator. Below approximately 150° engine coolant temperature, the TVS top nozzle is open, so that vacuum can operate the EFE actuator. Above 150° the top nozzle closes, shuts off the vacuum to the EFE actuator and vents it to atmosphere. The TVS bottom nozzle connects to the front vacuum break diaphragm on the carburetor. See Fast Idle Pulloff System, for a description of how it works.

Units in System, 6-Cylinder, One-Piece Head/Man, 1976

EFE valve
EFE actuator
Thermal vacuum switch, oil temperature, 2 nozzles, with vent

Units in System, 262, 305, 350, 400 V-8, 1976

EFE valve
EFE actuator
Thermal vacuum switch, 2 nozzles, with vent, coolant temperature

Units in System, 454 V-8, 1976

EFE valve
EFE actuator
Thermal vacuum switch, 3 nozzles, with vent, coolant temperature

Testing and Troubleshooting System, 1976

The 1976 2-nozzle (with vent) thermal vacuum switch used on the 305 V-8 is tested the same as the TVS on the 262, 350, and 400 V-8's.

FAST IDLE PULLOFF SYSTEM, 1976

Description of System

This is an entirely new system, completely different from the system used in 1975. The new system is only used on 4-bbl. carburetors with an electric choke.

The front vacuum diaphragm on the carburetor is made so that it will pull the throttle off the fast idle cam, from the high step down to the next lower step. This fast idle pulloff occurs only after the engine has warmed up to approximately 150° coolant temperature.

Vacuum to the diaphragm is controlled by the same thermal vacuum switch (TVS) that controls vacuum to the EFE actuator. Manifold vacuum comes through a hose to the middle nozzle on the TVS. The bottom nozzle is connected to the carburetor front vacuum diaphragm. The top nozzle is connected to the EFE valve actuator.

When the engine is cold (below 150° F.) the bottom nozzle on the TVS is shut off. At approximately 150° coolant temperature, the TVS allows vacuum to go through the bottom nozzle and operate the front vacuum diaphragm. The diaphragm then pulls the throttle off the high step of the fast idle cam.

If the car is driven normally while warming up, the pulloff system has no effect, because the normal movement of the throttle disengages the fast idle cam. But if the car is parked while it is warming up, the pulloff reduces the fast idle speed to prevent damage to the catalytic converter from excessive fast idle.

Units in System

Front vacuum pulloff diaphragm
3-nozzle thermo-vacuum switch

Testing and Troubleshooting System

If the engine is cold, set the choke by depressing the gas pedal to the floor. Start the engine and let it fast idle without touching the gas pedal. In a few minutes you should hear the engine reduce speed as the pulloff operates. If not, there is something wrong with the system.

If the engine is warm, set the choke with the engine off. Remove the air cleaner, open the throttle and close the choke by hand. Then let the throttle linkage go closed against the choke, so the throttle is on fast idle. Disconnect the hose to the front vacuum break and temporarily plug the hose. Start the engine and let it fast idle in Neutral. The speed should be approximately 1600 rpm. If not, the fast idle adjustment on the carburetor should be correctly set.

Unplug the hose from the TVS and connect it to the front vacuum diaphragm. The throttle should drop to the second step, and the engine should reduce speed. If not, the system is out of order.

Repairing System

The linkage connecting the front diaphragm to the choke is adjustable. The units themselves are not repairable if defective, and must be replaced.

Testing Thermal Vacuum Switch

Connect a vacuum gauge to the bottom nozzle on the thermal vacuum switch. With the engine at normal operating temperature, you should get full manifold vacuum on the gauge. If not, either the hoses are connected wrong, or the TVS is defective.

Repairing Thermal Vacuum Switch

Repairs are limited to replacement.
CAUTION: *Relieve the cooling system pressure before removing the switch.*

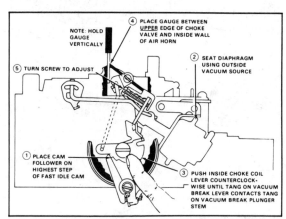

Front vacuum break adjustment

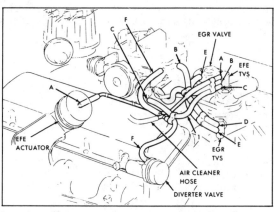

EFE and fast idle pulloff hoses, 454 V-8

Chevrolet 6 Cyl. — V-8 • Monza V-8

Testing Front Vacuum Diaphragm

Use a hand vacuum pump to apply 20″ Hg. vacuum to the diaphragm. The diaphragm should hold and not leak down. If it does leak down, the diaphragm is defective.

Repairing Front Vacuum Diaphragm

Repairs are limited to replacement. See the accompanying illustration for the diaphragm adjustment procedure.

ELECTRIC CHOKE 1976

Description of Choke

NOTE: *This choke is the same as the Cadillac electric choke, except for wire colors and connections.*

The choke cover has a dual-section heating element behind the coil spring. Whenever the engine is running, the choke receives current and the heater is in operation. A bimetal snap disc in the choke cover turns off the large section of the dual heating element below 50-70° F. Below that temperature, only the small section of the element provides heat. This gives a slower choke opening for better running in cold weather. Above 50-75° F. the bimetal snap disc turns on the large section of the heating element so that both sections provide heat for a quicker choke opening when the engine is warm.

Current to the choke is controlled by a three-terminal oil pressure switch. The tan wire is the hot lead. The dark blue wire is the ground circuit for the oil pressure light on the instrument panel. The light blue wire connects the choke to the oil pressure switch. Current flows to the choke only when there is oil pressure in the engine.

When the choke heater is receiving current, it is grounded to the carburetor to complete the circuit. A metal plate at the back of the choke cover touches the carburetor to make the ground.

CAUTION: *The choke cover must be installed without a gasket, so that the ground circuit will be complete. The gasket is not necessary, because there is no vacuum to the choke housing, as there is in older chokes.*

Testing Electric Choke

At an engine compartment temperature of approximately 70° F. (cold engine) the choke should open wide in approximately 1½ minutes. The easiest way to make this test is to move the air cleaner to one side, open the throttle to allow the choke to close, and run a hot wire from the positive battery post to the choke terminal. Then time how long it takes the choke to open.

Off the car, the choke coil can be tested by connecting a 12-volt hot wire to the coil terminal, and a ground wire to the choke cover grounding plate. Starting at a room temperature of 60-70° F. the choke coil should rotate 45 degrees within 54-90 seconds after the connection is made. If not, the choke coil is defective, and must be replaced.

If the choke coil is in good condition, but will not operate on the engine, it could be caused by:
1. Broken or disconnected light blue wire between oil pressure switch and choke coil.
2. Poor ground between choke cover grounding plate and housing.
3. Broken or disconnected wire in circuit from battery to oil pressure switch. See Repairing Electric Choke below, for circuit connections.
4. Burned out TCS SOL RDO fuse in fuse block.
5. Defective oil pressure switch.
6. No oil pressure to switch. This could be caused by blocked oil passageways or sludged oil. If the oil light on the instrument panel does not go off when the engine is running, there is no oil pressure, and the engine should be shut down until the cause is found.

Repairing Electric Choke

Repairs are limited to replacement. Current to the choke starts at the battery, goes through several splices and connections to the ignition switch. From the ignition switch accessory terminal the current goes to the TCS-SOL-RDO fuse in the fuse block, then through the firewall connector to the connector for the automatic transmission kickdown switch. From the kickdown switch connector a tan wire takes the current to the oil pressure switch, and from there to the choke when the switch is on.

EMISSION CONTROL SYSTEMS 1977

MANIFOLD CATALYTIC CONVERTER 1977

California 6-cylinder engines have a new manifold converter in addition to the usual underfloor converter. The manifold converter attaches to the exhaust manifold outlet in the engine compartment. Because the manifold converter is closer to the engine, it heats up faster after a cold start, and helps clean up the exhaust to meet the California requirement.

6-CYLINDER ELECTRIC CHOKE 1977

The 1-bbl. carburetor used on the 6-cylinder engines is now equipped with an electric choke. This is the same design choke that was used last year on the 454 V-8, and has also been used on Cadillac and Chevette.

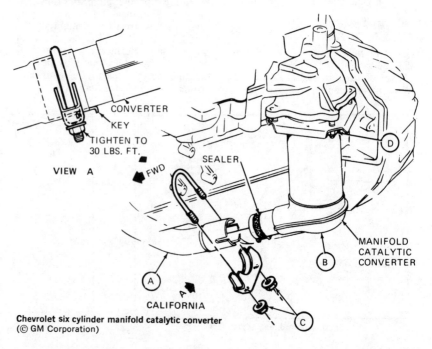

Chevrolet six cylinder manifold catalytic converter
(© GM Corporation)

Chevrolet 6 Cyl. — V-8 • Monza V-8

EMISSION CONTROL SYSTEMS

V-6 ENGINES 1978

You won't have any trouble recognizing the new 200-inch V-6 as a Chevrolet engine. It looks just like a Chevy small block with two cylinders lopped off. Everything about the engine is conventional Chevy V-8 design, except the crankshaft.

It is a fact of normal "V" engine design that if the firing impulses are going to be an equal number of degrees apart, the number of degrees between the crankshaft throws must be the same as the number of degrees between the banks. The number of degrees between the crankshaft throws is determined by the number of cylinders. An eight cylinder engine fires eight times in two revolutions, or 720 degrees. Divide the 720 degrees by eight, and the result is 90 degrees, the distance by which the crankshaft throws must be separated. Given the 90 degree separation, if the "V" block is then made with a 90 degree separation, the firing impulses will all come out evenly spaced.

A 6-cylinder engine fires 6 times in 720 degrees, or every 120 degrees. To make the firing impulses evenly spaced, the banks of a V-6 would have to be 120 degrees apart. But when a V-6 is made by cutting off two cylinders from a 90 degree V-8, it means that the firing impulses will not be even, with a normal 120 degree crankshaft. What happens is that the engine is an odd firing design, with firing impulses alternating between 90 degrees and 150 degrees apart.

Buick had the odd firing design for several years, but in 1977 they phased in a new crankshaft, with split or splayed pins. Instead of having only three throws, the new crank had six throws. This matched up each crank throw to its respective cylinder, and made the engine even firing, for a smoother idle.

Chevrolet had the benefit of all this Buick design work, and they were not really satisfied with the result. Through experimenting, they found out that the odd-firing design was actually smoother than even-firing, but with a different spacing than on the original Buick V-6. The new Chevrolet V-6 fires with alternating intervals of 108 degrees and 132 degrees. In order to do this, the Chevrolet V-6 has split crankpins, but separated by fewer degrees than the Buick engine.

When looking at a scope pattern of the new Chevrolet V-6, the timing between cylinders will be staggered, similar to the older Buick, but not as much. All of the General Motors V-6 engines have the same firing order, which is 1-6-5-4-3-2. And in the usual G.M. practice, the cylinders are numbered from front to back, with the odd numbers on the left bank.

The 196-inch V-6, used in the Monza, is a smaller version of the 231 V-6 made by Buick. The two Buick V-6 engines both have the distributor in the front, and look identical. The best way to tell them apart is by the code letter in the vehicle identification number.

The 200-inch V-6 is easily spotted, because it has the distributor in the rear.

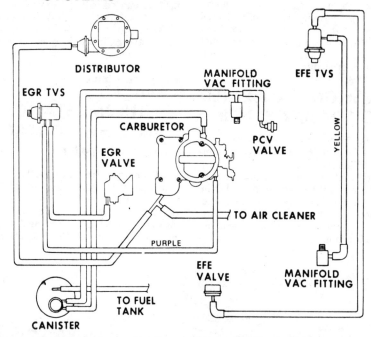

Vacuum hose routing Chevrolet 250 six cylinder—49 states
(© GM Corporation)

EMISSION CONTROL SYSTEMS

EMISSION CONTROLS POSITIVE CRANKCASE VENTILATION (PCV) 1978

The 1978 Chevrolets have a closed Positive Crankcase Ventilation System (PCV). This system provides a more complete scavenging of crankcase vapors. The valve must be replaced periodically to insure that proper crankcase ventilation is taking place, otherwise serious engine damage could result.

1979

All engines are equipped with a closed Positive Crankcase Ventilation System (PCV). This system is designed to provide a more complete dispersion of crankcase vapors.

An engine that is operated without any crankcase ventilation can be seriously damaged, therefore it is important to check and replace the PCV valve periodically.

EMISSION CALIBRATED CARBURETOR 1978

Carburetor calibration is important in maintaining proper emission levels. The carburetor operating systems is calibrated to provide the best possible combination of engine performance, fuel economy, and emission control. Adjustments and service must be performed to insure that engine exhaust emission levels remain within legislated limits.

1979

While the carburetor's main function is to provide the engine with a combustible air/fuel mixture, it's calibration is also very important to maintaining proper emission levels.

Chevrolet 6 Cyl. — V-8 • Monza V-8

The carburetor's idle, off-idle, main metering, power enrichment, and accelerating pump systems are calibrated in order to provide a combination of engine performance, fuel economy, and exhaust emission control. Adjustments and service must be performed according to recommended procedures.

EMISSION CALIBRATED DISTRIBUTOR 1978

Distributor calibration is a critical part of engine exhaust emission control. Therefore the initial timing centrifugal advance and vacuum advance are calibrated to provide the best engine performance and fuel economy at varying speeds and loads while remaining within exhaust emission limits.

1979

Distributor calibration is a critical part of engine exhaust emission control. Therefore the initial timing centrifugal advance and vacuum advance are calibrated to provide the best engine performance and fuel economy at varying speeds and loads while remaining within exhaust emission limits.

THERMOSTATIC AIR CLEANER (TAC) 1978

All engines are equipped with the Thermostatic Air Cleaner system (TAC). This system incorporates a damper assembly within the air cleaner inlet to allow pre-heated and non pre-heated air that is entering the air cleaner to maintain a controlled air temperature into the carburetor.

1979

The thermostatic air cleaner system (TAC) used on all 1979 engines, which controls the amount of heated air that enters the carburetor to maintain a controlled air temperature, is exactly the same as the system that was used on the 1978 engines.

EVAPORATIVE CONTROL SYSTEM (ECS) 1978

This system reduces the amount of escaping gasoline vapors. Float bowl emissions are controlled by in-

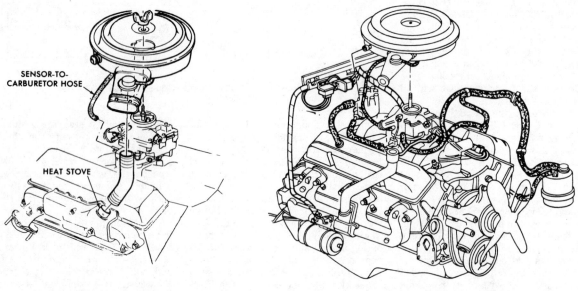

Chevrolet, TAC assembly V-8 engine   **Evaporative control system—typical V-8 engine**

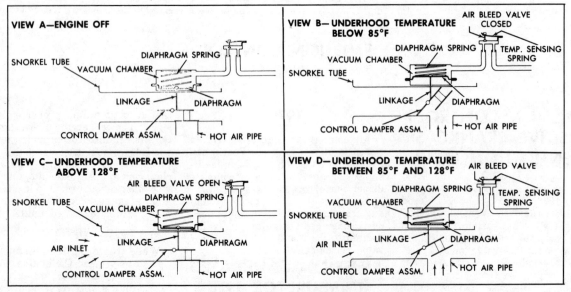

Chevrolet, thermostatic air cleaner operation

ternal carburetor modifications. Redesigned bowl vents, reduced bowl capacity, heat shields, and improved intake manifold-to-carburetor insulation serve to reduce vapor loss into the atmosphere. The venting of fuel tank vapors into the air has been eliminated. Fuel vapors are now directed through lines to a canister containing an activated charcoal filter. Unburned vapors are trapped here until the engine is running, the canister is purged by air drawn in by manifold vacuum. The air and fuel vapors are then directed into the engine to be burned. This system is designed to reduce fuel vapor emission.

1979

All vehicles are equipped with this system, which is designed to prevent the escape of fuel vapor into the atmosphere. Before this system was introduced the vapor generated by the evaporation of fuel in the gas tank was expelled into the atmosphere. With the inception of the evaporative control system (ECS) the vapor is transfered by an emissions line to the engine compartment. When the vehicle is operating vapors are fed directly into the engine for consumption. During periods of inoperation, an activated charcoal canister stores any vapor generated for consumption during the next period of vehicle operation.

EXHAUST GAS RECIRCULATION (EGR)
1978

All engines are equipped with exhaust gas recirculation (EGR). This system consists of a metering valve, a vacuum line to the carburetor, and cast-in exhaust gas passages in the intake manifold. The EGR valve is controlled by carburetor vacuum, and accordingly opens and closes to admit exhaust gases into the fuel/air mixture. The exhaust gases lower the combustion temperature, and reduce the amount of oxides of nitrogen (NO_x) produced. The valve is closed at idle between the two extreme throttle positions.

All California models and cars delivered in areas above 4000 ft. are equipped with back pressure EGR valves. The EGR valve receives exhaust back pressure through its hollow shaft. This exerts a force on the bottom of the control valve diaphragm, opposed by a light spring. Under low exhaust pressure (low engine load and partial throttle), the EGR signal is reduced by an air bleed. Under conditions of high exhaust pressure (high engine load and large throttle opening), the air bleed is closed and the EGR valve responds to an unmodified vacuum signal. At wide open throttle, the EGR flow is reduced in proportion to the amount of vacuum signal available.

The negative transducer backpressure EGR valve assembly has the same function as the positive transducer backpressure EGR valve except the transducer is designed to allow the valve to open with a negative exhaust backpressure. The flow of the valve is controlled by manifold vacuum, negative exhaust backpressure and the carburetor ported vacuum signal.

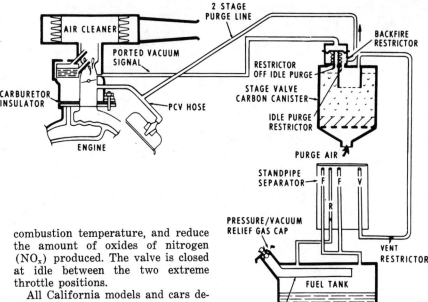

Schematic of evaporative emissions system

1979

The exhaust gas recirculation system (EGR) is used on all engines. This system regulates exhaust gas into the induction system for recirculation through the combustion cycle in order to reduce oxides of nitrogen (NO_x).

The EGR valve remains closed during engine idle and deceleration to prevent rough idle from excessive exhaust gas dilution in the idle air/fuel mixtures. The operation of the exhaust gas recirculation system for 1979 is the same as the operation of the system for 1978.

CATALYTIC CONVERTER
1978

All models are equipped with a

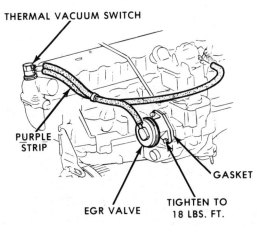

Exhaust gas recirculation system (EGR), 6-cyl. engine

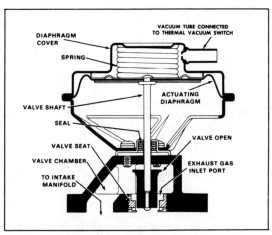

Exhaust gas recirculation valve (EGR)

Chevrolet 6 Cyl. — V-8 • Monza V-8

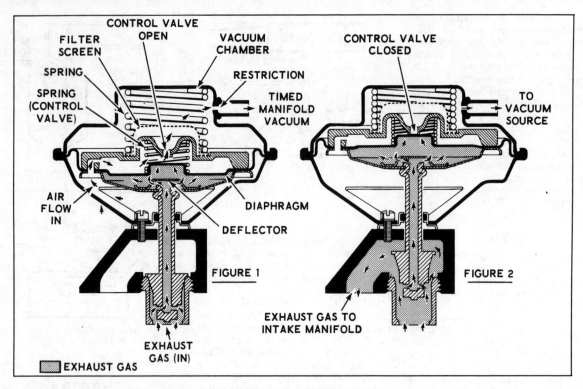

Positive back pressure EGR valve

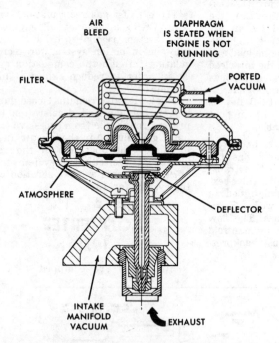

Negative transducer backpressure EGR valve

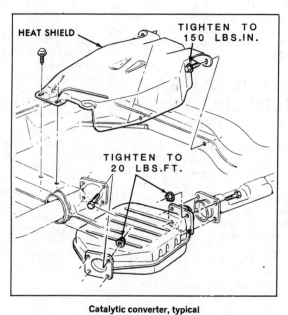

Catalytic converter, typical

catalytic converter. This unit is located midway in the exhaust system. The converter is stainless steel with an aluminized steel cover and a ceramic felt blanket to insulate the converter from the floor plan. The catalyst pellet bed inside the converter consist of noble metals which cause a reaction that converts hydrocarbons and carbon monoxide into water and carbon dioxide. The six cylinder engines are equipped with an additional catalytic converter located directly under the exhaust manifold.

1979

The catalytic converter, which is an emission control device, is located midway in the exhaust system. It is designed to reduce hydrocarbons (HC) and carbon monoxide (CO) pollutants from the exhaust gas stream. The catalytic converter contains beads that are coated with a

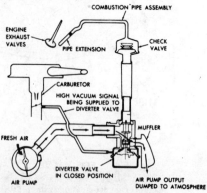

Operation of air diverter valve

Chevrolet 6 Cyl. — V-8 • Monza V-8

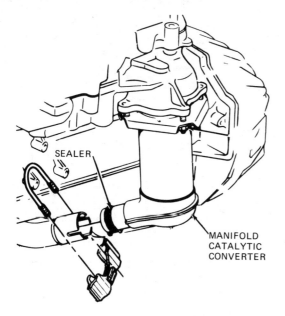

Manifold catalytic converter

Early fuel evaporation—6-cyl. engine

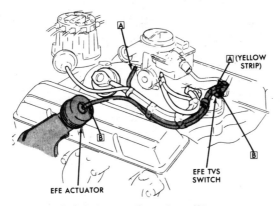

Early fuel evaporation system—V-8

material containing platinum and palladium. All six cylinder engines have an additional converter located directly under the exhaust manifold.

EARLY FUEL EVAPORATION (EFE) 1978

All models are equipped with this system to reduce engine warm up time, and improve driveability, and reduce emissions.

On start-up, a vacuum motor acts to close a heat valve in the exhaust manifold which causes exhaust gases to enter the intake manifold heat riser passages. Incoming fuel mixture is than heated and more complete fuel evaporation is provided during warm-up.

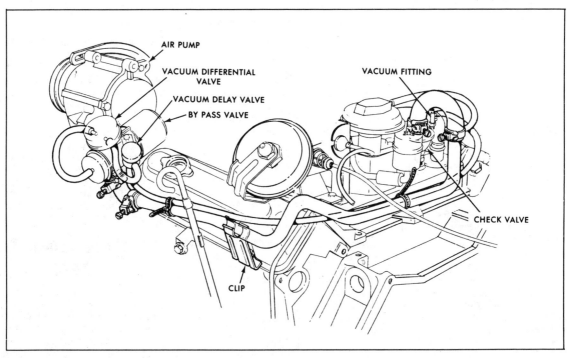

Air injector system—V-6 (231 CID) Calif.

111

Chevrolet 6 Cyl. — V-8 • Monza V-8

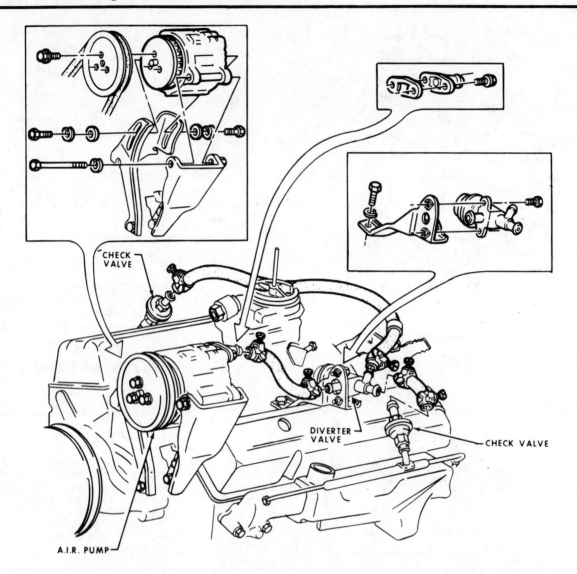

Air injection system and components

1979

The early fuel evaporation system (EFE) is used to provide a fast heat source to the engine induction system during cold driveway. This heat provides quick fuel evaporation and more uniform fuel distribution to aid cold vehicle operation. It also reduces the length of time carburetor choking is required making reductions in exhaust emission levels possible.

VACUUM DELAY VALVE 1978

The vacuum delay valve is nothing more than a retard delay valve. It allows the manifold vacuum to act on the distributor diaphragm without any restriction. But when the manifold vacuum drops to zero, as during heavy acceleration or if the engine should die at idle, the valve "traps" the vacuum and lets it bleed down slowly. This keeps the spark advanced for better performance, or for easier starting.

1979

The vacuum delay valve is nothing more than a retard delay valve. It allows the manifold vacuum to act on the distributor diaphragm without any restriction. But when the manifold vacuum drops to zero, as during heavy acceleration or if the engine should die at idle, the valve "traps" the vacuum and lets it bleed down slowly. This keeps the spark advanced for better performance, or for easier starting.

Above 120° F. coolant temperature, a thermal vacuum switch, that is connected in the system by hoses, opens and bypasses the vacuum delay valve so that the vacuum advance works normally.

SPARK DELAY VALVE 1978

The spark delay valve controls vehicle exhaust emissions by delaying vacuum advance during some vehicle acceleration modes. This system is used on all 305 V8 engines in California vehicles.

1979

The spark delay valve assembly is used on all California 305 V8 engines. This device controls vehicle exhaust emissions by delaying vacuum advance during vehicle acceleration modes. The valve is connected between the vaccum advance can of the distributor and the carburetor.

AIR PUMP 1978

The air injection reactor (AIR) injects compressed air into the exhaust system, close enough to the exhaust valves to continue the burning of the normally unburned segment of the exhaust gases. To do this it employs an air injection pump and

Chevrolet 6 Cyl. — V-8 • Monza V-8

a system of hoses, valves, tubes necessary to carry the compressed air from the pump to the exhaust manifolds.

A diverter valve is used to prevent backfiring. The valve senses sudden increases in manifold vacuum and ceases the injection of air during fuel-rich periods. During coasting, this valve diverts the entire air flow through the muffler and during high engine speeds, expels it through a relief valve. Check valves in the system prevent exhaust gases from entering the pump.

1979

The air injector reactor system (AIR) consists of an air injection pump, air diverter valve, a check valve, and an air manifold assembly. This system compresses the air and injects it through the air manifold into the exhaust system area of the exhaust valves. The diverter valve shuts off the injected air to the exhaust port and prevents backfiring when triggered by a sharp increase in manifold vacuum.

PULSE AIR SYSTEM
1979

The Pulse Air system consists of four pipes of equal length. Each pipe is to be secured to a check valve which is inserted into a grommet in the "pod" which is a plenum welded to the top of the rocker arm cover. The other end of the pipe routes to and is secured into an exhaust manifold passage in the head assembly using a tube nut. A separate pipe is installed between the pods as a fresh air supply source. A tee is part of this pipe onto which is to be slipped a hose which leads to a nipple on the clean side of the air cleaner.

A deceleration valve is used on California engines. This valve mounts on the forward leg of the inlet manifold and dumps air into the engine during deceleration.

ELECTRIC CHOKE
1978

The Electric Choke is used to assist in controlling the choke thermostatic coil for precise timing of the choke valve opening for good engine warm-up performance. This is accomplished with the use of a ceramic resistor with-in the choke unit. A small section resistor is used to gradually heat the coil, while a large section resistor is used to quickly heat the coil as the temperature dictates. The electric choke is not used on all models.

1979

The Electric Choke is used on select models of 1979, and operates in the same manner as the 1978 models. At temperatures below 50° F., the electrical current is directed to the small segment resistor of the choke to allow gradual opening of the choke valve. At temperatures of 70° F., and above, the electrical current is directed to both the small and the large segment resistors, to allow quicker opening of the choke valve.

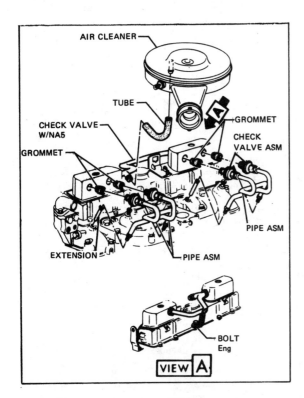

Pulse air system, 6-cyl. engines (49-States)

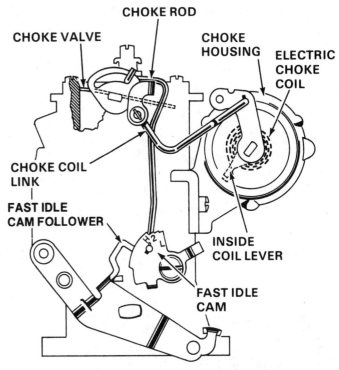

Electric choke assembly and carburetor components

Chevrolet 6 Cyl. — V-8 • Monza V-8

VACUUM CIRCUITS

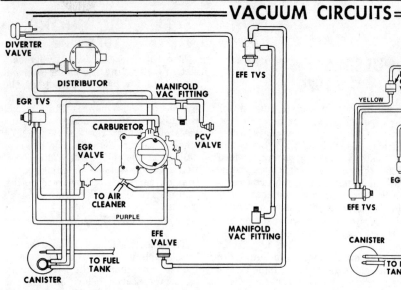

1977 Vacuum hose routing Chevrolet 250 six cylinder—California
(© GM Corporation)

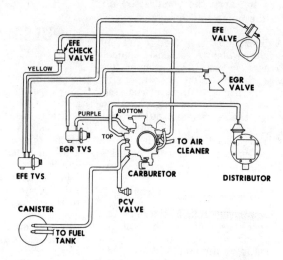

1977 Vacuum hose routing Chevrolet V-8 305 2bbl—49 states
(© GM Corporation)

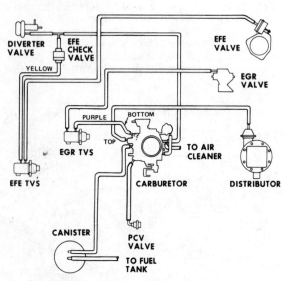

1977 Vacuum hose routing Chevrolet V-8 305 2bbl—Altitude
(© GM Corporation)

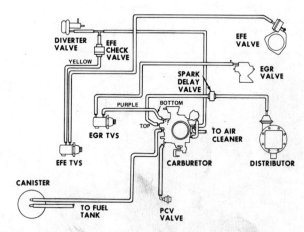

1977 Vacuum hose routing Chevrolet V-8 305 2bbl—California
(© GM Corporation)

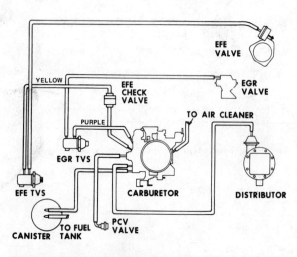

1977 Vacuum hose routing Chevrolet V-8 350 4bbl—49 states
(© GM Corporation)

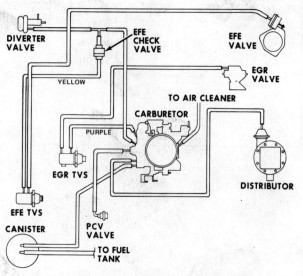

1977 Vacuum hose routing Chevrolet V-8 350 4bbl—Altitude
(© GM Corporation)

Chevrolet 6 Cyl. — V-8 • Monza V-8

VACUUM CIRCUITS

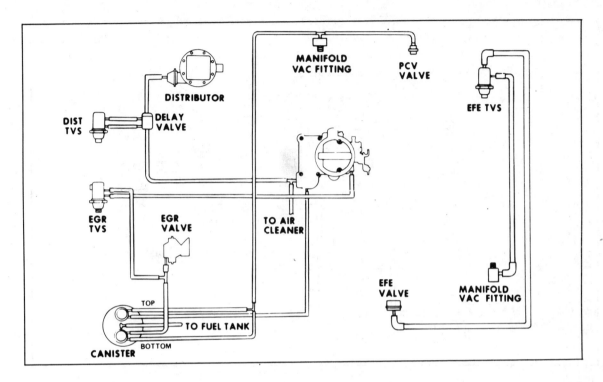

Vacuum hose routing, 1978 Chevrolet—6-cyl.—49-States and Altitude

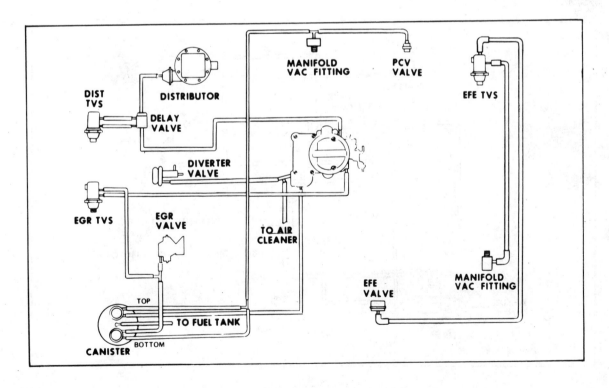

Vacuum hose routing, 1978 Chevrolet—6 cyl.—Calif.

Chevrolet 6 Cyl. — V-8 • Monza V-8

VACUUM CIRCUITS

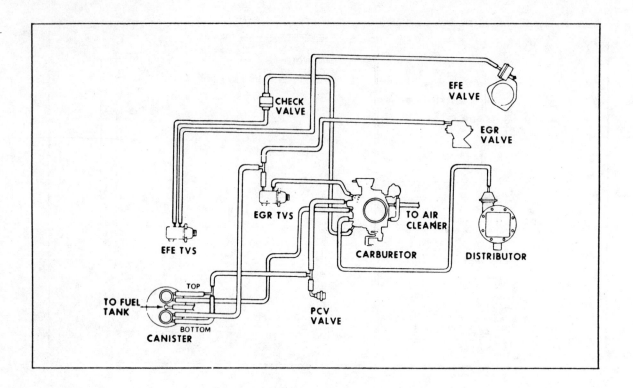

Vacuum hose routing, 1978 Chevrolet—350 V-8—49-States

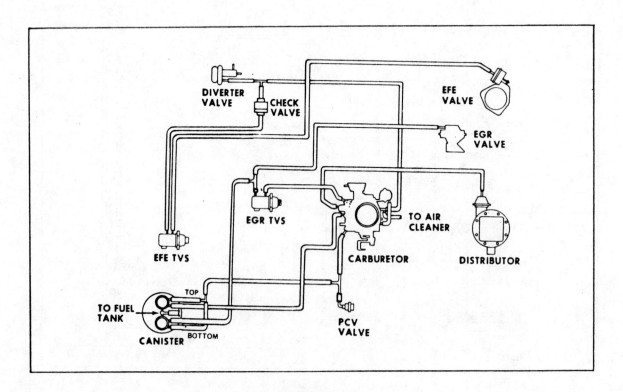

Vacuum hose routing, 1978 Chevrolet—V-8 305—Calif.

Chevrolet 6 Cyl. — V-8 • Monza V-8

VACUUM CIRCUITS

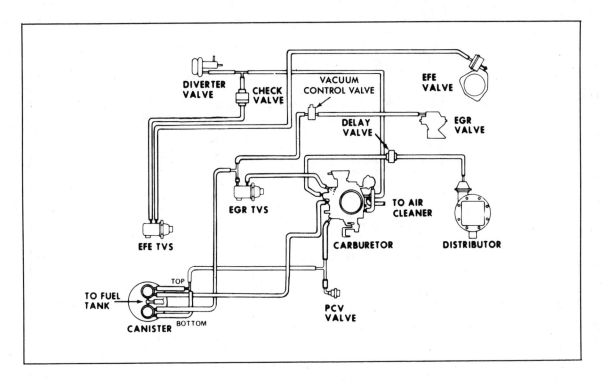

Vacuum hose routing 1978 Chevrolet—305 V-8—Calif.

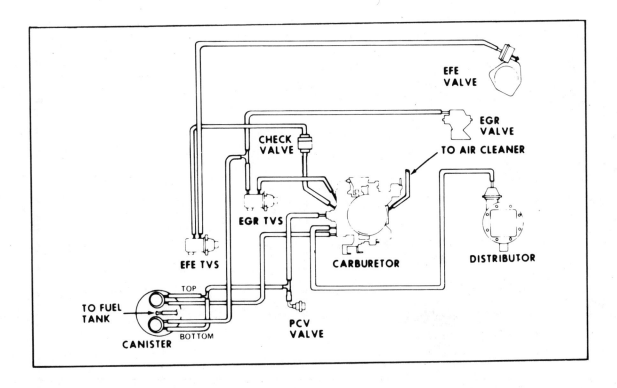

Vacuum hose routing 1978 Chevrolet—350 V-8—49-States

Chevrolet 6 Cyl. — V-8 • Monza V-8

VACUUM CIRCUITS

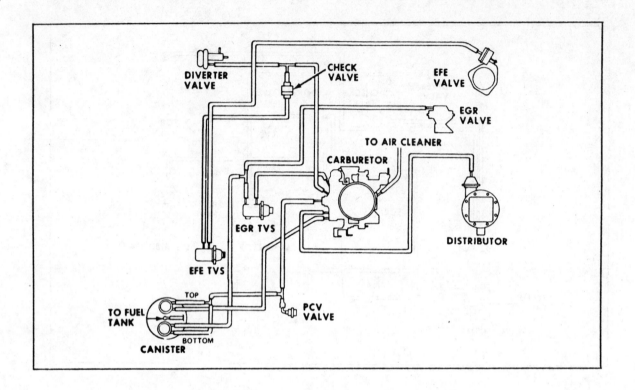

1978 Calif. and Altitude 350 V-8 vacuum hose routing

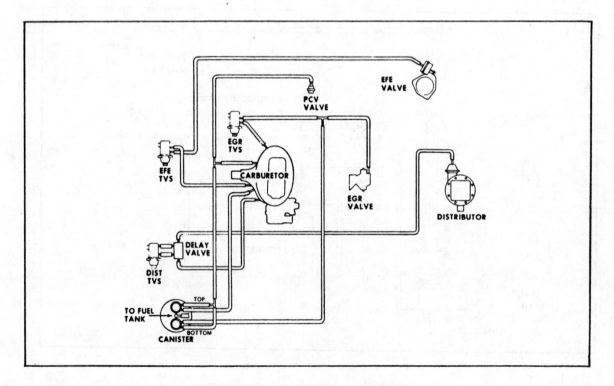

1978 49 States Chev. 200 V-6 vacuum hose routing

Chevrolet 6 Cyl. — V-8 • Monza V-8

VACUUM CIRCUITS

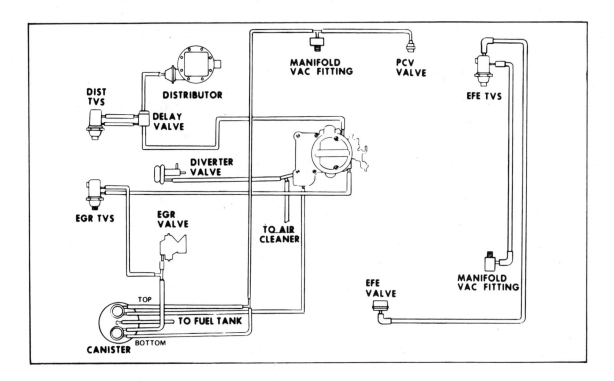

Vacuum hose routing 1978 Chevrolet—250 6-cyl.—Calif.
Vacuum hose routing, 1978 Chevrolet—6-cyl.—49-States and Altitude

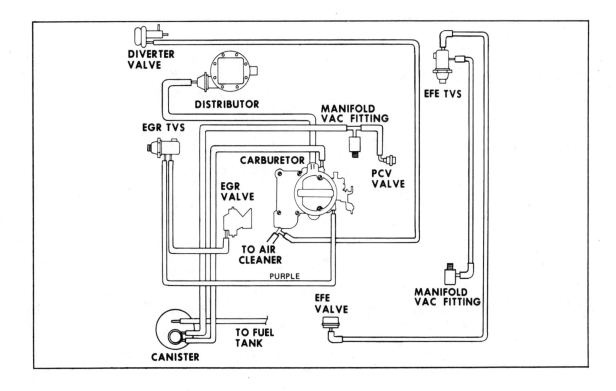

Typical 78-79 vacuum hose schematic, 151 CID, California with EGR

Chevrolet 6 Cyl. — V-8 • Monza V-8

VACUUM CIRCUITS

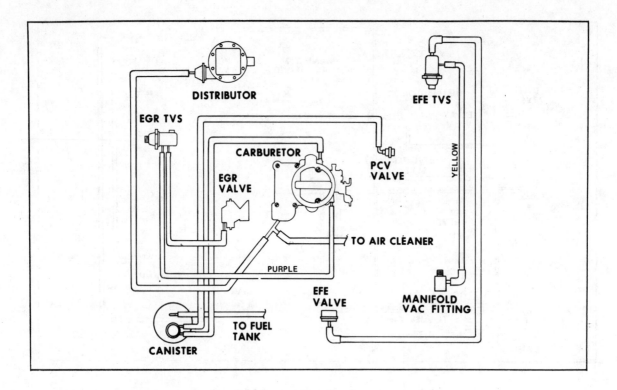

Vacuum hose routing 1978 Chevrolet—250 6-cyl.—49-States and Altitude

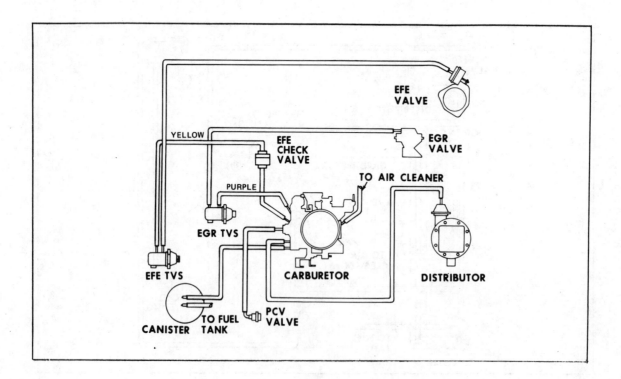

Vacuum hose routing, 1978 Chevrolet—350 V-8—49-States

Chevrolet 6 Cyl. — V-8 • Monza V-8

VACUUM CIRCUITS

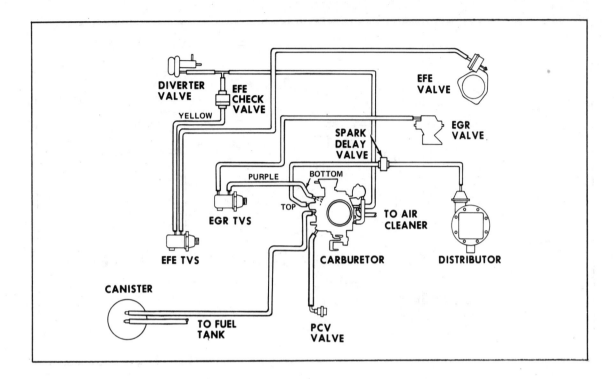

Vacuum hose routing 1978 Chevrolet—305 V-8—Calif.

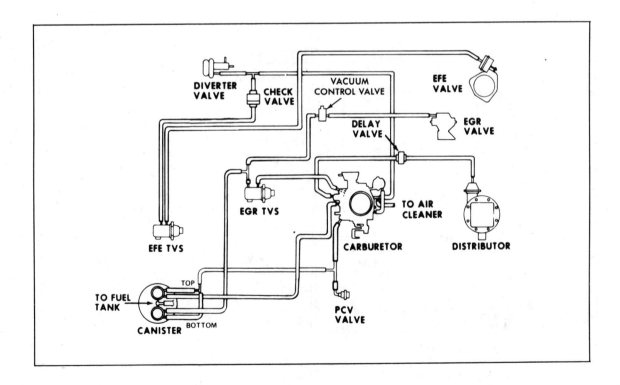

Vacuum hose routing, 1978 Chevrolet—V-8 350—Calif.

121

Chevrolet 6 Cyl. — V-8 • Monza V-8

VACUUM CIRCUITS

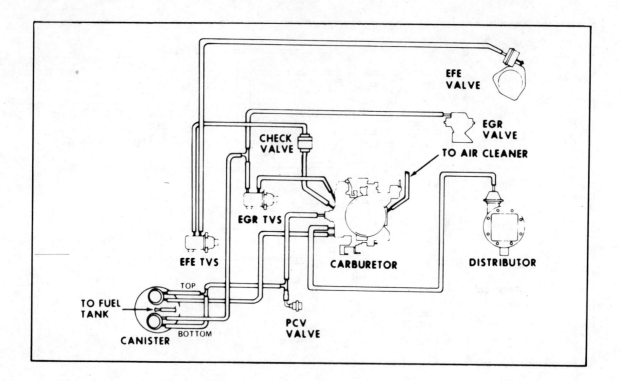

Vacuum hose routing 1978 Chevrolet—350 V-8—49-States

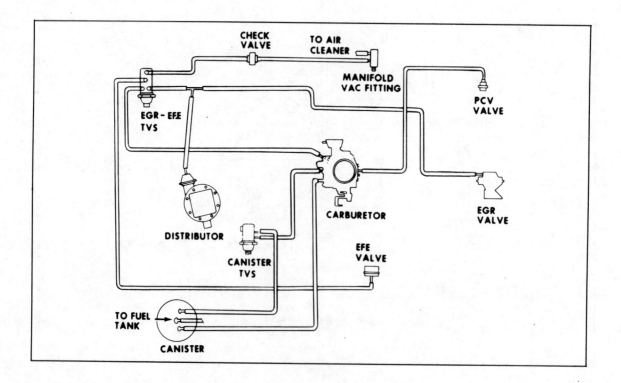

Vacuum hose routing, 1978 Chevrolet—V-6 231—49-States and auto. trans.

Vega • Astre • Monza & Sunbird 4 Cyl.

Astre • Monza • Skyhawk • Starfire • Sunbird • Vega

TUNE-UP SPECIFICATIONS

Engine Code, 5th character of the VIN number
Model Year Code, 6th character of the VIN number

Yr.	Eng. V.I.N. Code	Engine No. Cyl. Disp. (cu in)	Eng.■ Mfg.	Carb Bbl	H.P.	SPARK PLUGS Orig. Type	Gap (in)	DIST. Point Dwell (deg)	Point Gap (in)	IGNITION TIMING (deg BTDC) Man	Auto	VALVES Intake Opens (deg BTDC)	FUEL Pump Pres. (psi)	IDLE SPEED (rpm) Man Trans	Auto Trans In Drive
75	A	4-140	C	1v	75	R43TSX	.035	Electronic		8 @ 700	10	22	3.0-4.5	1200	750
	B	4-140	C	2v	85	R43TSX	.060	Electronic		10 @ 700	12	28	3.0-4.5	700	750
	C	6-231	B	2v	110	R44SX	.060	Electronic		12	12	17	3②	800	700
	F	8-260	O	2v	110	R46SX	.080	Electronic		16 @ 1100*	⑦	14	6	750	550⑩
	G	8-262	C	2v	110	R44TX	.060	Electronic		8	8	26	7.0-8.0	800	600
76	O	4-122	C	FI	120	R43T8X	.035	Electronic		12	—	38	—	1600	—
	A	4-140	C	1v	75	R43TS	.035	Electronic		8	10	22	3.0-4.5	1200	750
	B	4-140	C	2v	85	R43TS	.035	Electronic		10	12	28	3.0-4.5	700	750
	C	6-231	B	2v	110	R44SX	.060	Electronic		12	12	17	3②	800	600
	F	8-260	O	2v	110	R46SX	.060	Electronic		16 @ 1100	⑦	14	5.5-6.5	750	550
	G	8-262	C	2v	110	R45TS	.045	Electronic		6	8	26	7.0-8.0	800	600
	Q	8-305	C	2v	140	R45TS	.045	Electronic		6	8	26	7.0-8.5	800	600
77	B	4-140	C	2v	84	R43TS	.035	Electronic		TDC @ 700⑫	2	34	3.0-4.5	1250③⑨	850③⑪
	V	4-151	P	2v	87	R44TSX	.060	Electronic		14 @ 1000*	14(12)	33	4.5-5.0	500	650
	C	6-231	B	2v	105	R46TSX	.040	Electronic		12 @ 600	12	17	3②	600	550
	U	8-305	C	2v	145	R45TS	.045	Electronic		8	8	28	7.0-8.5	600	500
78	1	4-151	P	2v	85	R43TSX	.060	Electronic		14	14⑯	33	4-5.5	⑤	⑤
	V	4-151	P	2v	85	R43TSX	.060	Electronic		14	14	33	4-5.5	⑤	⑤
	C	6-196	B	2v	90	R46TSX	.060	Electronic		15	15	18	3-4.5⑬	⑤	⑤
	A	6-231	B	2v	105	R46TSX	.060	Electronic		15	15	17	3-4.5⑬	800	600
	U	8-305	C	2v	145	R45TS	.045	Electronic		4	6⑰	28	7.5-9.0	600	500⑮
79	1	4-151	P	2v	85	R43TSX	.060	Electronic		12	12	33	4-5.5	⑤	⑤
	V	4-151	P	2v	85	R43TSX	.060	Electronic		14	14	33	4-5.5	1000⑭	650⑭
	9	4-151	P	2v	85	R43TSX	.060	Electronic		14	14	33	4-5.5	⑤	⑤
	C	6-196	B	2v	90	R46TSX	.060	Electronic		⑤	⑤	18	3-4.5⑬	⑤	⑤
	A	6-231	B	2v	115	R46TSX	.060	Electronic		15	15	17	3-4.5⑬	800	600
	2	6-231	B	2v	115	R46TSX	.060	Electronic		15	15	17	3-4.5⑬	800	600
	G	8-305	C	2v	145	R45TS	.045	Electronic		4	4(4)	28	7.5-9.0	600	500⑮

Should the information provided in this manual deviate from the specifications on the underhood tune-up label, the label specifications should be used, as they may reflect production changes.

① THERE ARE TWO VERSIONS OF THE THIS ENGINE IN 1978, ONE FEATURES A SPECIAL CLOSED-LOOP EMISSION SYSTEM. THIS ENGINE CAN BE IDENTIFIED BY THE DEALER ORDER NUMBER UPC L36.
② 1975 SKYHAWK 3-4 PSI 1976-1977 SKYHAWK 3.-0-45 PSI.
• Figures in parenthesis are California specs.
③ ADJUST SPEEDUP SOLENOID TO OBTAIN SPECIFIED RPM, WITH AIR CONDITIONING UNIT TURNED ON AND COMPRESSOR CLUTCH WIRE DISCONNECTED.
④ AT 700 OR LESS RPM
⑤ SEE UNDERHOOD TUNE-UP DECAL.
⑥ –34
⑦ (14 @ 1100)
⑧ (1200)
⑨ (800)
⑩ (600)
⑪ (700)
⑫ 2ATDC @ 700
* Timing is with both automatic and manual transmissions.
⑬ AT 12.6 VOLTS
⑭ BASE IDLE 500 RPM.
⑮ HIGH ALT. BASE IDLE IS 600 RPM
⑯ WITH EGR VALVE: 12° WITHOUT EGR VALVE
⑰ HIGH ALT. IS 8°
■ B–Buick
C–Chevrolet
O–Oldsmobile
P–Pontiac

Vega • Astre • Monza & Sunbird 4 Cyl.

Astre • Monza • Skyhawk • Starfire • Sunbird • Vega

DISTRIBUTOR SPECIFICATIONS

Year	Distributor Identification	Centrifugal Advance Start Dist. Deg. @ Dist. RPM	Centrifugal Advance Finish Dist. Deg. @ Dist. RPM	Vacuum Advance Start @ In. Hg.	Vacuum Advance Finish Dist. Deg. @ In. Hg.
75	1112862	0 @ 810	11 @ 2400	5	12 @ 12
	1112880	0 @ 600	11 @ 2100	4	9 @ 12
	1112862	0 @ 810	11 @ 2400	5	12 @ 12
	1112863	0 @ 550	8 @ 2100	3-5	3.25-9.75 @ 11-12.5
	1110650	0 @ 550	8 @ 2100	3-5	6.75-8.25 @ 11.5-12.
	1110651	0 @ 540	8 @ 2050	5-7	9 @ 10
	1112951	0 @ 325	14 @ 2200	4	12 @ 15
	1112956	0 @ 325	14 @ 2200	NONE	NONE
76	1112862	0 @ 810	11 @ 2400	5	12 @ 12
	1112983	0 @ 600	11 @ 2000	4	7.5 @ 10
	1112862	0 @ 450	11 @ 2400	5	10 @ 14
	1110668	0 @ 638	8 @ 1600	6	12 @ 12
	1110661	0 @ 525	8 @ 2050	6	9 @ 10
	1112863	0 @ 550	8 @ 2100	3-5	12 @ 14.5
	1112863	0 @ 550	8 @ 2100	4	9 @ 12
	1110666	0 @ 500	10 @ 2100	4	12 @ 15
	1110668	0 @ 638	8 @ 1588	6	12 @ 11.5
	1110661	0 @ 530	8 @ 2050	6	9 @ 10
	1112994	0 @ 325	14 @ 2200	4.5	12 @ 10.5
	1112995	0 @ 455	13 @ 2233	4	15 @ 11
	1112996	0 @ 550	14 @ 2200	—	—
77	1110538	0 @ 425	85 @ 1000	5	12 @ 10
	1110539	0 @ 425	8.5 @ 1000	5	12 @ 10
	1103239	0 @ 600	10 @ 2100	14	7.5 @ 10
	1103244	0 @ 500	10 @ 1900	4	10 @ 10
	1103252	0 @ 500	10 @ 1900	4	9 @ 12
	1103229	0 @ 600	10 @ 2200	3.5	10 @ 12
	1103263	0 @ 600	10 @ 2200	3.5	10 @ 9
	1103231	0 @ 600	10 @ 2200	3.5	10 @ 12
	1103230	0 @ 600	10 @ 2200	3.5	10 @ 9
	1103303	0 @ 600	10 @ 2200	9	10 @ 16
	1110677	0 @ 700	10 @ 1800	4	12 @ 11
	1110686	0 @ 700	10 @ 1800	7	4 @ 9
78	1103281	0 @ 500	10 @ 1900	4	9 @ 12
	1103282	0 @ 500	10 @ 1900	4	10 @ 10
	1103326	0 @ 850	10 @ 2325	3.4	10 @ 10.7
	1103328	0 @ 600	10 @ 2200	3.5	10 @ 9
	1103329	0 @ 600	10 @ 2200	3.5	10 @ 12
	1103365	0 @ 850	10 @ 2325	5.5	7 @ 9.5
	1110695	0-2 @ 1000	6-9 @ 1800	6	8 @ 9
	1110731	0-2 @ 1000	6-9 @ 1800	6	8 @ 9
	1110732	0-2 @ 1000	6-9 @ 1800	9	7 @ 13

Vega • Astre • Monza & Sunbird 4 Cyl.

Astre • Monza • Skyhawk • Starfire • Sunbird • Vega

DISTRIBUTOR SPECIFICATIONS

Year	Distributor Identification	Centrifugal Advance Start Dist. Deg. @ Dist. RPM	Centrifugal Advance Finish Dist. Deg. @ Dist. RPM	Vacuum Advance Start @ In. Hg.	Vacuum Advance Finish Dist. Deg. @ In. Hg.
79	1103229	0 @ 600	10 @ 2200	3.5	10 @ 12
	1103231	0 @ 600	10 @ 2200	3.5	10 @ 12
	1103239	0 @ 600	10 @ 2100	4	5 @ 8
	1103244	0 @ 500	10 @ 1900	4	10 @ 10
	1103282	0 @ 500	10 @ 1900	4	10 @ 10
	1103285	0 @ 600	11 @ 2100	4	5 @ 8
	1103365	0 @ 850	10 @ 2325	5	8 @ 11
	1110726	0 @ 500	9 @ 2000	4	10 @ 10
	1110757	0 @ 600	9 @ 2000	4	10 @ 10
	1110766	0 @ 810	7.5 @ 1800	3	10 @ 9
	1110767	0 @ 840	7.5 @ 1800	4	12 @ 11

TORQUE SPECIFICATIONS

All readings in ft/lbs

Engine No. Cyl. Disp. (cu in)	Eng. V.I.N. Code	Make/Year	Cylinder Head Bolts	Rod Bearing Bolts	Main Bearing Bolts	Crankshaft Pulley Bolt	Flywheel to Crankshaft Bolts	MANIFOLD Intake	MANIFOLD Exhaust
4-140	A,B	73-77	60	35	65	80	60	30	30
V-6 231	C	75-77	85	42	①	200-310	60	40	25
8-260	F	75-77	85	42	120	160	40	25	25
8-262	G	All	60-70	45	75②	60	60	30	③
8-305	Q	76	65	45	80	60	60	30	20
4-151	V,	77-79	95	30	65	160	55	40	40
8-305	U	77-78	65	45	80	60	60	30	20
6-196	C	78-79	80	40	100	225	60	45	25
6-231	A	78-79	80	40	100	225	60	45	25
8-305	G	79	65	45	80	60	60	30	20
6-231	2	79	80	40	100	225	60	45	25
4-151	1	78-79	95	30	65	160	55	40	40
4-151	9	79	95	30	65	160	55	40	40

① 1, 2, 3, 4-80
   5-120
② WITH 4 BOLT MAINS, TORQUE OUTER BOLTS TO 65
③ CENTER BOLTS; 25-30; END BOLTS; 15-20

Vega • Astre • Monza & Sunbird 4 Cyl.

2GC, 2GV, 2GE CARBURETOR SPECIFICATIONS
CHEVROLET VEGA, MONZA
Rochester

Year	Carburetor Identification①	Float Level (in.)	Float Drop (in.)	Pump Rod (in.)	Idle Vent (in.)	Primary Vacuum Break (in.)	Secondary Vacuum Break (in.)	Automatic Choke (notches)	Choke Rod (in.)	Choke Unloader (in.)	Fast Idle Speed (rpm)
1975	7045101, 7045105	19/32	1 7/32	1 19/32	—	0.130	—	Index	0.375	0.350	—
	7045401, 7045405	21/32	1 7/32	1 19/32	—	0.130	—	Index	0.380	0.350	—
	7045102, 7045106	19/32	1 7/32	1 19/32	—	0.130	—	Index	0.375	0.350	—
	7045406	21/32	1 7/32	1 19/32	—	0.130	—	Index	0.380	0.350	—
1976	17056101	17/32	1 9/32	1 5/8	—	0.130	—	Index	0.260	0.325	—
	17056102	17/32	1 9/32	1 5/8	—	0.130	—	Index	0.260	0.325	—
	17056104	17/32	1 5/32	1 5/8	—	0.140	—	Index	0.260	0.325	—
	17056404	9/16	1 3/16	1 21/32	—	0.140	—	Index	0.260	0.325	—
1977	17057104, 17057105	1/2	1 9/32	1 21/32	—	0.150	—	Index	0.260	0.325	—
	17057107, 17057109	1/2	1 9/32	1 21/32	—	0.160	—	Index	0.260	0.325	—
	17057404, 17057405	1/2	1 9/32	1 21/32	—	0.160	—	1/2 Lean	0.260	0.325	—
1978	17058102	15/32	1 9/32	1 17/32	—	0.150	—	Index	0.260	0.325	—
	17058103	15/32	1 9/32	1 17/32	—	0.150	—	Index	0.260	0.325	—
	17058104	15/32	1 9/32	1 21/32	—	0.160	—	Index	0.260	0.325	—
	17058107	15/32	1 9/32	1 17/32	—	0.160	—	Index	0.260	0.325	—
	17058109	15/32	1 9/32	1 17/32	—	0.160	—	Index	0.260	0.325	—
	17058404	1/2	1 9/32	1 21/32	—	0.160	—	1/2 Lean	0.260	0.325	—
	17058405	1/2	1 9/32	1 21/32	—	0.160	—	Index	0.260	0.325	—
	17058447	7/16	1 5/32	1 5/8	—	0.110	0.150	1 Rich	0.080	0.140	—
	17058143	7/16	1 5/32	1 5/8	—	0.040	0.110	1 Rich	0.080	0.140	—
	17058147	7/16	1 5/32	1 5/8	—	0.100	0.140	1 Rich	0.080	0.140	—
	17058144	7/16	1 5/32	1 5/8	—	0.060	0.110	1 Rich	0.080	0.140	—

① The carburetor identification number is stamped on the float bowl, next to the fuel inlet nut.
② TCS disconnected for full vacuum advance.

MV, 1MV CARBURETOR SPECIFICATIONS
CHEVROLET VEGA, MONZA
Rochester Carburetor

Year	Carburetor Identification①	Float Level (in.)	Metering Rod (in.)	Pump Rod (in.)	Vacuum Break (in.)	Auxiliary Vacuum Break (in.)	Fast Idle Off Car (in.)	Choke Rod (in.)	Choke Unloader (in.)	Fast Idle Speed (rpm)
1975	Manual	1/8	—	—	0.100	0.450	—	0.080	0.375	2000
	Automatic	1/8	—	—	0.100	0.450	—	0.080	0.375	2000
1976	Manual	1/8	—	—	0.060	0.450	—	0.045	0.215	1200
	Automatic	1/8	—	—	0.060	0.450	—	0.045	0.215	750

① The carburetor identification number is stamped on the float bowl, next to the fuel inlet nut.
② TCS disconnected for full vacuum advance.
③ No vacuum to distributor.

Vega • Astre • Monza & Sunbird 4 Cyl.

Model 5210-C
CHEVROLET VEGA, MONZA — Holley Carburetor

Year	Carb. Part No. ① ②	Float Level (Dry) (in.)	Float Drop (in.)	Pump Position	Fast Idle Cam (in.)	Choke Plate Pulldown* (in.)	Secondary Vacuum Break (in.)	Fast Idle Setting (rpm)	Choke Unloader (in.)	Choke Setting
1975	348659, 348663,	0.420	1	#2	0.110	0.325	—	1600⑥	—	3 Rich
	348661, 348665	0.420	1	#2	0.110	0.275	—	1600⑥	—	3 Rich
	348660, 348664	0.420	1	#2	0.110	0.300	—	1600⑥	—	4 Rich
	348662, 348666	0.420	1	#2	0.110	0.275	—	1600⑥	—	4 Rich
1976	366829, 366831	0.420	1	#3	0.320	0.313	—	2200	0.375	2 Rich
	366833, 366841	0.420	1	#3	0.320	0.268	—	2200	0.375	2 Rich
	366830, 366832	0.420	1	#2	0.320	0.288	—	2200	0.375	3 Rich
	366834, 366840	0.420	1	#2	0.320	0.268	—	2200	0.375	3 Rich
1977	458103, 458105	0.420	1	#2	0.120	0.250	—	2500	0.350	3 Rich
	458107, 458109	0.420	1	#2	0.120	0.275	—	2500	0.400	3 Rich
	458102, 458104	0.420	1	#1	0.085	0.250	—	2500	0.350	3 Rich
	458106, 458108	0.420	1	#1	0.120	0.275	—	2500	0.400	3 Rich
	458110, 458112	0.420	1	#1	0.120	0.300	—	2500	0.400	3 Rich
1978	see notes	.520	1	—	.150	⑦	.400	2500	.350	⑧

Model 5210-C
Pontiac Astre, Sunbird, Ventura — Holley Carburetor

Year	Carb. Part No. ① ②	Float Level (Dry) (in.)	Float Drop (in.)	Pump Position	Fast Idle Cam (in.)	Choke Plate Pulldown* (in.)	Secondary Vacuum Break (in.)	Fast Idle Setting (rpm)	Choke Unloader (in.)	Choke Setting
1975	Manual	0.420	1	#3	0.140	0.300	—	2000⑥	—	2½ Rich
	Automatic	0.420	1	#2	0.140	0.400	—	2200⑥	—	3½ Rich
1976	Manual	0.410	1	#3	0.420	0.313③	—	2200⑥	0.375	2 Rich
	Automatic	0.410	1	#2	0.320	0.288③	—	2200⑥	0.375	3 Rich
1977	458102, 458103, 458104, 458105	0.420	1	④	0.085	0.250	—	2500	0.350	3 Rich
	458107, 458109	0.420	1	④	0.125	0.275	0.400	2500	0.350	3 Rich
	458110, 458112	0.420	1	④	0.120	0.300	0.400	2500	0.350	3 Rich
1978	see notes	.520	1	—	.150	⑫	—	⑬	.350	⑭

① Located on tag attached to the carburetor, or on the casting or choke plate
② Beginning 1974, GM identification numbers are used in place of the Holley numbers
③ 0.268 in California
④ #1 manual, #2 automatic
⑤ Not used
⑥ With no vacuum to the distributor
* Vacuum break initial choke valve clearance on AMC
⑦ Part #10001048, 10001050: .300
    #10001047, 10001049, 10001052, 10001054: .325
⑧ Part #10001047, 10001049: 1 Rich
    #10001048, 10001050, 10001052, 10001054: 2 Rich
⑨ Part #10001047, 10001049: .325
    #10004048, 10004049: .300
⑩ Part #10001047, 10001049: 2200
    #10004048, 10004049: 2400
⑪ Part #10001047, 10001049: 1 Rich
    #10004048, 10004049: 2 Rich
⑫ Part #10001047, 10001049: .325
    #10004048, 10004049: .300
⑬ Part #10001047, 10001049: 2200
    #10004048, 10004049: 2400
⑭ Part #10001047, 10001049: 1 Rich
    #10004048, 10004049: 2 Rich

Vega • Astre • Monza & Sunbird 4 Cyl.

CAR SERIAL NUMBER AND ENGINE IDENTIFICATION

1976

Mounted behind the windshield on the driver's side is a plate with the vehicle identification number. The sixth character is the model year, with 6 for 1976. The fifth character is the engine code, as follows.

A 140 1-bbl. 4-cyl. L-13
B 140 2-bbl. 4-cyl. L-11
O .... 122 Fuel Inj. 4-cyl. LY-3

Engines can be identified by a 3-letter code stamped on the top of the block below number 3 spark plug.

CAY 140 4-cyl. Man. L-11 49-States
CAZ 140 4-cyl. Man. L-11 Calif.
CBK 140 4-cyl. Man. L-11 Calif.
CBL 140 4-cyl. Auto. L-11 Calif.
CBS 140 4-cyl. Man. L-11 49-States
CBT 140 4-cyl. Auto. L-11 49-States
CBU 140 4-cyl. Auto. L-13 49-States
CBW 140 4-cyl. Man. L-13 49-States
CBX 140 4-cyl. Man. L-13 49-States
CBY 140 4-cyl. Man. L-11 Calif.
CBZ 140 4-cyl. Man. L-11 49-States
ZCB 122 4-cyl. Man. LY-3 49-States

NOTE: Astre and Sunbird codes are the same, except that the first letter ("C" for Chevrolet) is not used.

1977

The vehicle identification number is mounted on a plate behind the windshield on the driver's side. The sixth character is the model year, with 7 for 1977. The fifth character is the engine code, as follows.

B 140 2-bbl. 4-cyl. L-11

Engines can be identified by a 3-letter code stamped on the top of the block below number 3 spark plug.

CAA 140 4-cyl. Man. Altitude
CAB 140 4-cyl. Auto. Altitude
CAC 140 4-cyl. Man. Altitude
CAY 140 4-cyl. Man. 49-States
CAZ 140 4-cyl. Man. Calif.
CBK 140 4-cyl. Man. Calif.
CBL 140 4-cyl. Auto. Calif.
CBS 140 4-cyl. Man. 49-States
CBT 140 4-cyl. Auto. 49-States
CBY 140 4-cyl. Man. Calif.
CBZ 140 4-cyl. Man. 49-States

NOTE: Astre, Sunbird, and Starfire codes are the same as Vega and Monza, except that the first letter ("C" for Chevrolet) may not be used.

1978

The vehicle identification number is on a plate behind the windshield on the driver's side. The sixth character is the model year, with 8 for 1978. The fifth character is the engine code, as follows.
V .... 151 2-bbl. 4-cyl. LX-6 Pont.
1 .... 151 2-bbl. 4-cyl. LS-6 Pont.

NOTE: The LS-6 (code 1) has the Electronic Fuel Control, and is available only on Oldsmobile Starfire and Pontiac Sunbird.

1979

The vehicle identification number is on a plate behind the windshield on the driver's side. The sixth character is the model year, with 9 for 1979. The fifth character is the engine code, as follows.
V .... 151 2-bbl. 4-cyl. LX8, LS6/LS8 Pont.
C .... 196 2-bbl. V-6 LC9 Buick
A .... 231 2-bbl. V-6 LD5 Buick
G .... 305 2-bbl. V-8 LG3 Chev.

NOTE: LS8 to replace LS6 in January 1979.

EMISSION EQUIPMENT

1976
All Models
Closed positive crankcase ventilation
Emission calibrated carburetor
  On all models except,
    Not on 122 4-cyl.
Emission calibrated distributor
Heated air cleaner
  On all models except,
    Not on 122 4-cyl.
Vapor control, canister storage
Temperature controlled choke vacuum break
Exhaust gas recirculation
  On all models except,
    Not on 122 4-cyl.
Catalytic converter
  49-States
    On all models except,
      Not on 1-bbl. engines
  Calif.
    All models
Single diaphragm vacuum advance
  On all models except,
    Not on 122 4-cyl.
Air pump
  49-States
    All 1-bbl. engines
    Not used on 2-bbl.
    Not used on 122 4-cyl.
  Calif.
    All engines
Pulse air injection
  All 122 4-cyl.
  Not used on carburetor engines

Thermal check and delay valve
  49-States
    Not used
  Calif.
    All 2-bbl. engines
Transmission controlled spark
  49-States
    1-bbl. Man. Trans. only
    Not used on 2-bbl.
  Calif.
    Not used
Deceleration assist solenoid
  49-States
    1-bbl. Man. Trans. only
    Not used on 2-bbl.
  Calif.
    2-bbl. Man. Trans. only
Electronic fuel injection
  122 4-cyl. only.

1977
All Models
Closed positive crankcase ventilation
Emission calibrated carburetor
Emission calibrated distributor
Heated air cleaner
Vapor control, canister storage
Temperature controlled choke vacuum break
  49-States
    Not used
  Calif.
    All models
  Altitude
    All models

Exhaust gas recirculation
Catalytic converter
Air pump
  Not used
Pulse air injection
Trapped vacuum spark system

1978
All Models
Closed positive crankcase ventilation
Emission calibrated carburetor
Emission calibrated distributor
Heated air cleaner
Vapor control, canister storage
Exhaust gas recirculation
Electric choke
Catalytic converter
Choke vacuum delay valve
Early fuel evaporation
  All engines, except
    Not used with electronic fuel control
Vacuum delay valve, 4-nozzle
  All engines, except
    Not used with electronic fuel control
Electronic fuel control
  Used on engines with VIN code 1
Cold engine air bleed TVS
  Used on engines with VIN code 1

1979
All Models
Closed positive crankcase ventilation
Emission calibration carburetor
Emission calibrated distributor

Vega • Astre • Monza & Sunbird 4 Cyl.

EMISSION EQUIPMENT

1979
Heated air cleaner
Vapor control, canister storage
Exhaust gas recirculation
Electric choke
Catalytic converter
Choke vacuum delay valve
Early fuel evaporation
   All engines, except
   Not used with electronic fuel control
Vacuum delay valve, 4-nozzle
   All engines, except
   Not used with electronic fuel control
Electronic fuel control
Cold engine air bleed TVS

IDLE SPEED AND MIXTURE ADJUSTMENTS

1976
Air cleaner
   1-bbl. .... On
   2-bbl. .... Off
   Fuel Inj. .... On
Air cond. .... Off
Vapor hose to carb. .... Plugged
Dist. vac. hose
   1-bbl. .... Plugged
   2-bbl. .... Connected
Auto. Trans. .... Drive
Mixture adj. ... See rpm drop below

122 Fuel Inj. 49-States 1600
140 1-bbl. 49-States
   Auto. Trans.
     Solenoid connected 750
     Solenoid disconnected 550
     Mix. adj. ..... 75 rpm drop
        (825-750)
   Man. Trans.
     Solenoid disconnected 750
     Mix. adj. .... 75 rpm drop
        (825-750)

NOTE: *Curb idle and mixture settings on the 49-State 1-bbl. transmission engine are made with the solenoid disconnected. Idle speed is adjusted with the Allen screw in the center of the solenoid. The solenoid-connected speed is 1200 rpm.*

140 2-bbl. 49-States
   Auto. Trans.
     Solenoid connected 750
     Solenoid disconnected 600
     Mix. adj. .... 80 rpm drop
        (830-750)
   Man. Trans. 700
     Mix. adj. ...... 200 rpm drop
        (900-700)
140 2-bbl. Calif.
   Auto. Trans.
     Solenoid connected 750
     Solenoid disconnected 600
     Mix. adj. ...... 80 rpm drop
        (830-750)
   Man. Trans.
     Solenoid disconnected 700
     Mix. adj. ...... 200 rpm drop
        (900-700)

NOTE: *Curb idle and mixture settings on the Calif. 2-bbl. manual transmission engine are made with the solenoid disconnected. Idle speed is adjusted with the throttle screw. The solenoid-connected speed is 1000 rpm*

1977
Air cleaner .... Set aside with hoses connected
Air cond. .... Off
Other hoses .... See label
Auto. trans. .... Drive
Idle CO .... Not used
Mixture adj. ..See rpm drop below
140 4-bbl. 49-States
   Auto. trans. 650
     Mixture adj. ...... 30 rpm drop
        (680-650)
     A.C. idle speedup 850
   Man. trans. 700
     Mixture adj. ...... 80 rpm drop
        (780-700)
     A.C. idle speedup 1250
140 4-bbl. Calif.
   Auto. trans. 650
     Mixture adj. ...... 30 rpm drop
        (680-650)
     A.C. idle speedup 850
   Man. trans. 800
     Mixture adj. ...... 80 rpm drop
        (880-800)
     A.C. idle speedup 1250
140 4-bbl. Altitude
   Auto. trans. 700
     Mixture adj. ...... 30 rpm drop
        (730-700)
     A.C. idle speedup 850
   Man. trans. 800
     Mixture adj. ...... 80 rpm drop
        (880-800)
     A.C. idle speedup 1250

1978
Air cleaner .... In place
Air cond. .... Off
Hoses .... See label
Auto. trans. .... Drive
Idle speed
151 4-cyl. no air cond.
   Auto. trans.
     Solenoid connected 650
     Solenoid disconnected 500
     Propane enriched
       Code V 690
       Code 1 760
   Man. trans.
     Solenoid connected 1000
     Solenoid disconnected 500
     Propane enriched 1150
151 4-cyl. with air cond.
   Auto. trans. 650
   Propane enriched
     Code V 690
     Code 1 760
   A.C. idle speedup 850
   Man. trans. 1000
   Propane enriched 1150
   A.C. idle speedup 1200

NOTE: *All 151 4-cyl. engines have a carburetor solenoid. On engines without air conditioning, the solenoid is for anti-dieseling, and the idle speed is set with the solenoid adjustment. On air conditioned cars, the solenoid is for idle speedup, and is only energized when the air conditioning is on. The curb idle speed adjustment on air conditioned engines is made with the throttle screw.*

1979
CAUTION: *Emission control adjustment changes are noted on the Vehicle Emission Information Label by the manufacturer. Refer to the label before any adjustments are made.*
Air cleaner .... In place
Air cond. .... Off
Hoses .... See label
Auto. trans. .... Drive
Idle speed
151 4-cyl. no air cond.
   Auto. trans.
     Solenoid connected 650
     Solenoid disconnected 500
     Propane enriched 690
   Man. trans.
     Solenoid connected 1000
     Solenoid disconnected 500
     Propane enriched 1150
151 4-cyl. with air cond.
   Auto. trans 650
   Propane enriched 690
   A.C. idle speedup 850
   Man. trans. 1000
   Propane enriched 1150
   A.C. idle speedup 1200

NOTE: *All 151 4-cyl. engines have a carburetor solenoid. On engines without air conditioning, the solenoid is for anti-dieseling, and the idle speed is set with the solenoid adjustment. On air conditioned cars, the solenoid is for idle speedup, and is only energized when the air conditioning is on. The curb idle speed adjustment on air conditioned engines is made with the throttle screw.*

196 V6
   Auto. trans. 600
   Propane enriched ....
   Man. trans.
     Solenoid connected 800
     Solenoid disconnected 600
231 V-6 49-States
   Auto. trans. 600
   Propane enriched ....
   Man. trans.
     Solenoid connected 800

129

Vega • Astre • Monza & Sunbird 4 Cyl.

IDLE SPEED AND MIXTURE ADJUSTMENTS

1979
Solenoid disconnected 600
Propane enriched ....
231 V-6 Calif.
  Auto. trans. 600
    Propane enriched ....
  Man. trans.
    Solenoid connected 800
    Solenoid disconnected 600
    Propane enriched ....

231 V-6 Altitude
  Auto. trans. 600
    Propane enriched ....
  Man. trans.
    Solenoid connected 800
    Solenoid disconnected 600
    Propane enriched ....
305 2-bbl. V-8 49-States
  Auto. trans. 500
    Propane enriched 520-540
    AC idle speedup 600

Man. trans. 600
  Propane enriched 700-740
  AC idle speedup 700
305 2-bbl. V-8 Calif.
  Auto. trans. 500
    Propane enriched 520-540
    AC idle speedup 650
305 2-bbl. V-8 Altitude
  Auto. trans. 600
    Propane enriched 520-640
    AC idle speedup 700

INITIAL TIMING

1976
NOTE: *Distributor vacuum hose must be disconnected and plugged. Set timing at idle speed unless shown otherwise.*
122 4-cyl. .... 12° BTDC
140 4-cyl.
  49-States 1-bbl.
    Auto. Trans. ...... 10° BTDC
    Man. Trans. .... 8° BTDC
  All 2-bbl.
    Auto. Trans. ...... 12° BTDC
    Man. Trans. .... 10° BTDC

1977
140 4-cyl. 49-States
  Auto. trans. .... 2° BTDC

Man. trans. ....0° TDC
140 4-cyl. Calif.
  Auto. trans. ....0° TDC
  Man. trans. ....2° ATDC
140 4-cyl. Altitude
  Auto. trans. ....2° BTDC
  Man. trans. ....0° TDC

CAUTION: *California manual transmission engines are set at 2 degrees AFTER top dead center, as shown above.*

1978
NOTE: *Distributor vacuum hose must be disconnected and plugged. Set timing at 1000 rpm.*

151 4-cyl. Auto. trans.
  Starfire & Sunbird
    Code V .... 12° BTDC
    Code 1 .... 14° BTDC
  Monza .... 12° BTDC
  Phoenix .... 14° BTDC

1979
NOTE: *Vehicle timing information is noted on the Vehicle Emission Information Label by the manufacturer. Refer to this information before any adjustments are made.*
151 4-cyl. ....12° BTDC
196 V-6 .... 8° BTDC
231 V-6 ....15° BTDC
305 V-8 .... 4° BTDC

SPARK PLUGS

1976
122 4-cyl.—AC-R43LTS .... .035
140 4-cyl.—AC-R43TS .... .035

1977
140 4-cyl. ... AC-R43TS ... .035

1978
151 4-cyl. .... AC-R43TSX .. .060

1979
151 4-cyl. .... AC-R43TSX .. .060
196 & 231 V-6 ..AC-R46TSX .. .060
305 V-8 .... AC-R45TS .. .045

VACUUM ADVANCE

1976
122 4-cyl. .... None
140 4-cyl. .... Ported

1977
Diaphragm type ....Single
Vacuum source ....Manifold

1978
Diaphragm type .... Single
Vacuum source
  Code 1 .... Ported
  Code V
    With air cond. .... Ported
    No air cond. .... Manifold

NOTE: *The 151 4-cyl. "V" with air conditioning uses ported vacuum at normal operating temperature, but manifold vacuum when cold.*

1979
Diaphragm type ....Single
Vacuum source
  4-cyl.
    With air cond. ....Ported
    No air cond. ....Manifold
  196 V-6 ....Manifold
  231 V-6
    49-States ....Manifold
    Calif. ....Ported
    Altitude ....Ported
  305 V-8
    49-States ....Manifold
    Calif. ....Ported
    Altitude ....Manifold

Vega • Astre • Monza & Sunbird 4 Cyl.

EMISSION CONTROL SYSTEMS 1976

TRANSMISSION CONTROLLED SPARK 1976

Description of System 1976
The 1976 TCS system is identical to the 1973-74 system.

Units in System 1976
Vacuum solenoid (normally closed)
Temperature switch, cold override
Transmission switch (normally open)

Repairing System 1976
Repairs are limited to replacement of parts. There are no adjustments. Current to operate the system comes from the battery, to the ignition switch, then to the wiper fuse in the fuse block and finally to the solenoid in the engine compartment. Separate ground wires go from the solenoid to the transmission switch and the temperature switch.

COSWORTH VEGA

Chevrolet first introduced the Cosworth Vega in mid-1975. Except for slight changes in specifications, the 1976 models are the same. The Cosworth engine uses the same aluminum cylinder block as the Vega, with a bore of 3½ inches. Almost everything else on the engine is different. The stroke of the Vega is 3⅝ inches, which gives an engine size of 140 cubic inches. On the Cosworth the stroke is 3.16 inches, which results in the smaller size of 122 cubic inches.

The Cosworth cylinder head has 16 valves and two camshafts. Each cylinder has two exhaust valves and two intake valves. The camshafts are driven by a rubber cog belt, similar to what the Vega uses.

The Cosworth's emission controls are not complicated. Only four add-on systems are used, positive crankcase ventilation, vapor control, a catalytic converter, and pulse air injection. Carburetors are eliminated in favor of a Bendix electronic fuel injection system, similar to what the Cadillac Seville uses.

122 4-CYL. CLOSED POSITIVE CRANKCASE VENTILATION

Description of System, 122 4-Cyl.
The PCV valve is plugged into a hose that extends about 6 inches out to the right from the camshaft cover. The valve is connected to manifold vacuum by a hose from the top of the throttle body. Fresh air enters the engine from the air cleaner, through a hose connected to the crankcase at the right side of the engine just above the oil pan flange. There is no PCV filter on the Cosworth engine.

Units in System 122 4-Cyl.
PCV valve

Testing and Troubleshooting System 1224-Cyl.
1. Unplug the PCV valve from its hose. With the engine idling, check the vacuum at the valve with your finger or a vacuum gauge. If the vacuum is weak, renew the valve or hose, whichever is needed.
2. Plug the PCV valve back into the hose.
3. Remove the fresh air hose from the air cleaner and plug the hose.
4. Remove the oil filler cap and put a sheet of paper or a vacuum gauge over the oil filler opening to check crankcase vacuum. If there is not enough vacuum, check for an air leak at the cam cover gasket, oil pan gasket, or other engine gaskets. Tightening the bolts will usually eliminate leaks at the gaskets. The dipstick must be seated to prevent a vacuum leak.
5. Blow through the fresh air hose to be sure it is clear, then install the hose and the oil filler cap.

Repairing System, 122 4-Cyl.
The PCV valve should be replaced when dirty. Cleaning the valve is not recommended.

PULSE AIR INJECTION SYSTEM 122 4-CYL.

Description of System, 122 4-Cyl.
This system does not use an air pump. Air enters the exhaust ports because of the slight amount of suction at each exhaust valve. Air for the system starts at the air cleaner, goes through a hose to the shut off valve (also known as the air switch) then into a manifold on the right side of the engine. From the manifold, 4 hoses go to 4 check valves mounted on top of the engine. Each check valve is screwed onto a chrome pipe assembly that connects to the dual injection tubes at each exhaust port.

The system works because each injection tube ends close to an exhaust valve. When the exhaust valve is closed, the flow of exhaust gas in the exhaust pipe creates suction (low pressure) at the valve. This low pressure opens the pulse air check valve and draws in air through the injection tube. When the exhaust valve opens, the end of the injection tube changes from a low to a high pressure area. The high pressure goes through the pulse air pipe and closes the check valve to prevent backflow.

During deceleration, the shut off valve (air switch) turns off the air to prevent exhaust popping. The shut vacuum through a hose connected to the throttle body.

The Cosworth engine looks impressive, but most of the tubing you see is for the pulse air injection system

Vega • Astre • Monza & Sunbird 4 Cyl.

The pulse air shutoff valve (air switch) is mounted between the distributor and the cam cover

Units in System, 122 4-Cyl.

Shut off valve (air switch)
Check valves (4)
Air manifold and pipes
Exhaust manifold with injection tubes

Testing and Troubleshooting System, 122 4-Cyl.

Testing is done on each part of the system, rather than on the whole system at once. See below for tests on the individual units. About the only outward indication that anything is wrong will be backfiring in the exhaust during deceleration. This is caused by a shutoff valve that does not shut off the air during deceleration, or by an excessively rich air-fuel mixture.

Repairing System, 122 4-Cyl.

There are no adjustments. All units are replaced when worn out.

Testing Check Valve

The engine should be off, and cool enough so that you won't be burned. Disconnect each hose at the air manifold that runs along the right side of the engine, and use mouth pressure to blow through the hose toward the exhaust. You should be able to blow through the hose easily, but the valve should seal tightly when you suck back. If not, the check valve is defective.

Repairing Check Valve

Repairs are limited to replacement of the valves. After disconnecting the hose, each valve can be unscrewed from its pipe. Use two wrenches, and be careful to avoid putting any strain on the pipe.

Testing Shutoff Valve (Air Switch)

With the engine idling in Neutral, open the throttle and let it snap shut. If the engine does not backfire in the exhaust system, you can assume that the shutoff valve is working. The shutoff valve is a simple on-off valve that does not at any time allow air to exhaust to the atmosphere. There is no way to tell whether the valve is open or shut by looking at the outside of it or listening to it operate.

Testing of the valve is possible by removing it from the air manifold and connecting the small hose to manifold vacuum on any engine. When the engine is decelerated, you can see the plunger move if you look through the hose connections with a flashlight. If a new valve is available, the temporary substitution of a new valve will be easier and faster than trying to test the old one.

If a new valve is not available, disconnect the hose from the valve and securely plug both the hose and the valve opening. Then take a test run. If the backfiring stops during deceleration, you can be sure that a defective shutoff valve was the problem.

CAUTION: *A defective shutoff valve will cause backfiring only during deceleration. If backfiring occurs during idle, acceleration, or cruise, the problem is somewhere else.*

122 4-CYL. VAPOR CONTROL SYSTEM

This system is identical to that used on other Vegas, except that the canister hoses, instead of connecting to the carburetor, are connected to the fuel injection throttle body. The small hose from the canister valve is connected to a ported vacuum source on the throttle body. When the throttle is opened above idle, vacuum acts on the port and opens the canister valve. The larger hose from the canister connects to manifold vacuum, and purges the canister of vapors whenever the canister valve is open. The third outlet on the canister, which is connected to the bowl vent on some 140 4-cyl. Vegas, is capped off on the 122 4-cyl.

122 4-CYL. FUEL INJECTION

To adequately test this system, you must have the Kent-Moore tester, J-24706, for sale by Kent-Moore, 1501 S. Jackson St., Jackson, Mich. 49203. The price is not firmly established yet, but the cost of the Cadillac fuel injection analyzer is approximately $900, and that may be an indication of what the Cosworth analyzer will cost.

This is a high price to pay, but judging from a description of the Cosworth analyzer, it appears to be a well thought out piece of equipment, and even includes a built-in substitute electronic control unit that allows you to actually drive a car in which the electronic control unit has failed.

EMISSION CONTROL SYSTEMS 1977

TEMPERATURE CONTROLLED CHOKE VACUUM BREAK 1977

In 1975 and '76 the 1-bbl. carburetor used a secondary vacuum break controlled by a vacuum solenoid, a relay, and a temperature switch. The 2-bbl. carburetor used an electric solenoid to pull the choke open, and it was controlled by a relay and a temperature switch. For 1977 the 1-bbl. carburetor has been eliminated, and the 2-bbl. system has been greatly simplified.

Now the 2-bbl. carburetor uses a secondary vacuum break, mounted on the opposite corner of the carburetor from the choke, on California and altitude engines only. The vacuum supply is controlled by a red, white, and blue thermal vacuum switch (TVS). Below 105°F. coolant temperature the TVS is closed, and the vacuum break is inoperative. Above that temperature the TVS opens, and allows vacuum to the break unit, which then opens the choke more than it had already been opened by the primary vacuum break.

The TVS has three nozzles, and also controls the vacuum to the distributor vacuum advance. The hose hookup is complicated because the vacuum supply to the TVS runs through a 4-nozzle retard delay valve, but this does not affect the operation of the vacuum break, unless the hoses get hooked up incorrectly. Manifold vacuum connects to one nozzle on the black side of the 4-nozzle valve. The TVS connects to the other black nozzle. The numbers shown on some vacuum diagrams do not exist on the valves we have seen. Although the manifold vacuum runs through the black side of the 4-nozzle valve on the way to the TVS, this has no effect on the system because the two black nozzles are connected together. Vacuum passing from one black nozzle to the other does not go through the valve mechanism. The connection is made this way just for convenience. For testing, see Trapped Vacuum Spark System.

PULSE AIR INJECTION 1977

This system is nothing more than check valves connected to each exhaust port, and a hose from the air cleaner that supplies clean air to the check valves. The air pump and diverter valve used with previous systems have both been eliminated.

The pulse air system works because there is suction in the exhaust. The check valves allow the exhaust to pull in air, and this works almost as well as using a pump to push air in.

All four of the check valves are combined into one large housing called a Pulse Air Valve. A single hose from the pulse air valve connects to the air cleaner. The crankcase fresh air hose also connects to the air cleaner hose, but this is only for convenience, and does not affect the pulse air system. Maintenance is not required on the system, except to check the hoses and connections for deterioration or leaks.

Testing Pulse Air Valve

Remove the rubber hose and connect a hand vacuum pump to the center nozzle on the pulse air valve. The engine must be off. Pump up 20 in. Hg. of vacuum, then stop pumping and watch the gauge hand fall. If it takes less than two seconds to drop from 15 in. to 5 in. Hg. the pulse air valve is defective and must be replaced. If it takes two seconds or longer, the pulse air valve is okay.

NOTE: *Because it is directly connected to the air cleaner, a leaking pulse air valve can cause poor running by allowing exhaust gas to enter the engine.*

Repairing Pulse Air Valve

Repairs are limited to replacement.

TRAPPED VACUUM SPARK SYSTEM 1977

All 140 4-cylinder distributors operate their vacuum advance on intake manifold vacuum. The trapped vacuum spark system uses a 4-nozzle retard delay valve and a thermal vacuum switch (TVS) to "trap" the vacuum and hold the vacuum advance in the full advance position when the engine is cold. The retard delay valve has a black side and a white side. The black side connects to manifold vacuum and the bottom nozzle (nearest the threads) of the TVS. The white side connects to the distributor vacuum advance, and the top nozzle of the TVS. Vacuum can pass freely at all times from black nozzle to black nozzle, and from white to white. But the path between black and white is blocked by a check valve and a bleed. When the TVS is closed on a cold engine, below 115°F., the vacuum source is forced to operate the vacuum advance through the check valve, which opens and allows full vacuum advance. When the throttle is opened during acceleration, the vacuum falls off, and the check valve closes, trapping the vacuum and holding the advance unit in the full advance position. The bleed allows a small amount of air to bleed through, so that the advance slowly returns to the no-advance position. This is necessary just in case the engine has to be restarted while cold, because starting an engine in the full advance position is difficult.

Above a coolant temperature of 115°F. the TVS opens its inner and outer nozzles and bypasses the delay valve. Vacuum can then pass freely from the intake manifold to the distributor.

The 4-nozzle valve is sometimes called a spark delay valve or a check valve, but it actually is a retard delay valve. In some published vacuum diagrams the retard delay valve is shown with numbered hose connections. The valves that we have seen do not have any numbers. When hooking up the valve, the two black nozzles connect to manifold vacuum and the bottom nozzle on the TVS. It doesn't matter which is which, because both black nozzles are connected together inside the valve. The two white nozzles connect to the top of the TVS and the distributor. Again, it doesn't matter which way you put them because both white nozzles are connected together inside the valve.

Two different thermal vacuum switches are used with the system. The 49-State TVS has two nozzles. The California and Altitude TVS has three nozzles. The middle nozzle on the 3-nozzle TVS supplies manifold vacuum to the choke vacuum break, and has no effect on the trapped vacuum spark system as long as the hose is correctly hooked up and neither the diaphragm or hose is leaking.

Testing Thermal Vacuum Switch

On an engine at normal operating temperature, both 2-nozzle and 3-nozzle switches should have all the nozzles connected internally so that air blown into any one nozzle will come out the other one or two. Below 115°F. the 2-nozzle switch should be blocked, not allowing any air flow. On the 3-nozzle switch, all three nozzles are blocked below 105°F. As the switch warms up to 105°F. the bottom and middle nozzles are connected. At 115°F. or above, all three nozzles are connected. It is not necessary to test the precise opening temperature of the TVS. As long as it is closed when cold, and open when warm, the system will work correctly.

Repairing Thermal Vacuum Switch

Repairs are limited to replacement. Manifold vacuum connects to the inner nozzle (close to the threads).

Testing 4-nozzle Retard Delay Valve

Connect a hand vacuum pump to either black nozzle and plug the other black nozzle. Leave the white nozzles open. Operate the pump. You should not be able to pump up vacuum. Connect the pump to either white nozzle and plug the other white nozzle. Leave the black nozzles open. Operate the pump. You should be able to pump up 15 in. Hg. vacuum. When you stop pumping, the vacuum should fall off slowly, taking about 5 seconds to drop from 15 to 5 in. Hg. If the valve does not operate this way, it is either plugged or the check valve is faulty.

Repairing 4-Nozzle Retard Delay Valve

Repairs are limited to replacement. The black side of the valve connects to manifold vacuum, and the white side to the distributor. One white nozzle connects to the outer TVS nozzle, and one black nozzle to the inner TVS nozzle.

CARBURETOR SOLENOID DASHPOT 1977

Manual transmission cars equipped with air conditioning have a new idle speedup solenoid combined with a dashpot. The dashpot works at all times to slow down the closing of the throttle and prevent overrich mixtures during deceleration, which can cause backfiring in the exhaust. The screw on the throttle arm that bears against the solenoid stem is used to

Vega • Astre • Monza & Sunbird 4 Cyl.

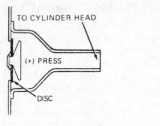

CLOSED POSITION   OPENED POSITION

Vega pair valve (© GM Corporation)

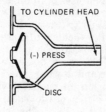

Vega retard delay valve (© GM Corporation)

set the idle speedup speed with the air conditioning on. The solenoid is energized only when the air conditioning is on. The dashpot part of the solenoid is not adjustable.

ELECTRONIC FUEL CONTROL SYSTEM (EFC) 1978

This system controls both the idle and main metering fuel mixtures according to how much oxygen there is in the exhaust. Most tune-up men have used an infra-red exhaust gas analyzer to measure the carbon monoxide (CO) and hydrocarbons (HC) in the exhaust. The infra-red measures the actual pollutants, except nitrogen oxides (NOx). Since these pollutants are what we want to control, it would seem that the best way to control them would be to install an analyzer on every car. The electronic fuel control system does just that, but instead of using an infra-red analyzer, it uses an oxygen sensor.

The easiest thing to analyze in the exhaust is the amount of oxygen. Because repair shops are not used to analyzing oxygen with their exhaust gas analyzers, most emission control technicians do not realize that the exhaust contains any oxygen. But oxygen is always present in the exhaust to some degree. If the mixture is rich, the extra fuel will combine with the oxygen in the air and the exhaust will have very little oxygen. If the mixture is lean, there won't be enough fuel to use up all the oxygen and more of it will pass through the engine.

As the oxygen content of the exhaust goes up, (lean mixture) the amount of CO goes down, because there is an excess of oxygen to combine with the CO and turn it into harmless CO2 (carbon dioxide). However, if the mixture gets too lean there is a tendency to form more NOx. The electronic fuel control solves all these problems by controlling the mixture within very narrow limits.

EMISSION CONTROL SYSTEMS 1978

The oxygen sensor has a zirconia element, and looks very much like a sparkplug. It screws into the exhaust pipe close to the exhaust manifold. Zirconia, when combined with heat and oxygen, has the peculiar property of being able to generate current. The amount of current is very small, but it is enough that it will pass through a wire and can be amplified by electronics.

A small black box under the hood is the electronic control unit. It receives the electric signals from the oxygen sensor and amplifies them enough to operate a vacuum modulator. The vacuum modulator is a vibrating electrical device. Engine manifold vacuum connects to the modulator, and then goes to the carburetor. The amount of vacuum that the carburetor receives is controlled by the modulator.

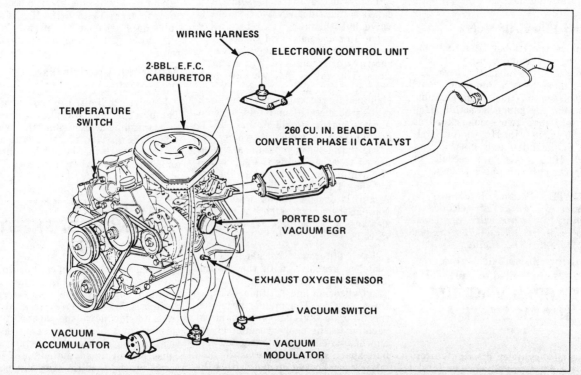

1978 electronic fuel control system—L-4 (Vin) (© CHEVROLET Div. GM Corp.)

Vega • Astre • Monza & Sunbird 4 Cyl.

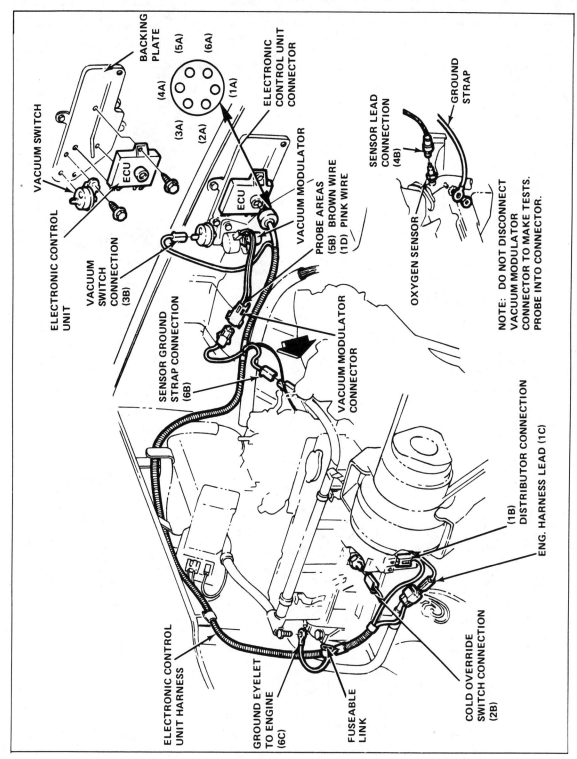

.1978 EFC electrical system (© Chevrolet Div. GM Corp.)

The carburetor is specially constructed with an idle air bleed valve and a main metering fuel valve. Both of the valves are operated by vacuum diaphragms. The idle air bleed gives a richer mixture at low vacuum because a spring pushes the regulating needle into the bleed. At high vacuum, the diaphragm compresses the spring and pulls the needle out of the bleed, letting more air into the idle system, which leans the mixture.

The main metering fuel valve actually takes the place of the power valve. It is separate from the main metering jet. It works the same as a power valve, in that it can add fuel to the system or shut off, but it cannot change the amount of fuel going through the main metering jet.

The spring tension on both the diagragms is adjustable with small screws. However, the main metering diaphragm is factory adjusted, and athere are no specifications for adjusting it in the field.

Both the idle diaphragm screw and the idle mixture screw on the carburetor are covered with press-in plugs. The idle mixture needle is covered by a cup plug that can be removed with a screw extractor. The screw extractor should fit the cup plug without drilling a hole. If a hole is drilled in the plug, the mixture needle will be damaged. The idle diaphragm plug is soft lead, and can be carefully pried out. Idle mixture settings are normally made only when the car-

135

Vega • Astre • Monza & Sunbird 4 Cyl.

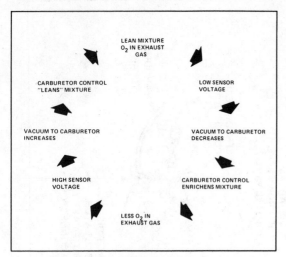

Cycle of Operations (© Chevrolet Div. GM Corp.)

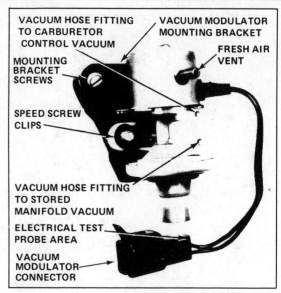

Vacuum modulator (© Chevrolet Div. GM Corp.)

buretor is overhauled. The necessary plugs are in the overhaul kit.

One of the first questions that comes up is, "How do you know it's working?" You can be sure the electronics are working if you hear a continuous clicking from the vacuum modulator. The vacuum part of the system can be checked with a vacuum gauge connected to a "T" inserted into the hose between the modulator and the carburetor. At a 650-rpm idle, the vacuum should be 2½ in. Hg. Pontiac recommends that the vacuum gauge should be a sensitive type with a maximum reading of 10 in. Hg. It is normal for the vacuum reading to fluctuate, but the needle should move an equal distance each side of 2½ in. Hg. To change the setting, remove the lead plug from the diaphragm adjusting screw and turn the screw until the average reading is 2½ in. Hg.

Two additional switches are used to send signals to the electronic control unit. A vacuum switch is connected to engine manifold vacuum. A wire goes from the switch to the electronic control unit. When engine vacuum is low, as it is under heavy load, the vacuum switch tells the ECU to give a slightly richer mixture.

The other switch is a temperature sensing switch in the cylinder head. When the engine is cold, it sends a signal to the ECU to allow a slightly richer mixture for better driveability.

The Electronic Fuel Control is not a complicated system. Except for the vacuum checks, which need to be done only at time of carburetor overhaul, no maintenance is required.

COLD ENGINE AIR BLEED THERMAL VACUUM SWITCH 1978

This switch, used only on code 1 engines with electronic fuel control, opens below 170°F. engine coolant temperature. When open, it allows air to bleed into the hose that connects between the PCV valve and the vapor control canister. The outer nozzle on the valve connects to the PCV-vapor canister hose. The inner nozzle connects to the top of the carburetor and supplies fresh air from the clean side of the air filter.

The reason for bleeding air into a cold engine is to put just enough oxygen into the exhaust to "fool" the oxygen sensor so it won't try to lean out the rich mixture caused by the choke. The small amount of oxygen added does not affect the running of the engine.

VACUUM DELAY VALVE 4-NOZZLE 1978

This new delay valve is a triumph of engineering ingenuity—and the despair of anyone who disconnects it and then tries to figure out how to hook it back up. The way the valve works is very simple. It is nothing more than a retard delay valve. It allows the manifold vacuum to act on the distributor diaphragm without

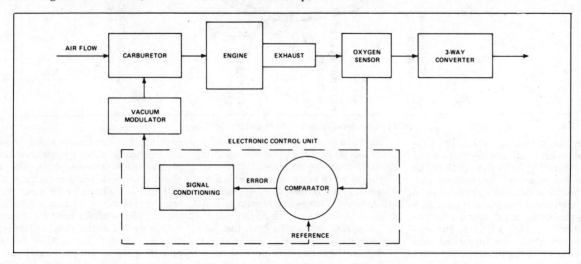

Electronic fuel control system schematic (© Chevrolet Div. GM Corp.)

Vega • Astre • Monza & Sunbird 4 Cyl.

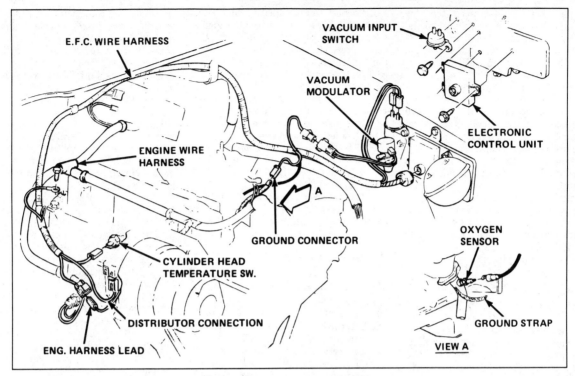

Wiring system (© Chevrolet Div. GM Corp.)

any restriction. But when the manifold vacuum drops to zero, as during heavy acceleration or if the engine should die at idle, the valve "traps" the vacuum and lets it bleed down slowly. This keeps the spark advanced for better performance, or for easier starting.

Above 120°F. coolant temperature, a thermal vacuum switch, that is connected in the system by hoses, opens and bypasses the vacuum delay valve so that the vacuum advance works normally.

The bypass system is what makes the hose hookup so complicated. Here's the way it works on the 4-cyl. without air conditioning. The MAN side of the valve has two nozzles. They are both connected together inside the valve, so it doesn't make any difference which one of them is connected to the manifold vacuum source. The other MAN nozzle connects to the thermal vacuum switch (TVS).

On the DIST side of the valve, one of the nozzles connects to the TVS, and the other connects to the distributor modular valve, or direct to the distributor. It dosen't make any difference which is which, because the two nozzles are connected together inside the valve.

When the TVS is open (above 120°F.) manifold vacuum goes in one MAN nozzle and right out the other MAN nozzle, without going through the valve. The vacuum then goes to the TVS and back to the DIST side of the valve. But all it does is go in one DIST nozzle and out the other one. In the bypass operation, the valve is only acting as a carrier for the vacuum, without actually doing anything.

But when the TVS is closed (below 120°F.) the vacuum can't bypass. It then goes in one MAN nozzle, through the valve, and out the DIST nozzle to the distributor. Then, the retard delay is in operation.

On the 4-cylinder with air conditioning, a different bypass system is used. Manifold vacuum connects to the MAN side of the valve, but the DIST side connects to the upper nozzle on a 3-nozzle TVS. The second nozzle on the DIST side of the vacuum delay valve is capped, and not used. The second nozzle on the MAN side is connected to another system, which doesn't affect the distributor. It is just a convenient way of hooking up manifold vacuum to the other system.

When the engine is cold (below 120°F.) the manifold vacuum goes through the vacuum delay valve and through the TVS to the distributor. When the engine warms up, the TVS shuts off the manifold vacuum, eliminating the vacuum delay valve entirely, and sends ported vacuum to the distributor.

Testing Vacuum Delay Valve

Connect a hand vacuum pump to the DIST side of the valve. Use a cap or a short length of hose and a bolt to plug the second nozzle on the DIST side. Plug both of the MAN nozzles with your finger. Pump up 15 in. Hg. vacuum and then remove your finger from the MAN side of the valve. The vacuum should drop slowly from 15 to 5 in. Hg. On the 151 4-cyl. it should take about 7 seconds on an automatic transmission car, or 3 seconds on a manual transmission car. If it takes longer, or shorter, or doesn't drop at all, the vacuum delay valve is defective.

Repairing Vacuum Delay Valve

Repairs are limited to replacement. The part numbers for the various applications are as follows.

Engine   Valve
151 4-cyl.
  Auto. trans. 527011
  Man. trans. 547833

When connecting the valve, the MAN side goes to manifold vacuum, and the DIST side goes to the distributor, through a TVS or a modulator valve.

NOTE: The vacuum delay valve is identical to that used on the Pontiac 151 4-cyl. in 1977.

CHOKE VACUUM BREAK VACUUM DELAY VALVE 1978

This valve has three nozzles. It connects between manifold vacuum and the choke vacuum break unit. When the engine starts, the delay valve restricts the vacuum so the vacuum break opens the choke slowly. If the engine is stopped, the vacuum is released through an internal check valve.

The operation of the valve is very simple, but the vacuum hookup is something else. At first glance it doesn't make any sense at all, because the manifold vacuum does not run directly from the intake manifold to the delay valve. For

137

Vega • Astre • Monza & Sunbird 4 Cyl.

convenience in manufacturing, the hose from manifold vacuum runs first to the distributor vacuum delay valve (DS-VDV), which has four nozzles. Then a second hose runs from the DS-VDV to the choke delay valve.

When the manifold vacuum goes through the DS-VDV it is not affected. The two nozzles on the MAN side are connected together inside the valve, without going through the valve mechanism. So the choke delay valve receives manifold vacuum just as if it were hooked up directly to the manifold.

Manifold vacuum connects to the 2-nozzle side of the choke delay valve. The second nozzle on the same side connects to a thermal vacuum switch again just for convenience. The single nozzle side of the delay valve connects to the choke vacuum break.

Testing Choke Vacuum Break Vacuum Delay Valve

Use a short length of hose and a "T" to insert a vacuum gauge between the delay valve and the vacuum break unit. Disconnect the second hose on the 2-nozzle side of the delay valve, and cap the nozzle. Leave the manifold vacuum hose hooked up. Start the engine and let it idle, then slip the manifold vacuum hose off the 2-nozzle side of the delay valve. The vacuum gauge should drop to zero immediately, without any delay. If it doesn't, the delay valve is defective, or the hose is restricted. When slipping the hose off the delay valve, hold your finger over the end of the hose to keep the engine from dying.

Reconnect the manifold vacuum hose to the delay valve. The vacuum gauge should rise slowly, taking about 40 seconds to reach full manifold vacuum. Note the highest reading you get, then disconnect the hose from the intake manifold and use a separate hose to attach the vacuum gauge to check manifold vacuum. Both vacuum readings should be within one inch of each other. If not, there is a leak somewhere.

NOTE: Because so many systems are connected together, a leak in one system can affect another. A single manifold vacuum connection supplies the distributor, the exhaust heat valve, and the vacuum break. When searching for leaks, always isolate each system before changing any parts.

Repairing Choke Vacuum Break Vacuum Delay Valve

Repairs are limited to replacement. This is a 3-nozzle valve, with part number 546801. Manifold vacuum connects to the 2-nozzle side. The 1-nozzle side connects to the vacuum break.

NOTE: The choke vacuum delay valve is identical to that used on the Pontiac 151 4-cyl. in 1977.

151 CUBIC INCH 4-CYLINDER ENGINE 1978

In the 1962 model year Chevrolet came out with the Chevy II compact car. It was designed as basic, thrifty transportation and came with a smaller version of Chevy's old reliable 6-cylinder engine. Also available, but seldom purchased, was a 153-cubic-inch 4-cylinder engine. It was simply a six with two cylinders lopped off. The four was available for nine model years, finally being discontinued when the 1971 Vega appeared.

From 1971 to 1976, the four was kept alive in some of General Motors remote outposts, mainly Brazil. When the energy crisis struck, the decision was made to bring back the four, and Pontiac is the first G.M. division to use it.

Although the new four will certainly look familiar to those who worked on the old engine, it has been extensively modified and brought up to date with all the modern G.M. features such as High Energy Ignition, bigger bore, and shorter stroke. Because of all the changes, it appears that very few, if any, of the parts will interchange with the old engine. In addition, the changes have made the new four a different engine from the current Chevy 6. You can't really say any more than it is a Chevy 6 with two cylinders lopped off, although some of the parts will interchange.

The carburetor on the 151 is the Holley-Weber, basically the same one that has been used for several years on the Vega, but with an electric choke. The rest of the engine is conventional, and about as easy to work on as a Chevy 6.

NOTE: The 151 4-cyl. engine, without electronic fuel control, is the same as the 151 4-cyl. engine used in the Pontiac Astre and Sunbird in 1977.

EMISSION CONTROL SYSTEMS 1979

ELECTRONIC FUEL CONTROL 1979

This system controls both the idle and main metering fuel mixture according to how much oxygen there is in the exhaust. Most tune-up men have used an infra-red exhaust gas analyzer to measure the carbon monoxide (CO) and hydrocarbon (HC) in the exhaust. The infra-red measures the actual pollutants, except nitrogen oxides (NOx). Since these pollutants are what we want to control, it would seem that the best way to control them would be to install an analyzer on every car. The electronic fuel control system does just that, but instead of using an infra-red analyzer, it uses an oxygen sensor.

The easiest thing to analyze in the exhaust is the amount of oxygen. Because repair shops are not used to analyzing oxygen with their exhaust gas analyzers, most emission control technicians do not realize that the exhaust contains any oxygen. But oxygen is always present in the exhaust to some degree. If the mixture is rich, the extra fuel will combine with the oxygen in the air and the exhaust will have very little oxygen. If the mixture is lean, there won't be enough fuel to use up all the oxygen and more of it will pass through the engine.

As the oxygen content of the exhaust goes up, (lean mixture) the amount of CO goes down, because there is an excess of oxygen to combine with the CO and turn it into harmless CO_2 (carbon dioxide). However, if the mixture gets too lean there is a tendency to form more NOx. The electronic fuel control solves all these problems by controlling the mixture within very narrow limits.

The oxygen sensor has a zirconia element, and looks very much like a spark plug. It screws into the exhaust pipe close to the exhaust manifold. Zirconia, when combined with heat and oxygen, has the peculiar property of being able to generate current. The amount of currents is very small, but it is enough that it will pass through a wire and can be amplified by electronics.

A small black box under the hood is the electronic control unit. It receives the electric signals from the oxygen sensor and amplifies them enough to operate a vacuum modulator. The vacuum modulator is a vibrating electrical device. Engine manifold vacuum connects to the modulator, and then goes to the carburetor. The amount of vacuum that the carburetor receives is controlled by the modulator.

The carburetor is specially constructed with an idle air bleed valve and a main metering fuel valve. Both of the valves are operated by vacuum diaphragms. The idle air bleed gives a richer mixture at low vacuum because a spring pushes the regulating needle into the bleed. At high vacuum, the diaphragm compresses the spring and pulls the needle out of the bleed, letting more air into the idle system, which leans the mixture.

The main metering fuel valve actually takes the place of the power valve. It is separate from the main metering jet. It works the same as a power valve, in that it can add fuel to the system or shut off, but it cannot change the amount of fuel going through the main metering jet.

The spring tension on both the diaphragms is adjustable with small screws. However, the main metering diaphragm is factory adjusted, and there are no specifications for adjusting it in the field. Both the idle diaphragm screw and the idle mixture screw on the carburetor are covered with press-in plugs. The idle mixture needle is covered by a cup plug that can be removed with a screw extractor. The screw extractor should fit the cup plug without drilling a hole. If a hole is drilled in the plug, the mixture needle will be damaged. The idle diaphragm plug is soft lead, and can be carefully pried out. Idle mixture settings are normally made only when the carburetor is overhauled. The necessary plugs are in the overhaul kit.

The Electronic Fuel Control is not a complicated system. Except for the vacuum checks, which need to be done only at time of carburetor overhaul, no maintenance is required.

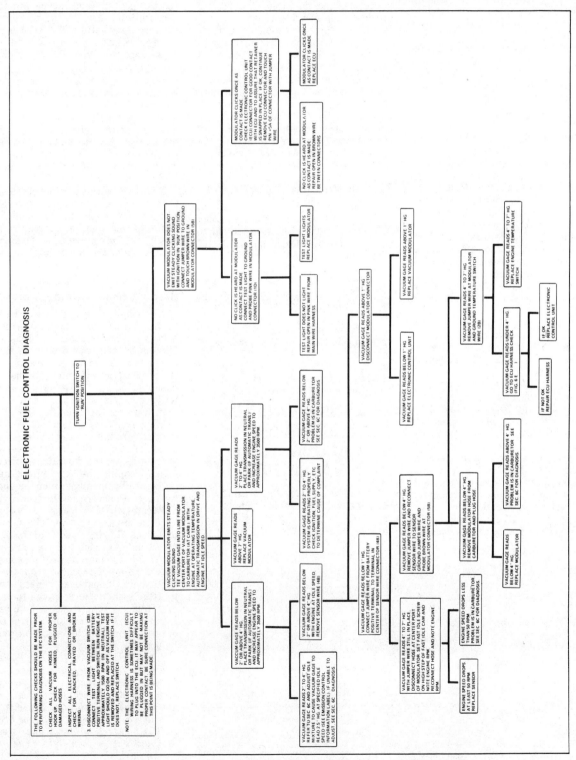

Electronic fuel control diagnosis chart (© CHEVROLET Div. GM Corp.)

Vega • Astre • Monza & Sunbird 4 Cyl.

VACUUM DELAY VALVE 4-NOZZLE 1979

The vacuum delay valve is nothing more than a retard delay valve. It allows the manifold vacuum to act on the distributor diaphragm without any restriction. But when the manifold vacuum drops to zero, as during heavy acceleration or if the engine should die at idle, the valve "traps" the vacuum and lets it bleed down slowly. This keeps the spark advanced for better performance, or for easier starting.

Above 120°F. coolant temperature, a thermal vacuum switch that is connected in the system by hoses, opens and bypasses the vacuum delay valve so that the vacuum advance works normally.

The bypass system is what makes the hose hookup so complicated. On the 4-cyl. without air conditioning, the MAN side of the valve has two nozzles. They are both connected together inside the valve, so it doesn't make any difference which one of them is connected to the manifold vacuum source. The other MAN nozzle connects to the thermal vacuum switch (TVS).

On the DIST side of the valve, one of the nozzles connects to the TVS, and the other connects to the distributor modular valve, or direct to the distributor. It doesn't make any difference which is which, because the two nozzles are connected together inside the valve.

When the TVS is open (above 120° F.) manifold vacuum goes in one MAN nozzle and right out the other MAN nozzle, without going through the valve. The vacuum then goes to the TVS and back to the DIST side of the valve. But all it does is go in one DIST nozzle and out the other one. In the bypass operation, the valve is only acting as a carrier for the vacuum, without actually doing anything.

But when the TVS is closed (below 120° F.) the vacuum can't bypass. It then goes in one MAN nozzle, through the valve, and out the DIST nozzle to the distributor. Then, the retard delay is in operation.

On the 4-cylinder with air conditioning, a different bypass system is used. Manifold vacuum connects to the MAN side of the valve, but the DIST side connects to the upper nozzle on a 3-nozzle TVS. The second nozzle on the DIST side of the vacuum delay valve is capped, and not used. The second nozzle on the MAN side is connected to another system, which doesn't affect the distributor. It is just a convenient way of hooking up manifold vacuum to the other system.

When the engine is cold (below 120° F.) the manifold vacuum goes through the vacuum delay valve and through the TVS to the distributor. When the engine warms up, the TVS shuts off the manifold vacuum, eliminating the vacuum delay valve entirely and sends ported vacuum to the distributor.

Testing Vacuum Delay Valve

Connect a hand vacuum pump to the DIST side of the valve. Use a cap or a short length of hose and a bolt to plug the second nozzle on the DIST side. Plug both of the MAN nozzles with your finger. Pump up 15 in. Hg. vacuum and then remove your finger from the MAN side of the valve. The vacuum should drop slowly from 15 to 5 in. Hg. On the 151 4-cyl. it should take about 7 seconds on an automatic transmission car, or 3 seconds on a manual transmission car. If it takes longer, or shorter, or doesn't drop at all, the vacuum delay valve is defective, and must be replaced.

CHOKE VACUUM BREAK VACUUM DELAY, VALVE 1979

This valve has three nozzles. It connects between manifold vacuum and the choke vacuum break unit. When the engine starts, the delay valve restricts the vacuum so the vacuum breaks opens the choke slowly. If the engine is stopped, the vacuum is released through an internal check valve.

The operation of the valve is very simple, but the vacuum hookup is something else. At first glance, it doesn't make any sense at all, because the manifold vacuum does not run directly from the intake manifold to the delay valve. For convenience in manufacturing, the hose from manifold vacuum runs first to the distributor vacuum delay valve (DS-VDV), which has four nozzles. Then a second hose runs from the DS-VDV to the choke delay valve.

When the manifold vacuum goes through the DS-VDV it is not affected. The two nozzles on the MAN side are connected together inside the valve, without going through the valve mechanism. So the choke delay valve receives manifold vacuum just as if it were hooked up directly to the manifold.

Manifold vacuum connects to the 2-nozzle side of the choke delay valve. The second nozzle on the same side connects to a thermal vacuum switch again just for convenience. The single nozzle side of the delay valve connects to the choke vacuum break.

Testing Choke Vacuum Break Vacuum Delay Valve

Use a short length of hose and a "T" to insert a vacuum gauge between the delay valve and the vacuum break unit. Disconnect the second hose on the 2-nozzle side of the delay valve, and cap the nozzle. Leave the manifold vacuum hose hooked up. Start the engine and let it idle, then slip the manifold vacuum hose off the 2-nozzle side of the delay valve. The vacuum gauge should drop to zero immediately, without any delay. If it doesn't, the delay valve is defective, or the hose is restricted. When slipping the hose off the delay valve, hold your finger over the end of the hose to keep the engine from dying.

Reconnect the manifold vacuum hose to the delay valve. The vacuum gauge should rise slowly, taking about 40 seconds to reach full manifold vacuum. Note the highest reading you get, then disconnect the hose from the intake manifold and use a separate hose to attach the vacuum gauge to check manifold vacuum. Both vacuum readings should be within one inch of each other. If not, there is a leak somewhere.

NOTE: *Because so many systems are connected together, a leak in one system can affect another. A single manifold vacuum connection supplies the distributor, the exhaust heat valve, and the vacuum break. When searching for leaks, always isolate each system before changing any parts.*

Repairs are limited to replacement.

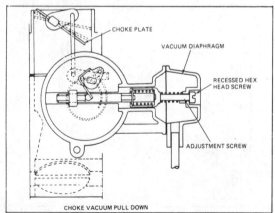

Vacuum Break

Vega • Astre • Monza & Sunbird 4 Cyl.
VACUUM CIRCUITS

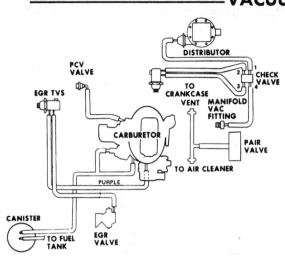

Vacuum hose routing Vega 140 4 cyl.—49 states (© GM Corporation)

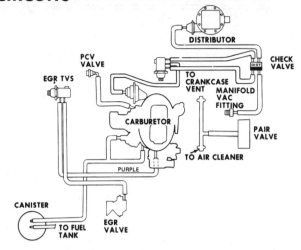

Vacuum hose routing Vega 140 4 cyl.—California and Altitude (© GM Corporation)

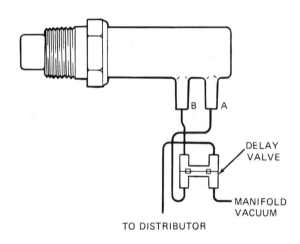

2-nozzle TVS switch—Vega (© GM Corporation)

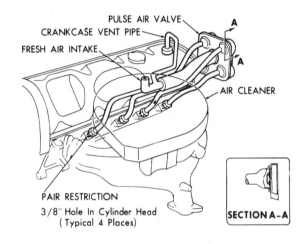

Pulse air system—Vega (© GM Corporation)

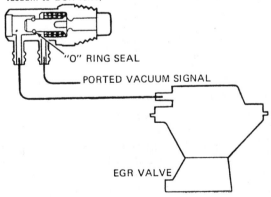

EGR cold override—Vega (© GM Corporation)

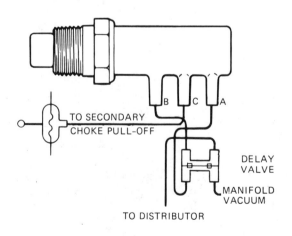

3-nozzle TVS switch—Vega (© GM Corporation)

141

Vega • Astre • Monza & Sunbird 4 Cyl.

VACUUM CIRCUITS

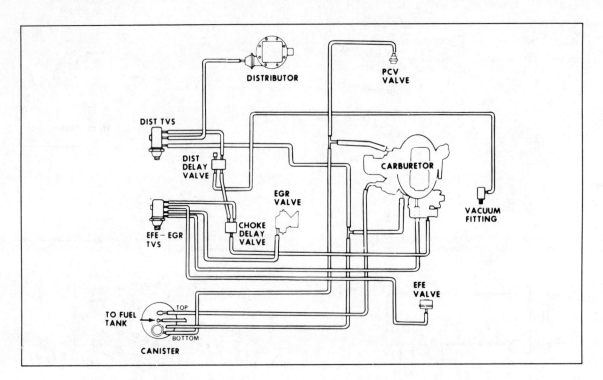

1978 49-States Monza, Phoenix, Starfire, Sunbird 151 4-cyl. auto. trans w/air cond.—
Vacuum hose routing

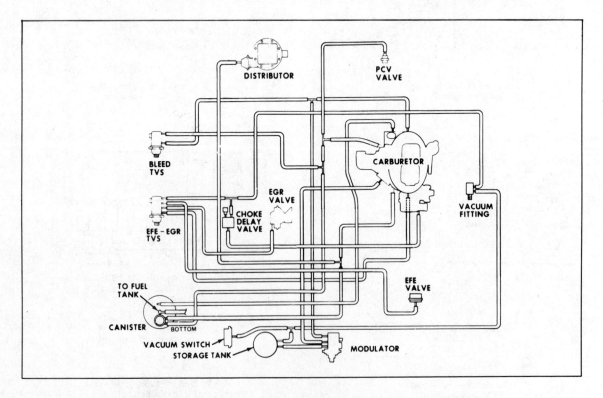

1978 California Monza, Phoenix, Starfire, Sunbird 151 4-cyl. w/ electonic fuel control—
Vacuum hose routing

Vega • Astre • Monza & Sunbird 4 Cyl.

VACUUM CIRCUITS

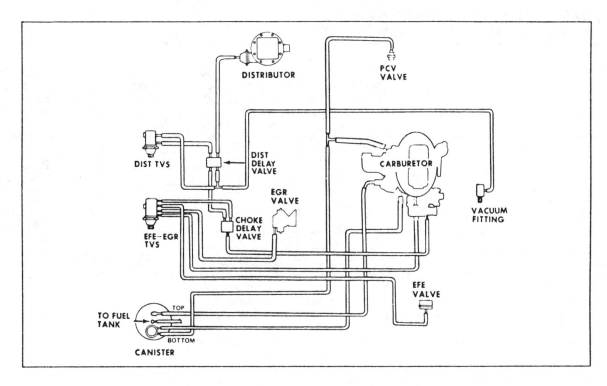

1978 49-States Monza, Phoenix, Starfire, Sunbird 151 4-cyl. manual trans.—Vacuum hose routing

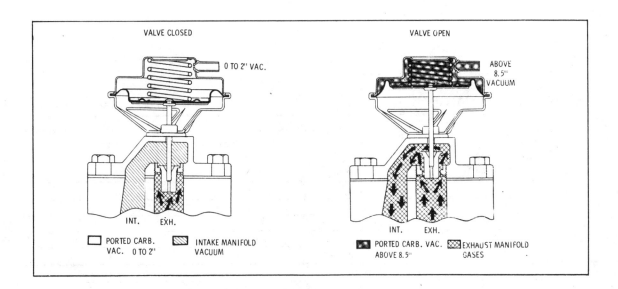

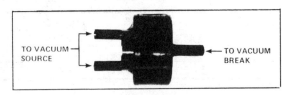

Vacuum Break Delay

143

Chevette

Chevette • Rally 1.6 • Scooter

TUNE-UP SPECIFICATIONS — CHEVROLET Chevette

Engine Code, 5th character of the VIN number
Model Year Code, 6th character of the VIN number

Year	Eng. V.I.N. Code	Eng. No. Cyl. Disp. (cu in)	Eng.* Mfg.	Carb Bbl	H.P.	SPARK PLUGS Orig. Type	SPARK PLUGS Gap (in)	DIST. Point Dwell (deg)	DIST. Point Gap (in)	IGNITION TIMING (deg BTDC) Man Trans	IGNITION TIMING (deg BTDC) Auto Trans	VALVES ■ Intake Opens (deg BTDC)	FUEL PUMP Pres. (psi)	IDLE SPEED ● (rpm) Man Trans	IDLE SPEED ● (rpm) Auto Trans In Drive
76	I	4-85	C	1	52	R43TS	.035	E.I.		10	10	32	5-6	①	②③
	E	4-97.6	C	1	60	R43TS	.035	E.I.		8	10	32	5-6	④	⑤⑥
77	I	4-85	C	1	52	R43TS	.035	E.I.		12	12	32	5-6	800/600	800/600
	E	4-97.6	C	1	60	R43TS	.035	E.I.		8	8	32	5-6	800/600	800/600⑦
78	E	4-97.6	C	1	63	R43TS	.035	E.I.		8	8	28	5-6.5	800/600	800/600⑦
	J	4-97.6	C	1	68	R43TS	.035	E.I.		8	8	31	5-6.5	800/600	800/600⑦
79	E	4-97.6	C	2		R43TS	.035	E.I.		⑧	⑧	28	5-6.5	⑧	⑧
	0	4-97.6	C	2		R43TS	.035	E.I.		⑧	⑧	31	5-6.5	⑧	⑧

Should the information in this manual deviate from the specifications on the underhood tune-up decal the decal specifications should be used as they may reflect production changes.

- ■ All readings in degrees Before Top Dead Center.
- ● Figures in parentheses indicate California usage. Where two figures appear separated by a slash the first is idle speed with solenoid energized, the second is idle speed with solenoid disconnected.
- * C—Chevrolet Division
- ① 85 C.I.D. W/MT-800N/600N (1000N/600N).
- ② 85 C.I.D. W/AT W/O AC-800DR/700DR (850DR/600DR).
- ③ 85 C.I.D. W/AT AND AC-950DR/800DR.
- ④ 97.6 C.I.D. W/MT-800N/600N (1000N/600N).
- ⑤ 97.6 C.I.D. W/AT W/O AC-800DR/700DR (850DR/600DR).
- ⑥ 97.6 C.I.D. W/AT AND AC-950DR/800DR (950DR/600DR).
- ⑦ 97.6 C.I.D. W/AT AND AC-950DR/800DR.
- ⑧ UNAVAILABLE AT TIME OF PUBLICATION. SEE UNDERHOOD DECAL.

DISTRIBUTOR SPECIFICATIONS — CHEVROLET Chevette

Year	Distributor Identification	Centrifugal Advance Start Dist. Deg. @ Dist. RPM	Centrifugal Advance Finish Dist. Deg. @ Dist. RPM	Vacuum Advance Start @ In. Hg.	Vacuum Advance Finish Dist. Deg. @ In. Hg.
'76	1110654	0 @ 600	10 @ 2400	4	7 @ 8
	1110655	0 @ 600	10 @ 2400	5	6 @ 12
	1110657	0 @ 600	10 @ 2400	5	12 @ 12①
	1110658	0 @ 600	10 @ 2400	5	13 @ 14.5
	1110659	0 @ 750	8 @ 2400	4	13 @ 12
'77	1110687	0 @ 600	10 @ 2400	4	13.5 @ 12
	1110693	0 @ 600	10 @ 2400	4	7 @ 8
	1110702	0 @ 600	10 @ 2400	4	15 @ 12
	1110703	0 @ 600	12 @ 2850	4	15 @ 12
'78	1110705	0 @ 600	10 @ 2400	4	15 @ 12
	1110707	0 @ 600	10 @ 2400	4	7 @ 8
	1110712	0 @ 600	10 @ 2400	4	13 @ 12
	1110713	0 @ 600	11 @ 2625	4	15 @ 12

Chevette

1ME CARBURETOR SPECIFICATIONS
CHEVROLET CHEVETTE

Year	Caburetor Identification① Number	Float Level (in.)	Metering Rod (in.)	Fast Idle Speed (rpm)	Fast Idle Cam (in.)	Vacuum Break (in.)	Choke Unloader (in.)	Choke Setting (notches)
1976	17056036, 17056030, 17056031, 17056037	5/32	0.072	2000②	0.065	0.070	0.165	3 Rich
	17056032, 17056034, 17056033, 17056035	5/32	0.073	2000③	0.045	0.070	0.200	3 Rich
	17056330, 17056331	5/32	0.072	2000	0.065	0.070	0.165	3 Rich
	17056332, 17056333, 17056334	5/32	0.073	2000	0.045	0.070	0.200	3 Rich
	17056335	5/32	0.073	2000	0.045	0.120	0.200	3 Rich
1977	17057016	3/8	0.070	2000	0.095	0.125	0.325	1 Lean
	17057013, 17057015	3/8	0.070	2000	0.100	0.125	0.325	1 Rich
	17057018	3/8	0.070	2000	0.085	0.120	0.325	1 Rich
	17057014, 17057020	3/8	0.070	2000	0.085	0.120	0.120	2 Rich
	17057310, 17057312	3/8	0.070	1800	0.100	0.100	0.110	Index
	17057314, 17057318	3/8	0.070	1800	0.100	0.110	0.110	Index
	17057042, 17057044, 17047045, 17057332, 17057334, 17057335	5/32	0.080	2300	0.050	0.080	0.200	2 Rich
	17057030, 17057031, 17057032, 17057034, 17057035	5/32	0.080	2300	0.050	0.080	0.200	3 Rich
1978	17058013	3/8	0.080	2000	0.180	0.200	0.500	Index
	17058014	5/16	0.100	2100	0.180	0.200	0.500	Index
	17058020	5/16	0.100	2100	0.180	0.200	0.500	Index
	17058314	3/8	0.100	2000	0.190	0.245	0.400	Index
	17058031	5/32	0.080	2400	0.105	0.150	0.500	2 Rich
	17058032	5/32	0.080	2400	0.080	0.130	0.500	3 Rich
	17058033	5/32	0.080	2400	0.080	0.130	0.500	2 Rich
	17058034	5/32	0.080	2400	0.080	0.130	0.500	3 Rich
	17058035	5/32	0.080	2300	0.080	0.130	0.500	3 Rich
	17058036	5/32	0.080	2400	0.080	0.130	0.500	3 Rich
	17058037	5/32	0.080	2400	0.080	0.130	0.500	2 Rich
	17058038	5/32	0.080	2400	0.080	0.130	0.500	3 Rich
	17058042	5/32	0.080	2400	0.080	0.160	0.500	2 Rich
	17058044	5/32	0.080	2400	0.080	0.160	0.500	2 Rich
	17058045	5/32	0.080	2300	0.080	0.160	0.500	2 Rich
	17058332	5/32	0.080	2400	0.080	0.160	0.500	2 Rich
	17058334	5/32	0.080	2400	0.080	0.160	0.500	2 Rich
	17058335	5/32	0.080	2300	0.080	0.160	0.500	2 Rich

Chevette

CAR SERIAL NUMBER AND ENGINE IDENTIFICATION

1976

Mounted behind the windshield on the driver's side is a plate with the vehicle identification number. The sixth character is the model year, with 6 for 1976. The fifth character is the engine code, as follows.

E 1.6 1-bbl. 4-cyl. LY5
I 1.4 1-bbl. 4-cyl. LX3

Engines can also be identified by a three-letter code stamped on the right side of the engine below number one spark plug.

CDD 1.4 4-cyl. Man. Calif.
CDS 1.4 4-cyl. Auto. 49-States
CDT 1.4 4-cyl. Auto. Calif.
CDU 1.4 4-cyl. Auto. 49-States
CNA 1.6 4-cyl. Man. 49-States
CNB 1.6 4-cyl. Man. 49-States
CVA 1.4 4-cyl. Man. 49-States
CVB 1.4 4-cyl. Man. 49-States
CYC 1.6 4-cyl. Auto. 49-States
CYD 1.6 4-cyl. Auto. 49-States
CYJ 1.6 4-cyl. Man. Calif.
CYK 1.6 4-cyl. Auto. Calif.
CYW 1.6 4-cyl. Man. Calif.
CYX 1.6 4-cyl. Man. Calif.

1977

The vehicle identification number is mounted on a plate behind the windshield on the driver's side. The sixth character is the model year, with 7 for 1977. The fifth character is the engine code, as follows.

E 1.6 1-bbl. 4-cyl. LY5
I 1.4 1-bbl. 4-cyl. LX3

Engine code on block

Engines can also be identified by a three-letter code stamped on the right side of the engine below number one spark plug.

CDS 1.4 4-Cyl. Auto. 49-States
CNA 1.6 4-Cyl. Man. 49-States
CNB 1.6 4-Cyl. Man. 49-States
CNC 1.6 4-Cyl. Man. Altitude
CND 1.6 4-Cyl. Man. Altitude
CNF 1.6 4-Cyl. Auto. Altitude
CNH 1.6 4-Cyl. Auto. Altitude
CNL 1.6 4-Cyl. Man. Calif.
CNN 1.6 4-Cyl. Man. Calif.
CNR 1.6 4-Cyl. Auto. Calif.
CNS 1.6 4-Cyl. Auto. Calif.
CVA 1.4 4-Cyl. Man. 49-States
CVB 1.4 4-Cyl. Man. 49-States
CYC 1.6 4-Cyl. Auto. 49-States
CYD 1.6 4-Cyl. Auto. 49-States
CYF 1.6 4-Cyl. Man. Calif.
CYH 1.6 4-Cyl. Man. Calif.
CYY 1.6 4-Cyl. Auto. Calif.
CYZ 1.6 4-Cyl. Auto. Calif.

1978

The vehicle identification number is mounted on a plate behind the windshield on the driver's side. The sixth character is the model year, with 8 for 1978. The fifth character is the engine code, as follows:

E .... 1.6 1-bbl. 4-cyl. LY-5 Chev.
J .... 1.6 1-bbl. 4 cyl. LW-5 Chev.

NOTE: 1.6 litres is the same as 98 cubic inches. The LW-5 is an optional high output engine with a bigger carburetor, different exhaust manifold, and a performance camshaft.

1979

The vehicle identification number is mounted on a plate behind the windshield on the driver's side. The sixth character is the model year, with 9 for 1979. The fifth character is the engine code, as follows.

E ....1.6 2-bbl. 4-cyl. L17 Chev.
J ....1.6 2-bbl. 4-cyl. L18 Chev.

NOTE: 1.6 litres is the same as 98 cubic inches.

EMISSION EQUIPMENT

1976

All Models

Closed positive crankcase ventilation
Emission calibrated carburetor
Emission calibrated distributor
Heated air cleaner
Vapor control, canister storage
Exhaust gas recirculation
Catalytic converter, single
Single diaphragm vacuum advance
Electric choke
Air pump (49-States: Not used. Calif.: All models)

1977

All Models

Closed positive crankcase ventilation
Emission calibrated carburetor
Emission calibrated distributor
Heated air cleaner
Vapor control, canister storage
Exhaust gas recirculation
Catalytic converter, single
Electric choke
  Calif. & Altitude only
Air pump
  Not used
Pulse air injection
  Calif. & Altitude only

1978

All Models

Closes positive crankcase ventilation
Emission calibrated carburetor
Emission calibrated distributor
Heated air cleaner
Vapor control, canister storage
Exhaust gas recirculation
Catalytic converter, single
Electric choke
  Calif. & Altitude only
Air pump
  Not used
Pulse air injection
  Calif. & Altitude only

1979

All Models

Closed positive crankcase ventilation
Emission calibrated carburetor
Emission calibrated distributor
Heated air cleaner
Vapor control, canister storage
Exhaust gas recirculation
Catalytic converter, single
Electric choke
  Calif. & Altitude only
Air pump
  Not used
Pulse air injection
  Calif. & Altitude only

Chevette

IDLE SPEED AND MIXTURE ADJUSTMENTS

1976

Air cleaner .... On
Air cond. .... Off
Vapor hose to carb. .... Plugged
Dist. vac. hose .... Plugged
Auto. trans. .... Drive
Mixture adj. .. See rpm drop below

1.4 and 1.6 49-States
  Auto. trans. no A.C.
    Solenoid connected 800
    Solenoid disconnected 700
    Mix. adj. .... 50 rpm drop
      (850-800)
  Auto. trans. with A.C.
    Solenoid off 800
    Mix. adj. .... 50 rpm drop
      (850-800)
    AC idle speedup 950
  NOTE: *Curb idle and mixture settings on the auto. trans. engine with air conditioning are made with the solenoid OFF (de-energized). Idle speed is adjusted with the Allen screw in the center of the solenoid. See Idle Speedup Solenoid in this section for a complete description.*
  Man. trans. no A.C.
    Solenoid connected 800
    Solenoid disconnected 600
    Mix. adj. .... 50 rpm drop
      (850-800)
  Man. trans. with A.C.
    Solenoid connected 800
    Solenoid disconnected 700
    Mix. adj. .... 50 rpm drop
      (850-800)

1.4 and 1.6 Calif.
  Auto. trans. no A.C.
    Solenoid connected 850
    Solenoid disconnected 600
    Mix. adj. .... 50 rpm drop
      (900-850)
  Auto. trans. with A.C.
    Solenoid off 850
    Mix. adj. .... 50 rpm drop
      (900-850)
    Idle speedup (A.C. on) 950
  NOTE: *Curb idle and mixture settings on the auto. trans. engine with air conditioning are made with the solenoid OFF (de-energized). Idle speed is adjusted with the Allen screw in the center of the solenoid. See Idle Speedup Solenoid in this section for a complete description.*
  Man. Trans.
    Solenoid connected 1000
    Solenoid disconnected ... 600
    Mix. adj. .... 100 rpm drop
      (1100-1000)

1977

Air cleaner .... Set aside with
    hoses connected
Air cond. .... Off
Other hoses .... See label
Auto. trans. .... Drive
Idle CO .... Not used
Mixture adj. .... See rpm drop below
1.4 4-cyl. Auto. trans.
  Solenoid connected 800
  Solenoid disconnected 600
  Mixture adj. .... 50 rpm drop
    (850-800)
1.4 4-cyl. Man. trans.
  Solenoid connected 800
  Solenoid disconnected 600
  Mixture adj. .... 75 rpm drop
    (875-800)
1.6 4-cyl. Auto. trans.
  With air cond. 800
    Mixture adj. ...... 50 rpm drop
      (850-800)
    A.C. idle speedup 800
  No air cond.
    Solenoid connected 800
    Solenoid disconnected 600
    Mixture adj. ...... 50 rpm drop
      (850-800)
1.6 4-cyl. Man. trans.
  49-States
    Solenoid connected 800
    Solenoid disconnected 600
    Mixture adj. ..... 100 rpm drop
      (900-800)
  Calif. 800
    Mixture adj. ...... 75 rpm drop
      (875-800)
  Altitude 800
    Mixture adj. ..... 100 rpm drop
      (900-800)

NOTE: *All 1.6 automatic transmission engines have a solenoid on the carburetor. On engines without air conditioning, the solenoid is used for anti-dieseling. It is energized whenever the ignition switch is on, and the curb idle speed is adjusted by turning the solenoid body.*

On engines with air conditioning, the solenoid is for A.C. idle speedup. It is energized only when the air conditioning is on, and the curb idle speed is adjusted with the ⅛ inch Allen screw in the end of the solenoid.

The A.C. idle speedup speed with the A.C. on is 800 rpm, the same as with the A.C. off. With the A.C. compressor operating, the engine would normally slow down at idle, but the solenoid opens the throttle enough to keep the engine idling at the same speed.

Some specifications show a "curb idle" speed of 950 rpm. This is the idle speed you will get if you have the A.C. on, but the compressor clutch wire disconnected. You will also get 950 rpm if you leave the A.C. off, but hot wire the solenoid to energize it.

Chevrolet instructions state that you must bottom the Allen screw in the solenoid before adjusting the "curb idle" to 950 rpm by turning the solenoid body. Then the Allen screw is backed out to adjust the "base idle" to 800 rpm. What they mean by "base idle" on this engine is the actual idle speed with the transmission in Drive.

Do not make the mistake of using the solenoid body to adjust the idle speed to 950 rpm without bottoming the Allen screw. If this is done, the engine will actually idle at 950 rpm, which is 150 rpm too fast.

Bottoming the Allen screw has the same effect as energizing the solenoid. At the carburetor plant they bottom the Allen screw instead of energizing the solenoid because they don't want the danger of sparks from making electrical connections where there are fuel vapors. Because of government regulations, servicing has to be done the same way that the original adjustment was made.

The term "curb idle" usually means the speed that an engine idles in normal operation. But Chevrolet uses the term "curb idle" for a setting that never happens in actual operation. Also, there are two meanings for the term "base idle." "Base idle" on an automatic transmission Chevette with air conditioning is the actual idle speed. But "base idle" on an automatic transmission Chevette without air conditioning is the idle speed with the solenoid disconnected, a condition which occurs only during testing.

1978

Air cleaner .... In place
Air cond. .... Off
Hoses .... See engine label
Auto trans. .... Drive
Idle CO .... Not used
Mixture adj. .. See propane rpm below
1.6 4-cyl. "E" auto. trans.
  With air cond. 800
    Propane enriched 850
    AC idle speedup 950
  No air cond.
    Solenoid connected 800
    Solenoid disconnected 600
    Propane enriched 850
1.6 4-cyl. "E" man. trans.
  49-States
    Solenoid connected 800
    Solenoid disconnected 600
    Propane enriched 900
  Calif. 800
    Propane enriched 875
  Altitude 800
    Propane enriched 900
NOTE: *California and Altitude "E" manual transmission engines use a device that mounts in the same position as a solenoid, but is not electric. The device is a dashpot, which slows the closing of the throttle. Correct adjustment is 5 turns in, after the plunger just touches the tang on the throttle lever.*
1.6 4-cyl. "J" auto. trans.
  With air cond. 800

Chevette

IDLE SPEED AND MIXTURE ADJUSTMENTS

1978

Propane enriched 875
AC idle speedup 950
No air cond.
  Solenoid connected 800
  Solenoid disconnected 600
  Propane enriched 875
1.6 4-cyl. "J" man. trans.
  Solenoid connected 800
  Solenoid disconnected 600
  Propane enriched 1000

NOTE: *All automatic transmission engines have a solenoid on the carburetor. On engines without air conditioning, the solenoid is used for anti-dieseling. It is energized whenever the ignition switch is on. Curb idle is adjusted by turning the solenoid body.*

On automatic transmission engines with air conditioning, the solenoid is for A.C. idle speedup. It is energized only when the air conditioning is on. Curb idle is adjusted with the 1/8 inch Allen screw in the end of the solenoid.

1979

CAUTION: *Emission control adjustment changes are noted on the Vehicle Emission Information Label by the manufacture. Refer to the label before any adjustments are made.*

Air Cleaner .... In place
Air cond. .... Off
Hoses .... See engine label
Auto. trans. .... Drive
Idle CO .... Not used
Mixture adj.
  ...... See propane rpm below
1.6 4-cyl. auto. trans.
  With air cond. 800
    Propane enriched 850
    AC idle speedup 950
  No air cond.
    Solenoid connected 800
    Solenoid disconnected 600
    Propane enriched 850
1.6 4-cyl. man. trans.
  49-States
    Solenoid connected 800
    Solenoid disconnected 600
    Propane enriched 900
  Calif. 800
    Propane enriched 875
  Altitude 800
    Propane enriched 900

NOTE: *California and Altitude manual transmission engines use a device that mounts in the same position as a solenoid, but is not electric. The device is a dashpot, which slows the closing of the throttle. Correct adjustment is 5 turns in, after the plunger just touches the tang on the throttle lever.*

1.6 4-cyl. auto. trans.
  With air. cond. 800
    Propane enriched 875
    AC idle speedup 950
  No air cond.
    Solenoid connected 800
    Solenoid disconnected 600
    Propane enriched 875
1.6 4-cyl. man. trans.
  Solenoid connected 800
  Solenoid disconnected 600
  Propane enriched 1000

NOTE: *All automatic transmission engines have a solenoid on the carburetor. On engines without air conditioning, the solenoid is used for anti-dieseling. It is energized whenever the ignition switch is on. Curb idle is adjusted by turning the solenoid body.*

On automatic transmission engines with air conditioning, the solenoid is for A.C. idle speedup. It is energized only when the air conditioning is on. Curb idle is adjusted with the 1/8 inch Allen screw in the end of the solenoid.

INITIAL TIMING

1976

NOTE: *Distributor vacuum hose must be disconnected and plugged. Set timing at idle speed unless shown otherwise.*

1.4 4-cyl. .... 10° BTDC
1.6 4-cyl.
  Auto. trans. .... 10° BTDC
  Man. trans. .... 8° BTDC

1977

1.4 4-cyl. .... 12° BTDC
1.6 4-cyl. .... 8° BTDC

1978

1.6 4-cyl. .... 8° BTDC

1979

1.6 4-cyl. .... 8° BTDC

SPARK PLUGS

1.4 4-cyl.—AC-R43TS .... .035
1.6 4-cyl.—AC-R43TS .... .035

VACUUM ADVANCE

Diaphragm type .... Single
Vacuum source .... Ported

EMISSION CONTROL SYSTEMS 1976

IDLE SPEEDUP SOLENOID, 1976

Description of Solenoid

This solenoid is the same one that has been used for many years on the one-barrel carburetor, but it is hooked up differently on automatic transmission cars. Instead of being energized whenever the ignition switch is on, this solenoid is energized only when the air conditioning is on. The solenoid speeds up the idle to make both engine cooling and the air conditioning more efficient, and to lessen the possibility of the engine dying.

Curb idle speed is adjusted with the Allen screw in the center of the solenoid. It is not necessary to disconnect the solenoid wire, because the idle speed is set with the air conditioning off, which also turns off the solenoid. Idle speedup rpm is set by turning the body of the solenoid.

If the solenoid on an automatic transmission car with A.C. comes on with the ignition switch, like an ordinary anti-dieseling solenoid, the wiring is probably hooked up wrong. The same wire loom is used for both automatic and manual transmission cars. A brown wire, that exits the harness near the solenoid, is used to connect the solenoid to the ignition switch on manual transmission cars. A green wire with a white stripe, that exits the harness near the firewall, is used

Chevette

to connect the solenoid to the air conditioning switch. You may find some cars that are hooked up incorrectly, because the person who worked on the car before you did not understand the system.

The brown wire should be connected on automatic transmission cars without A.C. and all manual transmission cars. The green-white-stripe wire should be connected on automatic transmission cars with A.C. The unused connector is left dangling in the engine compartment, but it should be curled out of the way so it won't get tangled in the throttle linkage. If you change the hookup, you must check the idle speeds and reset if necessary.

If the solenoid is connected for anti-dieseling (brown wire), curb idle speed is set with the solenoid energized, by turning the solenoid body. If the solenoid is connected as an air conditioning idle speedup (green wire with white stripe) curb idle speed is set with the solenoid not energized, using the Allen screw.

Testing and Adjusting Solenoid, 1976

With the air conditioning off, disconnect the wire at the solenoid and run a hot wire from the battery positive post to the solenoid. Start the engine and open the throttle slightly to allow the solenoid stem to extend. If the stem does not extend, the solenoid is defective.

Turn the body of the solenoid to adjust the idle speedup speed to 950 rpm in Drive.

CAUTION: *Turning the body of the solenoid to adjust the idle speedup speed will change the setting of the curb idle speed. After changing the idle speedup speed, you must reset the curb idle speed with the Allen screw in the center of the solenoid.*

Repairing Solenoid, 1976

Repairs are limited to replacement. Current to operate the solenoid comes from the battery to the ignition switch, to the Heater-A.C. fuse in the fuse block, then through the air conditioning wiring to the solenoid.

NOTE: *The air conditioning system has a ten second time delay relay, which prevents air conditioner operation in the first ten seconds after the ignition switch is turned on. If the air conditioning controls are in the ON position, with the ignition switch off, and the ignition switch is turned on, the idle speedup solenoid will not extend its stem until approximately ten seconds later.*

The PCV valve (arrow) is completely hidden under a tangle of hoses and wires

POSITIVE CRANKCASE VENTILATION (PCV) SYSTEM

Description of System

The PCV valve is plugged into a grommet at the left side of the camshaft cover, and connected by a hose to intake manifold vacuum. Fresh air enters from the air cleaner through a tube connected to the rear of the cam cover. There is no PCV filter in the Chevette system.

Units in System

PCV valve

Testing and Troubleshooting System

1. Unplug the PCV valve from its grommet. With the engine idling, check the vacuum at the valve with your finger or a vacuum gauge. If the vacuum is weak, renew the valve or hose, whichever is needed.
2. Plug the PCV valve back into the grommet.
3. Remove the fresh air tube from the rear of the rocker cover and plug the hole in the cover.
4. Remove the oil filler cap and put a sheet of paper or a vacuum gauge over the oil filler opening to check crankcase vacuum. Shield the paper from the fan blast so it will stay in place. If there is not enough vacuum, check for an air leak at the cam cover gasket, oil pan gasket, or other engine gaskets. Tightening the bolts will usually eliminate leaks at the gaskets. The dipstick must be seated to prevent a vacuum leak.
5. Blow through the fresh air tube to be sure it is clear, then install the tube and the oil filler cap.

Repairing System

The PCV valve should be replaced when dirty. Cleaning the valve is not recommended, because you can never be sure that it is clean.

AIR PUMP SYSTEM

Description of System

Fresh air enters the pump through a filter fan behind the pump pulley. From the pump, air under pressure flows through a large hose to a bypass valve at the right side of the engine compartment. From the bypass valve a large hose goes to the check valve on the right side of the cylinder head below the oil filler cap. From the check valve, air passages lead inside the head to each exhaust port.

A small hose from the bypass valve connects to the vacuum differential valve mounted nearby. The small hose then goes through a vacuum trap valve (not on all cars) to a fuel separator valve at the right front of the engine. The fuel separator valve is connected by tubing and hose to a fitting at the rear of the intake manifold. During deceleration, the sudden increase in intake manifold vacuum closes the vacuum differential valve, which shuts off the vacuum to the

Chevette

Removing the PCV valve is not difficult if you know where it is

bypass valve. The bypass valve closes off the outlet to the check valve, and opens the exhaust opening. All the pump air then goes into the atmosphere to prevent exhaust system backfire.

The bypass valve also opens from excess pressure caused by high engine rpm, and allows the pump air to go into the atmosphere.

The fuel separator valve keeps liquid fuel from running out of the intake manifold into the bypass valve. The vacuum trap valve maintains vacuum at the bypass valve during wide open throttle, so the system will continue to work when manifold vacuum is low.

Units in System
Air pump
Bypass valve
Vacuum differential valve
Vacuum trap valve (not on all cars)
Fuel separator valve
Check valve
Cylinder head with air passages

Testing and Troubleshooting System

Testing is done on each part of the system, rather than on the whole system at once. See below for tests on the individual units.

Backfire in the exhaust system can be caused by a bypass valve that is not working. However, it is more likely that exhaust popping or backfire is caused by the engine being out of tune. Misfiring, or a dirty carburetor, can cause unburned fuel to enter the exhaust system and explode when mixed with the air from the pump.

Any of the following can be a cause of low or no air supply to the exhaust ports.
Loose drive belt
Defective bypass valve
Pressure hose leaks
Vacuum hose leaks
Worn out pump
Plugged air filter on pump
Pump noise can be caused by:
Worn out pump
Pump mounting loose
Defective check valve
Defective diverter valve
Hose touching other engine parts
Loose or leaking pressure hose

Excessive belt noise can be caused by:
Slipping belt
Dragging or seized pump

New pumps may be noisy and require 10 to 15 miles of running at highest legal speeds before they quiet down. Do not get rid of a pump because it squeaks when turned by hand. Most of them do. If the driver complains of too much noise, temporarily disconnect the pump belt to see if the noise goes away. If the pump is definitely the source of excessive noise, replace it.

Sometimes the pump is blamed for poor idle or generally poor running of the engine. The pump is never in any way responsible for either of these and it must not be replaced because of them.

Repairing System
The only adjustment to the system is belt tension. The only repair possible on the pump is replacement of the pulley or filter fan.

Testing Air Pump
1. Disconnect the hose at the back of the pump or at the pump side of the bypass valve.
2. With the engine idling, check the flow of air by feeling it with your hand.
3. Increase the engine speed to approximately 1,500 rpm and check the flow of air again. If the flow increases as the engine is accelerated, the pump output is okay.
4. Inspect the filter fan for clogging with the engine off. If the fan is plugged or damaged, replace it. Do not try to clean it.

Repairing Air Pump
The only repair possible is replacement of the pulley or filter fan.

Testing for crankcase vacuum can be done with a piece of paper (arrow) over the oil fill opening, but you must remember to put a plug in the fresh air hole (arrow)

Fuel separator (1), vacuum differential valve (2), and bypass valve (3)

Testing Bypass Valve

With the engine idling in Neutral or Park, disconnect the small hose from the top of the bypass valve. The valve should immediately open, and all the pump air should go through the exhaust holes on the valve. If you can't hear the air escaping, speed the engine up a little and put the back of your hand next to the exhaust holes so you can feel the air coming out. If the air does not come out, the bypass valve is defective.

NOTE: *Do not attempt to test the bypass valve by pinching the hose and releasing it. This action will have no effect on the valve.*

To test the relief feature of the valve, pinch the large hose shut between the bypass valve and the check valve. Air should come out through the exhaust holes on the valve. If not, the valve is defective.

CAUTION: *Do not attempt to check the relief feature by revving the engine.*

Repairing Bypass Valve

Repairs are limited to replacement of the valve. There are no adjustments.

Testing Vacuum Differential Valve (VDV)

1. Disconnect the hose at the intake manifold side of the VDV and check to be sure there is full manifold vacuum at the hose with the engine running. If not, run a separate hose from the intake manifold to the VDV.
2. Reconnect the manifold vacuum hose to the VDV.
3. Remove the hose between the bypass valve and the VDV, and connect a vacuum gauge to the VDV.
4. With the engine idling, or running at part throttle above idle, you should see full manifold vacuum on the gauge. If not, the VDV is defective.
5. Open the throttle and let it snap shut. As the engine decelerates, the vacuum gauge should drop to zero. As the engine speed returns to normal, the vacuum gauge should rise and indicate full manifold vacuum again. If not, the VDV is defective.

NOTE: *If you do not get full manifold vacuum at the VDV hose, there is a restriction in either the fuel separator, the vacuum trap valve, or the hoses.*

Repairing Vacuum Differential Valve (VDV)

Repairs are limited to replacement of the valve.

Testing Vacuum Trap Valve

Connect a hand vacuum pump to the intake manifold side of the valve. You should not be able to pump up any vacuum. If you can pump up vacuum, the valve is plugged.

Connect the hand vacuum pump to the bypass valve side. You should be able to pump up several inches of vacuum and it should hold without leaking down. If not, the valve is defective.

Repairing Vacuum Trap Valve

Leaking valves can sometimes be cured by blowing out any dirt or foreign matter inside the valve. If not, the valve must be replaced.

Testing Fuel Separator Valve

The fuel separator valve is a sealed unit containing a check valve and an air bleed, protected by a filter. Whenever vacuum is applied to the separator (during idle, cruise, or deceleration) a small amount of air bleeds through the filter into the vacuum line. This constant flow of air purges the line of any fuel vapors and prevents them from reaching the vacuum differential valve or the bypass valve. The check valve inside the fuel separator opens when acted on by engine vacuum, but closes if high pressure should come through the vacuum line, as from an intake manifold flashback.

To test the fuel separator valve, connect a hand vacuum pump to the black (vacuum) nozzle, and plug the white nozzle with your finger. Pump up 15" Hg. of vacuum by operating the pump quickly. The vacuum gauge should leak down slowly, taking several seconds to reach zero. If it doesn't leak down, the air bleed is

Vacuum hoses on air pump engine. The vacuum trap valve has been discontinued, because the fuel separator has an internal check valve

151

Chevette

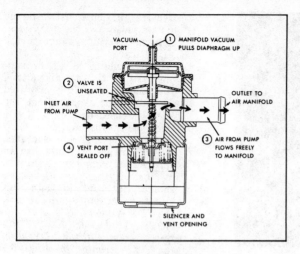

Bypass valve in normal running position

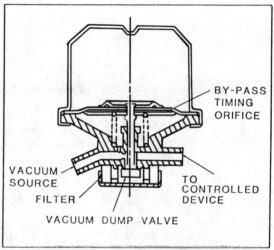

Bypass valve in pressure relief position

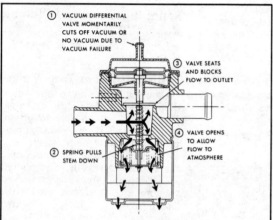

Bypass valve in deceleration position

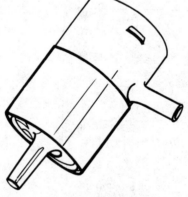

Vacuum differential valve

clogged. Do the test again, but remove your finger from the white nozzle when the gauge shows 15" Hg. of vacuum. The gauge should drop immediately to zero. If it doesn't, the valve is clogged.

Connect the vacuum pump to the white nozzle, and leave the black nozzle open. Pump up 15" Hg. vacuum. The gauge should fall slowly to zero, taking several seconds. If it falls immediately to zero, the check valve is leaking. If it doesn't fall at all, the air bleed is clogged.

Repairing Fuel Separator Valve

Repairs are limited to replacement of the valve. The part number of the valve is 373963. The black nozzle, marked VAC connects to engine manifold vacuum. The white nozzle, marked VDV, connects to the vacuum differential valve.

Testing Check Valve

Disconnect the hose at the bypass valve so you can blow through it into the check valve with the engine off. You should be able to blow through the check valve toward the cylinder

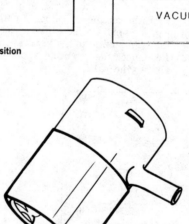

Fuel separator with internal check valve

head, but the valve should not leak when you suck back.

Repairing Check Valve

Repairs are limited to replacement of the valve, which can be unscrewed from the cylinder head. Use two wrenches, one to unscrew the check valve, and the other to hold the fitting that enters the head, so that the fitting does not unscrew.

VAPOR CONTROL SYSTEM

Description of System

A dome at the top of the tank prevents overfilling and provides an air pocket so that only vapors go out the vent line. The filler cap is a pressure-vacuum type with a ratchet tightening device that prevents overtightening.

A single 2-section vent line runs from the tank dome along the driveshaft tunnel to the carbon canister on the left side of the engine compartment. A rubber hose is used under the car at the rear of the driveshaft tunnel to connect the two sections of vent line. The canister collects the vapors from the tank and also acts as a tank vent, allowing air to enter the tank as the fuel is removed.

3-nozzle canisters are used, with a purge valve. The nozzle on the body of the canister receives the vapors from the tank. The nozzle on the top of the purge valve connects to ported carburetor vacuum, which is either of the two nozzles on the side of the carburetor. The nozzle at the bottom of the purge valve connects to manifold vacuum at a "T" in the PCV hose.

When the throttle is opened, ported vacuum goes through the hose to the purge valve and opens the valve, which allows full manifold vacuum to draw the vapors out of the canister and into the engine. Fresh air enters the bottom of the canister through a filter. At idle the purge valve is closed. This is known as the staged purge system.

Units in System
Domed fuel tank
Pressure-vacuum fuel tank cap
Carbon canister
Carburetor with purge connections

Testing and Troubleshooting System
No testing is required. The system should be inspected for fuel leaks or odors and deteriorated or damaged hoses and lines should be replaced. The tank cap can be tested with mouth pressure and suction to see if it opens. The only maintenance is replacement of the canister filter at the specified mileage intervals.

Repairing System
Both the fuel line and the vapor line enter the tank at the top of the dome through the fuel tank gauge unit. The usual cam lock ring and gasket is used to hold the gauge unit in place. For access to the gauge unit, the tank must be drained and removed from the car.

When replacing hoses, be sure that the replacement hose is marked or otherwise certified by the manufacturer for use in vapor control systems.

To replace the canister filter, remove the 10mm bolt on the side of the bracket. Spread the bracket and lift the canister out. It is not necessary to disconnect the hoses. Pull the filter out of the bottom of the canister. Be sure that the new filter is in place without any wrinkles.

HEATED AIR CLEANER

Description of Heated Air Cleaner
The air cleaner filter element is in a sealed "can" mounted on top of the carburetor. Air entry is through a removable duct, mounted to the cam cover so that the "can" fits down over the duct. Inside the duct is a flapper valve and thermostatic spring. When the temperature at the spring is below 50° F. the flapper will be up, shutting off the outside air and allowing only heated air from an exhaust manifold stove to go into the air

The air intake housing can be removed from the cam cover to reveal the thermostatic spring (arrow)

cleaner. When the temperature is above 110° F. the flapper drops down, which shuts off the heated air and allows only outside air to enter. Between 50 and 110° F. the flapper allows both heated and outside air to enter.

Testing Heated Air Cleaner
Remove the outside air tube by flipping the clamp loose. Use a screwdriver inside the opening to move the flapper valve and be sure it is free. At normal operating temperature the flapper should be down in the heat-off position, unless the outside air is very cold. When the engine is cold, the flapper should be up in the heat-on position. If necessary, the cold operation can be checked by putting the duct in a refrigerator for a few minutes, to see if it goes to the heat-off position.

Repairing Heated Air Cleaner
The duct assembly is not repairable. If the thermostatic spring and flapper valve do not work, replace the entire assembly.

When replacing the air cleaner, a gasket goes between the filter can and the duct, and a large ring gasket between the can and the carburetor.

EXHAUST GAS RECIRCULATION SYSTEM

Description of System
The exhaust gas recirculation (EGR) valve is mounted on the intake manifold between the carburetor and the cylinder head. An exhaust passageway through the engine carries exhaust heat to warm the fuel mixture and supply exhaust gas to the EGR valve. The valve is operated by ported vacuum through a hose connected to the carburetor. A spring in the valve keeps it closed when there is no ported vacuum to open it. The EGR valve is closed at idle, blocking off the exhaust gas recirculation. It is open at part throttle, which allows exhaust gas to enter the intake manifold.

California cars use a thermal vacuum switch mounted on the thermostat elbow that controls the ported vacuum to the EGR valve. Below 100-120° F. coolant temperature, the vacuum is shut off so there is no exhaust gas recirculating.

Units in System
EGR valve
Carburetor with EGR port
Intake manifold with EGR passages

Heated air cleaner takes cool outside air through a long tube connected to the front of the car. Note how easily the spark plugs can be reached

Chevette

EGR valve (arrow) is buried between the carburetor and the camshaft cover

EGR thermal vacuum switch (arrow) is mounted on the thermostat elbow

Thermal vacuum switch (Calif. only)

Testing and Troubleshooting System

1. Put the transmission in Park or Neutral with the parking brake on, and connect a tachometer to the engine.
2. Start the engine and set the throttle so that the engine runs at exactly 2,000 rpm. The engine must be at normal operating temperature with the choke fully open.
3. Disconnect the vacuum hose at the EGR valve. The engine speed should increase at least 100 rpm.
4. If the engine speed does not increase, use your finger to check for vacuum at the end of the hose. If there is no vacuum, look for a plugged hose or carburetor port.
5. If the vacuum is okay, hook the hose back up to the valve and watch or feel with your hand under the valve to find out if the diaphragm moves to the open position. If the diaphragm does not move, the valve is stuck or defective.
6. If the valve moves to the open position, but has no effect on engine speed at 2,000 rpm, either the valve or the passageway underneath it are plugged. The valve will have to be removed and either cleaned or replaced. See "Repairing EGR Valve" for the cleaning procedure.

Repairing System

Cleaning the system is recommended, particularly the EGR valve itself. See "Repairing EGR valve," below, for the cleaning procedure. None of the units are adjustable. They must be cleaned or replaced if defective.

Testing EGR Valve

Use a vacuum pump or manifold vacuum from a running engine to apply at least 9 in. Hg of vacuum to the valve. Use a mirror or your hand to check the diaphragm and stem. The valve should open fully, with the stem and diaphragm moving up from the mounting, and should not leak down.

If the valve does not open or is sluggish or jerks during opening, it must be removed and cleaned. If, after cleaning, the valve does not work right, it must be replaced. See "Repairing EGR Valve," below, for the cleaning procedure.

Repairing EGR Valve

Cleaning of the valve is recommended. Do not use any kind of solvent or cleaning fluids. They will damage the valve diaphragm. The base of the valve can be cleaned with a wire brush.

To clean the valve seat and plunger, use an ordinary spark plug sandblast cleaning machine. Insert the plunger into the spark plug opening on the machine and clean it. Use a vacuum pump to fully open the EGR valve, then hold the valve over the spark plug opening and clean it again. Use an air hose to blow all loose sand out of the valve and operate the valve several times to be sure it is not hanging up on a grain of sand.

Testing Thermal Vacuum Switch (TVS)

Connect a vacuum pump to the lower nozzle of the TVS. Leave the upper nozzle open. If the temperature of the TVS is below approximately 100° F. the vacuum passage should be closed, and you should be able to pump up several inches of vacuum. With 12" Hg. vacuum applied to the TVS, it should not lose more than 2" Hg. in 2 minutes. If not, the TVS is defective. Above 100-120° F. the TVS vacuum passage should be open. If not, the TVS is defective.

Repairing Thermal Vacuum Switch (TVS)

Repairs are limited to replacement of the TVS.

CAUTION: *Relieve the cooling system pressure before removing the switch.*

The vacuum hose from the carburetor connects to the lower nozzle, nearest the mounting threads. The EGR valve connects to the upper nozzle.

ELECTRIC CHOKE

Description of Choke

The choke has a dual-section heating element behind the coil spring. Whenever the engine is running, the choke receives current and the heater is in operation. A bimetal snap disc in the choke cover turns off the large section of the dual heating element below 50-70° F. Below that temperature, only the small section of the element provides heat. This gives a slower choke opening for better running in cold weather. Above 50-70° F. the bimetal snap disc turns on the large section of the heating element so that both sections provide heat for a quicker choke opening when the engine is warm.

Current to the choke is controlled by a three-terminal oil pressure switch. The pink-with-black-stripe

Chevette

To check the current to the electric choke, connect a 12-volt test light between the disconnected wire and the choke terminal. It should light when the engine is running

The Chevette choke is the same design used on the Chevrolet 454 V-8 and Cadillac 500 V-8

wire is the hot lead. The dark blue wire is the ground circuit for the oil pressure light on the instrument panel. The light blue wire (actually a light gray) connects the choke to the oil pressure switch. Current flows to the choke only when there is oil pressure in the engine.

Current to the choke starts at the battery, goes to the ignition switch, then to the direction signal-backup light fuse in the fuse block. From the fuse it goes through the firewall connector to the oil pressure switch.

The choke heater is grounded to the carburetor to complete the circuit. A metal plate at the back of the choke cover touches the carburetor to make the ground.

CAUTION: *The choke cover must be installed without a gasket, so that the ground circuit will be complete. The gasket is not necessary, because there is no vacuum to the choke housing, as there is in older chokes.*

Testing Electric Choke

At an engine compartment temperature of approximately 70° F. (cold engine) the choke should open wide in approximately 1½ minutes. The easiest way to make this test is to move the air cleaner to one side, open the throttle to allow the choke to close, and run a hot wire from the positive battery post to the choke terminal. Then time how long it takes the choke to open. The 1½ minutes time is only approximate. On a dead engine some chokes take 3 or 4 minutes to reach wide open. In actual operation this time is much less because engine vibration helps the linkage move.

Off the car, the choke coil can be tested by connecting a hot wire to the coil terminal, and a ground wire to the choke cover grounding plate. Starting at a room temperature of 60-70° F. the choke coil should rotate 45 degrees within 54-90 seconds after

Choke oil pressure switch is mounted dangerously close to the "hot" starter terminals

To remove the plug from the oil pressure switch, disconnect the battery ground cable, then flip the latch (arrow) up with a screwdriver

the connection is made. If not, the choke coil is defective, and the entire choke cover must be replaced.

If the choke coil is in good condition, but will not operate on the engine, it could be caused by:

1. Broken or disconnected light blue wire between oil pressure switch and choke coil.
2. Poor ground between choke cover grounding plate and housing.
3. Broken or disconnected wire in circuit from battery to oil pressure switch. See Description above, for circuit connections.
4. Burned out DIR SIG BU fuse in fuse block.
5. Defective oil pressure switch.
6. No oil pressure to switch. This could be caused by blocked oil passageways or sludged oil.

Repairing Electric Choke

Repairs are limited to replacement.

CHEVETTE DIAGNOSTIC PLUG

Mounted on the firewall on the left side of the engine compartment is an orange plastic connector that can be used to hook up engine diagnosis equipment. The engine diagnosis plug is mounted horizontally. Air conditioned cars have a second plug, mounted vertically, that can be used to diagnose the air conditioning.

If you have the adapter to connect to the plug, then all you have to do is hook it up. But even if you don't have the adapter, you can use the plug with an ordinary voltmeter to diagnose the engine. All the positions on the plug are numbered, and are hooked up to the following parts of the engine.

1. Starter "B" terminal (Battery)
2. Ignition switch 1 (Battery)
3. Blank
4. Ignition coil battery terminal

Chevette

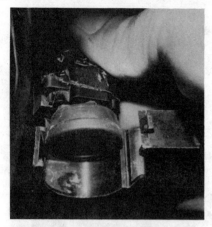

After the oil pressure switch plug latch is disengaged, it looks like this

Current to the oil pressure switch plug can be checked with the ignition switch on

This Chevette engine is without air conditioning, and you can see the distributor. With air conditioning, the distributor is almost inaccessible, but there are no points to replace in the High Energy Ignition. The white plastic next to the distributor is a splash shield for the ignition coil

5. Ignition switch 2 (Ignition coil)
6. Coil "C" terminal (Tachometer)
7. Blank
8. Starter "S" terminal
9. Ignition switch 3 (Starter)
G. Ground.

The diagnostic plug makes it very easy to check voltages at various points on the engine, just by hooking up to the plug.

CAUTION: *Do not use screwdrivers or other thick tools to make the connection, or you may run the plug. Only a thin, blade-type connector should be used.*

Following are examples of various tests that can be made.

Cranking voltage at coil: Connect a voltmeter between 4 and G.

Tachometer hookup: Connect a tachometer between 6 and G.

Remote cranking: Connect 1 and 8 with a jumper wire.

CAUTION: *When using a jumper wire, an incorrect connection can burn up the wiring or the ignition system.*

CLOSED POSITIVE CRANKCASE VENTILATION SYSTEM (PCV) 1977

A closed PCV system is used to remove the harmful blow-by gases and moisture from the crankcase. The PCV valve is positioned in the valve cover and routes the gases to the intake manifold. The air intake is from the air cleaner assembly.

1978

The closed PCV system is retained for use on the 1978 Chevette engines to reduce the hydrocarbon emission from the crankcase.

1979

The closed PCV system is used on the 1979 Chevette engines. Conditions indicative of malfunctioning PCV systems are rough idle, oil present in the air cleaner, engine oil leakage and

EMISSION CONTROL SYSTEMS 1977-79

excessive oil sludging or diluting.

The PCV system must be sealed so that the system operates properly. If the system is suspected of improper operation, check for the possible causes and correct so that the system functions as designed.

ENGINE MODIFICATIONS 1977-79

The major modifications to the engines consists of emission calibrated carburetors and distributors.

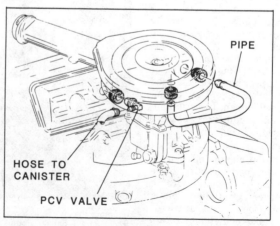

Positive crankcase ventilation system—1977-79 Chevette

Chevette

During each succeeding year, the calibrations for the carburetors and distributors are changed to conform with the changing emission laws and fuel requirements, as per the States and Federal mandated laws.

THERMOSTATIC AIR CLEANER (TAC) 1978

The Thermostatic Air Cleaner is used on all Chevette engines. A vacuum motor controls the damper assembly, located in the air cleaner inlet. The damper assembly controls the amount of preheated and non-preheated air entering the air cleaner, to maintain a controled air temperature into the carburetor. The damper assembly motor vacuum is modulated by a temperature sensor in the air cleaner.

1979

The Thermostatic Air Cleaner is used on all Chevette engines for 1979 and operate in the same manner as the 1978 system. The preheating of the inlet air to the carburetor results in lower emissions levels through leaner carburetors and choke calibrations and still maintains good drivability.

PULSE AIR INJECTION SYSTEM (PAIR) 1977

(High Altitude and California Models)

This system is basically the same as used on the Vega engines. The system consists of pulse air valve which has four check valves, one for each exhaust port. The firing operation of the engine creates a pulsating flow of gases, which have positive or negative pressures, due to the opening and closing of the exhaust valves. When positive pressure is present in the exhaust manifold, the check valve is forced closed and no exhaust gas will flow past the valve and into the fresh air supply line. When negative pressure is present, the fresh air supply line is opened to allow fresh air to enter the exhaust manifold to mix with the exhaust gases, and to further the burning of the gases.

1978

(High Altitude and California Models)

The Pulse Air Injection system is used on the Chevette engines sold in California and High Altitude areas. The operation is the same as in the 1977 models. The pulse air check valve must be maintained or replaced as necessary. If one or more check valves fail, the exhaust gases will enter the carburetor through the air cleaner assembly and cause surging or poor engine performance.

TAC system—1977-79 Chevette

1979

(High Altitude and California Models)

The Pulse Air Injection system is again used on the Chevette engines sold in California and High Altitude areas. The operation of the system remains the same as in previous years. Failure of the pulse air valve is indicated by paint being burned off the valve body by the hot exhaust gases, and by a hissing sound caused by deterioration of the rubber hose.

EVAPORATION CONTROL SYSTEM (ECS) 1977

The ECS system is designed to limit the release of fuel vapors into the atmosphere from the carburetor and fuel tank. The system consists of a

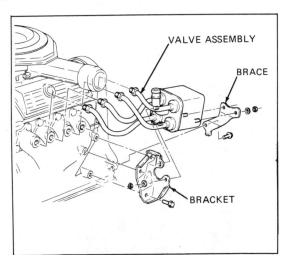

Pulse air valve mounting—1977-79 Chevette
(California and High Altitude models)

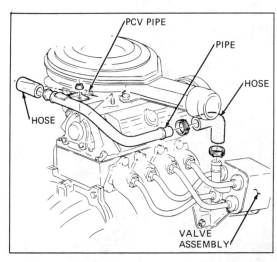

Pulse air pipe and hoses—1977-79 Chevette
(California and High Altitude models)

157

Chevette

charcoal canister to absorb the fuel vapors and to store them until the engine is in operation. Hoses and lines are used to direct the vapors from the fuel tank and carburetor, to the charcoal canister. A purge orfice is located in the carburetor and is connected by a hose to the canister, and during periods of part throttle, the vapors are routed to the PCV system, to be burned along with the crankcase gases.

1978

The ECS system is used on all Chevette models. The gas cap is equipped with a vacuum valve, to allow air to enter the fuel tank during periods of engine operation. The gasoline tank is domed to allow the vapors to escape to the canister without filling the system with fuel.

1979

The ECS system remains unchanged from the previous model years. The filter in the base of the canister should be replaced every two years or 3,000 miles. (More often when operating in dusty conditions).

If excessive gasoline fumes are present around the canister area, the canister may have to be replaced. Check the operation of the purge hose and valve.

EXHAUST GAS RECIRCULATION SYSTEM (EGR) 1977

The EGR system is used to reduce the combustion chamber temperatures by allowing metered amounts of exhaust gases into the induction system, to mix with the air/fuel mixture.

The federal vehicles have a ported vacuum signal at all times to the EGR valve, while the California and High Altitude vehicles have a thermal vacuum switch to prevent the EGR valve from operating until the engine temperature reaches 120° F.

1978

The EGR system is used on all 1978 Chevette engines, with the thermal vacuum switch again used on the California and High Altitude vehicles. Rough idling engines can be caused by a defective or broken EGR valve, vacuum hose disconnected or broken, or exhaust deposits lodged under the EGR valve, and should be checked before any major repairs are preformed on the engine.

1979

The EGR system is continued in use for the 1979 models, with no changes from the 1978 models. If the EGR valve must be changed, replacement must be made with the same numbered EGR valve to maintain the proper metering of the exhaust gas flow.

CATALYTIC CONVERTER 1977

A single catalytic converter is used as an emission control device and installed as part of the exhaust system. The purpose of the converter is to reduce the hydrocarbons and carbon monoxide pollutants from the exhaust gases, with the use of pellets coated with catalytic materials, containing Platinum and palladium. Only unleaded gasoline is to be used for fuel in vehicles equipped with the catalytic converter.

1978

The catalytic converter is used on all Chevette exhaust systems for 1978, to reduce the hydrocarbons and carbon monoxide levels of the exhaust gases.

The catalytic pellets can be changed in the converter unit. Special tools should be used in the performance of the repairs.

1979

The converter is used on all Chevette exhaust systems for 1979. The pellet replacement procedure remains the same. The complete converter replacement is necessary, if the general condition warrants. Only unleaded fuel can be used.

NOTE: *A bottom cover replacement is available for installation on the converter.*

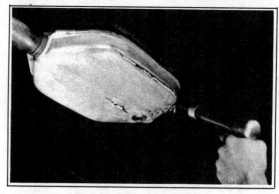
Removing catalytic converter bottom cover—1977-79 Chevette

EGR thermal vacuum switch—1977-79 Chevette (California and High Altitude models)

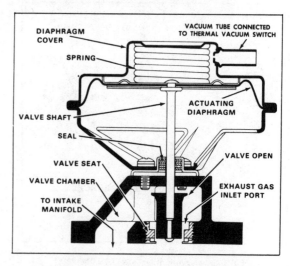
Cross section of EGR valve—1977-79 Chevette

Chevette

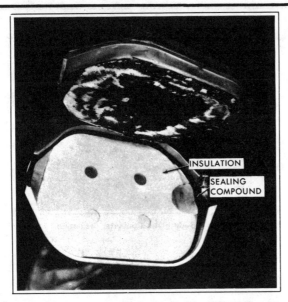

Installing replacement converter bottom cover—1977-79 Chevette

ELECTRIC CHOKE 1977

An electric choke is used to assist the heating of the choke coil to aid in quicker opening of the choke valve, while the engine is at higher temperatures. The electrical circuits to the choke is routed through the oil pressure switch, when the engine is operating. The current flows to a ceramic resistor that is divided into two sections. A small section is used for the gradual heating of the choke coil, while a large section is used for quick heating of the choke coil. The resistor warms the choke coil to provide proper choke valve opening for good engine warm up performance.

1978

The electric choke is used on the carburetors for the 1978 Chevettes.

At temperatures below 50° F., a temperature sensitive bi-metal disc causes a spring loaded contact to close, as the small section continues to heat, and changes the current to change from the small section to the large section, which increases the rate of heat flow to the choke coil for quicker opening of the choke valve.

At temperatures of 70° F. or above, the electrical current is directed to both the small section and the large section, to provide faster choke valve opening when leaner air/fuel mixtures are desired.

1979

The electric choke is continued in use on the Chevette carburetors for 1979. The operation of the choke and the current flow remains the same as 1978. When repairing or adjusting the choke mechanism, do not install a gasket between the electric choke assembly and the choke housing. To do so, would interrupt the grounding circuits.

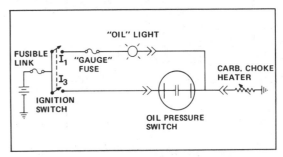

Electric choke circuit schematic—1977-79 Chevette

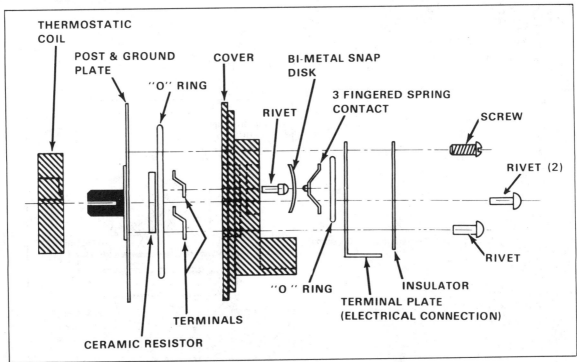

Exploded view of electric choke components—1977-79 Chevette

Chevette

VACUUM CIRCUITS

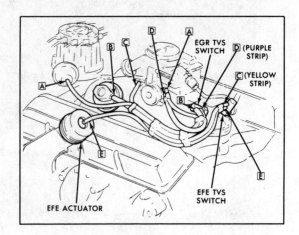

1976 Monza 262 V-8

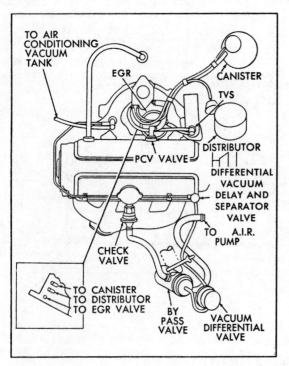

1976 Chevette 4-Cyl. with air pump

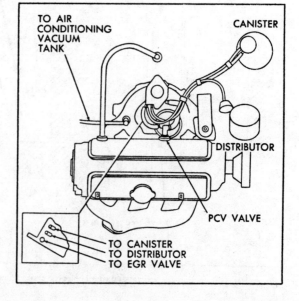

1976 Chevette 4-Cyl. no air pump

Chevette

VACUUM CIRCUITS

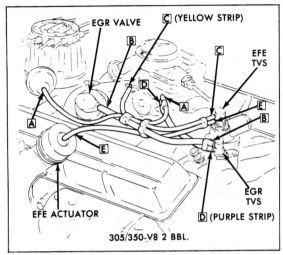

1976 305, 350 2-Bbl. V-8

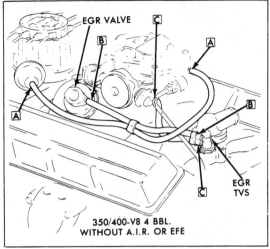

1976 350, 400 4-Bbl. V-8 no air pump, no EFE

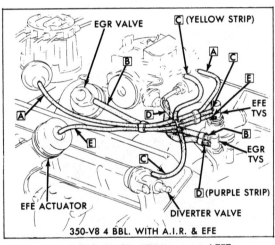

1976 350 4-Bbl. with air pump and EFE

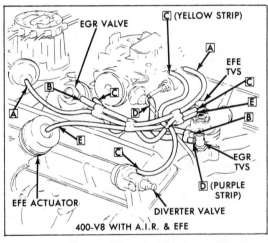

1976 400 4-Bbl. V-8 with air pump and EFE

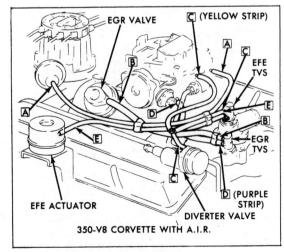

1976 350 4-Bbl. Corvette V-8 with air pump

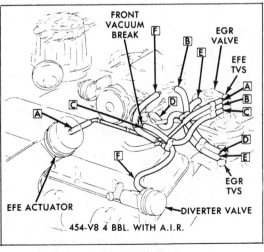

1976 454 4-Bbl. V-8

Chevette

VACUUM CIRCUITS

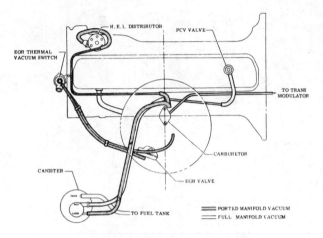

1976 49-State 250 two-piece head and manifold

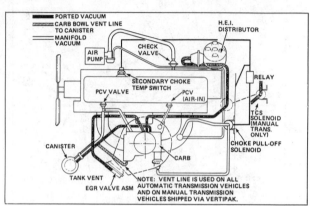

1976 49-State 140 1-Bbl.

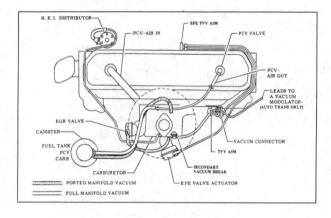

1976 49-State 250 one-piece head and manifold

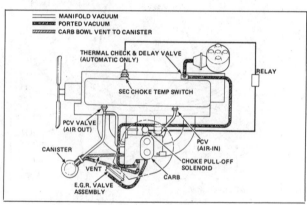

1976 49-State 140 2-Bbl.

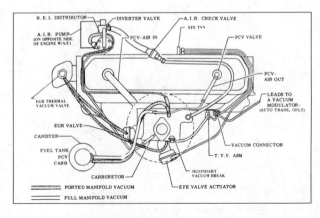

1976 Calif. 250 one-piece head and manifold

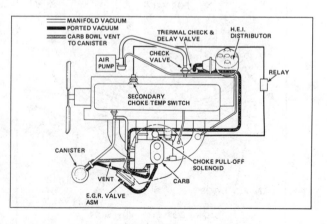

1976 Calif. 140 2-Bbl.

Chevette

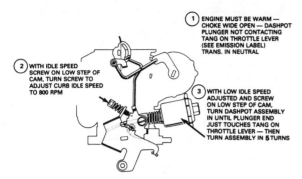

Idle dashpot adjustment 1.6L M.T. Altitude and California—Chevette
(© GM Corporation)

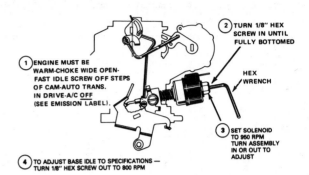

Idle speed adjustment 1.6L A.T. and Air Cond.—Chevette
(© GM Corporation)

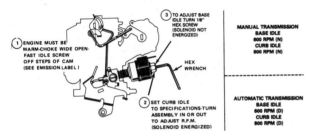

Idle speed adjustment 1.4L and 1.6L w/A.T. and wo/A.C. and 1.6L w/M.T.—49 states (© GM Corporation)

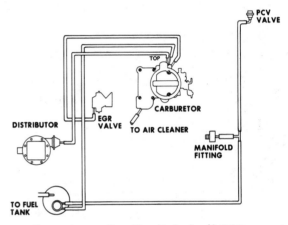

Vacuum hose routing—Chevette 4 cyl.—49 states
(© GM Corporation)

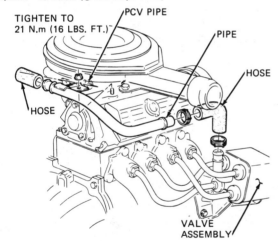

Pulse air pipe and hoses—Chevette
(© GM Corporation)

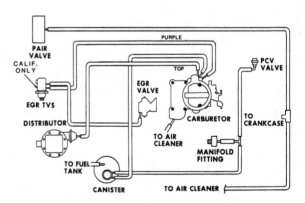

Vacuum hose routing—Chevette 4 cyl.—California and Altitude
(© GM Corporation)

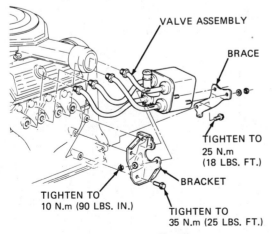

Pulse air valve—Chevette
(© GM Corporation)

163

Chevette

VACUUM CIRCUITS

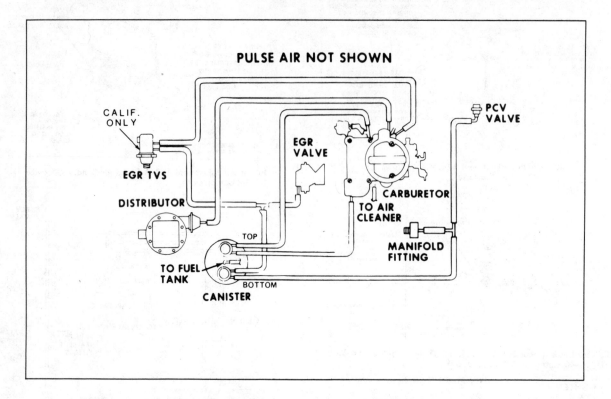

Vacuum hose routing—1978-79 Chevette (California and High Altitude models)

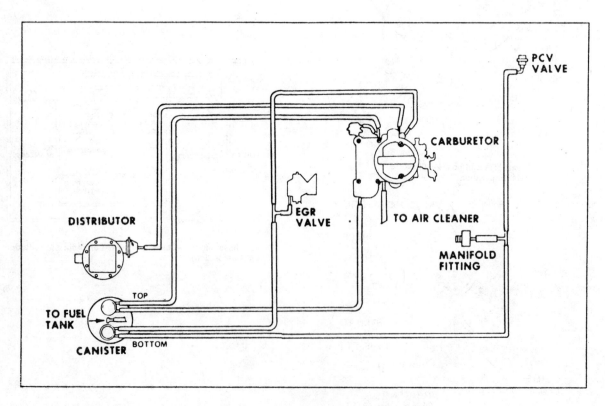

Vacuum hose routing—1978-79 Chevette (Low Altitude models)

Oldsmobile

Full-Size • Intermediate • Compact

TUNE-UP SPECIFICATIONS **OLDSMOBILE (EXCEPT STARFIRE)**

Engine Code, 5th character of the VIN number
Model Year Code, 6th character of the VIN number

Yr.	Eng. V.I.N. Code	Engine No. Cyl. Disp. (cu in)	Eng.*** Mfg.	Carb Bbl	H.P.	SPARK PLUGS Orig. Type	Gap (in)	DIST. Point Dwell (deg)	Point Gap (in)	IGNITION** TIMING (deg BTDC) Man	IGNITION** TIMING (deg BTDC) Auto	VALVES Intake Opens (deg BTDC)	FUEL PUMP Pres. (psi)	IDLE SPEED (rpm) Man Trans	IDLE SPEED (rpm) Auto Trans (In Drive)
73	D	6-250	C	1	100	R46T	.035	33	.019	6③	6③	16	4-5	700⑬	600⑬
	H	8-350	O	2	160	R46S	.040	30	.016	—	14④	22	5-6	—	700⑮
	K	8-350	O	4	180	R46S	.040	30	.016	8	12	22	5-6	1100⑯	650⑮
	M	8-350	O	4	220	R46S	.040	30	.016	—	12⑤	22	5-6	—	650⑮
	T	8-455	O	4	225	R45S	.040	30	.016	8	8	28	5-6	650⑯	650⑮
	U-V	8-455	O	4	250	R45S	.040	30	.016	10	8	28	5-6	1000⑯	650⑮
74	D	6-250	C	1	100	R46T	.040	33	.019	8	8(6)	16	4-5	850⑬	600⑬
	K	8-350	O	4	180	R46S	.040	30	.016	—	12	22	5-6	—	650⑮
	M	8-350	O	4	220	R46S	.040	30	.016	—	12	22	5-6	—	650⑮
	R	8-455	B	2	215	R46S	.040	30	.016	4	4	10	5-6	650⑯	650⑮
	T	8-455	O	4	210	R46S	.040	30	.016	—	8	22	5-6	—	650⑮
	U	8-455	O	4	230	R46SX	.080	E.I.		—	8	22	5-6	—	650⑮
	V	8-455	O	4	275	R46SX	.060	E.I.		—	14	22	5-6	—	650⑮
75	D	6-250	C	1	105	R46TX	.060	E.I.		10③	10③	16	4-5	850⑫	550⑫
	F	8-260	O	2	110	R46SX	.080	E.I.		16	18(16)⑥	22	5-6	750	650⑮
	H	8-350	B	2	145	R45TSX	.060	E.I.		—	12	19	3-6	—	600
	J	8-350	B	4	160	R45TSX	.060	E.I.		—	12③	19	3-6	—	600
	K	8-350	O	4	170	R46SX	.080	E.I.		—	20	22	5-6	—	650⑮
	R	8-400	P	2	170	R46TSX	.060	E.I.		—	16③	26	5-6	—	650
	S	8-400	P	4	185	R45TSX	.060	E.I.		—	16③	30	5-6	—	650
	T	8-455	O	4	190	R46SX	.080	E.I.		—	16	22	5-6	—	650⑮
	W	8-455	O	4	215	R46SX	.080	E.I.		—	12	22	5-6	—	650⑮
76	D	6-250	C	1	110	R46TS	.035	E.I.		6③	10③	16	4-5	850⑫	550⑳⑫
	F	8-260	O	2	110	R46SX	.080	E.I.		16⑦	18⑦	22	5-6	750	550
	H	8-350	B	2	145	R45TSX	.060	E.I.		—	12③	13.5	4-5	—	600
	J	8-350	B	4	165	R45TSX	.060	E.I.		—	12③	13.5	4-5	—	600
	R	8-350	O	4	170	R46SX	.080	E.I.		—	22(20)	22	5-6	—	650⑰⑳
	T	8-455	O	4	190	R46SX	.080	E.I.		—	16	22	5-6	—	650⑮
77	C-A	6-231	B	2	105	R46TS	.040	E.I.		12③	12③	17	4-5	800⑰	670⑰
	F	8-260	O	2	110	R46SZ	.060	E.I.		16	18	22	5-6	750	650⑮
	U	8-305	C	2	145	R45TS	.045	E.I.		8③	8③(6)	28	7-9	600	700⑭
	L	8-350	C	4	170	R45TS	.045	E.I.		—	8③(6)	28	7-9	—	650⑭
	R	8-350	O	4	170	R46SZ	.060	E.I.		—	20	13.5	5-6	—	650⑮
	K	8-403	O	4	185	R46SZ	.060	E.I.		—	20⑨	20	5-6	—	650⑮
78	A	6-231	B	2	105	R46TSX	.060	E.I.		15	15③	17	4-5	800	600
	F	8-260	O	2	110	R46SZ	.060	E.I.		—	20	14	5.5-6.5	800⑱	650
	U	8-305	C	2	145	R45TS	.045	E.I.		—	4(8)③	28	7.5-9.0	700⑰	600
	H	8-305	C	4	155	R45TS	.045	E.I.		—	4③	28	7.5-9.0	—	600⑭
	L	8-350	C	4	160	R45TS	.045	E.I.		—	8③	28	7.5-9.0	—	650
	R	8-350	O	4	160	R46SZ	.060	E.I.		—	20	16	5.5-6.5	—	650⑮
	K	8-403	O	4	190	R46SZ	.060	E.I.		—	20	16	5.5-6.5	—	650⑮
	N	8-350	O	Diesel	120*	—		—		—	—	16	㉒	—	650⑯
79	A	6-231	B	2	115	R46TSX	.060	E.I.		15③	15③	17	4.25-5.75	⑩	⑩
	2	6-231	B	2	㉒	⑩	⑩	E.I.		⑩	⑩	㉒	4.25-5.75	⑩	⑩
	F	8-260	O	2	110	R46SZ	.060	E.I.		—	20	14	5.5-6.5	⑩	⑩
	P	8-260	O	Diesel	㉒	—		—		—	—	㉒	㉒	—	565
	Y	8-301	P	2	130*	R46TSX	.060	E.I.		—	12③	27	7-8	—	650
	G	8-305	C	2	145	R45TS	.045	E.I.		4③	4③	28	7.5-9.0	600	500

165

Oldsmobile

Full-Size•• • Intermediate • Compact

TUNE-UP SPECIFICATIONS — OLDSMOBILE (EXCEPT STARFIRE)

Engine Code, 5th character of the VIN number
Model Year Code, 6th character of the VIN number

Yr.	Eng. V.I.N. Code	Engine No. Cyl. Disp. (cu in)	Eng.*** Mfg.	Carb Bbl	H.P.	SPARK PLUGS Orig. Type	Gap (in)	DIST. Point Dwell (deg)	Point Gap (in)	IGNITION** TIMING (deg BTDC) Man	Auto	VALVES■• Intake Opens (deg BTDC)	FUEL PUMP Pres. (psi)	IDLE SPEED • (rpm) Man Trans	Auto Trans (In Drive)
	H	8-305	C	4	160	R45TS	.045	E.I.		4③	4③	28	7.5-9.0	700	500
	L	8-350	C	4	170	R43TS	.045	E.I.		—	8③	28	7.5-9.0	—	⑩
	R	8-350	O	4	170	R46SZ	.060	E.I.		—	20	16	5.5-6.5	—	550
	N	8-350	O	Diesel	120*	—	—	—		—	—	16	㉒	—	575
	X	8-350	B	4	165*	R46TSX	.060	E.I.		—	15③	㉒	㉒	—	550
	K	8-403	O	4	190	R46SZ	.060	E.I.		—	20	16	5.5-6.5	—	550

Should the information provided in this manual deviate from the specifications on the underhood tune-up label, the label specifications should be used as they may reflect production changes.

** Set timing at 1100 rpm unless otherwise specified.
■ All figures are in degrees before top dead center.
* Chilton estimate
• Figure in parentheses indicates California usage. Where two figures appear separated by a slash the first is idle speed with solenoid energized, the second is idle speed with solenoid disconnected.
*** B—Buick Division
C—Chevrolet Division
O—Oldsmobile Division

P—Pontiac Division
① AT 850 RPM.
② MANUAL TRANSMISSION 44 DEGREES.
③ AT SLOW IDLE.
④ DELTA 88—12.
⑤ VISTA CRUISER—10.
⑥ CALIFORNIA OMEGA—14.
⑦ FEDERAL ONLY. FOR CALIFORNIA SEE UNDERHOOD TUNE-UP LABEL.
⑧ MANUAL TRANSMISSION—31.
⑨ NOT ADJUSTABLE ON MODELS EQUIPPED WITH ELECTRONIC SPARK TIMING (EST).
⑩ SEE UNDERHOOD TUNE-UP DECAL.
⑪ @ 850 RPM.
⑫ /425
⑬ /450

Starfire

TUNE-UP SPECIFICATIONS

Engine Code, 5th character of the VIN number
Model Year Code, 6th character of the VIN number

Yr.	Eng. V.I.N. Code	Engine No. Cyl. Disp. (cu in)	Eng.■ Mfg.	Carb Bbl	H.P.	SPARK PLUGS Orig. Type	Gap (in)	DIST. Point Dwell (deg)	Point Gap (in)	IGNITION• TIMING (deg BTDC) Man	Auto	VALVES Intake Opens (deg BTDC)	FUEL Pump Pres. (psi)	IDLE SPEED • (rpm) Man Trans	Auto Trans In Drive
75	A	4-140	C	1v	75	R43TSX	.035	Electronic		8 @ 700	10	22	3.0-4.5	1200	750
	B	4-140	C	2v	85	R43TSX	.060	Electronic		10 @ 700	12	28	3.0-4.5	700	750
	C	6-231	B	2v	110	R44SX	.060	Electronic		12	12	17	3②	800	700
	F	8-260	O	2v	110	R46SX	.080	Electronic		16 @ 1100*	⑦	14	6	750	550⑩
	G	8-262	C	2v	110	R44TX	.060	Electronic		8	8	26	7.0-8.0	800	600
76	O	4-122	C	FI	120	R43T8X	.035	Electronic		12	—	38	—	1600	—
	A	4-140	C	1v	75	R43TS	.035	Electronic		8	10	22	3.0-4.5	1200	750
	B	4-140	C	2v	85	R43TS	.035	Electronic		10	12	28	3.0-4.5	700	750
	C	6-231	B	2v	110	R44SX	.060	Electronic		12	12	17	3②	800	600
	F	8-260	O	2v	110	R46SX	.060	Electronic		16 @ 1100	⑦	14	5.5-6.5	750	550
	G	8-262	C	2v	110	R45TS	.045	Electronic		6	8	26	7.0-8.0	800	600
	Q	8-305	C	2v	140	R45TS	.045	Electronic		6	8	26	7.0-8.5	800	600
77	B	4-140	C	2v	84	R43TS	.035	Electronic		TDC @ 700⑫	2	34	3.0-4.5	1250③⑨	850③⑪
	V	4-151	P	2v	87	R44TSX	.060	Electronic		14 @ 1000*	14(12)	33	4.5-5.0	500	650
	C	6-231	B	2v	105	R46TSX	.040	Electronic		12 @ 600	12	17	3②	600	550
	U	8-305	C	2v	145	R45TS	.045	Electronic		8	8	28	7.0-8.5	600	500

Oldsmobile

Starfire

TUNE-UP SPECIFICATIONS

Engine Code, 5th character of the VIN number
Model Year Code, 6th character of the VIN number

Yr.	Eng. V.I.N. Code	Engine No. Cyl. Disp. (cu in)	Eng.■ Mfg.	Carb Bbl	H.P.	SPARK PLUGS Orig. Type	Gap (in)	DIST. Point Dwell (deg)	Point Gap (in)	IGNITION TIMING (deg BTDC) Man	Auto	VALVES Intake Opens (deg BTDC)	FUEL Pump Pres. (psi)	IDLE SPEED (rpm) Man Trans	Auto Trans In Drive
78	1	4-151	P	2v	85	R43TSX	.060	Electronic		14	14⑯	33	4-5.5	⑤	⑤
	V	4-151	P	2v	85	R43TSX	.060	Electronic		14	14	33	4-5.5	⑤	⑤
	C	6-196	B	2v	90	R46TSX	.060	Electronic		15	15	18	3-4.5⑬	⑤	⑤
	A	6-231	B	2v	105	R46TSX	.060	Electronic		15	15	17	3-4.5⑬	800	600
	U	8-305	C	2v	145	R45TS	.045	Electronic		4	6⑰	28	7.5-9.0	600	500⑮
79	1	4-151	P	2v	85	R43TSX	.060	Electronic		12	12	33	4-5.5	⑤	⑤
	V	4-151	P	2v	85	R43TSX	.060	Electronic		14	14	33	4-5.5	1000⑭	650⑭
	9	4-151	P	2v	85	R43TSX	.060	Electronic		14	14	33	4-5.5	⑤	⑤
	C	6-196	B	2v	90	R46TSX	.060	Electronic		⑤	⑤	18	3-4.5⑬	⑤	⑤
	A	6-231	B	2v	115	R46TSX	.060	Electronic		15	15	17	3-4.5⑬	800	600
	2	6-231	B	2v	115	R46TSX	.060	Electronic		15	15	17	3-4.5⑬	800	600
	G	8-305	C	2v	145	R45TS	.045	Electronic		4	4(4)	28	7.5-9.0	600	500⑮

Should the information provided in this manual deviate from the specifications on the underhood tune-up label, the label specifications should be used, as they may reflect production changes.

① THERE ARE TWO VERSIONS OF THE THIS ENGINE IN 1978, ONE FEATURES A SPECIAL CLOSED-LOOP EMISSION SYSTEM. THIS ENGINE CAN BE IDENTIFIED BY THE DEALER ORDER NUMBER UPC L36.
② 1975 SKYHAWK 3-4 PSI 1976-1977 SKYHAWK 3.0-45 PSI.
• Figures in parenthesis are California specs.
③ ADJUST SPEEDUP SOLENOID TO OBTAIN SPECIFIED RPM, WITH AIR CONDITIONING UNIT TURNED ON AND COMPRESSOR CLUTCH WIRE DISCONNECTED.
④ AT 700 OR LESS RPM
⑤ SEE UNDERHOOD TUNE-UP DECAL.
⑥ –34
⑦ (14 @ 1100)
⑧ (1200)
⑨ (800)
⑩ (600)
⑪ (700)
⑫ 2ATDC @ 700
* Timing is with both automatic and manual transmissions.
⑬ AT 12.6 VOLTS
⑭ BASE IDLE 500 RPM.
⑮ HIGH ALT. BASE IDLE IS 600 RPM
⑯ WITH EGR VALVE: 12° WITHOUT EGR VALVE
⑰ HIGH ALT. IS 8°
■ B–Buick
C–Chevrolet
O–Oldsmobile
P–Pontiac

Full-Size • Intermediate • Compact

DISTRIBUTOR SPECIFICATIONS — OLDSMOBILE (EXCEPT STARFIRE)

Year	Distributor Identification	Centrifugal Advance Start Dist. Deg. @ Dist. RPM	Finish Dist. Deg. @ Dist. RPM	Vacuum Advance Start @ In. Hg.	Finish Dist. Deg. @ In. Hg.
75	1110650	0 @ 550	8 @ 2100	3-5	6.75-8.25 @ 11.5-12.5
	1112500	0 @ 600	10 @ 2200	6-8	12.5 @ 11.5
	1112863	0 @ 550	8 @ 2100	3-5	3.25-9.75 @ 11-12.5
	1112896	0 @ 550	6 @ 2250	6.5-8.5	7 @ 11.5
	1112928	0 @ 600	8 @ 2200	6-8	12.5 @ 12
	1112936	0 @ 500	8 @ 2000	6.5	12 @ 16
	1112937	0 @ 500	6.5 @ 1800	8	18 @ 13
	1112951	0 @ 650	14 @ 2200	4	12 @ 7.5
	1112952	0 @ 500	7 @ 1800	8	9 @ 6.5
	1112953	0 @ 500	9.5 @ 2000	8	9 @ 8
	1112956	0 @ 650	14 @ 2200	—	—
	1112958	0 @ 600	8 @ 2200	5	12.5 @ 5.5

Oldsmobile

Full-Size • Intermediate • Compact

DISTRIBUTOR SPECIFICATIONS — OLDSMOBILE (EXCEPT STARFIRE)

Year	Distributor Identification	Centrifugal Advance Start Dist. Deg. @ Dist. RPM	Centrifugal Advance Finish Dist. Deg. @ Dist. RPM	Vacuum Advance Start @ In. Hg.	Vacuum Advance Finish Dist. Deg. @ In. Hg.
76	1103204	0 @ 325	14 @ 2200	6	10 @ 14.8
	1103208	0 @ 325	14 @ 2200	6	9 @ 10.2
	1103210	0 @ 500	8 @ 2000	6	12 @ 13.7
	1103211	0 @ 325	14 @ 2200	6	7 @ 9.2
	1103212	0 @ 500	6.5 @ 1800	8	14 @ 15
	1110666	0 @ 500	10 @ 2100	3-5	12 @ 14-15
	1112863	0 @ 550	8 @ 2100	6-8	12 @ 14-16
	1112936	0 @ 500	8 @ 2000	6.5	12 @ 15.8
	1112937	0 @ 500	6.5 @ 1800	8	9 @ 13
	1112952	0 @ 550	7 @ 1800	8	9 @ 13
	1112953	0 @ 500	8 @ 2000	8	9 @ 8
	1112956	0 @ 325	14 @ 2200	—	—
	1112988	0 @ 500	6.5 @ 1800	8	9 @ 13
	1112991	0-2 @ 872	8.7-11 @ 2500	6.9	9.25-10.75 @ 14.3
	1112994	0 @ 325	14 @ 2200	4.5	12 @ 10.5
	1112995	0 @ 455	13 @ 2232.5	4	15 @ 11
77	1103239	0 @ 600	10 @ 2100	14	7.5 @ 10
	1103246	0 @ 600	11 @ 2100	4	9 @ 12
	1103248	0 @ 600	10-12 @ 2100	3-5	4-6 @ 7.9
	1103259	0 @ 500	9.5 @ 2000	6	12 @ 13
	1103260	0 @ 500	6.5 @ 1800	6	12 @ 13
	1103262	0 @ 455	13 @ 2232.5	4	15 @ 11
	1103264	0 @ 500	6.5 @ 1800	5	8 @ 11
	1103266	0 @ 500	9.5 @ 2000	5	8 @ 11
	1110677	0-2.2 @ 764.5	8-11 @ 2500	6	11.25-12.75 @ 20
	1110686	0-2 @ 889.5	9-11 @ 2500	6	3.25-4.75 @ 20
78	1103281	0 @ 500	10 @ 1900	4	9 @ 12
	1103282	0 @ 500	10 @ 1900	4	10 @ 10
	1103285	0 @ 600	11 @ 2100	4	12 @ 8
	1103320	0 @ 455	13 @ 2232.5	4	15 @ 11
	1103322	0 @ 300	14.5 @ 2000	6	12 @ 13
	1103323	0 @ 500	9.5 @ 2000	5	8 @ 11
	1103324	0 @ 300	11.5 @ 1800	6	12 @ 13
	1103325	0 @ 500	6.5 @ 1800	5	8 @ 11
	1103346	0 @ 500	9.5 @ 2000	6	12 @ 13
	1103347	0 @ 500	6.5 @ 1800	6	12 @ 13
	1103353	0-1.5 @ 625	5.5-6 @ 2250	3-6	10 @ 9-12
	1103355	0 @ 455	13 @ 2232.5	4	15 @ 9
	1110695	0-3 @ 1000	6-9 @ 1800	7-9	12 @ 10-13
	1110731	0-2 @ 1000	6-9 @ 1800	4-6	8 @ 7-9
79	1103259	0 @ 500	9.5 @ 2000	6	12 @ 13
	1103260	0 @ 500	6.5 @ 1800	6	12 @ 13
	1103262	0 @ 455	13 @ 2232.5	4	15 @ 11
	1103264	0 @ 500	6.5 @ 1800	5	8 @ 11
	1103266	0 @ 500	9.5 @ 2000	5	8 @ 11
	1103320	0 @ 455	13 @ 2232.5	4	15 @ 11
	1103322	0 @ 3000	14.5 @ 2000	6	12 @ 13
	1103355	0 @ 455	13 @ 2232.5	4	15 @ 9

Oldsmobile

Full-Size • Intermediate • Compact

DISTRIBUTOR SPECIFICATIONS — OLDSMOBILE (EXCEPT STARFIRE)

Year	Distributor Identification	Centrifugal Advance Start Dist. Deg. @ Dist. RPM	Centrifugal Advance Finish Dist. Deg. @ Dist. RPM	Vacuum Advance Start @ In. Hg.	Vacuum Advance Finish Dist. Deg. @ In. Hg.
	1110695	0-3 @ 1000	6-9 @ 1800	7-9	12 @ 10-13
	1110731	0-2 @ 1000	6-9 @ 1800	4-6	8 @ 7-9
	1110766	0 @ 840	7.5 @ 1800	3.9	12 @ 10.9
	1110770	0 @ 810	7.5 @ 1800	3	10 @ 9
	1112995	0 @ 455	13 @ 2232.5	4	15 @ 11

Starfire

DISTRIBUTOR SPECIFICATIONS

Year	Distributor Identification	Centrifugal Advance Start Dist. Deg. @ Dist. RPM	Centrifugal Advance Finish Dist. Deg. @ Dist. RPM	Vacuum Advance Start @ In. Hg.	Vacuum Advance Finish Dist. Deg. @ In. Hg.
75	1112862	0 @ 810	11 @ 2400	5	12 @ 12
	1112880	0 @ 600	11 @ 2100	4	9 @ 12
	1112862	0 @ 810	11 @ 2400	5	12 @ 12
	1112863	0 @ 550	8 @ 2100	3-5	3.25-9.75 @ 11-12.5
	1110650	0 @ 550	8 @ 2100	3-5	6.75-8.25 @ 11.5-12.
	1110651	0 @ 540	8 @ 2050	5-7	9 @ 10
	1112951	0 @ 325	14 @ 2200	4	12 @ 15
	1112956	0 @ 325	14 @ 2200	NONE	NONE
76	1112862	0 @ 810	11 @ 2400	5	12 @ 12
	1112983	0 @ 600	11 @ 2000	4	7.5 @ 10
	1112862	0 @ 450	11 @ 2400	5	10 @ 14
	1110668	0 @ 638	8 @ 1600	6	12 @ 12
	1110661	0 @ 525	8 @ 2050	6	9 @ 10
	1112863	0 @ 550	8 @ 2100	3-5	12 @ 14.5
	1112863	0 @ 550	8 @ 2100	4	9 @ 12
	1110666	0 @ 500	10 @ 2100	4	12 @ 15
	1110668	0 @ 638	8 @ 1588	6	12 @ 11.5
	1110661	0 @ 530	8 @ 2050	6	9 @ 10
	1112994	0 @ 325	14 @ 2200	4.5	12 @ 10.5
	1112995	0 @ 455	13 @ 2233	4	15 @ 11
	1112996	0 @ 550	14 @ 2200	—	—
77	1110538	0 @ 425	85 @ 1000	5	12 @ 10
	1110539	0 @ 425	8.5 @ 1000	5	12 @ 10
	1103239	0 @ 600	10 @ 2100	14	7.5 @ 10
	1103244	0 @ 500	10 @ 1900	4	10 @ 10
	1103252	0 @ 500	10 @ 1900	4	9 @ 12
	1103229	0 @ 600	10 @ 2200	3.5	10 @ 12
	1103263	0 @ 600	10 @ 2200	3.5	10 @ 9
	1103231	0 @ 600	10 @ 2200	3.5	10 @ 12
	1103230	0 @ 600	10 @ 2200	3.5	10 @ 9
	1103303	0 @ 600	10 @ 2200	9	10 @ 16
	1110677	0 @ 700	10 @ 1800	4	12 @ 11
	1110686	0 @ 700	10 @ 1800	7	4 @ 9

Oldsmobile

Starfire

DISTRIBUTOR SPECIFICATIONS

Year	Distributor Identification	Centrifugal Advance		Vacuum Advance	
		Start Dist. Deg. @ Dist. RPM	Finish Dist. Deg. @ Dist. RPM	Start @ In. Hg.	Finish Dist. Deg. @ In. Hg.
78	1103281	0 @ 500	10 @ 1900	4	9 @ 12
	1103282	0 @ 500	10 @ 1900	4	10 @ 10
	1103326	0 @ 850	10 @ 2325	3.4	10 @ 10.7
	1103328	0 @ 600	10 @ 2200	3.5	10 @ 9
	1103329	0 @ 600	10 @ 2200	3.5	10 @ 12
	1103365	0 @ 850	10 @ 2325	5.5	7 @ 9.5
	1110695	0-2 @ 1000	6-9 @ 1800	6	8 @ 9
	1110731	0-2 @ 1000	6-9 @ 1800	6	8 @ 9
	1110732	0-2 @ 1000	6-9 @ 1800	9	7 @ 13
79	1103229	0 @ 600	10 @ 2200	3.5	10 @ 12
	1103231	0 @ 600	10 @ 2200	3.5	10 @ 12
	1103239	0 @ 600	10 @ 2100	4	5 @ 8
	1103244	0 @ 500	10 @ 1900	4	10 @ 10
	1103282	0 @ 500	10 @ 1900	4	10 @ 10
	1103285	0 @ 600	11 @ 2100	4	5 @ 8
	1103365	0 @ 850	10 @ 2325	5	8 @ 11
	1110726	0 @ 500	9 @ 2000	4	10 @ 10
	1110757	0 @ 600	9 @ 2000	4	10 @ 10
	1110766	0 @ 810	7.5 @ 1800	3	10 @ 9
	1110767	0 @ 840	7.5 @ 1800	4	12 @ 11

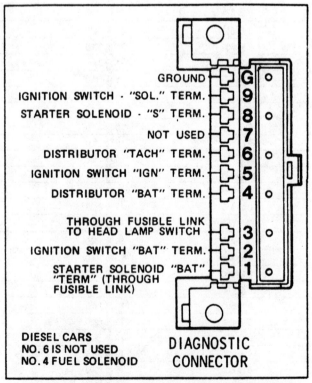

Engine electrical diagnostic connector terminals—1978 models

Oldsmobile

2MC, M2MC, M2ME CARBURETOR SPECIFICATIONS

OLDSMOBILE — Rochester Carburetor

Year	Carburetor Identification①	Float Level (in.)	Choke Rod (in.)	Choke Unloader (in.)	Vacuum Break Lean or Front (in.)	Vacuum Break Rich or Rear (in.)	Pump Rod (in.)	Choke Coil Lever (in.)	Automatic Choke (notches)
1975	7045297	3/16	0.130	0.300	0.300	0.150	9/32②	0.120	1 Rich
	7045354	3/16	0.130	0.300	0.300	0.150	5/16③	0.120	1 Rich
	7045358	3/16	0.130	0.300	0.300	0.150	5/16③	0.120	1 Rich
	7045156	5/32	0.130	0.300	0.300	0.150	9/32②	0.120	1 Rich
	7045598	5/32	0.130	0.300	0.300	0.150	3/16②	0.120	Index
	7045298	5/32	0.130	0.300	0.300	0.150	3/16②	0.120	1 Rich
	7045356	5/32	0.130	0.300	0.300	0.150	3/16②	0.120	Index
1976	17056156	1/8	0.105	0.210	0.175	0.110	9/32②	0.120	1 Rich
	17056157	1/8	0.105	0.210	0.175	0.110	3/16③	0.120	1 Rich
	17056158	1/8	0.105	0.210	0.175	0.110	9/32②	0.120	1 Rich
	17056454	1/8	0.105	0.210	0.210	0.110	3/16③	0.120	1 Rich
	17056455	1/8	0.120	0.210	0.210	0.130	9/32②	0.120	1 Rich
	17056456	1/8	0.105	0.210	0.210	0.110	3/16③	0.120	Index
	17056457	1/8	0.105	0.210	0.245	0.110	3/16③	0.120	Index
	17056458	1/8	0.105	0.210	0.210	0.110	3/16③	0.120	1 Rich
	17056459	1/8	0.105	0.210	0.210	0.110	3/16③	0.120	Index
1977	17057150, 17057151	1/8	0.085	0.190	0.160	0.090	11/32③	0.120	2 Rich
	17057157	1/8	0.090	0.190	0.190	0.100	3/8③	0.120	1 Rich
	17057156, 17057158	1/8	0.085	0.190	0.160	0.090	11/32③	0.120	1 Rich
1978	17058150	3/8	0.065	0.203	0.203	0.133	1/4②	0.120	2 Rich
	17058151	3/8	0.065	0.203	0.229	0.133	11/32③	0.120	2 Rich
	17058152	3/8	0.065	0.203	0.203	0.133	1/4②	0.120	2 Rich
	17058154	3/8	0.065	0.203	0.146	0.245	11/32③	0.120	2 Rich
	17058155	3/8	0.065	0.203	0.146	0.245	11/32③	0.120	2 Rich
	17058156	3/8	0.065	0.203	0.229	0.133	11/32③	0.120	2 Rich
	17058158	3/8	0.065	0.203	0.229	0.133	11/32③	0.120	2 Rich
	17058450	3/8	0.065	0.203	0.146	0.289	11/32③	0.120	2 Rich

MV, 1MV CARBURETOR SPECIFICATIONS

OLDSMOBILE

Year	Carburetor Identification①	Float Level (in.)	Metering Rod (in.)	Pump Rod	Idle Vent (in.)	Vacuum Break (in.)	Auxiliary Vacuum Break (in.)	Fast Idle Off Car (in.)	Choke Rod (in.)	Choke Unloader (in.)	Fast Idle Speed (rpm)
1975	Manual	11/32	0.080	—	—	0.350	0.312	—	0.275	0.275	1800②
	Automatic	11/32	0.080	—	—	0.200	0.215	—	0.160	0.275	1800②
1976	4-140 Man.	1/8	—	—	—	0.055	0.450	—	0.045	0.215	—
	4-140 Auto.	1/8	—	—	—	0.060	0.450	—	0.045	0.215	—
	6-250 Man.	11/32	—	—	—	0.165	0.320	—	0.140	0.265	—
	6-250 Auto.	11/32	—	—	—	0.140	0.265	—	0.100	0.265	—
	6-250 Calif.	11/32	—	—	—	0.150	0.260	—	0.135	0.265	—

① The carburetor identification number is stamped on the float bowl, next to the fuel inlet nut.
② Preset
③ Low step of cam.

Oldsmobile

Oldsmobile Starfire — Model 5210-C

Year	Carb. Part No. ①②	Float Level (Dry) (in.)	Float Drop (in.)	Pump Position	Fast Idle Cam (in.)	Choke Plate Pulldown* (in.)	Secondary Vacuum Break (in.)	Fast Idle Setting (rpm)	Choke Unloader (in.)	Choke Setting
1976	Manual	0.420	1	#3	0.320	0.313③	—	2200	0.375	2 Rich
	Automatic	0.420	1	#2	0.320	0.288③	—	2200	0.375	3 Rich
1977	458102, 458104	0.420	1	④	0.085	0.250	0.400	2500	0.350	3 Rich
	458103, 458105	0.420	1	④	0.120	0.250	0.400	2500	0.350	3 Rich
	458106, 458107, 458108, 458109	0.420	1	④	0.120	0.275	0.400	2500	0.400	3 Rich
	458110, 458112	0.420	1	④	0.120	0.300	0.400	2500	0.400	3 Rich
1978	see notes	.520	1	—	.150	⑨	—	⑩	.350	⑪

2GC, 2GV, 2GE CARBURETOR SPECIFICATIONS

OLDSMOBILE — Rochester Carburetor

Year	Carburetor Identification①	Float Level (in.)	Float Drop (in.)	Pump Rod (in.)	Idle Vent (in.)	Primary Vacuum Break (in.)	Secondary Vacuum Break (in.)	Automatic Choke (notches)	Choke Rod (in.)	Choke Unloader (in.)	Fast Idle Speed (rpm)
1975	7045143	15/32	1 9/32	1 19/32	—	0.140	0.120	1 Rich	0.080	0.080	Preset
	7045147	7/16	1 9/32	1 19/32	—	0.120	0.120	1 Lean	0.080	0.140	1800②
	7045149	7/16	1 9/32	1 19/32	—	0.120	0.120	1 Rich	0.080	0.140	1800②
	7045160	9/16	1 7/32	1 11/32	—	0.145	0.265	1 Rich	0.085	0.180	Preset
	7045161	9/16	1 7/32	1 11/32	—	0.145	0.265	1 Rich	0.085	0.180	Preset
	7045449	7/16	1 9/32	1 19/32	—	0.120	0.120	1 Lean	0.080	0.140	Preset
1976	17056143	15/32	1 5/32	1 11/32	—	0.140	0.100	1 Rich	0.080	0.180	—
	17056145	7/16	1 5/32	1 19/32	—	0.110	0.100	1 Rich	0.080	0.140	—
	17056149	7/16	1 5/32	1 19/32	—	0.120	0.100	1 Rich	0.080	0.140	—
	17056447	7/16	1 5/32	1 19/32	—	0.130	0.110	1 Rich	0.080	0.140	—
	17056449	7/16	1 5/32	1 19/32	—	0.130	0.110	1 Rich	0.080	0.140	—
1977	17057145, 17057146, 17057148	7/16	1 5/32	1 19/32	—	0.110	0.090	1 Rich	0.080	0.140	—
	17057143, 17057144, 17057447	7/16	1 5/32	1 19/32	—	0.130	0.100	1 Rich	0.080	0.140	—
	17057445	7/16	1 5/32	1 19/32	—	0.140	0.110	1 Lean	0.080	0.140	—
	17057446, 17057448	7/16	1 5/32	1 19/32	—	0.130	0.110	1 Rich	0.080	0.140	—
	17057104, 17057105	7/16	1 9/32	1 21/32	—	—	0.130	Index	0.260	0.325	—
	17057107, 17057109	7/16	1 9/32	1 5/8	—	—	0.130	Index	0.260	0.325	—
	17057112, 17057114	19/32	1 9/32	1 21/32	—	—	0.130	Index	0.260	0.325	—
	17057113, 17057123	19/32	1 9/32	1 5/8	—	—	0.130	Index	0.260	0.325	—
	17057404	1/2	1 9/32	1 21/32	—	—	0.140	1 Lean	0.260	0.325	—
	17057405	1/2	1 9/32	1 5/8	—	—	0.140	1 Lean	0.260	0.325	—
1978	17058102	15/32	1 9/32	1 17/32	—	0.130	—	Index	0.260	0.325	—
	17058103	15/32	1 9/32	1 17/32	—	0.130	—	Index	0.260	0.325	—
	17058104	15/32	1 9/32	1 21/32	—	0.130	—	Index	0.260	0.325	—
	17058105	15/32	1 9/32	1 21/32	—	0.130	—	Index	0.260	0.325	—
	17058107	15/32	1 9/32	1 17/32	—	0.130	—	Index	0.260	0.325	—
	17058108	19/32	1 9/32	1 21/32	—	0.130	—	Index	0.260	0.325	—
	17058109	15/32	1 9/32	1 17/32	—	0.130	—	Index	0.260	0.325	—
	17058110	19/32	1 9/32	1 21/32	—	0.130	—	Index	0.260	0.325	—

Oldsmobile

2GC, 2GV, 2GE CARBURETOR SPECIFICATIONS
OLDSMOBILE

Year	Carburetor Identification①	Float Level (in.)	Float Drop (in.)	Pump Rod (in.)	Idle Vent (in.)	Primary Vacuum Break (in.)	Secondary Vacuum Break (in.)	Automatic Choke (notches)	Choke Rod (in.)	Choke Unloader (in.)	Fast Idle Speed (rpm)
	17058111	19/32	1 9/32	1 17/32	—	0.130	—	Index	0.260	0.325	—
	17058113	19/32	1 9/32	1 17/32	—	0.130	—	Index	0.260	0.325	—
	17058121	19/32	1 9/32	1 17/32	—	0.130	—	Index	0.260	0.325	—
	17058123	19/32	1 9/32	1 17/32	—	0.130	—	Index	0.260	0.325	—
	17058126	19/32	1 9/32	1 17/32	—	0.130	—	Index	0.260	0.325	—
	17058128	19/32	1 9/32	1 17/32	—	0.130	—	Index	0.260	0.325	—
	17058140	7/16	1 5/32	1 19/32	—	0.070	0.110	1 Rich	0.080	0.140	—
	17058145	7/16	1 5/32	1 19/32	—	0.060	0.110	1 Rich	0.080	0.160	—
	17058147	7/16	1 5/32	1 19/32	—	0.100	0.140	1 Rich	0.080	0.140	—
	17058182	7/16	1 5/32	1 19/32	—	0.080	0.110	1 Rich	0.080	0.140	—
	17058183	7/16	1 5/32	1 19/32	—	0.080	0.110	1 Rich	0.080	0.140	—
	17058185	7/16	1 5/32	1 19/32	—	0.050	0.110	1 Rich	0.080	0.140	—
	17058187	7/16	1 5/32	1 19/32	—	0.080	0.110	1 Rich	0.080	0.140	—
	17058189	7/16	1 5/32	1 19/32	—	0.080	0.110	1 Rich	0.080	0.140	—
	17058404	1/2	1 9/32	1 21/32	—	0.140	—	1/2 Lean	0.260	0.325	—
	17058405	1/2	1 9/32	1 21/32	—	0.140	—	1/2 Lean	0.260	0.325	—
	17058408	21/32	1 9/32	1 21/32	—	0.140	—	1/2 Lean	0.260	0.325	—
	17058410	21/32	1 9/32	1 21/32	—	0.140	—	1/2 Lean	0.260	0.325	—
	17058444	7/16	1 5/32	1 19/32	—	0.100	0.140	1 Rich	0.080	0.140	—
	17058446	7/16	1 5/32	1 19/32	—	0.110	0.130	1 Rich	0.080	0.140	—
	17058447	7/16	1 5/32	1 19/32	—	0.110	0.150	1 Rich	0.080	0.140	—
	17058448	7/16	1 5/32	1 9/16	—	0.100	0.140	1 Rich	0.080	0.140	—

① The carburetor identification number is stamped on the float bowl, next to the fuel inlet nut.
② In Park
③ In Neutral

QUADRAJET CARBURETOR SPECIFICATIONS
OLDSMOBILE

Year	Carburetor Identification①	Float Level (in.)	Air Valve Spring (turn)	Pump Rod (in.)	Primary Vacuum Break (in.)	Secondary Vacuum Break (in.)	Secondary Opening (in.)	Choke Rod (in.)	Choke Unloader (in.)	Fast Idle Speed② (rpm)
1975	7045183	3/8	1/2	9/32	0.190	0.140	④	0.135	0.235	③
	7045250	3/8	1/2	9/32	0.250	0.180	④	0.170	0.300	③
	7045483	3/8	1/2	9/32	0.275	0.180	④	0.135	0.235	③
	7045550	3/8	1/2	9/32	0.275	0.180	④	0.135	0.235	③
	7045264	17/32	1/2	9/32	0.150	0.260	④	0.130	0.235	③
	7045184	3/8	3/4	9/32	0.190	0.140	④	0.135	0.235	③
	7045185	3/8	3/4	9/32	0.275	0.140	④	0.135	0.235	③
	7045251	3/8	3/4	9/32	0.190	0.140	④	0.135	0.235	③
	7045484	3/8	3/4	9/32	0.190	0.140	④	0.135	0.235	③
	7045485	3/8	3/4	9/32	0.190	0.180	④	0.160	0.235	③
	7045551	3/8	3/4	9/32	0.190	0.140	④	0.135	0.235	③
	7045246	5/16	3/4	3/8	0.130	0.115	④	0.095	0.240	③
	7045546	5/16	3/4	3/8	0.145	0.130	④	0.095	0.240	③

Oldsmobile

QUADRAJET CARBURETOR SPECIFICATIONS
OLDSMOBILE

Year	Carburetor Identification①	Float Level (in.)	Air Valve Spring (turn)	Pump Rod (in.)	Primary Vacuum Break (in.)	Secondary Vacuum Break (in.)	Secondary Opening (in.)	Choke Rod (in.)	Choke Unloader (in.)	Fast Idle Speed② (rpm)
1976	17056246	5/16	3/4	3/8	0.130	0.120	④	0.095	0.250	—
	17056250	13/32	1/2	9/32	0.190	0.140	④	0.130	0.230	—
	17056251	13/32	3/4	9/32	0.190	0.140	④	0.130	0.230	—
	17056252	13/32	3/4	9/32	0.190	0.140	④	0.130	0.230	—
	17056253	13/32	1/2	9/32	0.190	0.140	④	0.130	0.230	—
	17056255	13/32	3/4	9/32	0.190	0.140	④	0.130	0.230	—
	17056256	13/32	3/4	9/32	0.190	0.140	④	0.130	0.230	—
	17056257	13/32	3/4	9/32	0.190	0.140	④	0.130	0.230	—
	17056258	13/32	1/2	9/32	0.190	0.140	④	0.130	0.230	—
	17056259	13/32	1/2	9/32	0.190	0.140	④	0.130	0.230	—
	17056546	5/16	3/4	3/8	0.130	0.130	④	0.095	0.250	—
	17056550	13/32	1/2	9/32	0.190	0.140	④	0.130	0.230	—
	17056551	13/32	3/4	9/32	0.190	0.140	④	0.130	0.230	—
	17056552	13/32	3/4	9/32	0.200	0.140	④	0.130	0.230	—
	17056553	13/32	1/2	9/32	0.190	0.140	④	0.130	0.230	—
	17056556	13/32	3/4	9/32	0.190	0.140	④	0.130	0.230	—
1977	17057250, 17057252, 17057253, 17057255, 17057256	13/32	1/2	9/32	0.135	0.180	④	0.100	0.220	⑤
	17057257, 17057258, 17057550, 17057552, 17057553	13/32	1/2	9/32	0.135	0.225	④	0.100	0.220	⑤
	17057202, 17057204	15/32	7/8	9/32	0.160	—	④	0.325	0.280	⑤
	17057502, 17057504, 17057582, 17057584	15/32	7/8	9/32	0.175	—	④	0.325	0.280	⑤
1978	17058202	15/32	7/8	9/32	0.157	—	④	0.314	0.277	⑤
	17058204	15/32	7/8	9/32	0.157	—	④	0.314	0.277	⑤
	17058250	13/32	1/2	9/32	0.129	0.183	④	0.096	0.220	⑤
	17058253	13/32	1/2	9/32	0.129	0.183	④	0.096	0.220	⑤
	17058257	13/32	1/2	9/32	0.136	0.230	④	0.103	0.220	⑤
	17058258	13/32	1/2	9/32	0.136	0.230	④	0.103	0.220	⑤
	17058259	13/32	1/2	9/32	0.136	0.183	④	0.103	0.220	⑤
	17058502	15/32	7/8	9/32	0.164	—	④	0.314	0.277	⑤
	17058504	15/32	7/8	9/32	0.164	—	④	0.314	0.277	⑤
	17058553	13/32	1/2	9/32	0.136	0.230	④	0.103	0.220	⑤
	17058555	13/32	1/2	9/32	0.136	0.230	④	0.103	0.220	⑤
	17058582	15/32	7/8	9/32	0.179	—	④	0.314	0.277	⑤
	17058584	15/32	7/8	9/32	0.179	—	④	0.314	0.277	⑤

① The carburetor identification number is stamped on the float bowl, next to the secondary throttle lever.
② On low step.
③ 1800 rpm on Omega and 400 cu. in. engines with the cam follower on the highest step of the fast idle cam; 900 rpm on all others with the fast idle cam follower on the lowest step of the fast idle cam.
④ No measurement necessary on two point linkage; see text.
⑤ 3 turns after contacting lever for preliminary setting.

Oldsmobile

CAR SERIAL NUMBER AND ENGINE IDENTIFICATION

1976

Mounted behind the windshield on the driver's side is a plate with the vehicle identification number. The sixth character is the model year, with 6 for 1976. The fifth character is the engine code, as follows.

C	....	231 2-bbl. V-6 LD-7
D	....	250 1-bbl. 6-Cyl. L-22
F	....	260 2-bbl. V-8 LV 8
H	....	350 2-bbl. V-8 L-32
J	....	350 4-bbl. V-8 L-77
R	....	350 4-bbl. V-8 L-34
S	....	455 4-bbl. V-8 L-78
T	....	455 4-bbl. V-8 L-74

NOTE: *The L-32 and L-77 350 V-8's are Buick engines with the distributor in the front. The other V-8's are all Oldsmobile engines.*

Engines can be identified by a two or three-letter code. The 6-cylinder inline code is next to the distributor. The 260, 350 L-34, and 455 V-8 codes are on the oil filler tube. The 350 L-32 and L-77 V-8 codes are near No. 1 plug. The V-6 code is near the first spark plug on the right side.

CCC 250 6-cyl. Auto. Calif.
CCD 250 6-Cyl. Man. 49-States
CCF 250 6-cyl. Auto. 49-States
FH 231 V-6 Man. 49-States
FI 231 V-6 Auto. 49-States
FJ 231 V-6 Auto. Calif.
FO 231 V-6 Man. Calif.
PA 350 V-8 Auto. L-32 49-States
PB 350 V-8 Auto. L-32 49-States
PE 350 V-8 Auto. L-77 49-States
PF 350 V-8 Auto. L-77 49-States
PM 350 V-8 Auto. L-77 Calif.
PN 350 V-8 Auto. L-77 Calif.
QA 260 V-8 Man. 49-States
QB 260 V-8 Auto. 49-States
QC 260 V-8 Auto. 49-States
QD 260 V-8 Man. 49-States
QK 260 V-8 Man. 49-States
QN 260 V-8 Man. 49-States
QP 260 V-8 Auto. 49-States
QT 260 V-8 Auto. 49-States
Q2 350 V-8 Auto. L-34 49-States
Q3 350 V-8 Auto. L-34 49-States
Q5 350 V-8 Auto. L-34 49-States
Q6 350 V-8 Auto. L-34 49-States
TE 260 V-8 Auto. Calif.
TJ 260 V-8 Auto. Calif.
TL 350 V-8 Auto. L-34 Calif.
TO 350 V-8 Auto. L-34 Calif.
TP 260 V-8 Auto. Calif.
TT 260 V-8 Auto. Calif.
TW 350 V-8 Auto. L-34 Calif.
TX 350 V-8 Auto. L-34 Calif.
TY 350 V-8 Auto. L-34 Calif.
T6 350 Auto. L-34 Calif.
UB 455 V-8 Auto. 49-States
UC 455 V-8 Auto. 49-States
UD 455 V-8 Auto. 49-States
UE 455 V-8 Auto. 49-States
U3 455 V-8 Auto. 49-States
U4 455 V-8 Auto. 49-States
U5 455 V-8 Auto. 49-States
VB 455 V-8 Auto. Calif.
VD 455 V-8 Auto. Calif.
VE 455 V-8 Auto. Calif.
V3 455 V-8 Auto. Calif.
V4 455 V-8 Auto. Calif.
V5 455 V-8 Auto. Calif.

1977

The vehicle identification number is mounted on a plate behind the windshield on the driver's side. The sixth character is the model year, with 7 for 1977. The fifth character is the engine code, as follows.

C	....	231 2-bbl. V-6 LD-7	Buick
F	....	260 2-bbl. V-8 LV-8	Olds.
G	....	350 2-bbl. V-8 L-65	Chev.
K	....	403 4-bbl. V-8 L-80	Olds.
L	....	350 4-bbl. V-8 LM-1	Chev.
R	....	350 4-bbl. V-8 L-34	Olds.
U	....	305 2-bbl. V-8 LG-3	Chev.

Engines can be identified by a two or three-letter code. Oldsmobile-built engines have a 2-letter code on a paper sticker attached to the oil filler tube. Chevrolet engines have a 3-letter code stamped on the front face of the right bank, just below the head. The V-6 (a Buick-built engine) has the code on a paper sticker attached to the front of one rocker cover.

CHY 305 V-8 4-bbl. Auto. "L" 49-States
CKM 350 V-8 4-bbl. Auto. "L" Altitude
CKR 350 V-8 4-bbl. Auto. "L" Calif.
CLY 350 V-8 2-bbl. Auto. "G" 49-States
CPA 305 V-8 2-bbl. Man. 49-States
CPY 305 V-8 2-bbl. Auto. 49-States
CRL 305 V-8 2-bbl. Auto. 49-States
CRM 305 V-8 2-bbl. Auto. Calif.
  260 V-8 auto. trans. only
  403 V-8 except
  Not used on Toronado
CRS 305 V-8 2-bbl. Auto. Calif.
CRT 305 V-8 2-bbl. Auto. Altitude
CUB 350 V-8 4-bbl. Auto. "L" 49-States
Q2 350 V-8 4-bbl. Auto. "R" Altitude
Q3 350 V-8 4-bbl. Auto. "R" Altitude
Q6 350 V-8 4-bbl. Auto. "R" Altitude
Q7 350 V-8 4-bbl. Auto. "R" Altitude
Q8 350 V-8 4-bbl. Auto. "R" Altitude
Q9 350 V-8 4-bbl. Auto. "R" Altitude
QC 260 V-8 2-bbl. Auto. 49-States
QD 260 V-8 2-bbl. Auto. 49-States
QE 260 V-8 2-bbl. Auto. 49-States
QJ 260 V-8 2-bbl. Auto. 49-States
QK 350 V-8 4-bbl. Auto. "R" 49-States
QL 350 V-8 4-bbl. Auto. "R" 49-States
QN 350 V-8 4-bbl. Auto. "R" 49-States
QO 350 V-8 4-bbl. Auto. "R" 49-States
QP 350 V-8 4-bbl. Auto. "R" 49-States
QQ 350 V-8 4-bbl. Auto. "R" 49-States
QS 260 V-8 2-bbl. Man. 49-States
QT 260 V-8 2-bbl. Man. 49-States
QU 260 V-8 2-bbl. Auto. 49-States
QV 260 V-8 2-bbl. Auto. 49-States
SA 231 V-6 2-bbl. Man. 49-States
SB 231 V-6 2-bbl. Man. Calif.
SD 231 V-6 2 bbl. Auto. 49-States
SE 231 V-6 2 bbl. Auto. Calif.
SF 231 V-6 2-bbl. Auto. Altitude
SG 231 V-6 2-bbl. Man. Altitude
SI 231 V-6 2-bbl. Auto. 49-States
SK 231 V-6 2-bbl. Auto. Calif.
SL 231 V-6 2-bbl. Auto. Calif.
SM 231 V-6 2-bbl. Auto. Altitude
SN 231 V-6 2-bbl. Auto. Altitude
SU 231 V-6 2-bbl. Man. Calif.
SW 231 V-6 2-bbl. Auto. 49-States
SY 231 V-6 2-bbl. Auto. Calif.
TB 350 V-8 4-bbl. Auto. "R" Calif.
TC 350 V-8 4-bbl. Auto. "R" Calif.
TN 350 V-8 4-bbl. Auto. "R" Calif.
TO 350 V-8 4-bbl. Auto. "R" Calif.
TU 350 V-8 4-bbl. Auto. "R" Calif.
TV 350 V-8 4-bbl. Auto. "R" Calif.
TX 350 V-8 4-bbl. Auto. "R" Calif.
TY 350 V-8 4-bbl. Auto. "R" Calif.
U2 403 V-8 4-bbl. Auto. Altitude
U3 403 V-8 4-bbl. Auto. Altitude
U6 403 V-8 4-bbl. Auto. Altitude
UE 403 V-8 4-bbl. Auto. 49-States
UJ 403 V-8 4-bbl. Auto. 49-States
UK 403 V-8 4-bbl. Auto. 49-States
UL 403 V-8 4-bbl. Auto. 49-States
UN 403 V-8 4-bbl. Auto. 49-States
VA 403 V-8 4-bbl. Auto. Calif.
VB 403 V-8 4-bbl. Auto. Calif.
VE 403 V-8 4-bbl. Auto. Calif.
VJ 403 V-8 4-bbl. Auto. Calif.
VK 403 V-8 4-bbl. Auto. Calif.

1978

The vehicle identification number is mounted on a plate behind the windshield on the driver's side. The sixth character is the model year, with 8 for 1978. The fifth character is the engine code, as follows:

A	....	231 2-bbl. V-6 LD-5 Buick
F	....	260 2-bbl. V-8 LV-8 Olds
H	....	305 4-bbl. V-8 LG-4 Chev.
K	....	403 4-bbl. V-8 L-80 Olds
L	....	350 4-bbl. V-8 LM-1 Chev.
N	....	350 Diesel V-8 LF-9 Olds
R	....	350 4-bbl. V-8 L-34 Olds
U	....	305 2-bbl. V-8 LG-3 Chev.
V	....	151 2-bbl. 4-cyl. LX-6 Pont.
1	....	151 2-bbl. 4-cyl. LS-6 Pont.

The "V" 4-cylinder engine is the same as used last year on Pontiac, and this year on Monza, Phoenix, Starfire, and Sunbird. The "1" 4-cylinder engine has the electronic fuel control. Both engines are covered in the Monza chapter in this supplement.

1979

The vehicle identification number is mounted on a plate behind the windshield on the driver's side. The sixth character is the model year, with 9

175

Oldsmobile

CAR SERIAL NUMBER AND ENGINE IDENTIFICATION

1979

for 1979. The fifth character is the engine code, as follows:

- A .... 231 2-bbl. V-6 LD-5 Buick
- F .... 260 2-bbl. V-8 LV-8 Olds
- H .... 305 4-bbl. V-8 LG-4 Chev.
- K .... 403 4-bbl. V-8 L-80 Olds
- L .... 350 4-bbl. V-8 LM-1 Chev.
- N .... 350 Diesel V-8 LF-9 Olds
- R .... 350 4-bbl. V-8 L-34 Olds
- G .... 305 2-bbl. V-8 LG-3 Chev.
- V .... 151 2-bbl. 4-cyl. LX-8 Pont.
- 1 .... 151 2-bbl. 4-cyl. LS-6 Pont.
- Y .... 301 2-bbl. V-8 L-27 Pont.
- P .... 260 Disel V-8 LF-7 Olds
- X .... 350 2-bbl. V-8 L-77 Chev.

EMISSION EQUIPMENT

1976

All Models

Closed positive crankcase ventilation
Emission calibrated carburetor
Emission calibrated distributor
Heated air cleaner
Vapor control, canister storage
Exhaust gas recirculation
Catalytic converter
Single diaphragm vacuum advance
    49-States
        All engines
    Calif.
        All engines except,
        No vacuum advance on 260 V-8
Air pump
    49-States
        Not used
    Calif.
        250 6-Cylinder
Early fuel evaporation
    49-States
        250 6-Cyl.
        231 V-6
        350 2-bbl. Omega only
        350 4-bbl. Omega only
    Calif.
        250 6-Cyl.
        231 V-6
        260 V-8 codes TE, TJ, TP, and TT
        350 4-bbl. Omega only
Temperature controlled choke vacuum break
    250 6-Cyl.
    350 4-bbl. L-34 V-8
    455 4-bbl. V-8
Spark delay valve
    49-States
        Not used
    Calif.
        All 455 4-bbl. except
        Not used on Toronado
Vacuum reducer valve
    49-States
        350 Cutlass V-8 codes Q4, Q6, only
        455 V-8 except,
        Not used on Toronado
    Calif.
        Not used
Choke air modulator
    All 231 V-6
    All Omega 350 V-8
Air cleaner thermal control valve
    All 26 V-8
Spark advance vacuum modulator
    49-States
        260 V-8 auto. trans. only
    Calif.
        Not used

1977

All Models

Closed positive crankcase ventilation
Emission calibrated carburetor
Emission calibrated distributor
Heated air cleaner
Vapor control, canister storage
Exhaust gas recirculation
Catalytic converter
Single diaphragm vacuum advance
    All engines except
    Not used on Toronado
Air pump
    49-States
        Not used
    Calif.
        All models
    Altitude
        All models except,
        Not used on 403 4-bb.l V-8
        Not used on 350 4-bbl. V-8
            Codes Q2, Q3, Q6, Q7, Q8, Q9
Early fuel evaporation
    All V-6
    All 305 V-8 (Chevy)
    350 V-8 LM-1 only (Chevy) codes CHY, CKM, CKR, CLY, CUB.
    Not used on Olds-built engines
Temperature controlled choke vacuum break
    350 V-8 L-34 only. Codes: All 350 V-8 codes that begin with Q or T.
    All 403 V-8
Spark delay valve
    49-States
        Not used
    Calif.
        305 V-8 codes CRM, CRS (Chevy)
        350 V-8 L-34 only. Codes: All 350 V-8 codes that begin with T.
        All 403 V-8 except
        Not used on Toronado
    Altitude
        305 V-8 code CRT (Chevy)
Vacuum reducer valve
    49-States
        350 V-8 codes QK, QL, QN, QO, QP, QQ
    Calif.
        Not used
    Altitude
        Not used
Spark advance vacuum modulator
    49-States
    Calif.
        Not used

Distributor vacuum delay valve
    See spark delay valve
Electronic spark timing
    All Toronado
Electric choke
    231 V-6 only

1978

All Carburetor Models

Closed positive crankcase ventilation
Emission calibrated distributor
Emission calibrated carburetor
Heated air cleaner
Vapor control, canister storage
Exhaust gas recirculation
Catalytic converter
Air pump
    49-States
        Not used
    Calif.
        V-8 and V-6
    Altitude
        231 V-6
        260 V-8
        305 2-bbl. V-8 Starfire only
        350 V-8 "L"
Early fuel evaporation
    4-cyl.
    V-6
    V-8 Chev. only ("H", "L", "U")
    Not used on Olds-built engines
Temperature controlled choke vacuum break
    49-states
        231 V-6
        260 V-8
        350 V-8 "R"
        403 V-8
    Calif.
        231 V-6 auto. trans. only
        260 V-8
        350 V-8 "R"
        403 V-8
    Altitude
        231 V-6
        260 V-8
        350 V-8 "R"
        403 V-8
Distributor thermal control valve
    260 V-8 Calif. only
Spark delay valve
    305 V-8 altitude and Calif. only
Retard delay valve
    151 4-cyl. "V"
Distributor vacuum delay valve
    350 V-8 "R" Calif. only
    403 V-8 Calif. only

Diesel Models

Positive crankcase ventilation

Oldsmobile

EMISSION EQUIPMENT

1979
All Carburetor Models
Closed positive crankcase ventilation
Emission calibrated distributor
Emission calibrated carburetor
Heated air cleaner
Vapor control, canister storage
Exhaust gas recirculation
Catalytic converter
Air Pump
  49-States
    Not used
  Calif.
    V-8 and V-6
  Altitude
    231 V-6
    305 4-bbl. V-8
    350 V-8 "L"
Early fuel evaporation
  4-cyl. (Vin I)
  V-6
  V-8 Chev. only ("H" "L" "G" "X")
  V-8 Pont.
Choke thermal vacuum switch
  49-States
    260 V-8
    301 V-8
    350 V-8 "R"
    403 V-8
  Calif.
    260 V-8
    350 V-8 "R"
    403 V-8

Altitude
  260 V-8
  350 V-8 "R"
  403 V-8
Spark delay valve
  305 V-8 all except Calif. and altitude
Retard delay valve
  231 V-6 Calif.
  260 V-8 Calif.

Diesel Models
Positive crankcase ventilation

Emission Equipment Abbreviations:
AIR—Air Injection Reaction System
BP-EGR—Back Pressure Exhaust Gas Recirculation
CEAB-TVS—Cold Engine Airbleed—TVS
CP-TVS—Canister Purge Thermal Vacuum Switch
CTVS—Choke Thermal Vacuum Switch
CVB-VDV—Choke Vacuum Break-Vacuum Delay Valve
DCV—Distributor Check Valve
DTCV—Distributor Thermal Control Valve
DTVS—Distributor Thermal Vacuum Switch

DVDV—Distributor Vacuum Delay Valve
EEF—Evaporative Emission Control
EFE—Early Fuel Evaporation Valve and Actuator
EFE-CV—EFE Check Valve
EFE/TVS—EFE Thermal Vacuum Switch
EFE-DTVS—EFE Distributor Thermal Vacuum Switch
EGR—Exhaust Gas Recirculation
EGR/EFE-TVS—EGR/Early Fuel Evaporation-TVS
EGR/CP-TVS—EGR Canister Purge Thermal Vacuum Switch
EGR-DTVS—EGR Distributor Thermal Vacuum Switch
EGR-TCV—EGR Thermal Control Valve
EGR-TVS—EGR Thermal Vacuum Switch
OS—Oxygen Sensor
PCV—Postive Crankcase Ventilation
SAVM—Spark Advance Vacuum Modulator
SDV—Spark Delay Valve
SRDV—Spark Retard Delay Valve
TAC—Thermostatic Air Cleaner
TACS—Thermostatic Air Cleaner Sensor
VIS—Vacuum Input Switch
VM—Vacuum Modulator
VM-CV—Vacuum Modulator Check Valve

IDLE SPEED AND MIXTURE ADJUSTMENTS

1976

Air cleaner
  1-bbl. carb. .... On
  2-bbl. 350 V-8 .... On
  2-bbl. 260 V-8 .... Off
  All 4-bbl. .... Off
Vapor hose to carb. .... Plugged
EGR hose to carb.
  1-bbl. carb. .... Plugged
  2-bbl. 350 V-8 .... Plugged
  2-bbl. 260 V-8
    For speed adj.
      Man. trans. .... Plugged
      Auto. trans. ..... Connected
    For mixture adj. ..... Plugged
  4-bbl. carb. .... Plugged
Air cond. .... Off
Auto. trans. .... Drive
Mixture adj. .. See rpm drop below
231 V-6
  Auto. trans. 60
    Mixture adj. .... 80 rpm drop
    (680-600)
  Man. trans.
    Solenoid connected 800
    Solenoid disconnected 600
    Mixture adj. ... 100 rpm drop
    (900-800)
250 6-cyl. 49-States
  Auto. trans.
    Solenoid connected 550
    Solenoid disconnected 425

    Mixture adj. .... 25 rpm drop
    (575-550)
  Man. trans.
    Solenoid connected 850
    Solenoid disconnected 425
    Mixture adj. .... 350 rpm drop
    (1200-850)
250 6-cyl. Calif.
  Auto. trans.
    Solenoid connected 600
    Solenoid disconnected 425
    Mixture adj. .... 40 rpm drop
    (640-600)
260 V-8 49-States
  Auto. trans. 550
    Mixture adj. .... 60 rpm drop
    (610-550)
    AC idle speedup 650
  Man. trans. 750
    Mixture adj. .... 325 rpm drop
    (1075-750)
260 V-8 Calif.
  Auto. trans.
    Codes, TE, TJ, TP, TT 600
    Mixture adj. .. 100 rpm drop
    (700-600)
    Codes T2, T3, T4,
      T5, T7, T8 550
    Mixture adj. .. 100 rpm drop
    (650-550)
  Man. trans. 750
    Mixture adj. ... 325 rpm drop
    (1075-750)

350 V-8 Omega 2-bbl. & 4-bbl.
  Auto. trans. 600
    Mixture adj. .... 80 rpm drop
    (680-600)
350 V-8 Cutlass & 88
  49-States 550
    Mixture adj. .... 30 rpm drop
    (580-550)
    AC idle speedup 650
  Calif. 600
    Mixture adj. .... 25 rpm drop
    (625-600)
    AC idle speedup 650
455 V-8 49-States 550
  Mixture adj. ...... 30 rpm drop
  (580-550)
  AC idle speedup 650
455 V-8 Calif. 600
  Mixture adj. ...... 25 rpm drop
  (625-600)
  AC idle speedup 650

NOTE: *Plastic limiter caps on the mixture screws must be removed or the tabs cut off to make the idle mixture adjustment. On the 350 4-bbl. and 455 4-bbl. Oldsmobile engines (distributor in the rear) Oldsmobile recommends hooking up a vacuum gauge before setting the idle mixture. If the lean drop procedure reduces the vacuum reading more than 2" Hg. make*

Oldsmobile

IDLE SPEED AND MIXTURE ADJUSTMENTS

1976

the adjustment again but stop turning the mixture screws when the vacuum drops 2" Hg. even though the rpm may not have dropped by the full amount specified. See AC Idle Speedup Solenoid in this section for a full explanation of the difference between an idle stop solenoid and an idle speedup solenoid.

1977

231 V-6 & 305 V-8 (2GE or 2GC carbs.)
For Speed Adj.
  Air cleaner ....In place
  Air cond. ....Off
  Vapor hose to carb. .....Plugged
  Dist. vac. hose ....Plugged
  EGR hose (at EGR
    valve) ....Plugged
  Auto. trans. ....Drive
For Mixture Adj.
  Air cleaner ....Set aside
  Air cleaner vac. hose..Connected
  Level cont. comp. hose..Plugged
  Other hoses . See underhood label
  Air cond. ....Off
  Auto. trans. ....Drive
  Idle CO ....Not used
  Mixture adj. .....See rpm drop
               below

All Other V-8 (speed or mixture adj.)
  Air cleaner ....Removed
  Vac. fitting on
    manifold ....Plugged
  Air cond. ....Off
  Vapor hose to carb .....Plugged
  EGR hose (at EGR
    Valve) ....Plugged
  Auto. trans. ....Drive
  Idle CO ....Not used
  Mixture adj. .....See rpm drop
               below

231 V-6 49-States
  Auto. trans. 600
    Mixture adj. ......40 rpm drop
               (640-600)
    A.C. idle speedup 670
  Man. trans.
    Solenoid connected 800
    Solenoid disconnected 600
    Mixture adj. ......60 rpm drop
               (860-800)
    A.C. idle speedup 800

NOTE: *All 49-State manual transmission V-8 engines have solenoids. On air conditioning engines the solenoid is connected as an idle speedup, that only operates when the air conditioning is on. The curb idle speed is adjusted with the throttle screw.*

Manual transmission 49-State V-6 engines without air conditioning also have a solenoid, but it is connected as an anti-dieseling solenoid that comes on with the ignition switch. The curb idle speed is adjusted with the solenoid.

231 V-6 Calif.
  Auto. trans. 600
    Mixture adj. ......10 rpm drop
               (610-600)
    A.C. Idle speedup 670
  Man. trans.
    Solenoid connected 800
    Solenoid disconnected 600
    Mixture adj. ......10 rpm drop
               (810-800)

231 V-6 Altitude
  Auto. trans. 600
    Mixture adj. ......10 rpm drop
               (610-600)
    A.C. idle speedup 670

260 V-8 49-States
  Auto. trans. 550
    Mixture adj. .....60 rpm drop
               (610-550)
    A.C. idle speedup 650
  Man. trans. 750
    Mixture adj. .....325 rpm drop
               (1075-750)

305 2-bbl. V-8 49-States
  Auto. trans. 500
    Mixture adj. ......30 rpm drop
               (530-500)
    A.C. Idle speedup
      Starfire 700
      Omega 650
  Man. trans. 600
    Mixture adj. ......50 rpm drop
               (650-600)
    A.C. Idle speedup 650

305 2-bbl. V-8 Calif.
  Auto. trans. 500
    Mixture adj. ......30 rpm drop
               (530-500)
    A.C. Idle speedup 700

305 2-bbl. V-8 Altitude
  Auto. trans. 600
    Mixture adj. ......30 rpm drop
               (630-600)
    A.C. Idle speedup 700

350 2-bbl. V-8 49-States
  Auto. trans. 500
    Mixture adj. ......30 rpm drop
               (530-500)
    A.C. Idle speedup 650

350 4-bbl. V-8 "L" 49-States
  Auto. trans. 500
    Mixture adj. ......50 rpm drop
               (550-500)
    A.C. Idle speedup 650

350 4-bbl. V-8 "L" Calif.
  Auto. trans. 500
    Mixture adj. ......50 rpm drop
               (550-500)
    A.C. Idle speedup 650

350 4-bbl. V-8 "L" Altitude
  Auto. trans. 600
    Mixture adj. ......50 rpm drop
               (650-600)
    A.C. Idle speedup 650

350 4-bbl. V-8 "R" 49-States
  Auto. trans. 550
    Mixture adj. ......30 rpm drop
               (580-550)
    A.C. Idle speedup 650

350 4-bbl. V-8 "R" Calif.
  Auto. trans. 550
    Mixture adj. ......25 rpm drop
               (575-550)
    A.C. Idle speedup 650

350 4-bbl. V-8 "R" Altitude
  Auto. trans. 600
    Mixture adj. ......25 rpm drop
               (625-600)
    A.C. Idle speedup 700

403 4-bbl. V-8 49-States
  Auto. trans. 550
    Mixture adj. ......30 rpm drop
               (580-550)
    A.C. Idle speedup 650

403 4-bbl. V-8 Calif. except Toronado
  Auto. trans. 550
    Mixture adj. ......25 rpm drop
               (575-500)
    A.C. Idle speedup 650

403 4-bbl. V-8 Calif. Toronado
  Auto. trans. 600
    Mixture adj. ......25 rpm drop
               (625-600)
    A.C. Idle speedup 650

403 4-bbl. V-8 Altitude
  Auto. trans. 600
    Mixture adj. ......25 rpm drop
               (625-600)
    A.C. Idle speedup 700

1978

Air cleaner .... In place
Air cond. .... Off
Auto. trans. .... Drive
Idle CO .... Not used
Mixture adj. . See propane speed below
Vacuum hoses .... See engine label
151 4-cyl. with air cond.
  Auto. trans. 650
    Propane enriched 680-700
    AC idle speedup 850
  Man. trans. 1000
    Propane enriched 1140-1160
    AC idle speedup 1200
151 4-cyl. no. air cond.
  Auto. trans.
    Solenoid connected 650
    Solenoid disconnected 500
    Propane enriched 680-700
  Man. trans.
    Solenoid connected 1000
    Solenoid disconnected 500
    Propane enriched 1140-1160
NOTE: *The mixture on the 4-cyl. with Electronic Fuel Control (VIN 1) must be adjusted to 2-4 in. Hg. See procedure under Electronic Fuel Control in Monza section.*
231 V-6 49-States
  Auto. trans. 600
    Propane enriched 650
    AC idle speedup 670
  Man. trans.
    Solenoid connected 800
    Solenoid disconnected 600
    Propane enriched 940
231 V-6 Calif.
  Auto. trans. 600
    Propane enriched 615
    AC idle speedup 670

Oldsmobile

IDLE SPEED AND MIXTURE ADJUSTMENTS

1978

Man. trans.
  Solenoid connected 800
  Solenoid disconnected 600
  Propane enriched 880
231 V-6 Altitude
  Auto. trans. 600
    Propane enriched 615
    AC idle speedup 670
  Man. trans.
    Solenoid conneted 800
    Solenoid disconnected 600
    Propane enriched 880

NOTE: The air conditioning idle speedup solenoid is not used on some air conditioned cars. AC idle speedup figures above apply only to those cars equipped with the solenoid.

260 V-8 auto. trans. 500
  Propane enriched
    49-States 560-580
    Calif. 530-550
    Altitude ....
    AC idle speedup 650
260 V-8 man. trans. 650
  Propane enriched 780-800
  AC idle speedup 800
305 2-bbl. V-8 "U" 49-States
  Auto. trans. 500
    Propane enriched 520-540
    AC idle speedup
      Omega 600
      Starfire 700
  Man. trans. 600
    Propane enriched 700-740
    AC idle speedup 700
305 2-bbl. V-8 "U" Calif.
  Auto. trans. 500
    Propane enriched 520-540
    AC idle speedup
      Cutlass 650
      Starfire 700
305 2-bbl. V-8 "U" Altitude
  Auto. trans. 600
    Propane enriched 620-640
    AC idle speedup 700
305 4-bbl. V-8 "H" 49-States
  Auto. trans. 500
    Propane enriched 540-560
    AC idle speedup 600
350 4-bbl. V-8 "L" Calif.
  Auto. trans. 500
    Propane enriched 530-570
    AC idle speedup 600
350 4-bbl. V-8 "L" Altitude
  Auto. trans. 600
    Propane enriched 630-670
    AC idle speedup 650
350 4-bbl. V-8 "R" 49-States
  Auto. trans. 550
    Propane enriched 625-645
    AC idle speedup 650
350 4-bbl. V-8 "R" Calif.
  Auto. trans. 550
    Propane enriched 565-585
    AC idle speedup 650
350 4-bbl. V-8 "R" Altitude
  Auto. trans. 600
    Propane enriched ....
    AC idle speedup 700

350 Diesel V-8
  Auto. trans. 575
    Solenoid (fast idle) 650
    Mixture adj. .... None
403 V-8 49-States
  Auto. trans. 550
  Propane enriched
    Toronado 605-625
    All others 625-645
    AC idle speedup 650
403 V-8 Calif.
  Auto. trans.
    Toronado 600
    All others 550
  Propane enriched
    Toronado 605-625
    All others 565-585
  AC idle speedup 650
403 V-8 Altitude
  Auto. trans. 600
  Propane enriched ....
  AC idle speedup 700

1979

NOTE: *Refer to Vehicle Emission Information Label before making any adjustments.*

Air cleaner .... In place
Air cond. .... Off
Auto. trans. .... Drive
Idle CO .... Not used
Mixture adj. See propane speed below
Vacuum hoses ..... See engine label
151 4-cyl. 49-States Vin code (V)
  Auto. trans. and AC
    Curb or "on" idle 850
    Slow or "off" idle 650
    Propane enriched idle 695
  Manual trans. and AC
    Curb or "on" idle 1250
    Slow or "off" idle 900
    Propane enriched idle ... 1040
  Auto. trans. no AC
    Curb or "on" idle 650
    Slow or "off" idle 500
    Propane enriched idle 695
  Manual trans. no AC
    Curb or "on" idle 900
    Slow or "off" idle 500
    Propane enriched idle ... 1040
151 4-cyl. Vin code (1)
  Auto. trans. AC
    Curb or "on" idle 850
    Slow or "off" idle 650
    Propane enriched idle .....
  Manual trans. AC
    Curb or "on" idle 1200
    Slow or "off" idle 1000
    Propane enriched idle .....
  Auto. trans. no AC
    Curb or "on" idle 650
    Slow or "off" idle 500
    Propane enriched idle .....
  Manual trans. no AC
    Curb or "on" idle 1000
    Slow or "off" idle 500
    Propane enriched idle .....
231 V-6 49-States Vin code (A)
  Auto. trans.
    Curb or "on" idle 670
    Slow or "off" idle 550

    Propane enriched idle 575
  Manual trans.
    Curb or "on" idle 800
    Slow or "off" idle 600
    Propane enriched idle 1000
231 V-6 California Vin code (A)
  Auto trans.
    Curb or "on" idle .... none
    Slow or "off" idle 600
    Propane enriched idle 615
  Manual trans.
    Curb or "on" idle 800
    Slow or "off" idle 600
    Propane enriched idle 840
231 V-6 Altitude Vin code (A)
  Auto trans.
    Curb or "on" idle .... none
    Slow or "off" idle 600
    Propane enriched idle 615
260 V-8 49-States Vin code (F)
  Auto. trans.
    Curb or "on" idle 625
    Slow or "off" idle 500
    Propane enriched idle . .560-580
  Manual trans.
    Curb or "on" idle 800
    Slow or "off" idle 650
    Propane enriched idle . .780-800
260 V-8 California Vin code (F)
  Auto. trans.
    Curb or "on" idle 625
    Slow or "off" idle 500
    Propane enriched idle . .530-550
60 V-8 Altitude Vin code (F)
  Auto. trans.
    Curb or "on" idle 650
    Slow or "off" idle 550
    Propane enriched idle 575
260 Diesel All Vin code (P)
  Auto. trans.
    Curb or "on" idle 650
    Slow or "off" idle 590
  Manual trans.
    Curb or "on" idle 659
    Slow or "off" idle 575
301 V-8 All Vin code (Y)
  Auto. trans.
    Curb or "on" idle 650
    Slow or "off" idle 500
    Propane enriched idle 530
305 V-8 49-States Vin code (G)
  Auto trans.
    Curb or "on" idle 600
    Slow or "off" idle 500
    Propane enriched idle . .520-540
  Manual trans.
    Curb or "on" idle .... none/
                           700 Omega
    Slow or "off" idle ..... 600 (N)
    Propane enriched idle . .710-750
305 V-8 California Vin code (G)
  Auto. trans.
    Curb or "on" idle 650
    Slow or "off" idle ..... 600 (N)
    Propane enriched idle . .640-660
305 V-8 4-bbl. 49-States Vin (H)
  Auto. trans.
    Curb or "on" idle 600
    Slow or "off" idle 500
    Propane enriched idle . .530-570

179

Oldsmobile

IDLE SPEED AND MIXTURE ADJUSTMENTS

1979

Manual trans.
  Curb or "on" idle ....none
  Slow or "off" idle 700
  Propane enriched idle ..800-850
305 V-8 4-bbl. California Vin (H)
  Auto. trans.
  Curb or "on" idle 600
  Slow or "off" idle 600
  Propane enriched idle ..525-560
305 V-8 4-bbl. Altitude Vin code (H)
  Auto. trans.
  Curb or "on" idle 650
  Slow or "off" idle 600
  Propane enriched idle ..630-670
350 V-8 California Vin code (L)
  Auto. trans.
  Curb or "on" idle 600
  Slow or "off" idle 500
  Propane enriched idle ..525-560
350 V-8 Altitude Vin code (L)
  Auto. trans.
  Curb or "on" idle 600
  Slow or "off" idle 500
  Propane enriched idle ..630-670

350 Diesel Vin code (N)
  Auto. trans.
  Curb or "on" idle 650
  Slow or "off" idle 575
350 V-8 49-States Vin code (R)
  Auto. trans.
  Curb or "on" idle 650
  Slow or "off" idle 550
  Propane enriched idle .625-645/
    Toronado 585-590
350 V-8 California Vin code (R)
  Auto. trans.
  Curb or "on" idle 600
  Slow or "off" idle 500
  Propane enriched idle ..565-585
350 V-8 Altitude Vin code (R)
  Auto. trans.
  Curb or "on" idle 650
  Slow or "off" idle 550
  Propane enriched idle 590
403 V-8 49-States Vin code (K)
  Auto. trans.
  Curb or "on" idle 650
  Slow or "off" idle 550
  Propane enriched idle ..625-645

403 V-8 California Vin code (K)
  Auto. trans.
  Curb or "on" idle 600
  Slow or "off" idle 500
  Propane enriched idle ..565-585
403 V-8 Altitude Vin code (K)
  Auto. trans.
  Curb or "on" idle 650
  Slow or "off" idle 550
  Propane enriched idle 590

NOTE: *One cars equipped with automatic transmission the idle adjustments are made with the transmission selector lever in the (DR) range, unless otherwise noted.*

On cars equipped with manual transmission adjustments are made with transmission selector lever in the (N) range, unless otherwise noted.

When making Curb or "on" idle adjustments turn the idle solenoid screw in or out to adjust RPM (AC on) clutch wires disconnected and wheels blocked. w/o AC, idle stop solenoid energized—if used.

INITIAL TIMING

1976

NOTE: *Distributor vacuum hose must be disconnected and plugged. Set timing at idle speed unless shown otherwise. Set automatic transmission engines with the transmission in Neutral, unless shown otherwise.*

231 V-6 .... 12° BTDC
250 6-cyl.
  Man. trans. .... 6° BTDC
  Auto. trans. .... 10° BTDC
260 V-8 (At 1100 rpm)
  49-States
  Auto. trans. .... 18° BTDC
  Man. trans. .... 16° BTDC
  Calif.
  Auto Trans.
    Omega codes TE,
      TJ .... 14° BTDC
    Omega codes T2,
      T3, T7, T8 .... 16° BTDC
    Cutlass codes TP,
      TT .... 16° BTDC
    Cutlass codes T4,
      T5 .... 14° BTDC
  Man. trans. .... 14° BTDC
350 V-8 Omega .... 12° BTDC
350 V-8 Cutlass & 88 (1100 rpm)
  All except codes Q4,
    Q6 .... 20° BTDC
  Codes Q4, Q6 ...... 22° BTDC
455 V-8 49-States (1100 rpm)
  Cutlass, 88 .... 16° BTDC
  98 codes UC, U3, U5 . 16° BTDC
  98 codes U6, U7, U8 .. 18° BTDC
  Toronado .... 14° BTDC
455 V-8 Calif. (110 rpm)
  Cutlass, 88, 98 ..... 16° BTDC
  Toronado .... 12° BTDC

1977

NOTE: *Distributor vacuum hose must be disconnected and plugged. Set timing at idle speed unless shown otherwise. Set automatic transmission engines with the transmission in Neutral, unless shown otherwise.*

231 V-6 .... 12° BTDC
260 V-8 (At 1100 rpm)
  49-States auto. trans.
    Cutlass .... 18° BTDC
    Omega .... 20° BTDC
    88 .... 20° BTDC
  49-States man. trans. ...16° BTDC
305 2-bbl. V-8
  49-States .... 8° BTDC
  Calif. .... 6° BTDC
  Altitude .... 8° BTDC
350 2-bbl. V-8 "G" ...... 8° BTDC
350 4-bbl. V-8 "L"
  49-States .... 8° BTDC
  Calif. .... 6° BTDC
  Altitude .... 8° BTDC
350 4-bbl. V-8 "R" (At 1100 rpm)
  49-States ....20° BTDC
  Calif.
    Cutlass ....20° BTDC
    88 wagon ....20° BTDC
    98 ....20° BTDC
    Omega ....18° BTDC
    88 except wagon ....18° BTDC
  Altitude ....20° BTDC
403 4-bbl. V-8 (At 1100 rpm)
  49-States
    Cutlass wagon ......22° BTDC
    All others ....20° BTDC
  Calif. ....20° BTDC
  Altitude ....20° BTDC

NOTE: *A special procedure is used to set the timing on the Toronado. Do not turn the distributor.*

1978

NOTE: *Distributor vacuum hose must be disconnected and plugged. Set timing at idle speed unless shown otherwise. Set automatic transmission engines with the transmission in Neutral, unless shown otherwise.*

151 4-cyl. (at 1000 rpm) ..... 14° BTDC
231 V-6 .... 15° BTDC
260 V-8 (at 1100 rpm)
  49-States auto. trans. .... 20° BTDC
  49-States man. trans. .... 18° BTDC
  Calif. auto. trans. .... 18° BTDC
  Altitude auto. trans. ..... 18° BTDC
305 2-bbl. V-8 "U"
  49-States
  Calif.
    Starfire .... 6° BTDC
    Cutlass .... 8° BTDC
  Altitude .... 8° BTDC
305 4-bbl. V-8 "H"
  49-States .... 4° BTDC
350 4-bbl. V-8 "L" ...... 8° BTDC
350 4-bbl. V-8 "R"
  (at 1100 rpm) .... 20° BTDC
350 Diesel V-8 ..... Align pump marks
    with engine dead
403 V-8 (at 1100 rpm)
  49-States
    88 wagon .... 20° BTDC
    88 coupe & sedan ...... 18° BTDC
    98 .... 18° BTDC
    Toronado .... 20° BTDC
  Calif.
    Toronado .... 22° BTDC
    All others .... 20° BTDC
  Altitude .... 20° BTDC

Oldsmobile

INITIAL TIMING

1978
NOTE: A special procedure (different from 1977) is used to set the timing on the Toronado. See Toronado Electronic Spark Timing, in this section.

1979
NOTE: Engine timing information is noted on the Vehicle Emission Information Label by the manufacture. Refer to this information before any adjustments are made.

151 4-cyl.	14° BTDC
231 V-6	15° BTDC
260 V-8 (at 1100 rpm)	
49-States auto. trans.	20° BTDC
49-States man. trans.	18° BTDC
Calif. auto. trans.	18° BTDC
Altitude auto. trans.	18° BTDC
301 V-8	12° BTDC
305 2-bbl. V-8 "G"	
49-States	4° BTDC
Calif.	
Starfire	2° BTDC
305 4-bbl. V-8 "H"	
49-States	4° BTDC
Altitude	8° BTDC
350 4-bbl. V-8 "L"	8° BTDC
350 4-bbl. V-8 "R"	
(at 1100 rpm)	20° BTDC
260 & 350 Diesel V-8	Align pump marks with engine stopped
403 V-8	20° BTDC

SPARK PLUGS

1976

231 V-6	AC-R44SX	.060
250 6-Cyl.	AC-R46TS	.035
260 V-8	AC-R46SX	.080
	AC-R46SZ	.060
350 V-8 Omega	AC-R45TSX	.060
350 V-8 Cutlass & 88	AC-R46SX	.080
	AC-R46SZ	.060
455 V-8	AC-R46SX	.080
	AC-R46SZ	.060

NOTE: Oldsmobile-built engines may come with either "X" or "Z" plugs (last letter) as shown above, but not mixed on the same engine. The "X" plug must be gapped at .080", and the "Z" plug at .060". This does not apply to the 231 V-6, 250 6-cyl. and 350 V-8 Omega, which are not Olds-built engines.

1977

231 V-6	AC-R46TS	.040
231 V-6	AC-R46TSX	.060
260 V-8	AC-R46SZ	.060
305 2-bbl.	AC-R45TS	.045
350 2-bbl.	AC-R45TS	.045
350 4-bbl.		
"L"	AC-R45TS	.045
"R"	AC-R46SZ	.060
403 4-bbl.	AC-R46SZ	.060

NOTE: "L" and "R" refer to the engine code letter in the vehicle identification number. The V-6 engine was originally equipped with TS plugs, but later production used the TSX plugs.

1978

151 4-cyl.	AC-R43TSX	.060
231 V-6	AC-R46TSX	.060
260 V-8	AC-R46SZ	.060
305 2-bbl.	AC-R45TS	.045
305 4-bbl.	AC-R45TS	.045
350 4-bbl.		
"L"	AC-R45TS	.045
"R"	AC-R46SZ	.060
403 4-bbl.	AC-R46SZ	.060

1979

151 4-cyl.	AC-R43TSX	.060
231 V-6	AC-R46TSX	.060
260 V-8	AC-R46SZ	.060
301 2-bbl.	AC-R46TSX	.060
305 2-bbl.	AC-R45TS	.045
305 4-bbl.	AC-45TS	.045
350 4-bbl.		
"L"	AC-R45TS	.045
"R"	AC-R46SZ	.060
"X"	AC-R46TSX	.060
403 4-bbl.	AC-R46SZ	.060

VACUUM ADVANCE

1976

250 6-cyl.	
49-States	
Calif.	
231 V-6	Ported
260 V-8	
49-States	
Auto. trans.	Modulated
Man. trans.	Ported
Calif.	None
350 V-8 Omega	Ported
350 V-8 Cutlass & 88	
49-States	Manifold
Calif.	Ported
455 V-8 Cutlass, 88, 98	
49-States	Manifold
Calif.	Ported
455 V-8 Toronado	Manifold

NOTE: The California Toronado uses manifold vacuum for the vacuum advance only when the engine is overheated enough to open the distributor thermal vacuum switch. At normal operating temperatures there is no vacuum advance. This also applies to 1975 California Toronados.

1977

Diaphragm type .... Single
Vacuum source

231 V-6	
49-States	
Early production	Manifold
Late production	Ported
Calif.	Ported
Altitude	Ported
260 V-8 49-States	
Auto. trans.	Modulated
Man. trans.	Ported
305 V-8	Manifold
350 2-bbl. V-8	Manifold
350 V-8 "L"	Manifold
350 V-8 "R"	
49-States	Manifold
Calif.	Ported
Altitude	Ported
403 V-8 except Toronado	
49-States	Modulated
Calif.	Ported
Altitude	Ported
403 V-8 Toronado	
49-States	Manifold
Calif.	Ported
Altitude	Manifold

1978

Diaphragm type .... Single
Vacuum source

151 4-cyl.	Ported
231 V-6	
49-States	Manifold
Calif.	Ported
Altitude	Manifold
260 V-8 49-States	
Auto. trans.	Modulated
Man. trans.	Manifold
260 V-8 Calif.	Ported
260 V-8 Altitude	Modulated
305 2-bbl. V-8	
49-States	Manifold
Calif.	Ported
Altitude	Manifold
305 4-bbl. V-8	Manifold
350 4-bbl. V-8 "L"	Manifold
350 4-bbl. V-8 "R"	
49-States	Manifold
Calif.	Ported
Altitude	Ported
403 V-8 except Toronado	
49-States	Manifold
Calif.	Ported
Altitude	Ported
403 V-8 Toronado	
49-States	Manifold

Oldsmobile

VACUUM ADVANCE

1978
Calif. .... Ported
Altitude .... Manifold
NOTE: The advance vacuum hose on the Toronado connects to the EST controller.

1979
Diaphragm type .... Single
Vacuum source
  151 4-cyl. .... Ported

231 V-6
  49-State .... Manifold
  Calif. .... Ported
  Altitude .... Manifold
260 V-8 49-States
  Auto. trans. .... Modulated
  Man. trans. .... Manifold
260 V-8 Calif. .... Ported
260 V-8 Altitude ..... Modulated
305 2-bbl. V-8
  49-States .... Manifold
  Calif. .... Ported

Altitude .... Manifold
305 4-bbl. V-8 .... Manifold
350 4-bbl. V-8 "L" ..... Manifold
350 4-bbl. V-8 "R"
  49-States .... Manifold
  Calif. .... Ported
  Altitude .... Ported
403 V-8
  49-States .... Manifold
  Calif. .... Ported
  Altitude .... Ported

EARLY FUEL EVAPORATION 6-CYLINDER, 1976

See the Chevrolet chapter in this supplement.

AIR CONDITIONING IDLE SPEEDUP SOLENOID

Description of Solenoid

The solenoids used on carburetors for several years have been known as idle stop solenoids, or anti-dieseling solenoids. They are connected to the ignition switch so that they are energized whenever the switch is on. Curb idle speed with that type of solenoid is set with the solenoid adjustment.

The idle speedup solenoid is the same solenoid, mounted in the same place on the carburetor, but hooked up so that it is energized only when the air conditioning is on. When the idle speedup solenoid is used, the curb idle adjustment is made with the throttle screw.

To find out if a solenoid is hooked up as an anti-dieseling solenoid or as an idle speedup solenoid, all you have to do is open the throttle, hold your finger on the end of the solenoid plunger, and have someone else turn the ignition on, with the air conditioning off. If the plunger pops out as you relax the pressure with your finger, it's an ordinary anti-dieseling solenoid. On some cars you can make this test without even opening the hood, because the throttle linkage is sensitive enough to feel the solenoid open as you push on the gas pedal.

Testing Solenoid

With the ignition switch on, the solenoid stem should extend whenever the air conditioning is turned on. To check the actual idle speed with the air conditioning on, the compressor clutch wires must be disconnected at the compressor. This eliminates the problem of the compressor cycling on and off while you are adjusting the idle. With the solenoid energized, adjust the idle speedup speed to specifications, which is 650 rpm for all engines.

Repairing Solenoid

Repairs are limited to replacement.

SPARK ADVANCE VACUUM MODULATOR (SAVM) 1976

Description of Modulator

The SAVM has hoses connected to it from the distributor vacuum advance unit, manifold vacuum, and carburetor ported vacuum. The SAVM allows either ported vacuum, manifold vacuum, or 7" Hg. vacuum to operate the vacuum advance. At idle, when manifold vacuum in the engine is high, but ported vacuum is non-existent, the SAVM allows only 7" Hg. vacuum to go to the distributor. As the throttle is opened, and the ported vacuum increases, the distributor continues to receive 7" Hg. until the ported vacuum rises to 7" Hg. After the ported vacuum goes over 7" Hg. the SAVM switches the distribu-

1976 spark advance vacuum modulator

EMISSION CONTROL SYSTEMS 1976

tor over to ported vacuum only, whatever it may be. If both the ported vacuum and the manifold vacuum are below 7" Hg. as happens near wide open throttle, the distributor operates on manifold vacuum only.

In other words, at idle the distributor gets 7" Hg, at part throttle it gets ported vacuum, and near full throttle it gets manifold vacuum. This system gives more spark advance for better running, but not so much advance that emissions go up.

Testing Modulator

1. Connect a vacuum gauge to the nozzle marked "distributor," and a hand vacuum pump to the nozzle marked "intake manifold." Slowly pump up the vacuum. The gauge should equal the pump vacuum up to 7" Hg. As the pump goes on up to 15" Hg. or more, the gauge should stay at 7" Hg.

2. Switch the pump to the nozzle marked "carburetor" and apply vacuum. The gauge should stay at zero until the pump output reaches 7" Hg. At that point the gauge should show the same vacuum as the pump, and it should continue to show the same vacuum as the pump output rises to 15" Hg. and beyond.

3. Switch the hoses so the vacuum pump is connected to the distributor nozzle, and the vacuum gauge connected to the carburetor nozzle. Pump up several inches of vacuum. The vacuum gauge should stay at zero. If not, the modulator is leaking.

The modulator must pass all three tests.

Repairing Modulator

Repairs are limited to replacement. The modulator part number is 553952.

Oldsmobile

EMISSION CONTROL SYSTEMS 1977

DISTRIBUTOR VACUUM DELAY VALVE 1977

This is a spark delay valve. It is used on ported vacuum systems, and delays the application of vacuum advance for as much as 30 seconds, depending on throttle opening. When the ported vacuum drops off, as during idle or wide open throttle, an internal check valve opens to release the trapped vacuum and allow the distributor advance to return to the no-advance position.

Testing Distributor Vacuum Delay Valve

Remove the valve from the car and connect a vacuum gauge to the DTVS nozzle and a hand vacuum pump to the CARB nozzle. Pump up vacuum and there should be a slight delay in getting a reading on the separate vacuum gauge. Stop pumping and it should take 3 or 4 seconds for the pump gauge and the separate gauge to balance with equal readings. Remove the separate gauge and hold your finger over the DTVS nozzle. Pump up 15 in. Hg. vacuum. The vacuum should hold without leaking. Remove your finger and the pump gauge reading should drop slowly. If the valve does not work exactly as described, it is defective.

Repairing Distributor Vacuum Delay Valve

Repairs are limited to replacement. The CARB nozzle connects to the source of ported vacuum. The DTVS nozzle connects to the Distributor Thermal Vacuum Switch.

ELECTRIC CHOKE 1977

V-6 engines now use the 2GE car-

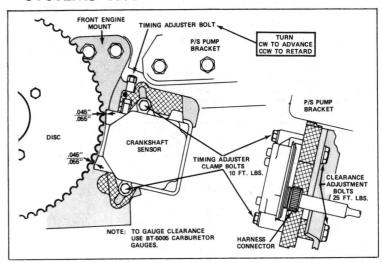

Sensor to disc clearance (© GM Corporation)

buretor, which has an electric choke. This choke is similar to the choke used on other GM cars, such as Chevette, Cadillac, and the discontinued 454 Chevrolet engine. The choke receives current from the oil pressure switch in all cars, so that it is only being heated while the engine is running. A minor difference occurs when the V-6 is used in the Starfire. The oil pressure switch supplies current to the electric fuel pump when the engine is either cranking or running. The choke is connected to the fuel pump circuit, so it also receives current when the engine is either cranking or running. Other than that, the only thing that changes between car models is the colors of the wires, in some cases.

TORONADO ELECTRONIC SPARK TIMING 1977

In the 1977 Toronado distributor, all the timing mechanism has been removed. There is no vacuum advance, no centrifugal advance, and no pick up coil or pole piece. The distributor is simply a rotor and cap. The timing of the spark is taken care of by an electronic controller assembly mounted under the glove box. The electronic controller knows the position of the engine at all times because it is connected to a crankshaft sensor mounted on the front of the block. The sensor is positioned next to a toothed wheel mounted behind the crankshaft front pulley. At four positions on the pulley, one of the teeth is a slightly different size, and the sensor knows whenever it goes by. In effect, the toothed wheel and sensor have taken the place of the pick up coil and pole piece that are inside the normal HEI distributor.

The controller also receives signals from a temperature sensor mounted on top of the engine. This is a special sensor that is not the same as the older HOT light sensor, but accomplishes the same thing. It sends temperature information to the controller, and the controller turns on the HOT light if the engine overheats.

Two hoses are connected to the controller. One senses the atmospheric pressure under the hood. The hose just ends in the harness so its open end is in the engine compartment. This is necessary so the controller is not sensitive to passenger compartment pressure, which can be considerably higher than atmospheric because of the movement of the car and the ventilation blowers.

The second hose connects to engine manifold vacuum, and senses engine load in the same way that the vacuum advance on a distributor would.

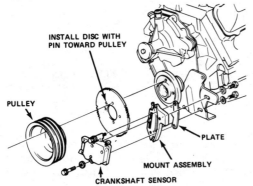

Crankshaft Position (CP) Sensor (© GM Corporation)

Oldsmobile

ENGINE TIMING ADVANCE IN CRANKSHAFT DEGREES

1. Engine **MUST BE** at operating temperature.
2. Connect tachometer and timing meter.
3. Disconnect controller assembly vacuum tube (white) from manifold vacuum "T". Plug "T" and connect vacuum pump and gauge to white tube. Use pump to get vacuum readings shown.

NOTE: 1400 RPM at 15" of vacuum gives maximum advance. If advance is less than specification, slowly increase vacuum to get maximum advance. If advance is now within specification, instruments/gauges used are inaccurate. If not within specification, replace controller assembly.

EXCEPT CALIFORNIA CARS			CALIFORNIA CARS		
ENGINE RPM	VACUUM (INCHES)	*CRANKSHAFT DEGREES	ENGINE RPM	VACUUM (INCHES)	*CRANKSHAFT DEGREES
600	16.5	29 to 34	600	13.5	17 to 20
600	15	27 to 31	600	12	17 to 20
1000	12	29 to 37	1000	12	19 to 22
1400	15	44 to 61	1400	15	44 to 61
1400	10.5	37 to 41	2000	18	31 to 49
1400	6	25 to 35	2000	0	31 to 34
2000	18	44 to 53			
2000	0	28 to 31			

*Advance specifications are approximate depending upon accuracy of tachometer, vacuum gauge and timing meter.

Engine vacuum timing—Toronado (© GM Corporation)

While the engine is running, the controller will set the spark advance on 49-State cars anywhere from 25 to 61 degrees, depending on the amount of engine vacuum and the rpm. Thus, the controller takes care of all engine conditions, the same as a vacuum and centrifugal advance, but has the advantage of being able to retard the spark if the engine overheats. It can also advance the spark at high altitude to help prevent loss of power.

Initial timing is not set by moving the distributor. The distributor has to be in a certain position so that the spark will go to the right cylinder, but this has nothing to do with the timing. When the engine is positioned with number 1 piston at the top of the compression stroke, a white line on the rotor should align with a pointer in the distributor. This insures that the rotor will be lined up with the segments in the cap.

Initial timing is set by moving the crankshaft position sensor at the front of the engine. Two bolts are loosened, and then the timing adjuster bolt is turned to move the sensor in relation to the toothed wheel. As the sensor is moved, a timing light will show the timing marks on the front pulley, the same as other engines.

Before setting the timing, it is necessary to disable the controller so it won't advance the spark. If the controller is not disabled, the timing will be off considerably, probably retarded. To prepare for setting the timing, you must find an open connector under the instrument panel a few inches from the controller. Connect a jumper wire from this connector to ground. If you have done it correctly, the Check Ignition light on the instrument panel will come on while the engine is running. Ordinarily this light only comes on when the ignition is turned to the Start position, or there is something wrong with the ignition. After setting the timing by moving the crankshaft sensor, the jumper must be removed from the connector, and the light will go out.

Reaching the sensor to make the timing adjustment can be difficult if you haven't done it before. With the engine off, use a mirror and flashlight to find the location of the clamp bolts and the adjuster bolt. The adjuster bolt can be turned with the engine running if a long extension is used. Do not attempt to reach the clamp bolts with the engine running. They are reached from the front of the engine.

4-NOZZLE THERMAL VACUUM SWITCH 1977

Toronado engines manufactured for high altitude use have a new 4-nozzle TVS. It connects to the exhaust gas recirculation system (EGR), to the electronic spark timing controller (EST), and to both ported and manifold vacuum. It is called the EGR-EST-TVS.

Below 120° F. coolant temperature, the TVS is closed, shutting off vacuum to both the EGR system and the EST controller. Above 120° F. coolant temperature, ported vacuum is allowed to go to the EGR system, and manifold vacuum goes to the EST controller.

The EGR system also has a thermal control valve (TCV) in the hose, mounted above the water pump. The TCV is closed below 61-76°F. and open above that temperature.

In most cases the coolant temperature TVS would open after the TCV. But if the air coming through the radiator is cold enough, the TCV would stay closed, blocking vacuum to the EGR valve, even though the TVS is open.

Testing 4-Nozzle TVS

With the engine at normal operating temperature, the EGR valve should receive ported vacuum above idle, and the electronic spark timing controller should receive manifold vacuum. Check it by removing the EGR hose at the EGR valve, or the controller hose at the TVS.

When the engine is cold, vacuum to the EGR system and the EST controller should be shut off. Check it by removing the hoses at the TVS, with the engine running above idle.

Repairing 4-Nozzle TVS

Repairs are limited to replacement. Relieve the cooling system pressure before removing the TVS.

Oldsmobile

TO SET REFERENCE TIMING

1. CONNECT OPEN REFERENCE TIMING CONNECTOR (TAPED TO HARNESS) TO GROUND USING ABOUT A 2 FT. JUMPER WIRE.
2. RUN ENGINE AT IDLE (NOTE: "CHECK IGNITION" LIGHT WILL REMAIN ON IF PROPERLY GROUNDED PER STEP 1.)
3. SET REFERENCE TIMING TO 20° BTC BY MOVING CRANK SENSOR IN SLOTS PROVIDED IN MOUNT. DO NOT ROTATE DISTRIBUTOR.
4. DISCONNECT GROUND CABLE. "CHECK IGNITION" LIGHT SHOULD GO OUT.

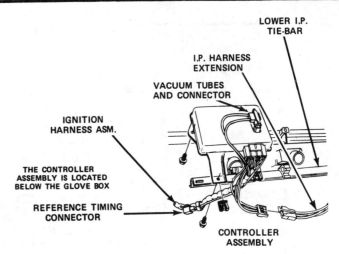

Controller and connector (© GM Corporation)

EMISSION CONTROL SYSTEMS

DIESEL ENGINE 1978

Oldsmobile developed the Diesel out of their 350 V-8. Very few of the parts are interchangeable with the gasoline 350, because of the necessity for beefing up the engine parts. The engine block, crankshaft, camshaft, connecting rods, and valve lifters were all strengthened to withstand the stress from the 22.3 to 1 compression ratio.

The Olds diesel, like other diesels, draws in only air through its induction system. The air is compressed in the cylinder, and then the fuel is injected through a high pressure nozzle which squirts the fuel directly into the combustion chamber. Because the air in the combustion chamber is so hot, after being compressed to a 22.3 to 1 ratio, the injected fuel ignites and burns instantly. This eliminates the need for any spark ignition system. The diesel has no spark plugs, wires, or distributor.

Some diesels have a throttle, but the Olds diesel has none. The induction system is wide open at all times. Power is varied by the length of time the injector nozzles stay open. The longer they stay open, the more fuel is injected, and the more power is produced.

The timing for the injectors is controlled by a high pressure pump, mounted in the valley below the intake manifold. The gas pedal, or accelerator pedal, that the driver pushes, is connected to the injection pump. There is no carburetor. When the air cleaner is removed, all that shows is a big hole where the air goes into the intake manifold.

Because there is no throttle, the engine creates very little vacuum. A vacuum pump is mounted in the former distributor drive hole, and driven by the camshaft. It provides vacuum to operate the heater controls, air conditioning, and cruise control. The power brakes are operated by a hydraulic booster, using hydraulic fluid under pressure from the power steering pump.

Because diesels are inherently clean, there is only one emission control on the

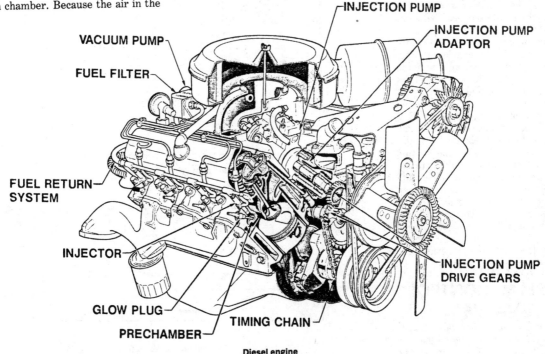

Diesel engine

Oldsmobile

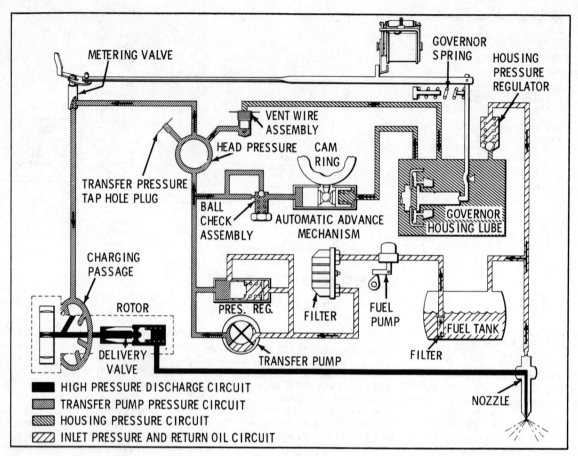

Pump fuel circuit

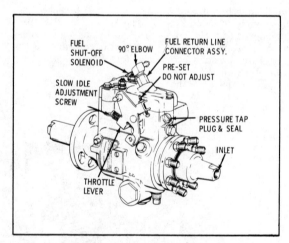

Injection pump slow idle screw

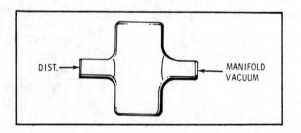

Distributor Thermal Control Valve (DTCV)

engine, a crankcase ventilation system. A filter on top of each rocker cover connects with tubes to a flow control valve on the air crossover. Fresh air enters through a check valve in the oil filler cap. Fumes from the crankcase move up to the rocker covers and go through the tubes into the air crossover. The slight amount of vacuum in the air crossover pulls in the fumes and keeps the crankcase ventilated. If the crankcase pressure should increase enough to force fumes out of the oil filler cap, the check valve in the cap closes and prevents this.

The flow control valve, mounted on the air crossover, is not an ordinary PCV valve. A diesel engine will actually run on crankcase fumes. If the engine air filter should become clogged, the vacuum inside the air crossover would increase. The increased vacuum might pull enough crankcase fumes into the engine to keep it running even though the injection pump was shut down. To prevent this, the flow control valve will close whenever the flow gets large enough that it might keep the engine running.

DISTRIBUTOR THERMAL CONTROL VALVE

This is the same design valve that is used at the front of the engine under a metal shield to control vacuum to the

Oldsmobile

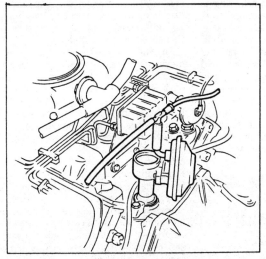
Diesel vacuum pump

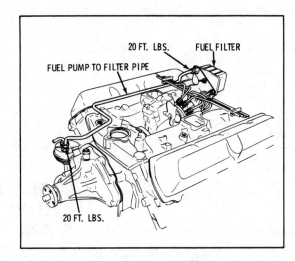
Diesel fuel lines and filter

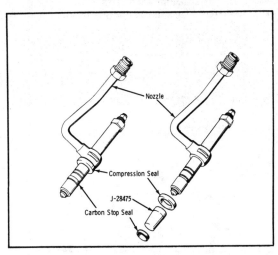
Installation of nozzle seals—Diesel engine

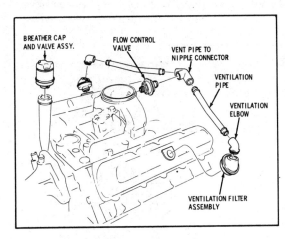
Diesel ventilation system

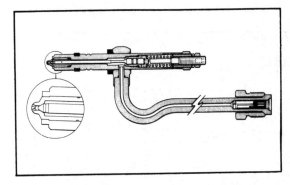
Nozzle cross section—Diesel engine

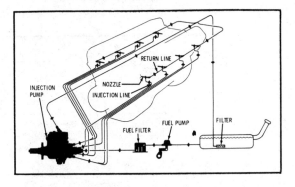
Diesel fuel system circuit

EGR system. In this application it is mounted at the rear of the engine, and is connected to intake manifold vacuum. Below 61-76°F. the Thermal Control Valve (TCV) is open. The distributor then receives full manifold vacuum, through the other two valves in the system, a check valve and the Exhaust Gas Recirculation-Distributor Thermal Vacuum Switch (EGR-DTVS). The check valve holds the highest vacuum applied to it, so once the engine has started the vacuum advance is held in the advanced position. If the engine is started when the TCV is open (below 61-76°F.) the check valve holds the vacuum advance in the advanced position until the EGR-DTVS switches at 120°F. Above that temperature, the distributor operates on ported vacuum.

If the engine is started with the TCV closed (above 61-76°F.) but below 120°F. coolant temperature, the TCV then blocks vacuum to the distributor. In this situation, the distributor will not get vacuum until the coolant temperature gets up to 120°F. and the EGR-DTVS switches. Any time the engine is started in that narrow range between the closing of the TCV (61-76°F.) and the switching of the EGR-DTVS (120°F. coolant temperature) there will be no vacuum to the distributor, and therefore no vacuum

Oldsmobile

advance. The lack of advance helps to heat up the exhaust and start the catalytic converter going.

When testing an engine it is common to stop and start it several times. While the engine is sitting, it could easily cool off to just the right temperature so that when it is restarted there will be no vacuum advance. If this happens, just wait until the engine warms up, and it will have ported advance.

When testing a cold engine, there will be full manifold vacuum to the distributor at idle. But after the engine warms up and the EGR-DTVS switches, there will be vacuum advance above idle only.

Testing of the valves is easily done by blowing through them at various temperatures to see if they are open or closed. The check valve is particularly important. If it is installed backwards, the system will not work. The best way to test the position of the check valve is to connect it to engine vacuum and start the engine. If the valve sucks air, it is connected properly. If not, it is backwards, or defective. To test its holding power, connect it to a vacuum gauge and apply full manifold vacuum. Then disconnect the vacuum source. The reading on the gauge should hold, with no more tha very slight leakage. When installing the check valve, the end that was connected to the vacuum gauge should face toward the distributor vacuum advance.

TORONADO ELECTRONIC SPARK TIMING 1978

In the 1977 system, the distributor had no timing mechanism inside it. Timing was taken care of by a crankshaft sensor at the front of the engine, and a disc that revolved with the front pulley. An adjustment bolt on the sensor was turned to set the initial timing.

In the 1978 system, the timing mechanism has been put back into the distributor. The timing module used with EST has three terminals. The timing module used with standard High Energy Ignition has four terminals. The modules are not interchangeable.

With the timing mechanism back in the distributor, initial timing is now changed by loosening the distributor clamp and turning the distributor body, the same as standard HEI. Everything else is the same as in 1977, including the necessity of grounding the reference timing connector at the controller assembly before setting the initial timing. Vacuum and centrifugal advance are still taken care of by the controller. There are no vacuum or centrifugal advance mechanisms in the distributor.

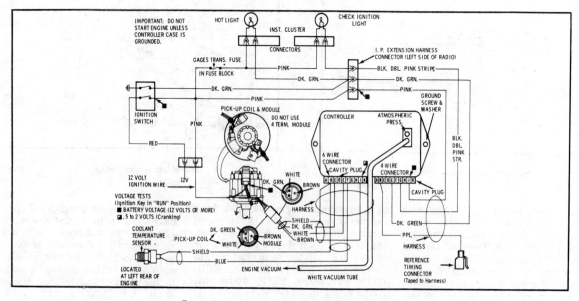

Reference timing connector—Toronado electronic spark timing system

Toronado electronic spark timing wiring schematic

Oldsmobile

CLOSED POSITIVE CRANKCASE VENTILATION SYSTEM 1978

All engines used by Oldsmobile are equipped with the closed PCV system. This system is designed to prevent crankcase vapors from entering the atmosphere. A vapor metering PCV valve is used to restrict the vapor flow when the intake manifold vacuum is high.

NOTE: *PCV Systems are used with the Diesel Engines.*

1979

The closed PCV system is continued for all engines installed by Oldsmobile Division. The operation of the system remains the same.

NOTE: *PCV Systems are used with the Diesel Engines.*

EMISSION CALIBRATED DISTRIBUTORS 1978-79

The emission calibrated distributors are used to provide the engine with the correct timing to fire the air/fuel mixture for good engine performance and fuel economy, and to remain with-in the allowable emission levels. Various type valves and switches are used with the vacuum advance units to assist in controlling the emissions and driveability.
The switches and valves are as follows:
Distributor Check Valve (DCV) Calif. Vin code F
Distributor Thermal Control Valve (DTCV) Vin code F
Distributor Thermal Vacuum Switch (DTVS) Vin code 5
Distributor Thermal Vacuum Switch (DTVS) Vin code R&K
Distributor Thermal Vacuum Switch (DTVS) Vin code V – Non/AC
Distributor Thermal Vacuum Switch (DTVS) Vin code V – W/AC
Distributor Vacuum Delay Valve (DVDW) Calif. Vin code – R & K
Spark Advance Vacuum Modulator (SAVM)
Spark Delay Valve (SDV)
Spark Retard Delay Valve (SRDV)
Electronic Spark Timing/Exhaust Gas Recirculation – Thermal Vacuum Switch (EST/EGR-TVS)

EMISSION CALIBRATED CARBURETORS 1978-79

Emission calibrated carburetors are used to provide the engine with the proper air/fuel mixture for good engine performance and fuel economy. Various switches and valves are used in conjunction with the fuel delivery system.

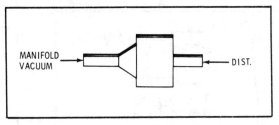

1978-79 Distributor Check Valve (DCV) Vin code F—Calif.

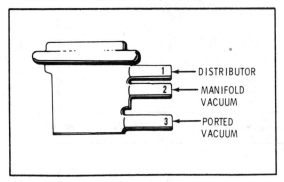

1978-79 Spark Advance Vacuum Modulator (SAVM)

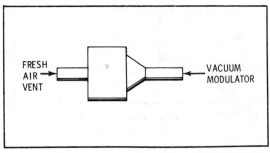

1978-79 Vacuum Modulator Check Valve (VMCV) Vin code 1

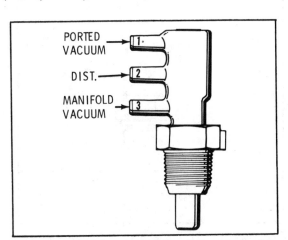

1978-79 Distributor Thermal Vacuum Switch (DTVS) Vin codes R and K

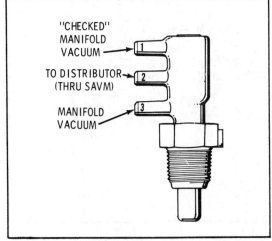

1978-79 Distributor Thermal Vacuum Switch (DTVS) Vin code F (w/integral check valve)

Oldsmobile

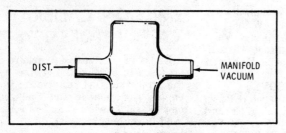

1978-79 Distributor Thermal Control Valve (DTVC) Vin code F—Calif.

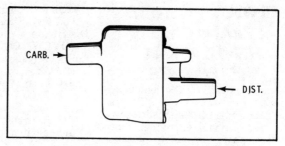

1978-79 Distributor Vacuum Delay Valve (DVDV)

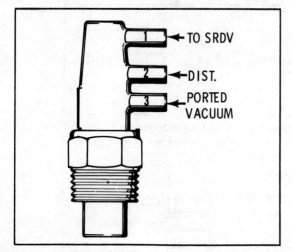

1978-79 Distributor Thermal Vacuum Switch (DTVS) Vin code V (with AC)

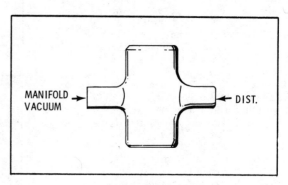

1978-79 Distributor Thermal Vacuum Switch (DTVS) Vin code V (without AC)

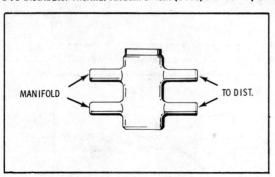

1978-79 Spark Retard Delay Valve (SRDV) Vin code V

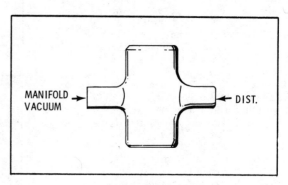

1978-79 Spark Delay Valve (SDV)

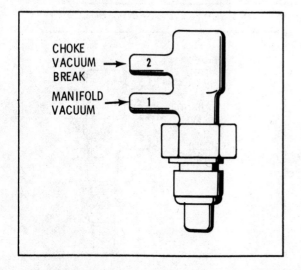

1978-79 Choke Thermal Vacuum Switch (CTVS) Vin code A

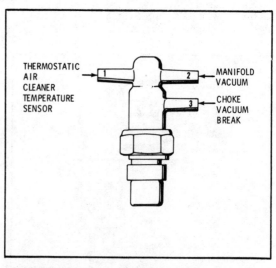

1978-79 Choke Thermal Switch (CTVS) Vin code F, R, and K

Oldsmobile

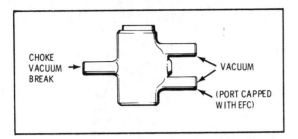

1978-79 Choke Vacuum Break-Vacuum Delay Valve (CVB-VDV) Vin code 1 and V

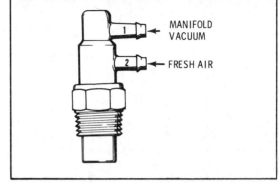

1978-79 Cold Engine Air Bleed-Thermal Vacuum Switch (CEAB-TVS) Vin code 1 (EFC)

The switches and valves are as follows;
Vacuum Modulator Check Valve (VMCV)
Choke Thermal Vacuum Switch (CTVS) three port – Vin code F, R, & K
Choke Thermal Vacuum Switch (CTVS) two port – Vin code A
Choke Vacuum Break – Vacuum Delay Valve (CVB-VDV)
Cold Engine Air Bleed – Thermal Vacuum Switch – (CEAB-TVS) Vin code 1

THERMOSTATIC AIR CLEANER (TAC) 1978

The TAC system is used to preheat the induction air into the carburetor, to allow leaner choke and carburetor calibrations, to lower the emission levels. A vacuum motor or thermostatic coil is used to control the inlet damper valve opening and closing, so that a controlled temperature of inlet air is directed to the carburetor.

1979

The TAC system is continued in use for the 1979 models with little or no change from the 1978 models.

EVAPORATION EMISSION CONTROL SYSTEM (EEC) 1978

The ECC system is used to control the emission of gasoline vapors from the fuel system. The system stores the evaporating fuel vapors when the engine is not operating, so that they may be burned during the combustion process, after the engine is started.

This process is accomplished by venting the fuel tank and carburetor bowl through a series of tubes, hoses and valves, to a canister containing activated charcoal. The canisters and valves differ from engine to engine and cannot be interchanged.

1979

The EEC system used is the same as that used on the 1978 models. Safeguards are built into the system to avoid the possibility of liquid fuel being drawn into the system.
The safeguards are;
1. A fuel tank overfill protector is installed on all models to provide adequate room for expansion of the fuel during temperature changes.
2. A fuel tank venting system is provided on all models to assure that the tank will be vented during any vehicle attitude. A domed fuel tank is used on sedans and coupes.
3. A pressure-vacuum relief valve, located in the gas cap, controls the fuel tank internal pressure.

EXHAUST GAS RECIRCULATION SYSTEM (EGR) 1978

The EGR system is used on all engines to meter exhaust gases into the induction air system, so that during the combustion cycle, the Oxides of Nitrogen (NO_x), emission can be reduced.

Ported and Back pressure types of EGR systems are used. The ported system uses a timed vacuum port in the carburetor to regulate the EGR valve, while the back pressure modulated system regulates the timed vacuum according to the exhaust back pressure level.

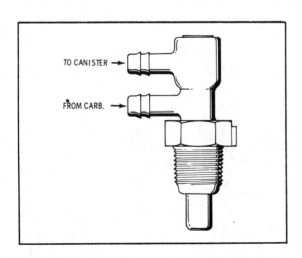

1978-79 Canister purge—Thermal Vacuum Switch (CP-TVS) Vin code A

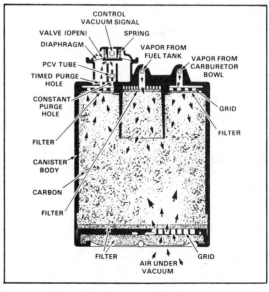

1978-79 Vapor canister—Vin code V and 1

Oldsmobile

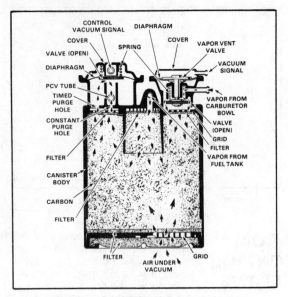

1978-79 Vapor canister—Vin code U, L, and H

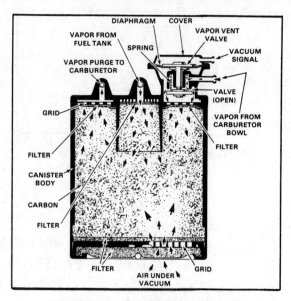

1978-79 Vapor canister—Vin code F, R, and K

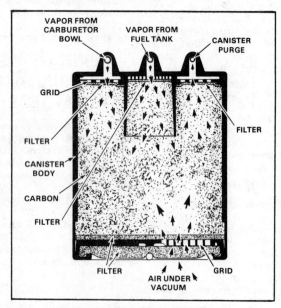

1978-79 Vapor canister—Vin code A

Two types of EGR valves are used, one without a transducer and one with a back pressure transducer, which is built into the internal components of the EGR valve. The transducer acts as a pressure regulator to control the flow of exhaust gases more evenly in certain engines.

NOTE: *Various Switches and Check Valves are used in the vacuum circuits.*

EXHAUST GAS RECIRCULATION SYSTEM (EGR) 1979

The EGR system remains in use on all 1979 engines and the system operation remains the same as used on the 1978 engine models. The EGR valve is designed to be closed during periods of deceleration and idle, to avoid rough idle and stalling from the dilution of the air/fuel mixture

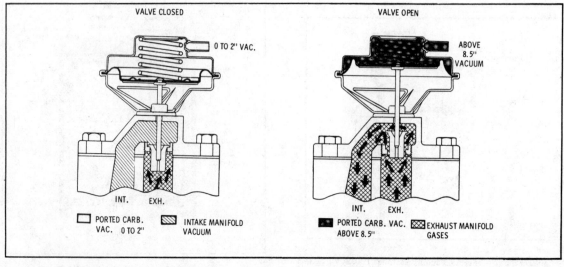

EGR valve—typical (open and closed position)

Oldsmobile

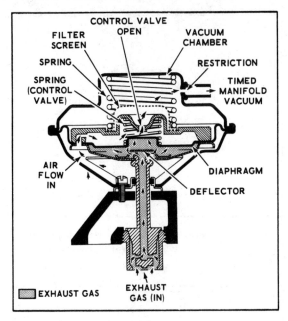

EGR valve with back pressure transducer—typical (Control valve open)

EGR valve with back pressure transducer—typical (Control valve closed)

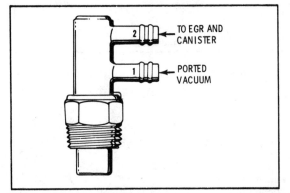

1978-79 Exhaust Gas Recirculation/Canister Purge—Thermal Vacuum Switch (EGR/CP-TVS) Vin code U, H and L

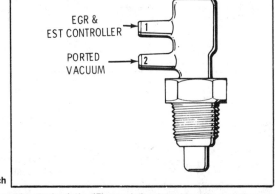

1978-79 Exhaust Gas Recirculation/Electronic Spark Timing—Thermal Vacuum Switch (EGR/EST-TVS)

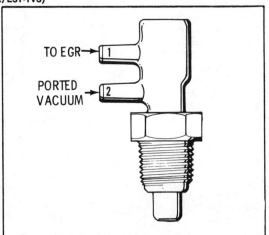

1978-79 Exhaust Gas Recirculation-Distributor Thermal Vacuum Switch (EGR-DTVS) Vin code R and K

by the exhaust gases.

NOTE: *Various Switches and Check Valves are used in the vacuum circuits.*

CATALYTIC CONVERTER (CC) 1978

The Catalytic Converter is used on all vehicles models to lower the levels of Hydrocarbons (HC) and Carbon Monoxide (CO) from the exhaust emission with the use of materials coated with Platinum and Palladium, to act as a catalyst.

A Phase II catalytic converter is used with the Electronic Fuel Control system. This converter uses Platinum and Rhodium plated material as the catalyst.

1979

The Catalytic Converter is used on the exhaust systems for 1979 models. The Phase II converter is again used with the Electronic Fuel Control System. An oxygen sensor is used in conjunction with the Phase II converter.

193

Oldsmobile

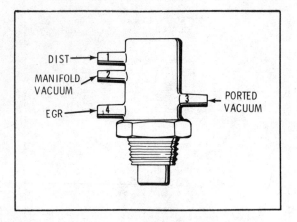

1978-79 Exhaust Gas Recirculation-Distributor Thermal Vacuum Switch (EGR-DTVS) Vin code F

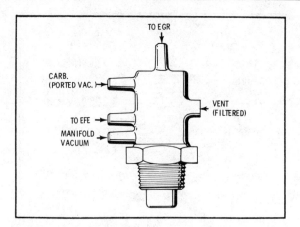

1978-79 Exhaust Gas Recirculation/Early Fuel Evaporation-Thermal Vacuum Switch (EGR/EFE-TVS) Vin code A (except Calif.)

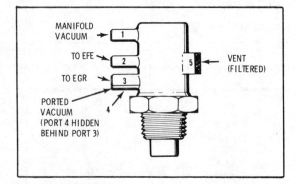

1978-79 Exhaust Gas Recirculation/Early Fuel Evaporation-Thermal Vacuum Switch (EGR/EFE-TVS) Vin code A (Calif.), V and 1

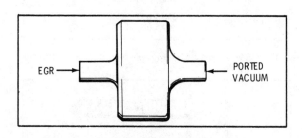

1978-79 EGR Thermal Control Valve (EGR-TCV)

AIR INJECTION REACTION SYSTEM (AIR) 1978

The Air Injection system is used on some engines to provide additional oxygen to the combustion process, after the exhaust gases leave the cylinders, to continue the combustion in the exhaust manifolds. The following components are used in the system:
1. Engine driven air pump
2. Vacuum differential valve
3. Air flow and control hoses
4. Air by-pass valve
5. Differential vacuum delay and separator valve
6. Exhaust check valve

1979

The Air Injection Reaction system is carried over to most of the 1979 models and operates in the same manner as the 1978 models.

EARLY FUEL EVAPORATION SYSTEM (EFE) 1978

The EFE system is designed to direct rapid heat to the engine air induction system to provide quick fuel evaporation and more uniform fuel distribution during cold driveaway. A vacuum operated valve is located between the exhaust manifold and the exhaust pipe is controlled by

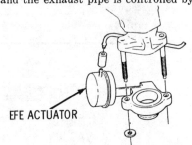

EFE actuator attached to valve—typical

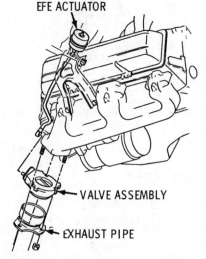

EFE actuator located above valve—typical

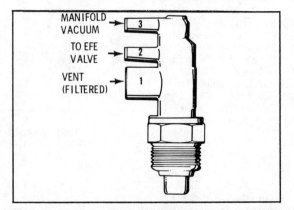

1978-79 Early Fuel Evaporation-Thermal Vacuum Switch (EFE-TVS)

Oldsmobile

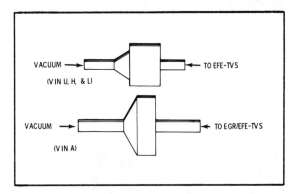

1978-79 Early Fuel Evaporation Check Valve (EFE-CV)

a thermal vacuum switch (TVS) to pass vacuum to the valve when temperatures are below a specific calibration point. As the vacuum is applied to the valve, the exhaust gas flow is increased under the intake manifold to preheat the manifold during cold engine operation. The system is used on most models.

1979

The Early Fuel Evaporation System is used on most 1979 models. The operation of the system remains the same as the 1978 models. Refer to the Emission Control Information label for usage clarification.

ELECTRIC CHOKE 1978

The Electric Choke is used to assist in controlling the choke thermostatic coil for precise timing of the choke valve opening for good engine warm-up performance. This is accomplished with the use of a ceramic resistor with-in the choke unit. A small section resistor is used to gradually heat the coil, while a large section resistor is used to quickly heat the coil as the temperature dictates. The electric choke is not used on all models.

1979

The Electric Choke is used on selected models of 1979, and operates in the same manner as the 1978 models. At temperatures below 50° F., the electrical current is directed to the small segment resistor of the choke to allow gradual opening of the choke valve. At temperatures of 70° F. and above, the electrical current is directed to both the small and the large segment resistors, to allow quicker opening of the choke valve.

ELECTRONIC FUEL CONTROL SYSTEM (EFC) 1978

The EFC system is new for 1978, and is designed to control the exhaust emissions by regulating the air/fuel mixture with the use of the following components;
1. Exhaust gas oxygen senor
2. Electric control unit
3. Vacuum modulator
4. Controlled air-fuel ratio carburetor
5. Phase II catalytic converter

The exhaust gas sensor is placed in the exhaust gas stream and generates a voltage which varies with the oxygen content of the exhaust gases. As the exhaust gas oxygen content rises, indicating a lean mixture, the voltage falls. As the exhaust gas oxygen content decreases, indicating a rich mixture, the voltage rises.

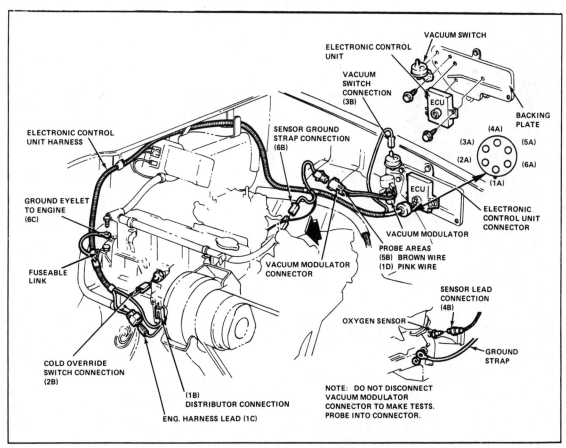

Electrical wiring connections—1978-79 Electronic Fuel Control system

Oldsmobile

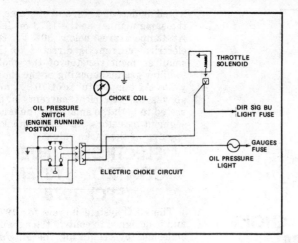

Electric choke circuit—typical

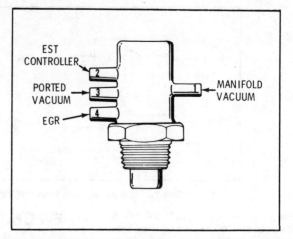

1978-79 Electronic Spark Timing/Exhaust Gas Recirculation-Thermal Vacuum Switch (EST/EGR-TVS)

The voltage signal is sent to the electronic control unit, where the voltage signal is monitored. The electronic control unit transmits a signal to the vacuum modulator. This signal is of a constant current and is continually cycling on and off.

To assist in maintaining proper electrical signal regulation during periods of unstable engine operation, a bi-metal switch is located in the cylinder head and provides an open circuit or a ground circuit for the electrical signal, when the temperature is above or below the calibration level value.

If below the calibration point, the electronic control unit uses this signal to restrict the amount of leanness that the carburetor can go lean. A temperature sensitive diode within the ECU, assists in the determination of the strength of the electrical signal. When the engine warms up, this feature drops out, indicating the engine is warm and the drivability will not be impaired while the carburetor is in the maximum lean range.

1979

The EFC system is continued for 1979 models. The operation remains the same as for the 1978 models. The EFC system should be included in any diagnosis of the following problems;

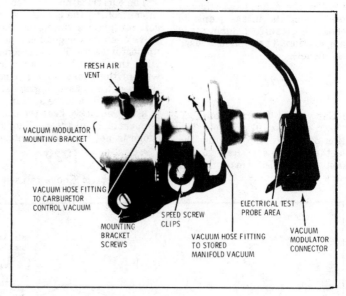

1978-79 Electronic Fuel Control vacuum modulator

1. Detonation
2. Stalls or rough idle – cold
3. Stalls or rough idle – hot
4. Missing
5. Hesitation
6. Surges
7. Sluggish or spongy operation
8. Poor gas mileage
9. Hard starting – cold
10. Hard starting – hot
11. foul exhaust odor
12. Engine cuts out

Oldsmobile

VACUUM CIRCUITS

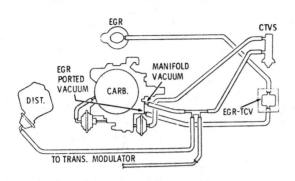

1976 49-State 350 4-Bbl. Cutlass & 88
1976 49-State 455 4-Bbl. Toronado

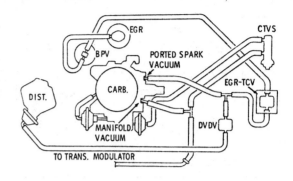

1976 Calif. 350 4-Bbl. Cutlass & 88

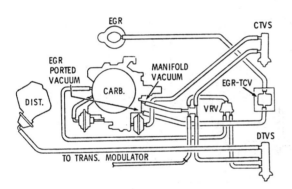

1976 Calif. 350 4-Bbl. Cutlass with 241 axle
1976 49-State 455 4-Bbl. Cutlass, 88, 98

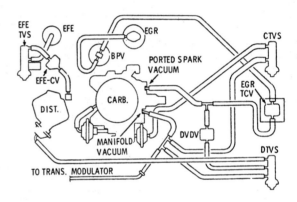

1976 Calif. 455 4-Bbl. Cutlass, 88, 98

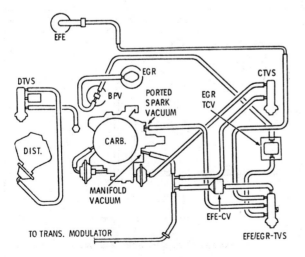

1976 Calif. 455 5-Bbl. Toronado

AIR	-	AIR INJECTOR REACTOR
BPV	-	BACK PRESSURE TRANSDUCER VALVE
CC	-	CATALYTIC CONVERTER
CTVS	-	CHOKE THERMAL VACUUM SWITCH
DTVS	-	DISTRIBUTOR THERMAL VACUUM SWITCH
DVDV	-	DISTRIBUTOR VACUUM DELAY VALVE
EEC	-	EVAPORATIVE EMISSION CONTROL
EFE	-	EARLY FUEL EVAPORATION VALVE AND ACTUATOR
EFECV	-	EFE CHECK VALVE
EFE-TVS	-	EARLY FUEL EVAPORATION TVS SWITCH
EFE/EGR-TVS	-	EARLY FUEL EVAPORATION/EXHAUST GAS RECIRCULATION TVS SWITCH
EFE-OTVS	-	EFE OIL THERMAL VACUUM SWITCH
EGR	-	EXHAUST GAS RECIRCULATION
EGRCV	-	EGR CHECK VALVE
EGR-TCV	-	EGR THERMAL CONTROL VALVE
EGR-TVS	-	EGR THERMAL VACUUM SWITCH
EGR/EFE-TVS	-	EXHAUST GAS RECIRCULATION/EARLY FUEL EVAPORATION TVS SWITCH
EGR-VDV	-	EXHAUST GAS RECIRCULATION/VACUUM DELAY VALVE
PCV	-	POSITIVE CRANKCASE VENTILATION SYSTEM
SAVM	-	SPARK ADVANCE VACUUM MODULATOR
TAC	-	THERMOSTATIC AIR CLEANER (CCS)
TAC-TCV	-	THERMOSTATIC AIR CLEANER TCV
VRV	-	VACUUM REDUCER VALVE

Abbreviations used in Oldsmobile vacuum circuits

Oldsmobile

VACUUM CIRCUITS

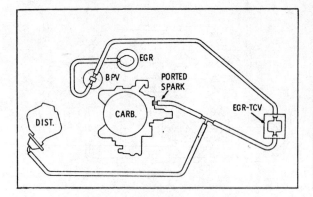

1976 49-State 260 2-Bbl. man. trans.

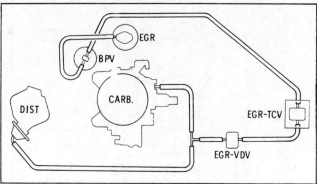

1976 Calif. 260 2-Bbl. man. trans.

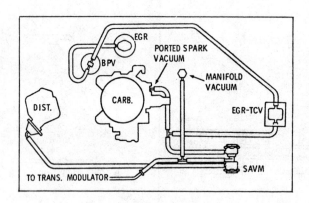

1976 49-State 260 2-Bbl. auto. trans.

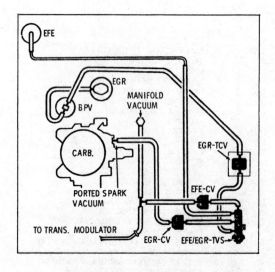

1976 Calif. 260 2-Bbl. auto. trans. 1st type

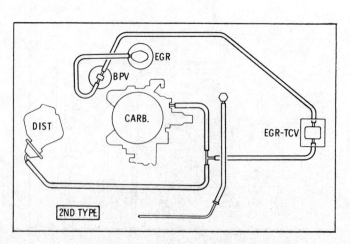

1976 Calif. 260 2-Bbl. auto. Trans. 2nd type

Oldsmobile

VACUUM CIRCUITS

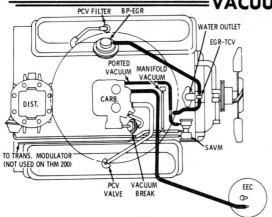

Vacuum hose routing 260 C.I.D. A.T. Omega-Cutlass-88 all except high altitude and California (© GM Corporation)

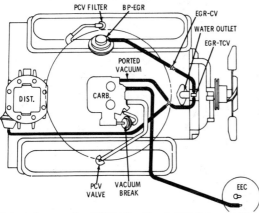

Vacuum hose routing 260 C.I.D. w/5 spd. M.T. Omega-Cutlass all except high altitude and California (© GM Corporation)

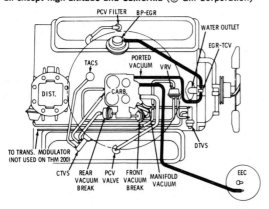

Vacuum hose routing 350-403 Cutlass-88-98 all except high altitude and California (© GM Corporation)

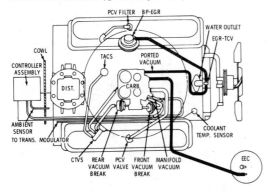

Vacuum hose routing 403 C.I.D. Toronado all except Calif. (© GM Corporation)

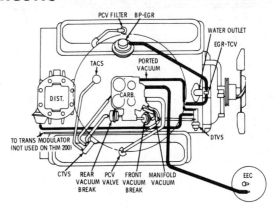

Vacuum hose routing 350 & 403 Cutlass-88-98 altitude (© GM Corporation)

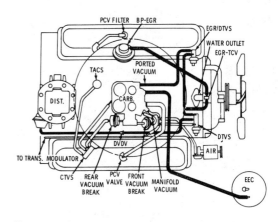

Vacuum hose routing 350 & 403 all California except Toronado (© GM Corporation)

AIR	— AIR INJECTOR REACTOR
BP-EGR	— BACK PRESSURE EXHAUST GAS RECIRCULATION
CTVS	— CHOKE THERMAL VACUUM SWITCH
DTVS	— DISTRIBUTOR THERMAL VACUUM SWITCH
DVDV	— DISTRIBUTOR VACUUM DELAY VALVE
EEC	— EVAPORATIVE EMISSION CONTROL
EFECV	— EARLY FUEL EVAPORATION CHECK VALVE
EGR	— EXHAUST GAS RECIRCULATION
EGRCV	— EGR CHECK VALVE
EGR/DTVS	— EXHAUST GAS RECIRCULATION DISTRIBUTOR THERMAL VACUUM SWITCH
EGR-TCV	— EGR THERMAL CONTROL VALVE
PCV	— POSITIVE CRANKCASE VENTILATION SYSTEM
SAVM	— SPARK ADVANCE VACUUM MODULATOR
TACS	— THERMOSTATIC AIR CLEANER SENSOR

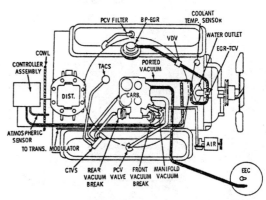

Vacuum hose routing—Toronado—California (© GM Corporation)

Oldsmobile

VACUUM CIRCUITS

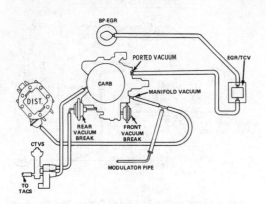

1978 49-States—350 Vin code R and 403 V-8

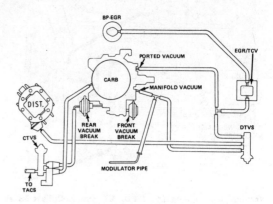

1978 Alt.—350 Vin code R and 403 V-8

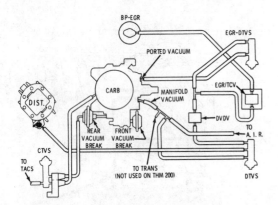

1978 Calif. 350 Vin code R and 403 V-8

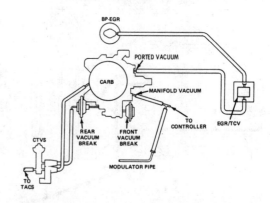

1978 49-States—Toronado 403 V-8

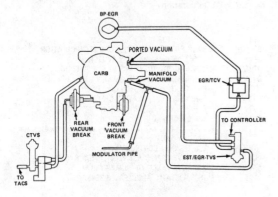

1978 Alt.—Toronado 403 V-8

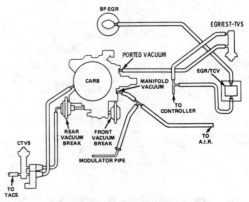

1978 Calif.—Toronado 403 V-8

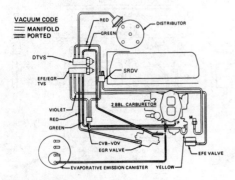

1978 49-States and Alt.—151 (with AC)

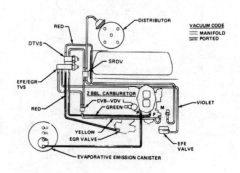

1978 49-States and Alt.—151 (without AC)

Oldsmobile

VACUUM CIRCUITS

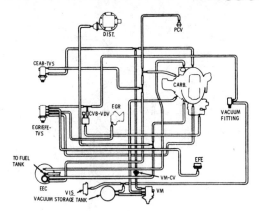

1978 Calif.—151 (with EFC)

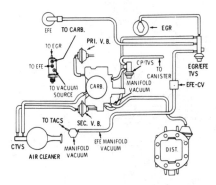

1978 49-States and Alt.—231 V-6

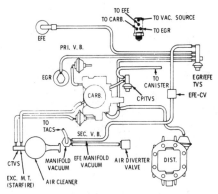

1978 Calif.—231 V-6

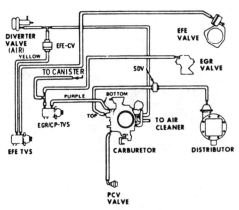

1978 Calif. and Alt.—305 V-8

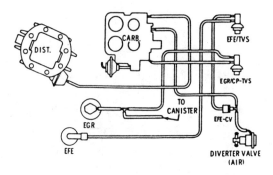

1978 49-States—305 V-8

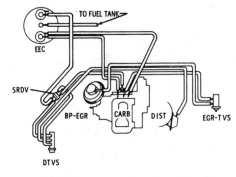

1979 151 4-cyl.—Except Calif. (w/o AC, MT)

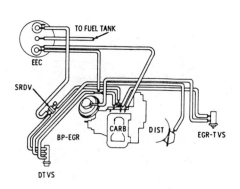

1979 151 4-cyl.—Except Calif. (with AC, AT)

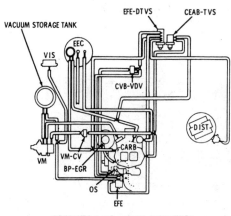

1979 151 4-cyl.—Calif. (with EFC)

201

Oldsmobile

VACUUM CIRCUITS

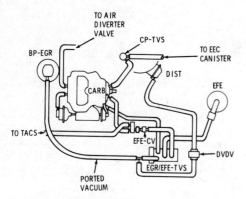

1979 231 V-6—Calif. Starfire (with MT)

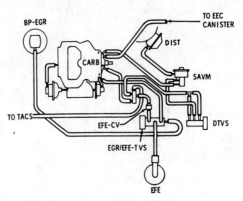

1979 231 V-6—Except Calif. and Alt. (with AT)

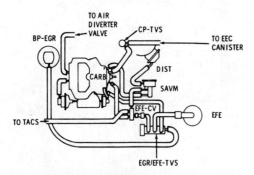

1979 231 V-6—Calif. Omega and 88 models, Alt. Starfire and Cutlass (with AT)

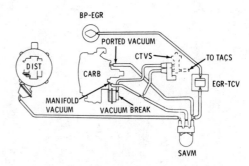

1979 260 V-8—Except Calif. and Alt. (with MT)

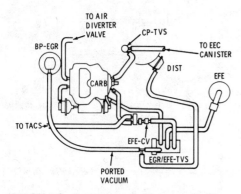

1979 231 V-6—Calif. Starfire and Cutlass (with AT)

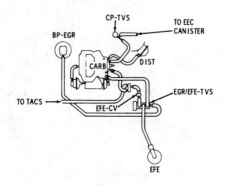

1979 231 V-6—Except Calif. and Alt. (with MT)

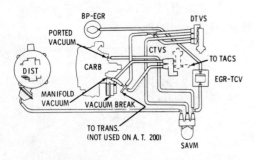

1979 260 V-8—Except Calif. and Alt. (with AT)

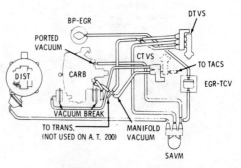

1979 260 V-8—Alt. (With AT)

Oldsmobile

VACUUM CIRCUITS

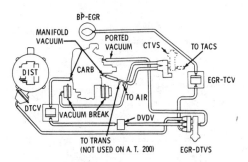

1979 260 V-8—Calif. (with AT)

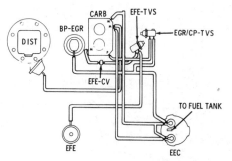

1979 305 V-8 Vin code G—Except Calif. and Alt. Starfire models

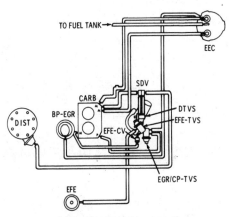

1979 305 V-8 Vin code G—Except Calif. and Alt. Omega and 88 models, except SW

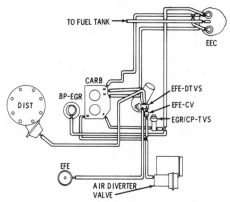

1979 305 V-8 Vin code G—Calif. 88 models, except SW

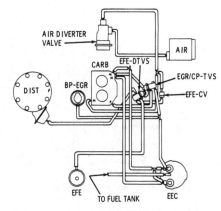

1979 305 V-8 Vin code G—Calif. Starfire models

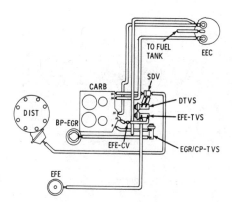

1979 305 V-8 Vin code H—Except Calif. and Alt.

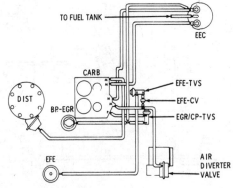

1979 305 V-8 Vin code H and 350 V-8 Vin code L—Calif. and Alt.

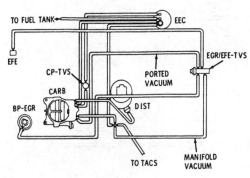

1979 350 V-8 Vin code X—except Calif. and Alt.

203

Oldsmobile

VACUUM CIRCUITS

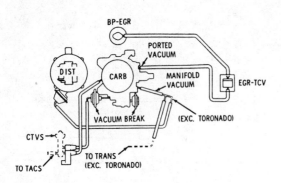

1979 350 V-8 Vin code R and 403 V-8 Vin code K—Except Calif. and Alt.

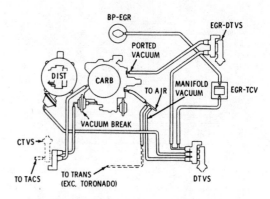

1979 350 V-8 Vin code R and 403 V-8 Vin code K—Calif.

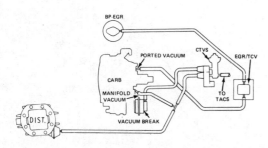

1978 49-States—260 V-8 (with 5 spd. trans.)

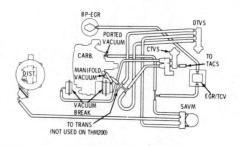

1978 Alt.—260 V-8 (with auto. trans.)

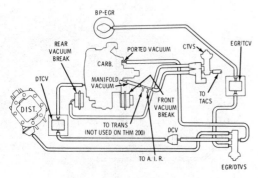

1978 Calif.—260 V-8

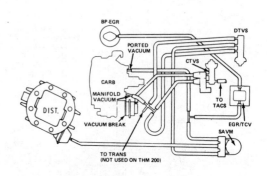

1978 49-States and Alt.—260 V-8

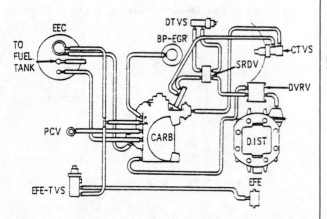

1979 301 V-8—Except Calif. and Alt. (with AC)

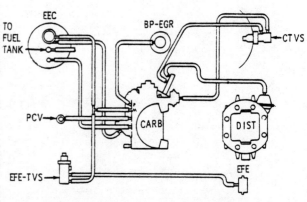

1979 301 V-8—Except Calif. and Alt. (without AC)

Pontiac

Full-Size • Intermediate • Compact

TUNE-UP SPECIFICATIONS — PONTIAC

Engine Code, 5th Character of the VIN number
Model Year Code, 6th character of the VIN number

Yr.	Eng. V.I.N. Code	Eng. No. Cyl. Disp. (cu in)	Eng.** Mfg.	Carb Bbl	H.P.	SPARK PLUGS Orig. Type	Gap (in)	DIST. Point Dwell (deg)	Point Gap (in)	IGNITION TIMING (deg BTDC) Man Trans	IGNITION TIMING Auto Trans	VALVES Intake Opens (deg BTDC)	FUEL PUMP Pres. (psi)	IDLE SPEED (rpm) Man Trans	IDLE SPEED Auto Trans In Drive
75	D	6-250	C	1	105	R46TX	.060	E.I.		10	10	25	3-5	850①①	550①①
	F	8-260	O	2	110	R46SX	.080	E.I.		16	18㉔	14	3-6	750	550⑰
	M	8-350	P	2	155	R46TSX	.060	E.I.		12	16㉓	26	3-6	—	600①①
	E	8-350	P	4	175	R45TSX	.060	E.I.		12	16㉓	26	3-6	775	650⑱
	H	8-350	B	2	145	R46TSX	.060	E.I.		12	12	19	3-6	—	600①①
	J	8-350	B	4	165	R45TSX	.060	E.I.		12	12	19	3-6	775	650⑱
	R	8-400	P	2	170	R46TSX	.060	E.I.		12	16㉓	26	3-6	—	650
	S	8-400	P	4	185	R45TSX	.060	E.I.		12	16㉓	23/30	3-6	775	650②⑰
	W	8-455	P	4	200	R45TSX	.060	E.I.		—	16㉒	23	3-6	—	650⑱
76	C	V6-231	B	2	110	R44SX	.060	E.I.		12	12	17	3-6	800	600
	D	6-250	C	1	110	R46TX	.060	E.I.		6	10	25	3-5	850	550③⑰
	F	8-260	O	2	110	R46SX	.080	E.I.		16㉔	18㉔	14	3-6	750	550⑰
	M	8-350	P	2	160	R46TSX	.060	E.I.		—	16	22	3-6	—	550
	E	8-350	P	4	165	R46TSX	.060	E.I.		—	16	26	3-6	—	600
	H	8-350	B	2	140	R45TSX	.060	E.I.		—	12	13	3-6	—	600
	J	8-350	B	4	155	R45TSX	.060	E.I.		—	12	13	3-6	—	600
	R	8-400	P	2	170	R46TSX	.060	E.I.		—	16	26	3-6	—	550
	S	8-400	P	4	185	R45TSX	.060	E.I.		12	16	23/30	3-6	775	575
	W	8-455	P	4	200	R45TSX	.060	E.I.		12	16㉓	23	3-6	775	550⑰
77	V	4-151	P	2	88	R44TSX	.060	E.I.		14	14(12)	27	4-5.5	1000	650㉘
	A	V6-231	B	2	105	R46TSX④	.060	E.I.		12	12	17	4-6	800⑮①	600
	C	V6-231	B	2	105	R46TSX④	.060	E.I.		12	12	17	4-6	800⑮①	600
	Y	8-301	P	2	135	R46TSX	.060	E.I.		16	12	31/27	7-8	750	550
	U	8-305	C	2	140	R45TS	.045	E.I.		—	⑤	28	7-9	—	500
	L	8-350	C	4	165	R45TS	.045	E.I.		8	8	28	7-9	—	600⑯
	P	8-350	P	4	170	R45TSX	.060	E.I.		16	16B	29	7-8	—	575
	R	8-350	O	4	170	R46SZ	.060	E.I.		20	20	29	5-6	—	600
	Z	8-400	P	4	180	R45TSX	.060	E.I.		18	16⑥	29	7-8	775	575⑧
	K	8-403	O	4	185	R46SZ	.060	E.I.		—	22㉓⑦	16	5-6	—	550⑨
78	V	4-151	P	2	85	R43TSX	.060	E.I.		—	14㉕	28	5-6.5	—	650㉘
	A	6-231	B	2	105	R46TSX	.060	E.I.		15㉘	15㉗	17	4.5-5.7	800	600
	Y	8-301	P	2	140	R46TSX	.060	E.I.		—	12	27	7-8.5	—	550㊳
	W	8-301	P	4	150	R45TSX	.060	E.I.		—	12㉙	14	7-8.5	—	550㊳
	U	8-305	C	2	145	R45TS	.045	E.I.		4	4(6)㉛	28	7.5-9	600㉜	500㉝
	H	8-305	C	4	160	R46SZ	.080	E.I.		—	4(4)	28	7.5-9	—	500(500)
	L	8-350	C	4	160	R45TS	.045	E.I.		6	(8)㉞	28	7.5-9	700	(500)㉞
	X	8-350	B	4	155	R46TSX	.060	E.I.		—	15	13.5	5.9-7.4㊵	—	550
	R	8-350	O	4	170	R46SZ	.060	E.I.		—	(20)㉟	16	5.5-6.5	—	(500)㊲
	Z	8-400	P	4	180	R45TSX	.060	E.I.		—	16	21㊴	7-8.5	—	575㊱
	Z	8-400	P	4	220	R45TSX	.060	E.I.		18	18	16	7-8.5	775	600㊱
	K	8-403	O	4	185	R46SZ	.060	E.I.		—	20(20)㉟	16	5.5-6.5	—	700(650)
79	V	4-151	P	2	85	R43TSX	.060	E.I.		—	12㉕	33	5-6.5	1000	750
	2	6-231	B	2	100	⑩	⑩	E.I.		⑩	⑩	17	4.5-5.7	⑩	⑩
	A	6-231	B	2	105	R46TSX㊶	.060	E.I.		15	15	17	4.5-5.7	600	600
	Y	8-301	P	2	140	R46TSX	.060	E.I.		—	12	16	7-8.5	—	550
	W	8-301	P	4	150	R45TSX	.060	E.I.		—	12	27㊸	7-8.5	—	750
	G	8-305	C	1	145	R45TS	.045	E.I.		4	4(6)	28	7.5-9	600	500(500)

Pontiac

Full-Size • Intermediate • Compact

TUNE-UP SPECIFICATIONS — PONTIAC

Engine Code, 5th Character of the VIN number
Model Year Code, 6th character of the VIN number

Yr.	Eng. V.I.N. Code	Eng. No. Cyl. Disp. (cu in)	Eng.** Mfg.	Carb Bbl	H.P.	SPARK PLUGS Orig. Type	Gap (in)	DIST. Point Dwell (deg)	DIST. Point Gap (in)	IGNITION TIMING (deg BTDC) Man Trans	IGNITION TIMING (deg BTDC) Auto Trans	VALVES Intake Opens (deg BTDC)	FUEL PUMP Pres. (psi)	IDLE SPEED (rpm) Man Trans	IDLE SPEED (rpm) Auto Trans In Drive
	H	8-305	C	4	150	R46SZ	.080	E.I.	—	—	4(4)	28	7.5-9	—	500(500)
	L	8-350	C	4	160	R46SZ	.080	E.I.	—	6	8	28	7.5-9	700	500
	R	8-350	O	4	170	R45TSX	.060	E.I.	—	—	(20)㊷	16	5.5-6.5	—	(550)
	X	8-350	B	4	155	R46TSX	.080	E.I.	—	—	15	13.5	5.9-7.4㊵	—	550
	Z	8-400	P	4	220	R45TSX	.060	E.I.	—	18	—	16	7-8.5	775	—
	K	8-403	O	4	185	R46SZ	.080	E.I.	—	—	(20)㊷	16	7-8.5	—	(550)

Should the information provided in this manual deviate from the specifications on the underhood tune-up label, the label specifications should be used, as they may reflect production changes.

- — Not applicable
- ** B Buick
  - C Chevrolet
  - O Oldsmobile
  - P Pontiac
- • Figure in parenthesis indicates California engine. All figures are in degrees before top dead center. Where two figures appear, the first represents timing with manual transmission, the second with automatic transmission.
- ① LOWER FIGURE INDICATES IDLE SPEED WITH SOLENOID DISCONNECTED.
- ② CATALINA AND GRAND SAFARI WAGONS 625 RPM
- ③ LEMANS WITH AIR COND. 575 RPM.
- ④ EARLY PRODUCTION USED R46TS AT .040.
- ⑤ TIMING IS 4B OR 8B. SEE UNDERHOOD LABEL.
- ⑥ W-72 ENGINE IS 18B.
- ⑦ 20B ON ALTITUDE ENGINES.
- ⑧ W-72 ENGINE IS 600 RPM.
- ⑨ ALTITUDE ENGINE IS 600 RPM.
- ⑩ SEE ENGINE DECAL. SPECIFICATIONS VARIES WITH CAR APPLICATIONS.
- ⑪ /425
- ⑫ /450
- ⑬ /500
- ⑭ /550
- ⑮ /600
- ⑯ (500)
- ⑰ (600)
- ⑱ (625)
- ⑲ (675)
- ⑳ (14)
- ㉑ (20)
- ㉒ (10)
- ㉓ (12)
- ㉔ (16)
- ㉕ SET TIMING AT 1000 RPM(N)
- ㉖ W/O AIR CONDITIONING—500 RPM(D). SLOW SCREW RPM—W/AIR COND.—850(D) W/O AIR COND.—650(D)
- ㉗ SLOW SCREW RPM—670(D)
- ㉘ SLOW SCREW RPM—800(N)
- ㉙ SET TIMING AT 750 RPM
- ㉚ SLOW SCREW RPM 650(D)
- ㉛ HIGH ALT. TIMING 8B @ 600(D), SLOW SCREW 700(D)
- ㉜ SLOW SCREW RPM—700(D)
- ㉝ SLOW SCREW RPM—600(D)(650(D))
- ㉞ W/DIST. NO. 1103310—R45TSX at .060 W/DIST. NO. 1103314—R46TSX AT .060
- ㉟ SET TIMING AT 1100 RPM
- ㊱ SET TIMING AT 700, SLOW SCREW RPM—700
- ㊲ SLOW SCREW RPM (650D)
- ㊳ CALIFORNIA SLOW SCREW RPM (600) HIGH ALTITUDE TIMING 8 @ 600(D) HIGH ALTITUDE SLOW SCREW 650(D)
- ㊴ W/AT—29°
- ㊵ W/AIR CONDITIONING 5.0-6.5 PSI
- ㊶ CALIFORNIA AND HIGH ALTITUDE R45TSX
- ㊷ TIME AT 1100 RPM
- ㊸ W/AUTO. TRANS. 27°

DISTRIBUTOR SPECIFICATIONS — PONTIAC

Year	Distributor Identification	Centrifugal Advance Start Dist. Deg. @ Dist. RPM	Centrifugal Advance Finish Dist. Deg. @ Dist. RPM	Vacuum Advance Start @ In. Hg.	Vacuum Advance Finish Dist. Deg. @ In. Hg.
'75	1110650	0 @ 600	7 @ 2100	4	8 @ 15
	1112495	0 @ 550	4 @ 2200	7	12.5 @ 12
	1112498	0 @ 600	8.5 @ 1800	6	12 @ 13
	1112500	0 @ 600	10 @ 2200	7	12.5 @ 12
	1112863	0 @ 365	8 @ 2200	4	9 @ 12
	1112896	0 @ 575	6 @ 2250	7	8 @ 12
	1112918	0 @ 500	7 @ 2200	7	10 @ 11
	1112928	0 @ 600	8 @ 2200	7	12.5 @ 12
	1112929	0 @ 500	10 @ 2200	7	10 @ 14
	1112930	0 @ 700	10 @ 2200	7	12.5 @ 12
	1112946	0 @ 500	11 @ 1800	7	12 @ 15

Pontiac

Full-Size • Intermediate • Compact

DISTRIBUTOR SPECIFICATIONS — PONTIAC

Year	Distributor Identification	Centrifugal Advance Start Dist. Deg. @ Dist. RPM	Centrifugal Advance Finish Dist. Deg. @ Dist. RPM	Vacuum Advance Start @ In. Hg.	Vacuum Advance Finish Dist. Deg. @ In. Hg.
	1112947	0 @ 600	10 @ 1900	8	10 @ 15
	1112949	0 @ 500	7 @ 2200	7	10 @ 14
	1112950	0 @ 600	10 @ 1900	7	10 @ 11
	1112951	0 @ 325	14 @ 2200	4	12 @ 15
	1112956	0 @ 325	14 @ 2200	②	②
'76	1103201	0 @ 600	10 @ 1900	6	12 @ 13
	1103205	0 @ 600	8 @ 2200	7	10 @ 11
	1103206	0 @ 600	8.5 @ 1800	7	10 @ 11
	1103207	0 @ 500	7 @ 2200	8	10 @ 15
	1110650	0 @ 600	7 @ 2100		8 @ 15
	1110661	0 @ 525	8 @ 2050	6	9 @ 10
	1110666	0 @ 500	10 @ 2100	4	12 @ 15
	1110668	0 @ 635	8 @ 1600	6	12 @ 12
	1112495	0 @ 550	4 @ 2200	7	12.5 @ 12
	1112497	0 @ 600	10 @ 1900	5	12.5 @ 11
	1112500	0 @ 600	10 @ 2200	7	12.5 @ 12
	1112862	0 @ 450	11 @ 2400	5	10 @ 14
	1112863	0 @ 385	8 @ 2200	4	9 @ 12
	1112896	0 @ 550	6 @ 2250	6.5	7 @ 12
	1112923	0 @ 500	7 @ 2200	7	12.5 @ 12
	1112928	0 @ 600	8 @ 2200	7	12.5 @ 12
	1112930	0 @ 700	5 @ 2200	7	12.5 @ 12
	1112950	0 @ 600	10 @ 1900	7	10 @ 11
	1112956	0 @ 325	14 @ 2200	②	②
	1112958	0 @ 600	8 @ 2200	5	12.5 @ 11
	1112960	0 @ 500	7 @ 2200	9	10 @ 16
	1112991	0 @ 715	10 @ 2215	7	10 @ 13
	1112992	0 @ 450	10 @ 2200	8	7 @ 11
	1112994	0 @ 325	14 @ 2200	5	12 @ 11
	1112995	0 @ 450	13 @ 2225	4	15 @ 11
'77	1103231	0 @ 600	10 @ 2200	3.5	10 @ 12
	1103239	0 @ 600	10 @ 2100	4.5	10 @ 10
	1103244	0 @ 500	10 @ 1900	4.5	10 @ 10
	1103246	0 @ 600	11 @ 2100	3.5	9 @ 11
	1103248	0 @ 600	10 @ 2100	3.5	5 @ 7
	1103257	0 @ 600	8.5 @ 1800	5	10 @ 10
	1103259	0 @ 500	9.5 @ 2000	6	12 @ 13
	1103260	0 @ 500	6.5 @ 1800	6	12 @ 13
	1103263	0 @ 600	10 @ 2200	3.5	10 @ 9
	1103264	0 @ 500	12.5 @ 1800	5	8 @ 11
	1103266	0 @ 500	9.5 @ 2000	5	8 @ 11
	1103269	0 @ 500	8.5 @ 2300	5	10 @ 10
	1103271	0 @ 500	10 @ 2200	5	12.5 @ 11
	1103272	0 @ 415	11.2 @ 1715	4	12.5 @ 12
	1103273	0 @ 500	9.5 @ 1800	4	12.5 @ 12
	1103276	0 @ 400	2 @ 500	5	10 @ 10
	1103278	0 @ 600	8 @ 2200	5	10 @ 10
	1110677	0 @ 700	10 @ 1800	4	12 @ 11
	1110686	0 @ 700	10 @ 1800	7	4 @ 9

② NOT EQUIPPED WITH VACUUM CONTROL

Pontiac

Full-Size • Intermediate • Compact

DISTRIBUTOR SPECIFICATIONS — PONTIAC

Year	Distributor Identification	Centrifugal Advance Start Dist. Deg. @ Dist. RPM	Centrifugal Advance Finish Dist. Deg. @ Dist. RPM	Vacuum Advance Start @ In. Hg.	Vacuum Advance Finish Dist. Deg. @ In. Hg.
'78	1103264	0 @ 500	6.5 @ 1800	5	8 @ 11
	1103266		9.5 @ 2000	5	8 @ 11
	1103281	0 @ 500	10 @ 1900	4	9 @ 12
	1103282	0 @ 500	10 @ 1900	4	10 @ 10
	1103285	0 @ 600	11 @ 2100	4	5 @ 8
	1103310	0 @ 500	7 @ 2200	4	12.5 @ 12
	1103314	0 @ 413	10.7 @ 1700	4	12.5 @ 12
	1103315	0 @ 500	10 @ 2200	5	12.5 @ 11
	1103316	0 @ 500	8.5 @ 2300	4	12.5 @ 12
	1103323	0 @ 500	9.5 @ 2000	5	8 @ 11
	1103325	0 @ 500	6.5 @ 1800	5	8 @ 11
	1103329	0 @ 600	10 @ 2200	3.5	10 @ 9
	1103337	0 @ 550	8 @ 1200	4	12 @ 10
	1103342	0-4 @ 1000	9.5 @ 2200	7	12 @ 13
	1103343	0 @ 400	8.25 @ 1820	4	12.5 @ 11
	1103346	0 @ 500	9.5 @ 2000	6	12 @ 13
	1103347	0 @ 500	6.5 @ 1800	6	12 @ 13
	1103359	0 @ 500	8.5 @ 2300	5	10 @ 10
	1110695	0-3 @ 1000	6-9 @ 1800	6	12 @ 13
	1110731	0-2 @ 1000	6-9 @ 1800	6	8 @ 9
'79	1103281	0 @ 500	10 @ 1900	4	9 @ 12
	1103282	0 @ 500	10 @ 1900	4	10 @ 10
	1103285	0 @ 600	11 @ 2100	4	10 @ 8
	1103310	0 @ 500	7 @ 2200	4	12.5 @ 12
	1103314	0 @ 413	10.7 @ 1700	4	12.5 @ 12
	1103315	0 @ 500	10 @ 2200	5	12.5 @ 11
	1103323	0 @ 500	9.5 @ 2000	5	8 @ 11
	1103325	0 @ 500	6.5 @ 1800	5	8 @ 11
	1103337	0 @ 550	8 @ 1200	4	12 @ 10
	1103346	0 @ 500	9.5 @ 2000	6	12 @ 13
	1103347	0 @ 500	6.5 @ 1800	6	12 @ 13
	1103353	0 @ 550	11 @ 2300	4	10 @ 10
	1103731	0 @ 840	7.5 @ 1800	6	8 @ 9
	1110695	0-3 @ 840	7.5 @ 1800	6	12 @ 13
	1110713	0-2 @ 1000	6-9 @ 1800	6	8 @ 9
	1110766	0 @ 840	7.5 @ 1800	3.9	12 @ 10.9

② NOT EQUIPPED WITH VACUUM CONTROL

Pontiac

Model 5210-C

Pontiac Astre, Sunbird, Ventura　　Holley Carburetor

Year	Carb. Part No. ① ②	Float Level (Dry) (in.)	Float Drop (in.)	Pump Position	Fast Idle Cam (in.)	Choke Plate Pulldown* (in.)	Secondary Vacuum Break (in.)	Fast Idle Setting (rpm)	Choke Unloader (in.)	Choke Setting
1975	Manual	0.420	1	#3	0.140	0.300	—	2000⑥	—	2½ Rich
	Automatic	0.420	1	#2	0.140	0.400	—	2200⑥	—	3½ Rich
1976	Manual	0.410	1	#3	0.420	0.313③	—	2200⑥	0.375	2 Rich
	Automatic	0.410	1	#2	0.320	0.288③	—	2200⑥	0.375	3 Rich
1977	458102, 458103, 458104, 458105	0.420	1	④	0.085	0.250	—	2500	0.350	3 Rich
	458107, 458109	0.420	1	④	0.125	0.275	0.400	2500	0.350	3 Rich
	458110, 458112	0.420	1	④	0.120	0.300	0.400	2500	0.350	3 Rich
1978	see notes	.520	1	—	.150	⑫	—	⑬	.350	⑭

① Located on tag attached to the carburetor, or on the casting or choke plate
② Beginning 1974, GM identification numbers are used in place of the Holley numbers
③ 0.268 in California
④ #1 manual, #2 automatic
⑤ Not used
⑥ With no vacuum to the distributor
* Vacuum break initial choke valve clearance on AMC
⑦ Part #10001048, 10001050: .300
　　#10001047, 10001049, 10001052, 10001054: .325
⑧ Part #10001047, 10001049: 1 Rich
　　#10001048, 10001050, 10001052, 10001054: 2 Rich
⑨ Part #10001047, 10001049: .325
　　#10004048, 10004049: .300
⑩ Part #10001047, 10001049: 2200
　　#10004048, 10004049: 2400
⑪ Part #10001047, 10001049: 1 Rich
　　#10004048, 10004049: 2 Rich
⑫ Part #10001047, 10001049: .325
　　#10004048, 10004049: .300
⑬ Part #10001047, 10001049: 2200
　　#10004048, 10004049: 2400
⑭ Part #10001047, 10001049: 1 Rich
　　#10004048, 10004049: 2 Rich

2MC, M2MC, M2ME CARBURETOR SPECIFICATIONS

PONTIAC　　Rochester Carburetor

Year	Carburetor Identification①	Float Level (in.)	Choke Rod (in.)	Choke Unloader (in.)	Vacuum Break Lean or Front (in.)	Vacuum Break Rich or Rear (in.)	Pump Rod (in.)	Choke Coil Lever (in.)	Automatic Choke (notches)
1975	7045156	5/32	0.130	0.275	0.230	0.150	9/32②	0.120	1 Rich
	7045297	3/16	0.130	0.275	0.275	0.180	9/32②	0.120	1 Rich
	7045298	5/32	0.130	0.275	0.275	0.150	9/32②	0.120	1 Rich
	7045598	5/32	0.160	0.275	0.230	0.150	9/32②	0.120	1 Rich
	7045356	5/32	0.160	0.275	0.275	0.180	9/32②	0.120	1 Rich
1976	8-260 Man.	1/8	0.105	0.210	0.175	0.110	3/16③	0.120	1 Rich
	8-260 Auto.	1/8	0.105	0.210	0.175	0.110	9/32②	0.120	1 Rich
	8-260 Calif.	1/8	0.105	0.210	0.210	0.110	3/16③	0.120	1 Rich④
1977	17057172	11/32	0.075	0.240	0.135	0.240	3/8③	0.120	2 Rich
	17057173	11/32	0.075	0.240	0.165	0.240	3/8③	0.120	2 Rich
1978	17058160	11/32	0.126	0.203	0.142	0.195	1/4②	0.120	2 Rich

① The carburetor identification number is stamped on the float bowl, next to the fuel inlet nut.
② Inner hole
③ Outer hole
④ Index on LeMans

Pontiac

MV, 1MV CARBURETOR SPECIFICATIONS
PONTIAC
Rochester Carburetor

Year	Carburetor Identification①	Float Level (in.)	Metering Rod (in.)	Pump Rod	Idle Vent (in.)	Vacuum Break (in.)	Auxiliary Vacuum Break (in.)	Fast Idle Off Car (in.)	Choke Rod (in.)	Choke Unloader (in.)	Fast Idle Speed (rpm)
1975	7045012	11/32	0.080	—	—	0.200	0.215	—	0.160	0.275	1800②
	7045013	11/32	0.080	—	—	0.350	0.312	—	0.275	0.275	1800②
	7045014	11/32	0.080	—	—	0.257	0.312	—	0.230	0.275	1800②
	Astre Man.	1/8	—	—	—	0.130	—	—	0.080	0.375	2000③
	Astre Auto.	1/8	—	—	—	0.130	—	—	0.080	0.375	2000③
1976	4-140 Man.	1/8	—	—	—	0.055	0.450	—	0.045	0.215	—
	4-140 Auto.	1/8	—	—	—	0.060	0.450	—	0.045	0.215	—
	6-250 Man.	11/32	—	—	—	0.165	0.320	—	0.140	0.265	—
	6-250 Auto	11/32	—	—	—	0.140	0.265	—	0.100	0.265	—
	6-250 Calif.	11/32	—	—	—	0.150	0.260	—	0.135	0.265	—

① The carburetor identification number is stamped on the float bowl, next to the fuel inlet nut.
② High step of cam.
③ No vacuum to distributor.

2GC, 2GV, 2GE CARBURETOR SPECIFICATIONS
PONTIAC
Rochester Carburetor

Year	Carburetor Identification①	Float Level (in.)	Float Drop (in.)	Pump Rod (in.)	Idle Vent (in.)	Primary Vacuum Break (in.)	Secondary Vacuum Break (in.)	Automatic Choke (notches)	Choke Rod (in.)	Choke Unloader (in.)	Fast Idle Speed (rpm)
1975	7045160	9/16	1 7/32	1 3/4	0.025	0.145	0.265	1 Rich	0.085	0.180	—
	7045162	9/16	1 7/32	1 13/16	0.025	0.145	0.260	1 Rich	0.085	0.180	—
	7045171	9/16	1 7/32	1 13/16	0.025	0.145	0.260	1 Rich	0.085	0.180	—
	7045143	15/32	1 7/32	1 13/16	0.025	0.140	0.120	1 Rich	0.080	0.180	—
1976	6-231 Man.	7/16	1 9/32	1 19/32	—	0.110	0.100	1 Rich	0.080	0.140	—
	6-231 Auto.	7/16	1 9/32	1 19/32	—	0.127	0.100	1 Rich	0.080	0.140	—
	6-231 Calif.	7/16	1 9/32	1 19/32	—	0.130	0.110	1 Rich	0.080	0.140	—
	8-350 Ventura	15/32	1 9/32	1 11/32	—	0.140	0.100	1 Rich	0.080	0.180	—
	8-350, 400 Auto.	9/16	1 9/32	1 11/32	—	0.165	0.285	1 Rich	0.085	0.180	—
1977	17057141, 17057147	7/16	1 5/32	1 5/8	—	0.110	0.090	1 Rich	0.080	0.140	—
	17057143, 17057144	7/16	1 5/32	1 19/32	—	0.130	0.100	1 Rich	0.080	0.140	—
	17057145	7/16	1 5/32	1 19/32	—	0.110	0.090	1 Rich	0.080	0.140	—
	17057446, 17057448	7/16	1 5/32	1 19/32	—	0.130	0.110	1 Rich	0.080	0.140	—
	17057447	7/16	1 5/32	1 19/32	—	0.130	0.100	1 Rich	0.080	0.140	—
	17057148	7/16	1 5/32	1 9/16	—	0.110	0.090	1 Rich	0.080	0.140	—
	17057149	7/16	1 5/32	1 9/16	—	0.110	0.040	1 Lean	0.080	0.140	—
	17057445	7/16	1 5/32	1 9/16	—	0.140	0.110	1 Lean	0.080	0.140	—
1978	17058102	19/32	1 9/32	1 17/32	—	0.130	—	Index	0.260	0.325	—
	17058103	19/32	1 9/32	1 17/32	—	0.130	—	Index	0.260	0.325	—
	17058108	19/32	1 9/32	1 21/32	—	0.130	—	Index	0.260	0.325	—
	17058110	19/32	1 9/32	1 21/32	—	0.130	—	Index	0.260	0.325	—
	17058111	19/32	1 9/32	1 5/8	—	0.130	—	Index	0.260	0.325	—
	17058112	19/32	1 9/32	1 21/32	—	0.130	—	Index	0.260	0.325	—

Pontiac

2GC, 2GV, 2GE CARBURETOR SPECIFICATIONS
PONTIAC
Rochester Carburetor

Year	Carburetor Identification①	Float Level (in.)	Float Drop (in.)	Pump Rod (in.)	Idle Vent (in.)	Primary Vacuum Break (in.)	Secondary Vacuum Break (in.)	Automatic Choke (notches)	Choke Rod (in.)	Choke Unloader (in.)	Fast Idle Speed (rpm)
	17058113	19/32	1 9/32	1 5/8	—	0.130	—	Index	0.260	0.325	—
	17058114	19/32	1 9/32	1 21/32	—	0.130	—	Index	0.260	0.325	—
	17058121	19/32	1 9/32	1 5/8	—	0.130	—	Index	0.260	0.325	—
	17058123	19/32	1 9/32	1 5/8	—	0.130	—	Index	0.260	0.325	—
	17058126	19/32	1 9/32	1 17/32	—	0.130	—	Index	0.260	0.325	—
	17058128	19/32	1 9/32	1 17/32	—	0.130	—	Index	0.260	0.325	—
	17058145	7/16	1 5/32	1 5/8	—	0.110	0.110	1 Lean	0.080	0.160	—
	17058147	7/16	1 5/32	1 5/8	—	0.140	0.140	1 Rich	0.080	0.140	—
	17058182	7/16	1 5/32	1 5/8	—	0.110	0.110	1 Rich	0.080	0.140	—
	17058183	7/16	1 5/32	1 5/8	—	0.110	0.110	1 Rich	0.080	0.140	—
	17058185	7/16	1 5/32	1 19/32	—	0.110	0.110	1 Rich	0.080	0.140	—
	17058187	7/16	1 5/32	1 19/32	—	0.110	0.110	1 Rich	0.080	0.140	—
	17058189	7/16	1 5/32	1 19/32	—	0.110	0.110	1 Rich	0.080	0.140	—
	17058408	21/32	1 9/32	1 21/32	—	0.140	0.140	½ Lean	0.260	0.325	—
	17058410	21/32	1 9/32	1 21/32	—	0.140	0.140	½ Lean	0.260	0.325	—
	17058412	21/32	1 9/32	1 21/32	—	0.140	0.140	½ Lean	0.260	0.325	—
	17058414	21/32	1 9/32	1 21/32	—	0.140	0.140	½ Lean	0.260	0.325	—
	17058444	7/16	1 5/32	1 5/8	—	0.140	0.140	1 Rich	0.080	0.140	—
	17058446	7/16	1 5/32	1 5/8	—	0.140	0.140	1 Rich	0.080	0.140	—
	17058447	7/16	1 5/32	1 5/8	—	0.150	0.150	1 Rich	0.080	0.140	—
	17058448	7/16	1 5/32	1 5/8	—	0.140	0.140	1 Rich	0.080	0.140	—

① The carburetor identification number is stamped on the float bowl, next to the fuel inlet nut.

QUADRAJET CARBURETOR SPECIFICATIONS
PONTIAC
Rochester Carburetor

Year	Carburetor Identification①	Float Level (in.)	Air Valve Spring (turn)	Pump Rod (in.)	Primary Vacuum Break (in.)	Secondary Vacuum Break (in.)	Secondary Opening (in.)	Choke Rod (in.)	Choke Unloader (in.)	Fast Idle Speed② (rpm)
1975	7045246	5/16	½	15/32	0.130	0.115	②	0.095	0.240	1800
	7045546	5/16	½	15/32	0.145	0.130	②	0.095	0.240	1800
	7045263	½	½	9/32	0.150	0.260	②	0.130	0.230	1800
	7045264	½	½	9/32	0.150	0.260	②	0.130	0.230	1800
	7045268	½	3/8	9/32	0.150	0.260	②	0.130	0.230	1800
	7045269	½	3/8	9/32	0.160	0.265	②	0.130	0.230	1800
	7045274	½	½	9/32	0.150	0.260	②	0.130	0.230	1800
	7045260	½	½	9/32	0.150	0.260	②	0.130	0.230	1800
	7045262	½	½	9/32	0.150	0.260	②	0.130	0.230	1800
	7045266	½	½	9/32	0.150	0.260	②	0.130	0.230	1800
	7045562	½	½	9/32	0.150	0.260	②	0.130	0.230	1800
	7045564	½	½	9/32	0.150	0.260	②	0.130	0.230	1800
	7045568	½	½	9/32	0.150	0.260	②	0.130	0.230	1800
	7045566	½	½	9/32	0.150	0.260	②	0.130	0.230	1800

Pontiac

QUADRAJET CARBURETOR SPECIFICATIONS
PONTIAC
Rochester Carburetor

Year	Carburetor Identification①	Float Level (in.)	Air Valve Spring (turn)	Pump Rod (in.)	Primary Vacuum Break (in.)	Secondary Vacuum Break (in.)	Secondary Opening (in.)	Choke Rod (in.)	Choke Unloader (in.)	Fast Idle Speed② (rpm)
1976	7045246	5/16	3/4	3/8	0.130	0.120	②	0.095	0.250	1800
	7045546	5/16	3/4	3/8	0.130	0.130	②	0.095	0.250	1800
	7045268	17/32	1/2	3/8	0.160	0.250	②	0.125	0.230	1800
	7045264, 7045274, 7045266	17/32	1/2	3/8	0.160	0.250	②	0.125	0.230	1800
	7045263	17/32	5/8	3/8	0.170	0.250	②	0.125	0.230	1800
	7045564	17/32	1/2	3/8	0.150	0.260	②	0.130	0.230	1800
	7045260, 7045262	17/32	1/2	3/8	0.160	0.250	②	0.125	0.230	1800
	8-455 Man.	17/32	1/2	3/8	0.160	0.250	②	0.125	0.230	1800
	7045562, 7045566	17/32	1/2	3/8	0.170	0.250	②	0.120	0.230	1800
1977	17057250, 17057253, 17057255, 17057256	13/32	1/2	9/32	0.120	0.170	②	0.095	0.205	900
	17057258	13/32	1/2	9/32	0.125	0.215	②	0.095	0.205	1000
	17057550, 17057553	13/32	1/2	9/32	0.125	0.215	②	0.095	0.200	1000
	17057262	17/32	1/2	3/8	0.150	0.240	②	0.130	0.220	1800
	17057263	17/32	5/8	3/8	0.165	0.240	②	0.130	0.220	1800
	17057266, 17057274	17/32	1/2	3/8	0.150	0.240	②	0.130	0.220	1800
1978	17058202	15/32	—	9/32	0.157	—	②	0.314	0.277	③
	17058204	15/32	—	9/32	0.157	—	②	0.314	0.277	③
	17058241	5/16	3/4	3/8	0.117	0.103	②	0.096	0.243	③
	17058250	13/32	1/2	9/32	0.119	0.167	②	0.088	0.203	③
	17058253	13/32	1/2	9/32	0.119	0.167	②	0.088	0.203	③
	17058258	13/32	1/2	9/32	0.126	0.212	②	0.092	0.203	③
	17058263	17/32	5/8	3/8	0.164	0.260	②	0.129	0.220	③
	17058264	17/32	1/2	3/8	0.149	0.260	②	0.129	0.220	③
	17058266	17/32	1/2	3/8	0.149	0.260	②	0.129	0.220	③
	17058272	15/32	5/8●	3/8	0.126	0.195	②	0.071	0.222	③
	17058274	17/32	1/2	3/8	0.149	0.260	②	0.129	0.220	③
	17058276	17/32	1/2	3/8	0.149	0.260	②	0.129	0.220	③
	17058278	17/32	1/2	3/8	0.149	0.260	②	0.129	0.220	③
	17058502	15/32	—	9/32	0.164	—	②	0.314	0.277	③
	17058504	15/32	—	9/32	0.164	—	②	0.314	0.277	③
	17058553	13/32	1/2	9/32	0.126	0.212	②	0.092	0.203	③
	17058582	15/32	7/8	9/32	0.179	—	②	0.314	0.277	③
	17058584	15/32	7/8	9/32	0.179	—	②	0.314	0.277	③

① The carburetor identification number is stamped on the float bowl, near the secondary throttle lever.

② No measurement necessary on two point linkage; see text.

③ 1½ turns after contacting lever for preliminary setting

Pontiac

CAR SERIAL NUMBER AND ENGINE IDENTIFICATION

1976

Mounted behind the windshield on the driver's side is a plate with the vehicle identification number. The sixth character is the model year, with 6 for 1976. The fifth character is the engine code, as follows.

C 231 2-bbl. V-6
D .... 250 1-bbl. 6-cyl.
F 260 2-bbl. V-8
H 350 2-bbl. V-8
M 350 2-bbl. V-8
E 350 4-bbl. V-8
J 350 4-bbl. V-8
R 400 2-bbl. V-8
S 400 4-bbl. V-8
W 455 4-bbl. V-8

NOTE: *The "H" and "J" 350 V-8's are Buick engines with the distributor in the front. The 260 V-8 is an Oldsmobile engine, which has an oil filler tube at the front of the block.*

Engines can be identified by a two or three-letter code. The 6-cylinder code is next to the distributor. The 260 V-8 code is on the oil filler tube. 350 V-8 Buick engines (distributor in front) have their codes near No. 1 plug. The 350 Pontiac engines and all 400 and 455 V-8's have the code on the front of the right bank.

CC 250-cyl. Auto. Calif.
CD 250 6-cyl. Man. 49-States
CF 250 6-Cyl. Auto. 49-States
CH 250 6-cyl. Auto. 49-States
CJ 250 6-cyl. Man. 49-States
FC 231 V-6 Man. Calif.
FH 231 V-6 Man. 49-States
FI 231 V-6 Auto. 49-States
FJ 231 V-6 Auto. Calif.
FK 231 V-6 Auto. Export
FM 231 V-6 Man. Export
FO 231 V-6 Man. Calif.
PA 350 V-8 2-bbl. Auto. 49-States
PB 350 V-8 2-bbl. Auto. 49-States
PE 350 V-8 4-bbl. Auto. 49-States
PF 350 V-8 4-bbl. Auto. 49-States
PM 350 V-8 4-bbl. Auto. Calif.
PN 350 V-8 4-bbl. Auto. Calif.
PO 350 V-8 2-bbl. Auto. Export
PP 350 V-8 4-bbl. Auto. Export
QA 260 V-8 2-bbl. Man. 49-States
QB 260 V-8 2-bbl. Auto. 49-States
QC 260 V-8 2-bbl. Auto. 49-States
QD 260 V-8 2-bbl. Man. 49-States
QK 260 V-8 2-bbl. Man. 49-States
QN 260 V-8 2-bbl. Man. 49-States
QP 260 V-8 2-bbl. Auto. 49-States
QT 260 V-8 2-bbl. Auto. 49-States
TA 260 V-8 2-bbl. Man. Calif.
TE 260 V-8 2-bbl. Auto. Calif.
TD 260 V-8 2-bbl. Man. Calif.
TJ 260 V-8 2-bbl. Auto. Calif.
TK 260 V-8 2-bbl. Man. Calif.
TN 260 V-8 2-bbl. Man. Calif.
TP 260 V-8 2-bbl. Auto. Calif.
TT 260 V-8 2-bbl. Auto. Calif.
T2 260 V-8 2-bbl. Auto. Calif.
T3 260 V-8 2-bbl. Auto. Calif.
T4 260 V-8 2-bbl. Auto. Calif.
T5 260 V-8 2-bbl. Auto. Calif.
WT 400 V-8 4-bbl. Man. 49-States
WX 455 V-8 4-bbl. Man. 49-States
XA 400 V-8 2-bbl. Auto. Export
YA 350 V-8 2-bbl. Auto. 49-States
YB 350 V-8 2-bbl. Auto. 49-States
YC 400 V-8 2-bbl. Auto. 49-States
YJ 400 V-8 2-bbl. Auto. 49-States
YK 350 V-8 2-bbl. Auto. 49-States
YL 350 V-8 2-bbl. Auto. 49-States
YP 350 V-8 2-bbl. Auto. 49-States
YR 350 V-8 2-bbl. Auto. 49-States
YS 400 V-8 4-bbl. Auto. 49-States
YT 400 V-8 4-bbl. Auto. 49-States
YY 400 V-8 4-bbl. Auto. 49-States
YZ 400 V-8 4-bbl. Auto. 49-States
Y3 455 V-8 4-bbl. Auto. 49-States
Y4 455 V-8 4-bbl. Auto. 49-States
Y6 400 V-8 4-bbl. Auto. 49-States
Y7 400 V-8 4-bbl. Auto. 49-States
Y8 455 V-8 4-bbl. Auto. 49-States
ZA 400 V-8 4-bbl. Auto. Calif.
ZB 455 V-8 4-bbl. Auto. Calif.
ZC 350 V-8 4-bbl. Auto. Calif.
ZK 400 V-8 4-bbl. Auto. Calif.
ZX 350 V-8 4-bbl. Auto. Calif.
Z3 455 V-8 4-bbl. Auto. Calif.
Z4 455 V-8 4-bbl. Auto. Calif.
Z6 455 V-8 4-bbl. Auto. Calif.
Z8 400 V-8 2-bbl. Auto. Export

1977

The vehicle identification number is mounted behind the windshield on the driver's side. The sixth character is the model year, with 7 for 1977. The fifth character is the engine code, as follows:

C .... 231 2-bbl. V-6 LD-7 Buick
K .... 403 4-bbl. V-8 L-80 Olds.
L .... 350 4-bbl. V-8 LM-1 Chev.
P .... 350 4-bbl. V-8 L-76 Pont.
R .... 350 4-bbl. V-8 L-34 Olds.
U .... 305 2-bbl. V-8 LG-3 Chev.
V .... 151 2-bbl. 4-cyl. LX-6 Pont.
Y .... 301 2bbl. V-8 L-27 Pont.
Z .... 400 4-bbl. V-8 L-78 Pont.

Engines can also be identified by a two or three-letter code. The V-6 has the code on a label at the front of one rocker cover. Olds V-8's have the code on the oil filler tube. Chevrolet and Pontiac V-8's have the code at the front of the right bank. The 151 4-cyl. has the code on the right side of the block behind the distributor.

CEB 305 2-bbl. V-8 Auto. 49-States
CED 305 2-bbl. V-8 Auto. 49-States
CKM 350 4-bbl. V-8 Auto. "L" Altitude
CKR 350 4-bbl. V-8 Auto. "L" Calif.
CPR 305 2-bbl. V-8 Auto. 49-States
CPY 305 2-bbl. V-8 Auto. 49-States
QP 350 4-bbl. V-8 Auto. "R" 49-States
QQ 350 4-bbl V-8 Auto. "R" 49-States
Q4 350 4-bbl. V-8 Auto. "R" Altitude
Q5 350 4-bbl. V-8 Auto. "R" Altitude
Q8 350 4-bbl. V-8 Auto. "R" Altitude
Q9 350 4-bbl. V-8 Auto. "R" Altitude
RE 231 V-6 Auto. 49-States
RG 231 V-6 Auto. 49-States
RH 231 V-6 Auto. 49-States
SA 231 2-bbl. V-6 49-States
SB 231 V-6 Man. Calif.
SD 231 V-6 Auto. 49-States
SI 231 V-6 Auto. 49-States
SJ 231 V-6 Auto. 49-States
SK 231 V-6 Auto. Calif.
SL 231 V-6 Auto. Calif.
SM 231 V-6 Auto. Altitude
SN 231 V-6 Auto. Altitude
SU 231 V-6 Man. Calif.
SX 231 V-6 Auto. Altitude
SY 231 V-6 Auto. Calif.
TK 350 4-bbl. V-8 Auto. "R" Calif.
TL 350 4-bbl. V-8 Auto. "R" Calif.
TN 350 4-bbl. V-8 Auto. "R" Calif.
TO 350 4-bbl. V-8 Auto. "R" Calif.
TS 350 4-bbl. V-8 Auto. "R" Calif.
TT 350 4-bbl. V-8 Auto. "R" Calif.
TX 350 4-bbl. V-8 Auto. "R" Calif.
TY 350 4-bbl. V-8 Auto. "R" Calif.
UA 403 4-bbl. V-8 Auto. 49-States
UB 403 4-bbl. V-8 Auto. 49-States
UZ 403 4-bbl. V-8 Auto. Altitude
U3 403 4-bbl. V-8 Auto. Altitude
VA 403 4-bbl. V-8 Auto. Calif.
VB 403 4-bbl. V-8 Auto. Calif.
VJ 403 4-bbl. V-8 Auto. Calif.
VK 403 4-bbl. V-8 Auto. Calif.
WA 400 4-bbl. V-8 Auto. 49-States
WB 301 2-bbl. V-8 Man. 49-States
WC 151 2-bbl. 4-cyl. Man. 49-States
WD 151 2-bbl. 4-cyl. Man. 49-States
WF 151 2-bbl. 4-cyl. Man. 49-States
WH 151 2-bbl. 4-cyl. Man. 49-States
XA 400 4-bbl. V-8 Auto. 49-States
XB 350 4-bbl. V-8 Auto. "P" 49-States
XC 350 4-bbl. V-8 Auto. "P" 49-States
XD 400 4-bbl. V-8 Auto. 49-States
XF 400 4-bbl. V-8 Auto. 49-States
XH 400 4-bbl. V-8 Auto. 49-States
XJ 400 4-bbl. V-8 Auto. 49-States
XK 400 4-bbl. V-8 Auto. 49-States
YA 350 4-bbl. V-8 Auto. "P" 49-States
YB 350 4-bbl. V-8 Auto. "P" 49-States
YH 301 2-bbl. V-8 Auto. 49-States
YK 301 2-bbl. V-8 Auto. 49-States
YL 151 2-bbl. 4-cyl. Auto. 49-States
YM 151 2-bbl. 4-cyl. Auto. 49-States
YR 151 2-bbl. 4-cyl. Auto. 49-States
YS 151 2-bbl. 4-cyl. Auto. 49-States
YU 400 4-bbl. V-8 Auto. 49-States
YW 301 2-bbl. V-8 Auto. 49-States
YX 301 2-bbl. V-8 Auto. 49-States
Y4 400 4-bbl. V-8 Auto. 49-States
Y6 400 4-bbl. V-8 Auto. 49-States
Y7 400 4-bbl. V-8 Auto. 49-States
Y9 350 4-bbl. V-8 Auto. "P" 49-States
ZH 151 2-bbl. 4-cyl. Auto. Calif.
ZJ 151 2-bbl. 4-cyl. Auto. Calif.
ZN 151 2-bbl. 4-cyl. Auto. Calif.
ZP 151 2-bbl. 4-cyl. Auto. Calif.

Pontiac

CAR SERIAL NUMBER AND ENGINE IDENTIFICATION

1978

The vehicle identification number is mounted behind the windshield on the driver's side. The sixth character is the model year, with 8 for 1978. The fifth character is the engine code, as follows:

A .... 231 2-bbl. V-6 LD-5 Buick
H .... 305 4-bbl. V-8 LG-4 Chev.
K .... 403 4-bbl. V-8 L-80 Olds.
L .... 350 4-bbl. V-8 LM-1 Chev.
R .... 350 4-bbl. V-8 L-34 Olds.
U .... 305 2-bbl. V-8 LG-3 Chev.
V .... 151 2-bbl. 4-cyl. LX-6 Pont.
W .... 301 4-bbl. V-8 L-37 Pont.
X .... 350 4-bbl. V-8 L-77 Buick
Y .... 301 2-bbl. V-8 L-27 Pont.
Z .... 400 4-bbl. V-8 L-78 Pont.
Z .... 400 4-bbl. V-8 W-72 Pont.
1 .... 151 2-bbl. 4-cyl. LS-6 Pont.

Code "V" is the normal 151 4-cyl. engine. Code "1" is the 151 4-cyl. with electronic fuel control. The "Z" engine comes in two versions. The L-78 is the standard engine with 7.7 to 1 compression ratio. The W-72 is the Trans Am version, with 8.1 to 1 compression ratio.

NOTE: The 4-cylinder engines are covered in the Monza chapter in this supplement.

1979

The vehicle identification number is mounted behind the windshield on the driver's side. The sixth character is the model year, with 9 for 1979. The fifth character is the engine code, as follows:

A .... 231 2-bbl. V-6 LD-5 Buick
H .... 305 4-bbl. V-8 LG-4 Chev.
K .... 403 4-bbl. V-8 L-80 Olds.
L .... 350 4-bbl. V-8 LM-1 Chev.
R .... 350 4-bbl. V-8 L-34 Olds.
V .... 151 2-bbl. 4-cyl. LX-8 Pont.
W .... 301 4-bbl. V-8 L-37 Pont.
X .... 350 4-bbl. V-8 L-77 Buick
Y .... 301 2-bbl. V-8 L-27 Pont.
Z .... 400 4-bbl. V-8 L-78 Pont.
   (w/Perf. Pkg. W-72)
Z .... 400 4-bbl. V-8 W-72 Pont.
1 .... 151 2-bbl. 4-cyl. LS-6 Pont.

Vin code "V" represents the 151 CID, 4-cyl. engine used in the 49-States, while the Vin code "1" represents the 151 CID engine used in the state of California.

EMISSION EQUIPMENT

1976

All Models
Closed positive crankcase ventilation
Emission calibrated carburetor
Emission calibrated distributor
Heated air cleaner
Vapor control, canister storage
Exhaust gas recirculation
Catalytic converter
Single diaphragm vacuum advance
  49-States
    All engines
  Calif.
    All engines except
    No vacuum advance on 260 V-8
Air pump
  49-States
    Not used
  Calif.
    250 6-Cylinder, all
    231 V-6 auto. trans. only
    350 4-bbl. V-8 in Firebird, Lemans, and Grand Prix.
    All 400 V-8
    455 V-8 code Z6 only
Early fuel evaporation
  49-States
    All engines except,
    Not on 260 V-8
  Calif.
    All 231 V-6
    All 250 6-Cyl.
    260 V-8 codes TE, TJ, TP, TT only
    All Ventura 350 V-8
    All 455 V-8
Temperature controlled choke vacuum break
  All engines except,
  Not on 260 V-8
  Not on Ventura 350 V-8
Spark delay orifice
  350 2-bbl. V-8 except Ventura
  350 4-bbl. V-8 auto. trans. only, except Ventura
  All 400 2-bbl. V-8
  400 4-bbl. V-8 auto. trans. only
  455 V-8 auto. trans. only
Distributor vacuum advance modulator valve
  49-States
    455 V-8 auto. trans. only
    260 V-8 auto. trans. only
  Calif.
    Not used
Choke air modulator
  All 231 V-6
  All Ventura 350 V-8
Air cleaner thermal control valve
  All 260 V-8

1977

Closed positive crankcase ventilation
Emission calibrated distributor
Emission calibrated carburetor
Heated air cleaner
Vapor control, canister storage
Exhaust gas recirculation
Catalytic converter
Air pump
  49-States
    Not used
  Calif.
    All V-6 and V-8 models
    Not used on 4-cyl.
  Altitude
    231 2-bbl. V-6
    350 4-bbl. V-8 "L"
Early fuel evaporation
  All models, except
  Not on Olds-built engines
Pulse air system
  49-States
    Not used
  Calif.
    151 4-cyl.
  Altitude
    Not used
Temperature controlled choke vacuum break
  On all V-8 engines except,
  Not on 305 2-bbl. V-8
  Not on 350 4-bbl. V-8 "L"

Vacuum delay valve, 4-nozzle
  151 4-cyl.
  49-State 301 V-8 auto. trans. with air cond.
Spark delay restrictor orifice
  350 4-bbl. V-8 "P"
  400 4-bbl. V-8
Distributor vacuum advance modulator valve
  49-State 301 2-bbl. V-8 auto. trans. with air cond.
Choke vacuum delay valve
  151 4-cyl.
Electric choke
  151 4-cyl.
  231 V-6
Air cleaner thermal control valve
  49-State 301 2-bbl. V-8
Transmission controlled spark
  49-States
    Not used
  Calif.
    231 V-6 man. trans.
  Altitude
    231 V-6 man. trans.

1978

All Models
Closed positive crankcase ventilation
Emission calibrated distributor
Emission calibrated carburetor
Heated air cleaner
Vapor control, canister storage
Exhaust gas recirculation
Catalytic converter
Air pump
  49-States
    Not used
  Calif.
    All V-6 and V-8 models
    Not used on 4-cyl.
  Altitude
    All models
Early fuel evaporation
  All models, except
  Not on Olds-built engines
Temperature controlled choke vacuum break

Pontiac

EMISSION EQUIPMENT

1978

49-States
  231 V-6
  260 V-8
  350 V-8 "R"
  403 V-8
Calif.
  231 V-6 auto. trans. only
  260 V-8
  350 V-8 "R"
  403 V-8
Altitude
  231 V-6
  260 V-8
  350 V-8 "R"
  403 V-8
Vacuum delay valve, 4-nozzle
  151 4-cyl. "V"
  301 2-bbl. V-8, engine codes XC, XD, YL, YN (on front of right bank.)
  301 4-bbl. V-8 with air cond.
Spark delay restrictor orifice
  400 V-8 auto. trans.
Distributor vacuum advance modulator valve
  301 2-bbl. V-8 engine codes XC, XD, YL, YN (stamped on front of right bank.
  301 4-bbl. V-8 with air cond.
  400 4-bbl. V-8 auto. trans. except W-72
Choke vacuum delay valve
  151 4-cyl.
Electric choke
  151 4-cyl.
  231 V-6

1979

All Models

Closed positive crankcase ventilation
Emission calibrated distributor
Emission calibrated carburetor
Heated air cleaner
Vapor control, canister storage
Exhaust gas recirculation
System w/back pressure transducer
Catalytic converter
Air pump
  49-States
    Not used
  Calif.
    All V-6 and V-8 models
    Not used on 4-cyl.
  Altitude
    Series H—231 CID—2-bbl. V-6 w/auto. trans.
    Series A & G—305 CID—4 bbl. V-8 w/auto. trans.
    Series A Wagon, F & X—350 CID—4-bbl. V-8 w/auto. trans.
Early fuel evaporation
  All models, except
    Olds-built engines and 151 CID 4-cyl. engines for 49-States use
Temperature Vacuum Switch (TVS)
Distributor Spark-Vacuum Modulated Valve (DS-VMV)
Distributor Spark-Thermal Vacuum Switch (DS-TVS)
Electric choke
Idle solenoid
Choke vacuum break

IDLE SPEED AND MIXTURE ADJUSTMENTS

1976

Air cleaner .... On
Vapor hose to carb .... Plugged
Air cond.
  V-6 .... Off
  V-8 .... Off
250 6-cyl.
  For idle speed adj.
    Man. trans. .... Off
    Auto. trans. .... On
  For mixture adj. .... Off
EGR hose to carb. .... Plugged
Distributor vac. hose. ... Connected
Auto. trans. .... Drive
Mixture adj. . See rpm drop below

NOTE: *Engines whose distributor vacuum advance operates on ported vacuum (see list below) should have their vacuum advance hose disconnected and plugged when setting idle speed and mixture. If, when reconnecting the hose after setting the idle, the engine speeds up, the throttle is open too far. This is usually caused by retarded initial timing.*

231 V-6
  Auto. trans. 600
    Mixture adj. .... 80 rpm drop (680-600)
  Man. trans. 800
    Mixture adj. ... 300 rpm drop (1100-800)
250 6-cylinder 49-States
  Auto. trans. Ventura and Firebird 550
    Mixture adj. .... 30 rpm drop (580-550)
  Auto. trans. LeMans
    No air cond. 550
      Mixture adj. .... 30 rpm drop (580-550)
    With air cond. 575
      Mixture adj. .. 30 rpm drop (605-575)
  Man. trans. 850
    Mixture adj. ... 350 rpm drop (1200-850)
250 6-cylinder, Calif.
  Auto. trans. 600
    Mixture adj. .... 40 rpm drop (640-600)
260 V-8 49-States
  Auto. trans. 550
    Mixture adj. .... 60 rpm drop (610-550)
  Man. trans. 750
    Mixture adj. ... 325 rpm drop (1075-750)
260 V-8 Calif.
  Auto. trans. 600
    Mixture adj. ... 100 rpm drop (700-600)
350 V-8 Ventura 600
  Mixture adj. ...... 80 rpm drop (680-600)
350 4-bbl. V-8 Calif.
350 2-bbl. V-8 49-States
  LeMans & Firebird
    Auto. trans. 550
      Mixture adj. .. 90 rpm drop (640-550)
  LeMans & Firebird
    Auto. trans. 600
      Mixture adj. .. 75 rpm drop (675-600)
400 2-bbl. V-8
  Auto. trans. 550
    Mixture adj. .... 90 rpm drop (640-550)
400 4-bbl. V-8
  Auto. trans. 575
    Mixture adj. .... 65 rpm drop (640-575)
  Man. trans. 775
    Mixture adj. ... 150 rpm drop (925-775)
455 4-bbl. V-8 49-States
  Auto. trans. 550
    Mixture adj. .... 90 rpm drop (640-550)
455 4-bbl. V-8 Calif.
  Auto. trans. 600
    Mixture adj. .... 40 rpm drop (640-600)

1977

For Idle Speed Adjustment

Air cleaner .... In place
Vapor hose to carb. .... Plugged
EGR hose to carb.
  V-6 and V-8 .... Plugged
  4-cyl. .... In place
Air cond. .... Off
Auto. trans. .... Drive

For Idle Mixture Adjustment

Air cleaner .... Set aside with hoses connected
Auto. level comp. hose ...... Plugged
Other hoses .... See underhood label
Air cond. .... Off
Auto. trans. .... Drive
Idle CO .... Not used
Mixture adj. ..... See rpm drop below
151 4-cyl. No air cond.
  Auto. trans.
    Solenoid connected 650
    Solenoid disconnected 500
    Mixture adj. ......35 rpm drop
  Man. trans.
    Solenoid connected 1000
    Solenoid disconnected 500
    Mixture adj. ..... 250 rpm drop (1250-1000)
151 4-cyl. with air cond.
  Auto. trans. 650

Pontiac

IDLE SPEED AND MIXTURE ADJUSTMENTS

1977

Mixture adj. ......35 rpm drop
     (685-650)
A.C. idle speedup 850
Man. trans. 1000
Mixture adj. .....250 rpm drop
     (1250-1000)
A.C. idle speedup 1250

NOTE: *All 151 4-cyl. engines have a carburetor solenoid. On engines without air conditioning, the solenoid is for anti-dieseling, and the idle speed is set with the solenoid adjustment. On air conditioned cars, the solenoid is for idle speedup, and is only energized when the air conditioning is on. The curb idle speed adjustment is made with the throttle screw.*

231 V-6 auto. trans. 600
  Mixture adj. ....40 rpm drop
     (640-600)
  A.C. idle speedup 670
231 V-6 man. trans.
  Solenoid connected 800
  Solenoid disconnected 600
  Mixture adj.
    49-States ....60 rpm drop
     (860-800)
    Calif. ....10 rpm drop
     (810-800)
301 2-bbl. V-8 auto. trans. 550
  Mixture adj. ....40 rpm drop
     (590-550)
  A.C. idle speedup 650
301 2-bbl. V-8 man. trans. 750
  Mixture adj. ....120 rpm drop
     (870-750)
  A.C. idle speedup 850
305 2-bbl. V-8 500
  Mixture adj. ....30 rpm drop
     (530-500)
  A.C. idle speedup 650
350 4-bbl. V-8 "P" 575
  Mixture adj. ....25 rpm drop
     (600-575)
  A.C. idle speedup 650
350 4-bbl. V-8 "L"
  Calif. 500
    Mixture adj. ......50 rpm drop
     (550-500)
    A.C. idle speedup 650
  Altitude 600
    Mixture adj. ......50 rpm drop
     (650-600)
    A.C. idle speedup 650
350 4-bbl. V-8 "R"
  49-States 550
    Mixture adj. ......30 rpm drop
     (580-550)
    A.C. idle speedup 650
  Calif. 550
    Mixture adj. ......25 rpm drop
     (575-550)
    A.C. idle speedup 650
  Altitude 600
    Mixture adj. ......25 rpm drop
     (625-600)
    A.C. idle speedup 700

400 V-8 49-States
  Auto trans. W-72 eng. (Y6) ..600
    Mixture adj. ......40 rpm drop
     (640-600)
    A.C. idle speedup 700
  Auto. trans. others 575
    Mixture adj. ......40 rpm drop
     (615-575)
    A.C. idle speedup 650
  Man. trans. 775
    Mixture adj. .....195 rpm drop
     (970-775)
403 V-8 49-States 500
  Mixture adj. ....80 rpm drop
     (580-500)
  A.C. idle speedup 650
403 V-8 Calif. 550
  Mixture adj. ....25 rpm drop
     (575-550)
  A.C. idle speedup 650
403 V-8 Altitude 600
  Mixture adj. ....25 rpm drop
     (625-600)
  A.C. idle speedup 650

1978

Air cleaner .... In place
Air cond. .... Off
Auto. trans. .... Drive
Vapor hose to carb. .... Plugged
EGR hose to carb. .... Plugged
Idle CO .... Not used
Mixture adj. . See propane speed below

NOTE: *Plug the distributor vacuum advance on the 400 4-bbl. man. trans. engine to keep the advance from affecting the idle speed.*

231 V-6 49-States
  Auto. trans. 600
    Propane enriched 650
    AC idle speedup 675
  Man. trans.
    Solenoid connected 800
    Solenoid disconnected 600
    Propane enriched 940
231 V-6 Calif.
  Auto. trans. 600
    Propane enriched 615
    AC idle speedup 670
  Man. trans.
    Solenoid connected 800
    Solenoid disconnected 600
    Propane enriched 880
231 V-6 Altitude
  Auto. trans. 600
    Propane enriched 615
    AC idle speedup 670
  Man. trans.
    Solenoid connected 800
    Solenoid disconnected 600
    Propane enriched 880
301 2-bbl. V-8 49-States
  Auto. trans. 550
    Propane enriched 580
    AC idle speedup 650
301 4-bbl. V-8 49-States
  Auto. trans. 550
    Propane enriched 590
    AC idle speedup 650

305 2-bbl. V-8 49-States
  Auto. trans. 500
    Propane enriched 530
    AC idle speedup 600
  Man. trans. 600
    Propane enriched 720
    AC idle speedup 700
305 2-bbl. V-8 Calif.
  Auto. trans. 500
    Propane enriched 530
    AC idle speedup 650
305 2-bbl. V-8 Altitude
  Auto. trans. 600
    Propane enriched 630
    AC idle speedup 700
350 4-bbl. V-8 "X" 49-States
  Auto. trans. 550
    Propane enriched 590
350 4-bbl. V-8 "L" 49-States
  Man. trans. 700
    Propane enriched 850-900
350 4-bbl. V-8 "L" Calif.
  Auto. trans. 500
    Propane enriched 550
    AC idle speedup 600
350 4-bbl. V-8 "L" Altitude
  Auto. trans. 600
    Propane enriched 650
    AC idle speedup 650
350 4-bbl. V-8 "R" Calif.
  Auto. trans. 550
    Propane enriched 575
    AC idle speedup 650
350 4-bbl. V-8 "R" Altitude
  Auto. trans. 550
    Propane enriched 635
    AC idle speedup 650
400 4-bbl. V-8 49-States
  Auto. trans. except W-72 575
    Propane enriched 615
    AC idle speedup 650
  Auto. trans. W-72 600
    Propane enriched 640
    AC idle speedup 700
  Man. trans. W-72 775
    Propane enriched 950
403 4-bbl. V-8 Calif.
  Auto. trans. 550
    Propane enriched 575
    AC idle speedup 650
403 4-bbl. V-8 Altitude
  Auto. trans. 600
    Propane enriched ....
    AC idle speedup 700

1979

The idle mixture screws have been preset and sealed at the factory. Idle mixture should only be adjusted in cases of major carburetor overhaul, throttle body replacement or high idle CO levels, as determined by state or local inspections, to obtain the correct idle speed with the use of the propane enrichment method. Adjustment of the idle speed by any other method, may violate emission laws of the Federal, States or Provincial agencies.

Refer to the Emission Control Information label to obtain the correct

Pontiac

IDLE SPEED AND MIXTURE ADJUSTMENTS

1979

idle speed, timing adjustment, disconnection of hoses or tubes and the location of the propane enrichment hose to the carburetor. The ECI Label reflects any manufacturing changes to the engine, concerning the emission levels.

Air cleaner .... In place
Air cond. .... Off
Auto. trans. .... Drive
Vapor hose to carb. ...... Plugged
EGR hose to carb. .... Plugged
Idle CO .... Not used
Mixture adj. See propane speed below

151 CID 4-cyl. Low Alt. Vin code V
  w/Auto. trans.
    Curb or "on" idle 850D
    Slow or "off" idle 650D
       (500 D—w/ AC)
    Propane enriched idle 695

  w/Manual trans.
    Curb or "on" idle 1250N
    Slow or "off" idle 900N
       (500 N—w/AC)
    Propane enriched idle 1040

151 CID 4-cyl. Calif. Vin code 1
  w/auto. trans.
    Curb or "on" idle 850D
    Slow or "off" idle 650D
       (500 D—w/AC)
    Propane enriched idle ... ②

  w/Manual trans.
    Curb or "on" idle 1000N
    Slow or "off" idle 500N
    Propane enriched idle ... ②

231 CID V-6 Low Alt. Vin code A
  w/Auto. trans.
    Curb or "on" idle 670D
    Slow or "off" idle 550D
    Propane enriched idle 575

  w/Manual trans.
    Curb or "on" idle 800N
    Slow or "off" idle 600N
    Propane enriched idle 1000

231 CID V-6 Calif. Vin Code A
  w/Auto. trans.
    Curb or "on" idle .... N/A
    Slow or "off" idle 600D
    Propane enriched idle 615

  w/Manual trans.
    Curb or "on" idle 800N
    Slow or "off" idle 600N
    Propane enriched idle 840

231 CID V-6 High Alt. Vin code A
  w/Auto. trans.
    Curb or "on" idle .... N/A
    Slow or "off" idle 600D
    Propane enriched idle 615

301 CID 2-bbl. V-8 Low Alt. Vin code Y
  w/Auto. trans.
    Curb or "on" idle 650D
    Slow or "off" idle 500D
    Propane enriched idle 530

301 CID 4-bbl. V-8 Low Alt. Vin code W
  w/Auto. trans.
    Curb or "on" idle 650D
    Slow or "off" idle 500D
    Propane enriched idle 540

  w/Manual trans.
    Curb or "on" idle 800N
    Slow or "off" idle 700N
    Propane enriched idle ... 810N

305 CID 2-bbl. V-8 Low Alt. Vin code G
  w/Auto. trans.
    Curb or "on" idle 600D
    Slow or "off" idle 500D
    Propane enriched idle .. 520-540

  w/Manual trans.
    Curb or "on" idle 700N
    Slow or "off" idle 600N
    Propane enriched idle .. 710-750

305 CID 2-bbl. V-8 Calif. Vin code G
  w/Auto. trans.
    Curb or "on" idle 650D
    Slow or "off" idle 600D
    Propane enriched idle .. 640-660

305 CID 4-bbl. V-8 Calif. Vin code H
  w/Auto trans.
    Curb or "on" idle 600D
    Slow or "off" idle 500D
    Propane enriched idle .. 540-560

305 CID 4-bbl. V-8 High Alt. Vin code H
  w/Auto. trans.
    Curb or "on" idle 600D
    Slow or "off" idle 500D
    Propane enriched idle .. 640-660

350 CID 4-bbl. V-8 Low Alt. Vin code X
  w/Auto. trans.
    Curb or "on" idle .... N/A
    Slow or "off" idle 550D
    Propane enriched idle 800

350 CID 4-bbl. V-8 Calif. Vin code R & L
  w/Auto. trans.
    Curb or "on" idle 600D
    Slow or "off" idle 500D
    Propane enriched idle ... R-①,
                       L-640-660

350 CID 4-bbl. V-8 High Alt. Vin code R & L
  w/Auto. trans.
    Curb or "on" idle 650D
    Slow or "off" idle 600D
    Propane enriched idle ... R-①,
                       L-640-660

400 CID 4-bbl. V-8 Low Alt. Vin code Z
  w/Manual trans.
    Curb or "on" idle .... N/A
    Slow or "off" idle 775N
    Propane enriched idle ... 800

403 CID 4-bbl. V-8 Low Alt. Vin code K
  w/Auto. trans.
    Curb or "on" idle 650D
    Slow or "off" idle 550D
    Propane enriched idle ... ①

403 CID 4-bbl. V-8 Calif. Vin code K
  w/Auto. trans.
    Curb or "on" idle 600D
    Slow or "off" idle 500D
    Propane enriched idle ... ①

403 CID 4-bbl. V-8 High Alt. Vin code K
  w/Auto. trans.
    Curb or "on" idle .... ①
    Slow or "off" idle .... ①
    Propane enriched idle ... ①

N/A—Not Applicable
① Refer to the Emission Control Information Label
② Refer to the 1979 Emission specifications

INITIAL TIMING

1976

231 V-6 .... 12° BTDC
250 6-cyl.
  Auto. trans. .... 10° BTDC
  Man. trans. .... 6° BTDC
260 V-8 (at 1100 rpm)
  49-States auto. trans. . 18° BTDC
  49-States man. trans. . 16° BTDC
  Calif. all .... 14° BTDC
350 V-8 Ventura .... 12° BTDC
350 V-8 except Ventura . 16° BTDC
400 V-8
  Auto. trans. .... 16° BTDC
  Man. trans. .... 12° BTDC
455 V-8 49-States
  Auto. trans. .... 16° BTDC
  Man. trans. .... 12° BTDC
455 V-8 Calif.
  Auto. trans. .... 12° BTDC

1977

NOTE: *Distributor vacuum hose must be disconnected and plugged. Set timing at idle speed unless shown otherwise.*

151 4-cyl. 49-States ..... 14° BTDC
151 4-cyl. Calif. .... 12° BTDC
231 V-6 .... 12° BTDC
301 2-bbl. V-8
  Auto trans. .... 12° BTDC
  Man. trans. .... 16° BTDC
305 2-bbl. V-8
  Codes CPY and CPR .. 8° BTDC
  Codes CEB and CED .. 4° BTDC
350 4-bbl. V-8 "P" ...... 16° BTDC
350 4-bbl. V-8 "R" (at 1100 rpm)
  49-States .... 20° BTDC
Calif.
  Firebird .... 18° BTDC
  Ventura .... 18° BTDC
  All others .... 20° BTDC
Altitude .... 20° BTDC
350 4-bbl. V-8 "L" ...... 8° BTDC
400 4-bbl. V-8 49-States
  Auto trans.
    Standard V-8 .... 16° BTDC
    W-72 Hi. Perf. ..... 18° BTDC
  Man. trans. .... 18° BTDC
403 4-bbl. V-8 (at 1100 rpm)
  49-States .... 22° BTDC
  Calif. .... 20° BTDC
  Altitude .... 20° BTDC

1978

NOTE: *Distributor vacuum hose must be disconnected and plugged. Set timing at idle speed unless otherwise.*

217

Pontiac

1978

- 231 V-6 .... 15º BTDC
- 301 2-bbl. V-8 .... 12° BTDC
- 301 4-bbl. V-8 .... 12° BTDC
- 305 2-bbl. V-8
  - 49-States .... 4° BTDC
  - Calif. .... 6° BTDC
  - Altitude .... 8° BTDC
- 350 4-bbl. V-8 "L"
  - 49-States .... 6° BTDC
  - Calif. .... 8° BTDC
  - Altitude .... 8° BTDC
- 350 4-bbl. V-8 "R" (at 1100 rpm)
  - Calif. .... 20° BTDC
  - Altitude .... 20° BTDC
- 350 4-bbl. V-8 "X"
  - 49-States .... 15° BTDC
- 400 4-bbl. V-8 49-States
  - Standard V-8 .... 16° BTDC
  - W-72 Hi. Perf. .... 18° BTDC
- 403 4-bbl. V-8 (At 1100 rpm)
  - Calif. .... 20° BTDC
  - Altitude .... 20° BTDC

1979

NOTE: *Distributor vacuum hose must be disconnected and plugged.*

INITIAL TIMING

Set timing at idle speed unless shown otherwise.

- 151 CID
  - Low Alt.
    - w/Auto. trans. ....12° @ 650D
    - w/Manual trans. ...12° @ 900N
  - Calif.
    - All ....14° @ 1000
- 231 CID
  - Low Alt.
    - w/Auto trans. ......15° @ 600
    - w/Manual trans. ....15° @ 800
  - Calif.
    - w/Auto trans. ......15° @ 600
    - w/Manual trans. ....15° @ 800
  - High Alt.
    - w/Auto. trans. ......15° @ 600
- 301 CID
  - Low Alt.
    - w/Auto. trans.
      - (2 & 4-bbl.) .....12° @ 650D
    - w/Manual trans.
      - (4-bbl.) ....14° @ 750N
- 305 CID
  - Low Alt.
    - w/Auto. trans.
      - (2-bbl.) ....4° @ 500D
    - w/Manual trans.
      - (2-bbl.) ....4° @ 600N
  - Calif.
    - w/Auto. trans.
      - (2-bbl., series F, X) 4° @ 500D
    - w/Auto. trans.
      - (2-bbl., series H) . 2° @ 600D
    - w/Auto. trans.
      - (4-bbl) ....4° @ 500D
  - High Alt.
    - w/Auto. trans. .... 4° @ 600D
- 350 CID
  - Low Alt.
    - w/Auto. trans.
      - (4-bbl.) ....15° @ 550
  - Calif.
    - w/Auto. trans.
      - (4-bbl., series X). 8° @ 500D
      - (Series B) ...20° @ 1100P
  - High Alt.
    - w/Auto. trans.
      - (4-bbl.) .... 8° @ 600D
- 400 CID
  - Low Alt.
    - w/Manual trans. ...18° @ 775N
- 403 CID
  - Low Alt.
    - w/Auto. trans. ...18° @ 1100P
  - Calif.
    - w/Auto. trans. ...20° @ 1100P

SPARK PLUGS

231 V-6	AC-R46TS	.040
231 V-6	AC-R46TSX	.060
301 V-8	AC-R46TSX	.060
305 V-8	AC-R45TS	.045
350 V-8 "P"	AC-R45TSX	.060
350 V-8 "R"	AC-R46SZ	.060
350 V-8 "L"	AC-R45TS	.045
400 V-8	AC-R45TSX	.060
403 V-8	AC-R46SZ	.060

1978

231 V-6	AC-R46TSX	.060
301 2-bbl.	AC-R46TSX	.060
301 4-bbl.	AC-R45TSX	.060
305 2-bbl.	AC-R45TS	.045
350 V-8 "L"	AC-R45TS	.045
350 V-8 "R"	AC-R46SZ	.060
350 V-8 "X"	AC-R46TSX	.060
400 V-8	AC-R45TSX	.060
403 V-8	AC-R46SZ	.060

1979

151 4-cyl.	AC-R46TSX	.060
231 V-6	AC-R46TSX	.060
301 2-bbl.	AC-R46TSX	.060
301 4-bbl.	AC-R45TSX	.060
305 2-bbl.	AC-R45TS	.045
350 V-8 "L"	AC-R45TSX	.045
350 V-8 "R"	AC-R46SZ	.060
350 V-8 "X"	AC-R46TSX	.060
400 V-8	AC-R45TSX	.060
403 V-8	AC-R46SZ	.060

1976

231 V-6	AC-R44SX	.060
250 6-Cyl.	AC-R46TX	.060
260 V-8	AC-R46SX	.080
350 V-8		
Ventura	AC-R45TSX	.060
Firebird, LeMans	AC-R46TSX	.060
400 V-8		
2-bbl.	AC-R46TSX	.060
4-bbl.	AC-R45TSX	.060
455 V-8	AC-R45TSX	.060

1977

151 4-cyl.	AC-R44TSX	.060

VACUUM ADVANCE

1976

- 231 V-6 .... Ported
- 250 6-cyl.
  - 49-States .... Manifold
  - Calif. .... Ported
- 260 V-8 49-States
  - Auto. trans. .... Modulated
  - Man. trans. .... Ported
- 260 V-8 Calif. .... None
- 350 V-8 Ventura .... Ported
- 350 V-8 except Ventura .. Manifold
- 400 V-8
  - 2-bbl. all .... Manifold
  - 4-bbl. auto. trans. ..... Manifold
  - 4-bbl. man. trans. ...... Ported
- 455 V-8 49-States
  - Auto. trans. .... Modulated
  - Man. trans. .... Ported

- 455 V-8 Calif.
  - Auto. trans. .... Manifold

1977

Diaphragm type ....Single
Vacuum source
- 151 4-cyl.
  - With air cond. .... Ported
  - No air cond. .... Manifold

NOTE: *The 151 4-cyl. with air conditioning uses ported vacuum at normal operating temperature, but manifold vacuum when cold.*

- 231 V-6
  - 49-States .... Manifold
  - Calif. .... Ported
  - Altitude .... Manifold

- 301 V-8 49-States
  - Man. trans. .... Ported
  - Auto. trans. no A.C. .....Manifold
  - Auto. trans. with A.C. .Modulated

- 305 V-8 .... Manifold
- 350 V-8 "P" .... Manifold

- 350 V-8 "R"
  - 49-States .... Manifold
  - Calif. .... Ported
  - Altitude .... Ported

- 350 V-8 "L" .... Manifold
- 400 V-8 .... Manifold

- 403 V-8
  - 49-States .... Manifold
  - Calif. .... Ported
  - Altitude .... Ported

Pontiac

VACUUM ADVANCE

1978

Diaphragm type .... single
Vacuum source
231 V-6 .... ported
301 2-bbl. V-8 .... manifold
301 4-bbl. V-8 .... manifold
305 2-bbl. V-8
  49-States .... manifold
  Calif. .... ported
  Altitude .... manifold
350 V-8 "L" .... manifold
350 V-8 "R" .... ported

350 V-8 "X" .... manifold
400 V-8
  Man. trans. .... ported
  Auto. trans.
    Single muffler .... modulated
    Dual mufflers .... manifold
403 V-8 .... ported

1979

Diaphragm Type .... Single
Vacuum Source
151 4-cyl.

Low Alt. (w/AC) ...... Ported
Calif. .... Ported
Low Alt. (without AC) . Manifold
231 V-6 .... Ported
301 2-bbl. V-8 .... Manifold
301 4-bbl. V-8 .... Manifold
305 2-bbl. V-8 .... Manifold
350 V-8 .... Manifold
400 V-8 .... Manifold
403 V-8
  Low and High Alt. .... Manifold
  Calif. .... Ported

EMISSION CONTROL SYSTEMS 1976

IDLE SPEEDUP SOLENOID 1976

On all Pontiac-built V-8 engines (distributor in rear) and the Oldsmobile 260 V-8, the carburetor is equipped with a solenoid when the car has air conditioning. This solenoid looks the same as the old idle stop solenoid, sometimes called an anti-dieseling solenoid. The difference is that the 1976 solenoid is connected to the air conditioning switch, and is only on when the air conditioning is on. It is strictly an idle speedup for better engine operation at idle when the air conditioning is on. The idle speedup rpm is set with the solenoid adjustment, but the curb idle speed is set with the throttle screw on the carburetor. The idle speedup speed for all Pontiac 1976 cars is 675 rpm in Drive. When making the setting, the air conditioning must be on, but the compressor should be disconnected by pulling the plug at the compressor. This eliminates any compressor cycling on and off that might change the idle speed while the setting is being made.

EXHAUST GAS RECIRCULATION 1976

Description of System

The basic system is the same as in 1975, but the thermal vacuum valve has been eliminated. In its place is a thermal control valve that senses engine temperature from a metal shield covering the valve. The switching temperature of the valve is approximately 53-68° F. Below that temperature the valve is closed, blocking the vacuum to the EGR valve. Above that temperature the valve opens, and al-

1976 Pontiac thermal control valve. It is hidden under a metal housing at the front of the engine

lows the ported vacuum to open the EGR valve above idle.

Testing Thermal Control Valve

Cool the valve in a refrigerator until its temperature is below 53° F. Connect a hand vacuum pump to the CARB side of the valve and pump up 15" Hg. of vacuum. The valve should hold the vacuum without leaking down.

Warm the valve until its temperature is 68° F. or more. The valve should be open and you should be able to blow through it. If not, the valve is defective.

Repairing Thermal Control Valve

Repairs are limited to replacement. The CARB side of the valve should be connected to the carburetor port, and the EGR side to the EGR valve or the back-pressure transducer, if a transducer is used.

SPARK DELAY RESTRICTOR 1976

Description of Restrictor

The restrictor is a small piece of plastic tube that inserts between two pieces of hose. It slows down the application of vacuum to the vacuum advance, and also slows down the loss of vacuum during acceleration. The restrictor has no moving parts. The restrictor takes the place of the spark delay valve that was used in former years.

Testing Restrictor

Remove the restrictor from the hose and look through it at the light to see if it is open. Blow it out if necessary.

Repairing Restrictor

Repairs are limited to replacement. The restrictor can be inserted into the hose in either direction.

DISTRIBUTOR VACUUM ADVANCE MODULATOR VALVE 1976

This valve is only used on the 260 V-8, and is identical to the Spark Advance Vacuum Modulator (SAVM) described in the Oldsmobile chapter, in this supplement.

231 V-6 ENGINE 1976

This is the Buick V-6 engine, available only in the Sunbird models. The emission controls are similar to what is on the 1975 Buick V-6.

Pontiac

151 CUBIC INCH 4-CYLINDER ENGINE 1977

In the 1962 model year Chevrolet came out with the Chevy II compact car. It was designed as basic, thrifty transportation and came with a smaller version of Chevy's old reliable 6-cylinder engine. Also available, but seldom purchased, was a 153-cubic-inch 4-cylinder engine. It was simply a six with two cylinders lopped off. The four was available for nine model years, finally being discontinued when the 1971 Vega appeared.

From 1971 to 1976, the four was kept alive in some of General Motors remote outposts, mainly Brazil. When the energy crisis struck, the decision was made to bring back the four, and Pontiac is the first G.M. division to use it.

Although the new four will certainly look familiar to those who worked on the old engine, it has been extensively modified and brought up to date with all the modern G.M. features such as High Energy Ignition, bigger bore, and shorter stroke. Because of all the changes, it appears that very few, if any, of the parts will interchange with the old engine. In addition, the changes have made the new four a different engine from the current Chevy 6. You can't really say any more that it is a Chevy 6 with two cylinders lopped off, although some of the parts will interchange.

The carburetor on the 151 is the Holley-Weber, basically the same one that has been used for several years on the Vega, but with an electric choke. The rest of the engine is conventional, and about as easy to work on as a Chevy 6.

301 V-8 1977

This engine is made by Pontiac, and appears to be just another Pontiac V-8. But it is a completely new engine, designed in an attempt to give 6-cylinder economy with V-8 smoothness. The engine has been made extremely light. In comparison with the Pontiac 350 V-8, the new 301 has a block that is 61 lbs. lighter. Each cylinder head is over seven pounds lighter. The intake manifold, available in a 2-bbl. design only, is a simple single-level type that saved 19 pounds.

The most surprising weight-saver is the crankshaft. If you are used to seeing counterweights alongside each throw, this shaft looks as if they left something off, and they did. There are only two counterweights, one at each end. The middle of the shaft is a plain crank, a daring, unconventional design for a modern engine, but it saved 23½ pounds.

Because of the lightweight block and cylinder walls, there is much more cylinder wall distortion when the heads are tightened. To avoid problems with rings and pistons, all 301 engines have a boring plate torqued to the head before they are bored. The plate simulates a tightened cylinder head. When the plate is removed, the bores distort, but go back to round holes when the actual heads are installed. Boring plates have been used for years by custom engine rebuilders as an extra, quality step. On the 301 they are a necessity.

Externally, the 301 is similar to all other Pontiac V-8's, with the same equipment. A unique feature of the 301, and an easy way to recognize it, is by the location of the engine oil dipstick. It is at the left rear of the engine.

PULSE AIR SYSTEM 1977

This system, used on 4-cylinder engines only, does away with the air pump and diverter valve. A simple check valve allows air to enter the exhaust, and keeps the exhaust from backing up. The air supply is from the clean side of the air cleaner. For details, see the Vega chapter in this supplement.

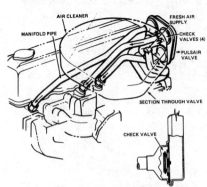

Pulse air system (© GM Corporation)

VACUUM DELAY VALVE 4-NOZZLE 1977

This new delay valve is a triumph of engineering ingenuity—and the despair of anyone who disconnects it and then tries to figure out how to hook it back up. The way the valve works is very simple. It is nothing more than a retard delay valve. It allows the manifold vacuum to act on the distributor diaphragm without any restriction. But when the manifold vacuum drops to zero, as during heavy acceleration or if the engine should die at idle, the valve "traps" the vacuum and lets it bleed down slowly. This keeps the spark advanced for better performance, or for easier starting.

Above 120°F. coolant temperature, a thermal vacuum switch, that is connected in the system by hoses, opens and bypasses the vacuum delay valve so that the vacuum advance works normally.

The bypass system is what makes the hose hookup so complicated. Here's the way it works on the 4-cyl. without air conditioning, and the V-8. The MAN side of the valve has two nozzles. They are both connected together inside the valve, so it doesn't make any difference which one of them is connected to the manifold vacuum source. The other MAN nozzle connects to the thermal vacuum switch (TVS).

On the DIST side of the valve, one of the nozzles connects to the TVS, and the other connects to the distributor modulator valve, or direct to the distributor. It doesn't make any difference which is which, because the two nozzles are connected together inside the valve.

When the TVS is open (above 120°F.) manifold vacuum goes in one MAN nozzle and right out the other MAN nozzle, without going through the valve. The vacuum then goes to the TVS and back to the DIST side of the valve. But all it does is go in one DIST nozzle and out the other one. In the bypass operation, the valve is only acting as a carrier for the vacuum, without actually doing anything.

But when the TVS is closed (below 120°F.) the vacuum can't bypass. It then goes in one MAN nozzle, through the valve, and out the DIST nozzle to the distributor. Then, the retard delay is in operation.

ELECTRIC CHOKE 1977

V-6 engines now use the 2GE carburetor, which has an electric choke. This choke is similar to the choke used on other GM cars, such as Chevette, Cadillac, and the discontinued 454 Chevrolet engine.

The choke receives current from

the oil pressure switch in all cars, so that it is only being heated while the engine is running. A minor difference occurs when the V-6 is used in the Sunbird. The oil pressure switch supplies current to the electric fuel pump when the engine is either cranking or running. The choke is connected to the fuel pump circuit, so it also receives current when the engine is either cranking or running. Other than that, the only thing that changes between car models is the colors of the wires, in some cases.

CHOKE VACUUM BREAK VACUUM DELAY VALVE 1977

This valve has three nozzles. It connects between manifold vacuum and the choke vacuum break unit. When the engine starts, the delay valve restricts the vacuum so the vacuum break opens the choke slowly. If the engine is stopped, the vacuum is released through an internal check valve.

The operation of the valve is very simple, but the vacuum hookup is something else. At first glance it doesn't make any sense at all, because the manifold vacuum does not run directly from the intake manifold to the delay valve. For convenience in manufacturing, the hose from manifold vacuum runs first to the distributor vacuum delay valve (DS-VDV), which has four nozzles. Then a second hose runs from the DS-VDV to the choke delay valve.

On the 4-cylinder with air conditioning, a different bypass system is used. Manifold vacuum connects to the MAN side of the valve, but the DIST side connects to the upper nozzle on a 3-nozzle TVS. The second nozzle on the DIST side of the vacuum delay valve is capped, and not used. The second nozzle on the MAN side is connected to another system, which doesn't affect the distributor. It is just a convenient way of hooking up manifold vacuum to the other system.

When the engine is cold (below 120°F.) the manifold vacuum goes through the vacuum delay valve and through the TVS to the distributor. When the engine warms up, the TVS shuts off the manifold vacuum, eliminating the vacuum delay valve entirely, and sends ported vacuum to the distributor.

Testing Vacuum Delay Valve

Connect a hand vacuum pump to the DIST side of the valve. Use a cap or a short length of hose and a bolt to plug the second nozzle on the DIST side. Plug both of the MAN nozzles with your finger. Pump up 15 in. Hg. vacuum and then remove your finger from the MAN side of the valve. The vacuum should drop slowly from 15 to 5 in. Hg. On the 301 V-8, it should take about 9 seconds. On the 151 4-cyl. it should take about 7 seconds on an automatic transmission car, or 3 seconds on a manual transmission car. If it takes longer, or shorter, or doesn't drop at all, the vacuum delay valve is defective.

Repairing Vacuum Delay Valve

Repairs are limited to replacement. The part numbers for the various applications are as follows.

Engine   Valve
151 4-cyl.
  Auto. trans. 527011
  Man. trans. 547833
301 V-8
  Auto. trans. with A.C. 546836

When connecting the valve, the MAN side goes to manifold vacuum, and the DIST side goes to the distributor, through a TVS or a modulator valve.

When the manifold vacuum goes through the DS-VDV it is not affected. The two nozzles on the MAN side are connected together inside the valve, without going through the valve mechanism. So the choke delay valve receives manifold vacuum just as if it were hooked up directly to the manifold.

Manifold vacuum connects to the 2-nozzle side of the choke delay valve. The second nozzle on the same side connects to a thermal vacuum switch, again just for convenience. The single nozzle side of the delay valve connects to the choke vacuum break.

Testing Choke Vacuum Break Vacuum Delay Valve

Use a short length of hose and a "T" to insert a vacuum gauge between the delay valve and the vacuum break unit. Disconnect the second hose on the 2-nozzle side of the delay valve, and cap the nozzle. Leave the manifold vacuum hose hooked up. Start the engine and let it idle, then slip the manifold vacuum hose off the 2-nozzle side of the delay valve. The vacuum gauge should drop to zero immediately, without any delay. If it doesn't, the delay valve is defective, or the hose is restricted. When slipping the hose off the delay valve, hold your finger over the end of the hose to keep the engine from dying.

Reconnect the manifold vacuum hose to the delay valve. The vacuum gauge should rise slowly, taking about 40 seconds to reach full manifold vacuum. Note the highest reading you get, then disconnect the hose from the intake manifold and use a separate hose to attach the vacuum gauge to check manifold vacuum. Both vacuum readings should be within one inch of each other. If not, there is a leak somewhere.

NOTE: *Because so many systems are connected together, a leak in one system can affect another. A single manifold vacuum connection supplies the distributor, the exhaust heat valve, and the vacuum break. When searching for leaks, always isolate each system before changing any parts.*

Repairing Choke Vacuum Break Vacuum Delay Valve

Repairs are limited to replacement. This is a 3-nozzle valve, with part number 546801. Manifold vacuum connects to the 2-nozzle side. The 1-nozzle side connects to the vacuum break.

EMISSION CONTROL SYSTEMS 1978-79

CLOSED POSITIVE CRANKCASE VENTILATION SYSTEM 1978

All engines used by Pontiac Motor Division, are equipped with the closed PCV system. The system is designed to prevent crankcase vapors from entering the atmosphere. A vapor metering PCV valve is used to restrict

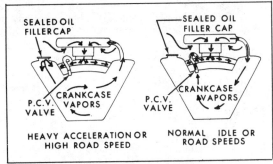

Typical PCV system air flow

Pontiac

the vapor flow when the intake manifold vacuum is high.

1979

The closed PCV system is continued on all engines used by Pontiac Motor Division for the 1979 model year. The operation of the system remains the same.

ENGINE MODIFICATION
1978

Emission calibrated distributors and carburetors are used to provide the engine with the proper air/fuel mixture and the correct timing to fire the mixture for good engine performance and fuel economy, and to remain with-in the allowable emission levels.

Vacuum delay valves are used with the vacuum advance units to assist in the controlling of emissions. They are as follows;
Thermal Vacuum Switches—TVS
Vacuum Delay valves—VDV
Check Valves—CV
Vacuum Modulator Valves—VMV
Vacuum Reducer Valve—VRV

Two types of vacuum advances are used, ported vacuum, resulting from a timed port in the throttle body above the throttle valve(s), and full manifold vacuum, taken from a manifold vacuum port on the carburetor or directly from the intake manifold.

1979

The engine modification system is basically the same for 1979 as was used on the 1978 models.

Minor carburetor calibration changes are made to comply with the Federal and States mandated changes in the emission standards.

Before any adjustments are made, check the emission control information label, attached to the vehicle, for up-to-date information.

THERMOSTATIC AIR CLEANER (TAC)
1978

The TAC system is used to preheat the induction air into the carburetor, to allow leaner choke and carburetor calibrations, to lower the emission levels. A vacuum motor or thermostatic coil is used to control the inlet damper valve opening and closing, so that a controlled temperature of inlet air is directed to the carburetor.

1979

The TAC system is continued in use for the 1979 modes with little or no change from the 1978 models.

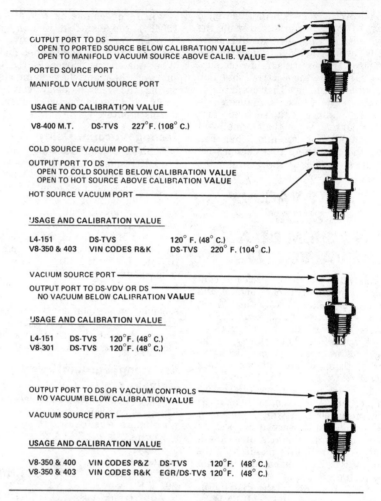

OUTPUT PORT TO DS
 OPEN TO PORTED SOURCE BELOW CALIBRATION VALUE
 OPEN TO MANIFOLD VACUUM SOURCE ABOVE CALIB. VALUE
PORTED SOURCE PORT
MANIFOLD VACUUM SOURCE PORT

USAGE AND CALIBRATION VALUE

V8-400 M.T.   DS-TVS   227°F. (108°C.)

COLD SOURCE VACUUM PORT
OUTPUT PORT TO DS
 OPEN TO COLD SOURCE BELOW CALIBRATION VALUE
 OPEN TO HOT SOURCE ABOVE CALIBRATION VALUE
HOT SOURCE VACUUM PORT

USAGE AND CALIBRATION VALUE

L4-151   DS-TVS   120°F. (48°C.)
V8-350 & 403   VIN CODES R&K DS-TVS 220°F. (104°C.)

VACUUM SOURCE PORT
OUTPUT PORT TO DS-VDV OR DS
 NO VACUUM BELOW CALIBRATION VALUE

USAGE AND CALIBRATION VALUE

L4-151   DS-TVS   120°F. (48°C.)
V8-301   DS-TVS   120°F. (48°C.)

OUTPUT PORT TO DS OR VACUUM CONTROLS
 NO VACUUM BELOW CALIBRATION VALUE
VACUUM SOURCE PORT

USAGE AND CALIBRATION VALUE

V8-350 & 400   VIN CODES P&Z   DS-TVS   120°F. (48°C.)
V8-350 & 403   VIN CODES R&K   EGR/DS-TVS   120°F. (48°C.)

Distributor spark thermal vacuum switch—Calibration points

EVAPORATION EMISSION CONTROL SYSTEM (EEC)
1978

The EEC system is used to control the emission of gasoline vapors from the fuel system. The system stores the evaporating fuel vapors when the engine is not operating, so that they may be burned during the combustion process, after the engine is started.

This process is accomplished by venting the fuel tank and carburetor bowl through a series of tubes, hoses and valves, to a canister containing activated charcoal. The canisters and valves differ from engine to engine and cannot be interchanged.

1979

The EEC system used is the same as that used on the 1978 models. Safeguards are built into the system to avoid the possibility of liquid fuel being drawn into the system.

The safeguards are;
1. A fuel tank overfill protector is installed on all models to provide adequate room for expansion of the fuel during temperature changes.
2. A fuel tank venting system is provided on all models to assure that the tank will be vented during any vehicle attitude. A domed fuel tank is used on sedans and coupes.
3. A pressure-vacuum relief valve, located in the gas cap, controls the fuel tank internal pressure.

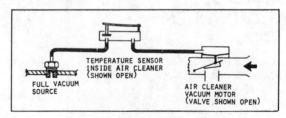

Thermostatically controlled air cleaner

Pontiac

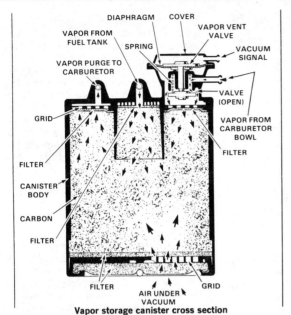

Vapor storage canister cross section
1978—Vin codes Y, W, X, R, K, Z   1979—Vin codes Y, W, X, R, K, Z, A

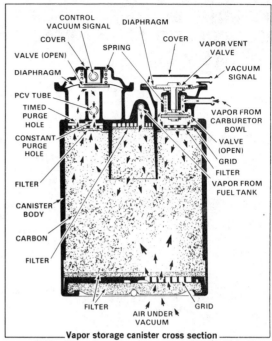

Vapor storage canister cross section
1978—Vin codes U, L   1979—Vin codes G, H, L, V

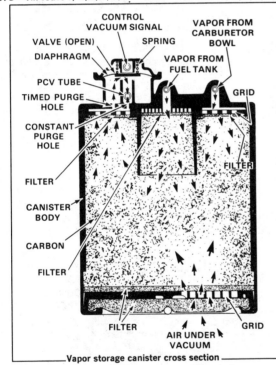

Vapor storage canister cross section
1978—Vin codes V, 1   1979—Vin code 1

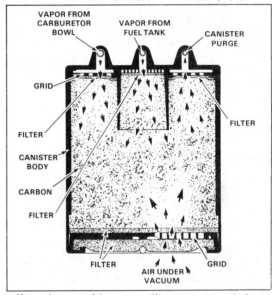

Vapor storage canister cross section 1978—Vin code A

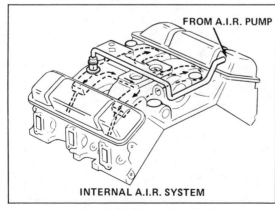

Internal AIR system—V-6 engines

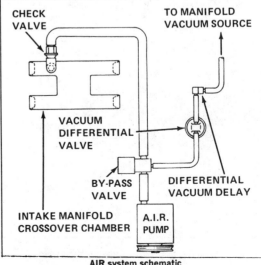

AIR system schematic

Pontiac

EXHAUST GAS RECIRCULATION SYSTEM (EGR) 1978

The EGR system is used on all engines to meter exhaust gases into the induction air system, so that during the combustion cycle, the Oxides of Nitrogen (NOx), emission can be reduced.

Ported and Back pressure types of EGR systems are used. The ported system uses a timed vacuum port in the carburetor to regulate the EGR valve, while the back pressure modulated system regulates the timed vacuum according to the exhaust back pressure level.

A separate back pressure transducer is used, or by the internal design of the EGR vave, to achieve the proper vacuum regulator.

1979

The EGR system remains in use on all 1979 engines and the system operation remains the same as used on the 1978 engine models. The EGR valve is designed to be closed during periods of deceleration and idle, to avoid rough idle and stalling from the dilution of the air/fuel mixture by the exhaust gases.

AIR INJECTION REACTION SYSTEM (AIR) 1978

The Air Injection system is used on some engines to provide additional oxygen to the combustion process, after the exhaust gases leave the cylinders, to continue the combustion in the exhaust manifolds. The following components are used in the system:
1. Engine driven air pump
2. Vacuum differential valve
3. Air flow and control hoses
4. Air by-pass valve
5. Differential vacuum delay and separator valve
6. Exhaust check valve

New for the 231 CID, V-6 engine, Vin code A, is an internal air distribution system, which eliminates much of the external piping.

1979

The Air Injection Reaction system is carried over to most of the 1979 models and operates in the same manner as the 1978 models.

EARLY FUEL EVAPORATION SYSTEM (EFE) 1978

The EFE system is designed to direct rapid heat to the engine air induction system to provide quick fuel evaporation and more uniform fuel distribution during cold driveaway. A vacuum operated valve is located between the exhaust manifold and the exhaust pipe and is controlled by a thermal vacuum switch (TVS) to pass vacuum to the valve when temperatures are below a specific calibration point. As the vacuum is applied to the valve, the exhaust gas flow is increased under the intake manifold to reheat the manifold during cold engine operation. The system is used on most models.

1979

The Early Fuel Evaporation System is used on most 1979 models. The operation of the system remains the same as the 1978 models. Refer to the Emission Control Information label for usage clarification.

ELECTRIC CHOKE 1978

The Electric Choke is used to assist in controlling the choke thermostatic coil for precise timing of the choke valve opening for good engine warm-up performance. This is accomplished with the use of a ceramic resistor with-in the choke unit. A small section resistor is used to gradually heat the coil, while a large sec-

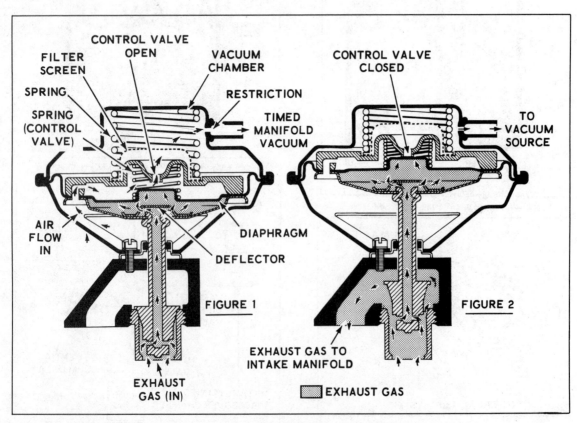

Cross section of ported EGR valve—Both on and off positions

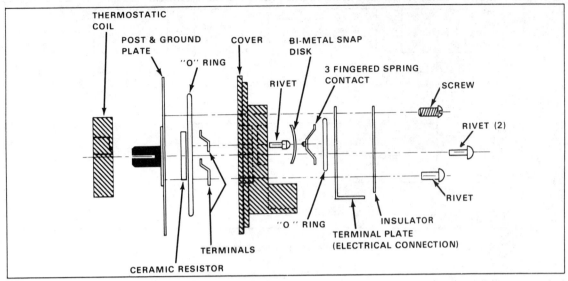

Exploded view of electric choke assembly

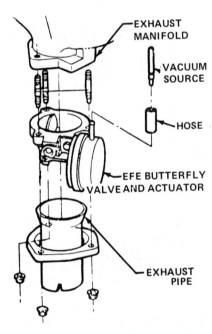

EFE systems—Typical

tion resistor is used to quickly heat the coil as the temperature dictates. The electric choke is not used on all models.

1979

The Electric Choke is used on select models of 1979, and operates in the same manner as the 1978 models.

At temperatures below 50° F., the electrical current is directed to the small segment resistor of the choke to allow gradual opening of the choke valve. At temperatures of 70° F. and above, the electrical current is directed to both the small and the large segment resistors, to allow quicker opening of the choke valve.

ELECTRONIC FUEL CONTROL SYSTEM (EFC) 1978

(L4-151 CID, Vin Code 1)

The EFC system is new for 1978, and is designed to control the exhaust emissions by regulating the air/fuel mixture with the use of the following components;
1. Exhaust gas oxygen sensor
2. Electric control unit
3. Vacuum modulator
4. Controlled air-fuel ratio carburetor
5. Phase II catalytic converter

The exhaust gas sensor is placed in the exhaust gas stream and generates a voltage which varies with the oxygen content of the exhaust gases. As the exhaust gas oxygen content rises, indicating a lean mixture, the voltage falls. As the exhaust gas oxygen content decreases, indicating a rich mixture, the voltage rises.

The voltage signal is sent to the electronic control unit, where the voltage signal is monitored. The electronic control unit transmits a signal to the vacuum modulator. This signal is of a constant current and is continually cycling on and off.

To assist in maintaining proper

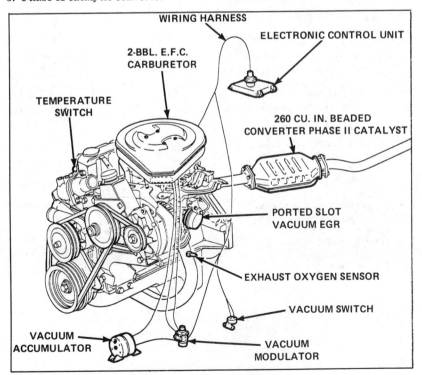

Electronic fuel control components

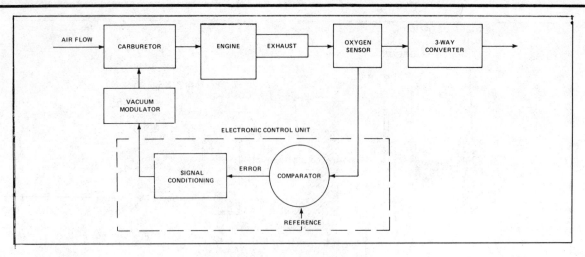

Schematic of electronic fuel control system

electrical signal regulation during periods of unstable engine operation, a bi-metal switch is located in the cylinder head and provides an open circuit or a ground circuit for the electrical signal, when the temperature is above or below the calibration level value.

If below the calibration point, the electronic control unit uses this signal to restrict the amount of leanness that the carburetor can go lean. A temperature sensitive diode with-in the ECU, assists in the determination of the strength of the electrical signal. When the engine warms up, this feature drops out, indicating the engine is warm and the drivability will not be impaired while the carburetor is in the maximum lean range.

1979

(L4-151 CID, Vin Code 1)

The EFC system is continued for 1979 models. The operation remains the same as for the 1978 models. The EFC system should be included in any diagnosis of the following problems;
1. Detonation
2. Stalls or rough idle—cold
3. Stalls or rough idle—hot
4. Missing
5. Hesitation
6. Surges
7. Sluggish or spongy operation
8. Poor gas milage
9. Hard starting—cold
10. Hard starting—hot
11. Foul exhaust odor
12. Engine cuts out

CATALYTIC CONVERTER (CC) 1978

The Catalytic Converter is used on all vehicles models to lower the levels of Hydrocarbons (HC) and Carbon Monoxide (CO) from the exhaust emission with the use of materials coated with Platinum and Palladium, to act as a catalyst.

A Phase II catalytic converter is used with the L-4, 151 CID, Vin code 1 engine, in the state of California, with the Electronic Fuel Control System. This converter uses Platinum and Rhodium plated material as the catalyst.

CATALYTIC CONVERTER (CC) 1979

The Catalytic Converter is used on the exhaust systems for 1979 models. The Phase II converter is again used on the L-4, 151 CID, Vin code 1 engine in California. An oxygen sensor is used in conjuction with the Phase II converter.

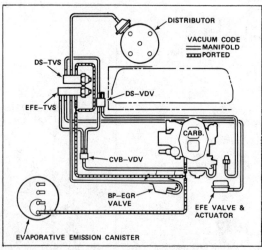

Vacuum hose routing 4 cyl. California w/auto. trans. and air cond.
(© GM Corporation)

Pontiac

VACUUM CIRCUITS

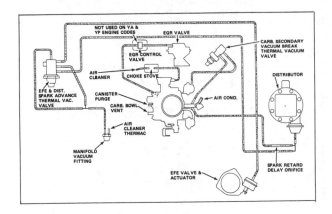

1976 49-State 350 2-Bbl.

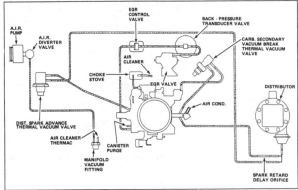

1976 49-State 400 4-Bbl. auto. trans.

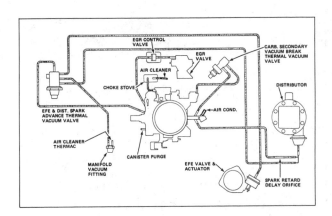

1976 Calif. 350, 400 4-Bbl.

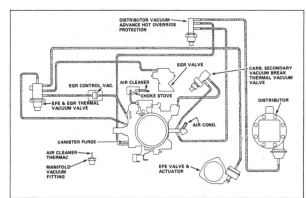

1976 49-State 400, 455, 4-Bbl. Man. Trans. Firebird

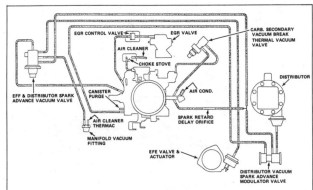

1976 49-State 455 4-Bbl. auto. trans.

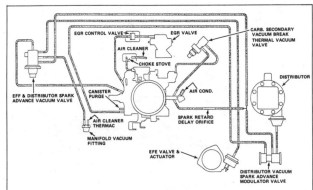

1976 Calif. 455 4-Bbl.

Pontiac

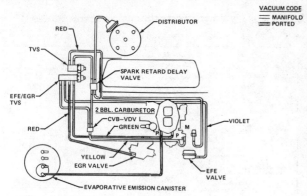

Vacuum hose routing 4 cyl. low altitude w/manual trans., w/auto. trans., wo/air cond. (© GM Corporation)

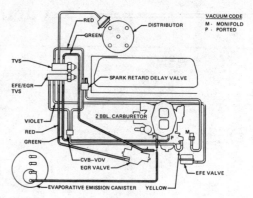

1977 Vacuum hose routing 4 cyl. low altitude w/auto. trans. and air cond. (© GM Corporation)

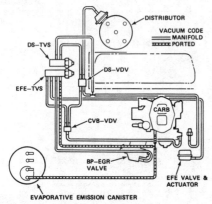

1977 Vacuum hose routing 4 cyl. California w/auto. trans., wo/air cond. (© GM Corporation)

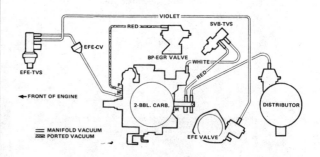

1977 Vacuum hose routing V-8 301 low altitude w/auto. trans., wo/air cond. (© GM Corporation)

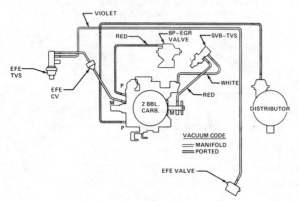

1977 Vacuum hose routing V-8 301 low altitude w/manual trans. (© GM Corporation)

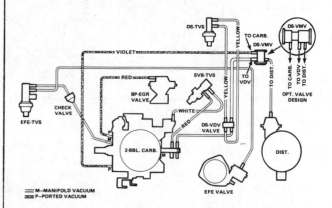

1977 Vacuum hose routing V-8 301 low altitude w/auto. trans. and air cond. (© GM Corporation)

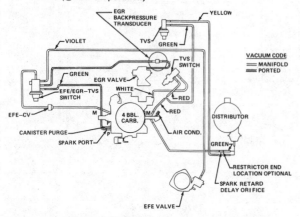

1977 Vacuum hose routing V-8 350 and 400 vin codes P & Z low altitude w/auto. trans. (© GM Corporation)

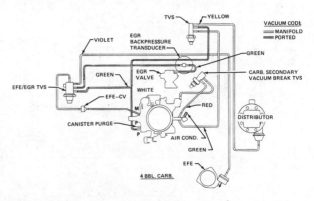

1977 Vacuum hose routing V-8 400 low altitude w/manual trans. (© GM Corporation)

Pontiac

VACUUM CIRCUITS

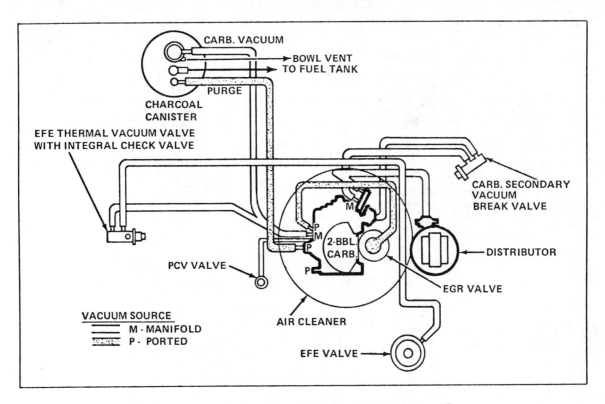

1978 49-States Pontiac 301 2-bbl. V-8 w/air cond. Engine codes YM, YP

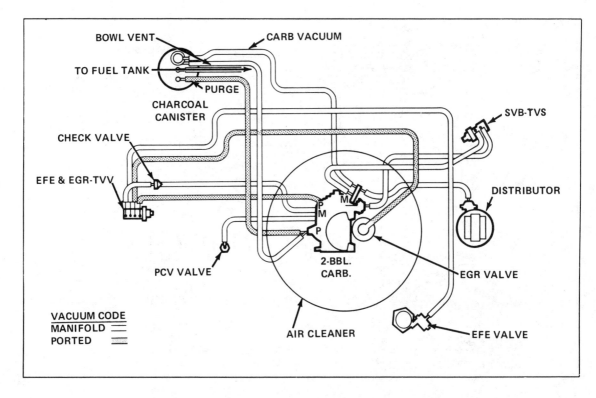

1978 49-States Pontiac 301 2-bbl. wo/air cond. Engine codes XA, XB

Pontiac

VACUUM CIRCUITS

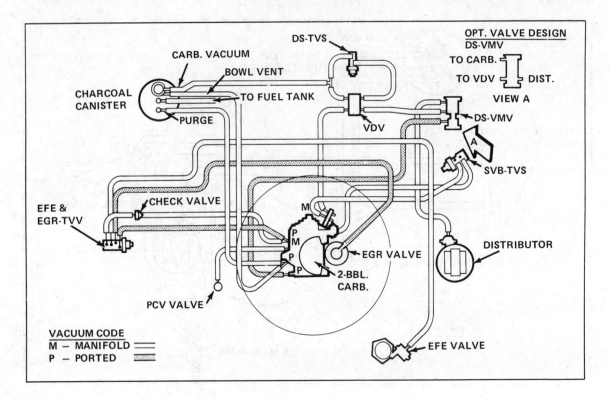

1978 49-States Pontiac 301 2-bbl. V-8 w/air cond. Engine codes XC, XD

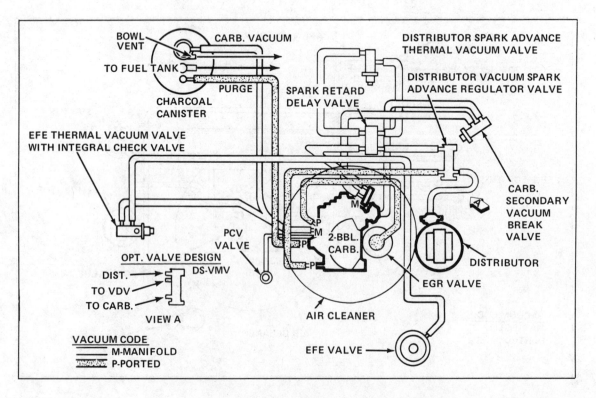

1978 49-States Pontiac 301 2-bbl. V-8 wo/air cond. Engine codes YL, YN

VACUUM CIRCUITS

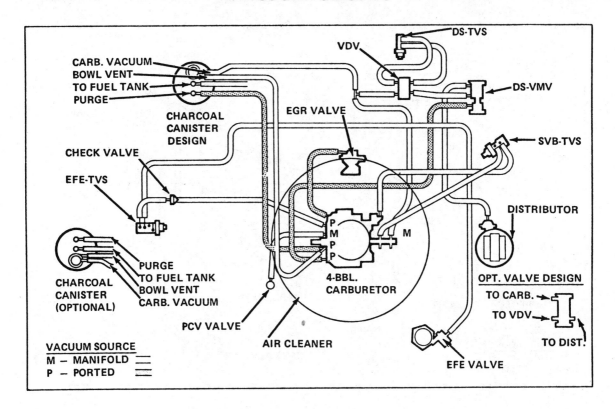

1978 49-States Pontiac 301 4-bbl. V-8 w/air cond.

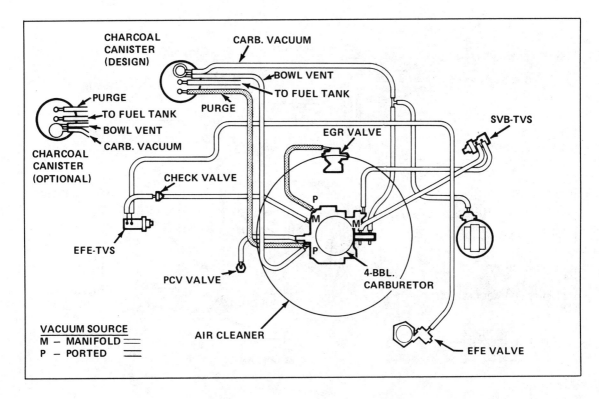

1978 49-States Pontiac 301 4-bbl. V-8 wo/air cond.

Pontiac

VACUUM CIRCUITS

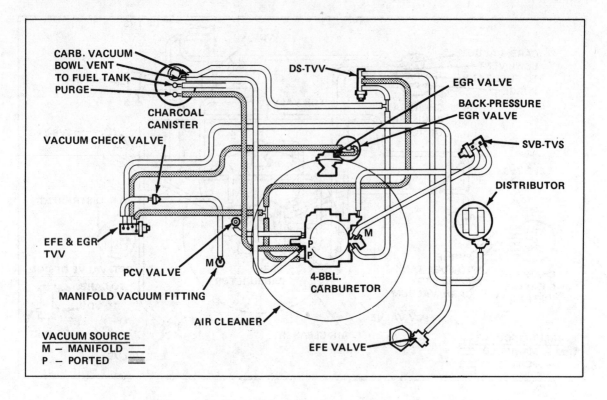

1978 Pontiac 400 4-bbl V-8 manual trans.

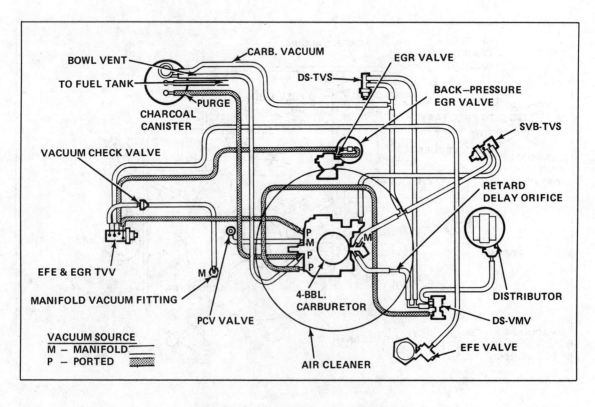

1978 Pontiac 400 4-bbl. V-8 auto. trans. single exhaust

Pontiac

VACUUM CIRCUITS

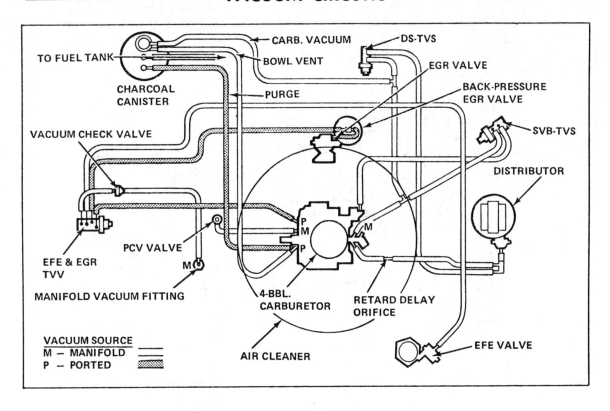

1978 Pontiac 400 4-bbl. V-8 auto. trans. dual exhaust

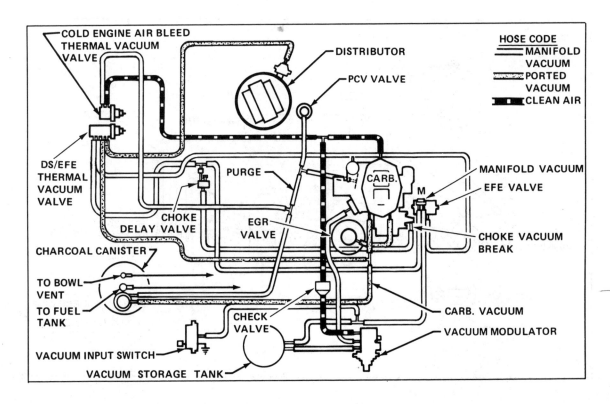

1979 151 CID 4 cyl.—Vin code 1—Calif.

Pontiac

VACUUM CIRCUITS

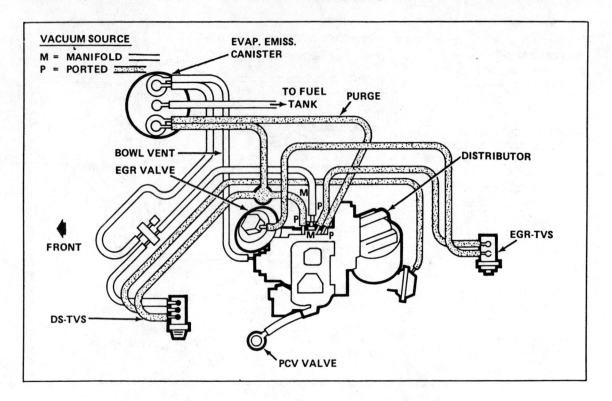

1979 151 CID 4 cyl.—Vin code V—Low Alt. (Auto. trans. w/AC)

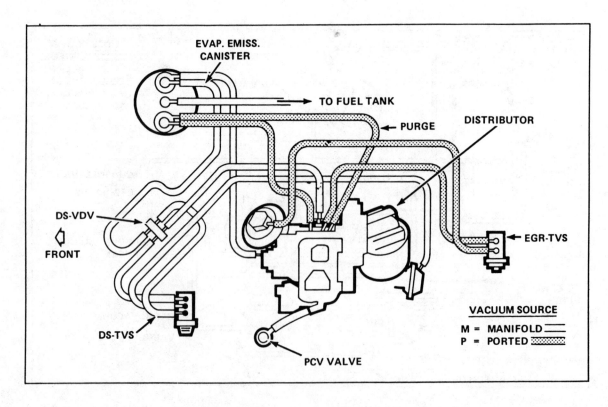

1979 151 CID 4 cyl.—Vin code V—Low Alt. (Auto. and Manual trans. without AC)

VACUUM CIRCUITS

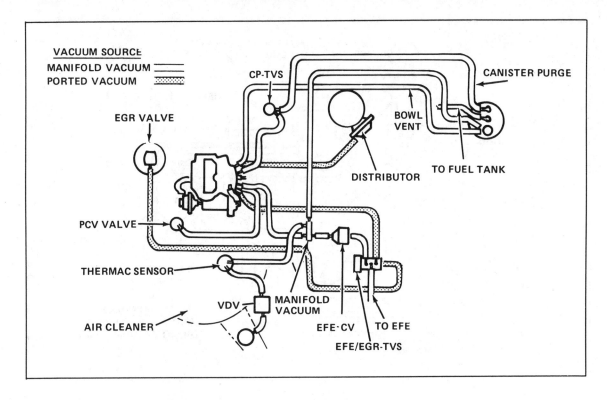

1979 231 CID V-6—Vin code A—Low Alt. (w/Manual trans.)

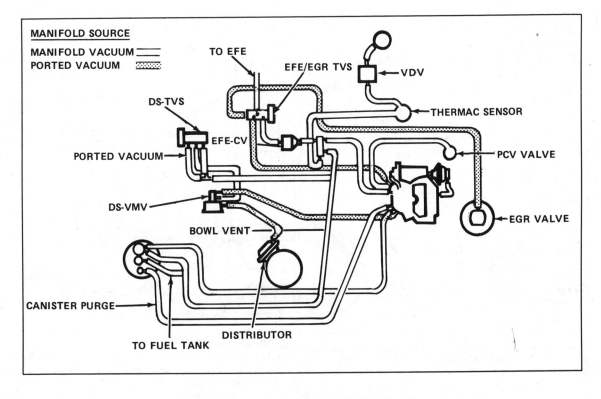

1979 231 CID V-6—Vin code A—Low Alt. (w/Auto. trans.)

Pontiac

VACUUM CIRCUITS

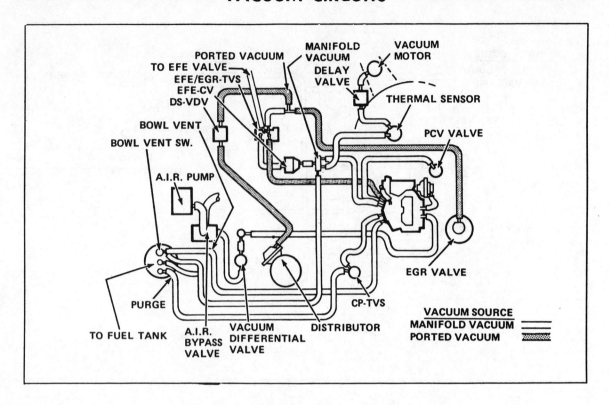

1979 231 CID V-6—Vin code A—Calif. (w/Manual trans.)

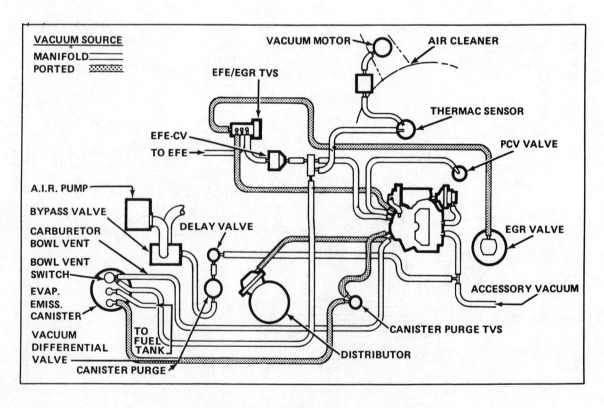

1979 231 CID V-6—Vin code A—Calif. and High Alt. (w/Auto. trans.)

Pontiac

VACUUM CIRCUITS

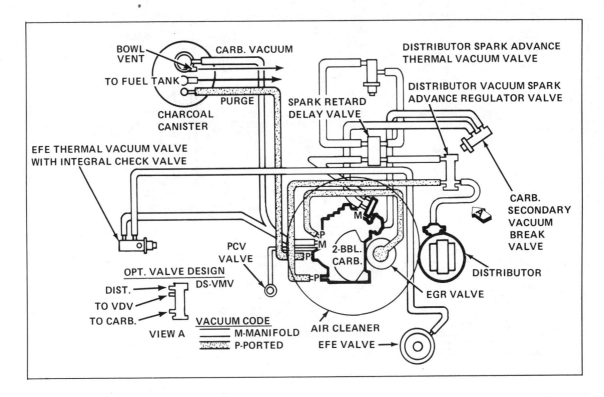

1979 301 CID V-8—Vin code Y (2-bbl. w/Auto. trans. and AC)

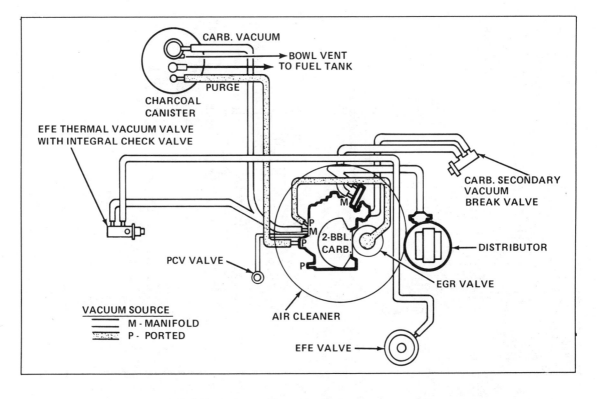

1979 301 CID V-8—Vin code Y (2-bbl. w/Auto. trans. and without AC)

Pontiac

VACUUM CIRCUITS

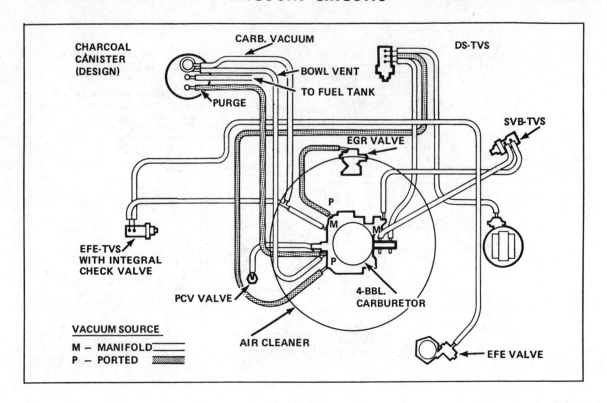

1979 301 CID V-8—Vin code W (4-bbl, w/Manual trans.)

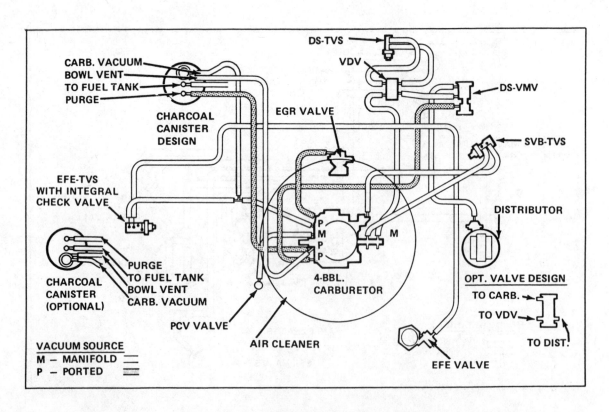

1979 301 CID V-8—Vin code W (4-bbl, w/Auto. trans. and AC)

Pontiac

VACUUM CIRCUITS

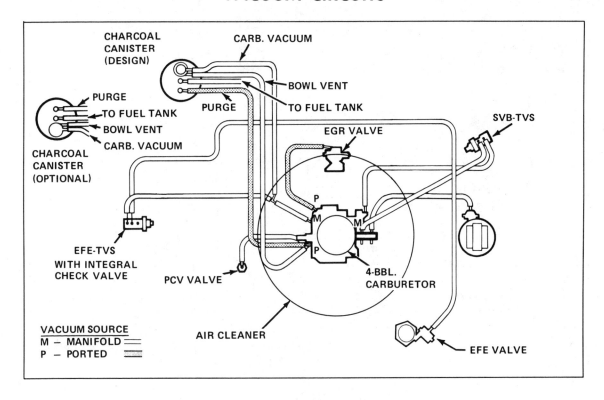

1979 301 CID V-8—Vin code W (4-bbl, w/Auto. trans and without AC)

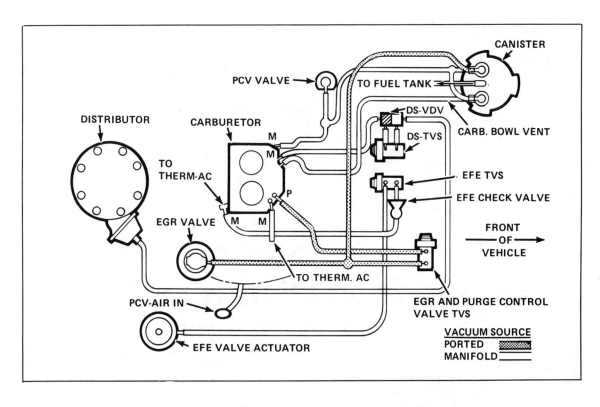

1979 305 CID V-8—Vin code G—Low Alt. (2-bbl. except Series H)

Pontiac

VACUUM CIRCUITS

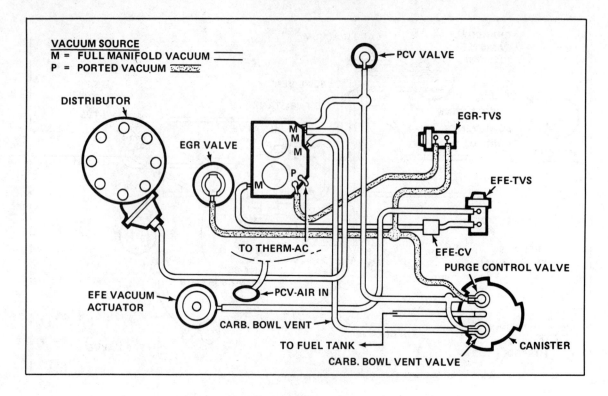

1979 305 CID V-8—Vin code G—Low Alt. (2-bbl, Series H only)

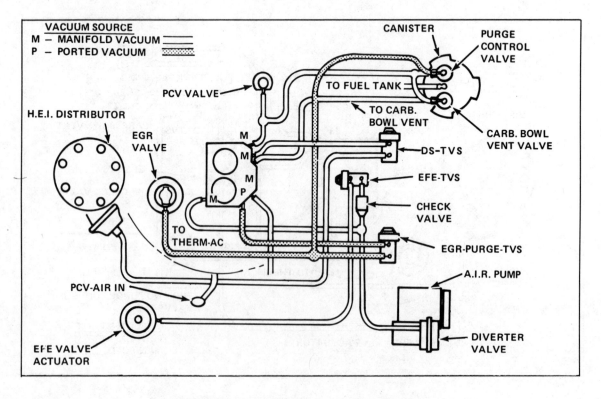

1979 305 CID V-8—Vin code G—Calif. (2-bbl. except Series H)

Pontiac

VACUUM CIRCUITS

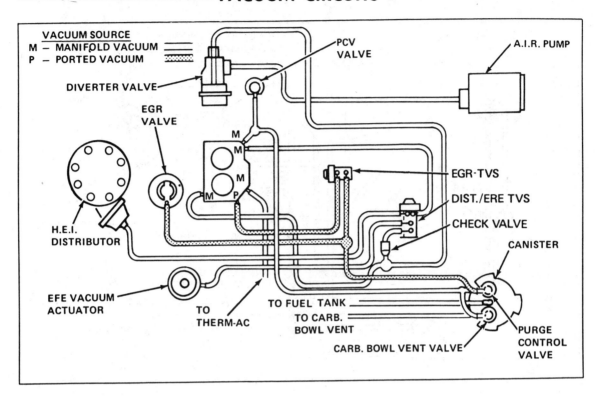

1979 305 CID V-8—Vin code G—Calif. (2-bbl, Series H only)

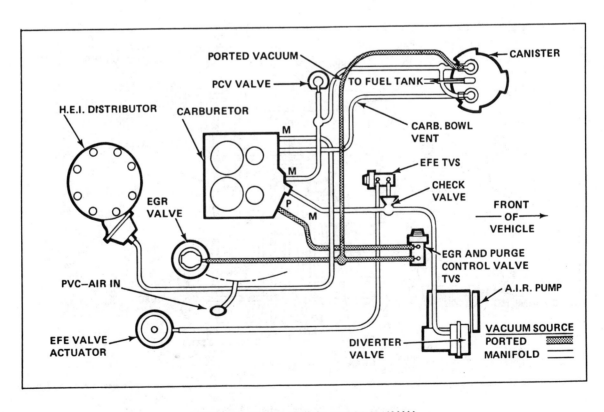

1979 305 CID V-8—Vin code H—Calif. and High Alt. (4-bbl.)
350 CID V-8—Vin code L—Calif. and High Alt. (4-bbl.)

241

Pontiac

VACUUM CIRCUITS

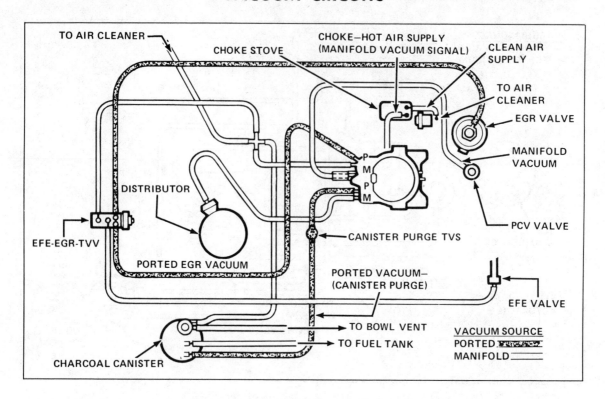

1979 350 CID V-8—Vin code X (4-bbl.)

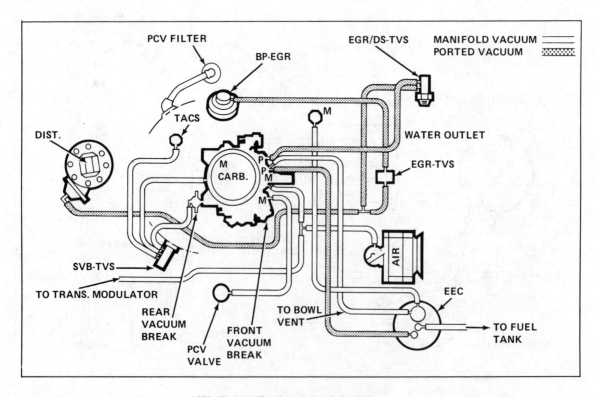

1979 350 CID V-8—Vin code R—Calif. (4-bbl.)
403 CID V-8—Vin code K—Calif. (4-bbl.)

Pontiac

VACUUM CIRCUITS

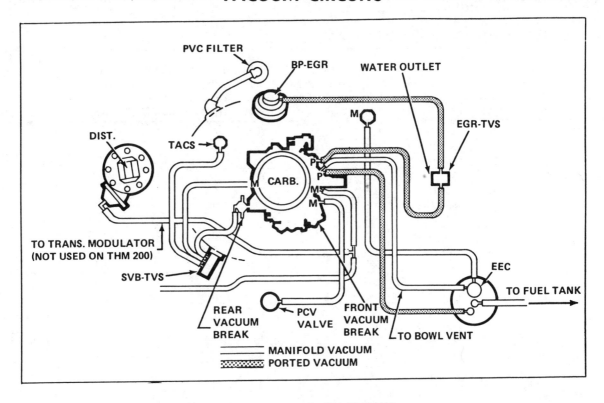

1979 350 CID V-8—Vin code R—High Alt. (4-bbl.)
403 CID V-8—Vin code K—High Alt. (4-bbl.)

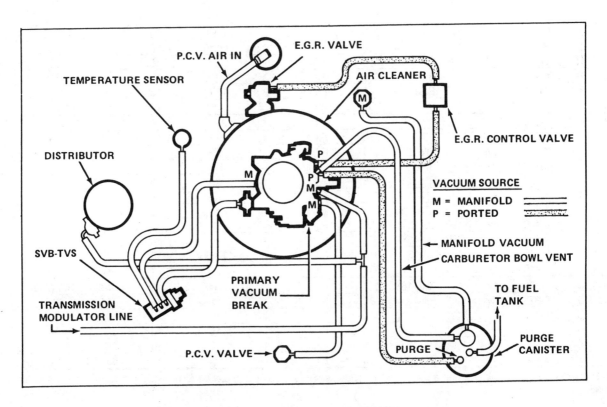

1979 403 CID V-8—Vin code K—Low Alt. (4-bbl.)

Chrysler Corporation

Full-Size • Intermediate • Compact

TUNE-UP SPECIFICATIONS — CHRYSLER CORP.

Engine Code, 5th character of the VIN number

Yr.	Eng. V.I.N. Code	Engine No. Cyl. Disp. (cu in)	Carb. Type	H.P.	SPARK PLUGS Orig. Type	Gap (in)	DIST. Point Dwell (deg)	DIST. Point Gap (in)	IGNITION TIMING (deg BTDC) Man Trans	IGNITION TIMING (deg BTDC) Auto Trans	VALVES Intake Opens (deg BTDC)	FUEL PUMP Pres. (psi)	IDLE SPEED (rpm) Man Trans	IDLE SPEED (rpm) Auto Trans (Neutral)
75	C	6-225	1v	95	BL-13Y	.035	E.I.		0	0	16	3½-5	800	750
	G	8-318	2v	150	N-13Y	.035	E.I.		2	0	10	5-7	750	750⑤
	K	8-360	2v	180	N-12Y	.035	E.I.		6	6	18	5-7	750	750
	J	8-360	4v	190	N-12Y	.035	E.I.		6	6	18	5-7	750	750
	L	8-360	4v	230	N-12Y	.035	E.I.		2	2	22	5-7	850	850
	M	8-400	2v	165	J-13Y	.035	E.I.		10	10	18	6-7½	750	750
	N	8-400	4v	190	J-13Y	.035	E.I.		8	8	21	3½-5	750	750
	P	8-400	4v	240	RJ-87P	.035	E.I.		8	8	21	3½-5	850	850
	T	8-440	4v	215	RJ-87P	.035	E.I.		8	8	18	4-6	750	750
	U	8-440	4v	260	J-11Y⑥	.035	E.I.		10 (8)	10 (8)	21	6-7½	750	750
76	C	6-225	1v	90	RBL-15Y	.035	E.I.		6 (4)	2	16	3½-5	700	700
	G	8-318	2v	140	N-12Y	.035	E.I.		0	0	10	5-7	700	700
	K	8-360	2v	170	N-12Y	.035	E.I.		6	6	18	5-7	700	700
	J	8-360	4v	175	N-12Y	.035	E.I.		6	6	18	5-7	700	700
	L	8-360	4v	220	N-12Y	.035	E.I.		2 (6)	2 (6)	22 (18)	5-7	850 (750)	850 (750)
	M	8-400	2v	175	J-13Y	.035	E.I.		10	10	18	5-7	700	700
	N	8-400	4v	185	J-13Y	.035	E.I.		8⑦	8	18	5-7	700	700
	P	8-400	4v	240	RJ-87P	.035	E.I.		6	6	18	5-7	850	850
	T	8-440	4v	200	RJ-87P	.035	E.I.		8	8	18	5-7	750	750
	U	8-440	4v	250	J-11Y	.035	E.I.		10 (8)	10 (8)	21	6-7½	750	750
77	C	6-225	1v	90	RBL-15Y	.035	E.I.		6⑧	2	16	3½-5	700	700
	D	6-225	2v	100	RBL-15Y	.035	E.I.		12	12	16	3½-5	700	700
	G	8-318	2v	135	RN-12Y	.035	E.I.		8 (2)	8 (2)	10	5-7	700	700 (850)
	K	8-360	2v	155	RN-12Y	.035	E.I.		10	10	18	5-7	700	700
	J	8-360	4v	170	RN-12Y	.035	E.I.		6	6	18	5-7	700	700
	L	8-360	4v	175	RN-12Y	.035	E.I.		10 (6)	10 (6)	18	5-7	700	700
	N	8-400	4v	190	RJ-13Y	.035	E.I.		10	10	18	5-7	750	750
	P	8-400	4v	240	RJ-13Y	.035	E.I.		10	10	18	5-7	750	750
	T	8-440	4v	195	RJ-13Y	.035	E.I.		12 (8)	12 (8)	18	5-7	750	750
	U	8-440	4v	230	RJ-11Y	.035	E.I.		8	8	21	6-7½	750	750
78	C	6-225	1v	100	RBL-16Y	.035	E.I.		12 (8)	12 (8)⑩	16	4-5½	700 (750)	700 (750)⑫
	D	6-225	2v	110	RBL-16Y	.035	E.I.		12	12	16	4-5½	700	700
	G	8-318	2v	140	RN-12Y	.035	E.I.		16	16	10	5-7	700	750
	H	8-318	4v	150	RN-12Y	.035	E.I.		—	10 (10)	10	5-7	—	750 (750)
	K	8-360	2v	155	RN-12Y	.035	E.I.		—	20	18	5-7	—	750
	J	8-360	4v	170	RN-12Y	.035	E.I.		—	6 (6)⑨	18	5-7	—	750 (750)
	L	8-360	4v	160	RN-12Y	.035	E.I.		—	16 (8)⑩	18	5-7	—	750
	N	8-400	4v	190	OJ-13Y	.035	E.I.		—	⑪	18	5-7	—	750
	P	8-400	4v	190	OJ-11Y	.035	E.I.		—	20	18	5-7	—	750
	T	8-440	4v	195	OJ-13Y	.035	E.I.		—	12 (8)⑩	18	5-7	—	750
	U	8-440	4v	255	OJ-11Y	.035	E.I.		—	16 (8)	21	5-7	—	750
79	C	6-225	1v	100*	RBL-16Y	.035	E.I.		12 (15)	12 (15)	16	4-5.5	700 (750)	700 (750)
	D	6-225	2v	110*	RN-12Y	.035	E.I.		12	12	16	4-5.5	750	750
	G	8-318	2v	140*	12N-12Y	.035	E.I.		—	16	10	5.8-7.3	—	750
	H	8-318	4v	150*	RN-12Y	.035	E.I.		—	16	10	5.8-7.3	—	750
	K	8-360	2v	155*	RN-12Y	.035	E.I.		—	16	18	5.8-7.3	—	750
	J	8-360	4v	170*	RN-12Y	.035	E.I.		—	16	18	5.8-7.3	—	750

NOTE: Should the information provided in this manual deviate from the specifications on the underhood tune-up label, the label specifications should be used, as they may reflect production changes.

Chrysler Corporation

① 30-34 WITH DISTRIBUTOR 3438517.
② 30-34 WITH DISTRIBUTORS 3438617, IBS-4018E.
③ DUAL POINTS—27-32 EACH SET.
④ 5 WITH DISTRIBUTORS 3656347, 3656350.
⑤ 900 FOR CARBURETORS 8001S, 8077S.
⑥ CALIFORNIA AND POLICE—BL-9Y.
⑦ E.L.B. SYSTEM—10.
   A.I.R. SYSTEM—6.
⑧ CALIFORNIA & HIGH ALT.—8.
   DISTRIBUTOR 3874876—12.
• Figures in Parentheses Apply to California Engines.
∗ Chilton Estimated Rating
⑨ CALIFORNIA AND HIGH ALT. SHOWN
⑩ W/HIGH ALT.—8BTDC
⑪ W/4500 LB. INERTIA WEIGHT—24BTDC
   W/5000 LB. INERTIA WEIGHT—20BTDC
⑫ HIGH ALT.—750 RPM
— Not applicable

DISTRIBUTOR SPECIFICATIONS — CHRYSLER CORP.

Year	Distributor Identification	Centrifugal Advance Start Dist. Deg. @ Dist. RPM	Centrifugal Advance Finish Dist. Deg. @ Dist. RPM	Vacuum Advance Start @ In. Hg.	Vacuum Advance Finish Dist. Deg. @ In. Hg.
1975	3874082	1–4.5 @ 600	11.5–14 @ 2,200	7	7–10 @ 11.5
	3874090	1.5–5.5 @ 550	11.5–14 @ 2,200	7	10–12 @ 12.5
	3874298	0.5–3.5 @ 700	13.5–16 @ 2,100	9	9.5–12.5 @ 15.5
	3874115	1–3.5 @ 600	10–12 @ 2,000	7	10–12 @ 12.5
	3874097	1–4.5 @ 650	13.5–16 @ 2,150	7	10–12 @ 12.5
	3874101	0.5–3 @ 600	10–12 @ 2,000	8	9–11 @ 14
	3874110	1–3.5 @ 600	10–12 @ 2,000	8	9–11 @ 14
	3874110	1–3.5 @ 600	10–12 @ 2,000	8	9–11 @ 14
	3874173	1–3.5 @ 600	8–10 @ 2,000	8	9–11 @ 14
	3874173	1–3.5 @ 600	8–10 @ 2,000	8	9–11 @ 14
1976	3874598	1.0–4.5 @ 600	9.5–11.5 @ 2,300	7	7.0–10.0 @ 11.5
	3874714	1.0–4.5 @ 450	7.0–9.0 @ 2,500	7	7.0–10.0 @ 11.5
	3874754	0–4.0 @ 500	11.5–13.5 @ 2,400	8	10.0–12.0 @ 13.5
	3874115	1.0–3.5 @ 600	10.0–12.0 @ 2,000	7	10.0–12.0 @ 12.5
	3874115	1.0–3.5 @ 600	10.0–12.0 @ 2,000	7	10.0–12.0 @ 12.5
	3874097	1–4.5 @ 650	13.5–16.5 @ 2,150	7	10.0–12.0 @ 12.5
	3874101	0.5–3.0 @ 600	10.0–12.0 @ 2,000	8	9.0–11.0 @ 14
	3874110	1.0–3.5 @ 600	10.0–12.0 @ 2,000	8	9.0–11.0 @ 14
	3874110	1.0–3.5 @ 600	10.0–12.0 @ 2,000	8	9.0–11.0 @ 14
	3874848	0.5–3.0 @ 550	5.5–9.0 @ 2,300	—	—
	3874596	0.5–4.0 @ 600	8.0–10.9 @ 2,000	8	9.0–11.0 @ 14
	3874173	1.0–3.5 @ 600	8.0–10.0 @ 2,000	7	9.0–12.0 @ 8.5
	3874795	1.5–3.5 @ 65	3.5–5.5 @ 2,300	7	9.0–12.0 @ 8.5
	3874119	1.0–3.5 @ 600	9.5–11.5 @ 2,400	7	10.0–12.0 @ 12.5
1977	3874115	1.3–3.1 @ 600	10.0–12.0 @ 2000	7	10.0–12.0 @ 12.5
	3874173	1.3–3.1 @ 600	8.0–10.0 @ 2000	8	9.0–11.0 @ 14
	3874714	0.3–2.4 @ 500	6.9–8.9 @ 2500	7	7.3–9.8 @ 11.5
	3874858	1.4–3.4 @ 600	6.5–8.5 @ 2400	7	10.0–12.5 @ 12.5
	3874876	0.2–2.2 @ 600	3.7–5.7 @ 2500	7	7.3–9.8 @ 11.5
	3874909	0.2–2.2 @ 600	7.6–9.6 @ 2300	8	10.0–12.0 @ 13.5
	3874913	1.7–4.1 @ 600	11.6–13.6 @ 2350	8	10.0–12.0 @ 13.5
	3874917	0.6–2.0 @ 500	8.0–10.0 @ 2000	7	10.0–12.5 @ 12.5
	3874929	1.4–3.4 @ 600	9.7–11.7 @ 2300	7	7.3–9.8 @ 11.5
	4091039	0.2–2.2 @ 600	3.7–5.7 @ 2500	7	7.3–9.8 @ 11.5
	4091101	1.0–1.2 @ 600	5.8–7.8 @ 2060	9	7.5–9.5 @ 12.5
1978①	3874115	1.3–3.1 @ 600	10.0–12.0 @ 2000	7	10.0–12.0 @ 12.5
	3874173	1.3–3.1 @ 600	8.0–10.0 @ 2000	8	9.0–11.0 @ 14
	3874858	1.4–3.4 @ 600	6.5–8.5 @ 2400	7	10.0–12.5 @ 12.5
	3874876	0.2–2.2 @ 600	3.7–5.7 @ 2500	7	7.3–9.8 @ 11.5
	3874929	1.4–3.4 @ 600	9.7–11.7 @ 2300	7	7.3–9.8 @ 11.5
	4091101	1.0–1.2 @ 600	5.8–7.8 @ 2060	9	7.5–9.5 @ 12.5
1979	3874876	.2–2.2 @ 600	3.7–5.7 @ 2500	7	7.3–9.8 @ 11.5

Chrysler Corporation

CARTER BBD SPECIFICATIONS
CHRYSLER PRODUCTS

Year	Model ④	Float Level (in.)	Accelerator Pump Travel (in.)	Bowl Vent (in.)	Choke Unloader (in.)	Choke Vacuum Kick ①	Fast Idle Cam Position ①	Fast Idle Speed (rpm)	Automatic Choke Adjustment
1975	8000S	¼	0.500③	—	0.280	0.130	0.070	1500	Fixed
	8064S	¼	0.500③	—	0.310	0.070	0.070	1500	Fixed
	8001S	¼	0.500③	—	0.310	0.110	0.070	1500	Fixed
	8003S	¼	0.500③	—	0.310	0.110	0.070	1500	Fixed
	8066S	¼	0.500③	—	0.280	0.130	0.070	1500	Fixed
	8062S	¼	0.500③	—	0.310	0.110	0.070	1500	Fixed
1976	8071S	¼	0.500③	—	0.280	0.130	0.070	1500	Fixed
	8069S	¼	0.500③	—	0.310	0.070	0.070	1200	Fixed
	8070S	¼	0.500③	—	0.310	0.110	0.070	1500	Fixed
	8077S, 8099S	¼	0.500③	—	0.280	0.110	0.070	1250	Fixed
	8072S	¼	0.500③	—	0.310	0.070	0.070	1500	Fixed
1977	8087S	¼	0.469③	—	0.280	0.100	0.070	1600	Fixed
	8089S	¼	0.469③	—	0.280	0.130	0.070	1600	Fixed
	8090S	¼	0.469③	—	0.280	0.130	0.070	1700	Fixed
	8127S	¼	0.469③	—	0.280	0.110	0.070	1500	Fixed
	8093S	¼	0.469③	—	0.310	0.130	0.070	1400	Fixed
	8094S	¼	0.469③	—	0.310	0.070	0.070	1400	Fixed
	8096S	¼	0.469③	—	0.310	0.110	0.070	1500	Fixed
	8126S	¼	0.469③	—	0.310	0.110	0.070	1500	Fixed
1978	8136S	¼	0.500③	0.080	0.280	0.110	0.070	1500	Fixed
	8137S	¼	0.500③	0.080	0.280	0.100	0.070	1600	Fixed
	8177S	¼	0.500③	0.080	0.280	0.100	0.070	1600	Fixed
	8175S	¼	0.500③	0.080	0.280	0.160	0.070	1400	Fixed
	8143S	¼	0.500③	0.080	0.280	0.150	0.070	1500	Fixed

Model 2245
Chrysler Corporation — Holley Carburetor

Year	Carb. ★ Part No.	Float Level (in.)	Accelerator Pump Adjustment (in.)	Bowl Vent Clearance (in.)	Fast Idle (rpm)	Choke Unloader Clearance (in.)	Vacuum Kick (in.)	Fast Idle Cam Position (in.)	Choke
1975	R-7226-A	.190	.250	.015	1600	.170	.150	.110	Fixed
	R-7211-A	.190	.250	.015	1600	.170	.150	.110	Fixed
	R-7027-A	.190	.250	.015	1600	.170	.150	.110	Fixed
1976	R-7364-A	.190	.265	.025	1600	.170	.150	.110	Fixed
	R-7366-A	.190	.265	.025	1600	.170	.150	.110	Fixed
1977	R-7671-A	.190	.265	.025	1700	.170	.110	.110	Fixed
1978	R-7991-A	.188	.265	.025	1600	.170	.110	.110	Fixed
	R-8326-A	.188	.265	.025	1600	.170	.110	.110	Fixed

★ Located on a tag attached to the carburetor.

Chrysler Corporation

CARTER TQ SPECIFICATIONS
CHRYSLER PRODUCTS

Year	Model ①	Float Setting (in.)	Secondary Throttle Linkage (in.)	Secondary Air Valve Opening (in.)	Secondary Air Valve Spring (turns)	Accelerator Pump (in.)	Choke Control Lever (in.)	Choke Unloader (in.)	Vacuum Kick (in.)	Fast Idle Speed (rpm)
1975	9004S	29/32	②	1/2	1 1/4	35/64	3 3/8	0.310	0.100	1600
	9002S	29/32	②	1/2	1 1/4	35/64	3 3/8	0.310	0.100	1600
	9046S	29/32	②	1/2	1 1/4	35/64	3 3/8	0.310	0.100	1800
	9008S	29/32	②	1/2	1 1/4	35/64	3 3/8	0.310	0.100	1800
	9053S	29/32	②	1/2	1 1/4	35/64	3 3/8	0.310	0.100	1800
	9009S	29/32	②	1/2	1 1/4	35/64	3 3/8	0.310	0.100	1600
	9010S	29/32	②	1/2	1 1/4	35/64	3 3/8	0.310	0.100	1600
	9011S	29/32	②	1/2	1 1/4	35/64	3 3/8	0.310	0.100	1600
	9012S	29/32	②	1/2	1 1/4	35/64	3 3/8	0.310	0.100	1800
1976	9002S	29/32	②	33/64	1 1/4	33/64	3 3/8	0.310	0.100	1700
	9055S	29/32	②	33/64	1 1/4	33/64	3 3/8	0.310	0.100	1700
	9074S	29/32	②	33/64	1 1/4	33/64	3 3/8	0.310	0.100	1600
	9057S	29/32	②	33/64	1 1/4	33/64	3 3/8	0.310	0.100	1600
	9054S	29/32	②	33/64	1 1/4	33/64	3 3/8	0.310	0.100	1800
	9058S	29/32	②	33/64	1 1/4	31/64	3 3/8	0.310	0.100	1600
	9059S	29/32	②	33/64	1 1/4	31/64	3 3/8	0.310	0.100	1600
	9066S	29/32	②	33/64	1 1/4	33/64	3 3/8	0.310	0.100	1600
	9062S	29/32	②	33/64	1 1/4	33/64	3 3/8	0.310	0.100	1600
	9052S	29/32	②	33/64	1 1/4	33/64	3 3/8	0.310	0.100	1600
1977	9076S	27/32	②	1/2	1 1/2	33/64	3 3/8	0.310	0.150	1700
	9077S	27/32	②	31/64	1 1/2	33/64	3 3/8	0.310	0.100	1400
	9078S	27/32	②	1/2	1 1/4	33/64	3 3/8	0.310	0.100	1400
	9080S	27/32	②	1/2	1 1/4	33/64	3 3/8	0.310	0.100	1200
	9081S	27/32	②	1/2	1 1/4	33/64	3 3/8	0.310	0.100	1600
	9093S	27/32	②	17/32	1 1/4	33/64	3 3/8	0.310	0.150	1500
	9101S	27/32	②	1/2	1 1/4	33/64	3 3/8	0.310	0.100	1600
1978	9147S	29/32	②	1/2	1 1/2	31/64	3 3/8	0.310	0.100	1600
	9137S	29/32	②	1/2	1 1/2	31/64	3 3/8	0.310	0.100	1600
	9134S	29/32	②	1/2	1 1/2	31/64	3 3/8	0.310	0.100	1500
	9104S	29/32	②	1/2	1 1/2	31/64	3 3/8	0.310	0.150	1500
	9140S	29/32	②	1/2	1 1/2	33/64	3 3/8	0.310	0.150	1500
	9108S	27/32	②	1/2	1 1/2	33/64	3 3/8	0.310	0.100	1400
	9109S	27/32	②	1/2	1 1/2	33/64	3 3/8	0.310	0.100	1400
	9110S	27/32	②	1/2	1 1/2	33/64	3 3/8	0.310	0.100	1600
	9111S	27/32	②	1/2	1 1/2	33/64	3 3/8	0.310	0.100	1400
	9112S	29/32	②	1/2	1 1/2	33/64	3 3/8	0.310	0.100	1200
	9148S	29/32	②	1/2	1 1/2	33/64	3 3/8	0.310	0.100	1600

① Model numbers located on the tag or on the casting
② Adjust link so primary and secondary stops both contact at same time

NOTE: All choke settings are fixed.

Chrysler Corporation

Model 2280
Chrysler Corporation — Holley Carburetor

Year	Carb. Part No. ⑦	Float Level (in.)	Accelerator Pump Adjustment (in.)	Bowl Vent Clearance (in.)	Fast Idle (rpm)	Choke Unloader Clearance (in.)	Vacuum Kick (in.)	Fast Idle Cam Position (in.)	Choke
1978	R-7990-A	.313	Flush	.030	1600	.310	.150	.070	Fixed

Model 1945
Chrysler Corporation — Holley Carburetor

Year	Carb. Part No. ②	Float Level (in.)	Accelerator Pump Adjustment (in.)	Bowl Vent Clearance (in.)	Fast Idle (rpm)	Choke Unloader Clearance (in.)	Vacuum Kick (in.)	Fast Idle Cam Position (in.)	Choke
1974	R-6721-A	.046	.680	—	1600	.250	.140	.080	Fixed
	R-6722-A	.046	.810	—	1800	.250	.090	.080	Fixed
	R-6723-A	.046	.680	—	1600	.250	.140	.080	Fixed
	R-6724-A	.046	.750	—	1800	.250	.080	.080	Fixed
	R-6725-A	.046	.750	—	1600	.250	.140	.080	Fixed
	R-6726-A	.046	.750	—	1800	.250	.090	.080	Fixed
1975	R-7329-A	.046	2.22	—	1700	.250	.130	.080	Fixed
	R-7017-A	.046	2.22	—	1600	.250	.130	.080	Fixed
	R-7018-A	.046	2.33	—	1700	.250	.090	.080	Fixed
	R-7019-A	.046	2.22	—	1600	.250	.130	.080	Fixed
	R-7020-A	.046	2.33	—	1700	.250	.090	.080	Fixed
	R-7029-A	.046	2.22	—	1600	.250	.130	.080	Fixed
	R-7210-A	.046	2.33	—	1700	.250	.090	.080	Fixed
1976	R-7356-A	①	2.22	.060	1600	.250	.110	.080	Fixed
	R-7357-A	①	2.65	.060	1700	.250	.100	.080	Fixed
	R-7360-A	①	2.22	—	1600	.250	.110	.080	Fixed
	R-7361-A	.046	2.65	—	1700	.250	.100	.080	Fixed
	R-7363-A	.046	2.65	—	1700	.250	.100	.080	Fixed
	R-7823-A	①	2.22	.070	1600	.250	.110	.080	Fixed
	R-7824-A	①	2.33	.105	1700	.250	.100	.080	Fixed
1977	R-7632-A	①	2.22	.060	1400	.250	.110	.080	Fixed
	R-7633-A	①	2.33	.060	1700	.250	.110	.080	Fixed
	R-7635-A	①	2.33	—	1700	.250	.110	.080	Fixed
	R-7744-A	①	2.33	.060	1700	.250	.130	.080	Fixed
	R-7745-A	①	2.22	.060	1600	.250	.150	.080	Fixed
	R-7746-A	①	2.33	.060	1700	.250	.110	.080	Fixed
	R-7764-A	①	2.22	.060	1700	.250	.110	.080	Fixed
	R-7765-A	①	2.33	.060	1700	.250	.110	.080	Fixed
1978	R-7988-A	①	2.22	.062	1400	.250	.110	.080	Fixed
	R-7989-A	①	2.33	.062	1600	.250	.110	.080	Fixed
	R-8008-A	①	2.33	.062	1700	.250	.110	.080	Fixed
	R-8010-A	①	2.33	.062	1500	.250	.130	.080	Fixed
	R-8394-A	①	2.33	.062	1700	.250	.110	.080	Fixed

① Flush with the top of the bowl cover gasket, plus or minus 1/32
② Located on a tag attached to the carburetor.

Chrysler Corporation

Model 5220
Chrysler Corporation

Year	Carb. Part No.	Accelerator Pump	Dry Float Level (in.)	Vacuum Kick (in.)	Curb Idle RPM (w/fan)	Fast Idle RPM (w/fan)	Throttle Position Transducer (in.)	Throttle Stop Speed RPM	Choke
1978	R-8376A, 8378A, 8384A, 8439A, 8441A, 8505A, 8507A	#2 hole	.480	.070	900	1100	.547	700	2 NR

CAR SERIAL NUMBER AND ENGINE IDENTIFICATON

1976

Mounted behind the windshield on the driver's side is a plate with the vehicle identification number. The sixth character is the model year, with 6 for 1976. The fifth character is the engine code, as follows.

```
C ............... 225 1-bbl. 6-cyl.
G ............... 318 2-bbl. V-8
J ............... 360 4-bbl. V-8
K ............... 360 2-bbl. V-8
L ............... 360 4-bbl. Hi. Perf.
M ............... 400 2-bbl.
N ............... 400 4-bbl.
P ............... 400 4-bbl. Hi. Perf.
T ............... 440 4-bbl.
U ............... 440 4-bbl. Hi. Perf.
```

Engines can be identified by a code, as follows.

225 6-cyl.: The code is stamped on the right side of the block next to number one cylinder. The first digit is the year, and the next three numbers are the cubic inch displacement, which is 225.

   6 225
 Year Engine

318 and 360 V-8: The code is stamped on the front of the left bank below the head. The first digit is the year, 6 for 1976, and the following letter is the manufacturing plant. The next three numbers are the cubic inch displacement, 318 or 360.

  6 M 318
 Year Plant Engine

400 V-8: The code is on a small pad forward of the distributor. The first digit is the year, 6 for 1976, and the following letter is the manufacturing plant. The next three numbers are the cubic inch displacement, 400.

  6 M 400
 Year Plant Engine

440 V-8: The code is either on a small pad forward of the distributor, or on a pad at the front of the left bank, forward of the intake manifold. The first digit is the year, 6 for 1976, and the following letter is the manufacturing plant. The next three numbers are the cubic inch displacement, 440.

  6 M 440
 Year Plant Engine

1977

The vehicle identification number is mounted on a plate behind the windshield on the driver's side. The sixth character is the model year, with 7 for 1977. The fifth character is the engine code, as follows.

```
C ............... 225 1-bbl. 6-cyl.
D ............... 225 2-bbl. 6-cyl.
G ............... 318 2-bbl. V-8
J ............... 360 4-bbl. V-8
K ............... 360 2-bbl. V-8
L ............... 360 4-bbl. Hi. Perf.
N ............... 400 4-bbl. V-8
P ............... 400 4-bbl. Hi. Perf.
T ............... 440 4-bbl. V-8
U ............... 440 4-bbl. Hi. Perf.
```

Engines can also be identified by a code, as follows.

225 6-cyl.: The code is stamped on the right side of the block next to number six cylinder. The first digit is the year, 7 for 1977, and the next three numbers are the cubic inch displacement, which is 225.

318 and 360 V-8: The code is stamped on the front of the left bank below the head. The first digit is the year, with 7 for 1977. The following letter is the manufacturing plant. The next three numbers are the cubic inch displacement, 318 or 360.

400 V-8: The code is on a small pad forward of the distributor. The first digit is the year, 7 for 1977, and the following letter is the manufacturing plant. The next three numbers are the cubic inch displacement, 400.

440 V-8: The code is in on a pad at the front of the left bank, forward of the intake manifold. The first digit is the year, with 7 for 1977, and the following letter is the manufacturing plant. The next three numbers are the cubic inch displacement, 440.

1978

The vehicle identification number is mounted on a plate behind the windshield on the driver's side. The sixth character is the model year, with 8 for 1978. The fifth character is the engine code, as follows:

```
C ............... 225 1-bbl. 6-cyl.
D ............... 225 2-bbl. 6-cyl.
G ............... 318 2-bbl. V-8
H ............... 318 4-bbl. V-8
J ............... 360 4-bbl. V-8
K ............... 360 2-bbl. V-8
L ............... 360 4-bbl. V-8 Hi. Perf.
N ............... 400 4-bbl. V-8
P ............... 400 4-bbl. V-8 Hi. Perf.
T ............... 440 4-bbl. V-8
U ............... 440 4-bbl. V-8 Hi. Perf.
```

Engines can also be identified by a code, as follows.

225 6-cyl.: The code is stamped on the right side of the block next to number six cylinder. The first digit is the year, 8 for 1978, and the next three numbers are the cubic inch displacement, which is 225.

318 and 360 V-8: The code is stamped on the front of the left bank below the head. The first digit is the year, with 8 for 1978. The following letter is the manufacturing plant. The next three numbers are the cubic inch displacement, 318 or 360.

400 V-8: The code is on a small pad forward of the distributor. The firist digit is the year, 8 for 1978, and following letter is the manufacturing plant. The next three numbers are the cubic inch displacement, 400.

440 V-8: The code is on a pad at the front of the left bank, forward of the intake manifold. The first digit is the year, with 8 for 1978, and the following letter is the manufacturing plant. The next three numbers are the cubic inch displacement, 440.

1979

The vehicle identification number is mounted on a plate behind the windshield on the driver's side. The sixth character is the model year, with 9 for 1979. The fifth character is the engine code, as follows:

```
C ............... 225 1-bbl. 6-cyl.
D ............... 225 2-bbl. 6-cyl.
G ............... 318 2bbl. V-8
H ............... 318 4-bbl. V-8
J ............... 360 4-bbl. V-8
K ............... 360 2bbl. V-8
L ............... 360 4-bbl. V-8 Hi. Perf.
```

Engines can also be identified by a code, as follows:

Chrysler Corporation

CAR SERIAL NUMBER AND ENGINE IDENTIFICATION

1978

225 6-cyl.: The code is stamped on th right side of the block next to number six cylinder. The first digit is the year, 9 for 1979, and the next three numbers are the cubic inch displacement, which is 225.

318 and 360 V-8: The code is stamped on the front of the left bank below the head. The first digit is the year, with 9 for 1979. The following letter is the manufacturing plant. The next three numbers are the cubic inch displacement, 318 or 360.

EMISSION EQUIPMENT

1976
All Models
Closed positive crankcase ventilation
Heated air cleaner
Vapor control, canister storage
Exhaust gas recirculation
Emission calibrated carburetor
Emission calibrated distributor
Single diaphragm vacuum advance
Electric assist choke
Orifice spark advance control
  6-cylinder 49-States
    All Man. Trans. except,
    Not used on Feather Duster
    and Dart Lite
    Not used on any Auto. Trans.
  6-cylinder Calif.
    All models
  V-8 49-States
    318 V-8 Auto. Trans. with air pump only
    All 360 V-8
    All 400 V-8 except,
    Not used on Lean Burn engine
    Not used on any 440 V-8
  V-8 Calif.
    All 318 V-8
    Not used on 360 V-8
    All 400 V-8
    All 440 V-8
Idle enrichment
  All V-8, Auto. Trans. only, except
  Not used on Lean Burn engine
  Not used on 6-Cyl.
Altitude compensation
  49-States
    Not used
  Calif.
    440 High Performance V-8
Catalytic Converter
  49-States
    All models except
    Not used on 360 4-bbl. V-8
    Not used on 400 4-bbl. V-8
    Not used on 318 V-8 with air pump
  Calif.
    All models
Air pump
  49-States
    All 360 4-bbl. V-8
    318 V-8 without converter
    All 400 4-bbl. V-8 except
    Not used on lean burn engine
  Calif.
    All models
Catalyst overheat protection
  Not used on 1976 models. Discontinued as a running change on late 1975 models.

1977
Closed positive crankcase ventilation
Heated air cleaner
Vapor control, canister storage
Exhaust gas recirculation
Emission calibrated carburetor
Emission calibrated distributor
Catalytic converter
Fresh air intake
Electric assist choke
Orifice spark advance control
  49-States
    225 1-bbl. man. trans. 4000 lb.
    All 225 2-bbl.
    All 318 2-bbl.
    Not used on Lean Burn engine
  Calif.
    All engines, except
    Not used on Lean Burn engine
  Altitude
    All engines
Idle enrichment
  49-States
    360 2-bbl. V-8
    360 4-bbl. Lean Burn
    440 4-bbl. Lean Burn
  Calif.
    All V-8 engines, except
    Not used on 318 2-bbl. Lean Burn
  Altitude
    Not used
Altitude compensation
  49-States
    Not used
  Calif.
    360 4-bbl.
    440 4-bbl. Lean Burn
  Altitude
    Not used
Air pump
  49-States
    Not used
  Calif.
    All models
  Altitude
    All models
Air aspirator system
  49-States
    All models
  Calif.
    Not used
  Altitude
    Not used
Air switching valve
  49-States
    Not used
  Calif.
    All models
  Altitude
    All models
Power heat control valve
  49-States
    Not used
  Calif.
    All V-8 engines
  Altitude
    Not used

1978
All Models
Closed positive crankcase ventilation
Heated air cleaner
Vapor control, canister storage
Exhaust gas recirculation
Emission calibrated carburetor
Emission calibrated distributor
Catalytic converter
Fresh air intake
Electric assist choke
Orifice spark advance control
  49-States
    225 1-bbl. 6-cyl.
    225 2-bbl. 6-cyl.
  Calif.
    225 1-bbl. 6-cyl.
    440 4-bbl. Hi. Perf.
  Altitude
    225 1-bbl. 6-cyl.
Idle enrichment
  49-States
    360 2-bbl. V-8
    360 4-bbl. V-8
    440 4-bbl. V-8
    440 4-bbl. V-8 Hi. Perf.
  Calif.
    360 4-bbl. V-8
  Altitude
    360 4-bbl. V-8
Altitude compensation
  49-States
    Not used
  Calif.
    All V-8 engines
  Altitude
    318 2-bbl. V-8
    360 4-bbl. V-8
Air pump
  49-States
    Not used, except
    440 4-bbl. standard engine was made with and without pump.
  Calif.
    All models
  Altitude
    225 1-bbl. 6-cyl.
    360 2-bbl. V-8
    360 4-bbl. V-8
    440 4-bbl. V-8
Air aspirator system
  49-States
    All engines except,
    Not used on 360 4-bbl. Hi. Perf.

Chrysler Corporation

EMISSION EQUIPMENT

1978

Not used on 440 4-bbl. Hi. Perf.
Calif.
  Not used
Altitude
  318 2-bbl. V-8
Air switching valve
  49-States
    Not used
  Calif.
    All engines except,
    Not used on 440 4-bbl. Hi. Perf.
Altitude
  225 1-bbl. 6-cyl.
  360 2-bbl. V-8
  360 4-bbl. V-8
Power heat control valve
  49-States
    Not used
  Calif.
    All V-8 engines
Altitude
  360 2-bbl. V-8
  360 4-bbl. V-8
  440 4-bbl. V-8

1979

All Models
Closed positive crankcase ventilation
Heated air cleaner
Vapor control, canister storage
Exhaust gas recirculation
Emission calibrated carburetor
Emission calibrated distributor
Catalytic converter
Fresh air intake
Electric assist choke
Orifice spark advance control
  49-States
    225 1-bbl. 6-cyl.
    225 2-bbl. 6-cyl.
  Calif.
    225 1-bbl. 6-cyl.
Idle enrichment
  49-States
    360 2-bbl. V-8
    360 4-bbl. V-8
  Calif.
    360 4-bbl. V-8
Air pump
  49-States
    Not used
  Calif.
    All models
Air aspirator system
  49-States
    All engines
  Calif.
    Not used
Air switching valve
  49-States
    Not used
  Calif.
    All engines
Power heat control valve
  49-States
    Not used
  Calif.
    All V-8 engines

IDLE SPEED AND MIXTURE ADJUSTMENTS

1976

Air cleaner .... On
Air cond. .... Off
Auto. trans.
  Lean Burn engine ...... Drive
  All others .... Park
Air pump .... Disconnected
Idle CO
  49-States
    Catalyst equipped
      (At probe) 0.3%
    No catalyst (at tailpipe)
      318 V-8 .... 0.5%
      360 4-bbl. V-8 .... 0.5%
      400 Lean Burn V-8 .... 0.1%
  Calif. (All at probe)
    225 6-Cyl. 1.0%
    318 V-8 .... 1.0%
    360 4-bbl. V-8 .... 2.0%
    400 4-bbl. V-8 .... 0.5%
    440 4-bbl. V-8 .... 0.3%
    440 4-bbl. Hi. Perf. V-8 .. 0.5%
  NOTE: *On engines equipped with an air pump, idle CO readings are taken with the air pump hose disconnected. "Probe" CO readings are taken in front of the converter.*
225 6-cyl.
  49-States
    Feather Duster 600
    Dart Lite 600
    All others 750
  Calif.
    Auto. trans. 750
    Man. trans. 800
318 V-8
  With catalyst 750
  No catalyst
    Solenoid connected 900
    Solenoid disconnected .. 1 turn
360 2-bbl. V-8 700
360 4-bbl. V-8
  49-States
    Standard and Hi. Perf.
      Solenoid connected 850
      Solenoid disconnected 1 turn
  Calif.
    Standard 750
400 2-bbl. V-8 700
400 4-bbl. V-8
  49-States
    With air pump
      Solenoid connected 850
      Solenoid disconnected 1 turn
    Lean Burn engine 700
  Calif. 750
440 4-bbl. V-8 750
NOTE: *To set the solenoid—disconnected idle speed on those engines equipped with a solenoid, adjust the solenoid-connected speed first. Then, with the engine idling and the solenoid still connected, adjust the throttle stop screw (on opposite side of carburetor from solenoid) so it just touches the stop. Then back the screw off exactly one turn.*

1977 CALIFORNIA CARS

Air cleaner .... In place
Air cond. .... Off
Auto. trans. .... Park
Air pump .... Disconnect for
               mixture setting
Mixture adj. .... Set idle CO
Idle CO (All at probe)
  225 1-bbl. 6-cyl. 0.3%
  318 2-bbl. V-8
    With lean burn ....
    No lean burn 0.5%
  360 4-bbl. V-8 .... 0.5%
  440 4-bbl. V-8
    With lean burn 0.3%
    Hi. Perf. 1.0%
Idle Speed
225 6-cyl. 750
318 2-bbl. V-8
  With lean burn ....
  No lean burn 850
360 4-bbl. V-8 750
440 4-bbl. V-8
  With lean burn 750
  Hi. Perf. 750

IDLE SPEED AND PROPANE ENRICHED MIXTURE ADJUSTMENTS 1977 49-STATE CARS

Air cleaner .... In place
Air cond. .... Off
Auto. trans. .... Park
Air pump .... Connected
Idle CO .... Not used
Mixture adj. .... See below
Lean burn engine ... Disconnect and plug vacuum advance hose and ground carb. switch with jumper.
225 1-bbl. 3500 lbs.
  Auto. trans. 700
    Propane enriched 890
  Man. trans. 700
    Propane enriched 835
225 1-bbl. 4000 lbs.
  Auto. trans. 700

Chrysler Corporation

1977

Propane enriched	830
Man. trans.	700
Propane enriched	805
225 2-bbl.	
Auto. trans.	750
Propane enriched	900
Man. trans.	750
Propane enriched	930
318 2-bbl. no lean burn	
Auto. trans.	700
Propane enriched	780
Man. trans.	700
Propane enriched	810
360 2-bbl.	700
Propane enriched	810
360 4-bbl. lean burn	750
Propane enriched	860
400 4-bbl. lean burn	750
Propane enriched	880
440 4-bbl. lean burn	750
Propane enriched	850
440 4-bbl. Hi. Perf.	750
Propane enriched	850

NOTE: *The 225 1-bbl. 3500-lb. cars are the Aspen, Aspen Custom, Volare, and Volare Custom without air conditioning. The 225 1-bbl. 4000 lb. cars are the Aspen, Aspen Custom, Volare, and Volare custom with air conditioning, and all Aspen Special Edition and station wagon, Volare Premier and station wagon, Fury, and Monaco.*

360 4-bbl.	750
Propane enriched	
Above 4000 ft.	850
Below 4000 ft.	830
440 4-bbl. Lean Burn	750
Propane enriched	
Above 4000 ft.	850
Below 4000 ft.	830

NOTE: *Curb idle speed on the Altitude 318 2-bbl. is 850, but the mixture screws are adjusted to bring the speed to 750 with the propane shut off, when doing the propane-assisted procedure. After adjusting the mixture screws to 750 rpm, the solenoid adjustment is used to raise the speed to 850.*

IDLE SPEED AND PROPANE ENRICHED MIXTURE ADJUSTMENTS 1977 ALTITUDE CARS

Air cleaner	In place
Air cond.	Off
Auto. trans.	Park
Air pump	Connected
Idle CO	Not used
Mixture adj	See below
Lean burn engines	Disconnect

and plug vacuum advance hose and ground carb. switch with jumper

225 1-bbl	750
Propane enriched	
Above 4000 ft.	850
Below 4000 ft.	820
318 2-bbl.	
Solenoid connected	See below
Solenoid disconnected	Not adjustable
Propane enriched	
Above 4000 ft.	930
Below 4000 ft.	900

IDLE SPEED AND MIXTURE ADJUSTMENTS 1978 EXHAUST ANALYZER METHOD

NOTE: *This is an alternate method for California cars only. The preferred method is to use propane enrichment.*

Air cleaner	In place
Air cond.	Off
Auto. trans.	Park
Air pump	Disconnect for mixture setting
Vac. adv. hose	Plugged
EGR vac. hose	Plugged
Lean burn engine	Ground carb. switch with jumper
Mixture adj.	Set idle CO
Idle CO	
225 1-bbl. 6-cyl.	Probe 0.3%
318 4-bbl. V-8	Tail 0.5%
360 4-bbl. V-8	Tail 0.5%
440 4-bbl. V-8	
With lean burn	Tail 0.3%
Hi. Perf.	Tail 1.0%
Idle Speed	
225 6-cyl	750
318 4-bbl. V-8	750
360 4-bbl. V-8	750
440 4-bbl. V-8	750

IDLE SPEED AND PROPANE ENRICHED MIXTURE ADJUSTMENTS 1978

NOTE: *This is the preferred method for adjusting all cars.*

Air cleaner	In place
Air cond.	Off
Auto. trans.	Park
Dist. vac. hose	Plugged
EGR vac. hose	Plugged
Lean burn engine	Ground carb. switch with jumper
Mixture adj.	See propane rpm below
Idle Speed	
225 1-bbl. 3500 lbs.	
49-States	
Auto. trans.	700
Propane enriched	790
Man. trans.	700
Propane enriched	835
Calif.	
Auto. trans.	750
Propane enriched	880
225 1-bbl. 4000 lbs.	
49-States	
Auto. trans.	700
Propane enriched	790
Calif.	
Auto. trans.	750
Propane enriched	820, 860 or 880 (See label)
Man. trans.	750
Propane enriched	860
Altitude	
Auto. trans.	750
Propane enriched	
Above 4000 ft.	910
Below 4000 ft.	880
225 2-bbl. 49-States	
Auto. trans.	750
Propane enriched	900
Man. trans.	750
Propane enriched	930
318 2-bbl.	
49-States	
Auto. trans.	750
Propane enriched	850
Man. trans.	700
Propane enriched	810
Altitude	
Auto. trans.	750
Propane enriched	
Above 4000 ft.	880
Below 4000 ft.	830
318 4-bbl.	
49-States auto. trans.	750
Propane enriched	825
Calif. auto. trans.	750
Propane enriched	825
360 2-bbl.	
49-States auto. trans.	750
Propane enriched	890
360 4-bbl.	
49-States auto. trans.	750
Propane enriched	900
Calif. auto. trans.	750
Propane enriched	830
400 4-bbl.	
49-States auto. trans.	750
Propane enriched	840
440 4-bbl.	
49-States auto. trans.	750
Newport & New Yorker	860
All others	825
Calif. auto. trans.	750
Propane enriched	860
Altitude auto. trans.	750
Propane enriched	
Above 4000 ft.	930
Below 4000 ft.	860

IDLE SPEED AND MIXTURE ADJUSTMENTS 1979 EXHAUST ANALYZER METHOD

NOTE: *This is an alternate method for California cars only. The pre-*

ferred method is to use propane enrichment.

Air cleaner .... In place
Air cond. .... Off
Auto. trans. .... Park
Air pump .... Disconnect for mixture setting
Vac. adv. hose .... Plugged
EGR vac. hose .... Plugged
Lean burn engine ..... Ground carb. switch with jumper
Mixture adj. .... Set idle CO

Idle CO
  225 1-bbl. 6-cyl. ..... Probe 0.3%
  318 4-bbl. V-8 .... Tail 0.5%
  360 4-bbl. V-8 .... Tail 0.5%

Idle Speed
  225 6-cyl. 750
  318 4-bbl. V-8 750
  360 4-bbl. V-8 750

IDLE SPEED AND PROPANE ENRICHED MIXTURE ADJUSTMENTS 1979

NOTE: *This is the preferred method for adjusting all cars.*

Air cleaner .... In place
Air cond. .... Off
Auto. trans. .... Park
Dist. vac. hose .... Plugged
EGR vac. hose .... Plugged
Lean burn engine ..... Ground carb. switch with jumper
Mixture adj. ..See propane rpm below

Idle Speed
225 1-bbl. 3500 lbs.
  49-States
    Auto. trans. 675
      Propane enriched 830
    Man. trans. 675
      Propane enriched 845
  Calif.
    Auto. trans. 750
      Propane enriched 925
225 1-bbl. 4000 lbs.
  49-States
    Auto. trans. 675
      Propane enriched 845
  Calif.
    Auto. trans. 750
      Propane enriched 925
    Man. trans. 750

      Propane enriched 925
225 2-bbl. 49-States
    Auto. trans. 725
      Propane enriched 890
318 2-bbl.
  49-States
    Auto. trans. 730
      Propane enriched 850
318 4-bbl.
  49-States
    Auto. trans. 750
      Propane enriched 850
  Calif.
    Auto. trans. 750
      Propane enriched 850
360 2-bbl.
  49-States
    Auto. trans. 750
      Propane enriched 850
360 4-bbl.
  49-States
    Auto. trans. 750
      Propane enriched 870
  Calif.
    Auto. trans. 750
      Propane enriched 870

INITIAL TIMING

1976

225 6-cyl.
  49-States
    Feather Duster ......10° BTDC
    Dart Lite ....12° BTDC
    Other auto. trans. ... 2° BTDC
    Other man. trans. ... 6° BTDC
  Calif.
    Auto. trans. .... 2° BTDC
    Man. trans. .... 4° BTDC
318 V-8 .... 2° BTDC
360 2-bbl. V-8 .... 6° BTDC
360 4-bbl. V-8
  49-States (Hi. Perf.) .. 2° BTDC
  Calif. (Standard) ...... 6° BTDC
400 2-bbl. V-8 ....10° BTDC
400 4-bbl. V-8
  49-States
    with air pump .... 6° BTDC
    Lean Burn engine ... 8° BTDC
  Calif. .... 8° BTDC
440 4-bbl. V-8
  49-States
    Standard .... 8° BTDC
    Hi. Perf. ....10° BTDC
  Calif. .... 8° BTDC

1977

225 1-bbl. 6-cyl. 49-States
  3500 lbs. ....12° BTDC

4000 lbs.
  Auto. trans. .... 2° BTDC
  Man. trans. .... 6° BTDC
225 1-bbl. 6-cyl. Calif. ... 8° BTDC
225 1-bbl. 6-Cyl. Altitude . 8° BTDC
225 2-bbl. 6-cyl. 49-States 12° BTDC
318 2-bbl. V-8 no lean burn
  49-States .... 8° BTDC
  Calif. .... 0° TDC
  Altitude .... 0° TDC
360 2-bbl. V-8 49-States ..10° BTDC
360 4-bbl. V-8 no lean burn
  Calif. .... 6° BTDC
  Altitude .... 6° BTDC
360 4-bbl. V-8 lean burn
  49-States ....10° BTDC
400 4-bbl. V-8 lean burn
  49-States ....10° BTDC
440 4-bbl. V-8 lean burn
  49-States ....12° BTDC
  Calif. .... 8° BTDC
  Altitude .... 8° BTDC
440 4-bbl. V-8 Hi. Perf.
  49-States .... 8° BTDC
  Calif. .... 8° BTDC

1978

225 1-bbl. 6-cyl.
  49-States ....12° BTDC
  Calif. .... 8° BTDC
  Altitude .... 8° BTDC

SPARK PLUGS

1976

225 6-cyl.—CH-RBL13Y ...... .035
318 V-8—CH-RN12Y .... .035
360 V-8—CH-RN12Y .... .035
400 2-bbl.—CH-RJ13Y .... .035

400 4-bbl.
  No catalyst—CH-RJ87P .... .035
  With catalyst—CH-RJ13Y .. .035
  Lean burn—CH-RJ13Y .... .035
440 4-bbl.
  Standard—CH-RJ87P ...... .035

225 2-bbl. 6 cyl. .... 12° BTDC
318 2-bbl. V-8 .... 16° BTDC
318 4-bbl. V-8 .... 10° BTDC
360 2-bbl. V-8 .... 20° BTDC
360 4-bbl. V-8
  49-States .... 16° BTDC
  Calif. .... 6° or 8° BTDC
    (See label)
  Altitude .... 6° or 8° BTDC
    (See label)
400 4-bbl. V-8 .... 20° or 24° BTDC
    (See label)
440 4-bbl. V-8
  49-States
    Newport &
    New Yorker .... 12° BTDC
    All others .... 16° BTDC
  Calif. .... 8° BTDC
  Altitude .... 8° BTDC

1979

225 1-bbl. 6-cyl.
  49-States ....12° BTDC
  Calif. .... 8° BTDC
225 2-bbl. 6-cyl. ....12° BTDC
318 2-bbl. V-8 ....16° BTDC
318 4-bbl. V-8 ....16° BTDC
360 2-bbl. V-8 ....12° BTDC
360 4-bbl. V-8
  49-States .... 16° BTDC
  Calif. ....16° BTDC

Hi. Perf.—CH-R11Y .... .035

NOTE: *High performance engines (dual exhaust) may use either resistor plugs shown above, or non-resistor, which are the same*

Chrysler Corporation

number without the "R." Other engines may also come from the factory with non-resistor plugs.

1977

225 6-cyl. ...... CH-BL15Y ...035
318 V-8 .... CH-N12Y 035
360 V-8 .... CH-N12Y 035
400 V-8 .... CH-J13Y 035

1976

49-States
  225 6-cyl.
    Feather Duster .... Ported
    Dart Lite .... Ported
    Other auto. trans. ... Manifold
    Other man. trans. ..... Ported
  318 V-8
    Auto. trans.
      With catalyst ..... Manifold
      No catalyst .... Ported
    Man. trans. .... Ported
All other engines .... Ported
Calif.
  All engines .... Ported

1977

Diaphragm type
  All engines .... Single

EMISSION CONTROLS NOTE, 1976

The controls described below are those that have been changed for 1976. All other control systems are the same as in 1975.

VAPOR CONTROL SYSTEM, 1976

In 1975, the fuel tank vent line, that runs from the fuel tank to the carbon canister, had an overfill limiting valve mounted in the engine compartment near the canister. In 1976, the overfill limiting valve has been discontinued, and a rollover valve has taken its place. The rollover valve performs the same overfill limiting job, but also prevents gas from running out of the tank when the vehicle is upside down. The rollover valve must be mounted vertically, the same as the overfill limiting valve, and all service procedures are the same.

All 1975 and 1976 models use a carbon canister with three hose connection nozzles. The purge valve that

SPARK PLUGS

440 V-8
  Lean burn ... CH-J13Y 035
  Hi. Perf. CH-J11Y 035

NOTE: *All engines may use the non-resistor plugs shown above, or resistor plugs, with an "R" in front of the basic number.*

1978

225 6-cyl. .... CH-RBL16Y ...035

VACUUM ADVANCE

Vacuum source
  49-States
    225 1-bbl. 3500 lbs. .....Ported
    225 1-bbl. 4000 lbs.
      Man. trans. ....Ported
      Auto. trans. ....Manifold
    225 2-bbl. ....Ported
    318 2-bbl. ....Ported
    All lean burn ....Manifold
  Calif.
    No lean burn ....Ported
    With lean burn ....Manifold
  Altitude
    No lean burn ....Ported

1978

Diaphragm type .... Single
Vacuum source
  225 1-bbl. 6-cyl. .... Ported

EMISSION CONTROL SYSTEMS 1976

used to be on top of the canister has been eliminated. Purging is done through a hose connected to ported vacuum at the carburetor. This eliminates all purging at idle, because the port is only exposed to vacuum above idle. Models with a carburetor external vent have a hose connecting the vent to the canister. Models without the external vent have the "Carb. Bowl" nozzle on the canister capped.

To control rollover fuel leakage from the tank cap, the 1976 cap has a slightly higher pressure relief. Also, a double tong arrangement makes it necessary to turn the cap an additional quarter turn to get it off the filler neck.

All other servicing procedures are the same as before.

EXHAUST GAS RECIRCULATION, 1976

All engines use the venturi vacuum system with an amplifier, except the 49-State 318 V-8, both manual and automatic, equipped with a catalytic converter. It uses the ported vacuum

318 V-8 .... CH-RN12Y ...035
360 V-8 .... CH-RN12Y ...035
400 V-8 .... CH-OJ13Y ...035
400 V-8 HP .... CH-OJ11Y ...035
440 V-8 .... CH-OJ13Y ...035
440 V-8 HP .... CH-OJ11Y ...035

1979

225 6-cyl. ..... CH-RBL16Y ...035
318 V-8 .... CH-RN12Y .. .035
360 V-8 .... CH-RN12Y .. .035

225 2-bbl. 6-cyl............. Ported
318 2-bbl. V-8 .... Ported
318 4-bbl. V-8 .... Ported
360 2-bbl. V-8 .... Ported
360 4-bbl. V-8 .... Ported
400 4-bbl. V-8 .... Ported
440 4-bbl. V-8 .... Manifold
440 4-bbl. Hi. Perf. V-8 .... Manifold

1979

Diaphragm type ....Single
Vacuum source
  225 1-bbl. 6-cyl. ....Ported
  225 2-bbl. 6-cyl. ....Ported
  318 2-bbl. V-8 ....Ported
  318 4-bbl. V-8 ....Ported
  360 2-bbl. V-8 ....Ported
  360 4-bbl. V-8 ....Ported

system, without any amplifier.

The 35-second delay timer is still used. The only engines that do not use it are the Calif. 360 4-bbl. V-8, and the 49-State 225 6-cylinder.

In late 1975 as a running change, the EGR maintenance reminder was discontinued, and is not used on any 1976 cars.

All 49-State V-8 cars with an amplifier continue to use a 3-nozzle vacuum solenoid that works with the EGR timer to control both the EGR system and the idle enrichment system.

This timer is mounted on the firewall, and controls a single solenoid or two separate solenoids that turn the vacuum on and off for both the EGR and idle enrichment systems.

Our efforts to develop a bench test have failed. Because of the electronic innards of the timer, it only works when it is hooked up to the car wiring.

We suggest you test the timer by making sure that everything else is okay. A simple check of the solenoids and wiring can be made by disconnecting the plug at the timer and con-

Chrysler Corporation

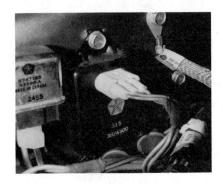

1975-76 EGR timer. This is a different design from the earlier timer

necting the gray wire (on the driver's side of the plug) to ground with the ignition switch on. If the solenoids click when the wire is connected to ground, you can be sure that the solenoid wiring is okay.

Also check to be sure you have battery voltage at the blue wire (in the center position), with the ignition switch on. The brown wire connects to the SOL terminal of the starter relay on the firewall. When the starter is not operating, the brown wire is a ground, and will light a test light connected between the battery and the wire. When the starter is operating, the brown wire becomes hot, and will light a test light connected between the wire and ground.

If the solenoids click, and the test lights show that the wiring is okay, a failure of the solenoids to operate in normal use indicates a bad timer. Substitute a new timer, but be sure it is mounted and connected before you turn on the ignition switch, to eliminate any accidental shorts.

Wide Open Throttle Kill Switch

A running change on 1976 Calif. 318 V-8's with automatic transmission is the addition of a wire open throttle kill switch. This switch is on the passenger side of the carburetor, and is connected to the throttle. The venturi vacuum hose that supplies vacuum to the EGR amplifier is connected to the switch so that the vacuum goes through the switch on its way to the amplifier. At wide open throttle, the switch shuts off the venturi vacuum to the EGR amplifier, so that there is no chance of having any EGR. EGR shutoff at wide open throttle is necessary so the engine will develop full power for maximum acceleration.

Ordinarily there is no venturi vacuum at wide open throttle, so the EGR system doesn't operate. But the 318 has a small 2-bbl. carburetor, and there can be enough vacuum at the venturi port to keep the EGR on at wide open throttle. The kill switch solves this problem by cutting off the vacuum.

IDLE ENRICHMENT SYSTEM, 1976

The system is the same, but the applications have changed. On April 7, 1975, the system was dropped from 49-State 6-cylinder engines as a running change. However, the system was not removed from the carburetor on many cars. The result is that it looks like someone forgot to hook it up, or that it was disconnected after it left the factory.

The 49-State 318 V-8 with a catalytic converter continues to use separate solenoids for the EGR and idle enrichment systems. The 318 with an air pump, including California, uses the single 3-nozzle solenoid which controls both the EGR and the idle enrichment systems. All other engines are the same as in 1975.

AUTOMATIC IDLE SPEED DIAPHRAGM 1976

The 1-bbl. Holley carburetor used on the 6-cylinder Feather Duster and Dart Lite cars is equipped with an automatic idle speed diaphragm. It is mounted in the same position as the old dashpot. The diaphragm is connected to manifold vacuum, and is made so that an internal spring is compressed by the vacuum. Any time a load is put on an idling engine, such as turning the air conditioner on, shifting into Drive, or operating the power steering, the vacuum will drop, and the engine will slow down. When the vacuum drops, the spring inside the automatic idle speed diaphragm pushes the stem out and speeds up the engine. Theoretically, any load that is put on the idling engine will be balanced by the diaphragm, and the idle speed will not change.

There are no adjustments to the diaphragm itself. Idle speed is adjusted with the throttle screw that pushes against the diaphragm stem. After adjusting the idle speed, the

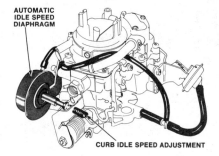

throttle should be opened wide and then dropped back to idle very quickly, so that the diaphragm will move. Then, if necessary, the idle speed should be reset to specifications.

ROLLOVER FUEL CONTROL 1976

All 1976 cars must have rollover fuel leakage control to prevent fuel from leaking out of the car if the car rolls over in an accident. Chrysler has added several valves to its cars to accomplish this. All cars have a rollover valve in the vent line from the fuel tank to the carbon canister. The rollover valve is mounted under the car. Cars equipped with a fuel return system also have a check valve in the return line near the fuel filter. Other changes include a slightly redesigned fuel pump, and a fuel tank cap with a higher pressure relief on some cars.

FUEL RETURN SYSTEM 1976

Many 1975 and 1976 cars use a 3-nozzle fuel filter. The 3rd nozzle connects to a return line that takes fuel and vapor back to the tank. This prevents the vapor from reaching the carburetor and causing rough running. Also, the constant flow of fuel back to the tank helps to keep the fuel lines cool and prevent vapor lock.

Two types of vapor separators are used. The 440 High Performance V-8 has used a filter with the return nozzle pointing down for several years. Other cars that use the system have a filter with the return nozzle pointing out to the side.

Regardless of which filter is used, the inlet to the filter must point down, and the outlet must point up. In other words, the hose from the fuel pump connects to the bottom of the filter, and the hose to the carburetor connects to the top of the filter. If the filter is installed upside down, or on its side, the vapors will not separate.

ORIFICE SPARK ADVANCE CONTROL 1975

A running change on 1975 49-State 440 High Performance V-8's is the elimination of the OSAC system. The OSAC valve is removed, and a cover plate installed over the mounting hole in the air cleaner. The hose from the distributor connects directly to the carburetor, or to the TIC valve, if used. This modification is legal and approved if done on the car assembly

Chrysler Corporation

line. The removal of the OSAC valve is not approved if it is done after the car leaves the factory. In addition, this modification is against federal law if done by any new car dealer, and may be against state law if done by independent shops.

The distributor also was changed, as follows.

1975 Distributor Numbers
49-State 440 High Perf. V-8
Early Production Late Production
With OSAC   No. OSAC
3874119   3874173

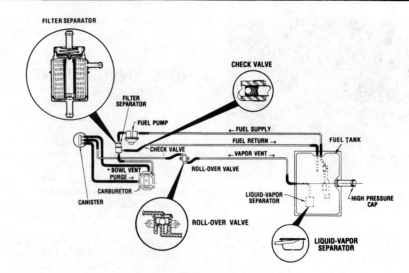

Fuel return system and rollover fuel control

EMISSION CONTROL SYSTEMS 1977

FRESH AIR INTAKE 1977

All 1977 engines use a tube that takes fresh, cool air from outside the engine compartment to the air cleaner snout. On a hot day this system can lower the intake air temperature as much as 30°F. The cooler air reduces the chance of detonation. The fresh air intake does not affect the operation of the heated air cleaner, which draws warm air from around the exhaust manifold when the intake air is cold.

AIR ASPIRATOR SYSTEM

The air aspirator valve is similar to the check valve used on air pump engines. The valve allows air flow to the exhaust ports, but blocks the flow of exhaust back out of the valve. There is no pump or diverter valve used with this system. The aspirator valve is connected to the exhaust manifold and to the clean side of the air cleaner. At idle, when there is suction in the exhaust, air is drawn in from the air cleaner and mixes with the exhaust to reduce tailpipe emissions. Above idle there is enough pressure in the exhaust to keep the aspirator valve closed.

Testing Air Aspirator Valve

The only test required is to be sure the valve is not stuck, or allowing exhaust gas to flow back towards the air cleaner. Remove the hose between the valve and the air cleaner with the engine idling. A piece of paper held against the valve will pulsate, indicating that air is being drawn into the valve. If there is no pulsation, or exhaust can be felt coming out of the valve, it is defective.

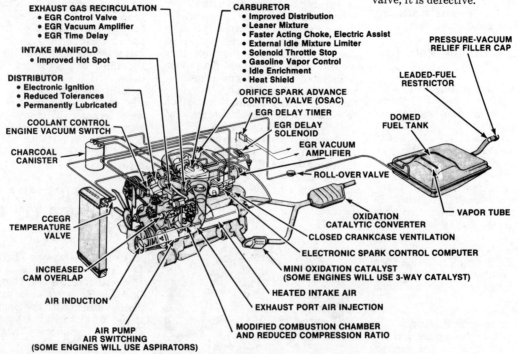

Chrysler Corporation 1977 emission control system (© Chrysler Corporation)

Chrysler Corporation

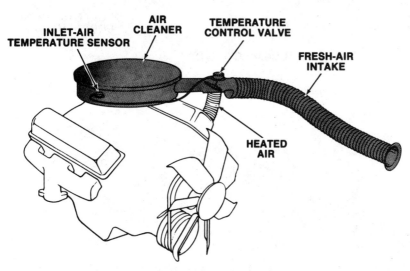

Chrysler Corporation fresh air intake (© Chrysler Corporation)

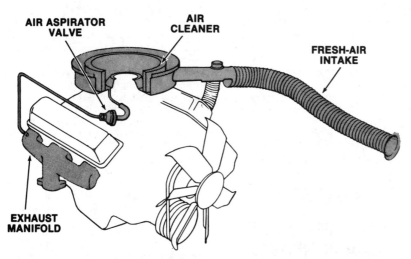

Chrysler Corporation air aspirator system (© Chrysler Corporation)

connection. 85% of the air is then injected downstream, but approximately 15% continues to flow to the exhaust ports.

On V-8's, the downstream entry point is on the right side manifold, above the heat control valve. On 6-cylinder engines, the entry is on the exhaust pipe, between the mini converter and the main converter.

The air switching system is used because it puts the air where it does the best job of burning up the pollutants. When the engine is cold, the air does its job best right at the engine exhaust ports, where there is enough heat. Also, the extra air helps to warm up the mini converter. After the engine is warm, the air is needed in the main underfloor converter. Also, getting the air away from the exhaust ports prevents it from recirculating with the exhaust gas into the EGR system, and leaning the mixture.

There is no change in testing procedures for the air pump or the diverter valve, because they operate the same as before.

Testing Air Switching Valve

Remove both large hoses from the air switching valve with the engine idling in PARK. If the engine is at normal operating temperature, the airflow should be toward the downstream connection, with a small amount of air from the other connection. Use a hand vacuum pump or a source of engine vacuum to apply 10 in. Hg. to the small vacuum hose connection on the valve. The airflow should switch so that all of the air comes out of the exhaust port connection. The valve should hold vacuum without leaking.

Repairing Air Switching Valve

Repairs are limited to replacement.

Repairing Air Aspirator Valve

Repairs are limited to replacement.

AIR SWITCHING VALVE 1977

Air output from the pump goes to the diverter valve, as in the past, but then goes to the air switching valve, which can send the air either to the exhaust ports or to a single entry point downstream. The air switching valve is operated by intake manifold vacuum, and this vacuum is controlled by a Coolant Control Engine Vacuum switch (CCEV).

When the engine coolant temperature is below 125°F. on the 318 V-8, or below 98°F. on any other engine, the CCEV is open, allowing manifold vacuum to operate the air switching valve and direct the air to the exhaust ports, the same as a normal air pump system. When the coolant temperature goes above 125°F. on the 318 V-8, or above 98°F. on any other engine, the CCEV shuts off the vacuum to the air switching valve. The spring in the valve then opens the tube that goes to the downstream

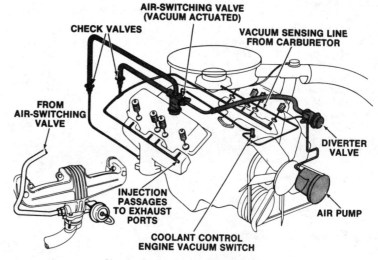

Chrysler Corporation air switching system (© Chrysler Corporation)

257

Chrysler Corporation

CAUTION: *The air switching valve looks just like a V-8 diverter valve, but they are not the same.*

Testing Coolant Control Engine Vacuum Switch

On any engine at normal operating temperature, the CCEVS must be closed, shutting off the manifold vacuum. If it is open at that temperature, it is defective. The valve should open below 125°F. on the 318 V-8, or below 98°F. on any other engine.

On some engines the CCEVS supplies vacuum to the idle enrichment system and the power heat control valve, in addition to the air switching valve. The idle enrichment system has a small air bleed in the hose about three inches from the carburetor. The idle enrichment valve, the power heat control valve, and the air switching valve all depend on the bleed to get rid of trapped vacuum when the CCEVS closes. If the bleed is missing or plugged, all three systems will be airlocked in one position when the CCEVS closes.

Repairing Coolant Control Engine Vacuum Switch

Repairs are limited to replacement.
CAUTION: *Relieve cooling system pressure before removing switch.*

POWER HEAT CONTROL VALVE 1977

This is similar to the vacuum operated heat riser valve used on other makes. It is installed only on Chrysler V-8 engines, in the right side exhaust manifold. A vacuum diaphragm is mounted near the valve, and connected to manifold vacuum through a Coolant Control Engine Vacuum Switch (CCEVS). When the engine coolant is below 125°F. on a 318 V-8, or below 98°F. on any other V-8, the CCEVS is open, allowing vacuum to operate the actuator and close the heat valve. The exhaust from the right bank is then forced through the crossover passage in the intake manifold and goes out the left side, warming the manifold as it goes.

When the engine warms up to above 125°F. on a 318 V-8, or above 98°F. on any other V-8, the CCEVS closes and shuts off the vacuum. The vacuum bleeds out through the idle enrichment bleed, and the spring in the actuator opens the heat valve. For testing on the CCEVS, see under Air Switching Valve, above.

Testing Power Heat Control Valve

When 6 in. Hg. or more vacuum is applied to the actuator, the valve

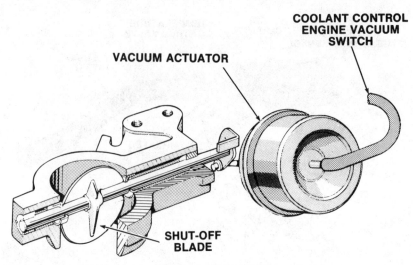

Chrysler power heat valve (© Chrysler Corporation)

should close. When the vacuum is removed, the spring in the actuator should open the valve. Also, the actuator should not leak when vacuum is applied to it.

If the actuator is in good condition, but the valve sticks, it may be possible to free up the valve by working it with solvent, the same as any other heat valve.

Repairing Power Heat Control Valve

Clean and lubricate the valve if necessary. If the actuator is bad, replace the entire assembly.

ANALYZER PROBE LOCATIONS 1977

49-State cars do not have the probe fitting in the catalytic converter. The factory recommended field method of setting idle mixture is the propane procedure, which does not use the HC-CO analyzer, so the probe fitting has been left out.

California cars are adjusted with an HC-CO analyzer, so the probe fitting is retained, but it is in a different place. California cars for 1977 use a mini-converter or "hybrid" converter welded into the exhaust pipe a few inches from the exhaust manifold. In all engines except the 440 4-bbl. dual exhaust, the probe fitting is in the mini or hybrid converter. On 6-cylinder engines the fitting is accessible from under the hood. On the 318 and 360 V-8's the mini converter is positioned so low that it is easier to reach it from under the car.

The 440 4-bbl. V-8 with dual exhaust is a special case. It uses a mini converter on the left bank only, and two underfloor converters, one on each side. The probe fitting is in the right side underfloor converter. The mini converter and the left side underfloor converter do not have a probe fitting.

PROPANE ASSISTED IDLE ADJUSTMENT 1977

Idle mixture settings on 49-State

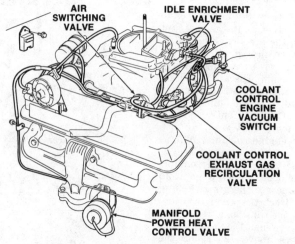

Chrysler vacuum control hoses (© Chrysler Corporation)

and Altitude cars for 1977 are adjusted with a propane-assisted procedure. The method is similar to what Ford Motor Co. started using in 1975, but the adjustment procedure is different.

When the carburetors are adjusted at the factory, the mixture screws are set lean to reduce emissions. The lean setting requires a greater throttle opening to keep the engine running. In effect, the engine is starving for fuel while it is idling. If propane is added to the engine, it will speed up. The amount that it will speed up depends on how lean the mixture screws were set. An engine can only take so much fuel, and the propane just makes up for the lack of fuel caused by the lean mixture screw setting.

The propane-assisted method accomplishes the same thing as the lean drop method used by General Motors. The advantage of the propane-assisted method is that it is not necessary to remove the idle mixture screws in most cases.

To make the adjustment, the engine must be thoroughly warmed up. Then with the engine at a normal curb idle, the air cleaner vacuum supply hose is disconnected and the propane rig is connected to the engine vacuum connection. The valve on the propane bottle should be slowly turned until the engine rises to its maximum idle speed and then starts to slow down. Reducing the amount of propane will bring the speed back to the maximum. Compare this with the specifications, and if it is within 10 rpm the mixture setting is okay, and should not be changed.

If the speed is too fast or too slow, the *idle speed screw* (not the mixture screws) should be adjusted to bring the speed to the "Enriched RPM" as shown in the specifications or on the engine label. This adjustment is made while the propane is still being fed at the rate that gave the fastest idle speed. After adjusting the throttle, check the propane feed rate again, to be sure the engine is idling at its fastest speed. If the engine speeds up, reset the idle speed screw to the "Enriched RPM."

NOTE: *The idle speed screw can be a throttle screw or a solenoid adjustment, depending on how the carburetor is equipped.*

Once the engine is idling at the "Enriched RPM" turn off the propane valve. The engine speed will drop. If the carburetor is adjusted correctly, it will drop to the correct idle speed. If the idle speed is too fast or too slow, change it by adjusting the mixture screws, not the idle speed screw. This will give the proper lean mixture for low emissions.

CAUTION: *The mixture screws should be used to change the idle speed only after following the entire propane-assisted procedure. To reset or change idle speed at any other time, use the idle speed screw.*

The 318 2-bbl. Altitude engine is adjusted with a special procedure. After adjusting the enriched rpm with the propane on, the propane is shut off and the mixture screws are adjusted to bring the speed to 100 rpm less than the specified idle rpm.

The idle rpm is 850, so the mixture screws are leaned evenly to bring the speed to 750. The final step is to raise the idle speed back to the 850 setting with the solenoid adjustment.

Although the purpose of the propane-assisted procedure is to allow setting the idle mixture without removing the limiter caps, there may be some engines that will not adjust within the range of the caps. If so, remove the caps to make the adjustment, but install the red "service" caps afterwards, with the tang positioned so that the screws can be adjusted leaner, but not richer.

CARBURETORS

Carburetors on 1978-79 vehicles are calibrated to control exhaust emissions. The Holley, Model 1945 carburetor has undergone some minor changes for 1979, which include an improved accelerator linkage and an improved accelerator pump.

The idle adjustment procedure for all vehicles will be propane assisted to determine the best idle. An exception to this would be some of the Carter, Model BBD carburetors which utilize tamperproof idle mixture screws.

All carburetor adjustments will be made after the carburetor has reached normal operating temperature, and with the transmission in neutral, headlights off, air conditioning compressor disconnected, idle stop switch grounded, and the vacuum hose to the EGR of spark control unit disconnected and plugged. Refer to the emission control label, located in the engine compartment, for further instructions.

HEATED INLET AIR SYSTEM

The purpose of the heated inlet air system is to control the temperature of the air entering the carburetor when ambient temperatures are low. By increasing the temperature of the air being introduced into the carburetor, a much leaner calibration is made possible, thereby reducing hydrocarbon emissions. In addition to a refined fuel air mixture, other benefits derived as a result of the heated air include smoother engine warm-up operation (a more volatile mixture as a result of heat) and a minimal amount of carburetor icing.

EMISSION CONTROL SYSTEMS 1978-79

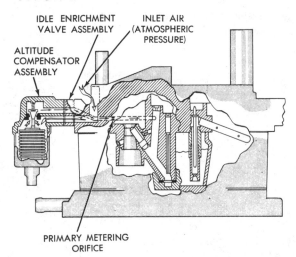

Altitude Compensator System—Thermo Quad Carburetor (California and High Altitude)

Chrysler Corporation

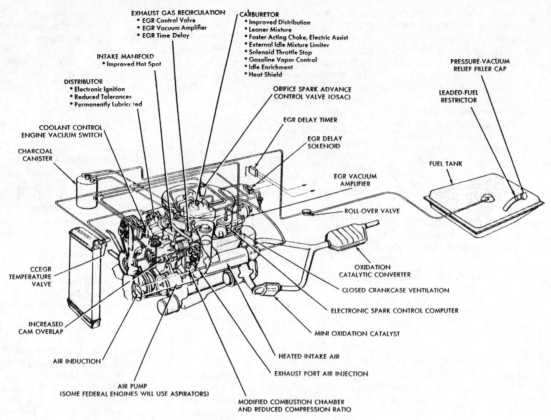

Emission Control Systems 1978-79

ELECTRIC ASSIST CHOKE SYSTEM

This system basically consists of an electric heating element, located next to a bimetal spring inside the choke housing, which assists engine heat to control choke duration. The electric current necessary to generate heat in the heater is routed through the oil pressure switch. A minimum pressure of 4 psi is required to close the switch before current can begin flowing through the switch and to the heater.

In addition to the electric heating element, two different electrical control units are used to regulate choke duration. A single stage control shortens only summer choke operation above approximately 80 degrees F. A duel stage control functions during summer temperatures in a manner similar to that of the single stage control, but also stabilizes choke duration in the winter by reducing current flow to the heater with the aid of a

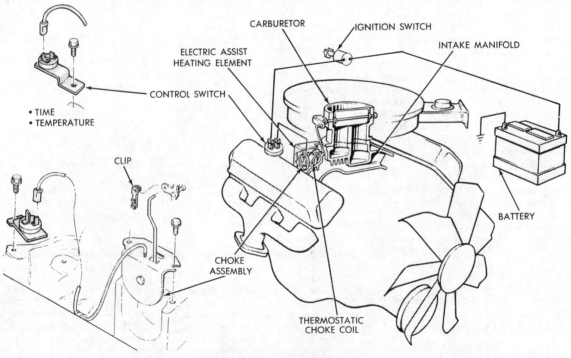

Electric Assist Choke System

Chrysler Corporation

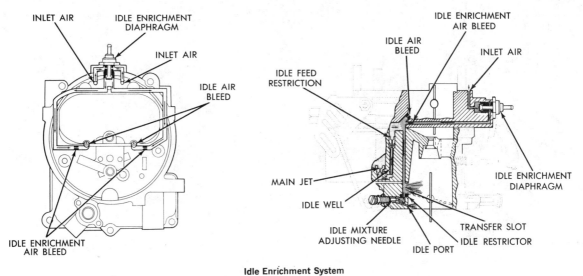

Idle Enrichment System

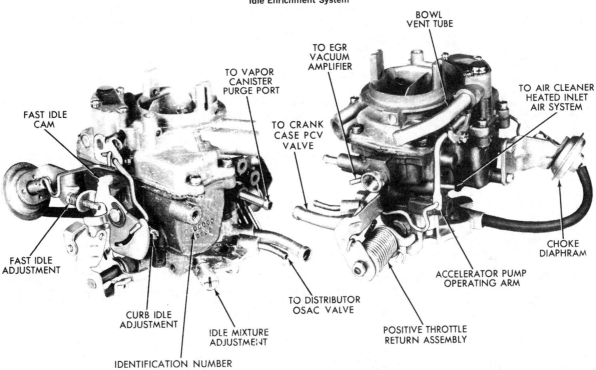

Carburetor—Model 1945

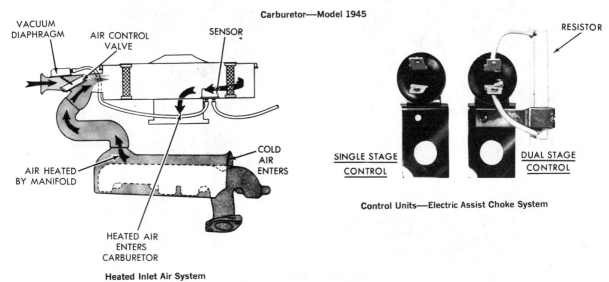

Heated Inlet Air System

Control Units—Electric Assist Choke System

Chrysler Corporation

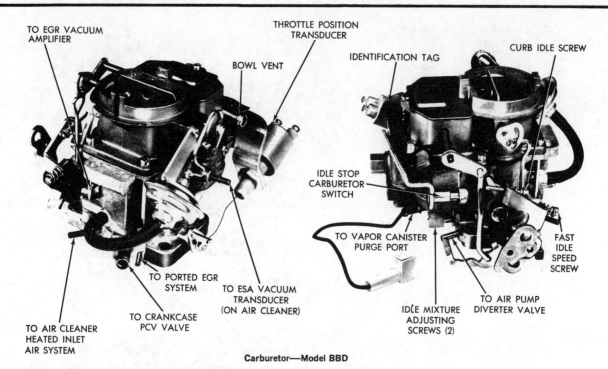

Carburetor—Model BBD

resistor permanently connected to both terminals of the control. The resistor is effective at temperatures below approximately 55 degrees F.

IGNITION SYSTEM

The purpose or design function of the ignition system is to allow a lean air fuel mixture to be burned under various operating conditions. Appropriately named the Electronic Lean Burn System (also known as Electronic Spark Control System – ESC), the ignition system is comprised of a spark control computer, five engine sensors, and a specially calibrated carburetor.

The spark control computer is capable of receiving several and various signals from the five engine sensors simultaneously, computing these signals within milliseconds, thereby monitoring engine performance, and advancing or retarding the ignition timing by signaling the ignition coil to produce the electrical impulses which fire the spark plugs.

ENGINE SENSORS

There are five engine sensors: vacuum transducer, pick-up coil, throttle position transducer, carburetor switch, and coolant switch.

The vacuum transducer, located on the spark control computer, senses intake manifold vacuum, and signals the computer which, in turn, regulates ignition timing.

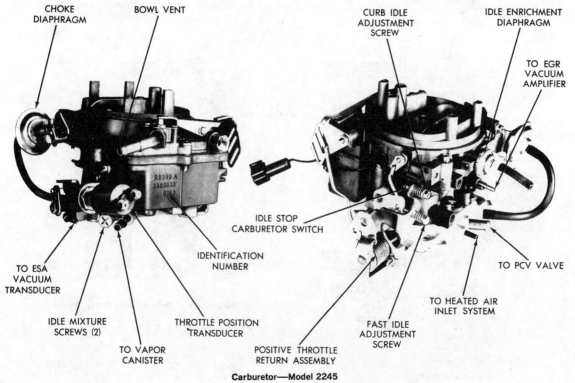

Carburetor—Model 2245

Chrysler Corporation

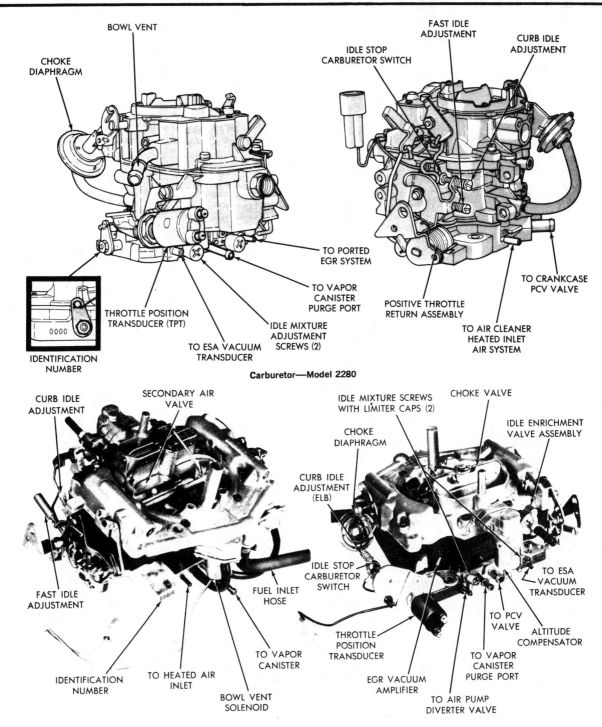

Carburetor—Model 2280

Thermo Quad Carburetor

A high vacuum reading will command an additional advance in timing. A low vacuum reading will command the timing to retard.

In order to achieve maximum advance for any inch of vacuum, the carburetor switch must remain open for a specified amount of time. During this state, ignition advance is not immediate; it is gradual, and develops at a slow rate. If the carburetor switch closes prematurely, the additional advance will be cancelled. This information is stored in the computer's memory, and the computer will slowly return the advance to 0. If the switch is opened before the computer can return the advance to 0, the advance will again gradually build from the point recorded in the computer's memory. If the switch is opened after the advance has been allowed to return to 0, ignition advance will again develop gradually.

The pick-up coil, located in the distributor, senses engine speed and crankshaft location, and transmits this information to the computer. The signal transmitted from the pick-up coil is reference signal. When the computer receives this signal, maximum timing advance is made available.

The function of the throttle position transducer is to sense the throttle position and rate of change. As the throttle plates start to open, regardless of position, additional advance will be directed on command from the computer.

The carburetor switch, located on the idle stop solenoid or air conditioning solenoid, signals the computer as to whether the engine is at idle or off idle.

263

Chrysler Corporation

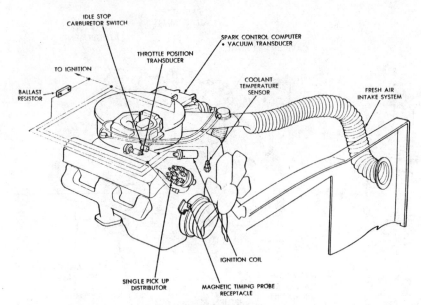

Electronic Lean Burn System

Valve Color Code	Opening Temperature (degrees F.)
Blue	75
Black	98
Yellow	125
Red	150

When the coolant reaches the calibration temperature, the valve opens thereby permitting vacuum to reach the EGR valve.

AIR INJECTION SYSTEM

A very effective way of reducing the level of carbon monoxide and hydrocarbons present in the exhaust gases is to inject a controlled amount of air directly into the exhaust system at a point very close to the exhaust valve. The injected air assists in further oxidation of the exhaust gases thereby holding the level of undesirable gases at a minimum. The system consists basically of a belt driven injection pump, a diverter valve, a switching valve, and a check valve.

Filtered air from the clean side of the air cleaner is drawn into the pump and delivered to an air switching valve. During engine warm-up, the air is directed to the exhaust ports to assist the oxidation process in the mini-catalyst. After the engine has reached normal operating temperature, the air switching valve will direct the air to a point in the exhaust pipe just behind the mini-catalyst on six cylinder engines, and to a point at the base of the right exhaust manifold on eight cylinder engines.

During periods of sudden deceleration, when abrupt throttle closing allows a rich air fuel mixture to reach the combustion chamber, the diverter valve will direct the injected air out into the atmosphere, thus preventing a backfire condition which would occur as a result of combining unburned fuel and injected air.

The coolant switch, located on the thermostat housing, senses engine coolant temperature. Its function is to signal the computer when the engine coolant temperature is below 150 degrees.

EXHAUST GAS RECIRCULATION (EGR)

Recirculated exhaust gas, when introduced into the combustion chamber, has the effect of minimizing the oxides of nitrogen emitted into the air by lowering combustion flame temperatures during engine operation. Only a modulated amount of exhaust gas is allowed to dilute the highly combustible air fuel mixture. The system responsible for regulating the volume of exhaust gas available during firing is called Exhaust Gas Recirculation (EGR).

During engine warm-up, when combustion flame temperature is relatively low, control vacuum is prevented from reaching the EGR valve by use of a coolant control exhaust gas recirculation valve (CCEGR). The CCEGR, mounted in the engine and/or radiator top tank, remains closed until the coolant reaches a specified temperature, plus or minus 5 degrees F.

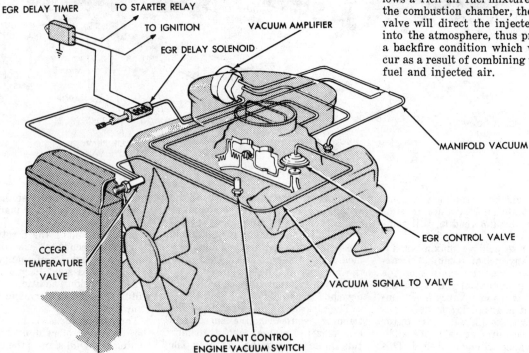

Exhaust Gas Recirculation

Chrysler Corporation

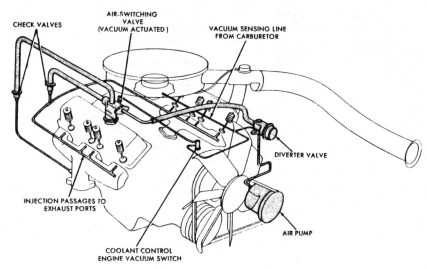

Air Injection System—Typical

repair when compared to the air injection system. An aspirator valve, located between the air cleaner and exhaust manifold, and connected by hoses, contains a spring loaded diaphragm which is sensitive to changes in pressure. Whenever a vacuum pulse occurs in the exhaust ports and manifold passages, the diaphragm opens to allow fresh air to mix with the exhaust gases. At high engine speeds, the aspirator valve remains closed.

To test if the valve has failed, disconnect the hose at the aspirator inlet, and, with the engine at idle and in neutral, check for escaping exhaust gases at the inlet. If the pulses can not be felt, and there is evidence of exhaust gases present at the inlet, the valve has failed and should be replaced.

A check valve, which is located in the injection tube assembly, will prevent hot exhaust gases from backing up into the hose and pump in the event of a system break down, such as would be caused by pump belt failure, excessively high exhaust system pressure, or a ruptured air hose.

ASPIRATOR AIR SYSTEM

The operating principle of the aspirator air system is similar to that of the air injection system in that both systems rely on the introduction of air into the exhaust manifold to reduce carbon monoxide and hydrocarbon emissions to an acceptable level. Though similar, both systems are distinctly different in design and operation.

The aspirator air system is a relatively simple system to diagnose and

Switching Valve Application

Word Code	Stamping Color	Engine
DO	Green	225 Calif. and high altitude.
BO	Yellow	318 and 360-4 Calif. and 360-4 high altitude.

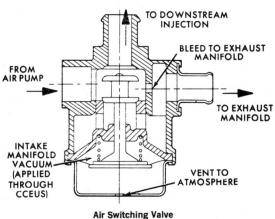

Air Switching Valve

Aspirator Valve

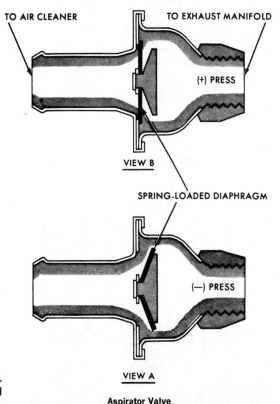

Diverter Valve

265

Chrysler Corporation

EVAPORATION CONTROL SYSTEM

Pollution can occur even when an engine is not running. This is because gasoline will continue to evaporate as long as the temperature of the fuel is high enough to permit evaporation. When the fuel is in a state of expansion, it will separate and rise away from the liquid fuel into the surrounding atmosphere. This constitutes pollution because gasoline fumes are hydrocarbons, an undesirable pollutant.

A method of controlling the emission of fuel vapors is to contain the fuel system in an evaporation control system. Fuel vapor leaving the carburetor bowl vent and fuel tank are trapped in vent hoses, and then routed to a charcoal cannister where the vapors are stored until they can be drawn into the intake manifold when the engine is running.

Some models still have the normal 3-hose carbon canister, with one hose bringing vapors from the tank, the second hose bringing vapors from the carburetor bowl vent, and the third hose for purging. The purge hose on 3-hose canisters is connected to a carburetor port that is above the throttle blade. At idle the port is not subject to engine vacuum, so the canister is not purged. Above idle the port is uncovered and purging takes place.

Some engines use a 4-hose canister. Two hoses bring vapors to the canister, one from the tank and the other from the carburetor bowl. The other two hoses are for purging. One of them is connected to ported vacuum at the carburetor. It turns on a purge valve on top of the canister. When the purge valve opens, a hose from the canister to manifold vacuum purges the canister.

Some of the larger V-8's use two carbon canisters. One canister is the 3-hose type. It connects to the carburetor bowl vent and to a carburetor port. The third outlet on the canister is capped. Purging takes place only above idle when the port is uncovered. The second canister is connected to the fuel tank. It is the purge valve 4-outlet type, but the outlet that normally connects to the bowl vent is capped off. With this system, one canister takes only the carburetor bowl vent vapors, and the other canister takes only the fuel tank vapors. This reduces the chance that any vapors will get into the atmosphere.

Both canisters have filters in the bottom, which must be replaced every 30,000 miles.

MINI CATALYTIC CONVERTER

In addition to the conventional type (underfloor converter) catalytic converter, a mini-converter is also used to initiate the exhaust gas oxidation before the exhaust gases reach the main underfloor catalytic converter.

CRANKCASE VENTILATION SYSTEM

This system, which utilizes a conventional PCV valve, is similar to the systems used on most vehicles marketed in the United States in that it directs crankcase vapors through the air cleaner and into the combustion area for complete burning.

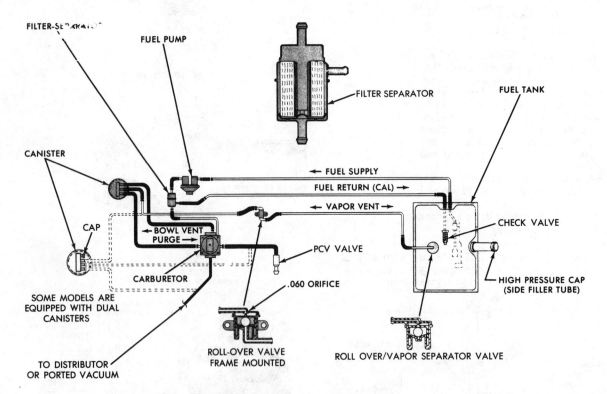

Evaporation Control System—1978-79 (Typical)

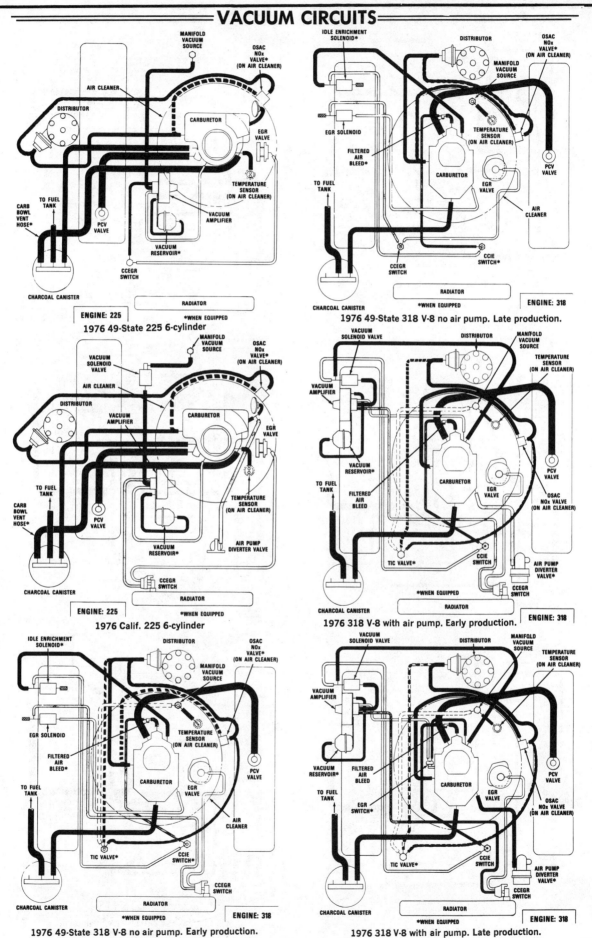

Chrysler Corporation

VACUUM CIRCUITS

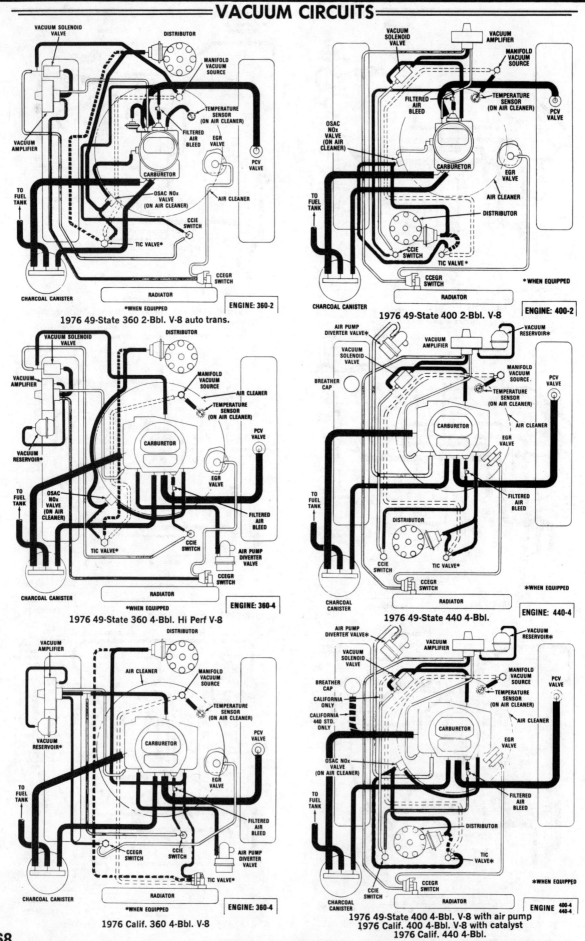

Chrysler Corporation

VACUUM CIRCUITS

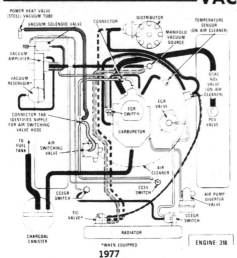

1977
Vacuum hose routing Chrysler 318-2bbl carb. high altitude air pump w/auto. trans. and California early—w/auto. trans. (© Chrysler Corporation)

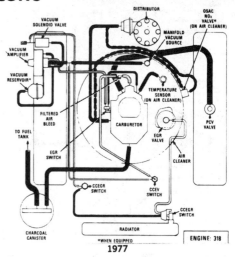

1977
Vacuum hose routing Chrysler 318-2bbl carb., Federal, Canada w/auto. trans.—early (© Chrysler Corporation)

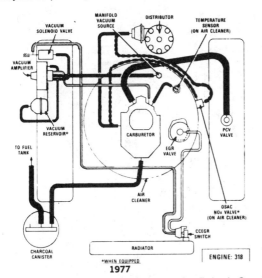

1977
Vacuum hose routing Chrysler 318-2bbl carb., Federal, Canada w/manual trans. (© Chrysler Corporation)

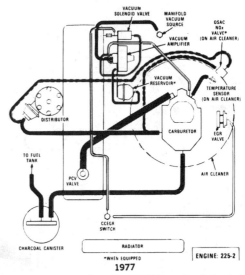

1977
Vacuum hose routing Chrysler 225-2bbl carb., Federal, Canada w/manual and auto. trans.

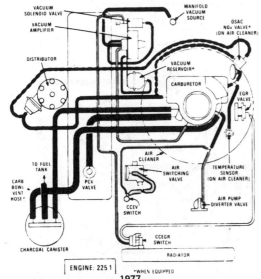

1977
Vacuum hose routing Chrysler 225-2bbl carb., high altitude air pump w/auto. trans. and California w/manual and auto. trans. (© Chrysler Corporation)

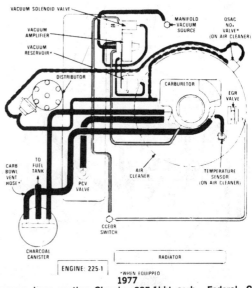

1977
Vacuum hose routing Chrysler 225-1bbl carb., Federal, Canada w/manual and auto. trans. (© Chrysler Corporation)

Chrysler Corporation

VACUUM CIRCUITS

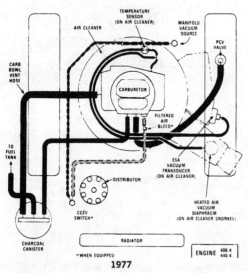

1977

Vacuum hose routing Chrysler 360-4bbl carb., high altitude air pump w/auto. trans. and 360-4bbl carb., H.P. eng., high altitude air pump and California w/auto. trans. (© Chrysler Corporation)

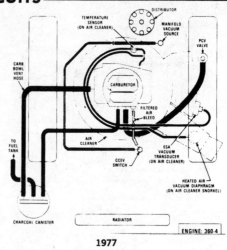

1977

Vacuum hose routing Chrysler 360-4bbl carb., Federal ELB w/catalytic converter, Canada ELB wo/catalytic converter w/auto. trans. (© Chrysler Corporation)

1977

Vacuum hose routing Chrysler 400-4bbl carb., or 440-4bbl carb., Federal, ELB w/catalytic converter, Canada, ELB wo/catalytic converter w/auto. trans. (© Chrysler Corporation)

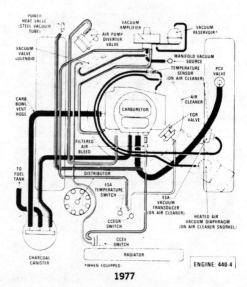

1977

Vacuum hose routing Chrysler 440-2bbl carb., Federal high altitude, ELB w/catalytic converter w/auto. trans., California, ELB w/catalytic converter, w/auto. trans. (© Chrysler Corporation)

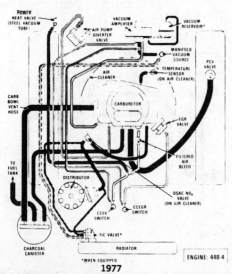

1977

Vacuum hose routing Chrysler 440-4bbl carb., H.P. California w/auto. trans. (© Chrysler Corporation)

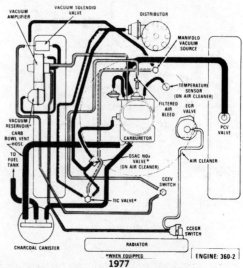

1977

Vacuum hose routing Chrysler 360-2bbl carb., Federal, Canada w/auto. trans.—early (© Chrysler Corporation)

Chrysler Corporation

VACUUM CIRCUITS

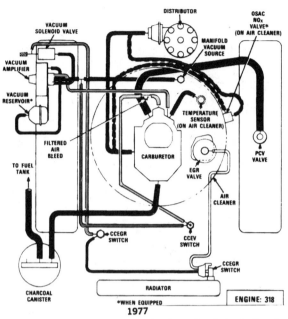

Vacuum hose routing Chrysler 318-2bbl carb., Federal, Canada w/auto. trans.—intermediate (© Chrysler Corporation)

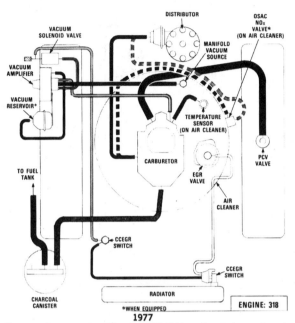

Vacuum hose routing Chrysler 318-2bbl carb., Federal, Canada w/auto. trans.—late (© Chrysler Corporation)

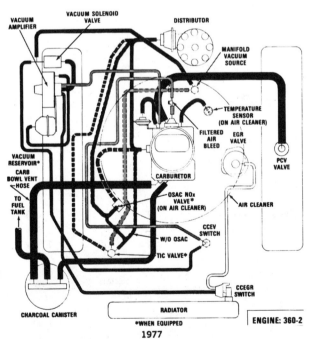

Vacuum hose routing Chrysler 360-2bbl Carb., Federal, Canada w/auto. trans.—late (© Chrysler Corporation)

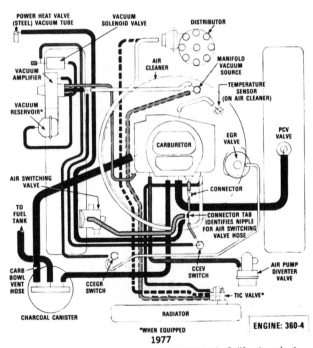

Vacuum hose routing Chrysler 360-4bbl std. California w/auto. trans. (© Chrysler Corporation)

271

Chrysler Corporation

VACUUM CIRCUITS

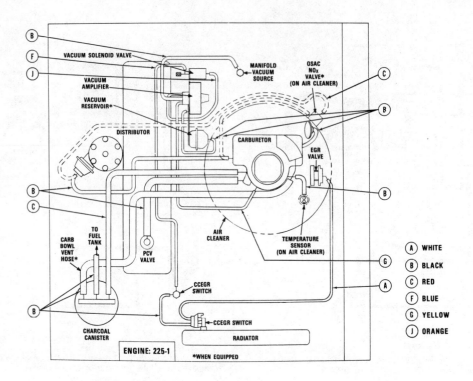

1978 Chrysler 225 1-bbl. 6-cyl., manual and auto. trans. with catalyst (49 states)

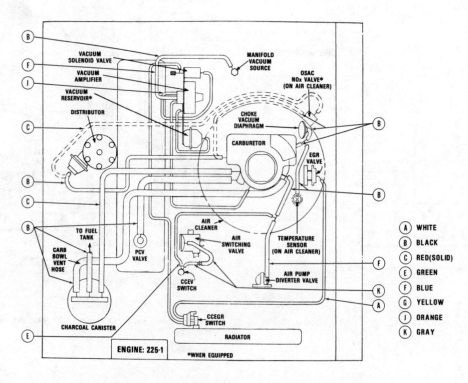

1978 Chrysler 225 1-bbl. 6-cyl., auto. trans. (Altitude and California)

Chrysler Corporation

VACUUM CIRCUITS

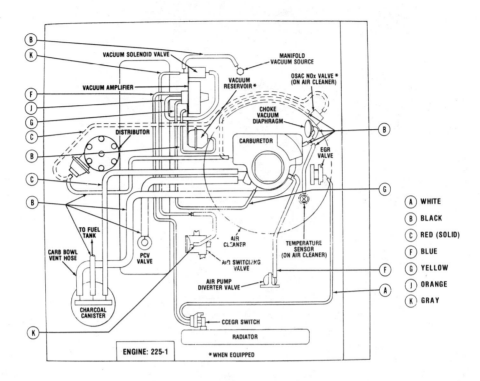

1978 Chrysler 225 1-bbl. 6-cyl., manual trans. (California)

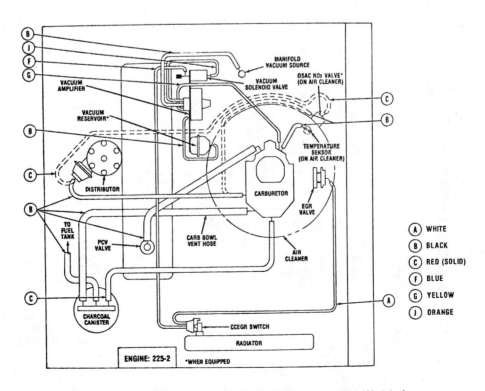

1978 Chrysler 225 2-bbl. 6-cyl., manual and auto. trans. with catalyst (49 states)

Chrysler Corporation

VACUUM CIRCUITS

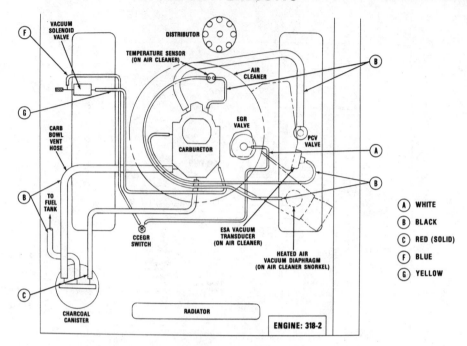

1978 Chrysler 318 2-bbl. V-8, early production (49 states and Altitude)

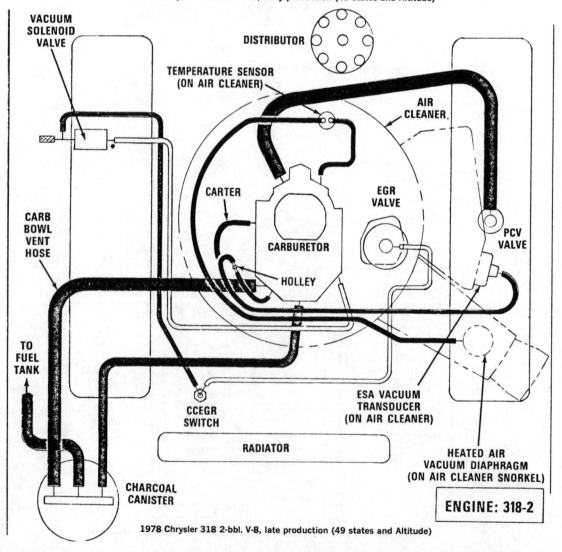

1978 Chrysler 318 2-bbl. V-8, late production (49 states and Altitude)

Chrysler Corporation

VACUUM CIRCUITS

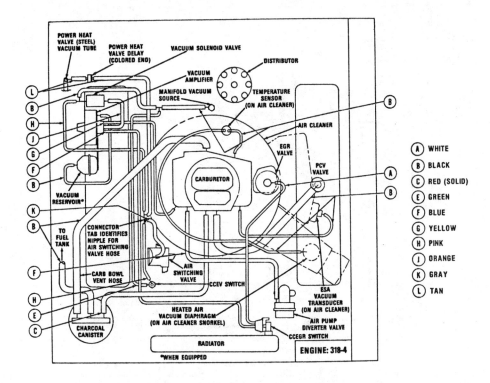

1978 Chrysler 318 4-bbl. V-8, auto. trans. (California)

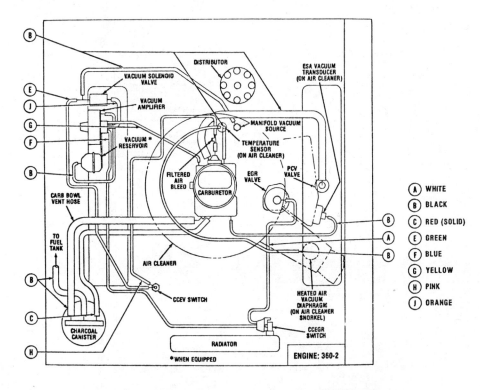

1978 Chrysler 360 2-bbl. V-8, lean burn and auto. trans. (49 states)

Chrysler Corporation

VACUUM CIRCUITS

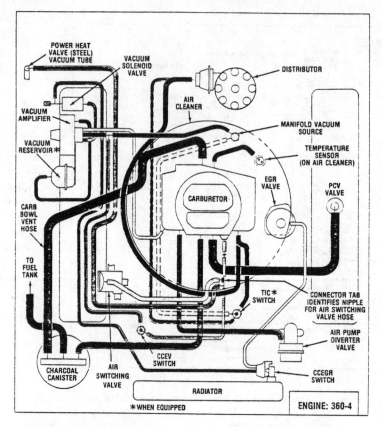

1978 Chrysler Aspen, Volare, Diplomat, LeBaron, 360 4-bbl. V-8, auto. trans./late production (Altitude)

1978 Chrysler 360 4-bbl. V-8, auto. trans./late production (California)

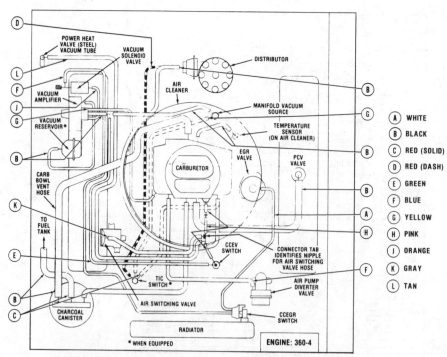

1978 Chrysler Aspen, Volare, Diplomat, LeBaron 360 4-bbl., V-8, auto. trans. early production (Altitude)

1978 Chrysler 360 4-bbl. V-8, auto. trans./early production (California)

Chrysler Corporation

VACUUM CIRCUITS

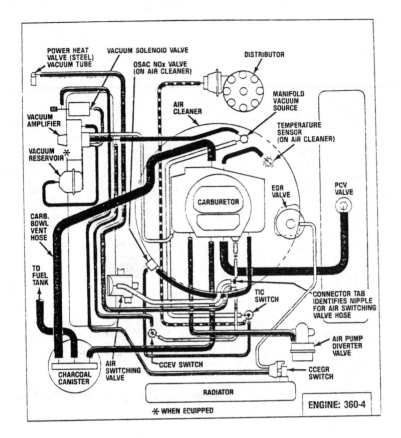

1978 Chrysler 360 4-bbl. V-8, auto. trans., for fleet states (California)

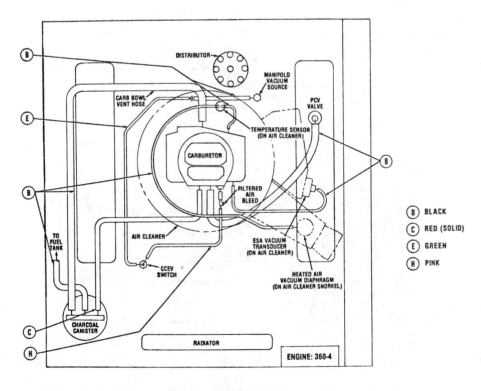

1978 Chrysler 360 4-bbl., auto. trans. with lean burn and catalyst (49 states)

Chrysler Corporation

VACUUM CIRCUITS

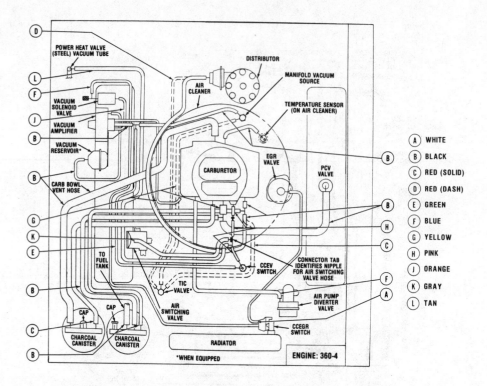

1978 Chrysler Fury, Monaco, Charger, Magnum, Cordoba, Newport, New York 360 4-bbl. V-8, auto. trans./early production (Altitude)

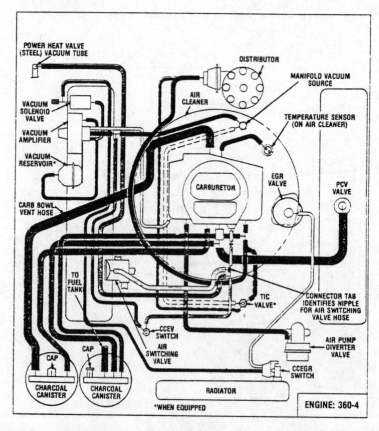

1978 Chrysler Fury, Monaco, Charger, Magnum, Cordoba, Newport, New York 360 4-bbl. V-8, auto. trans./late production (Altitude)

Chrysler Corporation

VACUUM CIRCUITS

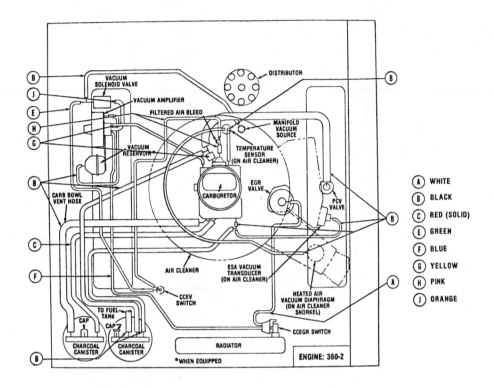

1978 Chrysler Newport and New Yorker 360 2-bbl., auto. trans./early production (49 states)

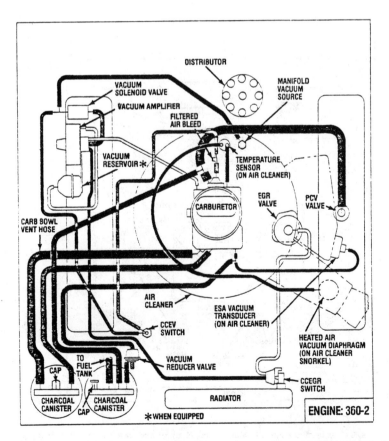

1978 Chrysler Newport and New Yorker 360 2-bbl., auto. trans./late production (49 states)

Chrysler Corporation

VACUUM CIRCUITS

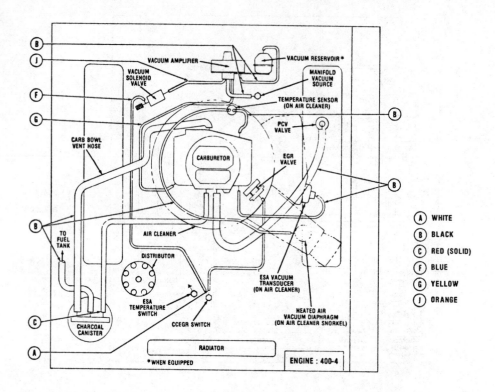

1978 Chrysler 400 4-bbl. V-8, auto. trans. except Newport and New Yorker (49 states)

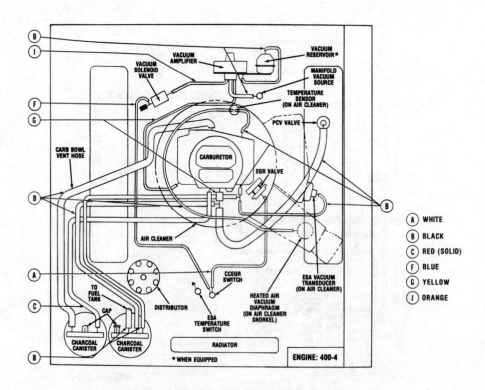

1978 Chrysler 400 4-bbl. V-8, auto. trans., Newport and New Yorker (49 states)

Chrysler Corporation

VACUUM CIRCUITS

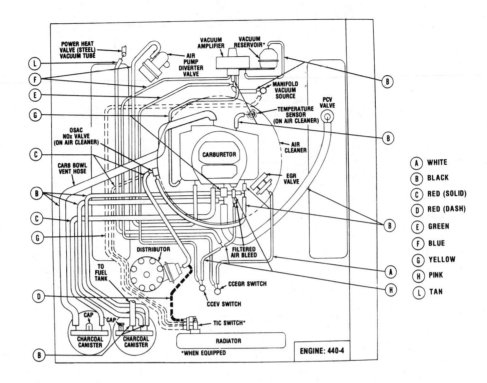

1978 Chrysler 440 4-bbl. V-8, auto. trans. with catalyst and air pump (California)

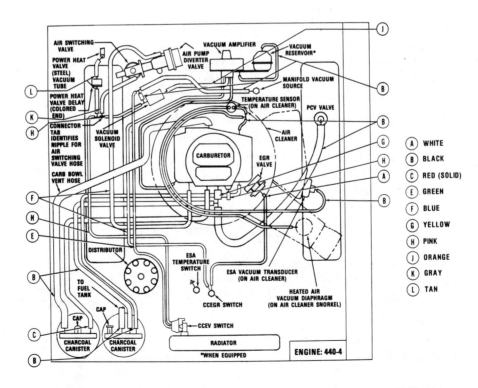

1978 Chrysler 440 4-bbl. V-8, auto. trans. with lean burn, catalyst, and air pump (California)

281

Chrysler Corporation

VACUUM CIRCUITS

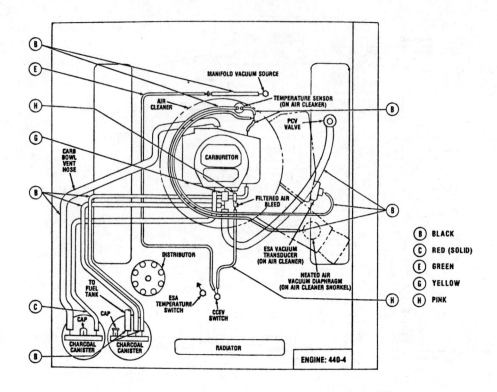

1978 Chrysler 440 4-bbl. V-8, auto. trans. with lean burn and catalyst (49 states)

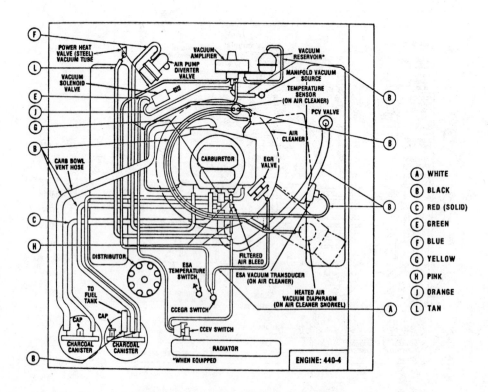

1978 Chrysler 440 4-bbl. V-8, auto. trans. (Altitude)

Chrysler Corporation

VACUUM CIRCUITS

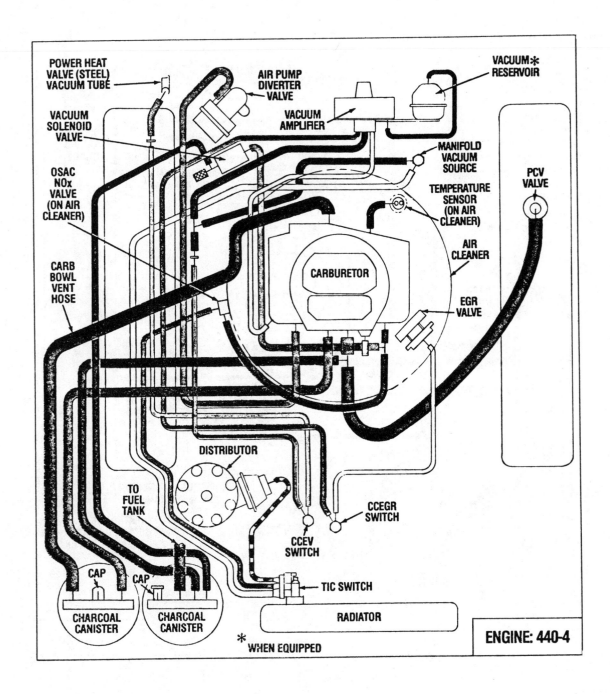

1978 Chrysler 440 4-bbl. V-8, auto. trans. Fury, Monaco, Charger, Magnum, and Cordoba (California)

Chrysler Corporation

VACUUM CIRCUITS

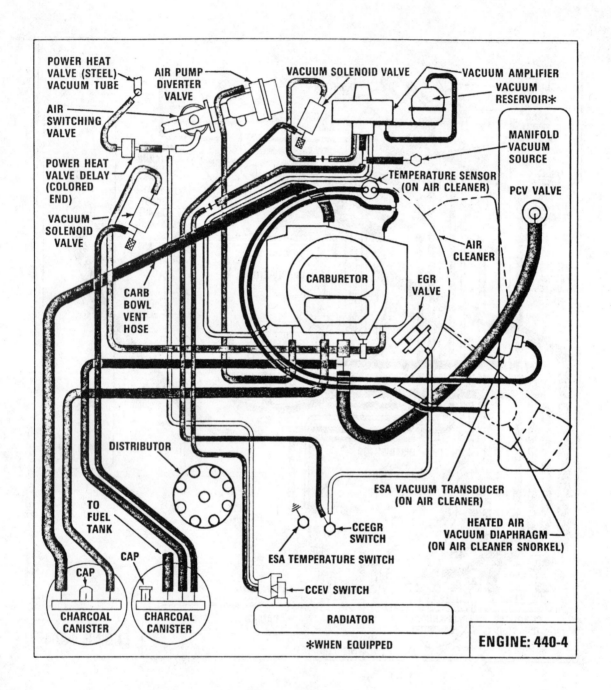

1978 Chrysler 440 4-bbl. V-8, auto. trans. Newport and New Yorker (California)

Chrysler Corporation

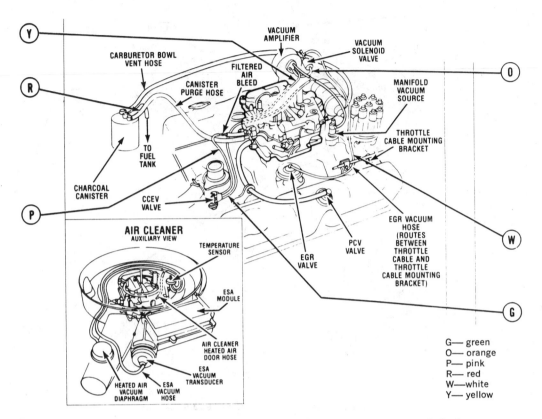

Vacuum house diagram, 360 4 bbl., 49 states, (automatic transmission)

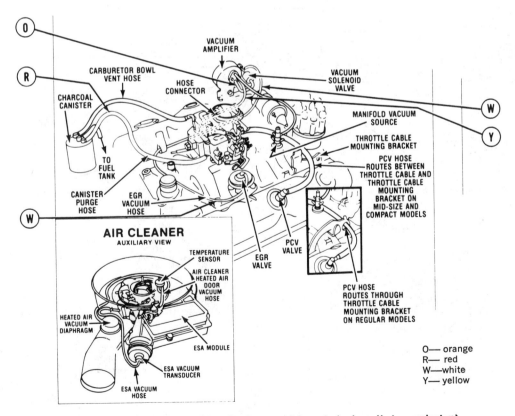

Vacuum hose diagram 318 2 bbl., 49 states and Canada (automatic transmission)

Chrysler Corporation

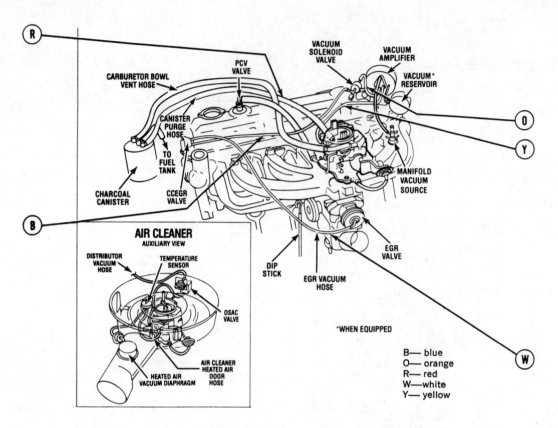

Vacuum hose diagram, 225 bbl., 49 states and Canada

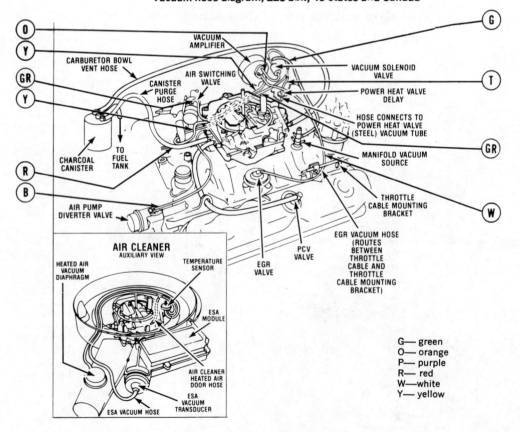

Vacuum hose diagram, 318 4 bbl., California (automatic transmission)

Chrysler Corporation

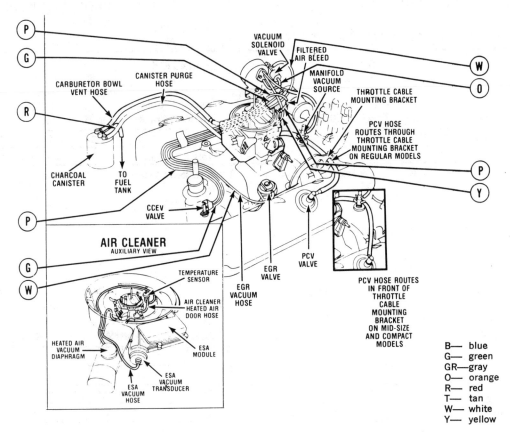

Vacuum hose diagram, 360 2 bbl., 49 states, (automatic transmission)

B— blue
G— green
GR—gray
O— orange
R— red
T— tan
W— white
Y— yellow

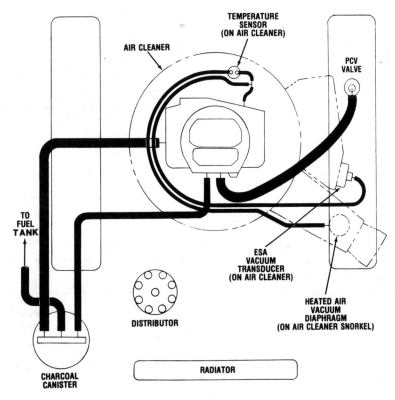

1976 49-State 400 4-Bbl. Electronic Lean Burn

Horizon & Omni

TUNE-UP SPECIFICATIONS — CHRYSLER HORIZON & OMNI

Engine Code, 5th character of the VIN number

Year	Eng. V.I.N. Code	ENGINE No. Cyl Disp. (cu in)	H.P.	SPARK PLUGS Orig. Type	Gap (in)	DIST. Point Dwell (deg)	Point Gap (in)	IGNITION TIMING (deg) Man Trans	Auto Trans	VALVES Intake Opens (deg)B.	FUEL PUMP Pres. (psi)	IDLE SPEED (rpm) Man Trans	Auto Trans in Drive
1978	A	104.7	75	RN-12Y	.035	ELB		15B①	15B	14B	4-6	900②	900②
1979	A	104.7	75	RN-12Y	.035	ELB		15B	15B	14B	4.4-5.9	900②	900②

Should the specifications in this manual deviate from the specifications on the engine compartment decal, the specifications on the decal should be used

ELB —ELECTRONIC LEAN BURN
① CANADA—10B
② W/AIR CONDITIONING—850 RPM; PROPANE ENRICHED IDLE SPEED FOR ALL MODELS—1075 RPM

ELECTRONIC LEAN BURN SPECIFICATIONS — CHRYSLER HORIZON & OMNI

Custom I.C. Spark Control Computer Part Number	5206467	5206501	5206516	5206525	5606526	5206666
Spark Timer Advance Schedule	8°	8°	8°	8°	8°	8°
Delay Time in Seconds	60	60	60	60	60	60
Throttle Advance Schedule	0°	0°	4°-6°@100°F	4°-6°@100°F	0°	4°-6°@100°F
Test Transducer Core Out 1 Inch	0°	0°	4°-6°@140°F	4°-6°@140°F	0°	4°-6°@140°F
Vacuum Advance Schedule						
(A) Operating Vacuum Range	0"-14"	0"-14"	0"-14"	0"-14"	0"-14"	0"-14"
(B) Advance Off Idle (Carb Switch Isolated With Paper)	None	6°-10°	2°-6°	2°-6°	2°-6°	6°-10°
(C) Accumulation Time (In Minutes)	8	7	8	8	8	7
(D) Advance After Accumulation Time	18°-22°	18°-22°	18°-22°	18°-22°	18°-22°	18°-22°
Speed Advance (Ground Carb Switch and Disconnect @2000 RPM	6°-10°	6°-10°	6°-10°	5°-9°	6°-10°	5°-9°
Throttle Transducer Before Checking) @4000 RPM	18°-22°	18°-22°	18°-22°	13°-17°	18°-22°	13°-17°

The above specifications are published from the latest information available at the time of publication. If anything differs from those on the Emission Control Information Label, use the specification on the label.

NOTE: *The underhood specifications sticker sometimes reflects tune-up specification changes made in production. Sticker figures must be used if they disagree with this information.*

HOLLEY 5220 2 BARREL

HOLLEY 5220 2 BARREL SPECIFICATIONS — OMNI/HORIZON

Year	Carb. Part No.	Accelerator Pump	Dry Float Level (in.)	Vacuum Kick (in.)	Curb Idle RPM (w/fan)	Fast Idle RPM (w/fan)	Throttle Position Transducer (in.)	Throttle Stop Speed RPM	Choke
1978	R-8376A, 8378A, 8384A, 8439A, 8441A, 8505A, 8507A	#2 hole	.480	.070	900	1100	.547	700	2 NR

Horizon & Omni

CAR SERIAL NUMBER AND ENGINE IDENTIFICATION

1978
Behind the windshield on the driver's side is the vehicle identification number. The sixth character is the model year, with 8 for 1978. The fifth character is the engine code, as follows:

A .... 1.7 liter 4-cyl.

1979
Behind the windshield on the driver's side is the vehicle identification number. The sixth character is the model year, with 9 for 1979. The fifth character is the engine code, as follows:

A .... 1.7 liter 4-cyl.

U.S. EMISSION EQUIPMENT

1978

All Models
Closed positive crankcase ventilation
Heated air cleaner
Vapor control, canister storage
Exhaust gas recirculation
Emission calibrated carburetor
Emission calibrated distributor
Catalytic converter
Fresh air intake
Full electric choke
Orifice spark advance control
   Not used
Idle enrichment
   Not used
Altitude compensation
   Not used
Air pump
   49-States
      Early models only (late models use air aspirator)
   Calif.
      All models
Air aspirator system
   49-States
      Late models only (early models use air pump)
   Calif.
      Not used

1979

All Models
Closed positive crankcase ventilation
Heated air cleaner
Vapor control, canister storage
Exhaust gas recirculation
Emission calibrated carburetor
Emission calibrated distributor
Catalytic converter
Fresh air intake
Full electric choke
Orifice spark advance control
   Not used
Idle enrichment
   Not used
Altitude compensation
   Not used
Air pump
   49-States
      Not used
   Calif.
      All models
Air aspirator system
   49-States
      All models
   Calif.
      Not used

IDLE SPEED AND MIXTURE ADJUSTMENTS EXHAUST ANALYZER METHOD

1978

NOTE: This is an alternate method for California cars only. The preferred method is to use propane enrichment

Air cleaner .... In place
Air cond. .... Off
Auto. trans. .... Park
Engine fan .... Operating
Air pump .... Disconnect for mixture setting
Vac. adv. hose .... Plugged
EGR vac. hose .... Plugged
Lean burn engine .... Ground carb. switch with jumper
Mixture adj. .... Set idle CO

NOTE: After disconnecting the air pump for mixture settings, the opening to the exhaust must be plugged to prevent air from entering the exhaust. Do not plug the pump output.

Idle CO
   1.7 4-cyl. .... Probe 0.5%
Idle Speed
   Auto. trans. with air cond. 900
      AC idle speedup 750
   Auto. trans. no air cond.
      Solenoid connected 900
      Solenoid disconnected 700
   Man. trans. with air cond. 900
      AC idle speedup 850
   Man. trans. no air cond.
      Solenoid connected 900
      Solenoid disconnected 700

NOTE: All engines have a solenoid on the carburetor. On engines without air conditioning, the solenoid is used for anti-dieseling. It is energized whenever the ignition switch is on. Curb idle speed is adjusted by turning the screw on top of the solenoid. The "solenoid disconnected" speed is set with the throttle stop screw.

On engines with air conditioning, the solenoid is used for air conditioning idle speedup. It is energized only when the air conditioning is on. The curb idle speed is adjusted by turning the screw on top of the solenoid. The "AC idle speedup" speed is adjusted with an Allen screw underneath the idle speed screw. To reach the Allen screw, the idle speed screw must be unscrewed and removed from the solenoid. There is no throttle stop screw on air conditioned engines.

1979

NOTE: This is an alternate method for California cars only. The preferred method is to use propane enrichment.

Air cleaner ....In place
Air cond. .... Off
Auto. trans. .... Park
Engine fan .... Operating
Air pump .... Disconnect for mixture setting
Vac. adv. hose .... Plugged
EGR vac. hose .... Plugged
Lean burn engine ..... Ground carb. switch with jumper
Mixture adj. .... Set idle CO

NOTE: After disconnecting the air pump for mixture settings, the opening to the exhaust must be plugged to prevent air from entering the exhaust. Do not plug the pump output.

Idle CO
   1.7 4-cyl. .... Probe 0.5%
Idle Speed
   Auto. trans. with air cond. 900
      AC idle speedup ....
   Auto. trans. no air cond.
      Solenoid connected 750
      Solenoid disconnected 700
   Man. trans. with air cond. 900
      AC idle speedup 850

Horizon & Omni

IDLE SPEED AND MIXTURE ADJUSTMENTS

Man. trans. no air cond.
  Solenoid connected 900
  Solenoid disconnected 700
  NOTE: All engines have a solenoid on the carburetor. On engines without air conditioning, the solenoid is used for anti-dieseling. It is energized whenever the ignition switch is on. Curb idle speed is adjusted by turning the screw on top of the solenoid. The "solenoid disconnected" speed is set with the throttle stop screw.

On engines with air conditioning, the solenoid is used for air conditioning idle speedup. It is energized only when the air conditioning is on. The curb idle speed is adjusted by turning the screw on top of the solenoid. The "AC idle speedup" speed is adjusted with an Allen screw underneath the idle speed screw. To reach the Allen screws, the idle speed screw must be unscrewed and removed from the solenoid. There is no throttle stop screw on air conditioned engines.

IDLE SPEED AND PROPANE ENRICHED MIXTURE ADJUSTMENTS

1978

NOTE: *This is the preferred method for adjusting all cars.*
Air cleaner .... In place
Air cond. .... Off
Auto. trans. .... Park
Engine fan .... Operating
Vac. adv. hose .... Plugged
EGR vac. hose .... Plugged
Lean burn engine .... Ground carb. switch with jumper
Mixture adj. .. See propane rpm below
Idle Speed
Auto. trans. with air cond. ....
  Propane enriched ....
  AC idle speedup ....
Auto. trans. no air cond.
  Solenoid connected ....

Solenoid disconnected ....
Propane enriched ....
Man. trans. with air cond. 900
  Propane enriched ....
  AC idle speedup 850
Man. trans. no air cond.
  Solenoid connected 900
  Solenoid disconnected 700
  Propane enriched ....

1979

NOTE: *This is the preferred method for adjusting all cars.*
Air cleaner .... In place
Air cond. .... Off
Auto. trans. .... Park
Engine fan .... Operating
Vac. adv. hose .... Plugged

EGR vac. hose .... Plugged
Lean burn engine ..... Ground carb. switch with jumper
Mixture adj. ..See propane rpm below
Idle Speed
Auto. trans. with air cond. .....
  Propane enriched ....
  AC idle speedup ....
Auto. trans. no air cond.
  Solenoid connected ....
  Solenoid disconnected ....
  Propane enriched ....
Man. trans. with air cond. 900
  Propane enriched ....
  AC idle speedup 850
Man. trans. no air cond.
  Solenoid connected 900
  Solenoid disconnected 700
  Propane enriched ....

INITIAL TIMING

NOTE: *Disconnect and plug vacuum advanced hose at spark control computer.*
1.7 4-cyl. .... 15° BTDC

SPARK PLUGS

1.7 4-cyl. ....CHRN12Y.. .035

VACUUM ADVANCE

Diaphragm type .... Single
Vacuum source .... Manifold

EMISSION CONTROL SYSTEMS

Exhaust Gas Recirculation	Controls NOx
Heated Inlet Air	Reduces HC
Initial Engine Timing	Controls exhaust HC, CO, and Nox
Positive Crankcase Ventilation	Controls Crankcase HC

CARBURETOR

Carburetion on the new OMNI and HORIZON is accomplished through the use of a Holly model 5220 carburetor, a staged, dual venturi carburetor on which the primary venturi is smaller than the secondary venturi.

The primary stage is comprised of an idle circuit, transfer system, diaphragm type accelerator pump system, main metering system, and a power enrichment system.

HEATED INLET AIR SYSTEM

The purpose of the heated inlet air system is to control the temperature of the air entering the carburetor when ambient temperatures are low. By increasing the temperature of the air being introduced into the carburetor, a much leaner calibration is made possible, thereby reducing hydrocarbon emissions. In addition to a refined fuel air mixture, other benefits derived as a result of the heated air include smoother engine warm-up operation (a more volatile mixture as a result of heat) and a minimal amount of carburetor icing.

EMISSION CONTROL SYSTEMS AND CORRESPONDING FUNCTIONS 1978-79

SYSTEM	FUNCTION
Air pumps (AIR)	Reduces exhaust HC and CO
Carburetor Calibration	Controls exhaust HC and CO
Catalytic Converter	Reduces exhaust HC and CO
Distributor Calibration	Controls exhaust HC and CO
Electric Choke	Reduces exhaust HC and CO
Evaporation Control System	Vapor control

Horizon & Omni

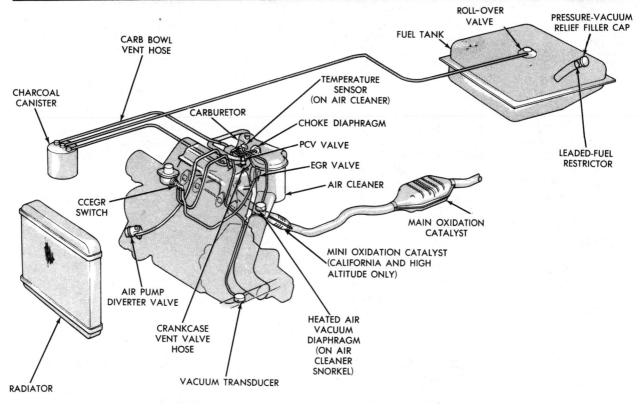

Emission control system—Omni and Horizon

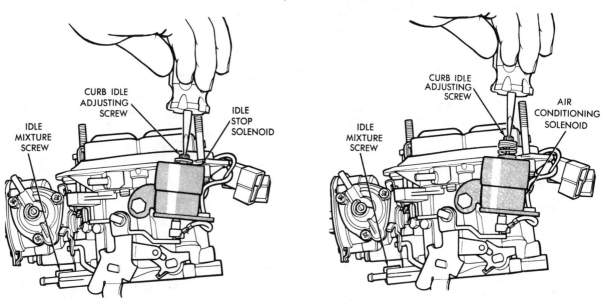

Propane assisted curb idle and mixture adjustment—without A/C

Propane assisted curb idle and mixture adjustment—with A/C

Idle mixture screw

FULL ELECTRIC CHOKE SYSTEM

A full-electric-choke is an electrically heated choke comprised of an electric heater and switch which are sealed within the choke housing. During cold weather the switch remains open to reduce heater output until the choke area reaches a sufficiently warm temperature. If the choke area is at a sufficiently warm temperature, then the switch closes and full heater output moves the choke to the open position.

Horizon & Omni

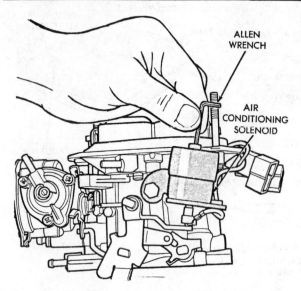

Air conditioning idle adjustment—propane assisted idle adjustment must be made first

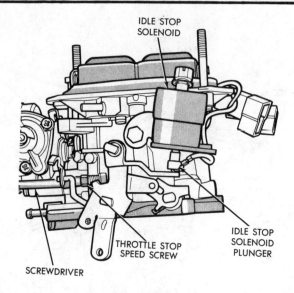

Idle stop adjustment—without A/C

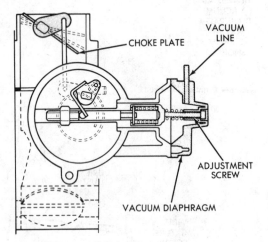

Full electric choke

IGNITION SYSTEM

The purpose or design function of the ignition system is to allow a lean air fuel mixture to be burned under various operating conditions. Appropriately named the Electronic Lean Burn System, the ignition system is comprised of a spark control computer, five engine sensors, and a specially calibrated Holley Carburetor.

The spark control computer is capable of receiving several and various signals from the five engine sensors simultaneously, computing these signals within milliseconds, thereby monitoring engine performance, and advancing or retarding the ignition timing by signaling the ignition coil to produce the electrical impulses which fire the spark plugs.

ENGINE SENSORS

There are five engine sensors: vacuum transducer, Hall Effect pickup assembly, throttle position transducer, carburetor switch, and coolant switch.

The vacuum transducer, located on the spark control computer, senses intake manifold vacuum, and signals the computer which, in turn, regulates ignition timing.

NOTE: *Ignition timing is not based on a fixed or constant curve. The number of curves can be without limit and variable.*

A high vacuum reading will command an additional advance in timing. A low vacuum reading will command the timing to retard.

NOTE: *In order to achieve maximum advance for any inch of vacuum, the carburetor switch must remain open for a specified amount of time. During this state ignition*

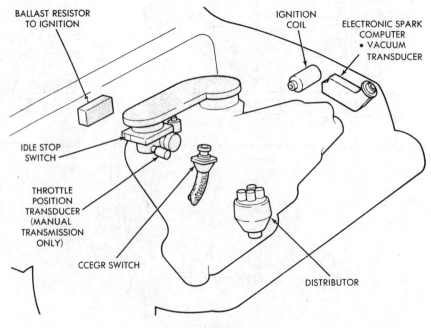

Ignition system

Horizon & Omni

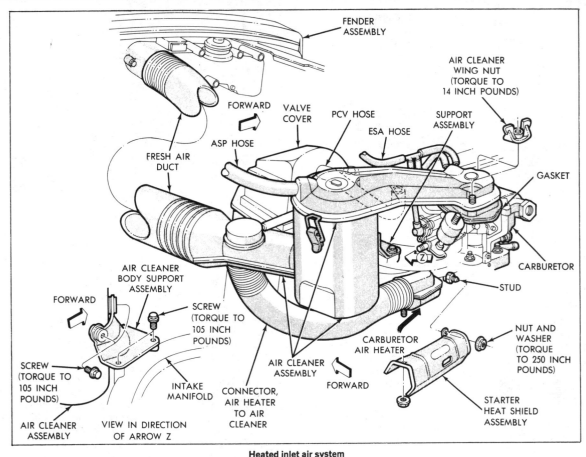

Heated inlet air system

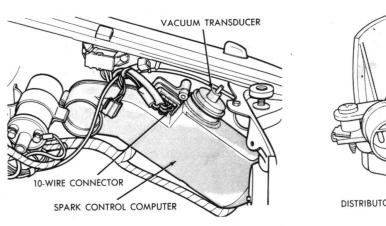

Spark control computer and vacuum transducer

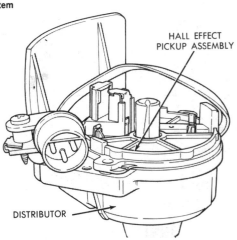

Hall effect pick-up assembly

advance is not immediate; it is gradual, and develops at a slow rate. If the carburetor switch closes prematurely, the additional advance will be cancelled. This information is stored in the computer's memory, and the computer will slowly return the advance to 0. If the switch is opened before the computer can return the advance to 0, the advance will again gradually build from the point recorded in the computer's memory. If the switch is opened after the advance has been allowed to return to 0, ignition advance will again develop gradually.

The Hall Effect pickup assembly, located in the distributor, senses engine speed and crankshaft location, and transmits this information to the computer. The signal transmitted from the Hall Effect pickup assembly is a reference signal. When the computer receives this signal, maximum timing advance is made available.

The function of the throttle position transducer is to sense the throttle position and rate of change. As the throttle plates start to open, regardless of position, additional advance will be directed on command from the computer.

The carburetor switch, located on the idle stop solenoid or air conditioning solenoid, signals the computer as to whether the engine is at idle or off idle.

The coolant switch, located on the thermostat housing, senses engine coolant temperature. It's function is to signal the computer when the engine coolant temperature is below 150 degrees.

EXHAUST GAS RECIRCULATION (EGR)

Recirculated exhaust gas, when in-

Horizon & Omni

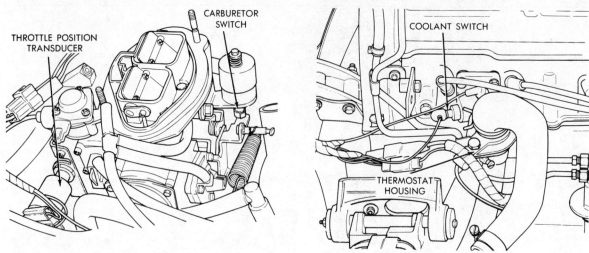

Throttle position transducer and carburetor switch

Coolant switch

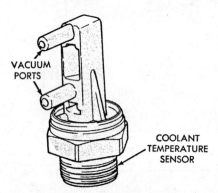

Coolant control EGR valve (CCEGR)

troduced into the combustion chamber, has the effect of minimizing the oxides of nitrogen emitted into the air by lowering combustion flame temperatures during engine operation. Only a modulated amount of exhaust gas is allowed to dilute the highly combustible air fuel mixture. The system responsible for regulating the volume of exhaust gas available during firing is called Exhaust Gas Recirculation (EGR).

During engine warm-up, when combustion flame temperature is relatively low, control vacuum is prevented from reaching the EGR valve by use of a coolant control exhaust gas recirculation valve (CCEGR). The CCEGR, mounted in the thermostat housing, remains closed until the coolant temperature exceeds 125 degrees F. When the coolant temperature reaches the calibration temperature of 125 degrees F., plus or minus 5 degrees, the valve opens thereby permitting vacuum to reach the EGR valve.

AIR INJECTION SYSTEM

NOTE: *This system is standard emission control equipment on all engines intended for sale in the 49 States and California, which were manufactured prior to and including January 10, 1978.*

One very effective way of reducing the level of carbon monoxide and hydrocarbons present in the exhaust gases is to inject a controlled amount of air directly into the exhaust system at a point very close to the exhaust valve. The injected air assists in further oxidation of the exhaust

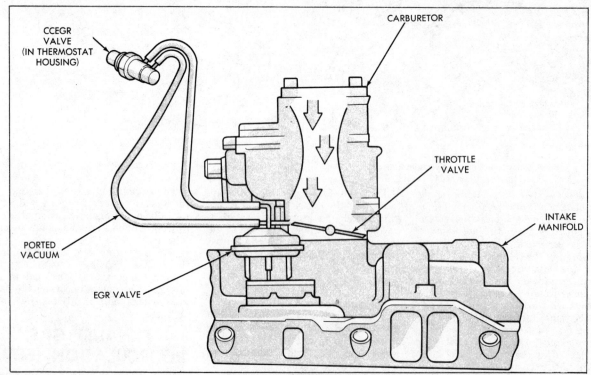

Exhaust gas recirculation system

Horizon & Omni

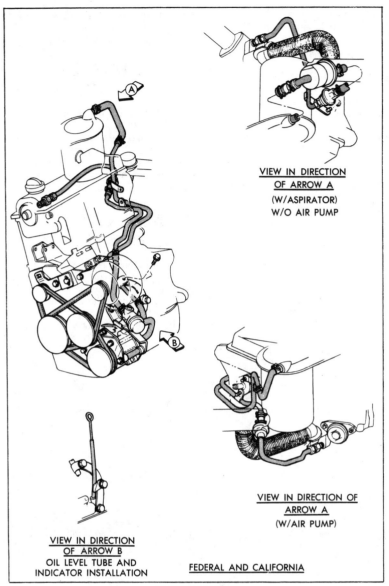

Air injection system

haust manifold. The Canadian air injection system injects air through the cylinder head at a point very close to the exhaust valve.

During periods of sudden deceleration, when abrupt throttle closing allows a rich air fuel mixture to reach the combustion chamber, the diverter valve will direct the injected air out into the atmosphere, thus preventing a backfire condition which would occur as a result of combining unburned fuel and injected air.

A check valve, which is located in the injection tube assembly, will prevent hot exhaust gases from backing up into the hose and pump in the event of a system break down, such as would be caused by pump belt failure, excessively high exhaust system pressure, or a ruptured air hose.

ASPIRATOR AIR SYSTEM

NOTE: *This system is standard emission control equipment on all Federal engines manufactured after January 10, 1978. Engines intended for sale in California will continue to utilize the Air Injection System.*

The operating principle of the aspirator air system is similar to that of the air injection system in that both systems rely on the introduction of air into the exhaust manifold to reduce carbon monoxide and hydrocarbon emissions to an acceptable level. Though similar, both systems are distinctly different in design and operation.

NOTE: *Refer to "Air Injection System."*

The aspirator air system is a relatively simple system to diagnose and repair when compared to the air injection system. An aspirator valve, located between the air cleaner and exhaust manifold, and connected by hoses, contains a spring loaded diaphragm which is sensitive to changes in pressure. Whenever a vacuum pulse occurs in the exhaust ports and

gases thereby holding the level of undesirable gases at a minimum. The system consists basically of a belt driven injection pump, a diverter valve, and a check valve.

Filtered air from the clean side of the air cleaner is drawn into the pump and delivered to an air injection manifold. Exactly where the air is injected depends on the emission control code. The air injection system, standard on vehicles manufactured for sale in the United States, injects air into the base of the ex-

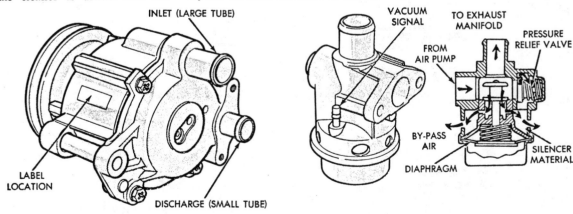

Air injection pump   **Diverter valve**

295

Horizon & Omni

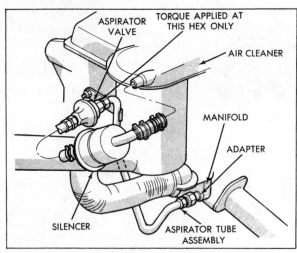
Aspirator air system

manifold passages, the diaphragm opens to allow fresh air to mix with the exhaust gases. At high engine speeds, the aspirator valve remains closed.

To test if the valve has failed, disconnect the hose at the aspirator inlet, and, with the engine at idle and in neutral, check for escaping exhaust gases at the inlet. If the pulses can not be felt, and there is evidence of exhaust gases present at the inlet, the valve has failed and should be replaced.

EVAPORATION CONTROL SYSTEM

Pollution can occur even when an engine is not running. This is because gasoline will continue to evaporate as long as the temperature of the fuel is high enough to permit evaporation. When the fuel is in a state of expansion, it will separate and rise away from the liquid fuel into the surrounding atmosphere. This constitutes pollution because gasoline fumes are hydrocarbons, an undesirable pollutant.

A method of controlling the emission of fuel vapors is to contain the fuel system in an evaporation control system. Fuel vapor leaving the carburetor bowl vent and fuel tank are trapped in vent hoses, and then routed to a charcoal cannister where the vapors are stored until they can be drawn into the intake manifold when the engine is running.

CRANKCASE VENTILATION SYSTEM

This system, which utilizes a conventional PCV valve, is similar to the systems used on most vehicles marketed in the United States in that it directs crankcase vapors through the air cleaner and into the combustion area for complete burning.

MINI CATALYTIC CONVERTER

In addition to the conventional type (underfloor converter) catalytic converter, a mini-converter is also used to initiate the exhaust gas oxidation before the exhaust gases reach the main underfloor catalytic converter.

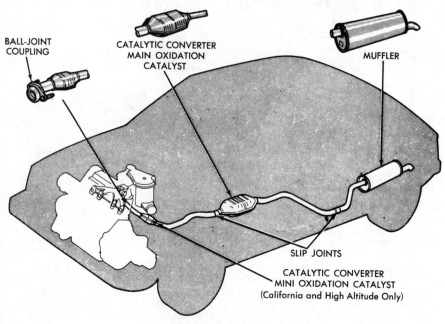
Exhaust system with mini-catalytic converter

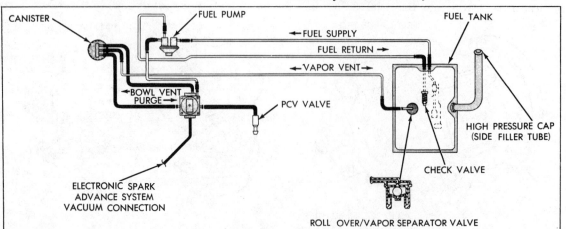
Evaporation control system

Horizon & Omni

VACUUM CIRCUITS

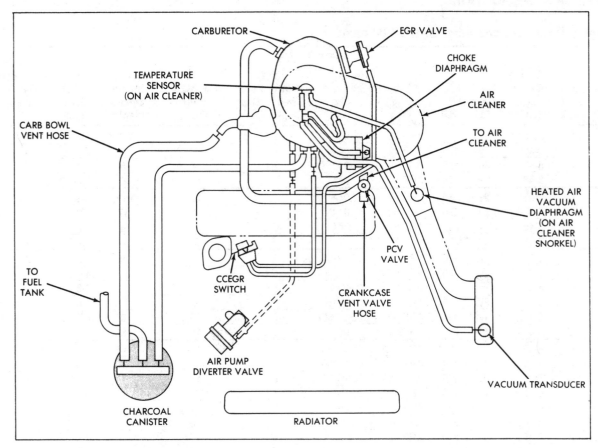

Vacuum hose routing diagram—Omni and Horizon

ENGINE FIRING ORDER

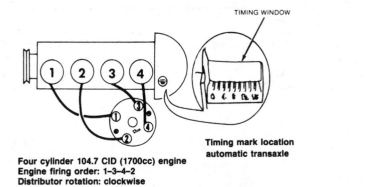

Four cylinder 104.7 CID (1700cc) engine
Engine firing order: 1-3-4-2
Distributor rotation: clockwise

Timing mark location automatic transaxle

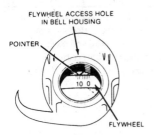

Timing mark location manual transaxle

VEHICLE IDENTIFICATION NUMBER (VIN)

Engine Code, 5th character of the VIN number

MODEL YEAR CODE		ENGINE CODE			
CODE	YEAR	CODE	DISP. (cu. in.)	CONFIG.	CARB.
8	1978	A	104.7 (1700cc)	L-4	2V
9	1979				

297

Ford Motor Company

Full-Size • Intermediate • Compact

TUNE-UP SPECIFICATIONS — FORD FULL-SIZE

Engine Code, 5th character of the VIN number

Yr.	Eng. V.I.N. Code	Engine No. Cyl Disp. (cu in)	Carb Bbl	H.P.	SPARK PLUGS Orig. Type	Gap (in)	DIST. Point Dwell (deg)	Point Gap (in)	IGNITION TIMING ■ (deg BTDC)	VALVES ● Intake Opens (deg)B	FUEL PUMP Pres. (psi)	IDLE SPEED ● (rpm) Man Trans	Auto Trans In Drive
75	Q	8-351	2v	154	ARF 42	.044	E.I.		⑥	19½°	5-7	1000	800
	S	8-400	2v	170	ARF 42	.044	E.I.		6°	17°	5-7	—	650(700)
	A	8-460	4v	195	ARF 52	.044	E.I.		14°	8°	5.7-7.7	—	650
	C	8-460	4v	275	ARF 52	.044	E.I.		14°	8°	5.7-7.7	—	650
76	Q	8-351	2v	152	ARF 52	.044	E.I.		12°	19½°	6-8	1000	800
	S	8-400	2v	180	ARF 52	.044	E.I.		10°	17°	6-8	—	625(650)
	A	8-460	4v	202	ARF 52	.044	E.I.		8°	8°	6.7-8.7	—	650
	C	8-460	4v	202	ARF 52	.044	E.I.		8°	8°	6.7-8.7	—	650
77	F	8-302	2v	134	ARF 52①	.050(.060)	E.I.		8	16	5.5-6.5	750	650
	H	8-351W	2v	149	ARF 52①	.050(.060)	E.I.		9	23	4.0-6.0	—	650
	Q	8-351	2v	161	ARF 52①	.050(.060)	E.I.		9(8)	19½°	6-8	1000	800
	S	8-400	2v	173	ARF 52①	.050(.060)	E.I.		8	17°	6-8	—	600(650)
	A	8-460	4v	197	ARF 52①	.050(.060)	E.I.		16	8°	6.7-8.8	—	600(650)
	C	8-460	4v	275*	AR52-6	.060	E.I.		16	8°	6.7-8.7	—	600(650)
78	F	8-302	2v	134	ARF 52①	.050(.060)	E.I.		14	16	5.5-6.5	—	650
	H	8-351W	2v	144	ARF 52①	.050(.060)	E.I.		4	23	4-6	—	650
	Q	8-351M	2v	145	ARF 52①	.050(.060)	E.I.		12	19½	6.5-7.5	—	650
	S	8-400	2v	160	ARF 52①	.050(.060)	E.I.		13(16)	17	6.5-7.5	—	650
	S	8-400	2v	166	ARF 52	.050	E.I.		13(16)	17	6.5-7.5	—	575(600)
	A	8-460	4v	202	ARF 52①	.050(.060)	E.I.		10	8	7.2-8.2	—	580
	A	8-460	4v	210	ARF 52	.050	E.I.		16⑦	8	7.2-8.2	—	580
	C	8-460	4v	255*	ARF 52①	.060	E.I.		16	18	7.2-8.2	—	580
79	F	8-302	vv	134*	ASF-52①	.050(.060)	E.I.		6	16	5.5-6.5	—	550
	H	8-351W	vv	144*	ASF-52①	.050(.060)	E.I.		⑥	23	4-6	—	650
	H	8-351W	2v	145*	ASF-52①	.050(.060)	E.I.		⑥	23	4-6	—	650
	S	8-400	2v	166*	ASF-52①	.050(.060)	E.I.		14⑧	17	6.5-7.5	—	600⑧

■ NOTE: Should the information provided in this manual deviate from the specifications on the underhood tune-up label, the label specifications should be used, as they reflect production changes
● Figures in parentheses indicates California specifications

① 52-6 IN CALIFORNIA
② MANUAL TRANSMISSION
③ EL AIR GAP SET AT FACTORY .017 NOT ADJUSTABLE
④ 3.00:1 AXLE, 6° BTDC
⑤ AIR CONDITIONING TURNED OFF.
⑥ SEE DECAL
⑦ AFTER JULY 15 ENGINE BUILD DATE 10° B
⑧ EXCEPT CALIFORNIA
VV Variable Venturi
E. I.=Electronic ignition
* Chilton estimate

TUNE-UP SPECIFICATIONS — FORD MID-SIZE

Engine Code, 5th character of the VIN number

Year	Eng. V.I.N. Code	Engine No. Cyl Disp. (cu in)	Carb Bbl	H.P.	SPARK PLUGS Orig. Type	Gap (in)	DIST. Point Dwell (deg)	Point Gap (in)	IGNITION TIMING ● (deg BTDC) Man Trans	Auto Trans	VALVES ● Intake Opens (deg)B	FUEL PUMP Pres. (psi)	IDLE SPEED ● (rpm) Man Trans	Auto Trans In Drive
75	H	8-351	2v	148	ARF 42	.044	E.I.			8°	15°	5-7		650(700)
	Q	8-351	2v	154	ARF 42	.044	E.I.			⑤	19½°	5-7	1000	800
	S	8-400	2v	170	ARF 42	.044	E.I.			6°	17°	5-7		650(700)
	A	8-460	4v	195	ARF 52	.044	E.I.			14°	8°	5.7-7.7		650
	C	8-460	4v	275	ARF 52	.044	E.I.			14°	8°	5.7-7.7		650

Ford Motor Company

ENGINE SPECIFICATIONS — FORD MID-SIZE

Engine Code, 5th character of the VIN number

Year	Eng. V.I.N. Code	Engine No. Cyl. Disp. (cu in)	Carb Type	TAX H.P.	Horsepower @ rpm	Torque @ rpm (ft lbs)	Bore and Stroke (in)	Comp. Ratio	Oil Pressure @ 2000 rpm
76	H	8-351	2v	154	ARF 52 .044	E.I.	12°	19½°	6-8	625(650)
	Q	8-351	2v	152	ARF 42 .044	E.I.	12°	19½°	6-8	1000	800
	S	8-400	2v	180	ARF 52 .044	E.I.	10°	17°	6-8	625(650)
	A	8-460	4v	202	ARF 52 .044	E.I.	8°	8°	6.7-8.7	650
	C	8-460	4v	202	ARF 52 .044	E.I.	8°	8°	6.7-8.7	650
77	T	6-200	1v	81	BRF 82 .050	E.I.	6° @ 750	20°	5-7	800	650
	L	6-250	1v	87	BRF 82 .050	E.I.	4° @ 750	18°	5-7	800	600
	F	8-302	2v	134	ARF 52① .050②	E.I.	2° @ 500	16°	6-8	800(850)	600(700)
	H	8-351	2v	152	ARF 52① .050②	E.I.	12° @ 500	23°	6-8	600(650)
	Q	8-351	2v	158	ARF 52① .050②	E.I.	⑤	19½°	6-8	1000	800
	S	8-400	2v	180	ARF 52① .050②	E.I.	6° @ 600	17°	6-8	600(650)
78	F	8-302	2v	134	ARF 52① .050②	E.I.	—	14	14	5.5-6.5	—	650
	H	8-351W	2v	144	ARF 52① .050②	E.I.	—	14	23°	4-6	—	650
	Q	8-351M	2v	152	ARF 52① .050②	E.I.	—	12(16)	19½°	6.5-7.5	—	650
	S	8-400	2v	166	ARF 52① .050②	E.I.	—	13(16)	17°	6.5-7.5	—	650
79	F	8-302	2v	134	ASF 52① .050②	E.I.	—	6⑦	16°	5.5-6.5	—	550⑧
	H	8-351W	2v	144	ASF 52 .050	E.I.	—	15	23°	4-6	—	⑨
	Q	8-351M	2v	152	ASF 52 .050	E.I.	—	12(14)	19½°	6.5-7.5	—	600

NOTE: Should the information provided in this manual deviate from the specifications on the underhood tune-up label, the label specifications should be used, as they may reflect production changes.

① 52-6 IN CALIFORNIA
② .060 IN CALIF.
③ EL AIR GAP SET AT FACTORY .017 NOT ADJUSTABLE.
④ (25-29) (32-35) (30-33)
⑤ SEE DECAL
⑦ 8 IN COUGAR, LTD II AND THUNDERBIRD
⑧ 600 IN COUGAR, LTD II AND THUNDERBIRD
⑨ 550 IN EEC EQUIPPED VEHICLES
• Figure in parentheses indicates California engine
E.I. Electronic Ignition
— Not applicable

TUNE-UP SPECIFICATIONS — FORD COMPACT AND INTERMEDIATE CARS

Engine Code, 5th character on the VIN number

Yr.	Eng. V.I.N. Code	Engine No. Cyl. Disp. (cu in)	Carb Bbl	H.P.	SPARK PLUGS Orig. Type	Gap (in)	DIST. Point Dwell (deg)	Point Gap (in)	IGNITION TIMING (deg BTDC) Man Trans	Auto Trans	VALVES Intake Opens (deg)B	FUEL PUMP Pres. (psi)	IDLE SPEED (rpm) Man Trans	Auto Trans In Drive
75	T	6-200	1v	75	BRF 82	.044	E.I.		6°	6	20°	4-6	800	650
	L	6-250	1v	91	BRF 82	.044	E.I.		6°	6	26°	4-6	800	600
	F	8-302	2v	140	ARF 42	.044	E.I.		⑦	⑦	20°	5-7	750(900)	650(700)
	H	8-351	2v	148	ARF 42	.044	E.I.		8°	8	15°	5-7		650(700)
76	T	6-200	1v	81	BRF 82	.044	E.I.		⑦	⑦	28°	5-7	800	650
	L	6-250	1v	87	BRF 82	.044	E.I.		⑦	⑦	26°	5-7	850	600
	F	8-302	2v	134	ARF 42	.044	E.I.		4°, 12°, 14°⑦		20°	6-8	750(800)	650(700)
	H	8-351	2v	154	ARF 52	.044	E.I.		12°	12	19½°	6-8		625(650)
77	T	6-200	1v	81	BRF 82	.050	E.I.		6	6	20°	5-7	800	650
	L	6-250	1v	87	BRF 82	.050	E.I.		4	6 (8)	18°	5-7	800	600
	F	8-302	2v	134	ARF 52	.050	E.I.		6	①④(12)	16°	6-8	800(850)	600(700)
	H	8-351	2v	152	ARF 52	.050	E.I.		—	4	23°	6-8		600(650)

299

Ford Motor Company

TUNE-UP SPECIFICATIONS — FORD COMPACT AND INTERMEDIATE CARS

Engine Code, 5th character on the VIN number

Yr.	Eng. V.I.N. Code	Engine No. Cyl Disp. (cu in)	Carb Bbl	H.P.	SPARK PLUGS Orig. Type	Gap (in)	Point Dwell (deg)	Point Gap (in)	IGNITION TIMING (deg BTDC) Man Trans	IGNITION TIMING (deg BTDC) Auto Trans	VALVES Intake Opens (deg)B	FUEL PUMP Pres. (psi)	IDLE SPEED (rpm) Man Trans	IDLE SPEED (rpm) Auto Trans In Drive
78	Y	4-140	2v	88	AWRF 42	.034	E.I.		6	20	22°	5.5-6.5	850	800
	T	6-200	1v	85	BRF 82	.050⑩	E.I.		10	10(6)	20°	5.5-6.5	800	650
	L	6-250	1v	97	BRF 82	.050	E.I.		4B	14(6)	18°	5.5-6.5	800	600
	F	8-302	2v	139	ARF 52⑧	.050	E.I.		TDC	6(12)⑫	16°	5.5-6.5	500	600
79	Y	4-140	2v	88*	AWSF42	.034	E.I.		6	20(17)⑬	22°	5.5-6.5	850	⑦
	T	6-200	1v	85*	BSF 82	.050	E.I.		8	10	20°	5.5-6.5	700	650(600)
	L	6-250	1v	97*	BSF 82	.050	E.I.		4	10(6)	18°	5.5-6.5	800	600
	F	8-302	2v⑨	139*	ASF 52⑧	.050	E.I.		⑦	⑦	16°	5.5-6.5	800	600

NOTE: Should the information provided in this manual deviate from the specifications on the underhood tune-up label, the label specifications should be used, as they may reflect production changes.

⑦ SEE UNDERHOOD TUNE-UP LABEL
⑧ ASF 52-6 @ .060 IN CALIF.
⑨ VARIABLE VENTURI IN CALF.
⑩ .060 CALIF.
⑪ MONARCH AND GRANADA 6B FEDERAL
⑫ HIGH ALT. 14BTDC
⑬ SET ALTITUDE AND CALIF. ENGINES IN NEUTRAL. NON AIR-CONDITIONED CALIF. ENGINE— 8B

● Figure in parentheses indicates California engine
* Chilton Estimate
— Does not apply
E.I. = Electronic Ignition

DISTRIBUTOR SPECIFICATIONS — FORD FULL-SIZE

Year	Distributor Identification	Centrifugal Advance Start Dist. Deg. @ Dist. RPM	Centrifugal Advance Finish Dist. Deg. @ Dist. RPM	Vacuum Advance Start In. Hg.	Vacuum Advance Finish Dist. Deg. @ In. Hg.
75	D5AE-12127BA	0-2 @ 900	8-11.5 @ 2,150	4	10.75-13.25 @ 13
	D5AE-12127BA	0-2 @ 900	8-11.5 @ 2,150	4	10.75-13.25 @ 13
	D5AE-12127DA	0-2.5 @ 975	8.25-10.75 @ 2,150	4	12.75-15.25 @ 13
	D4VE-12127CA	0-2 @ 825	9.75-12.5 @ 2,500	4	8.75-11.25 @ 14.6
76	D5OE-12127FA	0-3 @ 500	8.5-11.5 @ 2,500	5	10.5-13.25 @ 16
	D6AE-12127CA	0-2 @ 500	11.5-14 @ 2,500	4.3	12.5-15.25 @ 11.5
	D6AE-12127BA	0-3 @ 525	14.25-16.5 @ 2,500	5	10.5-13.25 @ 13
	D6AE-12127AA	0-2 @ 500	11.5-14 @ 2,500	3	12.5-15.25 @ 11.5
	D6VE-12127BA	0-3 @ 520	13.75-16.25 @ 4,000	3.8	10.75-13.25 @ 11
	D6VE-12127AA	0-2 @ 500	12.5-15 @ 2,500	5.5	10.75-13.25 @ 13.5
77	D6AE-12127AA	0-2 @ 500	11.5-14 @ 2,500	3.0-4.5	12.5-15.25 @ 11.5
	D6VE-12127CA	0-1 @ 450	11.25-14 @ 2,500	3.5-5.5	10.75-13.25 @ 15
	D7AE-12127BA	0-1 @ 425	11-13.5 @ 2,250	3.5-4.8	11.25-15.25 @ 12
	D7AE-12127CA	0-1 @ 700	8.25-11.75 @ 2,500	3-4.2	11.25-15.25 @ 11
	D7AE-12127DA	0-1 @ 450	11.25-14 @ 2,500	3.2-5.5	12.75-15.25 @ 14.5
	D7DE-12127CA	0-1 @ 450	12.5-15.5 @ 2,500	3-4	11.25-15.25 @ 11
	D7OE-12127CA	0-1 @ 450	13.5-16 @ 2,500	3.5-5.5	12.75-15.25 @ 14.5
78	D6VE-12127CA	0-1 @ 450	11.25-14 @ 2,500	3.5-5.5	10.75-13.25 @ 15
	D7AE-12127UA	0-1 @ 650	7.5-10.5 @ 2,500	4-5.2	10.75-13.5 @ 15
	D8AE-12127BA	0-1 @ 425	11.75-14.5 @ 2,500	3-4	14.75-17.25 @ 14.5
	D8AE-12127CA	0-1 @ 1100	6.9-9.5 @ 2,500	3-4	14.5-17.5 @ 14
	D8AE-12127GA	0-1 @ 450	5.75-8.25 @ 2,500	5.5-7.75	13.25-15.25 @ 16.5
	D8AE-12127HA	0-1 @ 450	11.25-14.25 @ 2,500	2.5-3.2	14.75-17.25 @ 14
	D8AE-12127JA	0-1 @ 700	8-10.5 @ 2,400	2.7-3.6	14.75-17.25 @ 14
	D8AE-12127LA	0-1 @ 775	6.75-9.25 @ 2,500	3-3.7	14.75-17.25 @ 13.7
	D8AE-12127NA	0-1 @ 490	9-12.75 @ 2,500	2.8-4	14.75-17.25 @ 15

DISTRIBUTOR SPECIFICATIONS — THUNDERBIRD 73-76

Year	Distributor Identification	Centrifugal Advance Start Dist. Deg. @ Dist. RPM	Centrifugal Advance Finish Dist. Deg. @ Dist. RPM	Vacuum Advance Start In. Hg.	Vacuum Advance Finish Dist. Deg. @ In. Hg.
73	D3MF12127DA	0-2 @ 650	8.3-10.5 @ 2,000	5	17.5-22.5 @ 20
	D3VF12127CA	0-2 @ 800	7.3-9.7 @ 2,000	4	17.5-22.5 @ 14
74	D4VE12127CA	0-.5 @ 500	8.3-10.5 @ 2,000	5	8.5-11.3 @ 20
75	D4VE-12127CA	0-2 @ 825	9.75-12.5 @ 2,500	4	8.75-11.25 @ 14.6
76	D6VE-12127BA	0-6 @ 1040	27.5-32.5 @ 4,000	3.8	21.5-26.5 @ 11
	D6VE-12127AA	0-4 @ 1000	25-30.5 @ 2,500	5.5	21.5-26.5 @ 13.5

DISTRIBUTOR SPECIFICATIONS — LINCOLN CONTINENTAL

Year	Distributor Identification	Centrifugal Advance Start Dist. Deg. @ Dist. RPM	Centrifugal Advance Finish Dist. Deg. @ Dist. RPM	Vacuum Advance Start In. Hg.	Vacuum Advance Finish Dist. Deg. @ In. Hg.
73	D3VF12127CA	0-2 @ 800	7.3-9.7 @ 2,000	4	17.5-22.5 @ 14
74	D4VE12127-CA	0-.5 @ 500	8.3-10.5 @ 2,000	5	8.5-11.3 @ 20
75	D4VE-12127CA	0-2 @ 825	9.75-12.5 @ 2,500	4	8.75-11.25 @ 14.6
76	D6VE-12127BA	0-6 @ 1040	27.5-32.5 @ 4,000	3.8	21.5-26.5 @ 11
	D6VE-12127AA	0-4 @ 1000	25-30.5 @ 2,500	5.5	21.5-26.5 @ 13.5
77	D6AE-12127AA	0-2 @ 500	11.5-14 @ 2,500	4.3	12.5-15.25 @ 11.5 ①
	D6VE-12127CA	0-1 @ 450	11.25-14 @ 2,500	3.5-5.5	10.75-13.25 @ 15
	D7AE-12127DA	0-1 @ 450	11.25-14 @ 2,500	3.2-5.5	12.75-15.25 @ 14.5
78	D6VE-12127CA	0-1 @ 450	11.25-14 @ 2,500	4.5	10.75-13.25 @ 15.5
	D7AE-12127UA	0-1 @ 650	7.5-10.5 @ 2,500	4	10.75-13.5 @ 15
	D7VE-12127CA	0-1 @ 450	7.25-10 @ 2,500	4.5	10.75-13.25 @ 15.5
	D8AE-12127BA	0-1 @ 450	11.25-14 @ 2,500	3.5	15.75-17.25 @ 14.5

DISTRIBUTOR SPECIFICATIONS — FORD COMPACT AND INTERMEDIATE CARS

Year	Distributor Identification	Centrifugal Advance Start Dist. Deg. @ Dist. RPM	Centrifugal Advance Finish Dist. Deg. @ Dist. RPM	Vacuum Advance Start In. Hg.	Vacuum Advance Finish Dist. Deg. @ In. Hg.
75	D5DE-12127UA	0-2 @ 700	9.25-12.25 @ 2,500	3	4.75-7.25 @ 6.5
	D5DE-12127VA	0-2 @ 700	10-12.5 @ 2,225	3	4.75-7.25 @ 6.5
	D5DE-12127DA	0-2 @ 700	8-11 @ 2,500	3	4.75-7.25 @ 5.25
	D5DE-12127FA	0-2 @ 580	10-12.75 @ 2,500	4.1	2.75-5.25 @ 8.5
	D5DE-12127AC	0-2 @ 575	11.75-14.25 @ 2,150	3.8	6.75-9.25 @ 12.75
	D5DE-12127HA	0-2 @ 550	11-13 @ 2,000	4.5	7.5-10.5 @ 13.0
	D5DE-12127BA	0-.2 @ 900	8-11.5 @ 2,150	4.0	6.75-9.75 @ 9
	D5AE-12127EA	0-2.5 @ 975	8-10.5 @ 2,150	3.0	6.5-9.5 @ 7.0
	D5AE-12127BA	0-2 @ 900	8-11.5 @ 2,150	4.0	6.75-9.75 @ 9
	D5AE-12127DA	0-2.5 @ 975	8-11 @ 2,150	4.0	8-11 @ 9
	D4VE-12127CA	0-2 @ 850	8-11 @ 2,000	4.0	5-8 @ 10.0
	D5DE-12127AA	0-2 @ 725	10.5-13.25 @ 2,500	3	4.75-7.25 @ 6.5
	D5DE-12127YA	0-2 @ 635	13.5-16.25 @ 2,500	5.25	6.75-9.25 @ 12.75

Ford Motor Company

DISTRIBUTOR SPECIFICATIONS — FORD COMPACT AND INTERMEDIATE CARS

Year	Distributor Identification	Centrifugal Advance Start Dist. Deg. @ Dist. RPM	Centrifugal Advance Finish Dist. Deg. @ Dist. RPM	Vacuum Advance Start In. Hg.	Vacuum Advance Finish Dist. Deg. @ In. Hg.
	D5DE-12127LA	0-2 @ 700	9.75-12.75 @ 2,500	4.3	8.75-11.25 @ 18
	D5DE-12127ZA	0-2 @ 660	8.25-10.75 @ 2,380	6	8.75-11.25 @ 14.25
	D5DE-12127AD	0-2 @ 650	8.75-11.25 @ 2,500	4.3	10.75-13.25 @ 12
	D5DE-12127SA	0-2 @ 525	14-16.5 @ 2,500	3	10.75-13.25 @ 14.5
	D5DE-12127MA	0-2 @ 710	8.5-11 @ 2,500	4.6	10.75-13.25 @ 19.8
76	D60E-12127AA	0-2 @ 550	14-16.5 @ 2,500	4.5	21.5-26.5 @ 19
	D5OE-12127FA	0-3 @ 500	8.5-11.5 @ 2,500	5	21-26.5 @ 16
	D6AE-12127CA	0-3 @ 500	8.4-11.1 @ 2,500	4	14-25.5 @ 11.3
	D6AE-12127BA	0-3 @ 525	14-16.5 @ 2,500	5	21-26.5 @ 23
	D6AE-12127AA	0-2 @ 500	11.5-14 @ 2,500	4.3	25-30.5 @ 11.5
	D6VE-12127BA	0-3 @ 520	13.75-16.25 @ 2,000	3.8	21.5-26.5 @ 11.5
	D6VE-12127AA	0-2 @ 500	12.5-15.2 @ 2,500	5	21.5-26.5 @ 13.5
	D6DE-12127BA	0-2 @ 550	12.25-15.25 @ 2,500	4.5	17.5-22.5 @ 14
	D6DE-12127KA	0-1 @ 500	10.2-13 @ 2,500	3	13.6-14.5 @ 6.7
	D6DE-12127AA	0-2 @ 700	12-15 @ 2,500	3	17.5 @ 11.2
	D5DE-12127ACA	0-2 @ 600	11.6-14.2 @ 2,140	6	13.5-18.5 @ 12.5
	D6DE-12127GA	—	—	—	—
	D5DE-12127AGA	0-2 @ 680	10.8-13.5 @ 2,500	5	25.5 @ 13.75
	D6DE-12127JA	0-3.5 @ 500	16.3-18.8 @ 2,200	5	21.5 @ 13
	D5DE-12127AFA	0-2.5 @ 550	10-12.4 @ 2,500	5.6	21.5 @ 15.3
	D6DE-12127LA	0-3.75 @ 550	15.5-18.25 @ 2,500	4	25.5 @ 12
	D6DE-12127CA	0-3 @ 600	11.1-13.8 @ 2,500	5	21.5 @ 27.2
	D6BE-12127BA	—	—	—	—
	D5DE-12127YA	—	—	—	—
	D6DE-12127AA	0-2 @ 700	12-15 @ 2,500	3	8.75 @ 11.2
	D5DE-12127NA	0-2 @ 550	8.75-11.5 @ 2,500	.4	8.75-11.25 @ 14.5
77	D5DE-12127AFA	0-2.5 @ 550	10-12.4 @ 2,500	5.6	13.25 @ 13.1
	D6AE-12127AA	0-2 @ 500	11.5-14 @ 2,500	4.3	12.5-15.25 @ 11.5
	D6DE-12127JA	0-3.5 @ 500	16.3-18.8 @ 2,200	5	13.25 @ 12.2
	D7AE-12127BA	0-1 @ 425	11-13.5 @ 2,250	3.5	12.75-15.25 @ 12
	D7AE-12127CA	0-1 @ 700	8.5-10.75 @ 2,500	3	12.75-15.25 @ 11
	D7AE-12127DA	0-1 @ 450	11.25-14 @ 2,500	3.2	12.25-15.25 @ 14.5
	D7BE-12127DA	0-1 @ 500	5.75-8.5 @ 2,500	3	8.75-11.25 @ 11
	D7BE-12127EA	0-1 @ 550	11.25-13.75 @ 2,450	3	8.75-11.25 @ 13.5
	D7BE-12127FA	0-1 @ 500	5.5-8.25 @ 2,500	3	4.75-7.25 @ 6.5
	D7DE-12127CA	0-1 @ 450	12.5-15.5 @ 2,500	3	12.75-15.25 @ 11
	D7DE-12127HA	(−)1 to ½ @ 550	3.5-6.5 @ 2,500	5	12.5-15 @ 14-15
	D7DE-12127FA	0-1 @ 425	11-13.75 @ 2,500	3	12.75-15.25 @ 13.5
	D7DE-12127GA	0-1 @ 525	4.5-7.25 @ 2,500	3	10.75-13.25 @ 8
	D7OE-12127CA	0-1 @ 450	13.5-16 @ 2,500	3.5	12.75-15.25 @ 14.5
	D7ZE-12127BA	0-1 @ 425	13.5-16 @ 2,500	3.5	10.75-13.25 @ 15.2
78	D7AE-12127UA	0-1 @ 650	7.5-10.5 @ 2,500	4	10.75-13.5 @ 15
	D7BE-12127GA	0-1 @ 500	6.25-8.75 @ 2,500	3	4.75-7.25 @ 7.0
	D7DE-12127AA	(−)1 to ½ @ 550	9.5-12.5 @ 2,500	3	13.25-15.25 @ 12
	D7DE-12127CA	(−)1 to ½ @ 450	12.5-15 @ 2,500	4	12.75-15.25 @ 11
	D7EE-12127CA	0-1 @ 800	5-7.5 @ 2,500	2.3	10.75-13.25 @ 15.75
	D7EE-12127DA	0-1 @ 510	11.5-14 @ 2,500	1.75	10.75-13.25 @ 12.4
	D7EE-12127EA	0-1 @ 525	11.5-14 @ 2,500	2	10.75-13.25 @ 15.75
	D8AE-12127BA	0-1 @ 450	11.25-14 @ 2,500	3.5	15.75-17.25 @ 14.5

DISTRIBUTOR SPECIFICATIONS — FORD COMPACT AND INTERMEDIATE CARS

Year	Distributor Identification	Centrifugal Advance Start Dist. Deg. @ Dist. RPM	Centrifugal Advance Finish Dist. Deg. @ Dist. RPM	Vacuum Advance Start In. Hg.	Vacuum Advance Finish Dist. Deg. @ In. Hg.
	D8AE-12127CA	0-1 @ 1100	6.9-9.5 @ 2,500	3	14.5-17.5 @ 14
	D8AE-12127GA	0-1 @ 450	5.75-8.25 @ 2,500	5.5	13.25-15.25 @ 16.5
	D8AE-12127HA	0-1 @ 450	11.25-14 @ 2,500	3.25	14.75-17.25 @ 13.5
	D8AE-12127JA	0-1 @ 700	8-10.5 @ 2,400	3.25	12.75-15.25 @ 13
	D8AE-12127LA	0-1 @ 775	6.25-9.25 @ 2,500	3.75	14.75-17.25 @ 6.75
	D8BE-12127CA	0-1 @ 1000	2.25-4.75 @ 2,500	3	8.75-11.25 @ 11.5
	D8BE-12127EA	0-1 @ 550	7-9.5 @ 2,500	4.5	8.75-11.25 @ 12.5
	D8BE-12127FA	0-1 @ 500	7-9.5 @ 2,500	3.5	8.75-11.25 @ 13.5
	D8BE-12127JA	0-1 @ 550	7.5-10.5 @ 2,500	3.5	6.75-9.25 @ 13
	D8DE-12127CA	0-1 @ 475	6.5-9 @ 2,500	2	10.75-13.25 @ 10.8
	D8DE-12127EA	(−)1 to ½ @ 450	9.5-12.25 @ 2,500	3	10.75-13.25 @ 13
	D8EE-12127EA	0-1 @ 550	7-9.5 @ 2,500	4.5	8.75-11.25 @ 12.5
	D8ZE-12127CA	0-1 @ 575	9.5-11.5 @ 2,500	2.5	9.75-12.25 @ 15.7

VEHICLE IDENTIFICATION NUMBER (VIN)

Typical vehicle identification number plate located on top left side of dash panel visible through windshield

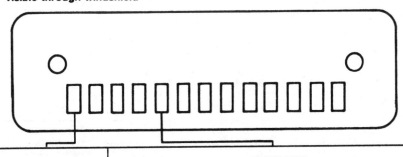

MODEL YEAR CODE		ENGINE CODE			
CODE	YEAR	CODE	DISP. (cu. in.)	CONFIG.	CARB.
3	1973	W	98(1600cc)	L-4	1V
4	1974	X	122(2000cc)	L-4	2V
5	1975	Y	140(2300cc)	L-4	2V
6	1976	U	170(2800cc)	V-6	1V
7	1977	Z	170(2800cc)	V-6	2V
8	1978	T	200	L-6	1V
9	1979	L	250	L-6	1V
		F	302	V-8	2V
		H	351	V-8	2V
		Q	351	V-8	4V
		R	351	V-8	4V
		S	400	V-8	2V
		N	429	V-8	4V
		CW	429	V-8	4V
		CX	429	V-8	4V
		O	429	V-8	4V
		A	460	V-8	4V
		C	460	V-8	4V

Ford Motor Company

CARTER YF, YFA SPECIFICATIONS
FORD MOTOR CO.

Year	Model ①	Float Level (in.)	Fast Idle Cam (in.)	Unloader (in.)	Choke
1975	D5DE-EA	3/8	0.140	0.250	2 Rich
	D5DE-MA	3/8	0.140	0.250	2 Rich
	D5DE-ZA	3/8	0.140	0.250	2 Rich
	D5DE-DA	3/8	0.140	0.250	2 Rich
	D5DE-GA	3/8	0.140	0.250	2 Rich
1976	D6BE-AA	25/32	0.140	0.250	1 Rich
	D6BE-BB	25/32	0.140	0.250	2 Rich
	D5DE-DB	25/32	0.140	0.250	2 Rich
	D5DE-MB	25/32	0.140	0.250	2 Rich
	D6DE-AB	25/32	0.140	0.250	Index
	D6DE-BB	25/32	0.140	0.250	Index
1977-78	D7BE-AA,AB,BA	25/32	0.140	0.250	Index
	D7BE-FA,HB,GB,GC	25/32	0.140	0.250	2 Rich
	D7BE-NA,DA	25/32	0.140	0.250	1 Rich

① Model number located on the tag or casting

Motorcraft Model 2700 VV Specifications
Ford Products

Year	Model	Float Level (in.)	Float Drop (in.)	Fast Idle Cam Setting (notches)	Cold Enrichment Metering Rod (in.)	Control Vacuum (in. H₂0)	Venturi Valve Limiter (in.)	Choke Cap Setting (notches)
1977-78	Pinto, Bobcat	1 3/64	1 15/32	4 Rich/2nd step	.125	5.0	13/32	Index
	All other	1 3/64	1 15/32	1 Rich/3rd step	.125	5.0	61/64	Index

Model 1946
Ford Motor Co. Fairmont & Zephyr

Year	Part Number	Float Setting	Choke Pulldown (in.)	Fast Idle Cam Slot	Accelerator Pump Stroke Slot	Fast Idle Clearance (in.)	Choke Setting
1978	D8BZ-9510R D8BZ-9510AA D8BZ-9510U D8BZ-9510A	see text	.026	#2	#2	.080	Fixed

Ford Motor Company

FORD, AUTOLITE, MOTORCRAFT MODELS 2100, 2150 SPECIFICATIONS
Ford Products

Year	(9510)* Carburetor Identification	Dry Float Level (in.)	Wet Float Level (in.)	Pump Setting Hole #①	Choke Plate Pulldown (in.)	Fast Idle Cam Linkage Clearance (in.)	Fast Idle (rpm)	Dechoke (in)	Choke Setting	Dashpot (in.)
1975	D5ZE-AC	3/8	3/4	2	0.145	②	1500	②	2 Rich	—
	D5ZE-BC	3/8	3/4	2	0.145	②	1500	②	2 Rich	—
	D5ZE-CC	3/8	3/4	3	0.145	②	1500	②	2 Rich	—
	D5ZE-DC	3/8	3/4	2	0.145	②	1500	②	2 Rich	—
	D5DE-AA	7/16	13/16	2	0.140	②	1500	②	3 Rich	—
	D5DE-BA	7/16	13/16	2	0.140	②	1500	②	3 Rich	—
	D5DE-JA	7/16	13/16	2	0.140	②	1500	②	3 Rich	—
	D5ZE-JA	7/16	13/16	2	0.140	②	1500	②	3 Rich	—
	D5OE-AA	7/16	13/16	2	0.140	②	1500	②	3 Rich	—
	D5OE-DA	7/16	13/16	2	0.140	②	1500	②	3 Rich	—
	D5DE-HA	7/16	13/16	3	0.140	②	1500	②	3 Rich	—
	D5DE-UA	7/16	13/16	2	0.140	②	1500	②	3 Rich	—
	D5OE-BA	7/16	13/16	3	0.125	②	1500	②	3 Rich	—
	D5OE-CA	7/16	13/16	3	0.125	②	1500	②	3 Rich	—
	D5OE-GA	7/16	13/16	2	0.125	②	1500	②	3 Rich	—
	D5AE-AA	7/16	13/16	3	0.125	②	1500	②	3 Rich	—
	D5AE-EA	7/16	13/16	3	0.125	②	1500	②	3 Rich	—
	D5ME-BA	7/16	13/16	2	0.125	②	1500	②	3 Rich	—
	D5ME-FA	7/16	13/16	2	0.125	②	1500	②	3 Rich	—
1976	D5ZE-BE	3/8	3/4	2	0.105	②	1600③	②	3 Rich	—
	D6ZE-AA	3/8	3/4	2	0.100	②	1600③	②	3 Rich	—
	D6ZE-BA	3/8	3/4	2	0.100	②	1600③	②	3 Rich	—
	D6ZE-CA	13/32	3/4	2	0.110	②	1600③	②	3 Rich	—
	D6ZE-DA	3/8	3/4	3	0.110	②	1600③	②	3 Rich	—
	D5DE-AEA	7/16	13/16	2	0.160	②	2000④	②	3 Rich	—
	D5DE-AFA	7/16	13/16	2	0.160	②	2000④	②	3 Rich	—
	D5WE-FA	7/16	13/16	2	0.160	②	2000④	②	3 Rich	—
	D6ZE-JA	7/16	13/16	2	0.160	②	2000④	②	3 Rich	—
	D6OE-AA	7/16	13/16	3	0.160	②	2000④	②	3 Rich	—
	D6OE-BA	7/16	13/16	3	0.160	②	2000④	②	3 Rich	—
	D6OE-CA	7/16	13/16	3	0.160	②	2000④	②	3 Rich	—
	D6WE-AA	7/16	13/16	2	0.160	②	1350⑤	②	3 Rich	—
	D6WE-BA	7/16	13/16	2	0.160	②	1350⑤	②	3 Rich	—
	D6AE-HA	7/16	13/16	2	0.160	②	1350⑤	②	3 Rich	—
	D6ME-AA	7/16	13/16	2	0.160	②	1350⑤	②	3 Rich	—
1977-78	D7YE-AA	0.375	0.750	3	0.122	0.142	1600	—	2 Rich	—
	D7YE-BA	0.375	0.750	3	0.122	0.142	1700	—	Index	—
	D7YE-EA	0.375	0.750	3	0.122	0.142	1600	—	2 Rich	—
	D7BE-JA	0.438	0.813	2	0.147	0.167	2100	—	1 Rich	—
	D7BE-LA	0.438	0.813	2	0.147	0.167	2100	—	1 Rich	—
	D7BE-MA	0.438	0.813	2	0.147	0.167	2000	—	1 Rich	—
	D7BE-PA	0.438	0.813	2	0.147	0.167	2100	—	1 Rich	—
	D7BE-YA	0.438	0.813	2	0.147	0.167	2100	—	1 Rich	—
	D7DE-KA	0.438	0.813	2	0.147	0.167	2100	—	1 Rich	—
	D7DE-LA	0.438	0.813	2	0.147	0.167	2000	—	1 Rich	—

Ford Motor Company

FORD, AUTOLITE, MOTORCRAFT MODELS 2100, 2150 SPECIFICATIONS
Ford Products

Year	(9510)* Carburetor Identification	Dry Float Level (in.)	Wet Float Level (in.)	Pump Setting Hole # ①	Choke Plate Pulldown (in.)	Fast Idle Cam Linkage Clearance (in.)	Fast Idle (rpm)	Dechoke (in)	Choke Setting	Dashpot (in.)
	D7WE-EA	0.438	0.813	2	0.147	0.167	2100	—	1 Rich	—
	D7WE-EB	0.438	0.813	2	0.147	0.167	2100	—	1 Rich	—
	D7AE-ADA	0.438	0.813	3	0.179	0.189	1400	—	2 Rich	—
	D7AE-AHA	0.438	0.813	3	0.179	0.189	1400	—	Index	—
	D7AE-CA	0.438	0.813	3	0.179	0.189	1400	—	Index	—
	D7AE-DA	0.438	0.813	3	0.179	0.189	1350	—	Index	—
	D7DE-RA	0.438	0.813	3	0.179	0.189	1400	—	3 Rich	—
	D7DE-RB	0.438	0.813	3	0.179	0.189	1400	—	3 Rich	—
	D7OE-CA	0.750	0.750	3	0.167	0.187	1350	—	2 Rich	—
	D7OE-LA	0.750	0.750	3	0.167	0.187	2000	—	2 Rich	—
	D7OE-NA	0.750	0.750	3	0.167	0.187	1350	—	2 Rich	—
	D7OE-RA	0.750	0.750	3	0.167	0.187	1350	—	2 Rich	—
	D7AE-ACA	0.438	0.813	2	0.156	0.170	1350	—	Index	—
	D7AE-AKA	0.438	0.813	3	0.179	0.189	1400	—	Index	—
	D7AE-GA	0.438	0.813	3	0.179	0.189	1350	—	Index	—
	D7OE-HA	0.438	0.813	3	0.185	0.205	1350	—	2 Rich	—
	D7OE-HB	0.438	0.813	3	0.185	0.205	1350	—	Index	—
	D7OE-MA	0.438	0.813	3	0.185	0.205	1400	—	Index	—
	D7OE-TA	0.438	0.813	3	0.185	0.205	1350	—	2 Rich	—

* Basic carburetor number for Ford products
① With link in inboard hole of pump lever
② Electric choke; see pulldown procedure in text
③ Figure given is for manual transmission; for automatics add 100 RPM.
④ Figure given is for 49 states Granada and Monarch; for Calif. Granada and Monarch and all Torino, Montego and Cougar models, figure is 1400 RPM.
⑤ Figure given is for 49 states model; Calif. specification is 1150 RPM.
⑥ 1600 with 360V8

FORD, AUTOLITE, MOTORCRAFT MODEL 5200 SPECIFICATIONS
Ford Products

Year	(9510)* Carburetor Identification ①	Dry Float Level (in.)	Pump Hole Setting	Choke Plate Pulldown (in.)	Fast Idle Cam Linkage (in.)	Fast Idle (rpm)	Dechoke (in.)	Choke Setting	Dashpot (in.)
1975	D52E-AA	0.460	2	0.200	0.100	1800	0.260	1 Lean	—
	D52E-BA	0.460	2	0.200	0.100	1800	0.260	1 Lean	—
	D52E-CA	0.460	2	0.200	0.100	1800	0.260	1 Lean	—
	D52E-DB	0.460	2	0.200	0.100	1800	0.260	1 Lean	—
	D5ZE-EA	0.460	2	0.200	0.100	1800	0.260	1 Lean	—
	D5ZE-FA	0.460	2	0.200	0.100	1800	0.260	1 Lean	—
	D5ZE-GA	0.460	2	0.200	0.100	1800	0.260	1 Lean	—
	D5ZE-HB	0.460	2	0.200	0.100	1800	0.260	1 Lean	—
1976	D6EE-BA	0.460	2	0.200	0.100	1500①	0.260	1 Lean	—
	D6EE-CA	0.460	2	0.270	0.160	1500①	0.260	1 Lean	—
	D6EE-DA	0.460	2	0.200	0.100	1500①	0.260	1 Lean	—
	D6ZE-EA	0.460	2	0.270	0.160	1500①	0.260	1 Lean	—

Ford Motor Company

FORD, AUTOLITE, MOTORCRAFT MODEL 5200 SPECIFICATIONS
Ford Products

Year	(9510)* Carburetor Identification①	Dry Float Level (in.)	Pump Hole Setting	Choke Plate Pulldown (in.)	Fast Idle Cam Linkage (in.)	Fast Idle (rpm)	Dechoke (in.)	Choke Setting	Dashpot (in.)
1977-78	D7EE-AAA	0.453	2	0.200	0.120	2000	0.180	Index	——
	D7EE-AB	0.453	2	0.240	0.120	1800	0.240	2 Rich	——
	D7EE-BDA	0.453	2	0.280	0.120	1500	0.240	2 Rich	——
	D7EE-BGA	0.453	2	0.240	0.120	1500	0.240	Index	——
	D7EE-BHA	0.453	2	0.240	0.120	1500	0.240	Index	——
	D7EE-BLA	0.453	2	0.240	0.120	2000	0.240	Index	——
	D7EE-BMA	0.453	2	0.240	0.120	2000	0.240	Index	——
	D7EE-DA	0.453	2	0.240	0.120	1500	0.240	2 Rich	——
	D7EE-EA	0.453	2	0.240	0.120	2000	0.240	Index	——
	D7EE-FA	0.453	2	0.240	0.120	1800	0.240	Index	——
	D7EE-GA	0.453	2	0.200	0.120	2000	0.200	Index	——
	D7EE-HA	0.453	2	0.240	0.120	1500	0.240	2 Rich	——
	D7EE-JA	0.453	2	0.240	0.120	2000	0.240	Index	——
	D7EE-KB	0.453	2	0.240	0.120	1800	0.240	2 Rich	——
	D7EE-LA	0.453	2	0.240	0.120	1800	0.240	Index	——
	D7EE-SA	0.453	2	0.240	0.120	1800	0.240	2 Rich	——
	D7EE-TA	0.453	2	0.240	0.120	1800	0.240	2 Rich	——
	D7EE-UA	0.453	2	0.240	0.120	1800	0.240	Index	——
	D7EE-VA	0.453	2	0.240	0.120	1800	0.240	Index	——

* Basic carburetor number
① Figure given is for all manual transmissions; for automatic trans. the figures are: (49 states) 2000 RPM; (Calif.) 1800 RPM.

FORD, AUTOLITE, MOTORCRAFT MODELS 4300, 4350 SPECIFICATIONS
Ford Products

Year	(9510)* Carburetor Identification	Dry Float Level (in.)	Pump Hole Setting	Choke Plate Pulldown (in.)	Fast Idle Cam Linkage (in.)	Fast Idle (rpm)	Dechoke (in.)	Choke Setting	Dashpot
1975	D5VE-AD	15/16	1	②	0.160	1600	0.300	2 Rich	——
	D5VE-BA	15/16	1	②	0.160	1600	0.300	2 Rich	——
	D5AE-CA	31/32	1	②	0.160	1600	0.300	2 Rich	——
	D5AE-DA	31/32	1	②	0.160	1600	0.300	2 Rich	——
1976	D6AE-CA	1.00	2	0.140③	0.140	1350	0.300	2 Rich	——
	D6AE-FA	1.00	2	0.140③	0.140	1350	0.300	2 Rich	——
	D6AE-DA	1.00	2	0.160④	0.160	1350	0.300	2 Rich	——
1977-78	D7AE-AAA	1.00	2	0.140	0.140	1350	0.300	Index	——
	D7AE-ANA	1.00	2	0.140	0.140	1350	0.300	Index	——
	D7AE-ZA	1.00	2	0.140	0.140	1350	0.300	Index	——
	D7PE-AA	1.00	2	0.140	0.140	1350	0.300	Index	——
	D7VE-KA	1.00	2	0.140	0.140	1350	0.300	2 Lean	——
	D7VE-SA	1.00	2	0.140	0.140	1350	0.300	Index	——

* Basic carburetor number for Ford products.
① The identification tag is on the bowl cover.
② Initial—0.160 in. Delayed—0.190 in.
③ Initial Figure given; delayed—0.190
④ Initial Figure given; delayed—0.210

Ford Motor Company

CAR SERIAL NUMBER AND ENGINE IDENTIFICATION

1976

Mounted behind the windshield on the driver's side is a plate with the vehicle identification number. The first character is the model year, with 6 for 1976. The fifth character is the engine code, as follows.

A .... 460 V-8 4-bbl.
C .... 460 V-8 4-bbl. Police
F .... 302 V-8 2-bbl.
H .... 351 V-8 2-bbl.
L .... 250 6-cyl. 1-bbl.
S .... 400 V-8 2-bbl.
T .... 200 6-cyl. 1-bbl.

NOTE: *351 V-8 engines come in two types, the 351W and the 351M. The basic specifications of the two engines are identical, but the design of many of the parts is different. The same code letter is used, regardless of which 351 V-8 is installed in the car. The easiest way to tell the two engines apart is by the fuel pump mounting bolts. The W engine has the bolts in a horizontal line. The M engine has the bolts in a vertical line.*

1977

The vehicle identification number is mounted on a plate behind the windshield on the driver's side. The first character is the model year, with 7 for 1977. The fifth character is the engine code, as follows.

A 460 4-bbl. V-8
C .... 460 4-bbl. V-8 Police
F 302 2-bbl. V-8
H 351W 2-bbl. V-8
L .... 250 1-bbl. 6-cyl.
Q 351M 2-bbl. V-8
S 400 2-bbl. V-8
T .... 200 1-bbl. 6-cyl.

1978

The vehicle identification number is mounted on a plate behind the windshield on the driver's side. The first character is the model year, with 8 for 1978. The fifth character is the engine code, as follows.

A 460 4-bbl. V-8
C .... 460 4-bbl. V-8 Police
F 302 2-bbl. V-8
H 351 2-bbl. V-8
L .... 250 1-bbl. 6-cyl.
S 400 2-bbl. V-8
T .... 200 1-bbl. 6-cyl.

351 engines come in two different designs, the "W" and the "M". Both engines use the "H" code letter. Although the bore and stroke of both engines is the same, the internal parts are not interchangeable. The two engines can be identified by looking at the fuel pump mounting bolts. The "W" bolts are on a horizontal line. The "M" bolts are on a vertical line. The "W" engine is the same general design as the 302 V-8. The "M" engine is the same general design as the 400 V-8.

1979

The vehicle identification number is mounted on the driver's side. The first character is the model year, with 8 for 1978. The fifth character is the engine code, as follows.

F 302 2-bbl. V-8
H 351 2-bbl. V-8
L .... 250 1-bbl. 6-cyl.
S 400 2-bbl. V-8
T .... 200 1-bbl. 6-cyl.

351 engines come in two different designs, the "W" and the "M". Both engines use the "H" code letter. Although the bore and stroke of both engines is the same, the internal parts are not interchangeable. The two engines can be identified by looking at the fuel pump mounting bolts. The "W" bolts are on a horizontal line. The "M" bolts are on a vertical line. The "W" engine is the same general design as the 302 V-8. The "M" engine is the same general design as the 400 V-8.

EMISSION EQUIPMENT

1976
All Models
Closed positive crankcase ventilation
Emission calibrated carburetor
Emission calibrated distributor
Heated air cleaner
Vapor control, canister storage
Air pump
Electric choke
Exhaust gas recirculation
Catalytic converter
Fresh air tube to air cleaner (Zip Tube)
Cold weather modulator
  Used on all models except
  Not on 351M and 400 V-8
  Not on 460 with high energy ignition
Delayed choke pulldown
  All 4-bbl. carburetors
Air cleaner vacuum delay valve
  351M V-8
Cold start spark advance
  49-States
    460 V-8
  Calif.
    351M V-8
    400 V-8
    460 V-8
Vacuum operated exhaust heat valve
  302 V-8
Computer controlled timing
  Continental Mark IV (Late production)

Altitude compensation carburetor
  Optional on 460 V-8 in Denver area only

1977
All Models
Closed positive crankcase ventilation
Emission calibrated carburetor
Emission calibrated distributor
Heated air cleaner
Vapor control, canister storage
Electric choke
Exhaust gas recirculation
Catalytic converter
Fresh air tube to air cleaner (Zip Tube)
Air pump
  49-States
    200 6-cyl.
    250 6-cyl. man. trans. only
    302 V-8
      All man. trans.
      Auto. trans. Cougar, LTD II, T-Bird, and Ranchero only.
    351M V-8
    400 V-8
    460 V-8
  Calif.
    All models
Delayed choke pulldown
  All 4-bbl. carburetors
Cold start spark advance
  49-States
    460 V-8
  Calif.
    Not used
Altitude compensation carburetor
  High altitude vehicles only
Distributor vacuum vent valve
  302 V-8 Calif. only.

1978
All Models
Closed positive crankcase ventilation
Emission calibrated carburetor
Emission calibrated distributor
Heated air cleaner
Vapor control, canister storage
Electric choke
Exhaust gas recirculation
Catalytic converter
Fresh air tube to air cleaner
Air pump
  49-States and Altitude
    200 6-cyl.
      All except calib. 8-7A-R0, 8-7B-R0
    250 6-cyl. man. trans. only
    302 V-8
      All except calib. 8-31A-R0, 8-31C-R0
    351M V-8
      All except calib. 8-34A-R00
    400 V-8
    460 V-8
  Calif.
    All models
Suction system
  49-States
    Only on calib. 8-7A-R0, 8-7B-R0

Ford Motor Company

EMISSION EQUIPMENT

1978

Calif.
  Not used
Altitude
  Not used
Delayed choke pulldown
  All 4-bbl. carburetors
Cold start spark advance
  All 460 V-8
Altitude compensation carburetor
  High altitude vehicles only

1979

All Models

Closed positive crankcase ventilation
Emission calibrated carburetor
Emission calibrated distributor
Heated air cleaner
Vapor control, canister storage
Electric choke
Exhaust gas recirculation
Catalytic converter
Fresh air tube to air cleaner

NOTE: *In addition to the above listed equipment some vehicles will also be equipped with an air pump system.*

Altitude compensation carburetors may be found on high altitude vehicles only.

IDLE SPEED AND MIXTURE ADJUSTMENTS

1976

Air cleaner .... On
Air cond. .... Off
Auto. trans. .... Drive
Solenoid .... Connected
  NOTE: *Idle speed with the solenoid disconnected is as follows.*
All engines except 460 V-8 500
460 V-8 600

200 6-cyl. 49-States
  Auto. trans. decals 6-000,
    6-200 650
  Man. Trans. decals 6-000,
    6-200 800
250 6-cyl. 49-States
  Auto. trans. decals 6-008,
    6-010 600
  Man. trans. decals 6-008,
    6-010 850
250 6-cyl. Calif.
  Auto. trans. decals 6-329,
    6-095, 6-063, 6-351 600
  Man. trans. decals 6-329,
    6-095, 6-063, 6-322 850
302 V-8 49-States
  Auto trans. decals 6-034,
    6-040, 6-041, 6-113, 6-114 .. 650
  Auto. trans. decals 6-030,
    6-004 700
  Man. trans. decal 6-225 800
  Man. trans. decals 6-034,
    6-040, 6-041, 6-113,
    6-114, 6-116, 6-117 750
302 V-8 Calif.
  Auto. trans. decals 6-064,
    6-089, 6-090, 6-103,
    6-302, 6-303, 6-304 700
351W V-8 49-States
  Auto. trans. decals 6-125,
    6-210 625
  Auto. trans. decals 6-001,
    6-032 650
351W V-8 Calif.
  Auto. trans. decals 6-094,
    6-308 650
351M V-8 49-States
  Auto. trans. decals 6-009,
    6-125, 6-226, 6-227 650
351M V-8 Calif.
  Auto. trans. decals 6-079,
    6-323 650
  Auto. trans. decal 6-087 675
400 V-8 49-States
  Auto. trans. decals 6-006,
    6-217, 6-218, 6-219, 6-252 .. 650

400 V-8 Calif.
  Auto. trans. decals 6-074,
    6-075 625
460 V-8 49-States
  Auto. trans. decals 6-033,
    6-124, 6-202, 6-236,
    6-250, 6-253 650
460 V-8 Calif.
  Auto. trans. decals 6-086,
    6-321 650

1977

Air cleaner .... In place
Air cond. .... Off
Auto. trans. .... Drive
200 6-Cyl.
  Cal. 7-6B-R1 800
250 6-cyl.
  Cal. 7-8A-R2 850
302 V-8
  Cal. 7-11E-RO
    Solenoid connected 600
    Solenoid disconnected 500
351W V-8
  Cal. 7-32B-R1
    Solenoid connected 625
    Solenoid disconnected 550
351M V-8
  Cal 7-14A-R1; 7-14B-R1 650
  Cal. 7-14N-RO
    Solenoid connected 600
    Solenoid disconnected 500
400 V-8
  Cal. 7-17B-R1
    Solenoid connected 600
    Solenoid disconnected 525
460 V-8
  Cal. 7-19A-R2
    Solenoid connected 650
    Solenoid disconnected 525

1978

Air cleaner .... In place
Air cond. .... Off
Auto. trans. .... Drive
200 6-cyl.
  Auto. trans. 650
  Man. trans. 800
250 6-cyl.
  Auto. trans. 600
  Man. trans. 800
302 V-8 auto. trans.
  Cal. 8-11D-R0, 8-11L-R0,
    8-11L-R11 625
  Cal. 8-11E-R0, 8-11E-R10
    Solenoid connected 600
    Solenoid disconnected 500
  Cal. 8-11G-R0, 8-11G-R17,
    8-11M-R16, 8-11M-R17,
    8-31B-R0, 8-31C-R1 600
    AC idle speedup 675
  Cal. 8-11M-R0, 8-11M-R10 ... 600
  Cal. 8-11N-R0 700
  Cal. 8-31A-R0 700
    AC idle speedup 775
  Cal. 8-11P-R0, 8-11Q-R0
    Solenoid connected 700
    Solenoid disconnected 600
  Cal. 8-11X-R0, 8-11Y-R0,
    8-11Z-R0 650
    AC idle speedup 725
NOTE: *The "solenoids" listed above may be vacuum operated throttle modulators, which have the same effect as a solenoid.*
302 V-8 man. trans.
  Cal. 8-10A-R0 900
    AC idle speedup 975
  Cal. 8-10A-R10 800
    AC idle speedup 875
  Cal. 8-10B-R0 800
NOTE: *"Cal." in these listings refers to the calibration, which is shown on a label attached to the engine rocker cover.*
351W V-8 auto. trans.
  Cal. 8-17X-R0 650
  All others 600
  AC idle speedup 675
351M V-8 auto. trans.
  49-State cal. 8-14X-R0 650
  All other 49-States 600
  AC idle speedup 675
  Calif. 600
    AC idle speedup 650
400 V-8 auto. trans.
  49-States 575
  AC idle speedup   625 or 650
                    (See decal)
460 V-8 auto. trans. 580
460 V-8 Police
  Solenoid connected 700
  Solenoid disconnected 550

IDLE MIXTURE SETTINGS 1978

NOTE: *When both a lean drop and a propane procedure are given, either may be used. Make all settings in Neutral or Drive, as shown.*

200 6-cyl. auto. trans. (in Drive)
  49-States lean drop .... 5 rpm

309

IDLE SPEED AND MIXTURE ADJUSTMENTS

1978

Calif. lean drop .... 10 rpm
Altitude lean drop .... 10 rpm
200 6-cyl. man. trans.
  49-States lean drop .... 30 rpm
250 6-cyl. auto. trans.
  49-States lean drop (Drive) .. 50 rpm
  Calif. lean drop
    In Drive .... 20 rpm
    In Neutral .... 70 rpm
250 6-cyl. man. trans.
  49-States lean drop .... 50 rpm
302 V-8 auto. trans. (In Neutral)
  Cal. 8-11D-R0
    Propane okay range .... 30-60 rpm
    Propane reset .... 60 rpm
  Cal. 8-11E-R0, 8-11E-R10
    Propane okay range ..... 5-55 rpm
    Propane reset .... 5-55 rpm
  Cal. 8-11G-R0, 8-11G-R17
    Propane okay range ... 10-100 rpm
    Propane reset .... 30-60 rpm
  Cal. 8-11L-R11
    Propane okay range .... 30-90 rpm
    Propane reset .... 60 rpm
    Lean drop .... 70 rpm
  Cal. 8-11M-R0, 8-11M-R10,
    8-11M-R16
    Propane okay range .. 100-160 rpm
    Propane reset .... 120-150 rpm
  Cal. 8-11N-R0
    Propane okay range ... 50-140 rpm
    Propane reset .... 70-110 rpm
    Lean drop .... 110 rpm
  Cal. 8-11P-R0, 8-11Q-R0
    Propane okay range ... 20-160 rpm
    Propane reset .... 60-120 rpm
    Lean drop .... 120 rpm
  Cal. 8-11X-R0, 8-11Z-R0
    Propane okay range
      Above 4000 ft. ..... 70-130 rpm
      Below 4000 ft. .... 180-240 rpm
    Propane reset
      Above 4000 ft. ...... 80-120 rpm
      Below 4000 ft ...... 190-230 rpm
  Cal. 8-31A-R0
    With std. flow PCV valve
      Propane okay range . 80-140 rpm
      Propane reset ..... 100-120 rpm
    With high flow PCV valve
      Propane okay range 160-250 rpm
      Propane reset ..... 170-230 rpm
  Cal. 8-31C-R1
    With std. flow PCV valve
    Propane okay
      range .... 90-160 rpm
      Propane reset .... 130 rpm
    With high flow PCV valve
      Propane okay range 200-320 rpm
      Propane reset .... 260 rpm
302 V-8 man. trans.
  Cal. 8-10A-R0, 8-10A-R10
    Propane okay range ...... 10 rpm
    Propane reset .... 30 rpm
  Cal. 8-10B-R0
    Lean drop .... 40 rpm
351W V-8 auto. trans. (Neutral)
  Cal. 8-32B-R0, 8-32B-R11, 8-32B-R13
    With std. flow PCV valve
      Propane okay range . 50-150 rpm
      Propane reset ..... 80-120 rpm
    With high flow PCV valve
      Propane okay range . 80-180 rpm
      Propane reset ..... 110-150 rpm
351M V-8 auto. trans.
  Cal. 8-14B-R0 (Drive)
    Propane okay range ..... 0-10 rpm
    Propane reset .... 0-5 rpm
  Cal. 8-14B-R13, 8-14B-R14 (Drive)
    Set to lean best idle
  Cal. 8-14B-R22 (Drive)
    Propane okay range ..... 0-10 rpm
    Propane reset .... 0-5 rpm
  Cal. 8-14N-R13 (Drive)
    Propane okay range ..... 0-30 rpm
    Propane reset .... 10-20 rpm
    Lean drop .... 20 rpm
  Cal. 8-14X-R0 (Drive)
    Propane okay range .. 240-360 rpm
    Propane reset .... 260-340 rpm
NOTE: After making setting on 8-14X-R0, install limiter caps and turn screws out to full rich setting, as far as limiter caps will allow.
  Cal. 8-34A-R00
    With std. flow PCV valve
    Propane okay range
      Neutral .... 75 rpm
      Drive .... 20-40 rpm
    Propane reset
      Neutral .... 70 rpm
      Drive .... 20 rpm
    With high flow PCV valve
    Propane okay range
      Neutral .... 125 rpm
      Drive .... 20-40 rpm
    Propane reset
      Neutral .... 100 rpm
      Drive .... 20 rpm
400 V-8 auto. trans.
  Cal. 8-17B-R0, 8-17C-R0 (Drive)
    Set to lean best idle
  Cal. 8-17P-R0 (Drive)
    Propane okay range ..... 0-30 rpm
    Propane reset .... 10-20 rpm
  Cal. 8-17X-R0 (Neutral)
    Propane okay range .. 80-200 rpm
    Propane reset .... 100-180 rpm
460 V-8 auto. trans.
  Cal. 8-19A-R1 (Drive)
    Propane okay range ..... 0-10 rpm
    Propane reset .... 0-10 rpm

1979

NOTE: Make all adjustments with the engine at normal operating temperature. Refer to the Engine Emission Label, located in the engine compartment, for additional instructions before making any adjustments.

Ford/Mercury 302 V-8 (F) 49 states

Calibration 9-11E-R1
  Auto. trans./air conditioned
    Fast idle 1750
    Curb idle
      AC/ on 625
      A/C off 550
    VOTM*
      A/C off 625
Calibration 9-11E-R1
  Auto. trans./ non A/C
    Fast idle 1750
    Curb idle 550
    VOTM* 625

Ford/Mercury 302 V-8 (F) Police station wgn. 49 states

Calibration 9-11F-R1
  Auto. trans./air conditioned
    Fast idle 1750
    Curb idle
      A/C on 625
      A/C off 550
    VOTM*
      A/C off 625
Calibration 9-11F-R1
  Auto. trans./non A/C
    Fast idle 1750
    Curb idle 550
    VOTM* idle 625

Ford/Mercury 302 V-8 (F) California only

Calibration 9-11Q-R01
  Auto. trans.
    Fast idle 1800
    Curb idle
      A/C on 625
      A/C off 550
      Non A/C 550
    VOTM*
      A/C 625
      Non A/C 625
Calibration 9-11Q-R-10
  Auto trans.
    Fast idle 1800
    Curb idle
      A/C on 625
      A/C off 550
      Non A/C 550
    VOTM*
      A/C 625
      Non A/C 625

Ford/Mercury 351 V-8 (H) 49 states

Calibration 9-12G-R0
  Auto. trans.
    Fast idle 2200
    Curb idle
      A/C on 650
      A/C off 600
      Non A/C 600
Calibration 9-12F-R0
  Auto. trans.
    Fast idle 2200
    Curb idle
      A/C on 650
      A/C off 600
      Non A/C 600

Mercury w/EEC II 351 V-8 (H) 49 states

Calibration 9-12A-R0
  Auto. trans.
    Fast idle 2000
    Curb idle
      A/C on 640
      A/C off 550
      Non A/C 550

Ford Motor Company

IDLE SPEED AND MIXTURE ADJUSTMENTS

1979

Calibration 9-12B-R0
  Auto. trans.
    Fast idle 2000
    Curb idle
      A/C on 640
      A/C off 550
      Non A/C 550

Ford/Mercury w/EEC II 351 V-8 (H) California only

Calibration 9-12N-R0
  Auto. trans.
    Fast idle 2100
    Curb idle
      A/C 620
      Non A/C 620

Calibration 9-12P-R0
  Auto. trans.
    Fast idle 2100
    Curb idle
      A/C 620
      Non A/C 620

Granada/Monarch 302 V-8 (F) 49 states

Calibration 9-11C-R0
  Auto. trans.
    Fast idle 2100
    Curb idle
      A/C on 675
      A/C off 600
      Non A/C 600

Calibration 9-11C-R1
  Auto. trans.
    Fast idle 2100
    Curb idle
      A/C on 675
      A/C off 600
      Non A/C 600

Calibration 9-10C-R0
  Manual trans.
    Fast idle 2300
    Curb idle
      A/C on 875
      A/C off 800
      Non A/C 800

Calibration 9-10C-R10
  Manual trans.
    Fast idle 2300
    Curb idle
      A/C on 875
      A/C off 800
      Non A/C 800

Granada/Monarch 302 V-8 (F) California only

Calibration 9-11S-R0
  Auto. trans.
    Fast idle 1800
    Curb idle
      A/C on 675
      A/C off 600
      Non A/C 600
    VOTM*
      A/C 700
      Non A/C 700

Granada/Monarch 250 I-6 (L) 49 states

Calibration 9-9A-R0
  Auto. trans.
    Fast idle 1700
    Curb idle
      A/C on 700
      A/C off 600
      Non A/C 600
    TSP off**
      A/C off 500
      Non A/C 500

Calibration 9-9A-R12
  Auto. trans.
    Fast idle 1700
    Curb idle
      A/C on 700
      A/C off 600
      Non A/C 600
    TSP off**
      A/C off 450
      Non A/C 450

Calibration 9-8A-R0
  Manual trans.
    Fast idle 1700
    Curb idle
      A/C 800
      Non A/C 800

Granada/Monarch 250 I-6 (L) California only

Calibration 9-9P-R0
  Auto. trans.
    Fast idle 2300
    Curb idle
      A/C on 700
      A/C off 600
      Non A/C 600
    Tsp off**
      A/C 450
      Non A/C 450

Cougar/LTD II/Thunderbird 302 V-8 (F) 49 states

Calibration 9-11J-R0
  Auto. trans.
    Fast idle 2100
    Curb idle
      A/C on 675
      A/C off 600
      Non A/C 600

Cougar/LTD II/Thunderbird 351 V-8 (H) 49 states

Calibration 9-12E-R0
  Auto. trans.
    Fast idle 2100
    Curb idle
      A/C on 650
      A/C off 600
      Non A/C 600

Calibration 9-14E-R0
  Auto. trans.
    Fast idle 2200
    Curb idle
      A/C on 650
      A/C off 600
      Non A/C 600

Cougar/LTD II/Thunderbird 351 V-8 (H) California only

Calibration R-140-R0
  Auto. trans.
    Fast idle 2300
    Curb idle
      A/C on 650
      A/C off 600
      Non A/C 600

Lincoln Continental/Mark V 400 V-8 (S) 49 states

Calibration 9-17F-R00
  Auto. trans.
    Fast idle 2200
    Curb idle
      A/C on 675
      A/C off 600

Lincoln Continental/Mark V 400 V-8 (S) California only

Calibration 9-17P-R0
  Auto. trans.
    Fast idle 2200
    Curb idle
      A/C on 650
      A/C off 650

Calibration 9-17Q-R0
  Auto. trans.
    Fast idle 2300
    Curb idle
      A/C on 650
      A/C off 600

Versailles 302 V-8 (F) 49 states

Calibration 8-11D-R0
  Auto. trans.
    Fast idle 1900
    Curb idle
      A/C 625

Calibration 8-11L-R11
  Auto. trans.
    Fast idle 3100
    Curb idle
      A/C 625
    VOTM*
      A/C 750

Versailles 302 V-8 (F) California only

Calibration 9-11P-R0/9-11P-R1
  Auto. trans.
    Fast idle 2900
    Curb idle
      A/C 625
    VOTM*
      A/C 800

Calibration 9-11V-R0
    Fast idle 1800
    Curb idle
      A/C on 675
      A/C off 600
    VOTM*
      A/C 700

* VOTM (Vacuum Operated Throttle Modulator)
** TSP (Throttle Solenoid Positioner)

NOTE: *All curb idle settings are made at the solenoid with the solenoid either energized or deenergized.*

Various throttle positioners are used depending on vehicle application. Curb idle adjustments are made at the throttle positioners and not at the "off idle" adjusting screw.

Ford Motor Company

IDLE SPEED AND MIXTURE AJDUSTMENTS

A combination of devices can be used including a solenoid-dashpot, a solenoid-diaphragm, a vacuum operated throttle solenoid, and a vacuum operated throttle kicker actuator.

When the A/C-heater control is positioned in the A/C mode the throttle kicker solenoid is energized thereby allowing intake manifold vacuum to reach the throttle kicker actuator which is positioned against the throttle lever. When a vacuum signal is applied to the throttle kicker actuator, it will increase the engine RPM for increased cooling and smoother idle. The throttle kicker actuator is also energized during engine warm-up or if an engine overheat condition exists.

IDLE MIXTURE ARTIFICIAL ENRICHMENT 1976

Air cleaner ....On
Vapor hose to air cleaner ....Off
PCV hose to air cleaner ....Off
PCV air cleaner connection ....Plugged
Air pump ....Disconnected
Auto. trans.............................See below
Vacuum retard hose (if used) ....Disconnected

Speed Gain (RPM)
NOTE: Check speed gain at idle speed. These figures can also be used for speed drop.

NOTE: Ford makes running changes throughout a model year, sometimes with considerable variation in engine timing and other engine specifications. Some engines may be identical except for the engine timing or idle speed specification. The only way to tell the engines apart is to read the number on the tune-up decal. If the decal is missing, a Ford parts man may be able to identify the engine by the identification tag or the part numbers of the carburetor or other parts. If later engines come out with different decal numbers, they will be in a later supplement.

IDLE MIXTURE ARTIFICIAL ENRICHMENT

Engine	Okay Range	Reset To
200 6-Cyl. 49-States		
Auto. Trans.		
Decal 6-000 (Drive)	10-80	30
Decal 6-200 (Drive)	0-20	10
Man. Trans.		
Decal 6-000	10-80	30
Decal 6-200	0-50	25
250 6-Cyl. 49-States		
Auto. Trans.		
Decal 6-008 (Drive)	0-20	0
Decal 6-010 (Drive)	0-20	10
Man. Trans.		
Decal 6-008 (At 750 rpm)	10-90	40
Decal 6-010	0-20	10
250 6-Cyl. Calif.		
Auto. Trans.		
Decal 6-095 (Drive)	0-40	20
Decal 6-329 (Drive)	20-80	50
Decal 6-063 (Drive)	0-20	10
Decal 6-351 (Neutral)	0-40	20
Man. Trans.		
Decal 6-063 (At 750 rpm)	50-200	125
Decal 6-095 (At 750 rpm)	50-200	125
Decal 6-322	0-20	10
302 V-8 49-States		
Auto. Trans.		
Decals 6-004, 6-040, 6-041, 6-113, 6-114 (Neutral)	0-100	0-100
Decal 6-030 (Neutral)		
Decal 6-034 (Neutral)	10-50	10-50
Man. Trans.		
Decals 6-034, 6-040, 6-113, 6-117, 6-225	0-10	0-10
Decals 6-041, 6-114, 6-116	0-20	0-20
302-V-8 Calif.		
Auto. Trans.		
Decal 6-064 (Drive)	10-60	20-60
Decals 6-089, 6-090, 6-103, 6-302, 6-303, 6-304 (Neutral)	10-60	20-60
351W V-8 49-States		
Auto. Trans.		
Decal 6-001 (Drive)	35-70	40-50
Decal 6-032 (Neutral)	30-90	20-50
Decals 6-125, 6-210 (Neutral)	20-100	30-80
351W V-8 Calif.		
Auto. Trans.		
Decal 6-094 (Neutral)	10-100	20-60
Decal 6-308 (Neutral)	10-100	20-80
351M V-8 49-States		
Auto. Trans.		
Decal 6-009 (Drive)	30 (min.)	50-70
Decals 6-215, 6-226, 6-227 (Drive)	0-100	40-60
351M V-8 Calif.		
Auto. Trans.		
Decals 6-079, 6-087, (Drive)	15 (min.)	30-50
Decal 6-323 (Drive)	30 (min.)	50-70
400 V-8 49-States		
Auto. Trans.		
Decal 6-006 (Drive)	15 (min.)	20-40
Decals 6-217, 6-218, 6-219, 6-252 (Drive)	0-60	20-40
400 V-8 Calif.		
Auto. Trans.		
Decals 6-074, 6-075, (Drive)	15 (min.)	20-40
460 V-8 49-States		
Auto. Trans.		
Decals 6-033, 6-124, 6-202 (Drive)	20-60	40
Decals 6-220, 6-236, 6-250, (Drive)	30-70	40
Decal 6-253 (Drive)	15-40	30
460 V-8 Calif.		
Auto. Trans.		
Decals 6-086, 6-321 (Drive)	15-40	30

Ford Motor Company

INITIAL TIMING 1976

NOTE: *Distributor vacuum hoses must be disconnected and plugged. Set timing with engine idling at speed shown. Set automatic transmission engines in Neutral unless shown otherwise.*

200 6-Cyl. 49-States
  Auto. Trans.
    Decal 6-000 (500 rpm) .... 6° BTDC
    Decal 6-200 (750 rpm max) .... 8° BTDC
  Man. Trans.
    Decal 6-000 (500 rpm) .... 6° BTDC
    Decal 6-200 (750 rpm max.) .... 10° BTDC

250 6-Cyl. 49-States
  Auto. Trans.
    Decal 6-008 (500 rpm) .... 6° BTDC
    Decal 6-010 (750 rpm max.) .... 12° BTDC
  Man. Trans.
    Decal 6-008 (650 rpm) .... 4° BTDC
    Decal 6-010 (750 rpm max.) .... 8° BTDC

250 6-Cyl. Calif.
  Auto. Trans.
    Decal 6-063 (500 rpm) .... 10° BTDC
    Decal 6-095 (750 rpm) .... 8° BTDC
    Decal 6-329 (500 rpm) .... 10° BTDC
    Decal 6-351 (750 rpm max.) .... 8° BTDC
  Man. Trans.
    Decal 6-063 (500 rpm) .... 6° BTDC
    Decal 6-095 (500 rpm) .... 6° BTDC
    Decal 6-322 (750 rpm max.) .... 6° BTDC

302 V-8 49-States
  Auto. Trans.
    Decal 6-004 (500 rpm) .... 8° BTDC
    Decal 6-113 (500 rpm) .... 12° BTDC
    Decal 6-114 (500 rpm) .... 8° BTDC
    Decal 6-030 (500 rpm) .... 6° BTDC
    Decal 6-040 (500 rpm) .... 8° BTDC
    Decal 6-041 (500 rpm) .... 12° BTDC
    Decal 6-034 (500 rpm) .... 6° BTDC
  Man. Trans.
    Decal 6-034 (500 rpm) .... 12° BTDC
    Decal 6-040 (500 rpm) .... 12° BTDC
    Decal 6-041 (500 rpm) .... 12° BTDC
    Decal 6-113 (500 rpm) .... 4° BTDC
    Decal 6-114 (500 rpm) .... 4° BTDC
    Decal 6-116 (500 rpm) .... 12° BTDC
    Decal 6-117 (500 rpm) .... 12° BTDC
    Decal 6-225 (500 rpm) .... 12° BTDC

302 V-8 Calif.
  Auto. Trans.
    Decal 6-064 (500 rpm) .... 8° BTDC
    Decal 6-089 (500 rpm) .... 8° BTDC
    Decal 6-090 (500 rpm) .... 12° BTDC

    Decal 6-103 (500 rpm) .... 8° BTDC
    Decal 6-302 (500 rpm) .... 4° BTDC
    Decal 6-303 (500 rpm) .... 8° BTDC
    Decal 6-304 (500 rpm) .... 12° BTDC

351W V-8 49-States
  Auto. Trans.
    Decal 6-001 (500 rpm) .... 10° BTDC
    Decal 6-125 (625 rpm in Drive) ...... 8° BTDC
    Decal 6-032 (650 rpm in Drive) ..... 12° BTDC
    Decal 6-210 (650 rpm in Drive) ..... 11° BTDC

351W V-8 Calif.
  Auto. Trans.
    Decal 6-094 (650 rpm in Drive) ..... 10° BTDC
    Decal 6-308 (650 rpm in Drive) ..... 10° BTDC

351M V-8 49-States
  Auto. Trans.
    Decal 6-009 (650 rpm) .... 8° BTDC
    Decal 6-215 (800 rpm) .... 12° BTDC
    Decal 6-226 (800 rpm) .... 12° BTDC
    Decal 6-227 (800 rpm) .... 10° BTDC

351M V-8 Calif.
  Auto. Trans.
    Decal 6-079 (650 rpm) .... 6° BTDC
    Decal 6-087 (650 rpm) .... 8° BTDC
    Decal 6-323 (650 rpm) .... 8° BTDC

400 V-8 49-States
  Auto. Trans.
    Decal 6-006 (650 rpm) .... 10° BTDC
    Decal 6-217 (650 rpm) .... 15° BTDC
    Decal 6-218 (650 rpm) .... 16° BTDC
    Decal 6-219 (650 rpm) .... 10° BTDC
    Decal 6-252 (650 rpm) .... 18° BTDC

400 V-8 Calif.
  Auto. Trans.
    Decal 6-075 (625 rpm) .... 12° BTDC
    Decal 6-074 (625 rpm) .... 10° BTDC

460-V-8 49-States
  Auto. Trans.
    Decal 6-033 (650 rpm in Drive) ..... 14° BTDC
    Decal 6-124 (650 rpm in Drive) ..... 14° BTDC
    Decal 6-202 (650 rpm in Drive) ..... 8° BTDC
    Decal 6-220 (650 rpm in Drive) ..... 10° BTDC
    Decal 6-236 (650 rpm in Drive) ..... 10° BTDC
    Decal 6-250 (650 rpm in Drive) ..... 8° BTDC
    Decal 6-253 (650 rpm in Drive) ..... 10° BTDC

460 V-8 Calif.
  Auto. Trans.
    Decal 6-086 (650 rpm in Drive) ..... 14° BTDC
    Decal 6-321 (650 rpm in Drive) ..... 10° BTDC

INITIAL TIMING

1977

NOTE: *Distributor vacuum hoses must be disconnected and plugged.*

200 6-cyl. .... 6° BTDC
250 6-cyl. .... 4° BTDC
302 V-8 .... 2° BTDC
351W V-8 .... 14° BTDC
351M 49-States .... 10° BTDC
351M Calif. .... 6° BTDC
400 V-8 .... 10° BTDC
460 V-8 .... 10° BTDC

1978

NOTE: *Distributor vacuum hoses must be disconnected and plugged.*

200 6-cyl. 49-States
  Auto. trans. .... 10° BTDC
  Man. trans. .... 10° BTDC
200 6-cyl. Calif.
  Auto. trans. .... 6° BTDC
  Man. trans. .... 10° BTDC
250 6-cyl. 49-States
  Auto trans. .... 14° BTDC
  Man. trans. .... 4° BTDC

250 6-cyl. Calif.
  Auto trans. .... 6° BTDC

302 V-8 auto. trans.
  Cal. 8-11D-R0 (EEC)
    At cranking speed ..... 10° BTDC
    At idle (625 rpm) ...... 30° BTDC
  Cal. 8-11E-R0,
    8-11E-R10, ... 14° BTDC
  Cal. 8-11G-R0, 8-11G-R17 .. 2° BTDC
  Cal. 8-11L-R0, 8-11L-R11 (EEC)
    At cranking speed ..... 10° BTDC
    At idle (625 rpm) ...... 15° BTDC

313

Ford Motor Company

1978

Cal. 8-11M-R0, 8-11M-R10,
  8-11M-R16 .... 6° BTDC
Cal. 8-11M-R17 .... 4° BTDC
Cal. 8-11N-R0 .... 10° BTDC
Cal. 8-11P-R0, 8-11Q-R0 .. 12° BTDC
Cal. 8-11X-R0, 8-11Y-R0,
  8-11Z-R0 .... 14° BTDC
Cal. 8-31A-R0 .... 4° BTDC
Cal. 8-31B-R0 .... 6° BTDC
Cal. 8-31C-R1 .... 2° BTDC

302 V-8 man. trans.
Cal. 8-10A-R0,
  8-10A-R10 .... 10° BTDC
Cal. 8-10B-R0 .... 12° BTDC

351W V-8
Cal. 8-12B-R0 .... 10° BTDC
Cal. 8-17P-R0, 8-17Q-R0 .. 16° BTDC
Cal. 8-17X-R0 .... 8° BTDC
Cal. 8-32B-R0, 8-32B-R11 . 14° BTDC

351M V-8
Cal. 8-14A-R0 .... 9° BTDC
Cal. 8-14A-R11, 8-14A-R12,
  8-14A-R13 .... 14° BTDC
Cal. 8-14A-R16 .... 12° BTDC
Cal. 8-14B-R0 .... 12° BTDC
Cal. 8-14B-R13, 8-14B-R14,
  8-14B-R22 .... 12° BTDC
Cal. 8-14E-R14 .... 14° BTDC
Cal. 8-14F-R16 .... 12° BTDC
Cal. 8-14N-R0,
  8-14N-R13 .... 16° BTDC
Cal. 8-14N-R14 .... 14° BTDC
Cal. 8-14Q-R14 .... 16° BTDC
Cal. 8-14X-R0 .... 12° BTDC

400 V-8
Cal. 8-17A-R0, 8-17B-R0 . 13° BTDC
Cal. 8-17B-R13,
  8-17B-R19 .... 13° BTDC
Cal. 8-17C-R0, 8-17C-R11 . 13° BTDC
Cal. 8-17E-R15,
  8-17F-R22 .... 13° BTDC
Cal. 8-17G-R15 .... 10° BTDC
Cal. 8-17N-R0 .... 14° BTDC

460 V-8
Cal. 8-19A-R1 .... 8° BTDC
Cal. 8-19B, R11,
  8-19C-R11 .... 10° BTDC

460 V-8 Police
Cal. 8-19D-R0 .... 16° BTDC

IGNITION TIMING AND SPARK PLUG APPLICATION

1979

Ford/Mercury 302 V-8 (F) 49 states
Calibration 9-11E-R1/Auto. trans./
with or without A/C
  6° BTDC @ 500 rpm
  ASF-52 .048-.052

INITIAL TIMING

Ford/Mercury Police station wgn. 302 V-8 (F) 49 states
Calibration 9-11F-R1/Auto. trans./
with or without A/C
  6° BTDC @ 500 rpm
  ASF-52 .048-.052

Ford/Mercury 302 V-8 (F) California only
Calibration 9-11Q-R01/9-11Q-R10
  6° BTDC @ 500 rpm
  ASF-52-6 .058-.062

Ford/Mercury 351 V-8 (H) 49 states
Calibration 9-12G-R0/Auto. trans.
  10° BTDC @ 800 rpm
  ASF-52 .048-.052
Calibration 9-12F-R0/Auto. trans.
  15° BTDC @ 800 rpm
  ASF-52 .048-052
Calibration 9-12A-R0/9-12B-R0/
Auto. trans.
  Timing is not adjustable (equipped with EEC)
  ASF-52 .048-.052

Ford/Mercury 351 V-8 (H) California only
Calibration 9-12N-R0/9-12P-R0/
Auto. trans.
  Timing is not adjustable (equipped with EEC)
  ASF-52 .048-.052

Granada/Monarch 302 V-8 (F) 49 states
Calibration 9-11C-R0/9-11C-R1/
Auto. trans.
  8° BTDC @ 500 rpm
  ASF-52 .048-.052
Calibration 9-10C-R0/Manual trans.
  12° BTDC @ 500 rpm
  ASF-52 .048-.052
Calibration 9-10C-R10/Manual trans.
  8° BTDC @ 500 rpm
  ASF-52 .048-.052

Granada/Monarch 302 V-8 (F) California only
Calibration 9-11S-R0/Auto. trans.
  12° BTDC @ 500 rpm
  ASF-52-6 .058-.062

Granada/Monarch 250 I-6 (L) 49 states
Calibration 9-9A-R0/9-9A-R12/Auto. trans.
  10° BTDC @ 750 rpm
  BSF-82 .048-.052

Calibration 9-8A-R0/Manual trans.
  4° BTDC @ 750 rpm
  BSF-82 .048-.052

Granada/Monarch 250 I-6 (L) California only
Calibration 9-9P-R0/Auto. trans.
  6° BTDC @ 750 rpm
  BSF-82 .048-.052

Cougar/LTD II/Thunderbird 302 V-8 (F) 49 states
Calibration 9-11J-R0/Auto. trans.
  8° BTDC @ 500 rpm
  ASF-52 .048-.052

Cougar/LTD II/Thunderbird 351 V-8 (H) 49 states
Calibration 9-12E-R0/Auto. trans.
  15° BTDC @ 600 rpm
  ASF-42 .048-.052
Calibration 9-14E-R0/Auto. trans.
  12° BTDC @ 800 rpm
  ASF-52 .048-.052

Cougar/LTD II/Thunderbird 351 V-8 (H) California only
Calibration 9-14Q-R0/Auto. trans.
  14° BTDC @ 800 rpm
  ASF-52 .048-.052

Lincoln Continental/Mark V 400 V-8 (S) 49 states
Calibration 9-17F-R00/Auto. trans.
  14° BTDC @ 800 rpm
  ASF-52 .048-.052

Lincoln Continental/Mark V 400 V-8 (S) California only
Calibration 9-17P-R0/Auto. trans.
  14° BTDC @ 800 rpm
  ASF-52-6 .058-.062
Calibration 9-17Q-R0°/Auto. trans.
  16° BTDC @ 800 rpm
  ASF-52 .048-.052

Versailles 302 V-8 (F) 49 states
Calibration 8-11D-R0/Auto. trans.
  30° BTDC @ 625 rpm (not adjustable)
  ASF-52 .048-.052
Calibration 8-11L-R11/Auto. trans.
  15° BTDC @ 625 rpm (not adjustable)
  ASF-52 .048-.052

Versailles 302 V-8 (F) California only
Calibration 9-11P-R0/9-11P-R1/Auto trans.
  Timing is not adjustable
  ASF-52 .048-.052
Calibration 9-11V-R0°/Auto. trans.
  12° BTDC @ rpm
  ASF-52-6 .058-.062

SPARK PLUGS

1976

200 6-cyl. ....AU-BRF-82 .... .044

250 6-cyl. ....AU-BRF-82 .... .044
302 V-8 ......AU-ARF-42 .... .044
  AU-ARF-52 .... .044

351W V-8 ....AU-ARF-42 .... .054
  AU-ARF-52 .... .044
  AU-ARF-52 .... .054

Ford Motor Company

SPARK PLUGS

1976

351M V-8 .... AU-ARF-42 .... .044
               AU-ARF-52 .... .044
400 V-8 ...... AU-ARF-42 .... .044
               AU-ARF-52 .... .044
460 V-8 ...... AU-ARF-52 .... .044

NOTE: *Above plug types and gaps were being used at the beginning of the model year. Later engines my use different types and gaps. Always check the decal of the engine you are working on.*

1976

Many Ford engines have been changed to manifold vacuum, instead of ported vacuum. Some of these are running changes, so you may find the same engine size with either ported or manifold vacuum to the distributor. We can't tell you which is which at this time, because Ford has not made their vacuum diagrams available yet.

1977

Diaphragm type
   200 6-cyl. .... Single
   250 6-cyl.
     49-States .... Single
     Calif. .... Dual
   302 V-8
     49-States .... Single
     Calif. .... Dual
   351W V-8 .... Single
   351M V-8 .... Single
   400 V-8 .... Single
   460 V-8 .... Single

AIR PUMP SYSTEM 1976

Description of System

The pump itself is unchanged from 1975, but many of the controls are new, and the plumbing is all new.

All V-8 engines and all 6-cylinder automatic transmission engines use the air bypass valve that was introduced last year on catalyst-equipped cars. This is the valve that has the small hose connection on the end. The big difference in 1976 is that the bypass valve is now connected to ported vacuum instead of manifold vacuum.

    CAUTION: *Some Calif. 6-cylinder engines use the 1975

1977

200 6-cyl. ..... AU-BRF82 ... .050
250 6-cyl. ..... AU-BRF82 ... .050
302 V-8 .... AU-ARF52 ... .050
351W V-8 .... AU-ARF52 ... .050
351M V-8
   49-States .. AU-ARF52 ... .050
   Calif. .... AU-ARF52 ... .060
400 V-8 .... AU-ARF52 ... .050
460 V-8 .... AU-ARF52 ... .050

1978

200 6-cyl. ..... AU-BSF82 .... .050
250 6-cyl. ..... AU-BSF82 .... .050

VACUUM ADVANCE

Vacuum source
   200 6-cyl. .... Manifold
   250 6-cyl.
     49-States .... Manifold
     Calif. .... Ported
   302 V-8
     49-States Auto. trans. .. Manifold
     49-States man. trans. ... Ported
     Calif. .... Ported
   351W V-8 .... Manifold
   351M V-8 .... Ported
   400 V-8 .... Ported
   460 V-8 .... Ported

1978

Diaphragm type
   200 6-cyl.
     49-States .... Single
     Altitude .... Single
     Calif. .... Dual
   250 6-cyl.
     49-States .... Single
     Altitude .... Single
     Calif. .... Dual
   302 V-8
     49-States .... Single
     Calif.
       Cal. 8-11N-R0 .... Single

EMISSION CONTROL SYSTEMS 1976

setup, with the bypass valve connected to manifold vacuum.

In the 1976 setup the pump air is dumped to atmosphere through the bypass valve during idle and deceleration, because there is no ported vacuum at those times. The vacuum differential valve used last year is eliminated, except on those engines such as some Calif. 6-cylinder engines that still use the 1975 setup.

The temperature control for vacuum to the bypass valve is also simplified. Instead of using an electric temperature switch with a separate solenoid, a mechanical temperature vacuum switch (TVS) in the air cleaner turns the vacuum off below approximately 60° F. air cleaner tem-

302 V-8
   49-States ... AU-ARF52 .... .050
   Calif. ...... AU-ARF52-6 ...... .060
351W V-8
   49-States ... AU-ARF52 .... .050
   Calif. ...... AU-ARF52-6 ...... .060
351M V-8
   49-States ... AU-ASF52 .... .050
   Calif. ...... AU-ASF52-6 ...... .060
400 V-8
   49-States ... AU-ASF52 .... .050
   Calif. ...... AU-ASF52-6 ...... .060
460 V-8
   49-States ... AU-ARF52 .... .050

   All others .... Dual
NOTE: *Versailles 302-V8 does not use any vacuum diaphragm.*
   351W .... Single
   351M .... Single
   400 .... Single
   460 .... Single
Vacuum source
   200 6-cyl. .... Ported
   250 6-cyl.
     49-States .... Manifold
     Calif. .... Ported
   302 V-8
     49-States
      Cal. 8-11G-R0, 8-11G-R17,
      8-11M-R0, 8-11M-R16,
      8-31C-R1 .... Manifold
     All others .... Ported
     Calif. .... Ported
   351W V-8 .... Ported
   351M V-8
     Cal. 8-14B-R0, 8-14B-R22 .. Manifold
     All others .... Ported
   400 V-8
     49-States .... Manifold
     Altitude .... Ported
     Calif. .... Ported
   460 V-8 .... Ported

perature on all except the 302 and 351W V-8 with automatic transmission. On those engines, a 2-nozzle coolant-sensitive PVS is threaded into the intake manifold coolant passage.

The EGR valve also receives its vacuum from either the air cleaner temperature valve or the coolant temperature valve, so the EGR valve is also shut off when the temperature valve is closed.

The reason for dumping the pump air during idle is to prevent overheating the catalytic converter. However, this overheating only becomes a problem during prolonged idle. Idle periods of one minute or less do not overheat the converter, and the car will have fewer emissions if the pump air

Ford Motor Company

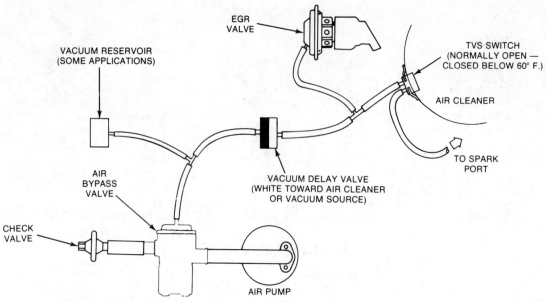

1976 air pump system, with the bypass valve connected to ported vacuum

is kept on during short idle periods. To solve this problem, a vacuum delay valve, similar to the old spark delay valve, is inserted in the hose between the TVS and the bypass valve. This vacuum delay valve traps the vacuum in the bypass valve, and takes about half a minute to a minute to bleed down. The result is that during short idle periods, the pump air keeps flowing to the exhaust, but if the idle is prolonged, pump air is dumped into the atmosphere through the bypass valve.

On some engines, a vacuum reservoir is connected into the hose between the vacuum delay valve and the bypass valve. This reservoir increases the amount of vacuum, so that it takes longer for the vacuum delay valve to bleed down, and insures that it will be a full minute before the pump air is dumped.

1976 49-State 6-cylinder manual transmission engines use a new type bypass valve, with two small hose connections. This is called a "Timed Air Bypass Valve with Vacuum Vent." Manifold vacuum connects to the hose connection on the body of the valve, and acts on the bottom side of a diaphragm. A calibrated hole in the diaphragm allows the vacuum to reach the top side of the diaphragm, so that vacuum equalizes on both sides. With the vacuum equalized, there is no force on the diaphragm, and the spring inside the bypass valve moves the plunger to the flow position, allowing the pump air to flow to the engine exhaust ports.

To get the bypass valve into the dump position, all that is necessary is to vent the top side of the diaphragm. With the top side vented, the manifold vacuum on the bottom side pulls the diaphragm and stem down, and the bypass valve goes into the dump position.

Venting of the top side of the diaphragm is taken care of on the 6-cylinder engines by an idle vacuum valve. This valve is simply a vacuum-operated vent. When vacuum is applied to the valve, the vent closes. If there is no vacuum on the valve, the vent opens. The vent is connected by a hose to the vent chamber on the air bypass valve. Vacuum does not pass through the idle vacuum valve at any time. All the vacuum does is open and close the vent. Because the idle vacuum valve is connected to ported vacuum, the vent is closed above idle, and open at idle and during deceleration.

Venting of the bypass valve is only necessary during prolonged idle. To delay the dumping for approximately half a minute to a full minute, a vacuum delay valve is inserted in the hose near the idle vacuum valve. The vacuum delay valve traps the vacuum, and it takes half a minute to a full minute to bleed down. This insures that the bypass valve only goes into the dump position during prolonged idle.

The 6-cylinder engines also use the

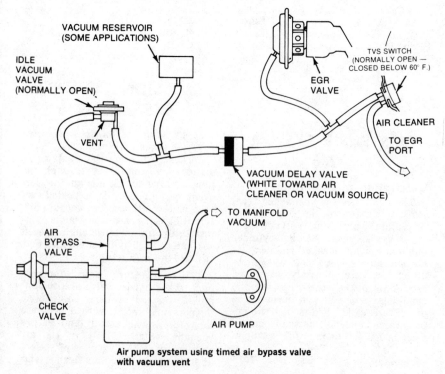

Air pump system using timed air bypass valve with vacuum vent

Ford Motor Company

same temperature control that the V-8's use. A thermal vacuum switch (TVS) in the air cleaner shuts off vacuum to the idle vacuum valve below approximately 60° F. air cleaner temperature. The TVS also shuts off the vacuum to the EGR valve.

Units in System, Single Nozzle Bypass Valve, 1976

Air pump with filter fan
Bypass valve with 1 small nozzle
Vacuum reservoir (some models)
Vacuum delay valve
Temperature vacuum switch in air cleaner
Check valves
Cylinder heads with air passages (V-8's)
Air manifolds and nozzles (6-cyl.)

Units in System, Dual Nozzle Bypass Valve, 1976

Air pump with filter fan
Bypass valve with 2 small nozzles
Idle vacuum valve
Vacuum reservoir (some models)
Vacuum delay valve
Temperature vacuum switch in air cleaner
Check valves
Cylinder heads with air passages (V-8's)
Air manifolds and nozzles (6-cyl. Calif.)
Single entry air tube (6-cyl. 49-States)

Testing and Troubleshooting System, 1976

Testing is done on each part of the system, rather than on the whole system at once. See below for tests on the individual units that have been changed for 1976.

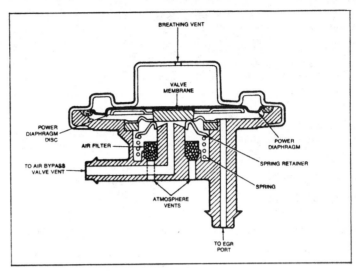

Air pump system idle vacuum valve

Repairing System, 1976

Belt tension is the only adjustment to the system. All units are replaced if defective.

Testing Single Nozzle Air Bypass Valve, 1976

This valve, with the single small hose connection nozzle on the end, is tested the same as in 1975, described on Page S-25 in Supplement No. 2. In 1975 the valve was connected to manifold vacuum. In 1976 it is connected to ported vacuum, but the valve is the same. When vacuum is applied to the small nozzle, air will flow through the valve. When the vacuum is shut off, air will exhaust out the bottom of the valve. If not, the valve is defective.

Repairing Single Nozzle Air Bypass Valve, 1976

Repairs are limited to replacement.

Testing Dual Nozzle Air Bypass Valve, 1976

This valve has two small hose connection nozzles, one on the side of the body, and the other on the side of the end cap. The nozzle on the body connects to manifold vacuum, and the nozzle on the end cap connects to the idle vacuum valve.

1. With the engine running at 1500 rpm, disconnect the small hose that goes to manifold vacuum and be sure you have full manifold vacuum at the end of the hose.
2. Reconnect the hose.
3. Disconnect the large hose between the bypass valve and the check valve, and position the hose so you can feel the blast of air coming out of it.
4. Disconnect the small hose from the end cap of the valve, and plug the opening in the valve with your finger. After a few

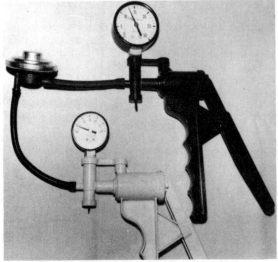

With vacuum applied to the bottom nozzle (white pump) the side nozzle (black pump) will hold vacuum also

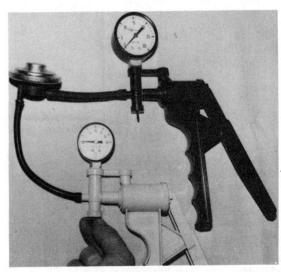

When vacuum is released from the bottom nozzle, (white pump) vacuum on the side nozzle (black pump) is also released

Ford Motor Company

seconds, air should flow through the bypass valve so that you can feel it coming out of the large hose.

5. Remove your finger from the bypass valve small hose nozzle, and the air should come out the exhaust holes on the valve. Recover the nozzle with your finger, and after a few seconds, air should stop coming out the exhaust holes and flow through the bypass valve again. If not, the bypass valve is defective.

Repairing Dual Nozzle Air Bypass Valve, 1976

Repairs are limited to replacement.

Testing Idle Vacuum Valve, 1976

1. Hold the valve with the hose connection nozzles on the bottom side. Connect a hand vacuum pump to the vent nozzle, which points out to the side, and another pump to the vacuum nozzle, which points down.
2. Pump up about 5" Hg. vacuum on the down pointing nozzle, and 15" Hg. on the side nozzle. Both pumps should hold vacuum without leaking down. If not, the idle vacuum valve is defective, but be sure you don't have any hose or pump leaks before you replace the valve.
3. Release the vacuum on the down-pointing nozzle. The side nozzle pump gauge should immediately drop to zero. If not, the idle vacuum valve is defective.

Repairing Idle Vacuum Valve, 1976

Repairs are limited to replacement of the valve. The nozzle that points down must connect to the vacuum source, and the side nozzle to the air bypass valve.

Testing Vacuum Delay Valve, 1976

1. Connect a hand vacuum pump to the colored side (opposite to the white side) of the valve. Leave the white side open. Work the pump in an attempt to pump up vacuum. If you can pump up vacuum, the valve is clogged or otherwise defective.
2. Connect a hand vacuum pump to the white side of the valve. Leave the colored side open. Operate the pump rapidly to pump up several inches of vacuum. The gauge should drop slowly to zero. If it drops instantly, or doesn't drop at all, the valve is defective.

Repairing Vacuum Delay Valve, 1976

Repairs are limited to replacement.

When installing the valve, the white side goes toward the vacuum source, and the colored or black side toward the idle vent valve or bypass valve.

Testing Temperature Vacuum Switch (TVS) 1976

1. Cool the valve in a refrigerator to 40° F. or less.
2. Connect a hand vacuum pump to the rim (smaller) nozzle on the valve.
3. Pump up at least 16" Hg. vacuum. The valve should hold without leaking.
   NOTE: *Because this valve is also the shutoff for the EGR valve, a slight leak may be enough to open the EGR valve when the engine is cold, and cause poor running, especially on those EGR systems that do not use a back pressure transducer.*
4. Warm the valve to 80° F. or more. It should open and not hold vacuum.

Repairing Temperature Vacuum Switch (TVS) 1976

Repairs are limited to replacement. The nozzle on the rim (smaller of the two) connects to the vacuum source. The center (larger) nozzle connects to the air bypass valve and EGR valve.

EXHAUST GAS RECIRCULATION, 1976, AND LATE 1975

Description of System

A running change in late 1975 was the introduction of a back pressure transducer. The transducer is installed in 1976 on almost all V-8 engines. The only one that doesn't use it is the 460 V-8 Police Interceptor. The transducer is a bleed installed in the vacuum hose that takes vacuum to the EGR valve. The base of the transducer is sandwiched between the EGR valve and its mounting pad, so that exhaust pressure can go through a small tube to the transducer. The exhaust pressure present in the exhaust system at part throttle closes a bleed against spring pressure so that the EGR valve receives whatever vacuum is available through its hose. When there is reduced exhaust pressure at idle or part throttle light loads, the exhaust pressure is weak, and the spring in the transducer opens the bleed so that the EGR operating vacuum is bled off. The result is that the EGR valve is only open at part throttle when it is needed to reduce NOx in the exhaust. Some 460 V-8 engines do not have enough back-pressure to operate the transducer, so a restrictor plate is inserted in the exhaust system.

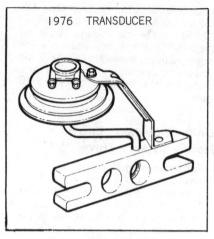

1976 EGR transducer

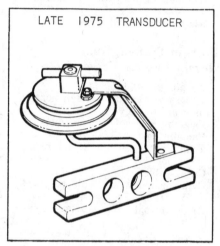

Late 1975 EGR transducer

The transducer has two hose connection nozzles, one of which connects to the vacuum source, and the other to the EGR valve. It is important that the nozzle with the small restriction be connected to the vacuum source. The mounting position of the transducer does not indicate which nozzle has the restriction. The only way to be sure is to use a small flashlight and look into each nozzle. Then connect the restricted nozzle to the vacuum source.

The temperature control on all 351M, 400, and 460 V-8's, 302 and 351W V-8 with manual transmission, and the 49-State 6-cylinder engines is changed from a switch that threads into the coolant passage to one that mounts in the air cleaner. This new switch, called the Temperature Vacuum Switch (TVS) also controls the vacuum to the air bypass valve. The switch is closed below approximately 60° F. so that exhaust gas will not recirculate on a cold engine.

302 and 351W V-8's, with automatic transmission, and Calif. 6-cylinder engines still use the PVS that threads into a coolant passage. This PVS also controls vacuum to both the

EGR valve and the air bypass valve.

Another new EGR control for 1976 is the Wide Open Throttle (WOT) valve, used only on the 400 V-8, to be sure that the EGR valve is closed at wide open throttle. On most engines, there isn't any vacuum at wide open throttle, so the EGR valve closes. But the 400 V-8 is under-carbureted with a small 2-bbl. for better mileage. At wide open throttle there is still several inches of vacuum at the EGR port, so the WOT is necessary to be sure exhaust gas does not recirculate at wide open throttle, for maximum power.

Vacuum from the temperature switch goes through the WOT, and then to the backpressure transducer and the EGR valve. A nozzle on the bottom of the WOT connects to venturi vacuum at the carburetor. The venturi vacuum opens a bleed against spring pressure inside the WOT. During idle, part throttle, and deceleration, venturi vacuum is not very strong, and the spring in the WOT keeps the bleed closed. But at wide open throttle, venturi vacuum is high, and moves the diaphragm inside the WOT to open the bleed. When the bleed is open, the vacuum supply to the EGR valve is weakened so that the EGR valve closes.

Testing and Troubleshooting System, 1976

The system test on transducer-equipped engines is the same as on older engines. To test for exhaust gas recirculation, apply manifold vacuum or hand pump vacuum directly to the EGR valve with the engine idling. If the engine dies or runs rough, you know the exhaust gas is recirculating. If not, the EGR valve or a passageway is plugged, and the valve will have to be removed and the blockage removed.

To check for system operation, make sure the engine is at normal operating temperature so that the temperature vacuum switch will be open. With the engine idling in Neutral, open the throttle to about the half-throttle position and watch for movement of the EGR valve stem. If the stem moves, the system is working okay.

CAUTION: *Open the throttle only enough to make the EGR valve stem move. Do not exceed 3000 rpm. If the stem does not move, check the amount of vacuum at the EGR valve hose. You should have at least 4" Hg. at half throttle. If not, check for a plugged EGR port in the carburetor throat, or a leaking or plugged hose. If the vacuum is okay, check the individual units as shown below.*

Units in System, 1976

EGR valve
Carburetor spacer
Temperature-controlled vacuum valve (PVS type) used on all 302 and 351W auto. trans. V-8's, and Calif. 6-cylinder engines
Temperature vacuum switch (in air cleaner) used on all 351M, 400, and 460 V-8's, 302 and 351W man. trans. V-8's, and 49-State 6-cylinder engines
Wide open throttle (WOT) valve used on 400 V-8 only.

Repairing System, 1976

Cleaning of the EGR valve, the transducer, and the passageways is recommended, if necessary. Solvent, or air pressure, must not be used on the diaphragm of either the EGR valve or the transducer to avoid damage. There are no adjustments, and all units are replaced when defective.

Testing Back Pressure Transducer

1. With the engine at normal operating temperature, run the engine at a fast idle by putting the throttle on the high step of the fast idle cam.
2. Remove the hose from the vacuum side of the transducer and check the vacuum at the hose with a gauge. Then reconnect the hose.
3. Disconnect the hose from the EGR valve and connect a vacuum gauge to the hose. The vacuum reading should be the same as at the vacuum side of the transducer, or within 2" Hg. If the vacuum is not within 2" Hg. there may be a clogged exhaust passage, a wrong hose connection, or a leaking transducer.

This restrictor is in the exhaust pipe on California 460 V-8's to create enough back pressure to operate the back pressure transducer

Off the car, the transducer can be checked by plugging one hose connection with your finger and using a hand vacuum pump on the other. The bleed valve should be open, and you should not be able to pump up any vacuum. If you can pump up vacuum, it means the bleed valve is clogged, and the transducer must be replaced.

To close the bleed, seal one side of the exhaust tube opening with your hand, and use mouth pressure on the other side. You should then be able to pump up vacuum, but the vacuum should drop to zero as soon as you remove the mouth pressure.

On the transducers we have tested, pressure applied to the exhaust tube opening does not close the bleed completely. There is still a little leakage. However, the leakage is so slight that it will allow you to pump up vacuum if you work the pump rapidly.

CAUTION: *Do not use air hose pressure on the transducer. It may rupture the diaphragm.*

If the bleed will not close from mouth pressure, the transducer is

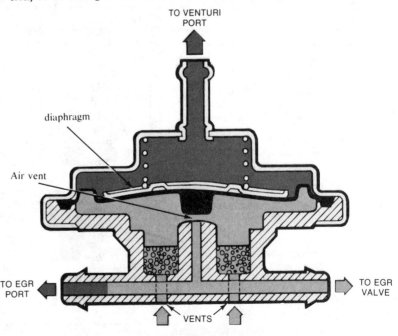

Wide open throttle valve

Ford Motor Company

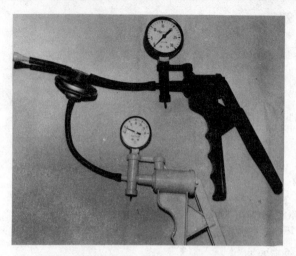

With vacuum applied to the venturi nozzle (white pump the vent is open and the black pump is unable to pump up any vacuum. Note the short hose and plug that must be used for this test

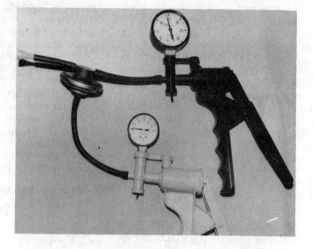

When vacuum on the venturi nozzle is released (white pump) the vent closes and the black pump can then get some vacuum

probably clogged, and must be replaced.

Repairing Back Pressure Transducer, 1976

Repairs are limited to replacement. One of the hose connection nozzles on the transducer has a restriction that you can see if you look in the end of the nozzle with a small flashlight. The nozzle with the restriction must connect to the vacuum source. The other nozzle connects to the EGR valve.

Testing Temperature Vacuum Switch (TVS) 1976

1. Cool the valve in a refrigerator to 40° F. or less.
2. Connect a hand vacuum pump to the rim (smaller) nozzle on the valve.
3. Pump up at least 16" Hg. vacuum. The valve should hold without leaking.
   NOTE: *Because this valve is also the shutoff for the EGR valve, a slight leak may be enough to open the EGR valve when the engine is cold, and cause poor running, especially on those EGR systems that do not use a back pressure transducer.*
4. Warm the valve to 80° F. or more. It should open and not hold vacuum.

Repairing Temperature Vacuum Switch (TVS) 1976

Repairs are limited to replacement. The nozzle on the rim (smaller of the two) connects to the vacuum source. The center (larger) nozzle connects to the air bypass valve and EGR valve.

Testing Wide Open Throttle (WOT) Valve, 1976

1. Connect a hand vacuum pump to the single (venturi) nozzle on

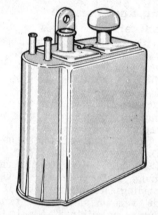

Carbon canisters for 1976 are made of plastic

one side of the valve.
2. Connect a second hand vacuum pump to one of the two nozzles that are in line on the other side of the valve.
3. Plug the other inline nozzle with a piece of hose and a golf tee.
4. Pump up 15" Hg. vacuum on the inline nozzle. The vacuum should hold without leaking. If not, the WOT is defective.
5. Pump up 5" Hg. vacuum on the venturi nozzle. The other pump should drop immediately to zero. If not, the WOT is defective

Repairing Wide Open Throttle (WOT) Valve, 1976

Repairs are limited to replacement. The single nozzle that points out from the valve must be connected to venturi vacuum. The other two nozzles are connected to the vacuum source (TVS), and to the backpressure transducer, in either direction.

COMPUTER CONTROLLED TIMING 1976

The CCT system provides an addi-

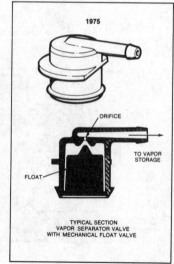
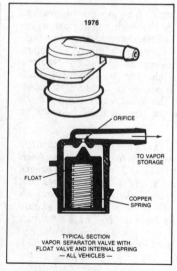

1975 and 1976 vapor separator valves

Ford Motor Company

← CALIBRATION NUMBER
← REVISION LEVEL

Ford vacuum circuits are listed in the manual under calibration number and revision level for 1973 and later engines. This is the label used on 1974 and later engines. See below for the label used on 1973 engines

sary to prevent detonation.

The CCT system was held up in late 1975, without ever having been installed on any 1976 cars. Introduction is planned for some time in the future.

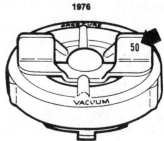

1976 Ford gas tank caps are a different color on the underside, and have the number "50" stamped either on the underside or the tang

This label, attached to the side of the valve cover, is used on 1973 Ford engines

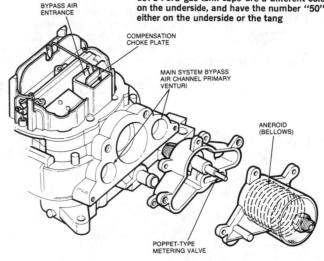

The altitude compensation is built onto the side of the 4-bbl. carburetor on 460 engines that are sold in the Denver, Colorado, high altitude area

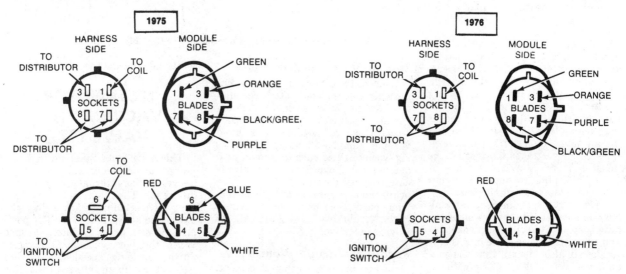

The wiring for the 1976 Ford ignition is changed from 1975, as shown here

321

tional six engine degrees advance or four degrees retard from the basic timing setting. This additional advance or retard is on top of whatever advance there happens to be in the centrifugal or vacuum unit. The additional advance comes from an extra magnetic pickup in the distributor. The retard comes from an electronic delay built into the control unit. The control unit is connected to sensors that sense engine temperature, manifold vacuum, and engine rpm. The control unit switch to the 6° advance pickup whenever the engine can use it to provide better running or improved gas mileage. It switches to the normal pickup at other times, and goes into 4° retard whenever necessary to prevent detonation.

The CCT system was held up in late 1975, without ever having been installed on any 1976 cars. Introduction is planned for some time in the future.

VAPOR CONTROL 1976

A new all-plastic carbon canister is used in 1976. The vapor separator now has a spring behind the float for a positive shutoff in case of a rollover. The vapor separator still plugs into a grommet on top of the fuel tank, but the separator with the spring is slightly longer than the older design.

Fuel tank caps are still the pressure-vacuum type, but it takes more pressure to open the relief. This helps to keep the fuel in the tank in case of a rollover. The underside of the 1975 caps was amber colored. The underside of the 1976 caps is silver colored. Also, the 1976 caps have the number "50" stamped into the cap metal.

ROLLOVER FUEL CONTROL 1976

All fuel pumps have a new valve design in the pump outlet to prevent fuel flow in case of a rollover. The pump is designed by Carter, and is a sealed unit that is discarded when worn out. The new valve is not removable or servicable. In addition, all vapor separator valves now have a spring behind the float, and all fuel tank caps now have a slightly higher pressure relief. All of these changes are made to keep fuel from leaking out if the car gets in a rollover accident.

2300 4-cyl. Calif.
  Auto. Trans.
    Decal 6-096 750
    Decal 6-313 750
  Man. Trans.
    Decal 6-096 850
    Decal 6-312 850
2800 V-6 49-States
  Auto. Trans.
    Decal 6-128 700
  Man. Trans.
    Decal 6-128 850
    Decal 6-216 850
    Decal 6-229 850
2800 V-6 Calif.
  Auto. Trans.
    Decal 6-066 700
    Decal 6-100 800
    Decal 6-325 800
  Man. Trans.
    Decal. 6-066 950
    Decal 6-100 850
    Decal 6-309 850

EMISSION CONTROL SYSTEMS 1977

AIR PUMP SYSTEM 1977

California 302 V-8 engines use a new system of dumping the vacuum to the air bypass valve so the pump air will divert to the atmosphere during idle or wide open throttle. A new Vacuum Vent Valve is used with a check valve and a delay valve. Above idle, ported vacuum from the carburetor keeps the vacuum vent valve closed. The vacuum can then pass through the check valve and act on the air bypass valve so that it functions normally.

At idle or wide open throttle there is not enough vacuum to keep the vacuum vent valve closed. It then opens, and allows air to enter the system, destroying the vacuum and letting the air bypass valve go into the bypass position. A delay valve makes the vacuum leak down slowly, so that the air pump does not bypass everytime the driver takes his foot off the throttle. It takes a few seconds of idling before the vacuum all leaks out and the system goes into bypass.

Because the vacuum vent valve allows outside air into the system every time it opens, it keeps the check valve and connecting hoses free of gasoline vapors. It also insures that the vacuum reservoir will not fill up with gasoline and reduce its capacity to store vacuum.

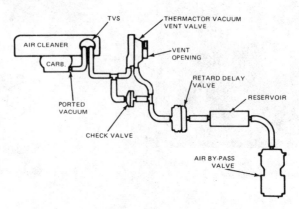

Typical air pump vent valve installation (© Ford Motor Co.)

Testing Vacuum Vent Valve

Connect the cover side of the valve, (usually colored black) to a continuous vacuum source, such as a running engine. Connect the vent side (usually colored white) to a hand vacuum pump, and pump up 10 in. Hg. vacuum. The vacuum should hold and not leak down. If not, the valve is defective. Disconnect the black side of the valve from the continuous vacuum source. The hand pump vacuum gauge should drop to zero. If not, the valve is defective.

Repairing Vacuum Vent Valve

Repairs are limited to replacement. The black nozzle should be connected to the source of vacuum, and the white side toward the bypass valve.

DISTRIBUTOR VACUUM VENT VALVE 1977

This valve looks the same as the air pump vacuum vent valve described above, but the internal construction is different. The distributor vacuum vent valve is connected in the hose between the spark port (or EGR port) and the distributor vacuum advance. When the port vacuum is high, as during cruising speed, the vacuum passes through an internal restrictor. It takes anywhere from 16 seconds to more than a minute for the vacuum

Ford Motor Company

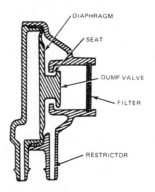

Air pump vacuum vent valve
(© Ford Motor Co.)

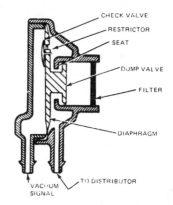

Distributor vacuum vent valve
(© Ford Motor Co.)

at the distributor to rise from zero to 8 in. Hg. When the port vacuum is low, as during heavy acceleration, deceleration, or idle, the valve opens a vent, which lets the vacuum in the system leak out, and the distributor advance returns immediately to the neutral or no-advance position.

The vent valve is positioned on a bracket so it is higher than the carburetor, to prevent fuel from accumulating in the valve. Any vapors that do build up in the valve when it is not under vacuum are blocked from traveling down the hose to the distributor by an internal check valve. When the valve does receive vacuum, the check valve opens and the gasoline vapors are drawn out.

Also in the system is a 2-nozzle ported vacuum switch, connected between the EGR port and the vacuum vent valve. The PVS blocks the vacuum below 95° F. engine coolant temperature. Thus, there is no vacuum advance at all when the engine is cold.

A 3-nozzle PVS is also used. It is connected between the vacuum vent valve and the distributor. When the engine coolant temperature goes over 225° F. the 3-nozzle PVS sends full manifold vacuum to the vacuum advance. At normal operating temperatures EGR port vacuum passes through the 3-nozzle PVS without being affected.

Testing Distributor Vacuum Vent Valve

Connect the cover side of the valve (the side with the code number) to a continuous source of vacuum, such as a distributor tester. The amount of vacuum applied to the valve must be held at a constant 10 in. Hg. If a distributor tester is not available, a running engine may be used, but some kind of regulator valve must be used to keep the vacuum at exactly 10 in. Hg.

Connect a vacuum gauge to the vent side of the valve, using a hose that is 24 inches long. As the vacuum gauge is connected start timing the number of seconds it takes for the gauge to reach 8 in. Hg. Check the time against the following chart.

Code No. on Valve	Minimum Time	Maximum Time
20	16 sec.	36 sec.
40	28 sec.	68 sec.

If the number of seconds does not fall within the correct range, the valve is defective.

Repairing Distributor Vacuum Vent Valve

Repairs are limited to replacement. There are two valves, code 20 and code 40. Be sure to replace with the same valve, or check the Ford parts book for the correct application.

CAUTION: *The distributor vacuum vent valve and the air pump vacuum vent valve have identical housings. Be sure to use the right valve. If in doubt, make the time test above The air pump vacuum vent valve will not pass the test.*

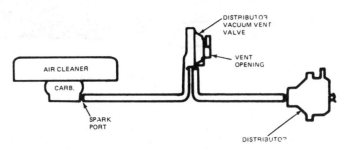

Typical distributor vacuum vent valve installation (© Ford Motor Co.)

POSITIVE CRANKCASE VENTILATION 1977

Some engines require that the PCV valve be changed from standard-flow to high-flow at either 22,500 or 30,000 miles. The engine decal identifies those engines that require the change, and gives the number on the valve. Most engines now use a PCV valve that connects to the hose through a plastic elbow. New valves come with the elbow separate from the valve. To avoid breaking the plastic elbow, it should be soaked in warm water for an hour or more, before attempting to push it onto the valve.

The color of the elbow is yellow for 6-cylinder engines, and black for the V-8's. High flow valve elbows are the same color, except that they have a white plastic collar.

VARIABLE VENTURI CARBURETOR 1977

A lot of carburetors are said to be "new," but are really nothing more than slightly modified versions of the same basic carburetor we have known for many years. The Motorcraft 2700 VV is certainly not in that category. This is the first really new and different carburetor that has been put on a production car in many years.

The 2700 VV does not have the usual venturi in the carburetor throats. The carburetor is a 2-barrel, and at the top of each throat is a black plastic venturi valve. A link from the valve is connected to a large diaphragm on the side of the carburetor. Vacuum to operate the diaphragm comes from the throat of the carburetor, between the throttle plates and the venturi valves. A spring keeps the venturi valves closed when there is no vacuum to open them.

Under actual running conditions both the throttle valves and the venturi valves are almost closed at idle. Just enough air enters to keep the engine running. The air that enters past the venturi valve picks up fuel from around a horizontal metering rod that is attached to the venturi valve. There is also an idle system that supplies a fuel-air mixture for idling.

When the throttle is opened, the area between the throttle plates and the venturi valves receives the high vacuum that is under the throttle plates. This high vacuum, called "control vacuum" acts on the control vacuum diaphragm and pulls the venturi valves open. As the valves open, they pull the tapered metering rod out so

323

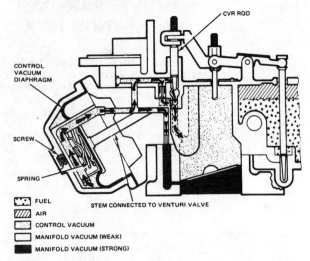

Carburetor control vacuum circuit (© Ford Motor Co.)

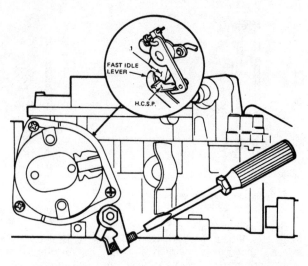

Fast idle speed adjustment (© Ford Motor Co.)

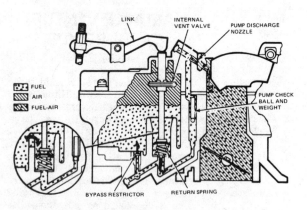

Accelerator pump circuit (© Ford Motor Co.)

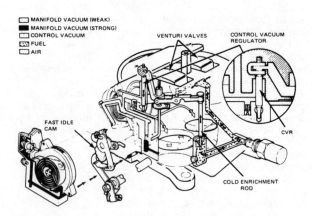

Cold enrichment circuit (© Ford Motor Co.)

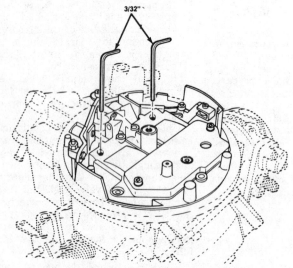

Idle trim adjustment (© Ford Motor Co.)

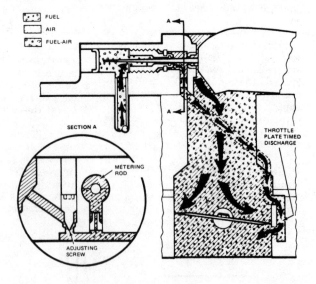

Idle trim system (© Ford Motor Co.)

Ford Motor Company

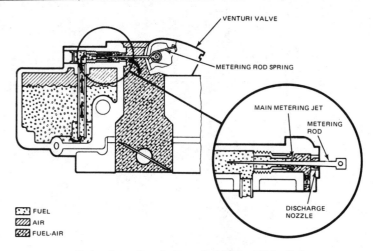

Carburetor main metering system (© Ford Motor Co.)

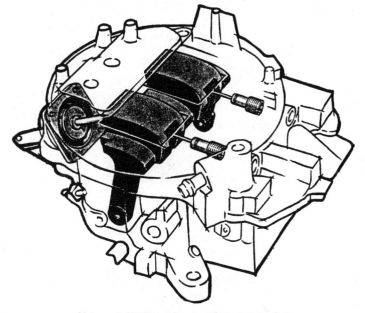

Motorcraft 2700VV carburetor (© Ford Motor Co.)

that more fuel can flow. Because the amount of venturi valve opening is controlled by vacuum, the engine always gets exactly the amount of air and fuel that it needs. It is possible to run a much leaner mixture without stumbling and dying, than would be possible in a normal carburetor.

The most important part of the new design is that this is a *solid fuel* carburetor. There are no air bleeds in the main metering system. As the fuel comes out of the passage around the metering rod and enters the venturi, it is what the engineers call "solid fuel" which has not had any air mixed in up to that point. All the mixing of air with fuel takes place in the venturi itself. The result is that there are no air bleeds to clog up and change the air-fuel ratio as the carburetor accumulates dust and dirt on the car.

Solid fuel carburetors are only possible if the air flow through the venturi is kept extremely high. Ideally, the airflow should exceed the speed of sound. It is not quite that high in the 2700 VV, but almost.

The 2700 VV is an interesting new design, but it is not a high performance carburetor. It was designed to solve the problem of emission control versus driveability. You can make any carburetor lean to reduce emissions, but you may not be able to drive the car. Hopefully, the 2700 VV will give good driveability and also meet the emission levels now and in the future.

DURA-SPARK I AND II 1977

The solid state ignition used in 1976 has been slightly modified, and is now called Dura Spark II. It is used on all 49-State and altitude engines. An all new system is called Dura Spark I. It is used only on California V-8 engines. The hardware on both of the systems looks much the same. The major differences are in the electronics. Dura Spark I has a different ignition coil, with primary connections that will only fit the matching connector. It is not possible to use the Dura Spark I coil in either Dura Spark II or the older systems.

The main reason for changing the systems was to increase the available voltage for more reliable spark plug firing. The older ignition would put out 26,000 volts. Dura Spark II will put out 36,000 volts, and Dura Spark I will go all the way to 42,000.

Dura Spark II uses a ballast resistor that has been reduced from the 1976 value of 1.35 ohms to 1.10 ohms. This raises the voltage in the system and thus increases the available secondary voltage. Dura Spark I does not use any ballast resistor at all. In addition, the coil primary resistance in the new coil is only 0.75 ohms. It was formerly 1.2 ohms in the 1976 system.

Special service procedures must be followed with both Dura Spark systems to prevent the spark from jumping to ground. Whenever a spark plug wire is removed, from either the cap, spark plug, or coil tower, the inside of the terminal boot must be greased with silicone grease before reconnecting. If possible, secondary wires should not be removed. Timing lights should be the induction type, using a connector that goes around the wire instead of connecting to the end. The direct connection type may false trigger the light, giving wrong timing.

Whenever a new rotor is installed, the brass parts of the rotor must be coated with silicone grease to approximately $\frac{1}{8}$ inch thick. As this silicone grease ages, it looks discolored and dirty, but this has no effect on performance, and the grease should not be removed.

If it is necessary to measure available secondary voltage by removing a plug wire, the following wires must not be removed.
V-8 engines: No. 1 or 8
6-Cyl. engines: No. 3 or 5

Typical timing pointer and probe receptacle (© Ford Motor Co.)

325

Ford Motor Company

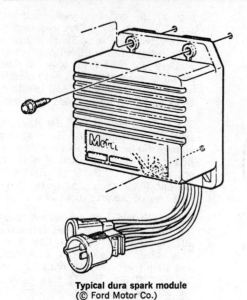

Typical dura spark module
(© Ford Motor Co.)

V-8 dura spark system (© Ford Motor Co.)

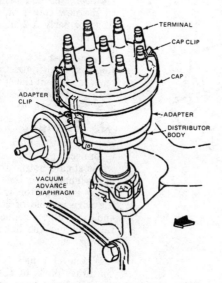

Typical dura spark distributor
(© Ford Motor Co.)

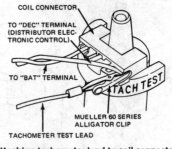

Attaching tachometer lead to coil connector
(© Ford Motor Co.)

ELECTRONIC ENGINE CONTROL SYSTEM 1978

This system, known as Electronic Engine Control, or EEC, is installed on all 1978 Versailles 302 V-8 engines. The system controls engine timing, exhaust gas recirculation, and air pump airflow. Seven sensors feed information into an electronic box mounted on the passenger side of the firewall near the steering column. The box, called an electronic control assembly, takes all the information, processes it, and then electrically controls the timing, EGR, and air pump airflow. First, let's look at the 7 sensors and how they do their job.

Inlet Air Temperature Sensor

Mounted in the air cleaner can, near the air inlet, the temperature sensor has

EMISSION CONTROL SYSTEMS 1978

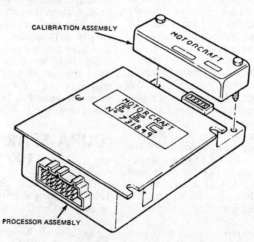

Electronic Control Unit (ECU)

Ford Motor Company

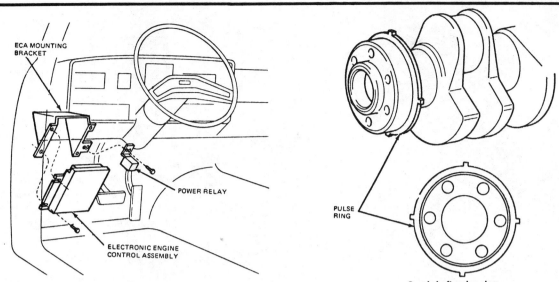

Power relay and ECA attachment

Crankshaft pulse ring

two wires connected to it, so current can flow from the electronic control assembly to the sensor and back. The sensor is a thermistor design, which changes resistance as the temperature changes. This affects the amount of current flowing, so that in effect the electronic control assembly knows at all times the temperature inside the air cleaner. The action of the sensor is opposite to temperature changes. In other words, as the temperature goes up, the sensor resistance goes down, and vice versa.

The information transmitted by the sensor is used by the electronic control assembly to determine the right setting for spark timing and air pump airflow. At high temperatures, the spark advance is less, to help prevent pinging.

The sensor is held in place by a clip inside the air cleaner. Resistance of the sensor should be between 6,500 and 45,000 ohms, as measured at the disconnected plug with an ohmmeter.

Throttle Position Sensor

This sensor is a variable resistor mounted on the carburetor and connected to the end of the throttle shaft. For every position of the throttle, there is a corresponding position of the sensor which gives a different current flow to the electronic control assembly. Using a throttle position sensor is almost the same as using a vacuum reading. The electronic control assembly combines the information on throttle position with information from other sensors, and then controls the spark advance, EGR flow rate, and air pump airflow.

The throttle position sensor is adjusted by loosening the mounting screws and rotating the sensor until the correct readings appear on a system analyzer. Although the sensor is a resistor, and could possibly be set using an ohmmeter, Ford does not recommend this. They want the sensor position adjusted with the analyzer to a specific voltage reading. This is more accurate than using an ohmmeter.

Crankshaft Position Sensor

Probably the most important sensor, it is mounted at the rear of the cylinder block, in front of the torque converter. If this sensor, or the wires to it are defective or damaged, the engine will not run. A pulse ring, permanently attached to the crankshaft, rotates just under the tip of the sensor. Every time one of the lobes on the ring pases by the sensor tip it creates a magnetic pulse in the electromagnet inside the sensor. With this signal, the elecronic control assembly knows the engine rpm and the position of the crank, so it can set the spark timing and EGR flow rate to exactly what the engine needs.

The signals from the crankshaft position sensor and the throttle position sensor are combined in the electronic control assembly to determine the load on the engine. This is exactly the same as taking a vacuum reading from the intake manifold, but it eliminates the need for any vacuum diaphragm or connecting hose.

A conventional engine gets its spark timing directly from the distributor, which is driven by the camshaft and timing chain. If the chain wears, the timing retards, so the timing must be periodically reset. This does not happen with the EEC system, because the timing is determined directly from the position of the crankshaft. There is never any need to adjust timing on the EEC system. An engine that is completely worn out will still have perfect timing.

A conventional engine also has problems with wear in the vacuum advance and centrifugal advance systems. The vacuum and centrifugal advance on the EEC engine is all controlled by the electronic control assembly. So it will also remain perfect over the life of the engine.

The crankshaft position sensor is held in place by a clamp and screw. Near the tip of the sensor is an "O" ring that keeps engine oil from escaping. There are no adjustments on the sensor.

Coolant Temperature Sensor

It's just an ordinary coolant temperature sensor with two wires connected to it. But it has a big effect on how the engine runs. It is mounted on the water passage at the left rear of the intake manifold. Its resistance varies with the coolant temperature, with high resistance at low temperatures, and vice versa. The sensor is connected to the electronic control assembly. When the engine is cold, the electronic control assembly cuts back on the amount of exhaust gas that is allowed to recirculate, or maybe eliminates the EGR entirely. If the engine is overheated because of long idling, the electronic control assembly advances the spark timing to increase the idle speed and help cool down the engine. If the coolant temperature sensor is damaged or defective, both EGR flow rate and spark timing may be affected.

The sensor is easily checked with an ohmmeter. At normal operating temperature, 160-220°F.) the resistance at the disconnected plug should be between 1,500 and 6,000 ohms. The sensor is removed by unscrewing it from the water passage with a wrench. Relieve the pressure first to avoid burns.

EGR Valve Position Sensor

This sensor is part of the EGR valve, which is mounted on a carburetor spacer. Wires connect the sensor to the electronic control assembly, to indicate the position of the EGR valve metering rod. This indirectly tells the electronic control assembly how much exhaust gas is recirculating. The electronic control assembly compares this information with what it receives from the other sensors, and then increases or decreases the exhaust gas recirculation rate with other controls that will be explained later. It is important to realize that the EGR valve position sensor is only a sensor. It does not actually move the metering rod or directly change the flow rate.

The sensor can be checked with an

Ford Motor Company

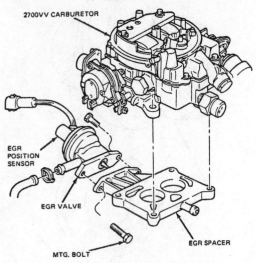

EGR valve and spacer

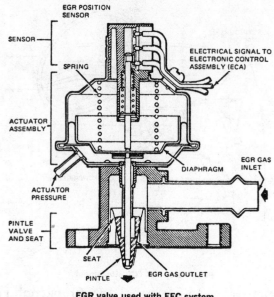

EGR valve used with EEC system

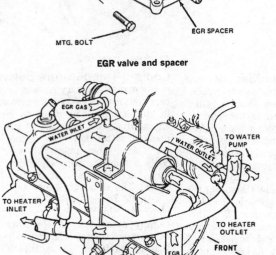

EGR gas cooler

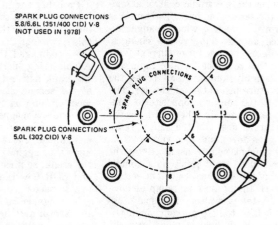

Spark plug wire connections

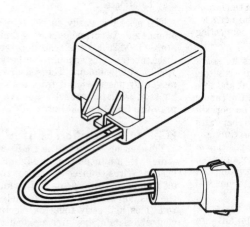

Barometric Pressure (BP) Sensor

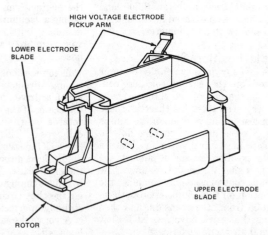

Distributor rotor

ohmmeter, after disconnecting the plug. There are three wires connecting to the sensor. The ohmmeter reading between the orange-white and black-white wires should be between 2800 and 5300 ohms, with the engine off. Between the orange-white and brown-light green wires, the ohmmeter should read 350 to 940 ohms, with the engine off. If the readings are outside these limits, the entire EGR valve must be replaced. There are no adjustments on the sensor or the valve.

Manifold Absolute Pressure Sensor

Manifold absolute pressure is defined as barometric pressure minus manifold vacuum. The sensor is connected to intake manifold vacuum. In effect, it simply measures intake manifold vacuum and sends a signal through wires to the electronic control assembly. The electronic control assembly uses this information, along with data from other sensors, to set the spark advance and EGR flow rate.

Ford Motor Company

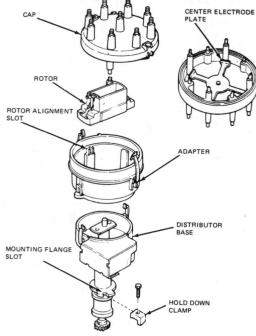
EEC distributor cap and rotor

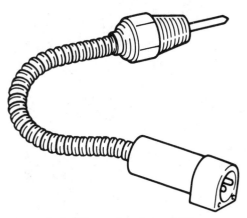
Coolant Temperature Sensor (ECT)

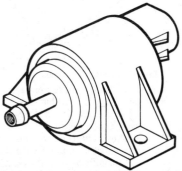
Manifold Absolute Pressure (MAP) Sensor

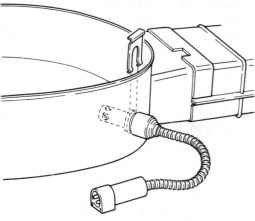
Inlet Air Temperature (IAT) Sensor

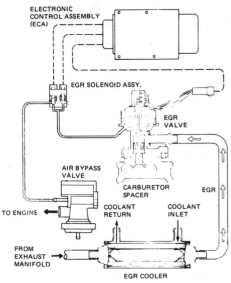
Typical EGR system used with EEC system

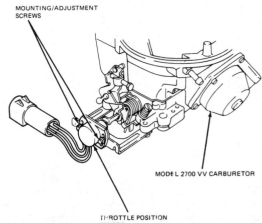
Throttle positioner Sensor

The MAP sensor is at the left rear of the intake manifold, next to the rocker cover. Ford does not give a simple ohmmeter testing procedure, preferring that the system analyzer be used. There are no adjustments to the MAP sensor.

Barometric Pressure Sensor

This sensor has only one purpose, to tell the electronic control assembly the altitude at which the car is operating.

At high altitude, the EGR flow rate is cut back to keep good engine performance. The sensor connects by wires to the electronic control assembly. A simple ohmmeter test is not provided for the barometric sensor. It must be tested with the system analyzer. There are no adjustments.

EEC Distributor

Because the electronic control assembly takes care of all timing and advance, the distributor has no vacuum advance, no centrifugal advance, and no timing mechanism. The distributor is nothing more than a rotor and some electrodes that distribute the spark to the correct spark plug wires.

Ford Motor Company

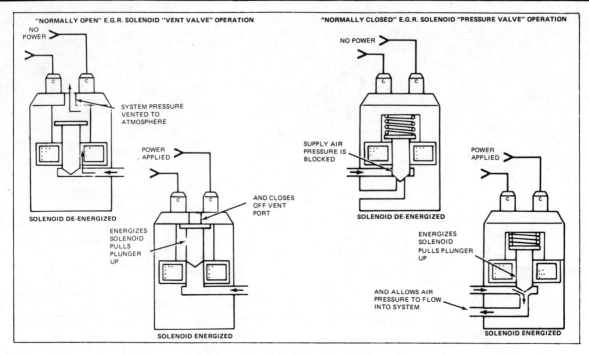

Vent and pressure solenoid

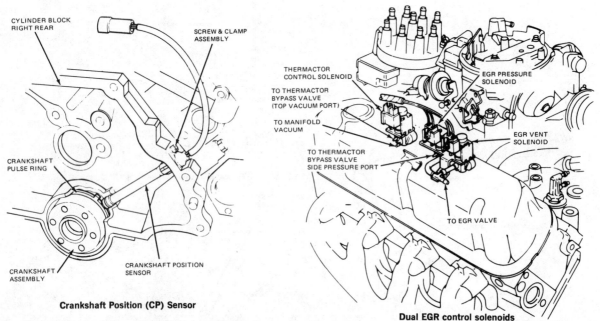

Crankshaft Position (CP) Sensor

Dual EGR control solenoids

The distributor cap and rotor are a completely new two-level design. Four of the plug wire electrodes are at one level, and the other four are a little bit lower. This puts more distance between the electrodes, and helps prevent crossfiring. In addition, a double-ended rotor is used, so that the spark jumps first from one end of the rotor, and then from the other end. This gives additional separation to prevent crossfiring. The construction of the rotor is such that the spark current actually has to jump twice to get to the plug wire electrode in the cap. It jumps once from the center tower to the pick up drum of the rotor, and a second time from the tip of the rotor to the cap electrode. When you look inside the cap, you will see a 4-spoke electrode plate connected to the center tower, and a rotor with two pickup arms. This plate is positioned so that when the correct rotor pickup arm is over one spoke, the other rotor pickup arm is between the spokes. This was also done to decrease the possibility of crossfire.

Because the distributor has nothing to do with timing, there are no adjustments for distributor position. The clamp and bolt at the base of the distributor fit into a slot, locking the distributor in one position. But there is a rotor position adjustment. A slot in the rotor must be opposite a slot in the housing, when number one cylinder is at top dead center of the compression stroke. This adjustment is made by removing the clamp bolt, lifting up the distributor, and meshing the drive gear teeth in a different position so that the slots line up. After the rotor is as close as possible this way, the rotor mounting screws can be loosened and the rotor moved to bring it into exact alignment with the slot. Ford has an official tool T78P-12200-A which fits into the rotor and housing slots at the same time and holds the rotor so it doesn't move while the screws are tightened. The tool is available from Owatonna Tool Company, Owatonna, Minnesota.

Because of the unique 2-level construction of the distributor cap and rotor, the plug wires do not fire in succession around the cap. The rotor fires first from one end, and then the other. This means that although the rotor only turns 45° between firings, the actual firing succession jumps back and forth across the cap. The numbers on the cap are the numbers of the cylinders, and must be connected to those cylinders. If you try to connect the wires in firing order succession around the cap, the engine will not run. Also, there are two sets of numbers on the cap. The inner ring of numbers is for the 302 V-8. The outer ring is for the 351 or 400 V-8, if it ever gets the EEC system. Because four of the numbers are different, the engine will not run if the plug wires are connected according to the wrong number ring.

A suprising thing about this EEC system is that there is no timing setting on the engine. The timing is done by the crankshaft positon sensor and the electronic engine control. At one end of the electronic engine control there is a small trim adjustment, but it only changes the timing a few degrees. The trim adjustment is sealed, and if the seal is broken, free replacement of a defective electronic engine control under the warranty may be refused.

Air Pump Airflow Control

A vacuum solenoid, mounted to the left of the intake manifold, is connected by wires to the electronic engine control. The solenoid vacuum path is a normally closed design. The electric coil in the solenoid is energized to open the vacuum passage by current from the electronic engine control. In the open position, the solenoid allows intake manifold vacuum to open the air bypass valve so that air from the pump can flow to each engine exhaust valve. When the solenoid is deenergized, the valve plunger moves to the closed position, shutting off the vacuum to the bypass valve, which then goes into the dump position. Air pump air is then diverted to the atmosphere.

The electronic engine control turns the solenoid on or off according to signals it receives from the inlet air temperature sensor, and the throttle position sensor.

The hoses should be connected to the solenoid with the lower outlet to manifold vacuum, and the upper outlet to the bypass valve.

Testing the solenoid can be done by idling the engine with a vacuum gauge connected to the upper outlet, after removing the bypass valve hose. When running the engine above idle in PARK, you should get a full manifold vacuum reading on the gauge. Let the engine come down to idle, and in a few seconds the solenoid should close, and the gauge should drop to zero. There is a delay built into the system so the air pump does not dump until the engine has been at idle for a few seconds. There are no adjustments on the solenoid. It must be replaced if it doesn't work right.

Exhaust Gas Recirculation System

The EGR valve is air-presure operated. A hose connects to the side of the air bypass valve and brings air pump air pressure to the EGR valve. A dual vacuum solenoid setup is in the hose between the bypass valve and the EGR valve. The two solenoids are mounted together on a bracket next to the left rocker arm cover, and connected by wires to the electronic control assembly. The rear solenoid is the vent solenoid. When the vacuum valve inside the solenoid is open, air pressure from the air pump is allowed to escape into the atmosphere, and the EGR valve closes.

The front solenoid is the pressure solenoid. When its vacuum valve is open, air pressure goes to the EGR valve and opens it. When the pressure solenoid is closed, the air from the pump is shut off. This also holds the pressure in the hose to the EGR valve, and makes the EGR valve stay in whatever position it was when the pressure solenoid was closed.

The electronic control assembly is continuously analyzing the information it receives from the sensors, and changing the positions of the solenoid valves to keep the flow of exhaust gas at the level the engine needs. Under certain conditions, there can be quite a bit of clicking from the solenoids, but this is normal.

On the pressure solenoid, the upper nozzle connects to the air pump bypass valve. The lower nozzle connects to a "T" that has hoses going to the vent solenoid and the EGR valve. The hose connection at the vent solenoid should be made to the upper nozzle.

The rear, vent solenoid is a normally open design. To test it for leaks, you must first disconnect the electric plug and apply battery voltage to one of the solenoid terminals, and ground the other. Then you can use a hand pressure pump to apply a few pounds pressure to the valve. These valves do not make a perfect seal, but as long as they do not leak any more than half a pound in 5 seconds, they can be considered okay.

The front, pressure solenoid is a normally closed design, so it can be tested by simply disconnecting the upper hose and applying a few pounds pressure from a hand pump. If you have the system analyzer, closing or opening the solenoids for testing is simply a matter of pushing a button, which makes it a lot easier.

Spacer plates under the carburetor have been used by Ford for years in EGR systems. A passage in the plate allows exhaust gas to travel from the exhaust crossover passage in the manifold to the EGR valve. From the valve the gas is metered back into the spacer and into the intake part of the manifold. The spacer on the EEC system is a different design. It does not connect to the exhaust crossover passage in the intake manifold. Exhaust gases are brought to the EGR valve and allow the gases to flow into the intake part of the manifold.

The tube that carries exhaust gas to the EGR valve is attached to a fitting at the front of the right exhaust manifold. Above the right rocker cover is an EGR cooler, which is connected to engine coolant. Engine coolant flows through the cooler and reduces the temperature of the exhaust gases. From the cooler, a tube takes the cooled exhaust gases to the EGR valve. The EGR valve has two hose connections. The large size hose is for the exhaust gases, and the small size is for pressure to operate the valve.

Diagnostic Tester

This analyzer, available from Owatonna Tool Company, is a plug in unit that will quickly analyze most of the EEC system. Additional equipment needed are an advance-reading timing light, a tachometer, and a vacuum-pressure gauge. The analyzer is made in two parts, and each part must be purchased separately. An advantage is that the voltmeter part can also be used with other Ford analyzers.

EEC Diagnostic Tester
T78L-50-EEC-1 $351.49
Digital Volt-Ohmmeter
T78L-50-DVOM-1 $331.24
Owatonna Tool Company
118 Eisenhower Drive
Owatonna, Minnesota 55060

HOLLEY MODEL 1946 1-BBL. CARBURETOR 1978

This carburetor is used only on Fairmont and Zephyr 200 inch. 6-cylinder engines. The carburetor is almost identical to the Holley Model 1945 used since 1978 on Chrysler Corporation slant-6 engines. Chrysler used a well-type choke, so there was no choke coil on the 1945. The 1946 has the choke coil mounted on the upper body in typical Ford fashion, and it is an electrically heated choke.

SUCTION AIR SYSTEM 1978

A hose runs from the clean side of the air cleaner to the exhaust manifold. Mounted about midway in the hose is a silencer and an air inlet valve, which works like a check valve. Suction in the exhaust manifold pulls in air from the air cleaner. This air oxidizes the hydrocarbons and carbon monoxide into harmless carbon monoxide and water vapor. The air inlet valve prevents backflow of the exhaust into the air cleaner. Because there is a lot of pulsating in the exhaust, it is normal for the air inlet valve to vibrate, expecially at idle. The silencer prevents these vibrations from reaching the air cleaner.

This system is used only on the 200 in. 6-cylinder engine in the Fairmont and Zephyr. To check the system, just remove the silencer and check the air inlet valve to be sure it is vibrating at idle. There should not be any backflow of exhaust out of the valve. The official test

Ford Motor Company

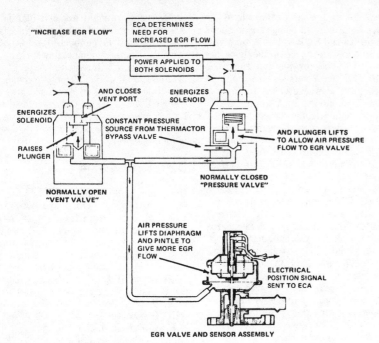

for the valve is to use a rubber bulb tester with the engine off. Squeeze the bulb, and if the bulb stays collapsed, it means the valve is okay. If the valve does not vibrate, the system may be clogged, or disconnected.

CATALYTIC CONVERTERS 1978

49-State 200 6-cylinder and 302 V-8 engines use a new clamshell type converter that mounts on a bracket attached to the rear engine mount. California 302 V-8 engines also use the clamshell converter, except for Granada and Monarch, which use the older design.

Calif. 200 6-cylinder engines use a new mini converter, in the exhaust pipe close to the exhaust manifold. The mini converter give better control of emissions during warmup, because it gets hot faster.

DUAL MODE IGNITION TIMING 1978

A special control assembly takes the place of the usual electronic ignition control. The control assembly retards the spark whenever the vacuum is low on 49-State cars, or the altitude is low on Altitude cars. You can recognize the special control because it has three plug connections, instead of the usual two. Also, there is a vacuum or barometric switch mounted next to the control, or in the engine compartment on the fender panel.

The vacuum switch used with 49-State calibrations is mounted on the fender panel and connected to a three-terminal plug on the control unit. When vacuum above 10 in. Hg. is applied to the switch,

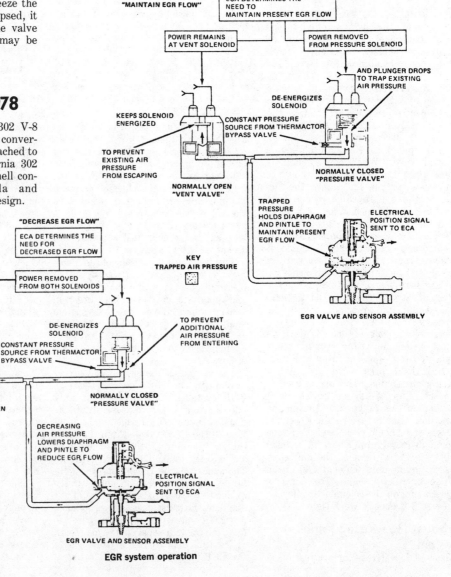

EGR system operation

Ford Motor Company

a signal goes through the wire to the ignition control, and the spark is retarded 3-6°. This means that during acceleration at wide open throttle, the spark will retard to prevent detonation.

A peculiar feature of the system is that some cars have the vacuum switches connected to manifold vacuum, and some to ported vacuum. With a ported vacuum connection, the system will go into 3-6° retard at idle. But with a manifold vacuum connection, there will be normal timing at idle.

When checking the initial timing on these engines, Ford recommends that the 3-terminal plug coming out of the ignition control be disconnected. This will insure that the engine is running at basic timing, without the retard.

The barometric pressure switch used with altitude calibrations does not connect to engine vacuum. It is mounted next to the ignition control. All it does is sense the altitude and tell the control when to give the 3-6° of retard. Below 2,400 ft. altitude, the switch is in the retard position. Between 2,400 and 4,300 ft. it can be in either position. Above 4,300 ft. the switch is in the basic timing position.

Testing the vacuum switch is easy, because you can apply vacuum to it and watch the timing change. But with the altitude switch, all you can do is check the timing. If the car is below 2,400 ft. check the timing with the switch plugged in, and it should be 3-6° retarded. Then disconnect the plug and the timing should go to the basic timing specification. There is no way to make the switch go into the altitude setting, unless you have a convenient 4,300-foot mountain nearby.

If you are checking the altitude switch with the car above 4,300 feet, it should be at the basic timing setting. If not, the switch is defective. Disconnecting the plug should not have any effect, because the switch should already be in the basic timing position. At an altitude of 4,300 feet or higher, there isn't any way of making the switch go into the retard position.

VAPOR CONTROLS 1978

The federal government imposed much stricter controls on the car makers 1978 and later models. In the past, emissions of gasoline vapor were measured by attaching hoses to certain parts of the car and measuring how much vapor came through the hose. Now, the entire car is placed in a sealed housing, and the new test is called Sealed Housing Evaporative Determination (SHED). This means that all the vapors coming from the car are trapped in the enclosure. A sample of the air in the

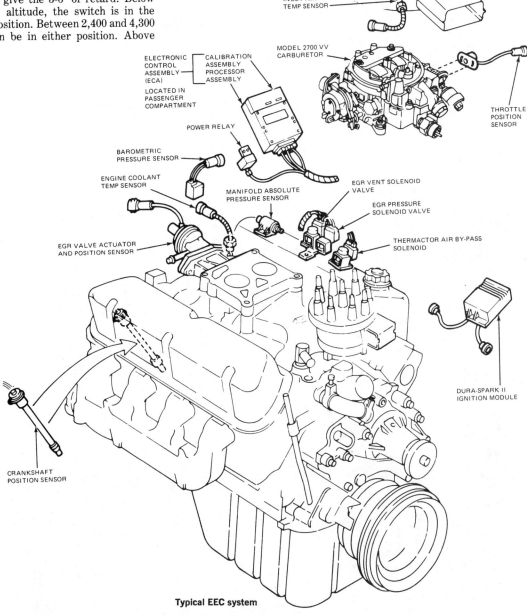

Typical EEC system

333

Ford Motor Company

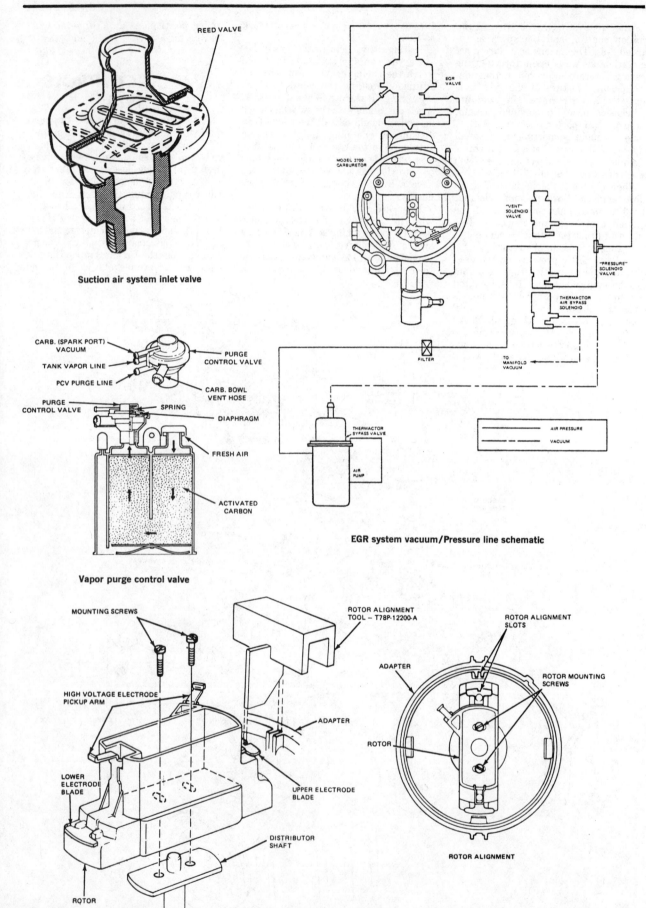

Suction air system inlet valve

Vapor purge control valve

EGR system vacuum/Pressure line schematic

Rotor alignment

Ford Motor Company

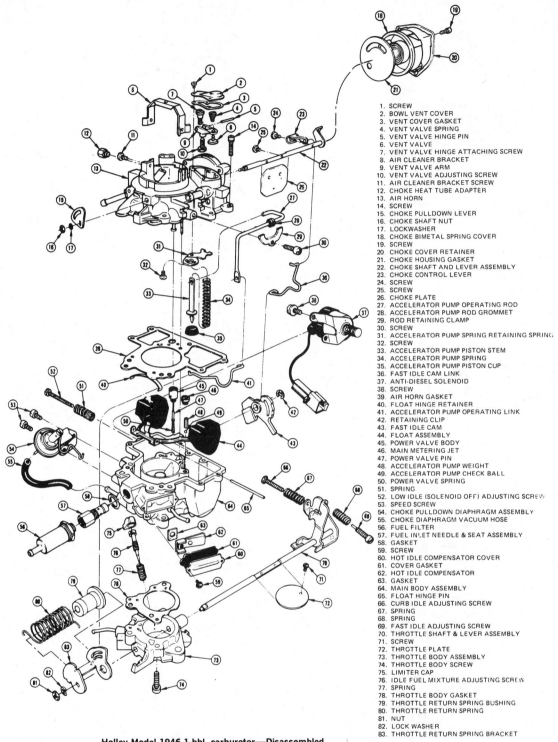

1. SCREW
2. BOWL VENT COVER
3. VENT COVER GASKET
4. VENT VALVE SPRING
5. VENT VALVE HINGE PIN
6. VENT VALVE
7. VENT VALVE HINGE ATTACHING SCREW
8. AIR CLEANER BRACKET
9. VENT VALVE ARM
10. VENT VALVE ADJUSTING SCREW
11. AIR CLEANER BRACKET SCREW
12. CHOKE HEAT TUBE ADAPTER
13. AIR HORN
14. SCREW
15. CHOKE PULLDOWN LEVER
16. CHOKE SHAFT NUT
17. LOCKWASHER
18. CHOKE BIMETAL SPRING COVER
19. SCREW
20. CHOKE COVER RETAINER
21. CHOKE HOUSING GASKET
22. CHOKE SHAFT AND LEVER ASSEMBLY
23. CHOKE CONTROL LEVER
24. SCREW
25. SCREW
26. CHOKE PLATE
27. ACCELERATOR PUMP OPERATING ROD
28. ACCELERATOR PUMP ROD GROMMET
29. ROD RETAINING CLAMP
30. SCREW
31. ACCELERATOR PUMP SPRING RETAINING SPRING
32. SCREW
33. ACCELERATOR PUMP PISTON STEM
34. ACCELERATOR PUMP SPRING
35. ACCELERATOR PUMP PISTON CUP
36. FAST IDLE CAM LINK
37. ANTI-DIESEL SOLENOID
38. SCREW
39. AIR HORN GASKET
40. FLOAT HINGE RETAINER
41. ACCELERATOR PUMP OPERATING LINK
42. RETAINING CLIP
43. FAST IDLE CAM
44. FLOAT ASSEMBLY
45. POWER VALVE BODY
46. MAIN METERING JET
47. POWER VALVE PIN
48. ACCELERATOR PUMP WEIGHT
49. ACCELERATOR PUMP CHECK BALL
50. POWER VALVE SPRING
51. SPRING
52. LOW IDLE (SOLENOID OFF) ADJUSTING SCREW
53. SPEED SCREW
54. CHOKE PULLDOWN DIAPHRAGM ASSEMBLY
55. CHOKE DIAPHRAGM VACUUM HOSE
56. FUEL FILTER
57. FUEL INLET NEEDLE & SEAT ASSEMBLY
58. GASKET
59. SCREW
60. HOT IDLE COMPENSATOR COVER
61. COVER GASKET
62. HOT IDLE COMPENSATOR
63. GASKET
64. MAIN BODY ASSEMBLY
65. FLOAT HINGE PIN
66. CURB IDLE ADJUSTING SCREW
67. SPRING
68. SPRING
69. FAST IDLE ADJUSTING SCREW
70. THROTTLE SHAFT & LEVER ASSEMBLY
71. SCREW
72. THROTTLE PLATE
73. THROTTLE BODY ASSEMBLY
74. THROTTLE BODY SCREW
75. LIMITER CAP
76. IDLE FUEL MIXTURE ADJUSTING SCREW
77. SPRING
78. THROTTLE BODY GASKET
79. THROTTLE RETURN SPRING BUSHING
80. THROTTLE RETURN SPRING
81. NUT
82. LOCK WASHER
83. THROTTLE RETURN SPRING BRACKET

Holley Model 1946 1-bbl. carburetor—Disassembled

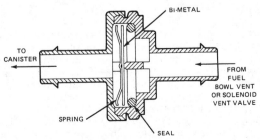

Fuel bowl thermal vent valve

Ford Motor Company

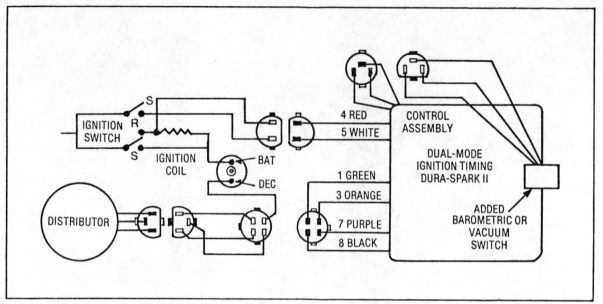

Dual mode ignition timing

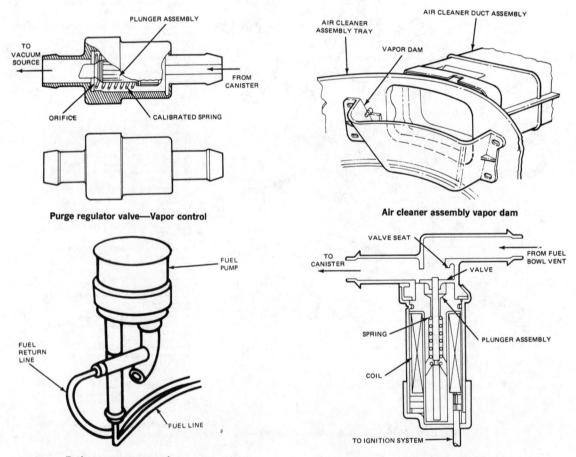

Purge regulator valve—Vapor control

Air cleaner assembly vapor dam

Fuel pump vapor separator

Fuel bowl solenoid vent valve

enclosure is then taken, and the amount of vapor given off by the car is calculated. To pass the test, it was necessary to redesign the vents on the carburetor and the carbon canister.

Carbon Canister

All canisters now have a purge control valve. The top of the valve is connected to ported carburetor vacuum. When the engine is idling, there is no vacuum from the port, and the valve is closed. Above idle, the port is uncovered, and vacuum goes to the top of the valve. This vacuum opens the valve and allows manifold vacuum through a PCV system connection to purge the vapors from the canister. Some cars use two canisters, arranged so that one canister absorbs the vapors from the fuel tank, and the other absorbs the vapors from the fuel bowl.

When the purge valve is closed, a small orifice allows some purging. Because of this constant purging, the valve may make a slight hissing sound. This is normal, and does not indicate that the valve is leaking. When the valve is open it will make a strong hissing sound if it is lifted off the canister. This hissing sound above idle means that it is working cor-

Ford Motor Company

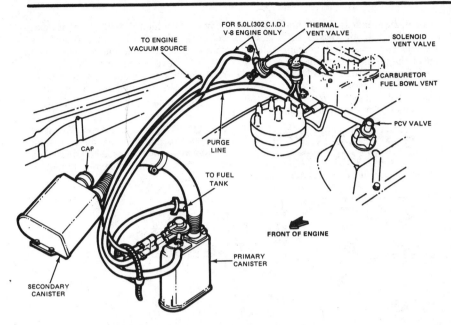

Typical dual canister carburetor Venting systems—Vapor control
5.8L/6.6L (351 M/400 C.I.D.) V-8 (EXCEPT ALTITUDE) – NON UNITIZED VEHICLES
(EXCEPT LINCOLN CONTINENTAL AND CONTINENTAL MARK V)

rectly. But if the strong hissing sound is heard at idle, it usually means the purge valve is stuck open, and should be replaced. An exception to this is those engines that have the top of the purge valve connected to intake manifold vacuum. On those engines the valve is open at idle and above idle, and the canister is purged continuously. To test the valve on those systems, disconnect the hose to the top of the purge valve, and the purging should stop.

Some engines use a ported vacuum switch mounted on a coolant passage. It shuts off the vacuum to the top of the purge valve when the engine is cold. This PVS may be a two-nozzle design, or a 4-nozzle design. If it has 4 nozzles, only two of them are used for the purge valve. The other two are for another emission control on the engine.

Some engines use a purge regulator valve in the purge hose between the carbon canister and intake manifold vacuum. This valve looks like a PCV valve. If intake manifold vacuum increases to where too much air would flow through the purge hose, the valve limits the flow to prevent leaning the mixture.

Some engines use a system of valves and plumbing with a distributor vacuum vent valve, a retard delay valve, a reservoir, and a thermactor idle vacuum valve. The names of these valves include words such as "distributor" and "thermactor" because they were originally designed for another use. Their use in the vapor control system does not mean that they are connected to any other system. The engineers used these valves because they had difficulty preventing rich or lean mixtures while purging the vapors from the canister.

The distributor vacuum vent valve has a 40-second delay built into it. The orifice is so small that it takes that long to build up vacuum and start purging the canister. This gives the engine a chance to settle down after starting without being upset by the purge vapors. During acceleration the vent valve opens and stops the purging, but the retard continues purging when the engine returns to cruising. The vacuum reservoir helps the retard delay valve maintain the vacuum.

When the engine is started warm, this delay system causes a rich mixture in the engine because it takes several seconds before the air from the canister passes through the delay into the engine. The thermactor idle vacuum valve was added to provide an air bleed. It is a normally open valve, so it bleeds air into the engine as soon as the engine starts. When the vacuum in the system is enough to open the purge valve, it also closes the thermactor idle vacuum valve, and the system purges normally.

Carburetor Bowl Vents

Several types of bowl vent valves are used. Some carburetors have a mechanical vent that opens from the throttle linkage when the engine idles. Others have a vapor operated vent that opens when the vapor pressure builds up. Those valves have been common for several years.

For 1978, Ford had added a thermal vent valve to some engines. This valve is in the hose connecting the bowl vent to the carbon canister. When the engine compartment is cold, the valve closes. This prevents canister vapors from traveling up the hose to the carburetor, and upsetting the mixture. When the engine compartmennt warms up, the valve opens, and then the vapors from the carburetor bowl are vented to the carbon canister. If there are any other valves in the system, such as a PVS, those valves also have to be open before the system does any purging.

An electrically operated vent valve, called a solenoid vent valve, is used on some engines. It is a normally open design, so that it vents the fuel bowl to the canister when the engine is not running. When the ignition switch is turned on, the valve closes and no purging takes

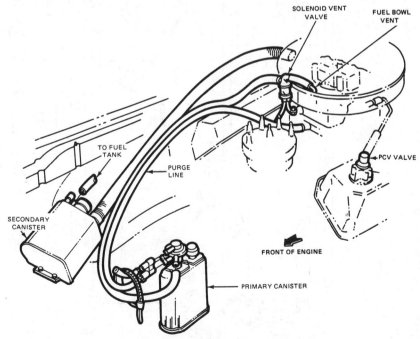

5.0L/5.8L (302/351 C.I.D.), 5.8L/6.6L (351 M/400 C.I.D.) ALTITUDE,
7.5L (460 C.I.D.) – NON UNITIZED VEHICLES
(EXCEPT LINCOLN CONTINENTAL AND CONTINENTAL MARK V)

Typical dual canister carburetor Venting systems—Vapor control

Ford Motor Company

place during engine operation. This prevents the purging vapors from upsetting the mixture.

With the vent to the canister shut off, some engines gave problems during running because of lack of venting. To solve this problem, the engineers added an auxiliary vent tube on some engines. This tube connects to a "T" in the bowl vent hose, and allows the vapors from the bowl to go into the air cleaner, even though the hose to the canister is shut off.

Additional escape of vapors was caused by the air intake on the air cleaner. To prevent vapor from spilling out of the air cleaner when the car was parked, a vapor dam was installed just inside the air cleaner can opposite the air intake. This keeps the vapors inside the air cleaner. They are purged as soon as the engine starts.

HIGH FLOW PCV VALVES 1978

The engine decal on certain engines shows that a high flow PCV valve must be installed according to the maintenance schedule. This means that when the PCV valve is due for replacement, a high flow valve should be installed, instead of the original low-flow valve.

After installing the valve, the idle speed and mixture should be checked, because the high flow valve might have changed it. This procedure of installing a high flow valve is not done on all engines. The instructions on the engine decal must be followed.

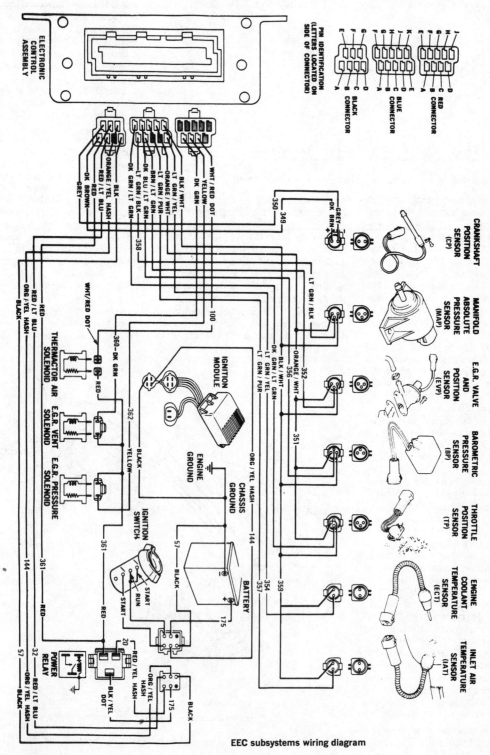

EEC subsystems wiring diagram

Ford Motor Company

ELECTRONIC ENGINE CONTROL SYSTEM 1979

This system, known as Electronic Engine Control, or EEC, is installed on all 1979 Versailles 302 V-8 engines. The system controls engine timing, exhaust gas recirculation, and air pump airflow. Seven sensors feed information into an electronic box mounted on the passenger side of the firewall near the steering column. The box, called an electronic control assembly, takes all the information, processes it and then electrically controls the timing, EGR, and air pump airflow. First, let's look at the 7 sensors and how they do their job.

Inlet Air Temperature Sensor

Mounted in the air cleaner can, near the air inlet, the temperature sensor has two wire connected to it, so current can flow from the electronic control assembly to the sensor and back. The sensor is a thermistor design, which changes resistance as the temperature changes. This affects the amount of current flowing, so that in effect the electronic control assembly knows at all times the temperature inside the air cleaner. The action of the sensor is opposite to temperature changes. In other words, as the temperature goes up, the sensor resistance goes down, and vice versa.

The information transmitted by the sensor is used by the electronic con-

EMISSION CONTROL SYSTEMS 1979

trol assembly to determine the right setting for spark timing and air pump airflow. At high temperatures, the spark advance is less, to help prevent pinging.

The sensor is held in place by a clip inside the air cleaner. Resistance of the sensor should be between 6,500 and 45,000 ohms, as measured at the disconnected plug with an ohmmeter.

Throttle Position Sensor

This sensor is a variable resistor mounted on the carburetor and connected to the end of the throttle shaft. For every position of the throttle, there is a corresponding position of the sensor which gives a different current flow to the electronic control assembly. Using a throttle position sensor is almost the same as using a vacuum reading. The electronic control assembly combines the information on throttle position with information from other sensors, and then controls the spark advance, EGR flow rate, and air pump airflow.

The throttle position sensor is adjusted by loosening the mounting screws and rotating the sensor until the correct readings appear on a system analyzer. Although the sensor is a resistor, and could possibly be set using an ohmmeter, Ford does not recommend this. They want the sensor position adjusted with the analyzer to a specific voltage reading. This is more accurate than using an ohmmeter.

Crankshaft Position Sensor

Probably the most important sensor, it is mounted at the rear of the cylinder block, in front of the torque converter. If this sensor, or the wires to it are defective or damaged, the engine will not run. A pulse ring, permanently attached to the crankshaft, rotates just under the tip of the sensor. Every time one of the lobes on the ring passes by the sensor tip it creates a magnetic pulse in the electromagnet inside the sensor. With this signal, the electronic control assembly knows the engine rpm and the position of the crank, so it can set the spark timing and EGR flow rate to exactly what the engine needs.

The signals from the crankshaft position sensor and the throttle position sensor are combined in the electronic control assembly to determine the load on the engine. This is exactly the same as taking a vacuum reading from the intake manifold, but it eliminates the need for any vacuum diaphragm or connecting hose.

A conventional engine gets its spark timing directly from the distributor, which is driven by the camshaft and timing chain. If the chain wears, the timing retards, so the timing must be periodically reset. This does not happen with the EEC system, because the timing is determined directly from the position of the crankshaft. There is never any need to adjust timing on the EEC system. An engine that is completely

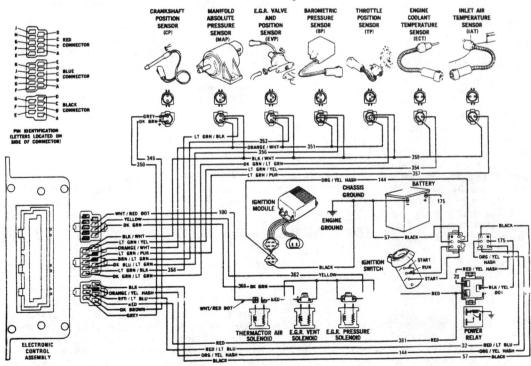

Wiring Diagram—EEC subsystems

Ford Motor Company

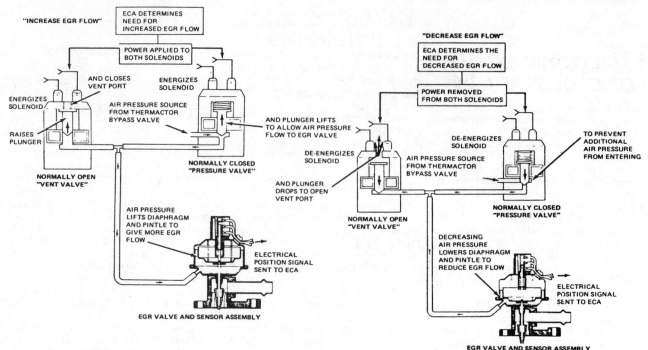

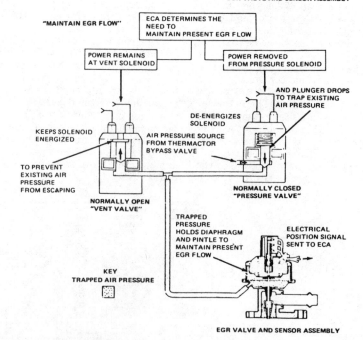

Operation of the EGR system (with EEC)

worn out will still have perfect timing.

A conventional engine also has problems with wear in the vacuum advance and centrifugal advance systems. The vacuum and centrifugal advance on the EEC engine is all controlled by the electronic control assembly. So it will also remain perfect over the life of the engine.

The crankshaft position sensor is held in place by a clamp and screw. Near the tip of the sensor is an "O" ring that keeps engine oil from escaping. There are no adjustments on the sensor.

Coolant Temperature Sensor

It's just an ordinary coolant temperature sensor with two wires connected to it. But it has a big effect on how the engine runs. It is mounted on the water passage at the left rear of the intake manifold. Its resistance varies with the coolant temperature, with high resistance at low temperatures, and vice versa. The sensor is connected to the electronic control assembly. When the engine is cold, the electronic control assembly cuts back on the amount of exhaust gas that is allowed to recirculate, or maybe eliminates the EGR entirely. If the engine is overheated because of long idling, the electronic control assembly advances the spark timing to increase the idle speed and help cool down the engine. If the coolant temperature sensor is damaged or defective both EGR flow rate and spark timing may be affected.

The sensor is easily checked with an ohmmeter. At normal operating temperature, (160-220°F.) the resistance at the disconnected plug should be between 1,500 and 6,000 ohms.

The sensor is removed by unscrewing it from the water passage with a wrench. Relieve the pressure first to avoid burns.

EGR Valve Position Sensor

This sensor is part of the EGR valve, which is mounted on a carburetor spacer. Wires connect the sensor to the electronic control assembly, to indicate the position of the EGR valve metering rod. This indirectly tells the electronic control assembly how much exhaust gas is recirculating. The electronic control assembly compares this information with what it receives from the other sensors, and then increases or decreases the exhaust gas recirculation rate with other controls that wil be explained later. It is important to realize that the EGR valve position sensor is only a sensor. It does not actually move the metering rod or directly change the flow rate.

The sensor can be checked with an ohmmeter, after disconnecting the plug. There are three wires connecting to the sensor. The ohmmeter reading between the orange-white and black-and-white wires should be between 2800 and 5300 ohms, with the engine off. Between the orange-white and brown-light green wires, the ohmmeter should read 350 to 940

Ford Motor Company

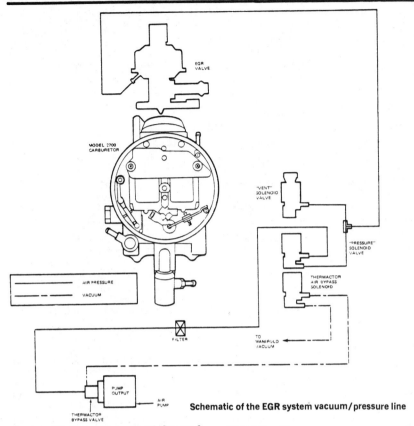
Schematic of the EGR system vacuum/pressure line

ohms, with the engine off. If the readings are outside these limits, the entire EGR valve must be replaced. There are no adjustments on the sensor or the valve.

Manifold Absolute Pressure Sensor

Manifold absolute pressure is defined as barometric pressure minus manifold vacuum. The sensor is connected to intake manifold vacuum. In effect, it simply measures intake manifold vacuum and sends a signal through wires to the electronic control assembly. The electronic control assembly uses this information, along with data from other sensors, to set the spark advance and EGR flow rate.

The MAP sensor is at the left rear of the intake manifold, next to the rocker cover. Ford does not give a simple ohmmeter testing procedure, preferring that the system analyzer be used. There are no adjustments to the MAP sensor.

Barometric Pressure Sensor

This sensor has only one purpose, to tell the electronic control assembly the altitude at which the car is operating.

At high altitude, the EGR flow rate is cut back to keep good engine performance. The sensor connects by wires to the electronic control assembly. A simple ohmmeter test is not provided for the barometric sensor. It must be tested with the sytem analyzer. There are no adjustments.

EEC Distributor

Because the electronic control assembly takes care of all timing and advance, the distributor has no vacuum advance, no centrifugal advance, and no timing mechanism. The distributor is nothing more than a rotor and some electrodes that distribute the spark to the correct spark plug wires.

The distributor cap and rotor are a completely new two-level design. Four of the plug wire electrodes are at one level, and the other four are a little bit lower. This puts more distance between the electrodes, and helps prevent crossfiring. In addition, a double-ended rotor is used, so that the spark jumps first from one end of the rotor, and then from the other end. This gives additional separation to prevent crossfiring. The construction of the rotor is such that the spark current actually has to jump twice to get to the plug wire electrode in the cap. It jumps once from the center tower to the pick up drum of the rotor, and a second time from the top of the rotor to the cap electrode. When you look inside the cap you will see a 4-spoke electrode plate connected to the center tower, and a rotor with two pickup arms. This plate is positioned so that when the correct rotor pickup arm is over one spoke, the other rotor pickup arm is between the spokes. This was also done to decrease the possibility of crossfire.

Because the distributor has nothing to do with timing, there are no adjustments for distributor position. The clamp and bolt at the base of the distributor fit into a slot, locking the distributor in one position. But there is a rotor position adjustment. A slot in the rotor must be opposite a slot in the housing, when number one cylinder is at top dead center of the compression stroke. This adjustment is made by removing the clamp bolt, lifting up the distributor, and meshing the drive gear teeth in a different position so that the slots line up. After the rotor is as close as possible this way, the rotor mounting screws can be loosened and the rotor moved to bring it into exact alignment with the slot. Ford has an official tool T78P-12200-A which fits into the rotor and housing slots at the same time and holds the rotor so it doesn't move while the screws are tightened. The tool is available from Owatonna Tool Company, Owatonna, Minnesota.

Because of the unique 2-level construction of the distributor cap and rotor, the plug wires do not fire in succession around the cap. The rotor fires first from one end, and then the other. This means that although the rotor only turns 45° between firings, the actual firing succession jumps back and forth across the cap. The numbers on the cap are the numbers of the cylinders, and must be connected to those cylinders. If you try to connect the wires in firing order succession around the cap, the engine will not run. Also, there are two sets of numbers on the cap. The inner ring of numbers is for the 302 V-8. The outer ring is for the 351 or 400 V-8, if it ever gets the EEC system. Because four of the numbers are different, the engine will not run if the plug wires are connected according to the wrong number ring.

A surprising thing about this EEC system is that there is no timing setting on the engine. The timing is done by the crankshaft position sensor and the electronic engine control. At one end of the electronic control there is a small trim adjustment, but it only changes the timing a few degrees. The trim adjustment is sealed, and if the seal is broken, free replacement of a defective electronic engine control under warranty may be refused.

Air Pump Airflow Control

A vacuum solenoid, mounted to the left of the intake manifold, is connected by wires to the electronic engine control. The solenoid vacuum path is a normally closed design. The electric coil in the solenoid is energized to open the vacuum passage by current from the electronic engine control. In the open position, the solenoid allows intake manifold vacuum to open the air bypass valve so that

341

Ford Motor Company

air from the pump can flow to each engine exhaust valve. When the solenoid is deenergized, the valve plunger moves to the closed position, shutting off the vacuum to the bypass valve, which then goes into the dump position. Air pump air is then diverted to the atmosphere.

The electronic engine control turns the solenoid on or off according to signals it receives from the inlet air temperature sensor, and the throttle position sensor.

The hose should be connected to the solenoid with the lower outlet to manifold vacuum, and the upper outlet to the bypass valve.

Testing the solenoid can be done by idling the engine with a vacuum gauge connected to the upper outlet, after removing the bypass valve hose. When running the engine above idle in PARK, you should get a full manifold vacuum reading on the gauge. Let the engine come down to idle, and in a few seconds the solenoid should close, and the gauge should drop to zero. There is a delay built into the system so the air pump does not pump until the engine has been at idle for a few seconds. There are no adjustments on the solenoid. It must be replaced if it doesn't work right.

Exhaust Gas Recirculation System

The EGR valve is air-pressure operated. A hose connects to the side of the air bypass valve and brings air pump air pressure to the EGR valve. A dual vacuum solenoid setup is in the hose between the bypass valve and the EGR valve. The two solenoids are mounted together on a bracket next to the left rocker arm cover, and connected by wires to the electronic control assembly. The rear solenoid is

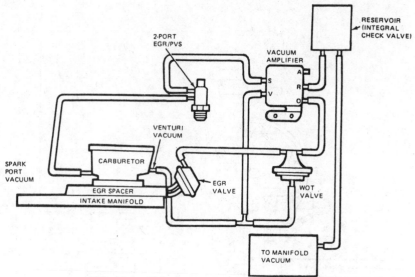

EGR system with Venturi vacuum amplifier—Typical

the vent solenoid. When the vacuum valve inside the solenoid is open, air pressure from the air pump is allowed to escape into the atmosphere and the EGR valve closes.

The front solenoid is the pressure solenoid. When its vacuum valve is open, air pressure goes to the EGR valve and opens it. When the pressure solenoid is closed, the air from the pump is shut off. This also holds the pressure in the hose to the EGR valve, and makes the EGR valve stay in whatever position it was when the pressure solenoid was closed.

The electronic control assembly is continuously analyzing the information is receives from the sensors, and changing the positions of the solenoid valves to keep the flow of exhaust gas at the level the engine needs. Under certain conditions, there can be quite a bit of clicking from the solenoids, but this is normal.

On the pressure solenoid, the upper nozzle connects to the air pump bypass valve. The lower nozzle connects to a "T" that has hoses going to the vent solenoid and the EGR valve. The hose connection at the vent solenoid should be made to the upper nozzle.

The rear, vent solenoid is a normally open design. To test it for leaks, you must first disconnect the electric plug and apply battery voltage to one of the solenoid terminals, and ground the other. Then you can use a hand pressure pump to apply a few pounds pressure to the valve. These valves do not make a perfect seal, but as long as they do not leak any more than half a pound in 5 seconds, they can be considered okay.

The front pressure solenoid is a normally closed design, so it can be tested by simply disconnecting the upper hose and applying a few

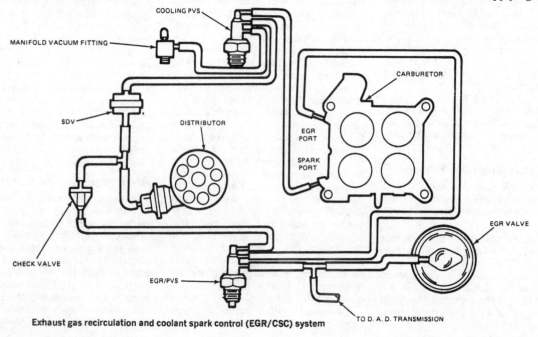

Exhaust gas recirculation and coolant spark control (EGR/CSC) system

Ford Motor Company

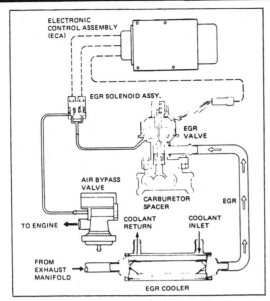

EGR system (with EEC system)—Typical

pounds pressure from a hand pump. If you have the system analyzer, closing or opening the solenoids for testing is simply a matter of pushing a button, which makes it a lot easier.

Spacer plates under the carburetor have been used by Ford for years in EGR systems. A passage in the plate allows exhaust gas to travel from the exhaust crossover passage in the manifold to the EGR valve. From the valve the gas is metered back into the spacer and into the intake part of the manifold. The spacer on the EEC system is a different design. It does not connect to the exhaust crossover passage in the intake manifold. Exhaust gases are brought to the EGR valve and allow the gases to flow into the intake part of the manifold.

The tube that carries exhaust gas to the EGR valve is attached to a fitting at the front of the right exhaust manifold. Above the right rocker cover is an EGR cooler, which is connected to engine coolant. Engine coolant flows through the cooler and reduces the temperature of the exhaust gases. From the cooler, a tube takes the cooled exhaust gases to the EGR valve. The EGR valve has two hose connections. The large size hose is for the exhaust gases, and the small size is for pressure to operate the valve.

Diagnostic Testor

This analyzer, available from Owatonna Tool Company, is a plug in unit that will quickly analyze most of the EEC system. Additional equipment needed are an advance-reading timing light, a tachometer, and a vacuum-pressure gauge. The analyzer is made in two parts, and each part must be purchased separately. An advantage is that the voltmeter part can also be used with other Ford analyzers.

EEC Diagnostic Tester
T78L-50-EEC-1
Digital Volt-Ohmmeter
T78L-50-DVOM-1
Owatonna Tool Company
118 Eisenhower Drive
Owatonna, Minnesota 55060

SUCTION AIR SYSTEM 1979

A hose runs from the clean side of the air cleaner to the exhaust manifold. Mounted about midway in the hose is a silencer and an air inlet valve, which works ilke a check valve. Suction in the exhaust manifold pulls in air from the air cleaner. This air oxidizes the hydrocarbons and carbon monoxides into harmless carbon monoxide and water vapor. The air inlet valve prevents backflow of the exhaust into the air cleaner. Because there is a lot of pulsating in the exhaust, it is normal for the air inlet valve to vibrate, especially at idle. The silencer prevents these vibrations from reaching the air cleaner.

This system is used only on the 200 in. 6-cylinder engine in the Fairmont and Zephyr. To check the system, just remove the silencer and check the air inlet valve to be sure it is vibrating at idle. There should not be any backflow of exhaust out of the valve. The official test for the valve is to use a rubber bulb tester with the engine off. Squeeze the bulb, and if the bulb stays collapsed, it means the valve is okay. If the valve does not vibrate, the system may be clogged, or disconnected.

CATALYTIC CONVERTERS 1979

49-State 200 6-cylinder and 302 V-8 engines use a new clamshell type converter that mounts on a bracket attached to the rear engine mount. California 302 V-8 engines also use the clamshell converter except for Granada and Monarch, which use the older design.

Calif. 200 6-cylinder engines use a new mini converter, in the exhaust pipe close to the exhaust manifold. The mini converter gives better control of emissions during warmup, because it gets hot faster.

DUAL MODE IGNITION TIMING 1979

A special control assembly takes the place of the usual electronic ignition control. The control assembly retards the spark whenever the vacuum is low on 49-State cars, or the altitude is low on Altitude cars. You can recognize the special control because it has three plug connections, instead of the usual two. Also, there is a vacuum or barometric switch mounted next to the control, or in the engine compartment on the fender panel.

The vacuum switch used with 49-State calibrations is mounted on the fender panel and connected to a three-terminal plug on the control unit. When vacuum above 10 in. Hg. is applied to the switch, a signal goes through the wire to the ignition control, and the spark is retarded 3-6°. This means that during acceleration at wide open throttle, the spark will retard to prevent detonation.

A peculiar feature of the system is that some cars have the vacuum switches connected to manifold vacuum, and some to ported vacuum. With a ported vacuum connection, the system will go into 3-6° retard at idle. But with a manifold vacuum connection, there will be normal timing at idle.

When checking the initial timing on these engines, Ford recommends that the 3-terminal plug coming out of the ignition control be disconnected. This will insure that the engine is running at basic timing, without the retard.

The barometric pressure switch used with altitude calibrations does not connect to engine vacuum. It is mounted next to the ignition control. All it does is sense the altitude and tell the control when to give the 3-6° of retard. Below 2,400 ft. altitude, the switch is in the retard position. Between 2,400 and 4,300 ft. it can be in either position. Above 4,300 ft. the switch is in the basic timing position.

Testing the vacuum switch is easy, because you can apply vacuum to it and watch the timing change. But with the altitude switch, all you can do is check the timing. If the car is below 2,400 ft. check the timing with

Ford Motor Company

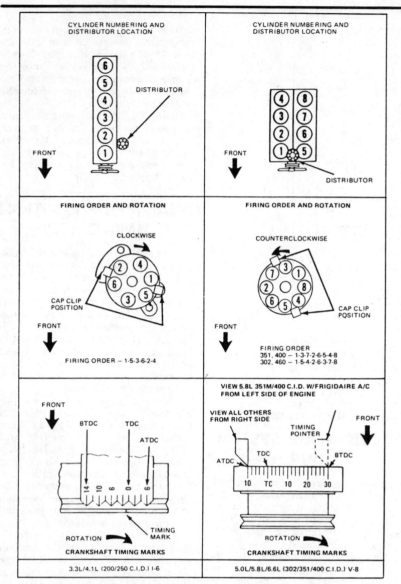

Cylinder firing order and engine timing—I-6 cyl. and V-8 cyl. engines

the switch plugged in, and it should be 3-6° retarded. Then disconnect the plug and the timing should go to the basic timing specification. There is no way to make the switch go into the altitude setting, unless you have a convenient 4,300-foot mountain nearby.

If you are checking the altitude switch with the car above 4,300 feet, it should be at the basic timing setting. If not, the switch is defective. Disconnecting the plug should not have any effect, because the switch should already be in the basic timing position. At an altitude of 4,300 feet or higher, there isn't any way of making the switch go into the retard position.

VAPOR CONTROLS 1979

The federal government imposed much stricter controls on the car makers 1978 and later models. In the past, emissions of gasoline vapor were measured by attaching hoses to certain parts of the car and measuring how much vapor came through the hose. Now, the entire car is placed in a sealed housing, and the new test is called Sealed Housing Evaporative Determination (SHED). This means that all the vapors coming from the car are trapped in the enclosure. A sample of the air in the enclosure is then taken, and the amount of vapor given off by the car is calculated. To pass the test, it was necessary to redesign the vents on the carburetor and the carbon canister.

Carbon Canister

All canisters now have a purge control valve. The top of the valve is connected to ported carburetor vacuum. When the engine is idling, there is no vacuum from the port, and the valve is closed. Above idle, the port is uncovered, and vacuum goes to the top of the valve. This vacuum opens the valve and allows manifold vacuum through a PCV system connection to purge the vapors from the canister. Some cars use two canisters, arranged so that one canister absorbs the vapors from the fuel tank, and the other absorbs the vapors from the fuel bowl.

When the purge valve is closed, a small orifice allows some purging. Because of this constant purging, the valve may make a slight hissing sound. This is normal, and does not indicate that the valve is leaking. When the valve is open it will make a strong hissing sound if it is lifted off the canister. This hissing sound above idle means that it is working correctly. But if the strong hissing sound is heard at idle, it usually means the purge valve is stuck open, and should be replaced. An exception to this is those engines that have the top of the purge valve connected to intake manifold vacuum. On those engines the valve is open at idle and above idle, and the canister is purged continuously. To test the valve on those systems, disconnect the hose to the top of the purge valve, and the purging should stop.

Some engines use a ported vacuum switch mounted on a coolant passage. It shuts off the vacuum to the top of the purge valve when the engine is cold. This PVS may be a two-nozzle design, or a 4-nozzle design. If it has 4 nozzles, only two of them are used for the purge valve. The other two are for another emission control on the engine.

Some engines use a purge regulator valve in the purge hose between the carbon canister and intake manifold vacuum. This valve looks like a PCV valve. If intake manifold vacuum increases to where too much air would flow through the purge hose, the valve limits the flow to prevent leaning the mixture.

Some engines use a system of valves and plumbing with a distributor vacuum vent valve, a retard delay valve, a reservoir, and a thermactor idle vacuum valve. The names of these valves include words such as "distributor" and "thermactor" because they were originally designed for another use. Their use in the vapor control system does not mean that they are connected to any other system. The engineers used these valves because they had difficulty preventing rich or lean mixtures while purging the vapors from the canister.

The distributor vacuum vent valve has a 40-second delay built into it. The orifice is so small that it takes that long to build up vacuum and start purging the canister. This gives the engine a change to settle down after starting wihout being upset by the purge vapors. During acceleration the vent valve opens and stops the purging, but the retard continues purging when the engine returns to

Ford Motor Company

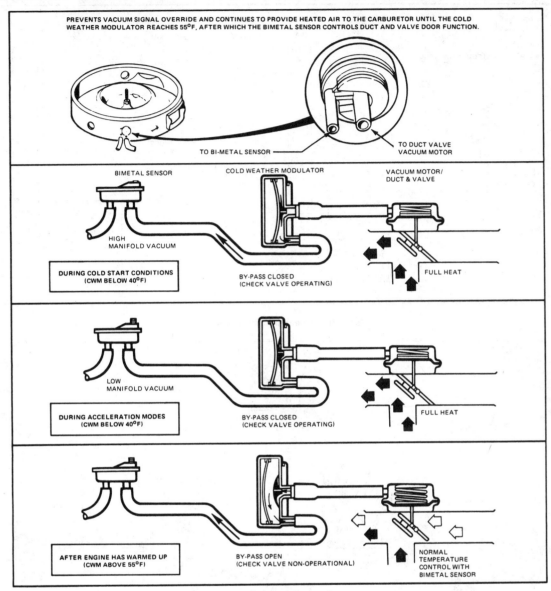

Operation of the cold weather modulator

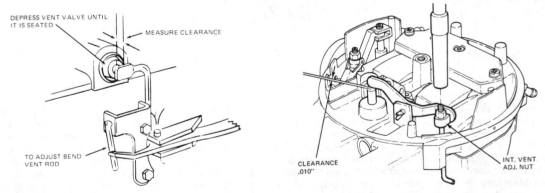

Bowl vent adjustment—Model 2150 2 bbl. carburetor

Internal vent system—Model 2700 VV carburetor

cruising. The vacuum reservoir helps the retard delay valve maintain the vacuum.

When the engine is started warm, this delay system causes a rich mixture in the engine because it takes several seconds before the air from the canister passes through the delay into the engine. The thermactor idle vacuum valve was added to provide an air bleed. It is a normally open valve, so it bleeds air into the engine as soon as the engine starts. When the vacuum in the system is enough to open the purge valve, it also closes the thermactor idle vacuum valve, and the system purges normally.

Carburetor Bowl Vents

Several types of bowl vent valves are used. Some carburetors have a mechanical vent that opens from the throttle linkage when the engine idles.

Ford Motor Company

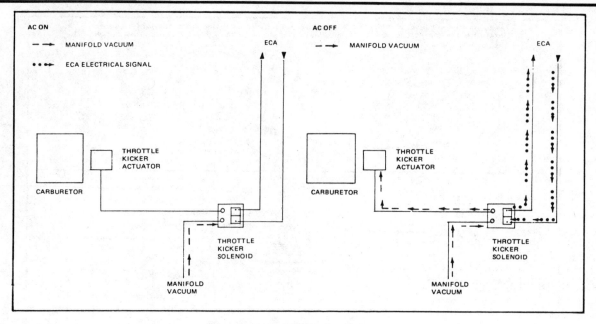

Schematic of the throttle kicker system

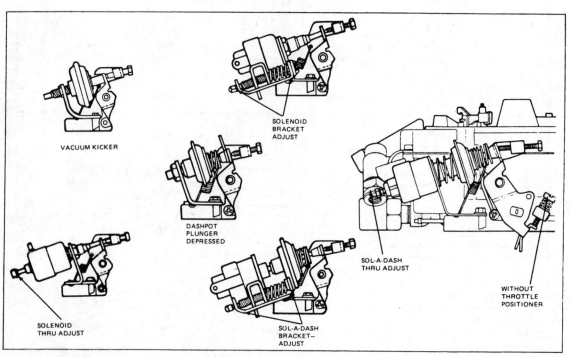

Throttle positioners

Others have a vapor operated vent that opens when the vapor pressure builds up. Those valves have been common for several years.

Ford has added a thermal vent valve to some engines. This valve is in the hose connecting the bowl vent to the carbon canister. When the engine compartment is cold, the valve closes. This prevents canister vapors from traveling up the hose to the carburetor, and upsetting the mixture. When the engine compartment warms up, the valve opens, and then the vapors from the carburetor bowl are vented to the carbon canister. If there are any other valves in the system, such as a PVS, those valves also have to be open before the system does any purging.

An electrically operated vent valve, called a solenoid vent valve, is used on some engines. It is a normally open design, so that it vents the fuel bowl to the canister when the engine is not running. When the ignition switch is turned on, the valve closes and no purging takes place during engine operation. This prevents the purging vapors from upsetting the mixture.

With the vent to the canister shut off, some engines gave problems during running because of lack of venting. To solve this problem, the engineers added an auxiliary vent tube on some engines. This tube connects to a "T" in the bowl vent hose, and allows the vapors from the bowl to go into the air cleaner, even though the hose to the canister is shut off.

Additional escape of vapors was caused by the air intake on the air cleaner. To prevent vapor from spilling out of the air cleaner when the car was parked a vapor dam was installed just inside the air cleaner can opposite the air intake. This keeps the vapors inside the air cleaner. They are purged as soon as the engine starts.

Ford Motor Company

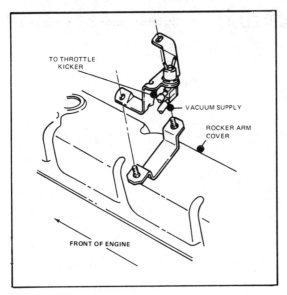

Throttle kicker solenoid

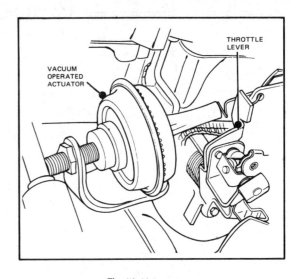

Throttle kicker actuator

HIGH FLOW PCV VALVES 1979

The engine decal on certain engines shows that a high flow PCV valve must be installed according to the maintenance schedule. This means that when the PCV valve is due for replacement, a high flow valve should be installed, instead of the original low-flow valve. After installing the valve, the idle speed and mixture should be checked, because the high flow valve might have changed it. This procedure of installing a high flow valve is not done on all engines. The instructions on the engine decal must be followed.

═══ VACUUM CIRCUITS ═══

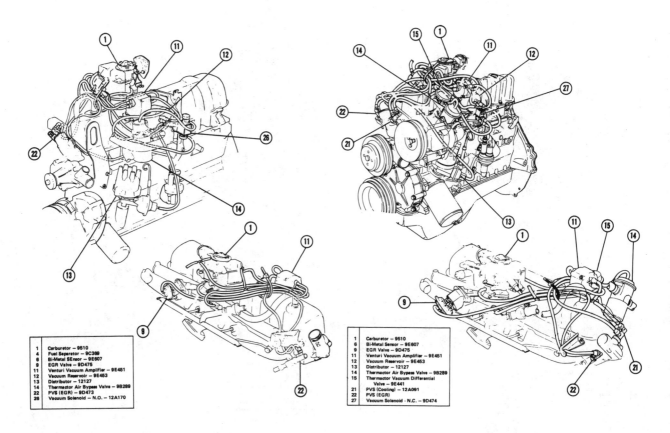

1976 200 calib. 5-6G, R-2

1976 200 calib. 5-7G, R-4 with air cond.

347

Ford Motor Company

VACUUM CIRCUITS

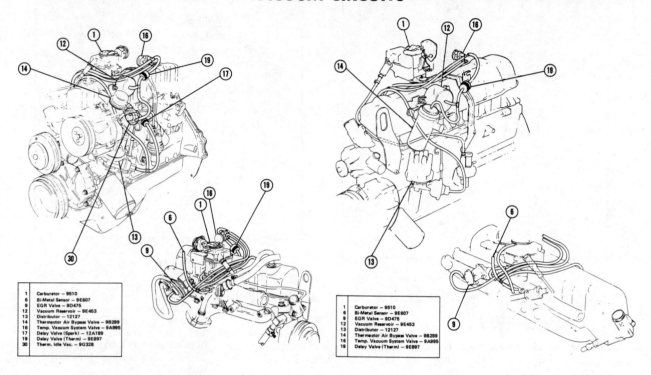

1976 200 calib. 6-6G, R-22

1976 200 calib. 6-7G, R-23

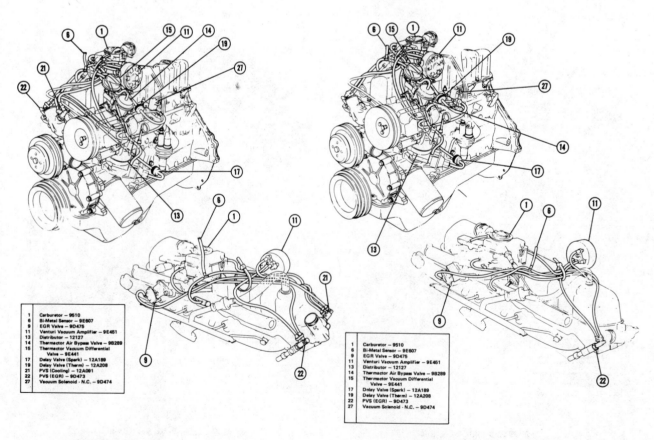

1976 250 calib. 5-8N, R-3S, with cooling PVS

1976 250 calib. 5-8N, R-3S, no cooling PVS

Ford Motor Company

VACUUM CIRCUITS

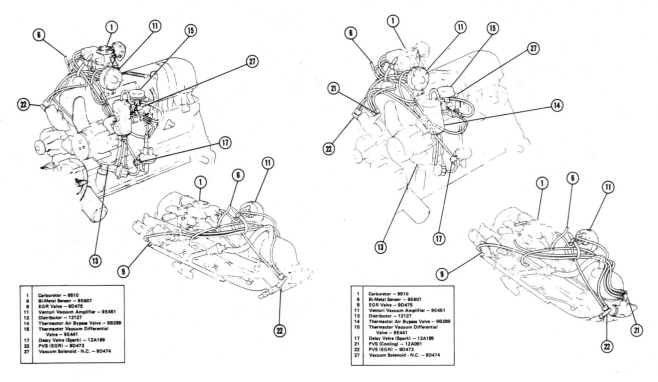

1976 250 calib. 5-9N, R-7S, no cooling PVS

1976 250 calib. 5-9N, R78, with cooling PVS

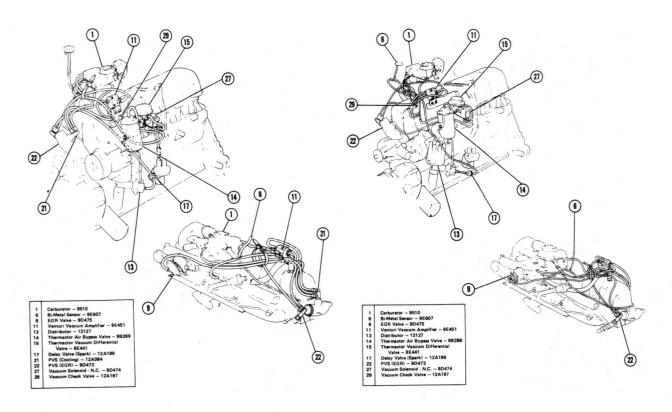

1976 250 calib. 5-9N, R-8S, with cooling PVS

1976 250 calib. 5-9N, R-8S, no cooling PVS

Ford Motor Company

VACUUM CIRCUITS

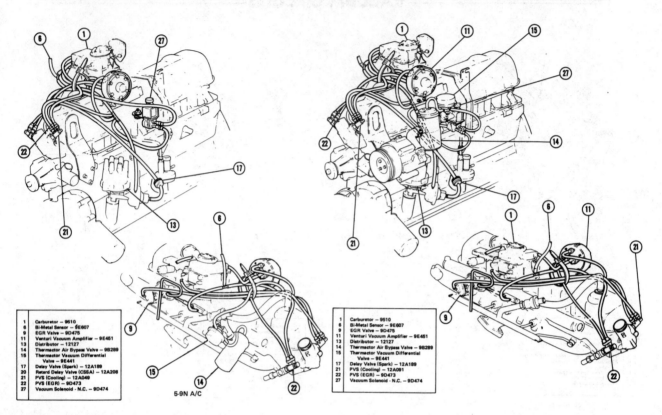

1976 250 calib. 5-9N, R-9, with cooling PVS, Granada and Monarch

1976 250 calib. 5-9N, R-9, with cooling PVS, Maverick & Comet

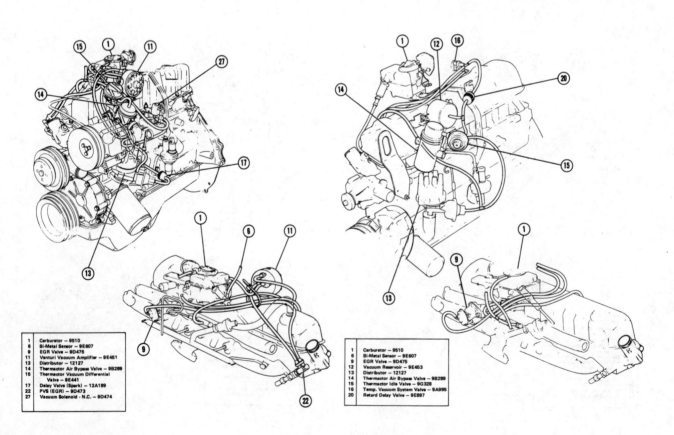

1976 250 calib. 5-9N, R-9, no cooling PVS

1976 250 calib. 6-8G, R-22, 1st type

Ford Motor Company

VACUUM CIRCUITS

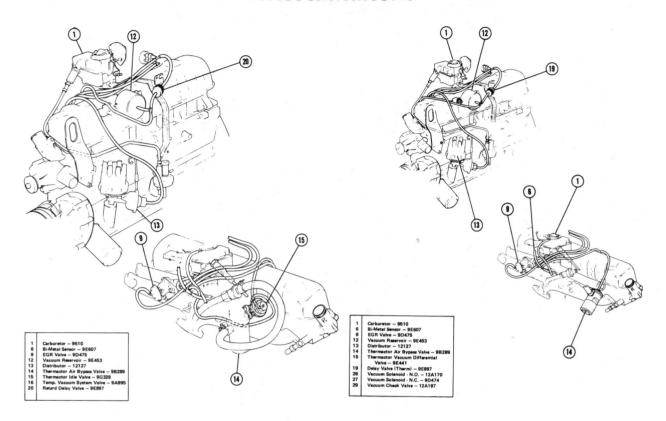

1976 250 calib. 6-8G, R-22, 2nd type

1976 250 calib. 6-9G, R-23, 1st type

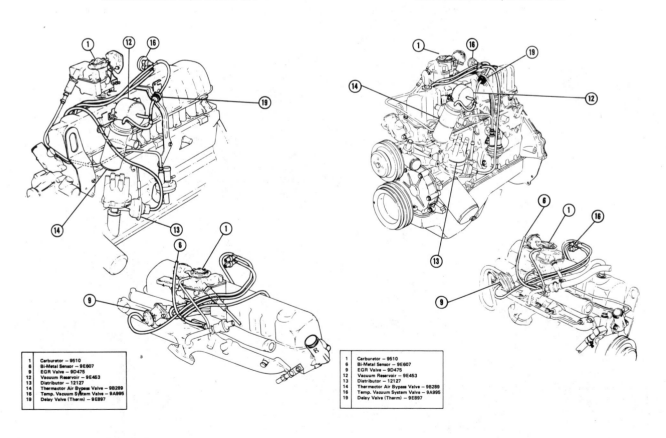

1976 calib. 6-9G, R-23, 2nd type

1976 250 calib. 6-9N, R-20, no cooling PVS.

Ford Motor Company

VACUUM CIRCUITS

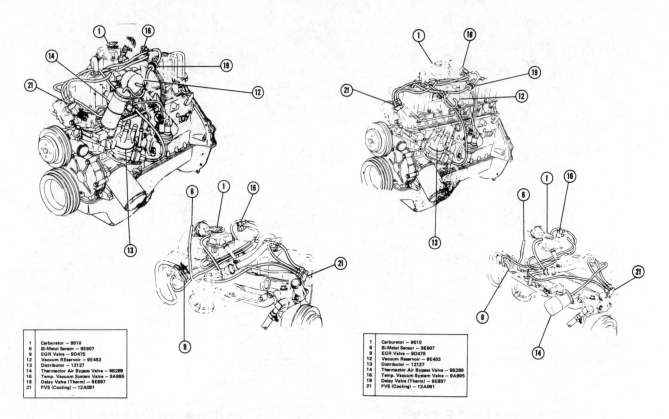

1976 250 calib. 6-9N, R-20, with cooling PVS, type 1

1976 250 calib. 6-9N, R-20, with cooling PVS type 2

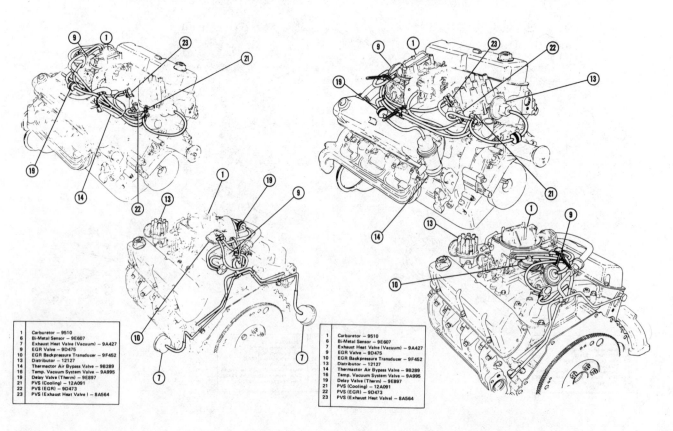

1976 302 calib. 5-10G, R-5

1976 302 calib. 5-10H, R-1, 5-11M, R-2

Ford Motor Company

VACUUM CIRCUITS

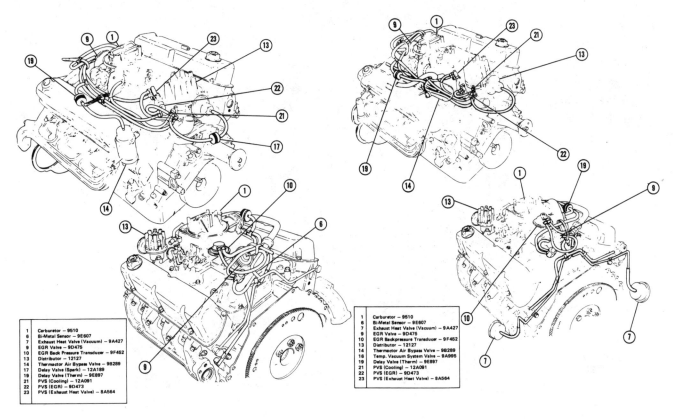

1976 302 calib. 5-11A, R-3

1976 302 calib. 5-11G, R-7

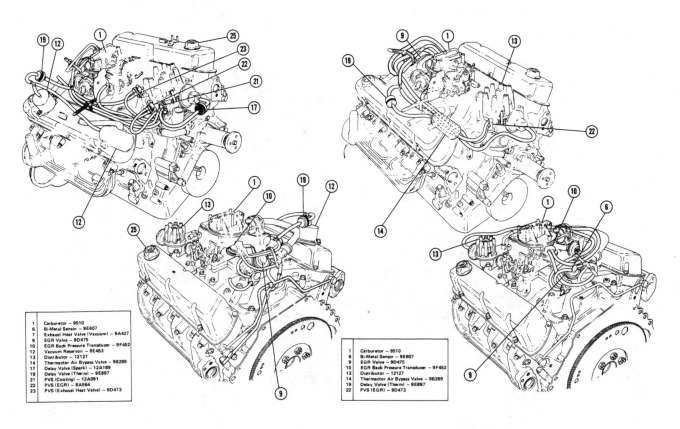

1976 302 calib. 5-11P, R-5

1976 302 calib. 6-11G, R-12

Ford Motor Company

VACUUM CIRCUITS

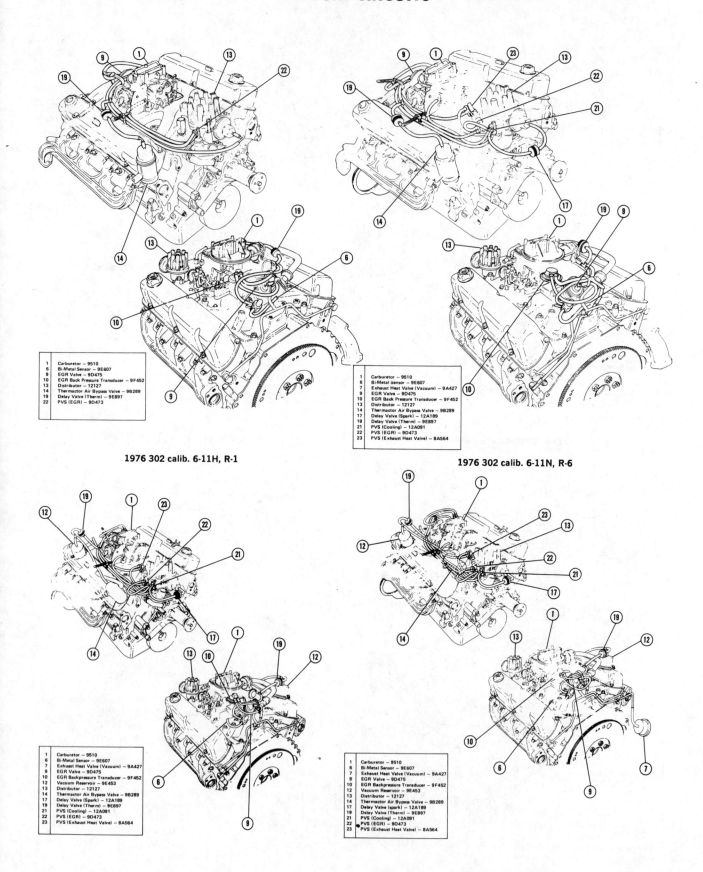

Ford Motor Company

VACUUM CIRCUITS

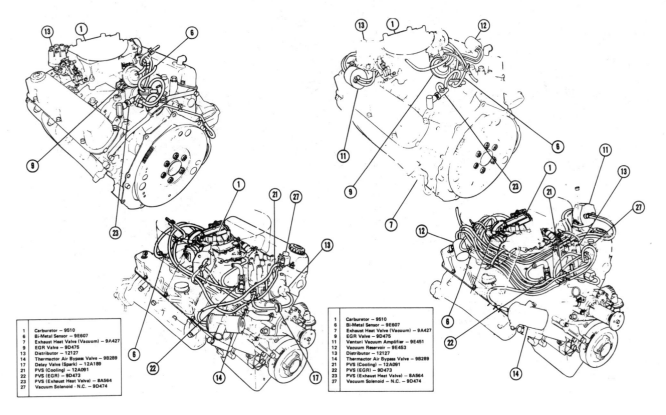

1976 351 w calib. 5-12G, R-75R

1976 351 w calib. 5-12N, R-5s

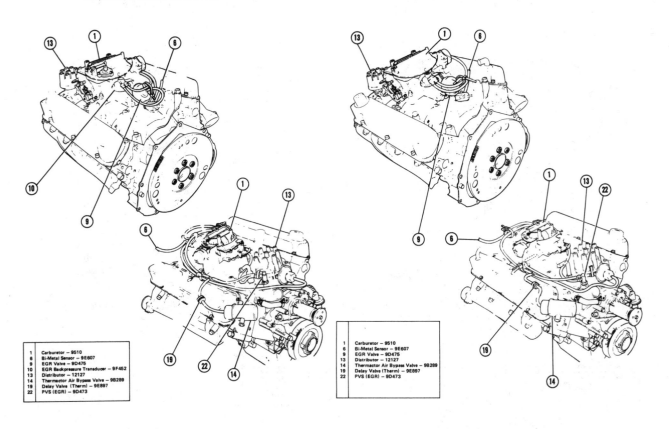

1976 351 w calib. 6-12H, R-O, R-1

1976 351 w calib. 6-12N, R-7

Ford Motor Company

VACUUM CIRCUITS

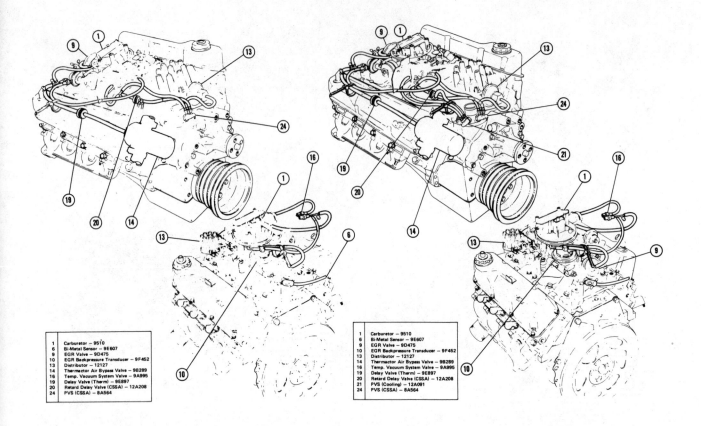

1976 351 M calib. 5-14H, R-61, no cooling PVS

1976 351 M calib. 5-14H, R-61 with cooling PVS

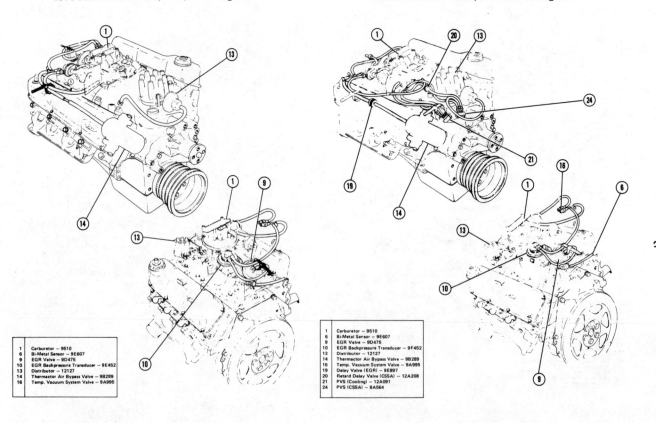

1976 351 M calib. 6-14K, R-11, R-12

1976 351 M calib. 6-14P, R-59, R-60

Ford Motor Company

VACUUM CIRCUITS

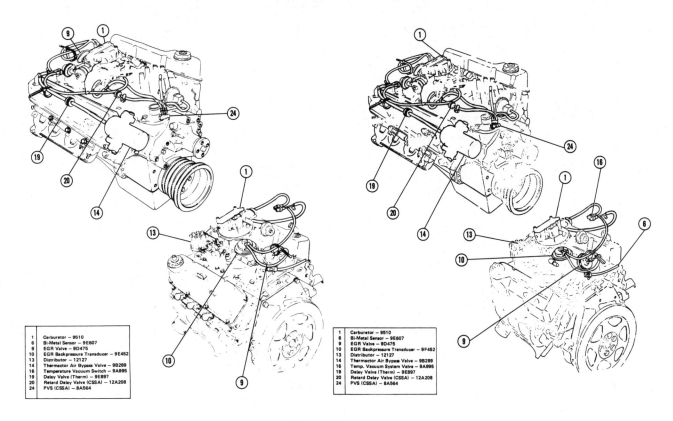

1976 351 M calib. 6-14P, R-64

1976 351 M calib. R-65, no cooling PVS

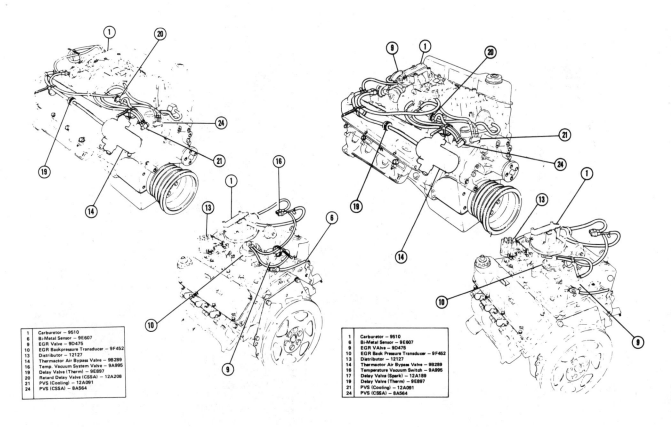

1976 351 M calib. 6-14P, R-65 with cooling PVS

1976 400 calib. 5-17G, R-22, Type 1

357

Ford Motor Company

VACUUM CIRCUITS

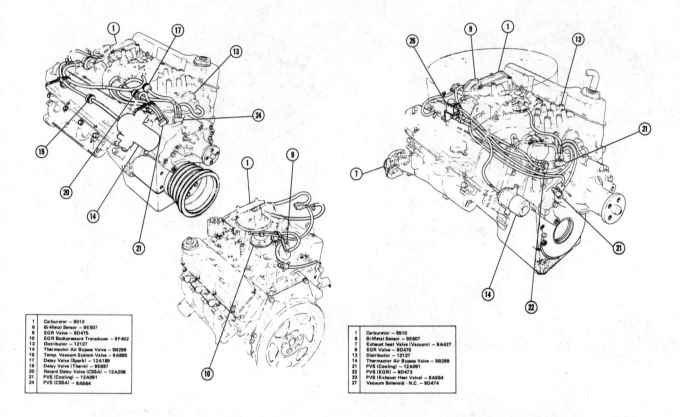

1976 400 calib. 5-17G, R-22, type 2, R-23

1976 400 calib. 5-17N, R-21S

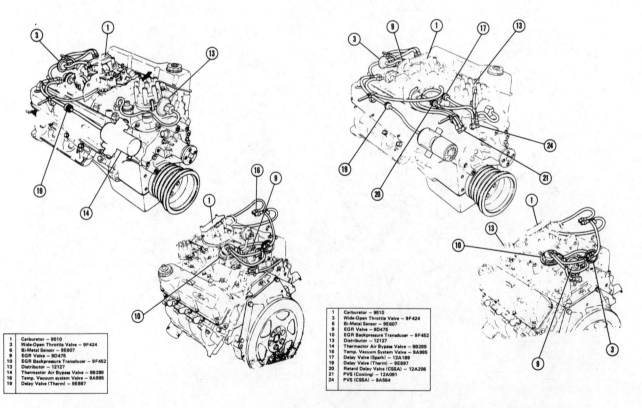

1976 400 calib. 6-17J, R-2, R-14, R-21

1976 400 calib. 6-17N, R-25, R-26, R-27, R-32, R-34

Ford Motor Company

VACUUM CIRCUITS

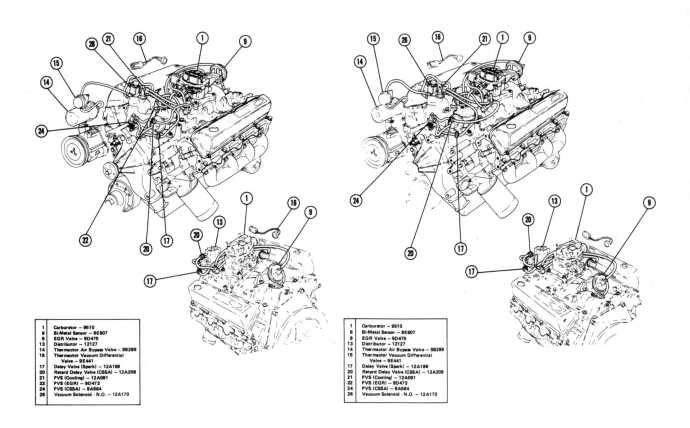

1976 460 calib. 5-19J, R-6, 5-19G, R-6X

1976 460 calib. 5-19K, R-6, 5-19G, R-6

1976 40 calib. 6-19L, R-15

1976 460 calib. 6-19L, R-20

Ford Motor Company

VACUUM CIRCUITS

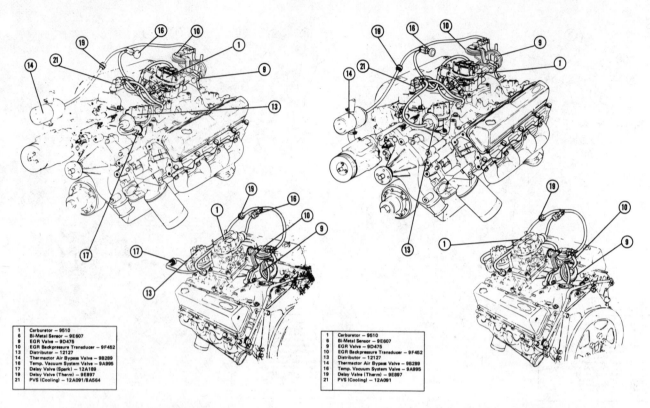

1976 460 calib. 6-19N, R-15, 1st type

1976 460 calib. 6-19N, R-15, 2nd type

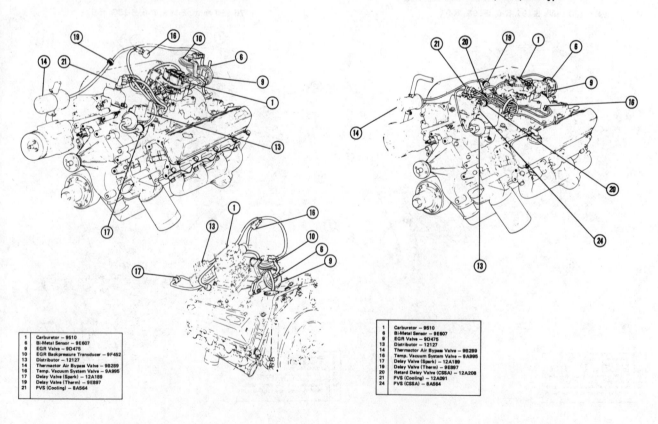

1976 460 calib. 6-19P, R-17

1976 460 Police Interceptor calib. 5-19G, R-6, R-14

Pinto • Mustang & Bobcat 4 Cyl. & V-6

TUNE-UP SPECIFICATIONS — FORD SMALL CARS

Engine Code, 5th character of the VIN number

Yr.	Eng. V.I.N. Code	Engine No. Cyl Disp. (cu in)	Carb Bbl	H.P.	SPARK PLUGS Orig. Type	Gap (in)	DIST. Point Dwell (deg)	Point Gap (in)	IGNITION TIMING (deg • BTDC) Man Trans	Auto Trans	VALVES • Intake Opens (deg)B	FUEL PUMP Pres. (psi)	IDLE SPEED • (rpm) Man Trans	Auto Trans In Drive
1975	Y	4-140	2v	88	AGRF 52	.035	E.I.			③	22°	3.5-5.5	850	750
	Z	6-170.8	2v	105	AGR 42	.035	E.I.			③	20°	3.5-5.5	850 (950)	700
	F	8-302	2v	140	ARF 42	.044	E.I.			③	20°	5-7	750 (900)	650 (700)
1976	Y	4-140	2v	92	AGRF 52	.035	E.I.			③	22°	5-7	850	750
	Z	6-170.8	2v	103	AGR 42	.035	E.I.			③	20°	3.5-5.8	850 (950)	700
	F	8-302	2v	134	ARF 42	.044	E.I.		4°, 12°, 14°③		20°	6-8	750 (800)	650 (700)
1977	Y	4-140	2v	92	AGRF 52	.035	E.I.			③	22°	5-7	850	750 (800)
	Z	6-170.8	2v	103	AGR 42	.035	E.I.			③	20°	3.5-5.8		700 (750)
	F	8-302	2v	134	ARF 52②	.050	E.I.		2° @ 500		16°	6-8	800 (850)	600 (700)
1978	Y	4-140	2v	88	AWRF 42	.034	E.I.		6B	20B		5.5-6.5	850	800 (750)
	Z	6-170.8	2v	90	AWSF 42	.034	E.I.		—	12B(6B)		3.5-5.8	700	650 (600)
	F	8-302	2v	139	ARF 52⑤	.050	E.I.		6B	4B(12B)④		5.5-6.5	900	700
1979	Y	4-140	2v	88*	AWSF 42	.034	E.I.		6B	20B(17B)⑥	22°	5.5-6.5	850	③
	W	4-140TC	2v	150*	⑨	⑨	E.I.			⑨	⑨	⑨	⑨	⑨
	Z	6-170.8	2v	90*	AWSF 42	.034	E.I.		10B	9(6)	20°	3.5-5.8	850	650 (700)
	F	8-302	2v	139*	ASF-52⑤	.050⑧	E.I.		12B	8B(12B)	16°	5.5-6.5	800	600

Note: Should the information provided in this manual deviate from the specifications on the underhood tune-up label, the label specifications should be used, as they may reflect production changes.
• Figure in parentheses indicates California engine
B BEFORE TOP DEAD CENTER
* ESTIMATED
② MANUAL TRANSMISSION
③ SEE DECAL
④ 16° BTDC IN HIGH ALT FORM
⑤ ASF-52-6 IN CALIF
⑥ SET ALTITUDE AND CALIF. ENGINES IN NEUTRAL. NON-AIR CONDITIONED. CALIF. ENGINE —8B
⑧ .060 IN CALIF.
⑨ INFORMATION NOT AVAILABLE AT PUBLICATION TIME

DISTRIBUTOR SPECIFICATIONS — FORD SMALL CARS

Year	Distributor Identification	Centrifugal Advance Start Dist. Deg. @ Dist. RPM	Centrifugal Advance Finish Dist. Deg. @ Dist. RPM	Vacuum Advance Start In. Hg.	Vacuum Advance Finish Dist. Deg. @ In. Hg.
1975	D52E-12127EA	0-2 @ 675	11.5-14 @ 2,250	4	2.75-5.25 @ 7
	D52E-12127FA	0-2 @ 715	11.5-14 @ 2,150	4	2.75-5.25 @ 7.5
	D5ZE-12127AA	0-2 @ 600	10.25 @ 2,500	4.8	2.75-5.25 @ 7.5
	D5ZE-12127CA	0-2.35 @ 600	10.25-12.75 @ 2,400	4.75	13-15.75 @ 15.5
	D5TF-12100NA	0-2 @ 650	9-12 @ 2,500	4.25	5-7 @ 9.5
	D5TF-12100EA	0-2 @ 650	8-10 @ 2,000	4.25	2-3 @ 6.75
	D5TF-12100MA	0-2 @ 660	6.5-9.5 @ 2,500	4.25	5-7 @ 9.5
	D5DE-12127KA	0-2 @ 775	8.25-10.75 @ 2,350	6	8.75-11.25 @ 14.5
	D5ZE-12127BA	0-2 @ 540	8.25-11.50 @ 2,500	5	8.25-11.25 @ 7.5
76	76TF-12100EA	.5 @ 550	10.5-12.5 @ 2,500	5	5 @ 8.5
	76TF-12100FA	0-2 @ 650	8-10 @ 2,000	5.5	4.5-7 @ 9.5
	76TF-12100GA	0-2 @ 650	8-10 @ 2,000	5.6	8-10 @ 12.5
	76TF-12100JA	0-2 @ 680	9.8-11.8 @ 2,000	4	2 @ 7.2
	D5DE-12127AFA	0-2.5 @ 550	10-12.4 @ 2,500	5.5	10.8-13.3 @ 15
	D6DE-12127JA	0-3.5 @ 500	16.3-18.8 @ 2,200	5	10.8 @ 13
	D6EE-12127AA	0-2.3 @ 650	13-15.5 @ 2,500	4	6.5-9.8 @ 7.2
	D6EE-12127BA	0-.5 @ 650	5-7.5 @ 2,500	5	10.8-13.3 @ 15.8
	D6EE-12127DA	0-2.5 @ 600	11.4-14 @ 2,500	4	10.8-12.3 @ 12.5

Pinto • Mustang & Bobcat 4 Cyl. & V-6

DISTRIBUTOR SPECIFICATIONS — FORD SMALL CARS

Year	Distributor Identification	Centrifugal Advance Start Dist. Deg. @ Dist. RPM	Centrifugal Advance Finish Dist. Deg. @ Dist. RPM	Vacuum Advance Start In. Hg.	Vacuum Advance Finish Dist. Deg. @ In. Hg.
77	77TF-12100AA	0-1 @ 625	8-10.5 @ 2,100	4	8-10 @ 12
	77TF-12100CA	0-1 @ 625	8-10.5 @ 2,100	4.5	5-7 @ 10
	77TF-12100DA	0-1 @ 510	11.5-14 @ 2,500	1.75	10.8-13.3 @ 12.4
	D0EA-12127GA	0-1 @ 425	13.5-16 @ 2,500	3	10.8-13.3 @ 16
	D7EE-12127CA	0-1 @ 800	5-7.5 @ 2,500	2.3	10.8-13.3 @ 15.75
	D7EE-12127DA	0-1 @ 525	11.5-14 @ 2,500	1.75	10.8-13.8 @ 12.4
	D7EE-12127EA	0-1 @ 525	11.5-14 @ 2,500	2	10.8-13.3 @ 15.75
	D7EE-12127GA	0-1 @ 525	11.5-14 @ 2,500	2.25	10.8-13.3 @ 15.75
	D7EE-12127HA	0-1 @ 775	5-7.5 @ 2,500	2	10.8-13.3 @ 15.75
	D7ZE-12127BA	0-1 @ 425	13.5-16 @ 2,500	3.5	10.8-13.3 @ 15.2
	D7ZE-12127CA	0-1 @ 575	9.3-12 @ 2,500	2.2	12.8-15.3 @ 16
78	77TF-12100AA	0-1 @ 625	8-10.5 @ 2,100	4	8-10 @ 12
	77TF-12100CA	0-1 @ 625	8-10.5 @ 2,100	4.5	5-7 @ 10
	77TF-12100HA	0-1 @ 600	11-12 @ 2,100	4.5	5-7 @ 10
	D7DE-12127HA	0-1 @ 550	9.5-12.5 @ 2,500	3	12.8-15.3 @ 11
	D7DE-12127JA	0-1 @ 550	8.8-10.8 @ 2,500	3	10.8-13.3 @ 16
	D7EE-12127CA	0-1 @ 800	5-7.5 @ 2,500	2.3	10.8-13.3 @ 15.75
	D7EE-12127DA	0-1 @ 525	11.5-14 @ 2,500	1.75	10.8-13.8 @ 12.4
	D7EE-12127EA	0-1 @ 525	11.5-14 @ 2,500	2	10.8-13.3 @ 15.75
	D8ZE-12127BA	0-1 @ 425	13.5-16 @ 2,500	3.5	10.8-13.3 @ 14
	D8ZE-12127CA	0-1 @ 575	9.5-12.5 @ 2,500	2.5	9.8-12.3 @ 15.7

VEHICLE IDENTIFICATION NUMBER (VIN) COMPOSITION

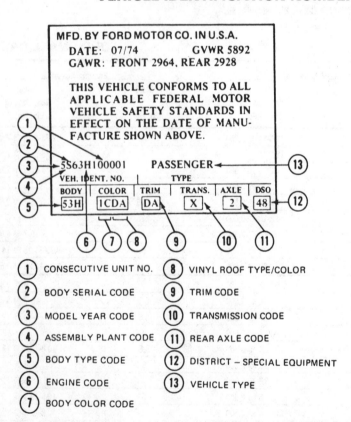

1. CONSECUTIVE UNIT NO.
2. BODY SERIAL CODE
3. MODEL YEAR CODE
4. ASSEMBLY PLANT CODE
5. BODY TYPE CODE
6. ENGINE CODE
7. BODY COLOR CODE
8. VINYL ROOF TYPE/COLOR
9. TRIM CODE
10. TRANSMISSION CODE
11. REAR AXLE CODE
12. DISTRICT – SPECIAL EQUIPMENT
13. VEHICLE TYPE

VIN TAG LOCATION

The official vehicle identification number is stamped on a plate attached to the forward left hand corner of the upper instrument panel. It can be seen and read through the windshield.

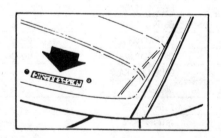

Pinto • Mustang & Bobcat 4 Cyl. & V-6

CAR SERIAL NUMBER AND ENGINE IDENTIFICATION

1976

Mounted behind the windshield on the driver's side is a plate with the vehicle identification number. The first character is the model year, with 6 for 1976. The fifth character is the engine code, as follows.

Y .... 140 (2300 cc) 4-cyl. 2-bbl.
Z .... 171 (2800 cc) V-6 2-bbl.

NOTE: *When the numbers 2300 cc and 2800 cc are converted to cubic inches, the 2300 comes out 140.3 cu. in. and the 2800 equals 170.8 cu. in. In some publications these engines have been called the 139 and the 169, but Ford is now referring to the 2300 as a 140, and to the 2800 as a 171. This does not mean the engines have been changed. They are the same bore and stroke they have always been.*

1977

The vehicle identification number is mounted on a plate behind the windshield on the driver's side. The first character is the model year, with 7 for 1977. The fifth character is the engine code, as follows.

Y .... 2300 2-bbl. 4-cyl.
Z 2800 2-bbl. V-6

1978

The vehicle identification number is mounted on a plate behind the windshield on the driver's side. The first character is the model year, with 8 for 1978. The fifth character is the engine code, as follows:

Y .... 2300 2-bbl. 4-cyl.
Z 2800 2-bbl. V-6

1979

The vehicle identification number is mounted on a plate behind the windshield on the driver's side. The first character is the model year, with 9 for 1979. The fifth character is the engine code, as follows:

Y ...... 139 2-bbl. 4-cyl. (all models)
Z ..... 169 2-bbl. V6 (Pinto, Bobcat, Mustang, Capri)
T ... 200 1-bbl. 6-cyl. (Fairmont and Zephyr)
F .. 302 2-bbl. V8 (Fairmont, Zephyr, Mustang, Capri)
W .... 139 Turbo Charged 4-cyl. (Mustang and Capri)

EMISSION EQUIPMENT

1976
All Models
Closed positive crankcase ventilation
Emission calibrated carburetor
Emission calibrated distributor
Heated air cleaner
Vapor control, canister storage
Electric choke
Catalytic converter
Spark delay valve (some engines)
Air pump
Exhaust gas recirculation, ported vacuum system
Decel enrichment valve, not used
Exhaust heat valve, not used

1977
All Models
Closed positive crankcase ventilation
Emission calibrated carburetor
Emission calibrated distributor
Heated air cleaner
Vapor control, canister storage
Exhaust gas recirculation
Electric choke
Catalytic converter
Air pump
  49-States
    Manual trans. only
  Calif.
    All models
Altitude
    All models
Decel enrichment valve idle speedup
  49-States
    Not used
  Calif.
    Man. trans. with A.C. calibrations 7-2N-R1, and 7-2T-R0
  Altitude
    Not used

1978
All Models
Closed positive crankcase ventilation
Emission calibrated carburetor
Emission calibrated distributor
Heated air cleaner
Vapor control, canister storage
Exhaust gas recirculation
Electric choke
Catalytic converter
Air pump
  All engines except
    Not used on 4-cyl. calibs. 8-21B-R0, 8-21B-R11, 8-21D-R0
Feedback carburetor
  Calif. 4-cyl. Pinto & Bobcat

1979
All Models
Closed positive crankcase ventilation
Emission calibrated carburetor
Emission calibrated distributor
Heated air cleaner
Vapor control, canister storage
Exhaust gas recirculation
Electric choke
Catalytic converter
Air pump
  All engines except:
  (Y) Fairmont and 6-cyl. (T) Fairmont
  Not used on 4-cyl.
Feedback carburetor
  California 4-cyl. Pinto and Bobcat

IDLE SPEED AND MIXTURE AJDUSTMENTS

1976

Air cleaner .... On
Air cond. .... Off
Fuel decel hose .... Plugged
Auto. Trans. .... Drive
2300 4-cyl. 49-States
  Auto. Trans.
    Decal 6-127 750
    Decal 6-211 750
  Man. Trans.
    Decal 6-127 850
    Decal 6-209 850
    Decal 6-231 850
    Decal 6-235 850

2300 4-cyl. Calif.
  Auto. Trans.
    Decal 6-096 750
    Decal 6-313 750
  Man. Trans.
    Decal 6-096 850
    Decal 6-312 850
2800 V-6 49-States
  Auto. Trans.
    Decal 6-128 700
  Man. Trans.
    Decal 6-128 850
    Decal 6-216 850
    Decal 6-229 850

2800 V-6 Calif.
  Auto. Trans.
    Decal 6-066 700
    Decal 6-100 800
    Decal 6-325 800
  Man. Trans.
    Decal. 6-066 950
    Decal 6-100 850
    Decal 6-309 850

IDLE MIXTURE ARTIFICIAL ENRICHMENT

Air cleaner .... On

Pinto • Mustang & Bobcat 4 Cyl. & V-6

IDLE SPEED AND MIXTURE ADJUSTMENTS

1976
Vapor hose to air cleaner ...... Off
PCV hose to air cleaner .... Off
PCV air cleaner connection . Plugged
Air pump .... Disconnected
Auto. Trans. .... See below

Speed Gain (RPM)
NOTE: *Check speed gain at idle speed. These figures can also be used for speed drop.*

Engine	Range	Reset
2300 4-cyl. 49-States		
Auto. Trans. (Drive)		
Decal 6-127	10-30	20
Decal 6-211	0-30	10
Man. Trans.		
Decal 6-127	10-30	20
Decal 6-209	0-30	10
Decal 6-231	0-30	10
Decal 6-235	0-30	10
2300 4-Cyl. Calif.		
Auto. Trans. (Drive)		
Decal 6-096	10-30	20
Decal 6-313	0-30	10
Man. Trans.		
Decal 6-096	10-30	20
Decal 6-312	0-20	10
2800 V-6 49-States		
Auto. Trans. (Neutral)		
Decal 6-128	0-80	20
Man. Trans.		
Decal 6-128	0-50	10
Decal 6-216	0-50	10
Decal 6-229	0-50	10
2800 V-6 Calif.		
Auto. Trans. (Drive)		
Decal 6-066	70-110	85
Auto. Trans. (Neutral)		
Decal 6-100	40-130	90
Decal 6-325	0-130	50
Man. Trans.		
Decal 6-066	0-20	0
Decal 6-100	0-75	10
Decal 6-309	0-75	10

1977
Air cleaner .... In place
Air cond. .... Off
Auto. trans. .... Drive
2300 4-cyl. 49-States
  Auto. trans.
    Solenoid connected 800
    Solenoid disconnected 600
  Man. trans. 850
2300 4-cyl. Calif.
  Auto. trans.
    Solenoid connected 750
    Solenoid disconnected 600
  Man. trans.
    Solenoid connected 850
    Solenoid disconnected 800
2300 4-cyl. Altitude
  Auto. trans.
    Solenoid connected 750
    Solenoid disconnected 550
2800 V-6 49-States
  Auto. trans. 700
  A.C. idle speedup 750
  Man. trans. 850
2800 V-6 Calif.
  Auto. trans.
    Solenoid connected 750
    Solenoid disconnected 600

NOTE: *On engines with a combination vacuum-electric solenoid, curb idle speed is set with the screw that contacts the solenoid stem, with the electric part of the solenoid energized. When the air conditioning is ON, the vacuum part of the solenoid is activated to give a faster idle speed, which is not adjustable.*

1978
Air cleaner .... In place
Air cond. .... Off
Auto. trans. .... Drive
2300 4-cyl. 49-States
  Auto. trans. 800
  Man. trans. 850
2300 4-cyl. Calif.
  Auto. trans. 750
  Man. trans.
2800 V-6 49-States
  Auto. trans. 650
  AC idle speedup 700
  Man. trans. 700
2800 V-6 Calif.
  Auto. trans. 600

IDLE MIXTURE SETTINGS 1978

NOTE: *When both a lean drop and a propane procedure are given, either may be used. Make all settings in Neutral or Drive, as shown. Figures given are in revolutions per minute.*

2300 4-cyl. auto. trans.
  Cal. 8-1A-R10, 8-1D-R10 (Drive)
    Propane okay range 5-40
    Propane reset 20
  Cal. 8-1N-R10 (Drive)
    Propane okay range 0-20
    Propane reset 10
  Cal. 8-1P-R1 (Drive)
    Propane okay range 20-40
    Propane reset 30
  Cal. 8-1X-R0 (Neutral)
    Propane okay range 10-40
    Propane reset 20
  Cal. 8021B-R0, 8-21B-R11,
  8-21D-R0 (Drive)
    Propane okay range 40-60
    Propane reset 50
2300 4-cyl. man. trans.
  Cal. 8-2A-R0, 8-2A-R10,
  8-2B-R0, 8-2D-R0
    Propane okay range 0-30
    Propane reset 10
  Cal. 8-2M-R0
    Propane okay range 0-30
    Propane reset 15
    Lean drop 25
  Cal. 8-2M-R10, 8-2N-R0
    Propane okay range 0-30
    Propane reset 15
  Cal. 8-2P-R0
    Propane okay range 20-40
    Propane reset 30
  Cal. 8-2T-R0
    Propane okay range 20-40
    Propane reset 30
    Lean drop 40
  Cal. 8-2X-R0
    Propane okay range 10-40
    Propane reset 20
2800 V-6 auto. trans.
  Cal. 8-4A-R0, 8-4N-R0,
  8-4N-R10 (Drive) ...... Set to lean best idle
2800 V-6 man. trans.
  Cal. 8-3A-R0 .... Set to lean best to idle

1979
NOTE: *The underhood specifications sticker sometimes reflects tune-up changes made in production. Sticker figures must be used if they disagree with this information.*

Air cleaner .... In place
Air Cond. .... Off
Auto. trans ..Drive unless otherwise noted
Headlights .... Off

NOTE: *On cars equipped with Vacuum Operated Throttle Modulator (VOTM), which provides intermediate fast idle speed during cold engine operation, the vehicle emissions label should be checked before any adjustments are made. Some models also use Throttle Solenoid Positioner (TSP) for curb idle adjustment. Refer to the vehicle emissions label to see if vehicle is so equipped and for the adjustment procedure. Refer to Ford Section for further explanation.*

139 2-bbl. 4-cyl. Vin code (Y) 49 States Pinto and Bobcat
  Auto. trans.
    Fast idle .... 2000 (N)
    Curb idle
      A/C 800
      Non A/C 800
    TSP off
      A/C 600
  Manual trans.
    Fast idle 1800
    Curb idle
      A/C on 1300
      A/C off 850
      Non A/C 850
139 2-bbl. 4-cyl. Vin code (Y) Calif. Pinto and Bobcat
  Manual trans.
    Fast idle 1850
    Curb idle
      Non A/C 850
139 2-bbl. 4-cyl. Vin code (Y) Altitude Pinto and Bobcat
  Auto. trans.
    Fast idle .... 2000 (N)
    Curb idle
      A/C 800

Pinto • Mustang & Bobcat 4 Cyl. & V-6

IDLE SPEED AND MIXTURE ADJUSTMENTS

1979

Non A/C 800
TSP off
  A/C 550
  Non A/C 550
Manual trans.
  Fast idle 1800
  Curb idle
    A/C 850
    Non A/C 850
  TSP off
    A/C 550
    Non A/C 550

169 2-bbl. V6 Vin code (Z) 49 States Pinto and Bobcat
Auto. trans.
  Fast idle .. 1600, (N), 1800 (N) Bobcat
  Curb idle
    A/C on 750
    A/C off 650
    Non A/C 650

169 2-bbl. V6 Vin code (Z) Calif. Pinto and Bobcat
Auto. trans.
  Fast idle .... 1750 (N)
  Curb idle
    A/C on 700
    Non A/C 700
  TSP off
    A/C .... 600 (N)
    Non A/C .... 600 (N)

139 2-bbl. 4-cyl. Vin code (Y) 49 States Fairmont and Zephyr
Auto. trans.
  Fast idle .... 2000 (N)
  Curb idle
    A/C 800
    Non A/C 800
  TSP off
    A/C 600
    Non A/C 600
Manual trans.
  Fast idle .... 1600 Fairmont, 1800 Zephyr
  Curb idle
    A/C 1300
    A/C off 850
    Non A/C 850

139 4-cyl. Vin code (Y) Calif. Fairmont and Zephyr
Auto. trans.
  Fast idle ... 1800 (N), 1300 (N) non A/C
  Curb idle
    A/C 750
    Non A/C 850
  TSP off
    A/C 600
    Non A/C 600
Manual trans.
  Fast idle 1800
  Curb idle
    A/C 850
    Non A/C 850
  TSP off
    A/C 600
    Non A/C 600

200 1-bbl. 6-cyl. Vin code (T) 49 States Fairmont and Zephyr
Auto. trans.
  Fast idle .... 1700 (N)
  Curb idle
    A/C 650
    Non A/C 650
  TSP off
    A/C .... 700 (N)
    Non A/C .... 700 (N)
Manual trans.
  Fast idle 1600
  Curb idle
    A/C on 850
    A/C off 700
    Non A/C 700

200 1-bbl. 6-cyl. Vin code (T) Calif. Fairmont and Zephyr
Auto. trans.
  Fast idle .... 1850 (N)
  Curb idle
    A/C on 650
    A/C off 600
    Non A/C 600

200 1-bbl. 6-cyl. Vin code (T) Altitude Fairmont and Zephyr
Auto. trans.
  Fast idle .... 1700 (N)
  Curb idle
    A/C 650
    Non A/C 650
  TSP off
    A/C 700
    Non A/C 700

302 2-bbl. V8 Vin code (F) 49 States Fairmont and Zephyr
Auto. trans.
  Fast idle .... 2100 (N)
  Curb idle
    A/C on 675
    A/C off 600
    Non A/C 600
Manual trans.
  Fast idle 2300
    A/C on 875
    A/C off 800
    Non A/C 800

302 2-bbl. V8 Vin code (F) Calif. Fairmont and Zephyr
Auto. trans.
  Fast idle .... 1800 N)
  A/C on 675
  A/C off 600
  Non A/C 600
  Non A/C 600
VOTM
  A/C 700
  Non A/C 700

302 2-bbl. V8 Vin code (F) Altitude Fairmont and Zephyr
Auto. trans.
  Fast idle .... 2100 (N)
  Curb idle
    A/C on 675
    A/C off 600
    Non A/C 600

139 2-bbl. 4-cyl. Vin code (Y) 49 States Mustang and Capri
Auto. trans.
  Fast idle .... 2000 (N)
  Curb idle
    A/C 800
    Non A/C 800
  TSP off
    A/C 600
    Non A/C 600
Manual trans.
  Fast idle 1600
  Curb idle
    A/C on 1300
    A/C off 850
    Non A/C 850

139 2-bbl. 4-cyl. Vin code (Y) Calif. Mustang and Capri
Auto. trans.
  Fast idle .... 1800 (N)
  Curb idle
    A/C 750
    Non A/C 850
  TSP off
    A/C 600
    Non A/C 600
Manual trans.
  Fast idle 1800
  Curb idle
    A/C 850
    Non A/C 850
  TSP off
    A/C 600
    Non A/C 600

139 2-bbl. 4-cyl. Vin code (Y) Altitude Mustang and Capri
Manual trans.
  Fast idle 1800
  Curb idle
    A/C 850
    Non A/C 850
  TSP off
    A/C 550
    Non A/C 550

169 2-bbl. V6 Vin code (Z) 49 States Mustang and Capri
Auto. trans.
  Fast idle ... 1800 (N) Mustang, 1600 (N) Capri
  Curb idle
    A/C on 750
    A/C off 650
    Non A/C 650

169 2-bbl. V6 Vin code (Z) Calif. Mustang and Capri
Auto. trans.
  Fast idle .... 1750 (N)
  Curb idle
    A/C 700
    Non A/C 700
  TSP off
    A/C .... 600 (N)
    Non A/C .... 600 (N)

302 2-bbl. V8 Vin code (F) 49 States Mustang and Capri
Auto. trans.
  Fast idle .... 2100 (N)
  Curb
    A/C on 675
    A/C off 600
    Non A/C 600
Manual trans.
  Fast idle 2300
  Curb idle
    A/C on 875
    A/C off 800

Pinto • Mustang & Bobcat 4 Cyl. & V-6

IDLE SPEED AND MIXTURE AJDUSTMENTS

1979

Non A/C 800

302 2-bbl. V8 Vin code (F) Calif.
Mustang and Capri
Auto. trans.
  Fast idle .... 1800 (N)

Curb idle
  A/C on 675
  A/C off 600
  Non A/C 600
VOTM
  A/C 700
  Non A/C 700

139 Turbo Charged 4-cyl. Vin code (W) 49 States Mustang and Capri
Manual trans.
  Fast idle 1800
Curb idle
  A/C on 1300
  A/C off 900
  Non A/C 900

INITIAL TIMING

1976

NOTE: *Distributor vacuum hoses must be disconnected and plugged. Set automatic transmission engines in Neutral unless shown otherwise. Set timing with engine idling at speed shown.*

2300 4-cyl. 49-States
  Auto. Trans.
    Decal 6-127 (550 rpm) .... 20° BTDC
    Decal 6-211 (550 rpm) .... 20° BTDC
  Man. Trans.
    Decal 6-127 (550 rpm) .... 6° BTDC
    Decal 6-209 (550 rpm) .... 6° BTDC
    Decal 6-231 (550 rpm) .... 6° BTDC
    Decal 6-235 (550 rpm) .... 6° BTDC
2300 4-cyl. Calif.
  Auto. Trans.
    Decal 6-096 (550 rpm) .... 20° BTDC

Dscal 6-313 (550 rpm) .... 20° BTDC
Man. Trans.
  Decal 6-096 (550 rpm) .... 20° BTDC
  Decal 6-312 (550 rpm) .... 6° BTDC
2800 V-6 49-States
  Auto. Trans.
    Decal 6-128 (700 rpm in Drive) .... 12° BTDC
  Man. Trans.
    Decal 6-128 (700 rpm) .... 10° BTDC
    Decal 6-229 (700 rpm) .... 10° BTDC
    Decal 6-216 (700 rpm) .... 10° BTDC
2800 V-6 Calif.
  Auto. Trans.
    Decal 6-066 (700 rpm in Drive) .... 8° BTDC
    Decal 6-100 (650 rpm) .... 6° BTDC
    Decal 6-325 (650 rpm) .... 6° BTDC

1977

NOTE: *Distributor vacuum hoses must be disconnected and plugged.*
2300 4-cyl.
  Auto. trans. .... 20° BTDC
  Man. trans. .... 6° BTDC
2800 V-6 49-States
  Auto. trans. .... 12° BTDC
  Man. trans. .... 10° BTDC
2800 V-6 Calif.
  Auto. trans. .... 6° BTDC

1978

NOTE: *Distributor vacuum hoses must be disconnected and plugged.*
2300 4-cyl.
  Auto. trans. .... 20° BTDC

Man. trans. .... 6° BTDC
2800 V-6 49-States
  Auto. trans. .... 12° BTDC
  Man. trans. .... 10° BTDC
2800 V-6 Calif.
  Auto. trans. .... 6° BTDC

1979

NOTE 1: *Before making any adjustments on the ignition timing refer to the Vehicle Emissions Label located in the engine compartment of the vehicle.*
2. *Distributer vacuum hoses must be disconnected and plugged.*

139 2-bbl. 4-cyl.
  Auto. trans. .... 20° BTDC

Manual trans. .... 6° BTDC
169 2-bbl. V6 49 States
  Auto. trans. .... 9° BTDC
  Manual trans. .... 6° BTDC
169 2-bbl. V6 California
  Auto. trans. .... 6° BTDC
200 1-bbl. 6 cyl. 49 States
  Auto. trans. .... 10° BTDC
  Manual trans. .... 8° BTDC
200 1-bbl. 6-cyl. California
  Auto. trans. .... 10° BTDC
302 2-bbl. V8 49 States
  Auto. trans. .... 8° BTDC
  Manual trans. .... 12° BTDC
302 2-bbl. V8 California
  Auto. trans. .... 12° BTDC
139 Turbo Charged 4-cyl. 49 States
  Manual trans. .... 2° BTDC

VACUUM ADVANCE

Diaphragm type
  2300 4-cyl.
    49-States .... Single
    Calif .... Dual and Single
  2800 V-6 .... Single

SPARK PLUGS

1976

2300 4-Cyl. ...AU-AGRF-52....034
2800 V-6 .....AU-AGRF-42....034

1977

2300 4-cyl. .. AU-AWRF-42.. .034
2800 V-6 .. AU-AWRF-42.. .034

NOTE: *The 1977 engines use tapered-seat plugs.*

1978

2300 4-cyl. ...... AU-AWSF-42 ...034
2800 V-6 .... AU-AWSF-42 ...034

1979

139 4-cyl. ...... AU-AWSF-42 ...034
169 V6 .... AU-AWSF-42 ...034
200 6-cyl. .... AU-BSF-82 ...050
302 V8 .... AU-ASF-52 ...050
139 Turbo Charged
  4-cyl .... AU-AWST-32 ...034

Pinto • Mustang & Bobcat 4 Cyl. & V-6

EMISSION CONTROL SYSTEMS 1976

AIR PUMP SYSTEM 1976

Description of System

The air bypass valve is a new type with two small hose connection nozzles. It is called a "Timed Air Bypass Valve with Vacuum Vent." Manifold vacuum connects to the hose nozzle on the body of the valve, and acts on the bottom side of a diaphragm. A calibrated hole in the diaphragm allows the vacuum to reach the top side of the diaphragm, so that the vacuum equalizes on both sides. With the vacuum equalized, there is no force on the diaphragm, and the spring inside the bypass valve moves the plunger to the flow position, allowing the pump air to flow to the engine exhaust ports.

To get the bypass valve into the dump position, all that is necessary is to vent the top side of the diaphragm. With the top side vented, the manifold vacuum on the bottom side pulls the diaphragm and stem down, and the bypass valve goes into the dump position, which sends the pump air into the atmosphere.

Venting of the top side of the diaphragm is done in different ways on 4-cylinder and V-6 engines. The 4-cylinder engines have the bypass valve vent (nozzle on side of valve end cap) connected to a vacuum solenoid mounted on the fender pan in the engine compartment. The solenoid is connected to an electric Thermo Actuated Valve (TAV) in the air cleaner. Below approximately 65° F. air cleaner temperature, the TAV is closed, which allows current to go to the vacuum solenoid and open it. With the vacuum solenoid open, the bypass valve is vented, and the pump air is dumped into the atmosphere.

As the 4-cylinder engine warms up, above an air cleaner temperature of 65° F. the TAV opens, which cuts off the current to the vacuum solenoid, allowing it to close. With the vacuum solenoid closed, the vent on the bypass valve is closed. The vacuum equalizes on both sides of the diaphragm through the bleed hole, and the spring in the bypass valve moves it to the flow position. The bypass valve will also dump during deceleration. Manifold vacuum increases on one side of the diaphragm faster than it can pass through the diaphragm hole. After a few seconds, the vacuum equalizes on both sides of the diaphragm, and the dumping stops.

On V-6 engines, venting is controlled by an idle vacuum valve. This valve is simply a vacuum-operated vent. When vacuum is applied to the valve, the vent closes. If there is no vacuum on the valve, the vent opens. The vent is connected by a hose to the vent chamber on the air bypass valve. Vacuum does not pass through the idle vacuum valve at any time. All the vacuum does is open and close the vent. Because the idle vacuum valve is connected to ported vacuum, the vent is closed above idle, and open at idle and during deceleration.

Venting of the bypass valve is only necessary during prolonged idle. To delay the dumping for approximately half a minute to a full minute, a vacuum delay valve is inserted in the hose near the idle vacuum valve. The vacuum delay valve traps the vacuum, and it takes half a minute to a full minute to bleed down. This insures that the bypass valve only goes into the dump position during prolonged idle. Some engines have a vacuum reservoir inserted between the delay valve and the bypass valve. This provides an additional amount of vacuum to leak down through the delay valve, and lengthens the time it takes the bypass valve to get into the dump position.

Vacuum from the carburetor port runs through a temperature vacuum switch (TVS) in the air cleaner. Below approximately 60° F. air cleaner temperature, the valve is closed, shutting off the vacuum to the idle vacuum valve, so that the bypass valve stays in the dump position. When the air cleaner warms up to above 60° F. the TVS opens and allows the ported vacuum to close the vent in the idle vacuum valve. The bypass valve then goes to the flow position.

Vacuum to the EGR valve is also controlled by the TVS in the air cleaner. The result is that the EGR valve is also off below an air cleaner temperature of less than approximately 60° F.

Units in System with Vacuum Solenoid

Air pump with filter fan
Bypass valve with 2 small nozzles
Vacuum solenoid
TAV switch in air cleaner
Check valves
Cylinder heads with air passages

Units in System with Idle Vacuum Valve

Air pump with filter fan
Bypass valve with 2 small nozzles
Idle vacuum valve
Vacuum reservoir (some models)
Vacuum delay valve
Temperature vacuum switch in air cleaner
Check valve
Cylinder heads with air passages

Testing and Troubleshooting System, 1976

Testing is done on each part of the system, rather than on the whole system at once. See below for tests on the individual units that have been changed for 1976.

Repairing System, 1976

Belt tension is the only adjustment to the system. All units are replaced if defective.

Testing Dual Nozzle Air Bypass Valve, 1976

This valve has two small hose connection nozzles, one on the side of the body, and the other on the side of the end cap. The nozzle on the body connects to manifold vacuum, and the nozzle on the end cap connects to either the idle vacuum valve on the V-6 or the vacuum solenoid on the 4-cylinder.

1. With the engine running at 1500 rpm, disconnect the small hose from the body of the bypass valve, and check to be sure you have full manifold vacuum at the end of the hose.
2. Reconnect the hose.
3. Disconnect the large hose between the bypass valve and the check valve, and position the hose so you can feel the blast of air coming out of it.
4. Disconnect the small hose from the end cap of the valve, and plug the opening in the valve with your finger. After a few seconds, air should flow through the bypass valve so you can feel it coming out of the large hose.
5. Remove your finger from the bypass valve small hose nozzle, and the air should come out the exhaust holes on the valve. Recover the nozzle with your finger, and after a few seconds, air should stop coming out the exhaust holes and flow through the large hose again. If not, the bypass valve is defective.

Repairing Dual Nozzle Air Bypass Valve, 1976

Repairs are limited to replacement.

Testing Vacuum Solenoid, 1976

This solenoid is normally closed. With the engine off and the electric plug disconnected from the solenoid, remove the hose and connect a hand vacuum pump to the solenoid. The solenoid should hold at least 15" Hg. vacuum without leaking down.

Pinto • Mustang & Bobcat 4 Cyl. & V-6

Connect a hot wire to one terminal of the solenoid and a ground wire to the other terminal, to energize the solenoid. When energized, the solenoid should open, and you should be able to blow through it easily. If not, the solenoid is clogged or defective.

Repairing Vacuum Solenoid, 1976

Repairs are limited to replacement.

Testing Thermo Actuated Valve (TAV) Switch, 1976

Disconnect the electric plug from the valve in the air cleaner. Use a penlight powered test light on the switch terminals. Above 65° F. the switch should be open, and the test light should not light. Below 65° F. the switch should be closed, lighting the test light.

The 65° F. temperature is approximate, and you should not throw away a TAV that is a few degrees off. As long as the TAV closes when cold and opens when hot, it should be considered okay.

Testing Thermo Actuated Valve (TAV) Switch, 1976

Repairs are limited to replacement.

Testing Temperature Vacuum Switch (TVS) 1976

1. Cool the valve in a refrigerator to 40° F. or less.
2. Connect a hand vacuum pump to the rim (smaller) nozzle on the valve.
3. Pump up at least 16" Hg. vacuum. The valve should hold without leaking.
   NOTE: *Because this valve is also the shutoff for the EGR valve, a slight leak may be enough to open the EGR valve when the engine is cold, and cause poor running, especially on those EGR systems that do not use a back pressure transducer.*
4. Warm the valve to 80° F. or more. It should open and not hold vacuum.

Repairing Temperature Vacuum Switch (TVS) 1976

Repairs are limited to replacement. The nozzle on the rim (smaller of the two) connects to the vacuum source. The center (larger) nozzle connects to the air bypass valve and EGR valve.

Testing Vacuum Delay Valve, 1976

1. Connect a hand vacuum pump to the colored side (opposite to the white side) of the valve. Leave the white side open. Work the pump in an attempt to pump up vacuum. If you can pump up vacuum, the valve is clogged or otherwise defective.
2. Connect a hand vacuum pump to the white side of the valve. Leave the colored side open. Operate the pump rapidly to pump up several inches of vacuum. The gauge should drop slowly to zero. If it drops instantly, or doesn't drop at all, the valve is defective.

Repairing Vacuum Delay Valve, 1976

Repairs are limited to replacement. When installing the valve, the white side goes toward the vacuum source, and the colored or black side toward the idle vent valve or bypass valve.

Testing Idle Vacuum Valve, 1976

1. Hold the valve with the hose connection nozzles on the bottom side. Connect a hand vacuum pump to the vent nozzle, which points out to the side, and another pump to the vacuum nozzle, which points down.
2. Pump up about 5" Hg. vacuum on the down pointing nozzle, and 15" Hg. on the side nozzle. Both pumps should hold vacuum without leaking down. If not, the idle vacuum valve is defective, but be sure you don't have any hose or pump leaks before you replace the valve.
3. Release the vacuum on the down-pointing nozzle. The side nozzle pump gauge should immediately drop to zero. If not, the idle vacuum valve is defective.

Repairing Idle Vacuum Valve, 1976

Repairs are limited to replacement of the valve. The nozzle that points down must connect to the vacuum source, and the side nozzle to the air bypass valve.

EXHAUST GAS RECIRCULATION 1976

Description of System

California manual transmission engines use a back pressure transducer. The transducer is a bleed installed in the vacuum hose that takes vacuum to the EGR valve. The base of the transducer is sandwiched between the EGR valve and its mounting flange, so that exhaust pressure can go through a small tube to the transducer. The exhaust pressure present in the exhaust system at part throttle closes a bleed against spring pressure so that the EGR valve receives whatever vacuum is available through its hose. When there is reduced exhaust pressure at idle or part throttle light loads, the exhaust pressure is weak, and the spring in the transducer opens the bleed so that the EGR operating vacuum is bled off. The result is that the EGR valve is only open at part throttle when it is needed to reduce NOx in the exhaust.

The transducer has two hose connection nozzles, one of which connects to the vacuum source, and the other to the EGR valve. It is important that the nozzle with the small restriction be connected to the vacuum source. The mounting position of the transducer does not indicate which nozzle has the restriction. The only way to be sure is to use a small flashlight and look into each nozzle. Then connect the restricted nozzle to the vacuum source.

On 49-State cars, the temperature control (PVS type) may not be used on some calibrations.

Units in System, 1976

EGR valve
Carburetor spacer
Temperature controlled vacuum valve (PVS type)
Vacuum delay valve (some models only)
Back pressure transducer (49-States: Not used. Calif.: 2.8 V-6 man. trans. only.)

Testing and Troubleshooting System, 1976

The system test on transducer-equipped engines is the same as on older engines. To test for exhaust gas recirculation, apply manifold vacuum or hand pump vacuum directly to the EGR valve with the engine idling. If the engine dies or runs rough, you know the exhaust gas is recirculating. If not, the EGR valve or a passageway is plugged, and the valve will have to be removed and the blockage removed.

To check for system operation, make sure the engine is at normal operating temperature so that the temperature vacuum switch will be open. With the engine idling in Neutral, open the throttle to about the half-throttle position and watch for movement of the EGR valve stem. If the stem moves, the system is working okay.

CAUTION: *Open the throttle only enough to make the EGR valve stem move. Do not exceed 3000 rpm. If the stem does not move, check the amount of vacuum at the EGR valve hose. You should have at least 4" Hg. at half throttle. If not, check for a plugged EGR port in the carburetor throat, or a leaking or plugged hose. If the vacuum is okay, check the individual units as shown below.*

Pinto • Mustang & Bobcat 4 Cyl. & V-6

Repairing System, 1976

Cleaning of the EGR valve, the transducer, and the passageways is recommended, if necessary. Solvent, or air pressure, must not be used on the diaphragm of either the EGR valve or the transducer to avoid damage. There are no adjustments, and all units are replaced when defective.

Testing Back Pressure Transducer

1. With the engine at normal operating temperature, run the engine at a fast idle by putting the throttle on the high step of the fast idle cam.
2. Remove the hose from the vacuum side of the transducer and check the vacuum at the hose with a gauge. Then reconnect the hose.
3. Disconnect the hose from the EGR valve and connect a vacuum gauge to the hose. The vacuum reading should be the same as at the vacuum side of the transducer, or within 2″ Hg. If the vacuum is not within 2″ Hg. there may be a clogged exhaust passage, a wrong hose connection, or a leaking transducer.

Off the car, the transducer can be checked by plugging one hose connection with your finger and using a hand vacuum pump on the other. The bleed valve should be open, and you should not be able to pump up any vacuum. If you can pump up vacuum, it means the bleed valve is clogged, and the transducer must be replaced.

To close the bleed, seal one side of the exhaust tube opening with your hand, and use mouth pressure on the other side. You should then be able to pump up vacuum, but the vacuum should drop to zero as soon as you remove the mouth pressure.

On the transducers we have tested, pressure applied to the exhaust tube opening does not close the bleed completely. There is still a little leakage. However, the leakage is so slight that it will allow you to pump up vacuum if work the pump rapidly.

CAUTION: *Do not use air hose pressure on the transducer. It may rupture the diaphragm.*

If the bleed will not close from mouth pressure, the transducer is probably clogged, and must be replaced.

Repairing Back Pressure Transducer, 1976

Repairs are limited to replacement. One of the hose connection nozzles on the transducer has a restriction that you can see if you look in the end of the nozzle with a small flashlight. The nozzle with the restriction must connect to the vacuum source. The other nozzle connects to the EGR valve.

EMISSION CONTROL SYSTEMS 1977

4 CYLINDER CARBURETOR 1977

The 5200 Holley-Weber 2-bbl. carburetor does not use the water hose connections that were used last year. All of the heat for the choke now comes from the electric heating unit, which is powered from the alternator so that it only receives current when the engine is running.

Three different throttle opening devices are used on the 4-cylinder engine. The Throttle Kicker is a simple vacuum-operated idle speedup device. It is used on air conditioned cars, with a vacuum solenoid controlling the vacuum to the kicker. When the air conditioning is ON, the solenoid is energized, allowing manifold vacuum to go to the kicker and increase the idle speed. When the air conditioning is OFF, the solenoid is not energized, which shuts off the vacuum to the kicker. The engine speed with the throttle kicker in operation is adjusted by changing the position of the kicker in its bracket. The idle speed with the air conditioning on and the kicker in operation should be the same as with the air conditioning off. The kicker increases the throttle opening just enough to maintain the idle speed under the load of the air conditioning compressor.

The Solenoid Kicker is a combination of an air conditioning idle speedup device and an anti-dieseling solenoid. The idle speedup part of the kicker is operated by vacuum, and the anti dieseling part is operated by electricity. In operation, the electric part of the solenoid comes on with the ignition switch. The curb idle speed is set with the screw that bears against the end of the solenmid stem. When the ignition switch is turned off, the solenoid is de-energized and the throttle closes to prevent dieseling.

The vacuum part of the solenoid opens the throttle an additional amount whenever the air conditioning is turned ON. The extra throttle opening makes up for the load of the compressor, and keeps the idle speed constant. The idle speed with the vacuum part of the solenoid in operation is not adjustable. You should not attempt to adjust the speed with the vacuum part in operation. If you do, then the curb idle speed will be incorrect when the air conditioning is OFF.

The vacuum control for the solenoid is a separate solenoid that allows the vacuum to pass whenever the air conditioning is ON.

The third device is a Solenoid Dashpot. This is an anti-dieseling solenoid with a dashpot built in. The electric part of the solenoid comes on with the ignition switch, and acts as an anti-dieseling solenoid. The dash-

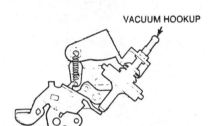

Throttle kicker (© Ford Motor Co.)

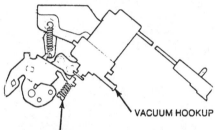

Solenoid kicker (© Ford Motor Co.)

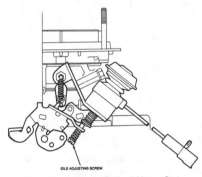

Solenoid dashpot (© Ford Motor Co.)

pot part is a cushioned stem that lets the throttle close slowly on a rubber diaphragm against a cushion of air. The dashpot is in operation at all times. It takes the place of the fuel deceleration valve, and helps control hydrocarbon emissions.

MANUAL ALTITUDE CONTROL 1977

The altitude valve is built into the fuel bowl on the 5200 2-bbl. carburetor used on the 4-cylinder engine.

Pinto • Mustang & Bobcat 4 Cyl. & V-6

When the valve is open, it allows extra fuel through a passage into the main metering system for normal sea level running. When the car is driven above 4000 feet, the valve should be closed to lean the mixture.

The control for the valve is a dash operated cable. Because there is a bell crank on the carburetor, the action of the cable is reversed. Pushing the cable in pulls the valve out to close it for altitude running. The cable should be pulled out for sea level running, which opens the valve.

The altitude control affects only the main metering system of the carburetor. It will not affect idle. No maintenance or adjustments are required. The valve should be cleaned or repaired if necessary, whenever the carburetor is overhauled.

The valve is either on or off. It doesn't do any good to move the valve through only part of the stroke in an attempt to partially lean or richen the mixture.

DURA-SPARK II 1977

The solid state Dura-Spark ignition used in 1976 has been slightly modified, and is now called Dura-Spark II. It is used on all 4 cylinder and V-6 altitude engines.

The main reason for changing the systems was to increase the available voltage for more reliable spark plug firing. The older ignition would put out 26,000 volts. Dura-Spark II will put out 36,000 volts.

Dura-Spark II uses a ballast resistor that has been reduced from the 1976 value of 1.35 ohms to 1.10 ohms. This reduction raises the voltage in the system and thus increases the available secondary voltage.

Special service procedures must be followed with the Dura-Spark system to prevent the spark from jumping to ground. Whenever a spark plug wire is removed, from either the cap, spark plug, or coil tower, the inside of the terminal boot must be greased with silicone grease before reconnecting. If possible, secondary wires should not be removed. Timing lights should be the induction type, using a connector that goes around the wire instead of connecting to the end. The direct connection type may false trigger the light, giving wrong timing.

Whenever a new rotor is installed, the brass parts of the rotor must be coated with silicone grease to approximately 1/8 inch thick. As this silicone grease ages, it looks discolored and dirty, but this has no effect on performance, and the grease should not be removed.

If it is necessary to measure available secondary voltage by removing a plug wire, the following wires must not be removed.

V-6 engines: No. 1 or 4
4-cyl. engines: No. 1 or 3

The distributor cap segments for the above wires are close to ground inside the distributor. When the wire is removed, they can easily arc over to ground and give a false indication of low available voltage. The other wires have enough distance so that they won't do this. Both Dura Spark systems have the capacity to jump great distance in open air. Any metal that might ground the system must be kept at least 3/4 inch from all terminal boots, and the rim of the distributor cap.

DECELERATION ENRICHMENT VALVE IDLE SPEEDUP 1977

The late style round type decel valve now works only when the air conditioning is ON. The valve speeds up the idle to take care of the added load of the air conditioning compressor.

The decel valve has two hose nozzles, one large and one small. The large nozzle connects to the carburetor. When the decel valve is open, a fuel-air mixture comes from the carburetor, goes through the decel valve and into the intake manifold to speed up the idle. The opening and closing of the decel valve is controlled by vacuum applied to the small nozzle. Intake manifold vacuum goes through a solenoid which is normally closed. When the air conditioning is turned ON, the solenoid is energized and opens, allowing vacuum to open the decel valve, which speeds up the idle.

A normal decel valve will not open unless it gets 20 in. Hg. or more of vacuum. The new decel valve has weaker spring. It will open at 6 in. Hg. or more. The result is that any time the air conditioning is turned ON, the decel valve is in operation. This includes idle, cruising, or deceleration. The only time the decel valve will not work is during wide open throttle, when the engine vacuum is too low to operate it.

Testing Decel Enrichment Valve

With the engine idling, and the air conditioning off, remove the small hose from the decel valve. Use a hand vacuum pump to apply over 6 in. Hg. vacuum to the valve. The engine should speed up. If not, either the valve is defective, or the fuel is not feeding from the carburetor to the valve.

Repairing Decel Enrichment Valve

Diaphragm kits are available for repairing the valve.

CAUTION: *This is a special valve, with a weaker spring. If the normal spring is used, the valve will not work to speed up the idle.*

Testing Decel Enrichment Valve Solenoid

Connect a vacuum gauge to one side of the solenoid, and vacuum from a running engine to the other side. You should not get a reading on the gauge, because this is a normally closed solenoid. Use a hot and ground wire to energize the solenoid. Vacuum should then pass through and read on the gauge.

Repairing Decel Enrichment Valve Solenoid

Repairs are limited to replacement. Be sure to replace with a normally closed solenoid.

4-CYLINDER DISTRIBUTOR VACUUM 1977

Several unique designs are used on

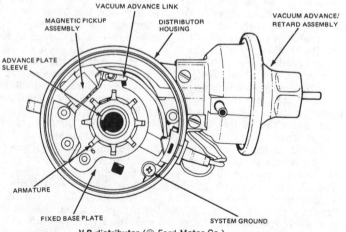

V-8 distributor (© Ford Motor Co.)

Pinto • Mustang & Bobcat 4 Cyl. & V-6

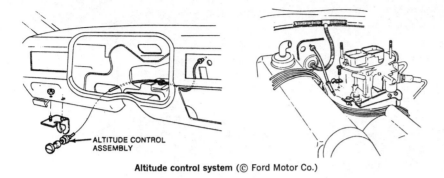

Altitude control system (© Ford Motor Co.)

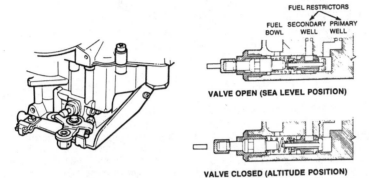

Manual altitude control (© Ford Motor Co.)

the California 2300 4-cylinder engine for 1977. There are both single and dual diaphragm distributors, with the vacuum controlled by a coolant temperature switch and an air cleaner temperature switch. And some engines use manifold vacuum, while others use ported vacuum. The systems break down basically into either single or dual diaphragm, described separately below.

Single Diaphragm System

Vacuum advance on some engines is operated by ported vacuum, and on some by manifold vacuum. Whichever is used, the system of temperature switches works exactly the same.

Vacuum to the advance is controlled by two temperature switches, a ported vacuum switch (PVS) that is sensitive to engine coolant temperature, and a Temperature Vacuum Switch (TVS) that is sensitive to air cleaner temperature. The vacuum advance unit can receive vacuum from either the PVS or the TVS.

The PVS opens when the engine coolant temperature is 95°F. or higher. Below 95°F. the PVS is closed. The TVS opens below 50°F. air cleaner temperature. As the TVS warms up, it closes above 65°F.

In actual operation, here's how it works. Let's say that you are starting an engine that is colder than 50°F. The TVS in the air cleaner is open, so you will have spark advance for better running. As the air cleaner warms up to 65°F. the TVS closes, shutting off the vacuum to the advance unit. There is no vacuum advance during this period, until the PVS warms up. Once the PVS reaches 95°F. coolant temperature, it opens and there is vacuum advance during normal operating temperature.

If you start an engine that is between 65 and 95°F. then there is no vacuum advance, because the TVS and the PVS are both closed. But as soon as the PVS reaches 95°F. there will be vacuum advance.

The TVS used with this system is the same shape as the one used with other systems on the V-8 engines. However, this TVS closes as it gets hotter, while the other TVS closes as it gets colder. The TVS for the system described above must be red, or have a red mark. If it is all white, it is the wrong TVS, and the system will not work.

Dual Diaphragm Distributor

Vacuum advance on some engines is operated by ported vacuum, and on others by manifold vacuum. Both diaphragms are operated by the manifold vacuum or by ported vacuum, from a single source. This is entirely different from the usual dual diaphragm system in which the retard diaphragm is operated by manifold vacuum and the advance diaphragm by ported vacuum.

Automatic transmission engines use ported vacuum to operate both diaphragms. Manual transmission engines use manifold vacuum. The ported system does not retard the spark at idle. There is no vacuum on either diaphragm, so the vacuum unit goes to the neutral or no-advance position. The manifold vacuum system retards the spark at idle whenever the advance vacuum is shut off by the temperature switches.

Both a PVS and a TVS are used, exactly the same as in the single diaphragm system described above, but only in the advance hose. There is no temperature regulation for the vacuum to the retard diaphragm. It gets vacuum (ported or manifold) whenever the engine is running.

If you start an engine that has cooled off all night and is below 50°F. the TVS is open, so you will have vacuum advance. As soon as the TVS warms up to 65°F. air cleaner temperature, the advance vacuum is shut off. This allows the advance diaphragm to rest against the retard diaphragm, which is in the retard position. Until the PVS warms up to 95°F. you will be running with a retarded spark. On the manifold vacuum system it will be retarded all the time, but on the ported system there will be no retard at idle because there is no ported vacuum.

After the PVS warms up to 95°F. the engine will have normal vacuum advance at all times. Of course there is vacuum to the retard diaphragm but it only retards the spark when there is no advance vacuum. On the manifold vacuum system there is advance vacuum all the time, so the retard diaphragm has no effect. On the ported vacuum system, the vacuum to both diaphragms is shut off at idle, so the advance diaphragm goes to the neutral or no advance position.

All of this complicated system was necessary because the 1977 catalytic converter is larger than last year's. To make it heat up faster, spark retard is needed when the engine is started. The PVS accomplishes this. But spark retard when the engine is extremely cold makes it run terrible. So the TVS was added to bring back the advance under cold starting conditions.

EMISSION CONTROL SYSTEMS 1978

FEEDBACK CARBURETOR

Pintos and Bobcats made for California are the only cars to use this system, and it is only on the 2300 4-cylinder engine. The system is called feedback, or closed loop, because there is a sensor in the exhaust that indirectly controls the fuel mixture at the carburetor. Whatever the carburetor does, the exhaust sensor detects, and this information goes

Pinto • Mustang & Bobcat 4 Cyl. & V-6

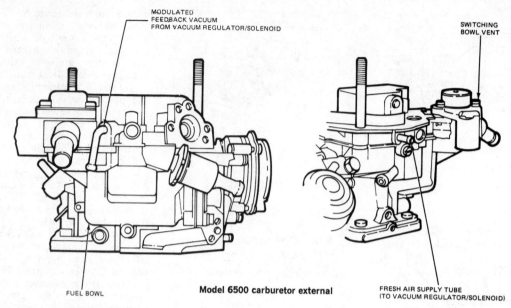

Model 6500 carburetor external

back to the carburetor. Thus, the system forms a complete loop, from carburetor to exhaust and back to carburetor. A similar system, using some of the same hardware, is used by General Motors on the Pontiac 151 4-cyl. engine, and is described in the Monza section of this supplement.

The carburetor used with this system is the familiar Holley-Weber 5200 2-bbl. It has been modified, and is called the 6500. The modifications consist of changing the power valve so it reacts to an external vacuum source, and renaming it the feedback metering valve. The rest of the carburetor is essentially the same as the 5200.

Because the feedback metering valve adds fuel to the main metering system, it does not affect the fuel mixture at idle. It only richens or leans the above-idle or cruising mixture. The design of the system is such that when the feedback metering valve is fully open, the mixture at cruising is slightly rich. When the valve is closed, the mixture is slightly lean. By varying the position of the metering valve, the mixture can be tailored to exactly what the engine needs.

The feedback metering valve is operated by intake manifold vacuum, which goes through a vacuum-solenoid regulator. The vacuum-solenoid regulator is operated electrically by an electronic control unit. The electronic control unit receives signals from an oxygen sensor

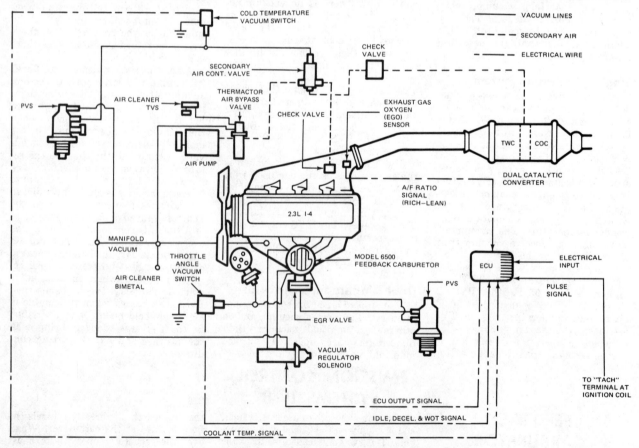

Feedback carburetor electronic engine control system—Schematic

Pinto • Mustang & Bobcat 4 Cyl. & V-6

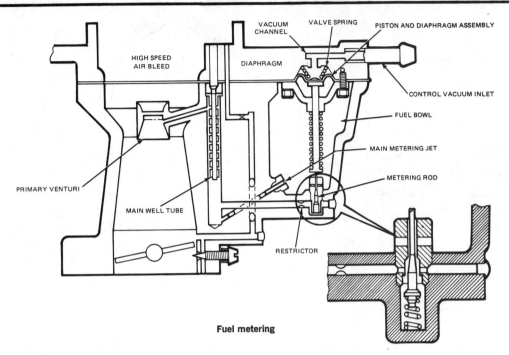

Fuel metering

in the exhaust, which tells whether the mixture is rich or lean.

The electronic control unit also receives signals from the tach terminal of the coil, to tell it how fast the engine is turning. And there is a cold temperature vacuum switch to signal engine temperature to the electronic control unit. A throttle angle switch indicates when the carburetor throttle is in the closed position. Now let's take a look at each one of the controls and how they work.

Vacuum Solenoid Regulator

This regulator is mounted on the left fender panel. It has 3 hoses connected to it. The top hose, near the electrical connection, goes to the air cleaner. All it does is let clean air into the regulator. The bottom hose connects to intake manifold vacuum. The middle hose supplies regulated vacuum from the regulator to the feedback metering valve in the carburetor.

Inside the regulator, a spring-operated regulator keeps the vacuum supply at 5 in. Hg. as long as intake manifold vacuum is 5 in. Hg. or higher. This regulated vacuum is then channeled up to the top part of the unit where there is a double-ended needle valve. The needle valve is moved up or down by the electric solenoid. When the valve is in the up position, it shuts off the vent and allows full regulated vacuum to go to the feedback metering valve. When the valve is in the down position it shuts off the vacuum and vents the hose to the feedback metering valve.

The solenoid valve never stays in one position. It is constantly vibrating up and down in response to a pulsating current from the electronic control unit. By varying the length of time that the valve is in the vent position or the vacuum position, the electronic control unit precisely regulates the amount of vacuum

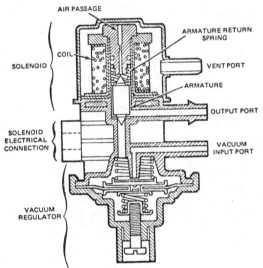

Vacuum solenoid regulator—Feedback system

going to the feedback metering valve in the carburetor. The range of vacuum that is supplied to the feedback metering valve is 0-5 in. Hg. At 5 inches it is in the full lean position. At zero it is in the full rich position.

The electronic control unit does not always continuously vary the vacuum to the feedback metering valve. During idle, or deceleration, the throttle angle switch signals the electronic control unit and the vacuum solenoid regulator goes into what is known as the open loop position. In this position the feedback metering valve receives a constant 2½ in. Hg. of vacuum, and the metering rod stays in the mid position. This is called open loop because the oxygen sensor is not telling the electronic control unit what to do.

Open loop also takes place when the engine coolant is below 125°F. In this case the signal comes from the cold temperature vacuum switch. The reason for going to open loop at these times is that the feedback metering valve would not have any effect. When the engine is cold, the choke is controlling the mixture. And at idle, or deceleration, the feedback metering valve would have no effect.

The system also goes into open loop control during wide open throttle, but because there is no engine vacuum at that time, the system cannot maintain 2½ in. Hg. at the feedback metering valve. So the springs push the valve down to the full rich position. Thus, the feedback metering valve acts just like an old fashioned power valve at wide open throttle.

Pinto • Mustang & Bobcat 4 Cyl. & V-6

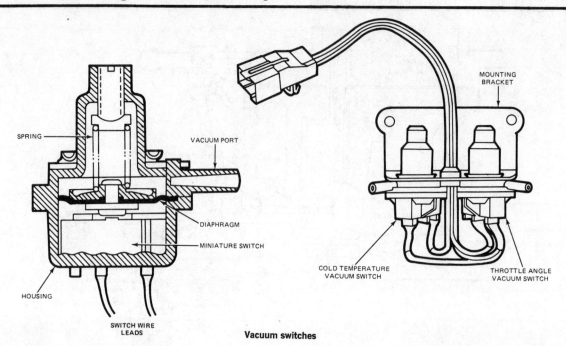

Vacuum switches

Cold Temperature Vacuum Switch and PVS Switch

There are two vacuum switches mounted together on the left fender apron. The rear switch is the cold temperature vacuum switch. It is connected by vacuum hose to a 3-nozzle ported vacuum switch (PVS) mounted at the front of the engine on a coolant passage. When the engine coolant is below 125°F. the PVS sends full manifold vacuum to the cold temperature vacuum switch. The contacts in the cold temperature vacuum switch are a normally open design. They close when the vacuum hits the switch, and this grounds a wire from the electronic control unit. The electronic control unit then goes into the open loop control, which maintains the feedback metering valve vacuum at a steady 2½ in. Hg.

When the engine warms up to more than 125°F. the PVS shuts off the manifold vacuum and connects the cold temperature vacuum switch to a vent.

This allows the spring inside the vacuum switch to open the contacts, and the electronic control unit then goes into the closed loop control.

On the PVS, the top nozzle must be connected to manifold vacuum, the middle nozzle to the cold temperature vacuum switch, and the bottom nozzle to the vent filter. Testing of the PVS is simply a matter of seeing if you get vacuum when you are supposed to. The cold temperature vacuum switch can be tested by disconnecting the wires and checking with a test light to see if the contacts close when vacuum is applied.

Throttle Angle Vacuum Switch

This switch is mounted on the same bracket, on the left fender panel, as the cold temperature vacuum switch. The throttle angle vacuum switch is the front switch. It is connected by vacuum hose to the carburetor spark port. When the throttle is closed, as during idle or deceleration, the spark port is above the throttle plate, and therefore does not receive vacuum. Above idle, the port is subject to vacuum, which then goes to the throttle angle switch. This switch is a normally closed design. With vacuum, the contacts stay closed, and the wire from the electronic control unit is grounded, telling the electronic control unit to stay in the open loop position. When vacuum acts on the switch above idle, the contacts open, and the electronic control unit then goes into the closed loop control.

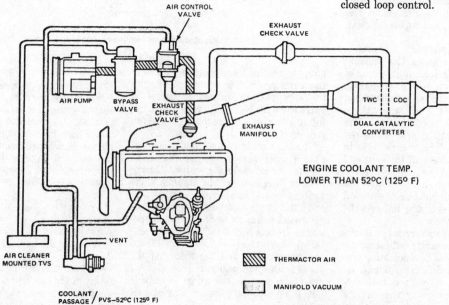

Thermactor system—Coolant temperature below 52 degrees C. (125 degrees F.)

Pinto • Mustang & Bobcat 4 Cyl. & V-6

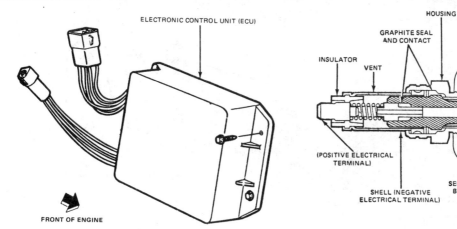

Electronic Control Unit (ECU)

Exhaust Gas Oxygen (EGO) Sensor

If the hoses should be switched between the throttle angle vacuum switch and the cold temperature vacuum switch, the system would be in closed loop control during idle and cold engine operation, and in open loop control during above-idle and cruising. In open loop, the feedback metering valve stays in the mid position, so the effect would be that the system was out of operation.

Electronic Control Unit

It is mounted on the left fender apron, and is about the same size as an ignition control. However, the connections are different, and it is labeled as the feedback carburetor control. No service or adjustments are possible on the control, except to change it if it doesn't work.

Diagnostic Tester

The tester is available from Owatonna Tool Company. It plugs into the electrical harness and analyzes most of the system. Other equipment needed are a tachometer, vacuum gauge, and hand vacuum pump. The tester is made in two parts, and each part must be purchased separately. The voltmeter part is the same as that used with the EEC system on the Versailles. So if you have already purchased the voltmeter for use on the Versailles, you don't have to buy it again.

Feedback Control Tester
T78L-50-FBC $398.12
Digital Volt-Ohmmeter
T78L-50-DVOM $331.24

Exhaust Gas Oxygen Sensor

This sensor is screwed into the exhaust manifold near the exhaust pipe connection. It is about the same size as an ordinary sparkplug. The sensor is made from a compound of zirconium, which has the unusual property of being able to generate small amounts of electricity when exposed to oxygen. The amount of voltage generated is less than one volt, but it is enough to send a signal to the electronic control unit. If the mixture in the exhaust is rich, it contains less oxygen, and the sensor sends a voltage of about ½ to 1 volt. When the exhaust mixture is lean, it contains more oxygen, and the sensor then sends a voltage of less than ½ volt.

Since the oxygen sensor generates voltage that can be read on a voltmeter, it might be possible to develop a test procedure with an ordinary voltmeter. But the only really practical and efficient way to test the sensor is with the diagnostic tester sold by Owatonna Tool Co.

Air Pump System and Dual Bed Converter

A special air pump system is used with the feedback carburetor. The pump and bypass valve are the same as before, but there is a secondary air control valve added to the system. The secondary air control valve receives the air under pressure from the pump and bypass valve, and sends it either to the exhaust valves, as before, or directly into the catalytic converter.

The converter used with this system has two sections. The front section is a 3-way catalyst, with a rhodium-platinum coating on the honeycomb. It will not only oxydize hydrocarbons and carbon monoxide into harmless carbon dioxide, but it will also reduce nitrogen oxides into harmless nitrogen and oxygen. The rear section is a conventional oxidation catalyst with a platinum coating. Air that enters the catalytic converter from the pump comes in behind the front section, so it doesn't disturb the reducing capability. It increases the oxydizing action in the rear section.

If the converter was warm when the car was started cold, there would be no need to switch the air from the converter to the exhaust manifold. When the car starts cold, the only warm area where oxidation can take place is at the exhaust valves. The system pumps the air into the exhaust valve area so that hydrocarbons and carbon monoxide can be kept down. The extra air introduced in front of the converter lowers the efficiency of the front converter section, but this is not a problem because nitrogen oxides are not produced in great quantities on a cold engine.

After the engine warms up, pumping the air into the exhaust valve area would make it difficult for the front section of the converter to reduce NOx. So the sys-

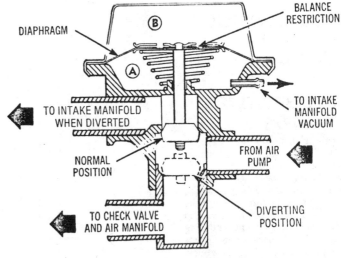

Backfire suppressor "gulp" valve

Pinto • Mustang & Bobcat 4 Cyl. & V-6

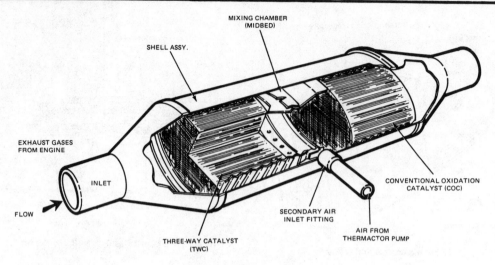

Dual catalytic converter

tem switches the air from the exhaust valve to the converter rear half.

Switching of the secondary air control valve is done by vacuum from a PVS switch. This is the same PVS that sends vacuum to the cold temperature vacuum switch. On a cold engine the PVS sends vacuum to the secondary air control valve, and it directs the air to the engine exhaust valves. When the engine warms up, the PVS shuts off the vacuum, and the secondary air control valve sends the air to the rear half of the converter.

Testing of the secondary air control valve can be done by disconnectig the outlet hoses and applying hand pump vacuum to the small hose with the engine running. With vacuum the air should come out the hose the exhaust manifold hose. Without vacuum, it should come out the hose to the converter. There are no adjustments on the secondary air control valve. If it doesn't work right, it must be replaced.

When testing the airflow, the air cleaner temperature must be above 49° F. A sensor in the air cleaner shuts off the air pump air below that temperature and diverts it to the atmosphere, the same as on past models.

VAPOR CONTROLS 1978

See the Ford section for information on vapor controls.

AIR PUMP SYSTEM 1978

California Fairmonts and Zephyrs with 2300 4-cyl. engines and manual transmissions are using the old "gulp" valve, which hasn't been seen on Ford products since the late 60's. Originally, the gulp valve was the forerunner of the bypass valve. Both the gulp valve and bypass valves divert the pump air during deceleration. The bypass valve diverts through a noise muffler into the atmosphere. The gulp valve diverts through a hose to the intake manifold. the additional air entering the intake manifold helps to lean out the rich mixture caused by the high vacuum during deceleration.

The Fairmont and Zephyr setup uses both a bypass valve and a gulp valve. The bypass valve diverts the air to the atmosphere in the usual way. The gulp valve allows air to enter the intake manifold and lean the mixture.

The gulp valve was not used alone because it has a tendency to stay open too long. A small shot of air is all that is needed to lean the mixture. The gulp valve opens first, gives the manifold the needed air, and then the bypass valve opens. When the bypass valve opens, it not only diverts the pump air, but also shuts off the air going to the gulp valve.

This insures that the intake manifold only gets a small amount of air, not enough to make the car drive down the road as if the throttle were still open.

MONOLITHE TIMING SYSTEM 1979

Some California engines are equipped with a "monolithe" timing system. The system is designed to accept an electronic probe that is connected to digital readout equipment. The probe receptacle is located at the front of all engines, except the 139 4-cyl. engine, which has a boss for monolithic timing in the left rear of the cylinder block.

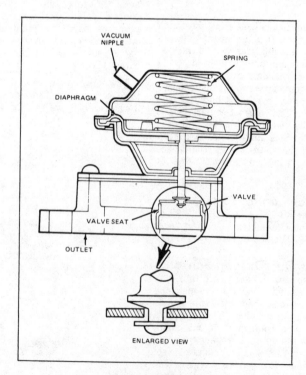

Typical EGR valve, poppet type

Pinto • Mustang & Bobcat 4 Cyl. & V-6

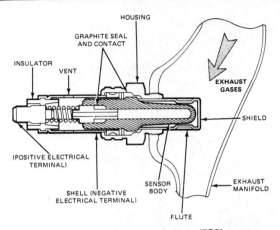

Feedback carburetor, exhaust gas oxygen (EGO) sensor

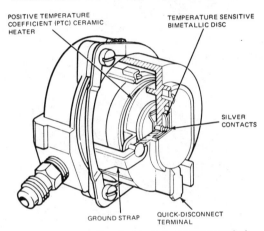

Temperature sensitive electric choke assembly, typical

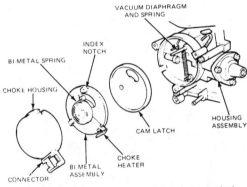

Constant operating electric assisted choke, typical

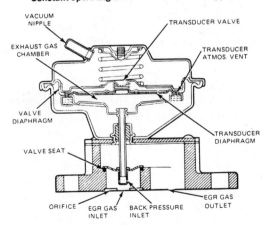

Typical EGR valve, integral back-pressure type

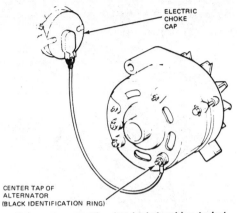

Temperature sensitive electric choke wiring, typical

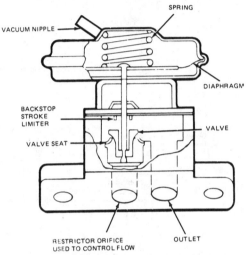

Typical EGR valve, internal tapered type

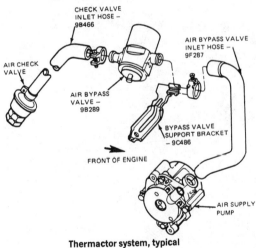

Thermactor system, typical

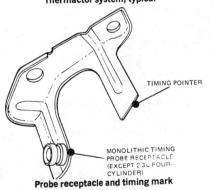

Probe receptacle and timing mark

377

Pinto • Mustang & Bobcat 4 Cyl. & V-6

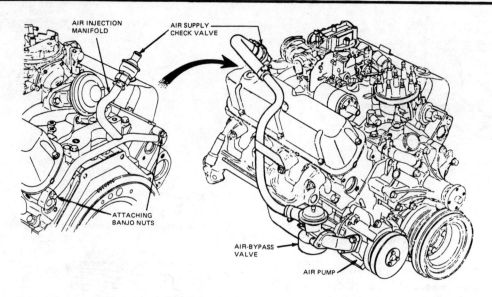

Typical thermactor system (Mustang and Capri)

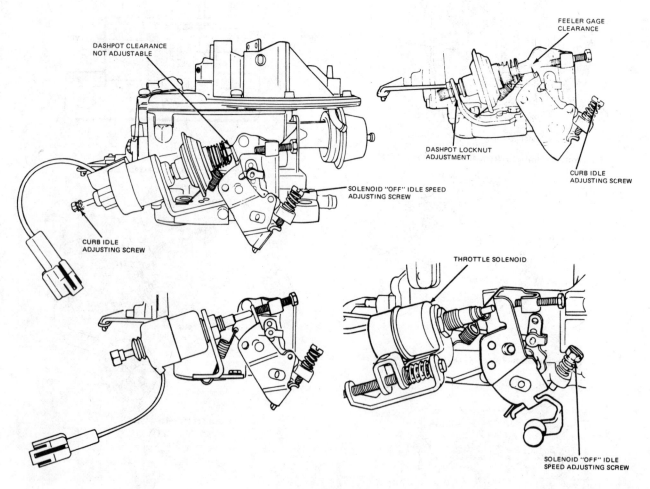

Typical throttle positioners

Pinto • Mustang & Bobcat 4 Cyl. & V-6
VACUUM CIRCUITS

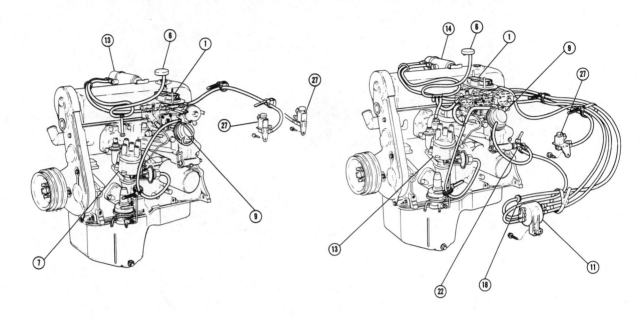

1	Carburetor – 9510
6	Bi-Metal Sensor – 9E607
9	EGR Valve – 9D475
13	Distributor – 12127
14	Thermactor Air Bypass Valve – 9B289
27	Vacuum Solenoid - N.C. – 9D474

1976 2300 cc. calib. 5-1G, R-3 no air cond.

1	Carburetor – 9510
6	Bi-Metal Sensor – 9E607
9	EGR Valve – 9D475
11	Venturi Vacuum Amplifier – 9E451
13	Distributor – 12127
14	Thermactor Air Bypass Valve – 9B289
18	Delay Valve (EGR) – 12A189
22	PVS (EGR) – 9D473
27	Vacuum Solenoid - N.C. – 9D474

1976 2300 cc. calib. 5-2G, R-6 no air cond. Pinto & Bobcat

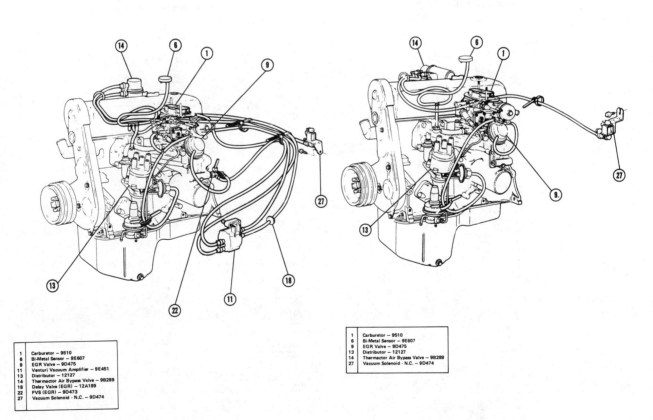

1	Carburetor – 9510
6	Bi-Metal Sensor – 9E607
9	EGR Valve – 9D475
11	Venturi Vacuum Amplifier – 9E451
13	Distributor – 12127
14	Thermactor Air Bypass Valve – 9B289
18	Delay Valve (EGR) – 12A189
22	PVS (EGR) – 9D473
27	Vacuum Solenoid - N.C. – 9D474

1976 2300 cc. calib. 5-2G, R-6 no air cond. Mustang

1	Carburetor – 9510
6	Bi-Metal Sensor – 9E607
9	EGR Valve – 9D475
13	Distributor – 12127
14	Thermactor Air Bypass Valve – 9B289
27	Vacuum Solenoid - N.C. – 9D474

1976 2300 cc. calib. 6-1G, R-1

Pinto • Mustang & Bobcat 4 Cyl. & V-6

VACUUM CIRCUITS

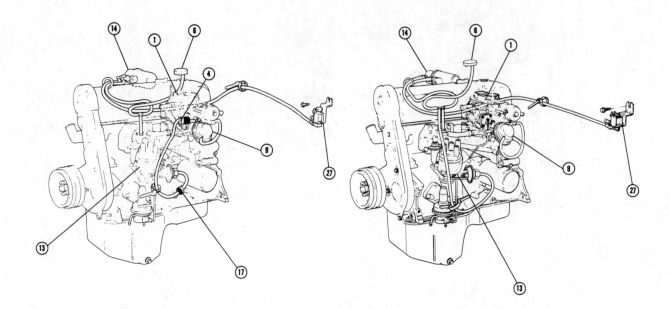

1976 2300 cc. calib. 6-1P, R-1

1976 2300cc. calib. 6-2G, R-1

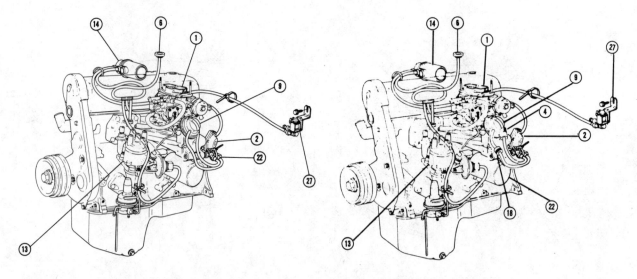

1976 2300 cc. calib. 6-2K, R-0, R-2, no air cond.

Pinto • Mustang & Bobcat 4 Cyl. & V-6

VACUUM CIRCUITS

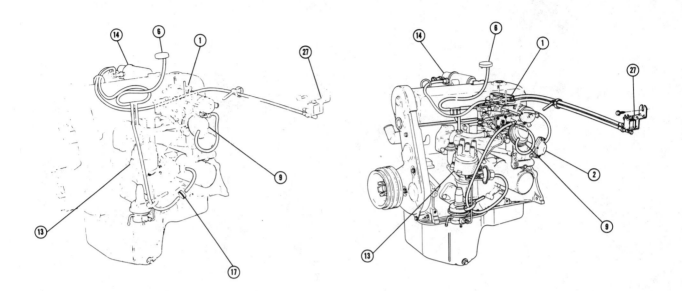

1 Carburetor — 9510 6 Bi-Metal Sensor — 9E607 9 EGR Valve — 9D475 13 Distributor — 12127 14 Thermactor Air Bypass Valve — 9B289 15 Thermactor Vacuum Differential Valve — 9E441 16 Temp. Vacuum System Valve — 9A995 17 Delay Valve (Spark) — 12A189 27 Vacuum Solenoid - N.C. — 9D474	1 Carburetor — 9510 2 Fuel Decel Valve — 9K793 6 Bi-Metal Sensor — 9E607 9 EGR Valve — 9D475 13 Distributor — 12127 14 Thermactor Air Bypass Valve — 9B289 27 Vacuum Solenoid - N.C. — 9D474

1976 2300 cc. calib. 6-2M, R-6   **1976 2300 cc. calib. 6-2N, R-1 no air cond.**

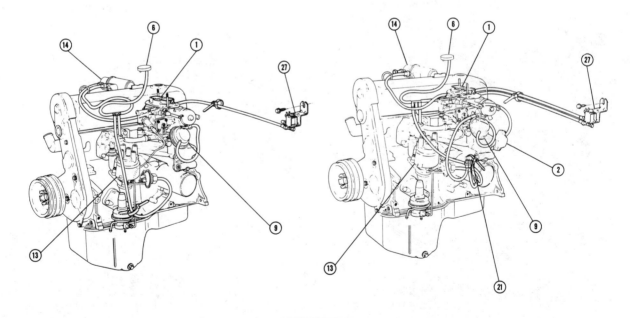

1 Carburetor — 9510 6 Bi-Metal Sensor — 9E607 9 EGR Valve — 9D475 13 Distributor — 12127 14 Thermactor Air Bypass Valve — 9B289	1 Carburetor — 9510 2 Fuel Decel Valve — 9K793 6 Bi-Metal Sensor — 9E607 9 EGR Valve — 9D475 13 Distributor — 12127 14 Thermactor Air Bypass Valve — 9B289 21 PVS (Cooling) — 12A091 27 Vacuum Solenoid - N.C. — 9D474

1976 2300 cc. calib. 6-2M, R-O, R-2   **1976 2300 cc. calib. 6-2N, R-1 with air cond.**

Pinto • Mustang & Bobcat 4 Cyl. & V-6

VACUUM CIRCUITS

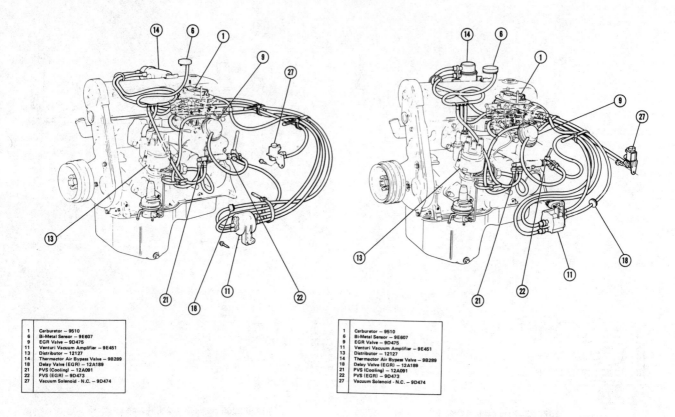

1976 2300 cc. calib. 5-2G, R-6 with air cond.

1976 2300 cc. calib. 5-2G, R-1

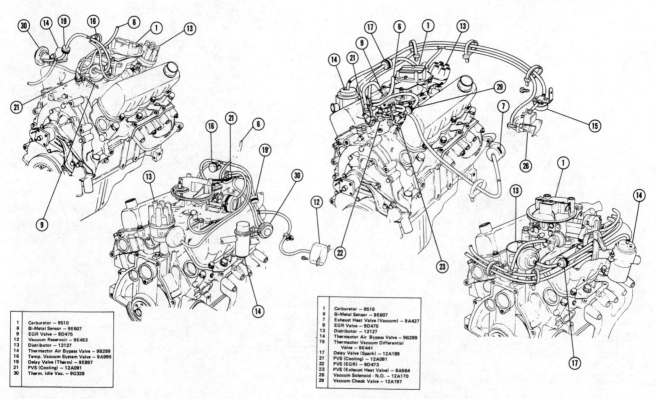

1976 2800 cc. calib. 6-4G, R-21

1976 2800 cc. calib. 5-4G, R-40

Datsun

TUNE-UP SPECIFICATIONS

Year	Model	SPARK PLUGS Type	SPARK PLUGS Gap (in.)	DISTRIBUTOR Point Dwell (deg)	DISTRIBUTOR Point Gap (in.)	IGNITION TIMING (deg) MT	IGNITION TIMING (deg) AT	Fuel Pump Pressure (psi)	IDLE SPEED (rpm) MT	IDLE SPEED (rpm) AT①	VALVE CLEARANCE (in.) (Hot) In	VALVE CLEARANCE (in.) (Hot) Ex	Percentage of CO at idle
1975	280 Z Federal	BP-6ES	0.028-0.031	Electronic	⑦	7B⑧	7B⑧	36.3	800	800	0.010	0.012	2.0
1975	280 Z (California)	BP-6ES	0.028-0.031	Electronic	⑦	10B	10B	36.3	800	800	0.010	0.012	2.0
1975	610	BP-6ES	0.031-0.035	49-55	0.017-0.022	12B	12B	3.8	750	650	0.010	0.012	2.0
1975	710	BP-6ES	0.031-0.035	49-55	0.017-0.022	12B	12B	3.8	750	650	0.010	0.012	2.0
1975	B210 (California)	BP-6ES	0.031-0.035	Electronic	⑦	10B	10B	3.8	750	650	0.014	0.014	2.0
1975	710, 610 (California)	BP-6ES	0.031-0.035	Electronic	⑦	12B	12B	3.8	750	650	0.010	0.012	2.0
1976	B-210 (Federal)	BP-5ES	0.031-0.035	49-55	0.017-0.022	10B	10B	3.8	700	650	0.014	0.014	2.0
1976	B-210 (California)	BP-5ES	0.031-0.035	Electronic	⑦	10B	10B	3.8	700	650	0.014	0.014	2.0
1976	610, 710 (Federal)	BP-6ES	0.031-0.035	Electronic	0.018-0.022	12B	12B	3.8	750	650	0.010	0.012	2.0
1976	610, 710 (California)	BP-6ES	0.031-0.035	Electronic	⑦	12B	12B	3.8	750	650	0.010	0.012	2.0
1976	620 (Federal)	BP-6ES	0.031-0.035	49-55	0.018-0.022	12B	12B	3.8	750	650	0.010	0.012	2.0
1976	620 (California)	BP-6ES	0.039-0.043	Electronic	⑦	10B	12B	3.8	750	650	0.010	0.012	2.0
1976	280 Z (Federal)	BP-6ES	0.028-0.031	Electronic	⑦	7B⑧	7B⑧	36.3	800	700	0.010	0.012	2.0
1976	280 Z (California)	BP-6ES	0.028-0.031	Electronic	⑦	10B	10B	36.3	800	700	0.010	0.012	2.0
1977	B-210 (Federal)	BP-5ES	0.039-0.043	49-55	0.018-0.022	10B	8B	3.8	700	650	0.014	0.014	2.0
1977	B-210 (California)	BP-5ES	0.039-0.043	Electronic	⑦	10B	10B	3.8	700	650	0.014	0.014	2.0
1977	610, 710 (Federal)	BP-6ES	0.031-0.035	49-55	0.018-0.022	12B	12B	3.8	750	650	0.010	0.012	2.0
1977	610, 710 (California)	BP-6ES	0.039-0.043	Electronic	⑦	12B	12B	3.8	750	650	0.010	0.012	2.0
1977	620 (Federal)	BP-6ES	0.031-0.035	49-55	0.018-0.022	12B	12B	3.8	750	650	0.010	0.012	2.0
1977	620 (California)	BPR-6ES	0.031-0.035	Electronic	⑦	10B	12B	3.8	750	650	0.010	0.012	2.0
1977-79	280 Z 280 ZX	BP-6ES	0.039-0.043	Electronic	⑦	10B	10B	36.3	800	700	0.010	0.012	2.0 ⑨
1977	F-10 (Federal)	BP-5ES	0.039-0.043	49-55	0.018-0.022	10B	10B	3.8	700	700	0.014	0.014	2.0
1977-78	F-10 (California) (1978 Federal)	BP-5ES	0.039-0.043	Electronic	⑦	10B	10B	3.8	700	700	0.014	0.014	2.0
1978-79	620	BP-6ES-11	0.039-0.043	Electronic	⑦	12B	12B	3.0-3.9⑩	600	600	0.010	0.012	1.0

Datsun

TUNE-UP SPECIFICATIONS—(Continued)

Year	Model	Spark Plugs Type	Spark Plugs Gap (in.)	Distributor Point Dwell (deg)	Distributor Point Gap (in.)	Ignition Timing (deg) MT	Ignition Timing (deg) AT	Fuel Pump Pressure (psi)	Idle Speed (rpm) MT	Idle Speed (rpm) AT①	Valve Clearance (in.) (Hot) In	Valve Clearance (in.) (Hot) Ex	Percentage of CO at idle
1978-79	510	BP-6ES-11	0.039-0.043	Electronic	⑦	12B	12B	3.0-3.9	600	600	0.010	0.012	1.0
1977-79	810	BP-6ES-11	0.039-0.043	Electronic	⑦	10B	10B	36.3 EFI	700	650	0.010	0.012	1.0 ⑨
1977	200 SX	BP-6ES	0.044	49-55	.020	10B	12B	3.0-3.9	600	600	0.010	0.012	0.3-2.0
1978-79	200 SX	BP-6ES-11	0.039-0.043	Electronic	⑦	12B	12B	3.0-3.9	600	600	0.010	0.012	1.0
1978	B210	BP-5ES-11	0.039-0.043	Electronic	⑦	10B	8B	3.9	700	700	0.010	0.012	2.0 ⑪
1979	210	BP-5ES-11	0.039-0.043	Electronic	⑦	10B	10B	3.9	700	700	0.010	0.012	1.0

NOTE: The underhood specifications sticker sometimes reflects tune-up specification changes made in production. Sticker figures must be used if they disagree with this chart.

① In Drive
② Air pump disconnected
③ Automatic—10B @ 600 below 30°F
④ Automatic—15B @ 600 below 30°F
⑤ Reluctor Gap 0.012-0.016
⑥ 10B—California
⑦ Reluctor Gap 0.008-0.016 in.
⑧ 13 BTDC—Advanced
⑨ 1978-79 1.0%—49 States
      0.5%—California
⑩ W/Electric Fuel Pump—4.6 lbs
⑪ Fu Models—1.0%
⑫ Fu Models—5°B @ 700
⑬ California Models—10B @ 650

Air Flow Meter Resistance Specifications

Air temperature °C (°F)	Resistance (kΩ)
—30 (—22)	20.3 to 33.0
—10 (—14)	7.6 to 10.8
10 (50)	3.25 to 4.15
20 (68)	2.25 to 2.75
50 (122)	0.74 to 0.94
80 (176)	0.29 to 0.36

Water Temperature Sensor Resistance Specifications

Cooling water temperature °C (°F)	Resistance (kΩ)
—30 (—22)	20.3 to 33.0
—10 (—14)	7.6 to 10.8
10 (50)	3.25 to 4.15
20 (68)	2.25 to 2.75
50 (122)	0.74 to 0.94
80 (176)	0.29 to 0.36

TORQUE SPECIFICATIONS
All readings in ft lbs

Engine Model	Cylinder Head Bolts	Main Bearing Bolts	Rod Bearing Bolts	Crankshaft Pulley Bolt	Flywheel to Crankshaft Bolts
L16	40	33-40	20-24	116-130	69-76
L24	47	33-40	20-24	116-130	101
A12	33-35	36-38	25-26	108-116	47-54
L18	47-62	33-40	33-40	87-116	101-116
A13	54-58	36-43	23-27	108-145	54-61
L20B	47-61	33-40	33-40	87-116	101-116
L26	54-61	33-40	27-31	94-108	94-108
A14, A15	51-54	36-43	23-27	108-145	54-61
L28	54-61	33-40	33-40	94-108	94-108

Datsun

CAR SERIAL NUMBER AND ENGINE IDENTIFICATION

The car identification number plate is located directly behind the windshield on the right side of the instrument panel. The I.D. number consists of the model and serial numbers.

The engine number is located on the left side of the engine block and consists of the model and serial number.

Engine Identification

Number of Cylinders	Displacement cu. in. (cc)	Type	Engine Model Code
4	97.3 (1,595)	OHC	L16
6	146.0 (2,393)	OHC	L24
4	71.5 (1,171)	OHV	A12
4	108.0 (1,770)	OHC	L18
4	78.59 (1,288)	OHV	A13
4	119.1 (1,952)	OHC	L20B
6	156.5 (2,655)	OHC	L26
4	85.24 (1,397)	OHV	A14
6	168.0 (2,753)	OHC	L28
4	90.80 (1,488)	OHV	A15

EMISSION CONTROLS

Various systems are used to control crankcase vapors, exhaust emissions, and fuel vapors. The accompanying chart shows the systems used with various models and engines.

Emission Control Equipment Applications Table

Year	Model	Engine	Emission Control Systems
1976-1977	610	L20B	1,3,4,5,6,7,8,9,10,11
1976-1977	710	L20B	1,3,4,5,6,7,8,9,10,11
1976-1979	B210	A14, A15	1,3,4,5,6,7,8,10,11
	280Z	L28	1,5,6,7,9,11,12
1976-1979	620	L20B	1,3,4,5,6,7,8,9,10
1977-1979	280Z	L28	1,5,6,7,9,12
1977-1978	F-10	A14	1,3,4,5,6,7,8,10,11
1978-1979	510	L20B	1,3,4,5,6,7,8,9,

Year	Model	Engine	Emission Control Systems
1978-1979①	810	L24	1,3,5,6,7,9,13
1978-1979	200SX	L20B	1,3,4,5,6,7,8,9

1. Closed Crankcase Ventilation System
2. Not used
3. Air Pump System
4. Engine Modification System
5. Fuel Vapor Control System
6. Exhaust Gas Recirculation System
7. Catalytic Converter California Cars Only
8. Early Fuel Evaporation System
9. Boost Controlled Deceleration Device
10. High Altitude Compensator—California Option
11. TCS—Manual Transmission exc. California
12. Floor Temperature Sensing Device
13. Spark Timing Control

① 1979—All 810 models are equipped with catalytic converter

CRANKCASE VENTILATION SYSTEM

The closed crankcase ventilation system is used to route the crankcase vapors (blow-by gases) to the intake manifold (carburetor equipped) or throttle chamber (EPI), to me mixed and burned with the air/fuel mixture.

An air intake hose is connected between the air cleaner assembly or the cover. A return hose is connected between a steel net baffle on the side throttle chamber, to the top engine of the crankcase to the intake manifold or throttle chamber, with a metering Positive Crankcase Valve mounted in the hose.

During periods of partial throttle, air is drawn through the air cleaner or throttle chamber, into the top engine cover and through the engine by the engine-developed vacuum. The air mixes with the crankcase vapors and is drawn through the steel net baffle which separates the heavy particles of oil from the vapors. The vapors are then metered through the PCV valve and directed into the intake manifold of the throttle chamber to be burned with the air/fuel mixture.

Under full throttle conditions, when the engine developed vacuum is insufficient to draw the vapors through the PCV valve, the vapors reverse in direction and are drawn into the air cleaner or throttle chamber from the top engine cover by the rush of induction air.

As the engine vacuum is raised, the vapors again flow through the PCV valve and into the combustion chambers. Therefore, no crankcase vapors are allowed to enter the atmosphere.

Air Pump System

In this system, an air injection pump, driven by the engine, compresses, distributes, and injects filtered air into the exhaust port of each cylinder. The air combines with unburned hydrocarbons and carbon monoxide to produce harmless compounds. The system includes an air cleaner, the belt driven air pump, a check valve, and an anti-backfire valve.

The air pump draws air through a hose connected to the carburetor air cleaner or to a separate air cleaner.

5. P.C.V. valve
6. Steel net
7. Baffle plate

1. Oil level gauge
2. Baffle plate
3. Flame arrester
4. Filter

⇨ Fresh air
➡ Blow-by gas

Crankcase ventilation system air flow (PCV)

Datsun

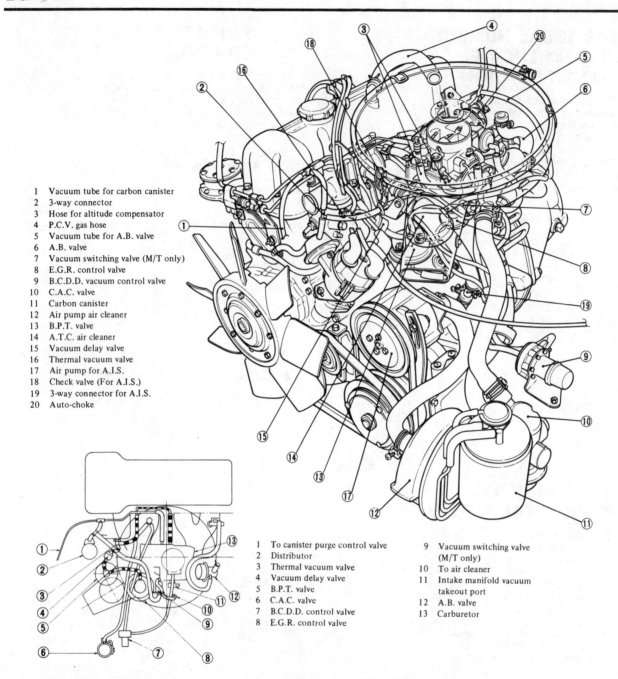

1. Vacuum tube for carbon canister
2. 3-way connector
3. Hose for altitude compensator
4. P.C.V. gas hose
5. Vacuum tube for A.B. valve
6. A.B. valve
7. Vacuum switching valve (M/T only)
8. E.G.R. control valve
9. B.C.D.D. vacuum control valve
10. C.A.C. valve
11. Carbon canister
12. Air pump air cleaner
13. B.P.T. valve
14. A.T.C. air cleaner
15. Vacuum delay valve
16. Thermal vacuum valve
17. Air pump for A.I.S.
18. Check valve (For A.I.S.)
19. 3-way connector for A.I.S.
20. Auto-choke

1. To canister purge control valve
2. Distributor
3. Thermal vacuum valve
4. Vacuum delay valve
5. B.P.T. valve
6. C.A.C. valve
7. B.C.D.D. control valve
8. E.G.R. control valve
9. Vacuum switching valve (M/T only)
10. To air cleaner
11. Intake manifold vacuum takeout port
12. A.B. valve
13. Carburetor

Emission control system—Calif. model—Typical (Model 710 illustrated)

The pump is a rotary vane unit with an integral pressure regulating valve. The pump outlet pressure passes through a check valve which prevents exhaust gas from entering the pump in case of insufficient pump outlet pressure. An anti-backfire valve admits air from the air pump into the intake manifold on deceleration to prevent backfiring in the exhaust manifold. In 1976 California models utilized a secondary system consisting of an air control valve which limits injection of secondary air and an emergency relief valve which controls the supply of secondary air. This system protects the converter from overheating. In 1977 the function of these two valves was taken by a single combined air control (C.A.C.) valve.

All engines with the air pump system have a series of minor alterations to accommodate the system. These are:

1. Special close-tolerance carburetor. Most engines, except the L16, require a slightly rich idle mixture adjustment.
2. Distributor with special advance curve. Ignition timing is retarded about 10° at idle in most cases.
3. Cooling system changes such as larger fan, higher fan speed, and thermostatic fan clutch. This is required to offset the increase in temperature caused by retarded timing at idle.
4. Faster idle speed.
5. Heated air intake on some engines.

The only periodic maintenance required on the air pump system is replacement of the air filter element and adjustment of the drive belt.

Engine Modification System

Engine modifications used on vehicles with the L16 and L18 and L20B, are:

1. A distributor with a secondary set of contact points which are retarded 5°. These secondary points are operational only when cruising or accelerating with a partially open throttle in third gear with manual

transmission, or over 13 mph with automatic transmission. A speed sensor is located at the speedometer on automatic transmission models. A temperature sensor in the engine compartment allows retarded timing only when the temperature inside the car is 50°F or above.

2. A solenoid valve in the carburetor opens to supply a lean fuel and air mixture, bypassing the throttle valve, in third gear or over 13 mph as above. The solenoid valve will not open if overridden by a closed throttle switch, a wide open throttle switch, a neutral switch, or a clutch disengaged switch. This arrangement is operational primarily during deceleration, when high intake manifold vacuum is present. For 1976-1979, this system is replaced with a vacuum controlled device in the carburetor to perform the same function.

The engine modification system used with the A12, A13, A14, A15 engines are relatively simple. It requires only a throttle positioner which holds the throttle slightly open on deceleration. A vacuum control valve connected to the intake manifold causes a vacuum servo to hold the throttle open slightly during the high vacuum condition of deceleration. The control valve is compensated for the effects of altitude and atmospheric pressure. The carburetor and distributor are specially calibrated for this engine.

The engine modification system for the 240Z sport coupe with the L24 engine and the 260Z with the L26 engine is quite similar to that for the A12, A13 engine, using a vacuum control valve, vacuum servo, and throttle positioner.

The 1976 280Z non-California models use the dual pick-up coil distributor while the California models use the single pick-up coil unit.

Beginning with the 1977 model year, all 280Z models were equipped with the single pick-up coil distributors.

Beginning with the 1979 models, the IC ignition unit has been miniaturized and mounted on the side of the distributor. The pick-up coil has been changed from an arm type of coil to a ring type of coil which eliminates dispersion of the signal waveform.

For 1976-79 engines, an exhaust gas recirculation (E.G.R.) system is used. This system uses vacuum to actuate a valve which allows a smaller amount of exhaust gases to be drawn into the intake manifold. This results in a decrease in oxides of nitrogen in the exhaust gases. The vacuum required to operate the system is not available at idle or wide throttle openings. A thermostatic switch inside the car shuts off the vacuum to the system when the temperature is below 30°F., thus allowing good cold starting and driveability.

Early Fuel Evaporation System (E.F.E.)

1976-79 Cars exc. 280Z

In this system, a control valve is welded to the valve shaft and installed on the exhaust manifold through bushing. This heat control valve is actuated by a coil spring, thermostatic spring and counterweight which are assembled on the valve shaft projecting at the rear outside of the manifold. The counterweight is secured to the shaft with a key, bolt and snap-ring. A chamber between the intake and exhaust mani-

1 Air control valve
2 E.G.R. control valve
3 Air relief valve
4 A.B. valve
5 B.C.D.D. solenoid valve
6 B.C.D.D.
7 Auto-choke
8 P.C.V. hose
9 Check valve (for A.I.S.)
10 3-way connector (M/T)
11 Thermal vacuum valve
12 B.P.T. valve
13 A.T.C. air cleaner
14 Air pump for A.I.S.
15 Air pump air cleaner
16 Canister

1 To canister purge control
2 Distributor
3 Thermal vacuum valve
4 B.P.T. valve
5 Air control valve
6 Carburetor
7 A.B. valve
8 Intake manifold vacuum takeout port (Idle compensator)
9 To air cleaner
10 Vacuum switching valve (M/T only)
11 E.G.R. control valve

Emission control system—Non-Calif. models—Typical (Model 710 illustrated)

Datsun

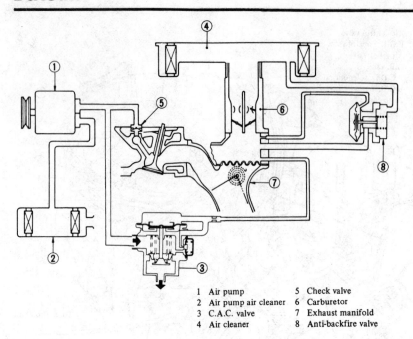

1. Air pump
2. Air pump air cleaner
3. C.A.C. valve
4. Air cleaner
5. Check valve
6. Carburetor
7. Exhaust manifold
8. Anti-backfire valve

Air injection system—Calif. models—Typical

folds above the manifold stove heats the air-fuel mixture by means of exhaust gases. This results in better atomization and lower HC content.

Boost Controlled Deceleration Device (B.C.D.D.)

All 1976-79 Cars exc. B210, F-10

The B.C.D.D. is installed under the throttle chamber as a part of it. It supplies additional air to the intake manifold during coasting to maintain manifold vacuum at the proper operating pressure.

There are two diaphragms in the device. Diaphragm I detects the manifold vacuum and opens the vacuum control valve when vacuum exceeds operating pressure. Diaphragm II operates the air control valve by way of the vacuum transmitted through the vacuum control valve. The air control valve regulates the amount of additional air so that the manifold vacuum can be kept at operating pressure.

On manual transmission models, in addition to the B.C.D.D., the system consists of a vacuum control solenoid valve, speed detecting switch and amplifier.

On automatic transmission models, in addition to the B.C.D.D., the system consists of vacuum control solenoid and inhibitor switch.

FUEL VAPOR CONTROL SYSTEM

The fuel vapor control system is used on all vehicles sold in the U.S. It has four major components:

1. A sealed gas tank filler cap to prevent vapors from escaping at this point.
2. A vapor separator which returns liquid fuel to the fuel tank, but allows vapors to pass into the system.
3. A vapor vent line connecting the vapor separator to a flow guide valve.
4. A flow guide valve which allows air into the fuel tank and prevents vapors from the crankcase ventilation system from passing into the vapor vent line and fuel tank.

When the engine is not running, fuel vapors accumulated in the fuel tank, vapor separator, and vapor vent line. When the vapor pressure exceeds 0.4 in. (10 mm) Hg, the flow guide valve opens to allow the vapors to pass into the crankcase ventilation system. Fuel vapors are thus accumulated in the crankcase. When the engine starts, the vapors are disposed of by the crankcase ventilation system. When enough fuel has been used to create a slight vacuum in the fuel tank and fuel vapor control system, the flow guide valve opens to let fresh air from the carburetor air cleaner into the tank.

On engines with sidedraft carburetors, float bowl vapors are routed through the float bowl overflow tubes to the carburetor air cleaner.

Catalytic Converter

All California Cars except Pick-Up

In addition to the air injection sys-

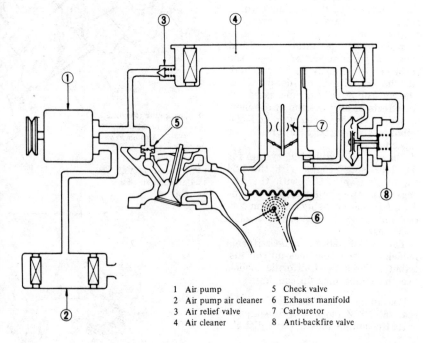

1. Air pump
2. Air pump air cleaner
3. Air relief valve
4. Air cleaner
5. Check valve
6. Exhaust manifold
7. Carburetor
8. Anti-backfire valve

Air injection system—Non-Calif. models—Typical

tem, EGR and the engine modifications, the catalyst further reduces pollutants. Through catalytic action, it changes residual hydrocarbons and carbon monoxide in the exhaust gas into carbon dioxide and water before the exhaust gas is discharged into the atmosphere.

NOTE: *Only unleaded fuel must be used with catalytic converters; lead in fuel will quickly pollute the catalyst and render it useless.*

The emergency air relief valve is used as a catalyst protection device. When the temperature of the catalyst goes above maximum operating tem-

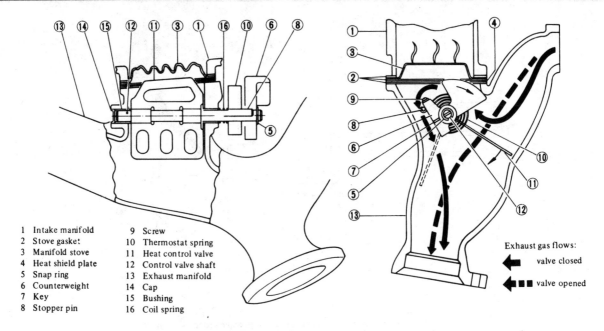

1	Intake manifold	9	Screw
2	Stove gasket	10	Thermostat spring
3	Manifold stove	11	Heat control valve
4	Heat shield plate	12	Control valve shaft
5	Snap ring	13	Exhaust manifold
6	Counterweight	14	Cap
7	Key	15	Bushing
8	Stopper pin	16	Coil spring

Early fuel evaporative system—Typical

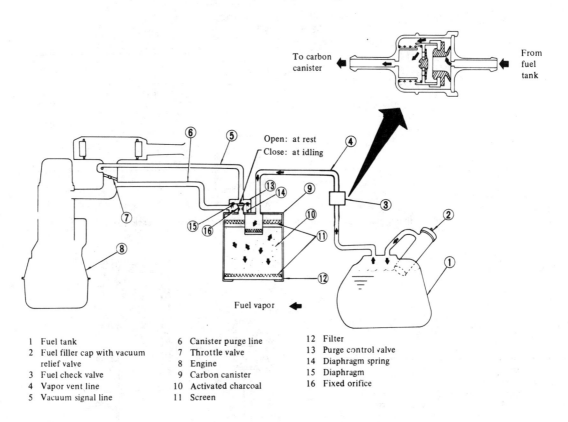

1	Fuel tank	6	Canister purge line	12	Filter
2	Fuel filler cap with vacuum relief valve	7	Throttle valve	13	Purge control valve
		8	Engine	14	Diaphragm spring
3	Fuel check valve	9	Carbon canister	15	Diaphragm
4	Vapor vent line	10	Activated charcoal	16	Fixed orifice
5	Vacuum signal line	11	Screen		

Evaporative emission control system operation with engine stopped or idling—Typical

perature, the temperature sensor signals the switching module to activate the emergency air relief valve. This stops air injection into the exhaust manifold and lowers the temperature of the catalyst.

NOTE: *The catalytic converter has been added to all 810 models for 1979 and the floor temperature warning light eliminated.*

Floor Temperature Warning System

1976-78 280Z

This system employs temperature sensors to warn of impending catalytic converter overheating. The system consists of a floor sensor located in the luggage compartment, a floor sensor relay located under the front passenger seat and a warning lamp located on the left side of the instrument panel.

Datsun

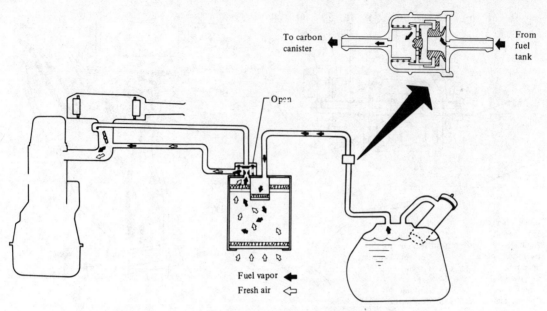

Evaporative emission control system operation with engine running—Typical

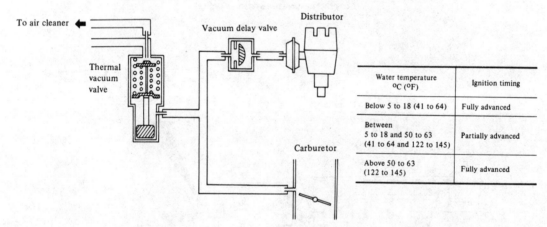

Spark timing control system—Non-Calif. models w/automatic transmission—Typical

Water temperature °C (°F)	Ignition timing
Below 5 to 18 (41 to 64)	Fully advanced
Between 5 to 18 and 50 to 63 (41 to 64 and 122 to 145)	Partially advanced
Above 50 to 63 (122 to 145)	Fully advanced

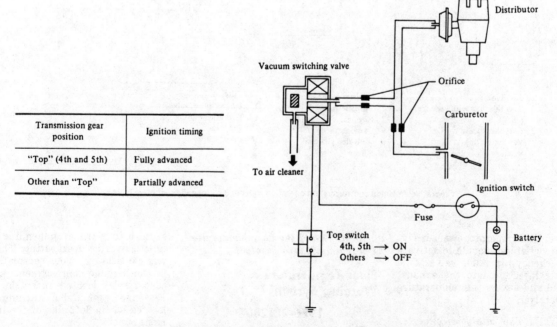

Transmission gear position	Ignition timing
"Top" (4th and 5th)	Fully advanced
Other than "Top"	Partially advanced

Spark timing control system—FU models w/manual transmission—Typical

Datsun

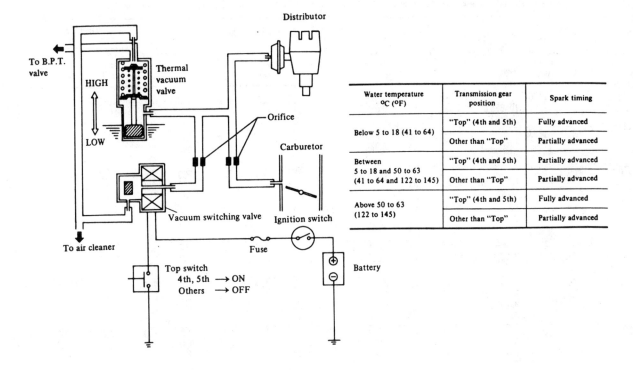

Spark timing control system—USA models w/manual transmission—Typical

Water temperature °C (°F)	Transmission gear position	Spark timing
Below 5 to 18 (41 to 64)	"Top" (4th and 5th)	Fully advanced
	Other than "Top"	Partially advanced
Between 5 to 18 and 50 to 63 (41 to 64 and 122 to 145)	"Top" (4th and 5th)	Partially advanced
	Other than "Top"	Partially advanced
Above 50 to 63 (122 to 145)	"Top" (4th and 5th)	Fully advanced
	Other than "Top"	Partially advanced

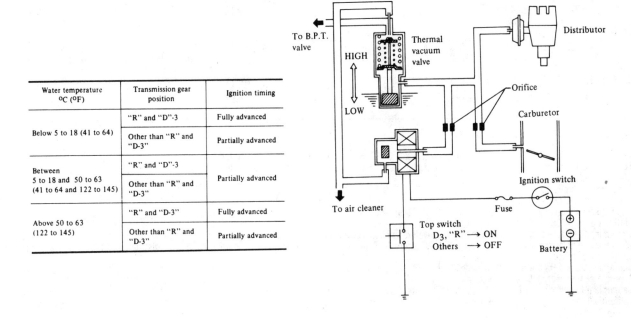

Spark timing control system—Calif. models w/automatic transmission—Typical

Water temperature °C (°F)	Transmission gear position	Ignition timing
Below 5 to 18 (41 to 64)	"R" and "D"-3	Fully advanced
	Other than "R" and "D-3"	Partially advanced
Between 5 to 18 and 50 to 63 (41 to 64 and 122 to 145)	"R" and "D"-3	Partially advanced
	Other than "R" and "D-3"	Partially advanced
Above 50 to 63 (122 to 145)	"R" and "D-3"	Fully advanced
	Other than "R" and "D-3"	Partially advanced

Honda

TUNE-UP SPECIFICATIONS

When analyzing compression test results, look for uniformity among cylinders, rather than specific pressures.

Year	Model	Engine Displacement (cc)	Original Equipment Spark Plugs Type	Gap (in.)	Distributor Point Dwell (deg)	Distributor Point Gap (in.)	Basic Ignition Timing (deg) MT	Basic Ignition Timing (deg) AT	Intake Valve Fully Opens (deg)	Fuel Pump Pressure (psi)	Idle Speed (rpm) MT	Idle Speed (rpm) AT	Valve Clearance (in.) Intake (cold)	Valve Clearance (in.) Auxiliary (cold)	Valve Clearance (in.) Exhaust (cold)
1973	Civic	1170	BP-6ES or W-20EP	0.028-0.031	49-55	0.018-0.022	TDC ③⑧	TDC ③⑧	10A	2.56	750-850 ④	700-800 ⑤	0.005-0.007	—	0.005-0.007
1974	Civic	1237	BP-6ES or W-20EP	0.028-0.031	49-55	0.018-0.022	5B ⑧	5B ⑧	10A	2.56	750-850 ④	700-800 ⑤	0.004-0.006	—	0.004-0.006
1975-76	Civic	1237	BP-6ES or W-20EP	0.028-0.032	49-55	0.018-0.022	7B ⑧	7B ⑧	10A	2.56	750-850 ④	700-800 ⑤	0.004-0.006	—	0.004-0.006
1977	Civic	1237	BP-6ES or W-20EP	0.028-0.032	49-55	0.018-0.022	TDC ⑧	TDC ⑧	10A	2.56	700-800 ④	700-800 ⑤	0.004-0.006	—	0.004-0.006
1978-79	Civic	1237	BP-6ES or W-20EP	0.028-0.032	49-55	0.018-0.022	2B ⑧	2B ⑧	10A	2.56	650-750 ④	650-750 ⑤	0.004-0.006	—	0.004-0.006
1975	Civic CVCC	1487	BP-6ES or W-20ES ②	0.028-0.032	49-55	0.018-0.022	TDC ⑨	3A ⑨	10A	1.85-2.56	800-900 ④	700-800 ⑤	0.005-0.007	0.005-0.007	0.005-0.007
1976	Civic CVCC	1487	B-6ES or W-20ES ②	0.028-0.032	49-55	0.018-0.022	2B ⑧⑨	2B ⑦⑨	10A	1.85-2.56	800-900 ④	700-800 ⑤	0.005-0.007	0.005-0.007	0.005-0.007
1977	Civic CVCC	1487	B6EB or W-20ES-L	0.028-0.032	49-55	0.018-0.022	6B ⑩	6B ⑩⑦	10A	1.85-2.56	750-850 ④	650-750 ⑤	0.005-0.007	0.005-0.007	0.005-0.007
1978-79	Civic CVCC	1487	B6EB or W-20ES-L	0.028-0.032	49-55	0.018-0.022	6B ⑩	6B ⑩	10A	1.85-2.56	650-750 ④	600-700 ⑤	0.005-0.007	0.005-0.007	0.007-0.009
1976-77	Accord CVCC	1600	B-6ES or W-20ES ①②	0.028-0.032	49-55	0.018-0.022	2B ⑨	TDC ⑨	10A	1.85-2.56	750-850 ④	630-730 ⑤	0.005-0.007	0.005-0.007	0.005-0.007
1978	Accord	1600	B6EB or W-20ES-L	0.028-0.032	49-55	0.018-0.022	6B ⑨⑩	6B ⑨⑩	10A	2.13-2.84	750-850 ④	650-750 ⑤	0.005-0.007	0.005-0.007	0.007-0.009
1979	Accord	1751	B7EB	0.028-0.032	Electronic		6B ⑨⑫	4B ⑬⑪	10A	2.13-2.84	650-750 ④	650-750 ⑤	0.005-0.007	0.005-0.007	0.010-0.012

① For continuous highway use over 70 mph, use cooler NGK B-7ES, Nippon Denso W-22ES or equivalent
② For continuous low-speed use under 30 mph, use hotter NGK B-5ES, Nippon Denso W-16ES or equivalent
③ Static ignition timing—5B
④ In neutral, with headlights on
⑤ In drive range, with headlights on
⑥ 5-speed sedan (hatchback) from engine number 2500001-up—6B
⑦ Station wagon—TDC
⑧ Aim timing light at red notch on crankshaft pulley with distributor vacuum hose(s) connected at specified idle speed
⑨ Aim timing light at red mark (yellow mark, 1978-79 Accord M/T) on flywheel or torque converter drive plate with distributor vacuum hose connected at specified idle speed
⑩ California (KL) and High Altitude (KH) models: 2B
⑪ Aim light at blue mark (49 States models)
⑫ California (KL) and High Altitude (KH) models: TDC (white mark)
⑬ California (KL) and High Altitude (KH) models: 2 ATDC (black mark)
TDC—Top Dead Center
B—Before Top Dead Center
A—After Top Dead Center
— Not Applicable
N.A. Not Available

Note: The underhood specifications sticker often reflects tune-up specification changes made in production. Sticker figures must be used if they disagree with those in this chart.

Honda

SERIAL NUMBER IDENTIFICATION CHART

Year	Model	VIN (Chassis Number)	Engine Number
1975	Civic	3300001—4000000	EB2-2000001—2025158
1975	Civic CVCC	1000001—2000000	ED1-1000001—1999999
1975	Civic CVCC Wagon	1000001—2000000	ED2-1000001—1999999
1976	Civic	4000001—5000000	EB2-2025159—2999999
1976	Civic CVCC	2000001—3000000	ED3-2000001—2499999
1976	Civic CVCC①	2000001—3000000	ED3-2500001—2999999
1976	Accord	1000001—2000000	EF1-1000001—2000000
1977	Civic	5000001—6000000	EB2-3000001—4000000
1977	Civic CVCC KL	3000001—4000000	EB3-3000001—3499999
1977	Civic CVCC KA	3000001—4000000	ED3-3500001—3899999
1977	Civic CVCC KH	3000001—4000000	ED3-3900001—4000000
1977	Accord KL	2000001—3000000	EF1-2000001—2499999
1977	Accord KA	2000001—3000000	EF1-2500001—2899999
1977	Accord KH	2000001—3000000	EF1-2900001—3000000
1978	Civic	6000001—7000000	EB3-1000001—1500000
1978	Civic CVCC KL	4000001—5000000	ED3-4000001—4499999
1978	Civic CVCC KA	4000001—5000000	ED3-4500001—4899999
1978	Civic CVCC KH	4000001—5000000	ED3-4900001—5000000
1978	Accord KL	3000001—4000000	EF1-3000001—3499999
1978	Accord KA	3000001—4000000	EF1-3500001—3899999
1978	Accord KH	3000001—4000000	EF1-3900001—4000000
1979	Civic	7000001—8000000	EB3-1500001—2000000
1979	Civic CVCC KL	5000001—6000000	ED3-5000001—5499999
1979	Civic CVCC KA	5000001—6000000	ED3-5500001—5899999
1979	Civic CVCC KH	5000001—6000000	ED3-5900001—6000000
1979	Accord KL	4000001—5000000	EF1-4000001—4499999
1979	Accord KA	4000001—5000000	EF1-4500001—4899999
1979	Accord KH	4000001—5000000	EF1-4900001—5000000

① 5 speed/49 States
KL: California
KH: High Altitude
KA: 49 States

Note: Beginning 1976, Civic CVCC Wagons have engine serial numbers prefixed ED4. KL, KH, and KA designations still apply.

TORQUE SPECIFICATIONS
All readings are given in ft lbs

Year	Engine Displacement (cc)	Cylinder Head Bolts	Main Bearing Bolts	Rod Bearing Bolts	Crankshaft Pulley Bolts	Flywheel to Crankshaft Bolts	Manifold In	Manifold Ex	Spark Plugs	Oil Pan Drain Bolt
1973-79	1170, 1237	30-35① 37-42②	27-31	18-21	34-38	34-38	13-17	13-17③	9-12	29-36
1975-79	1487, 1600, 1751 CVCC	40-47	30-35	18-21	58-65	34-48	15-17	15-17	11-18	29-36

① To engine number EB 1-1019949
② From engine number EB 1-1019950
③ 1975-76 models w/AIR—22-33 ft lbs

Honda

EMISSION CONTROLS

CRANKCASE EMISSION CONTROL SYSTEM

All engines are equipped with a "Dual Return System" to prevent crankcase vapor emissions. Blow-by gas is returned to the combustion chamber through the intake manifold and carburetor air cleaner. When the throttle is partially opened, blow-by gas is returned to the intake manifold through the breather tubes leading to the tee orifice, located on the outside of the intake manifold. When the throttle is opened wide and the vacuum in the air cleaner rises, blow-by gas is returned to the intake manifold through an additional passage in the air cleaner case.

EVAPORATIVE EMISSION CONTROL SYSTEM

Fuel vapor is stored in the expansion chamber, in the fuel tank, and in the vapor line up to the one-way valve. When the vapor pressure becomes higher than the set pressure of the one-way valve, the valve opens and allows vapor into the charcoal canister. While the engine is stopped or idling, the idle cut-off valve in the canister is closed and the vapor is absorbed by the charcoal.

At partially opened throttle, the idle cut-off valve is opened by manifold vacuum. The vapor that was stored in the charcoal canister and in the vapor line is purged into the intake manifold. Any excessive pressure of vacuum which might build up in the fuel tank is relieved by the two-way valve in the filler cap (Civic) or in the engine compartment (Accord).

EXHAUST EMISSION CONTROL SYSTEM

Special control devices are used with engine modifications. Improvements to the combustion chamber, intake manifold, valve timing, carburetor, and distributor comprise the engine modifications. The special control devices consist of the following:
 a. Intake air temperature control;
 b. Throttle opener;
 c. Ignition timing retard unit;
 d. Transmission and temperature controlled spark advance (TCS) for the transmission;
 e. Temperature controlled spark advance for Hondamatic automatic transmission.

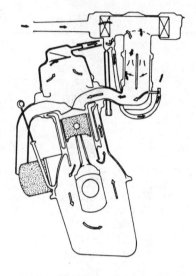

Crankcase ventilation system operation—Typical

1976-79 Models (Except 1976-79 CVCC)

Intake Air Temperature Control

When the temperature in the air cleaner is below approximately 100° F. the air bleed valve, which consists of a bimetallic strip and a rubber seal, remains closed. Intake manifold vacuum is then led into a vacuum motor, on the snorkel of the air cleaner, which moves the air control valve door, allowing only pre-heated air to enter the air cleaner.

When the temperature in the air cleaner becomes higher than approx. 100° F., the air bleed valve opens and the air control valve door returns to the open position allowing only unheated air through the snorkel.

Throttle Opener

The throttle opener is designed to prevent misfiring during deceleration by causing the throttle valve to remain slightly open, allowing better mixture control. The control valve is set to allow the passage of vacuum to the throttle opened diaphragm when the engine vacuum is equal to or greater than the control valve preset vacuum (21.6 ± 1.6 in. Hg) during acceleration.

Under running conditions, other than fully closed throttle deceleration, the intake manifold vacuum is less than the control valve set vacuum; therefore the control valve is not actuated. The vacuum remaining in the throttle opener and control valve is returned to atmospheric pressure by the air passage at the valve center.

Ignition Timing Retard Unit

When the engine is idling, the vacuum produced in the carburetor retarder port is communicated to the spark retard unit and the ignition timing, at idle, is retarded.

Transmission Controlled Spark Advance

When the coolant temperature is approximately 120° or higher, and the transmission is in First, Second, or Third gear, the solenoid valve cuts off the vacuum to the spark advance unit.

On later models, the vacuum is cut off to the spark advance unit regardless of temperature when First, Second, or Third gear is selected. Vacuum advance is restored when Fourth gear is selected.

Temperature controlled spark advance on cars equipped with Hondamatic transmission is designed to re-

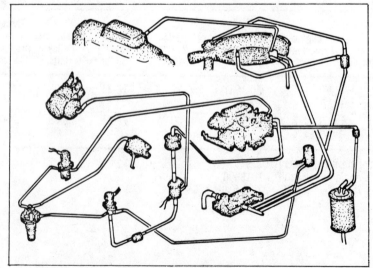

Emission Control system schematic (Accord with manual transmission)

Honda

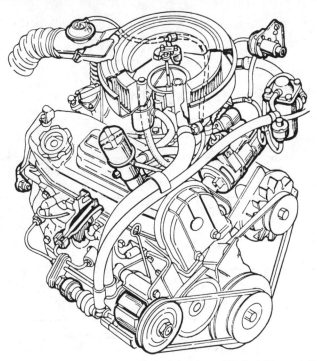
Emission control system components—Typical of engines with manual transmissions

duce Nox emissions. When the coolant temperature is approximately 120° or higher, the solenoid valve is energized, cutting off vacuum to the advance unit.

Ignition Timing Retard Unit

The ignition timing retard unit is used only on Hondamatic models and has no vacuum advance mechanism.

Air Injection System

Beginning with the 1975 model year, an air injection system is used. A belt-driven air pump delivers filtered air under pressure to injection nozzles located at each exhaust port. Here, the additional oxygen supplied by the vane-type pump reacts with any uncombusted fuel mixture, promoting an afterburning effect in the hot exhaust manifold. To prevent a reverse flow in the air injection manifold when exhaust gas pressure exceeds air supply pressure, a non-return check valve is used. To prevent exhaust afterburning or backfiring during deceleration, an anti-afterburn valve delivers air to the intake manifold instead. When manifold vacuum rises above the preset vacuum of the air control valve and/or below that of the air bypass valve, air pump air is returned to the air cleaner.

1976-79 CVCC Models

Throttle Controls

This system controls the closing of the throttle during periods of gear shifting, deceleration, or anytime the gas pedal is released. This system has two main parts, a dashpot system and

Start control solenoid valve

a throttle positioner system. The dashpot diaphragm and solenoid valve act to dampen or slow down the throttle return time to 1-4 seconds. The throttle positioner part consists of a speed sensor, a solenoid valve, a control valve and an opener diaphragm which will keep the throttle open a predetermined minimum amount any time the gas pedal is released when the car is traveling 15 mph (Civic) 20 mph (Accord) or faster, and closes it when the car slows to 10 mph.

Ignition Timing Controls

This system uses a coolant temperature sensor to switch distributor vacuum ignition timing controls on or off to reduce hydrocarbon and oxides of nitrogen emissions. The coolant switch is calibrated at 149° F.

Hot Start Control

This system is designed to prevent an overrich mixture condition in the intake manifold due to vaporization of residual fuel when starting a hot engine. This reduces hydrocarbon and carbon monoxide emissions.

CVCC Engine Modifications

By far, the most important part of the CVCC engine emission control system is the Compound Vortex Controlled Combustion (CVCC) cylinder head itself. Each cylinder has three valves: a conventional intake and conventional exhaust valve, and a smaller auxiliary intake valve. There are actually *two* combustion chambers per cylinder: a precombustion or auxiliary chamber, and the main chamber. During the intake stroke, an extremely lean mixture is drawn in to the main combustion chamber. Simultaneously, a very rich mixture is drawn into the smaller precombustion chamber via the auxiliary intake valve. The spark plug, located in the precombustion chamber, easily ignites the rich premixture, and this combustion spreads out into the main combustion chamber where the lean mixture is ignited. Due to the fact that the volume of the auxiliary chamber is much smaller than the main chamber, the overall mixture is very lean (about 18 parts air to one part fuel).

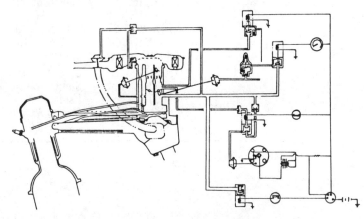
Emission control system schematic (Civic CVCC with manual transmission)

Subaru

TUNE-UP SPECIFICATIONS

Year	Engine Displacement (cc)	SPARK PLUGS Type	Gap (in.)	DISTRIBUTOR Point Dwell (deg)	Point Gap (in.)	Ignition Timing (deg)	Intake Valve Opens (deg)	Cranking Compression Pressure (psi)	IDLE SPEED (rpm)	VALVE CLEARANCE (in) In	Ex
1973-74	1400	BP-6ES	0.032	49-55	0.020	6B @ 800	24B	178	800	0.011-0.013	0.011-0.013
1975	1400	BP-6ES	0.030	49-55	0.020	8B @ 800M, 8B @ 900A	24B	178	①	0.012	0.014
1976	1400	BP-6ES	0.032	49-55	0.018	8B @ 900	24B	156	①	0.011	0.015
	1600	BP-6ES	0.032	49-55	0.018	8B @ 900	24B	156	①	0.011	0.015
1977-79	1600	BP-6ES	0.032	49-55	0.018	8B @ 850②	24B	156	①	0.010	0.014

B   Before Top Dead Center
TDC Top Dead Center
M—Manual, A—Automatic

① See Engine Compartment Sticker
② Calif. 900

NOTE: The underhood specifications sticker often reflects tune-up specification changes made in production. Sticker figures must be used if they disagree with those in this chart.

VEHICLE IDENTIFICATION

Year	Model (displacement)	Body Style	Vehicle Identification Number Code
1973-75	1400 GL	Sedan, Coupe, Hardtop	A22L
	1400 DL	Station Wagon	A62L
1975	4 wheel drive	Station Wagon	A64L
1976	1400, 1600 DL	Sedan, Coupe (M.T.)	A22L
	1400, 1600 DL	Sedan (A.T.)	A26L
	1400, 1600 DL	Station Wagon (M.T.)	A62L
	1400, 1600 DL	Station Wagon (A.T.)	A66L
	1400, 1600 GF	Hardtop (M.T.)	A22L
	1400, 1600 GF	Hardtop (A.T.)	A26L
	4 wheel drive	Station Wagon	A64L
1977-79	1600 DL	Sedan, Coupe	A26L
	1600 DL	Station Wagon	A66L
	1600 GF	Hardtop	A26L
	4WD	Station Wagon	A67L
	Brat	4 WD	—

Subaru

BREAKERLESS DISTRIBUTOR

Beginning with the 1977 model year, vehicles offered for sale in California were equipped with a breakerless distributor. The centrifugal advance, vacuum advance and retard units are the same as used with the conventional distributor.

The air gap between the reluctor and pick-up coil is adjustable and should be 0.008 to 0.016 inch for the standard transmission equipped vehicles and 0.012 to 0.016 inch for automatic transmission equipped vehicles. The ignition timing should be checked if any changes are made to the air gap.

CARBURETOR ADJUSTMENTS

(with or without CO Meter)

1976

With CO meter
(with air injector system connected)
Idle speed—850-950 RPM
CO %—0.15-0.55%
(w/o air injector connected)
Idle speed—850-950 RPM
CO %—0.5-1.5 %

Without CO meter
1. Adjust the engine idle speed to attain 980 RPM.
2. Turn the air/fuel mixture screw clockwise until the engine speed is 900 RPM.
3. The emitted CO should be within the proper range.

1977-78

With CO meter
(with air injected system connected)
49 states and high altitude
Idle speed—850 ± 50 RPM
CO %—0.5 ± 1.5%
California
Idle speed—900 ± 50 RPM
CO %—0.75 ± 0.25%
(w/o air injected system connected)
49 states and high altitude
Idle speed—850 ± 50 RPM
CO %—1.5 ± 0.5%
California
Idle speed—900 ± 50 RPM
CO % 0.75 ± 0.25%

Without CO meter
1. Adjust the engine speed to attain 930 RPM (990 RPM California) by adjusting the air/fuel mixture screw and the throttle screw.
2. Adjust the air/fuel mixture screw clockwise until the engine speed is changed to 850 RPM (900 RPM California).
3. The emitted CO should be within the proper range.

1979

Follow the instructions on the Emission Control Information Label, located within the engine compartment.

EMISSION CONTROLS CRANKCASE EMISSION CONTROL SYSTEM

The sealed crankcase emission control system takes blow-by gas emitted from the crankcase and routes the gas through the air cleaner and into the intake manifold for recombustion.

The system consists of a sealed oil filler cap, a rocker cover with an outlet pipe, an air cleaner with an inlet pipe to receive the connecting hoses and the connecting hoses and clamps.

There are no tests other than making sure that the system is kept clean.

EMISSION CONTROLS

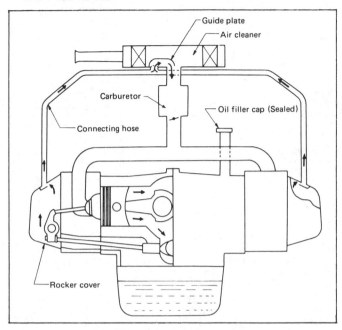

Cross section of crankcase emission control system

EVAPORATIVE EMISSION CONTROL SYSTEM

1976

Evaporative gas from the fuel tank is not discharged into the atmosphere but conducted to the air cleaner unit and then burned in the combustion chamber. No absorbent is used.

The system consists of a sealed fuel tank and filler cap, two reservoir tanks on the station wagon, an air breather valve or a restriction, breather hoses, breather pipe and the air cleaner.

There is an air breather valve located at the filler cap. When the flap (door) is opened, a spring exerts pressure on the rubber breather hose and pinches it shut.

The vacuum relief valve filler cap relieves any vacuum condition that might arise in the gas tank.

1977-79

This system includes a canister, a check valve, two orifices and on station wagons and 4-wheel drive vehicles, two reserve tanks.

Gasoline vapor evaporated from the fuel in the fuel tank is introduced into the canister located in the engine compartment and absorbed by the activated charcoal particles. As engine speed increases a purge valve is opened and fresh air is sucked in through the bottom filter of the canister purging the absorbed vapor from the activated charcoal.

AIR SUCTION SYSTEM

1976-79

The Air Suction System is used to reduce the exhaust emissions by supplying secondary air into the exhaust ports without the use of an air pump.

The negative pressure caused by the exhaust gas pulsation and the intake manifold pressure during the valve overlap period, is utilized for

Subaru

the further oxidation of HC and CO and to avoid the use of a catalytic converter or thermal reactor in the exhaust line.

A double reed valve assembly is used to direct fresh air to flow into the exhaust ports under negative pressure and to block the return of the exhaust gas under positive pressure. A reed valve controls air to each side of the engine.

An air suction silencer is used to muffle the sound of the air being admitted into the Air Suction System.

HOT AIR CONTROL SYSTEM

The hot air control system consists of the air cleaner, the air stove on the exhaust pipe and the air intake hose connecting the air cleaner and air stove. The air cleaner is equipped with an air control valve which maintains the temperature of the air being drawn into the carburetor at 100°-127° F. to reduce HG emission when the underhood temperature is below 100° F. This system should be inspected every 12,000 miles.

COASTING BY-PASS SYSTEM

To control the HC emissions while the vehicle is in the coasting or decelerating mode, a controlled amount of air/fuel mixture is channelled through the coasting by-pass passage in the carburetor to improve the combustion in the cylinders at the periods of high engine vacuum.

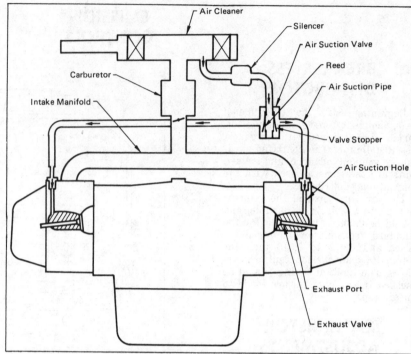

Cross section of air suction system

The high engine vacuum reacts on a by-pass valve diaphragm, opening a vacuum passage to the servo valve, located on the carburetor, which in turn opens a metered passage for air and fuel, from the carburetor air horn to the section of the throttle bore below the secondary throttle plate.

As the engine vacuum changes to a lower value due to acceleration, the by-pass valve closes the passage to the servo valve and the carburetor returns to its normal function.

ELECTRICALLY ASSISTED AUTOMATIC CHOKE

A vacuum-operated automatic choke replaces the manual chock previously used. The automatic choke uses a chock cap containing a heating element to speed up choke valve opening and reduce CO emissions during warm-up. The heating element gets its power from a special tap on the voltage regulator, when the ignition is on and the engine running.

EXHAUST GAS RECIRCULATION (EGR) SYSTEM

An exhaust gas recirculation (EGR) system is used on 1976 California and all 1977 and later models to reduce NO (oxides of nitrogen) emissions by lowering peak flame temperature during combustion. A small portion of the exhaust gases are routed into the intake manifold via a vacuum-operated EGR control valve.

A solenoid vacuum valve controls the flow of vacuum from a port on the carburetor (above the primary throttle valve) to the EGR valve vacuum diaphragm. The solenoid, in turn, is operated by a coolant temperature switch.

When the coolant temperature is above 122°F., the temperature switch breaks the current flow to the vacuum solenoid valve. The valve closes, permitting the throttle port vacuum to operate the EGR valve diaphragm.

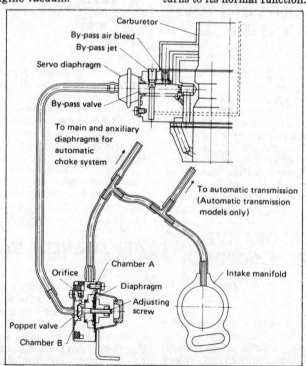

Coasting by-pass system components

Subaru

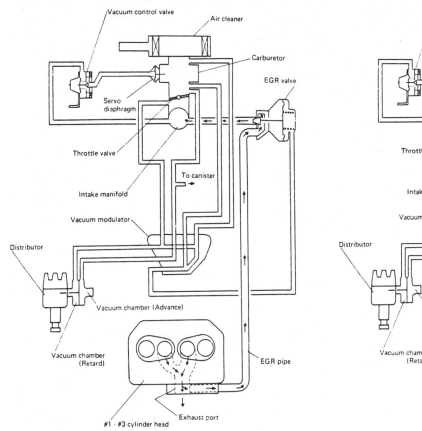

EGR system—49-states model (except with auto. trans. or 4WD)

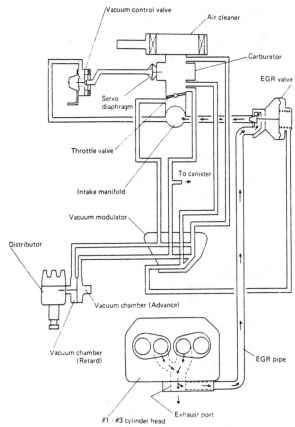

EGR system—49-states model (with auto. trans. or 4WD)

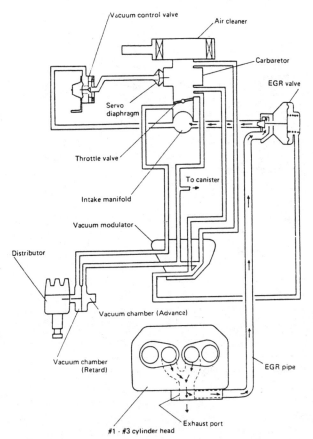

EGR system—Calif. models (except with auto. trans.), 49-states and high altitude models (except with auto. trans. or 4WD)

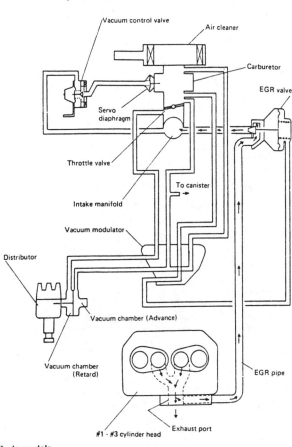

EGR system—Calif. models (with auto. trans.)

399

Subaru

This causes the EGR valve to open under conditions other than idle or wide-open throttle.

Below 122°F., the vacuum solenoid valve is energized to vent the vacuum from the throttle port into the atmosphere through a filter. By preventing exhaust gas recirculation from occurring before the engine has warmed-up, cold driveability is greatly improved.

ENGINE MODIFICATION SYSTEM

The vacuum is routed to the distributor vacuum retard unit. The vacuum unit retards the ignition spark in order to promote complete combustion in the cylinders.

There is an anti-dieseling solenoid mounted opposite the float bowl on the carburetor. This prevents the engine from dieseling when the ignition switch is turned off. When the ignition switch is turned off, an electromagnet in the switch is also cut off. A spring inside the housing forces a plunger into position, blocking the fuel passages leading to the opening below the throttle plates. When the ignition switch is turned on, it energizes the electromagnet in the switch and pulls the plunger out of the fuel passage, allowing fuel to reach the opening below the throttle plates.

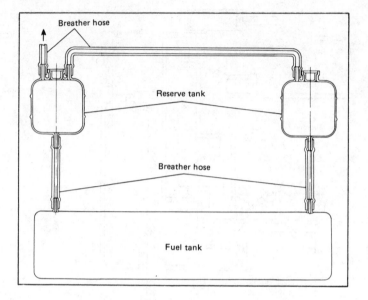

Vapor separators mounted on station wagon—typical

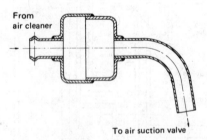

Cross section of air suction system silencer

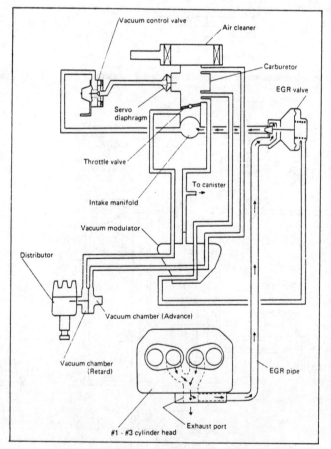

EGR system—49-states and high altitude (with auto. trans.)

Toyota

TUNE-UP SPECIFICATIONS

Year	Engine Type	SPARK PLUGS Type (ND)	SPARK PLUGS Gap (in.)	DISTRIBUTOR Point Dwell (deg)	DISTRIBUTOR Point Gap (in.)	Ignition Timing (deg) ▲ MT	Ignition Timing (deg) ▲ AT	Compression Press.	Fuel Pump Press. **	IDLE SPEED (rpm) ▲ MT	IDLE SPEED (rpm) ▲ AT	VALVE CLEARANCE (in.) Intake	VALVE CLEARANCE (in.) Exhaust
1975-77	2T-C	W16EP	0.030	52⑤	0.018	10B⑥	10B⑥	171	2.8-4.3⑦	850	850	0.008	0.013
	20R	W16EP	0.030	52	0.018⑩	8B	8B	156	2.2-4.2	850	850	0.008	0.012
	4M③	W16EP	0.030	41	0.018	10B	10B	156	4.2-5.4	800	750	0.007	0.010
	4M④	W16EP	0.030	41	0.018	5B	5B	156	4.2-5.4	800	750	0.007	0.010
	2F	W14EX	0.037	41	0.018	7B	—	149	3.4-4.7	650	—	0.008	0.014
	3K-C	W20EP	0.031	52	0.018	5B	—	156	2.8-4.3	750	—	0.008	0.012
1978-79	3K-C	BPR5EA-L	0.031	Electronic		8B	8B	156	3.0-4.5	750	750	0.008	0.012
	2T-C	BP5EA-L	0.031	Electronic		10B⑪	10B⑪	171	3.0-4.5	850	850	0.008	0.013
	4M	BPR5EA-L	0.031	Electronic		10B⑪	10B⑪	156	4.2-5.4	750	750	0.011	0.014
	20R	BP5EA-L⑫	0.031	Electronic		8B	8B	156	2.2-4.2	800	850	0.008	0.012

NOTE: If the information given in this chart disagrees with the information on the engine tune-up decal, use the specifications on the decal—they are current for the engine in your car.

▲ With manual transmission in Neutral and automatic transmission in Drive (D) (1973) or Neutral (1974-79).
③ Except Calif.
④ California only
⑤ Dual point—main 57°; sub 52°
⑩ California model Celica GT equipped with transistorized ignition
⑪ Calif.: 8B
⑫ Celica: BPR5EA-L
MT Manual transmission
AT Automatic transmission
TDC Top Dead Center
B Before top dead center
A After top dead center

TIMING MARK LOCATIONS

Engine Type	Location	Type of mark
3K-C and 2T-C	Crankshaft pulley	Notch and number scale
18R-C, 20R	Crankshaft pulley	Pointer and painted slot
4M	Crankshaft pulley	Slot and number scale
F, 2F	Flywheel	Ball and pointer

FAST IDLE ADJUSTMENT

Engine	Throttle Valve to bore clearance (in.)	Primary throttle angle (deg)	To adjust fast idle:
3K-C	0.040①	9②	Bend the fast idle lever
2T-C	0.032③	7	Turn the fast idle adjusting screw
18R-C	0.041	13—from closed	Turn the fast idle adjusting screw
20R	0.047	—	Turn the fast idle screw
4M	—	16—from closed④	Turn the fast idle adjusting screw
F	—	30—from closed	Bend the fast idle lever
2F	0.051	30—from closed	Bend the fast idle lever

— Not available
① 0.051 in 1976; 0.056 in 1977; 0.037 in 1978-79
② 20° open
③ 1976-79: 0.043
④ 1977-79: 9°

Toyota

CAR SERIAL NUMBER AND ENGINE IDENTIFICATION

ENGINE IDENTIFICATION

All models have the vehicle identification number (VIN) stamped on a plate which is attached to the left side of the instrument panel. This number consists of model and serial numbers.

The engine number is stamped on the right side of the block and consists of the model and serial number. On the 78 and 79 4M engine the identification number is stamped on the left side of the block.

Model	Year	Displacement Cu. in. (cm³)	No. of Cylinders	Type	Engine Series Identification
Corolla					
1200	1976-79	71.2 (1166)	4	OHV	3K-C
1600	1976-79	96.9 (1588)	4	OHV	2T-C
Corona					
2200	1976-79	133.6' (2189)	4	OHC	20R
Mark II & Cressida					
2600	1976-79	156.4 (2563)	6	OHC	4M

Model	Year	Displacement Cu. in. (cm³)	No. of Cylinders	Type	Engine Series Identification
Celica					
2200	1976-79	133.6' (2189)	4	OHC	20R
Hi-Lux					
2200	1976-79	133.6' (2189)	4	OHC	20R
Land Cruiser					
	1976-79	256.00 (4200)	6	OHV	2F

OHV—Overhead valve
OHC—Overhead cam

EMISSION CONTROL SYSTEMS

EMISSION CONTROLS
POSITIVE CRANKCASE VENTILATION (PCV) SYSTEM

A (PCV) valve is used in the line to prevent the gases in the crankcase from being ignited in case of a backfire. The amount of blow-by gases entering the mixture is also regulated by the PCV valve, which is spring-loaded and has a variable orifice.

The valve is ether mounted on the valve cover or in the line which runs from the intake manifold to the crankcase.

AIR INJECTION SYSTEM

A belt-driven air pump supplies air to an injection manifold which has nozzles in each exhaust port. Injection of air at this point causes combustion of unburned hydrocarbons in the exhaust manifold rather than allowing them to escape into the atmosphere. An antibackfire valve controls the flow of air from the pump to prevent backfiring which results from an overly rich mixture under closed throttle conditions.

A check valve prevents hot exhaust gas back-flow into the pump and hoses, in case of a pump failure, or when the antibackfire valve is working.

In addition newer engines have an air switching valve. On engines without catalytic converters, the ASV is used to stop air injection under a constant heavy engine load.

On engines with catalytic converters the ASV is used to protect the catalyst from overheating, by blocking the air necessary for the reaction.

On all 1976-79 engines, except the 2F, the relief valve is built into the ASV.

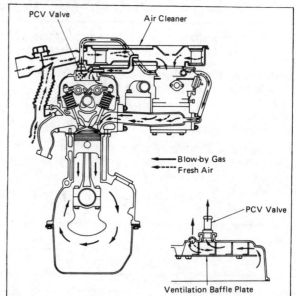

Positive crankcase ventilation system (PCV)—Typical

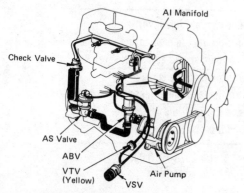

Air injection system (AI)—Typical Calif.

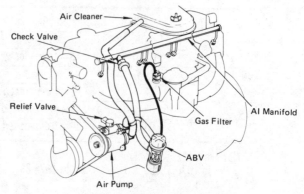

Air injection system (AI)—Typical

Toyota

AIR INJECTION SYSTEM DIAGNOSIS CHART

Problem	Cause	Cure
1. Noisy drive belt	1a Loose belt	1a Tighten belt
	1b Seized pump	1b Replace
2. Noisy pump	2a Leaking hose	2a Trace and fix leak
	2b Loose hose	2b Tighten hose clamp
	2c Hose contacting other parts	2c Reposition hose
	2d Diverter or check valve failure	2d Replace
	2e Pump mounting loose	2e Tighten securing bolts
	2g Defective pump	2g Replace
3. No air supply	3a Loose belt	3a Tighten belt
	3b Leak in hose or at fitting	3b Trace and fix leak
	3c Defective anti-backfire valve	3c Replace
	3d Defective check valve	3d Replace
	3e Defective pump	3e Replace
4. Exhaust backfire	4a Vacuum or air leaks	4a Trace and fix leak
	4b Defective anti-backfire valve	4b Replace
	4c Sticking choke	4c Service choke
	4d Choke setting rich	4d Adjust choke

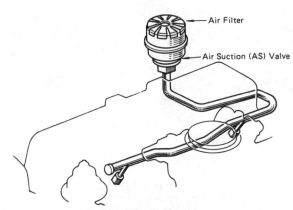

Air suction system (AS)—Typical

Air suction system inspection

AIR SUCTION SYSTEM

The Air Suction System, available on the 3K-C and 2T-C engines, brings fresh, filtered air into the exhaust ports to promote better burning of hydrocarbons. It also supplies the air necessary for the oxidizing reaction in the catalytic converter.

There are no adjustments on the system and, should it malfunction, the unit must be replaced as a whole.

To check the system, look over all lines for cracks or damage. If checks indicate no problems, start the engine and put a thin sheet of paper over the inlet port of the filter. If it is drawn to the opening, the unit is operating. If not, remove the filter and test the valve opening the same way. Replace the filter, if necessary.

EVAPORATIVE EMISSION CONTROL SYSTEM

Toyota vehicles use evaporative emission control (EEC) systems. All

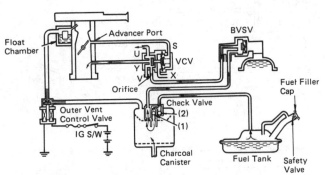

Fuel evaporative emission control system (EVAP) 4M engine—Typical

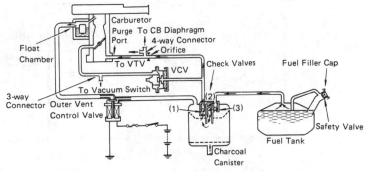

Fuel evaporative emission control system (EVAP) 3K-C engine—Typical

403

Toyota

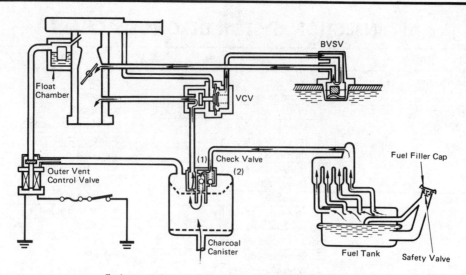

Fuel evaporative emission control system (EVAP) 2F engine

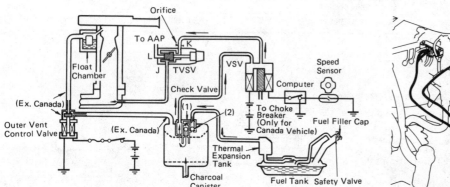

Fuel evaporative emission control system (EVAP) 2OR engine—Typical

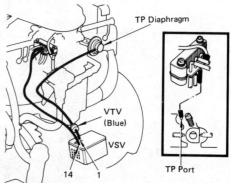

Throttle positioner system (TP)—Typical (2F engine)

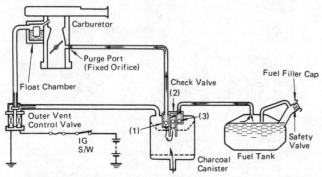

Fuel evaporation emission control system (EVAP) 2T-C engine—Typical

models use a "charcoal canister" storage system.

The charcoal canister storage system stores the fuel vapors in a canister filled with activated charcoal. All models use a vacuum switching valve to purge the system. The air filter is an integral part of the charcoal canister.

California models equipped with the 4M six-cylinder engine have a canister-mounted purge control valve. The purge control valve is connected to a carburetor port, which is located above the throttle control valve.

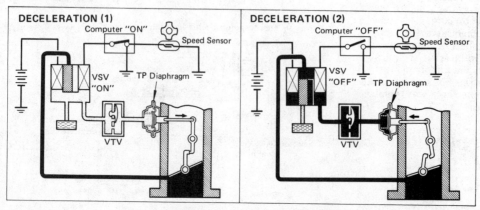

Throttle positioner operation—Typical (2F engine)

404

Toyota

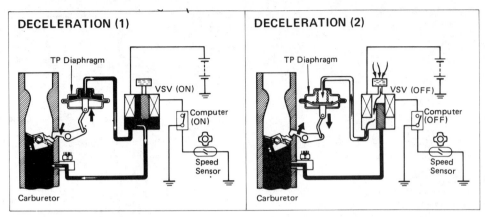

Throttle positioner operation Calif. typical 20R engine (manual trans.)

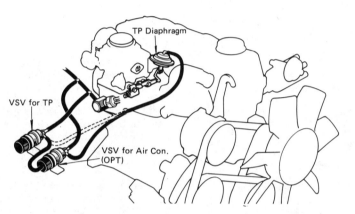

Throttle positioner system (TP) Calif. typical 20R engine (manual trans.)

When the engine is stopped or idling, there is no vacuum signal at the purge control valve so that it remains closed. When the throttle valve opens, the carburetor port is uncovered and a vacuum signal is sent to the purge control valve, which opens and allows the vapors stored in the canister to be pulled into the carburetor.

THROTTLE POSITIONER

On Toyotas with an engine modification system, a throttle positioner is included to reduce exhaust emissions during deceleration. The positioner prevents the throttle from closing completely. Vacuum is reduced under the throttle valve which, in turn, acts on the retard chamber of the distributor vacuum unit (if so equipped). This compensates for the loss of engine braking caused by the partially opened throttle.

Once the vehicle drops below a predetermined speed the vacuum switching valve provides vacuum to the throttle positioner diaphragm; the throttle positioner retracts allowing the throttle valve to close completely. The distributor also is returned to normal operation.

MIXTURE CONTROL SYSTEM

The mixture control valve, used on

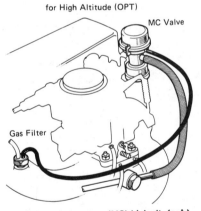

Mixture control system (MC) high alt. (opt.)—Typical

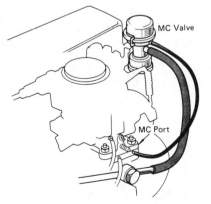

Mixture control system (MC)—Typical

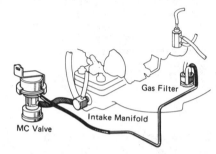

Mixture control system (MC) Calif.—Typical

all 1978 and later 3K-C and 2T-C engines aids in combustion of unburned fuel during periods of deceleration. The mixture control valve is operated by the vacuum switching valve during periods of deceleration to admit additional fresh air into the intake manifold. The extra air allows more complete combustion of the fuel, thus reducing hydrocarbon emissions.

EXHAUST GAS RECIRCULATION (EGR)

The Exhaust Gas Recirculation System (EGR) is used on all U.S. engines, except the 2T-C engine used in the Corolla. The Corolla did not adapt to the EGR system until 1977. The EGR valve is controlled by the same computer and vacuum switching valve that is used to operate other emis-

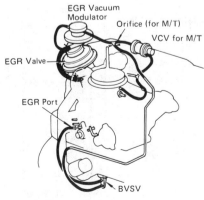

Exhaust gas recirculation system (EGR)—Typical

Toyota

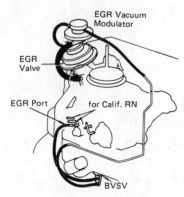

Exhaust gas recirculation system (EGR) Calif.—Typical

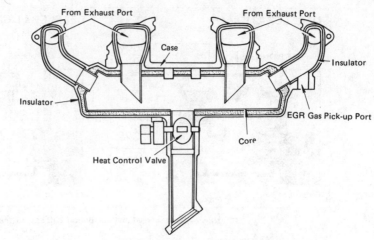

Thermal reactor system Calif. 2F engine only

sions control system components.

On all engines there are several conditions, determined by the computer and vacuum switching valve, which permit exhaust gas recirculation to take place.
1. Vehicle speed
2. Engine coolant temperature
3. EGR valve temperature, (F engine)

2F Thermal Reactor System and Heat Control Valve

Installed in place of the exhaust manifold on the Land Cruiser 2F engine for California, is the Thermal Reactor System. It collects the exhaust gases in a common area in order to keep their temperatures higher for a longer period of time to increase the efficiency of the exhaust gas and secondary air, restricting the release of unburned emissions.

Choke Return System

Because of the chance of seriously damaging the catalytic converter by operating the automobile with the choke out for long periods of time, the 3K-C engine is equipped with a choke return system.

Utilizing a holding coil, a holding plate and a return spring, the system generates a magnetic force when the coolant temperature is below 104°F, holding the choke plate open.

However, when the coolant exceeds 104°F., the thermo switch opens, cut-

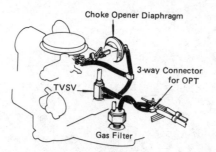

Choke opener system—Typical

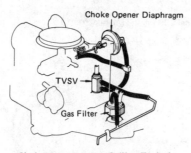

Choke opener system Calif.—Typical

ting off the current flow and allows the return spring to pull the choke plate open.

Most problems in this system will be electrical. Should the system malfunction, check for continuity in all circuits and replace the part not operating.

Choke Opener System

If a cold engine is driven soon after starting, the automatic choke system will close the choke plate, resulting in high levels of emissions. To combat that situation 1976-1977 California and High Altitude 2T-C engines and all 1978-79 engines are equipped with a system that forcibly holds the choke plate open.

When the coolant is below 140° F, the thermo wax in the TVSV closes the valve and prohibits any vacuum from acting on the choke diaphragm. This keeps the choke open. Above 140°F, the wax expands, opening the valve, and allows the choke plate to operate normally. Should the system malfunction, repace the TVSV.

When the engine is cold, an auxiliary enrichment circuit in the carburetor is operated to squirt extra fuel into the acceleration circuit in order to prevent the mixture from becoming too lean.

DUAL-DIAPHRAGM DISTRIBUTOR

Some Toyota models are equipped with a dual-diaphragm distributor unit. This distributor has a retard diaphragm, as well as a diaphragm

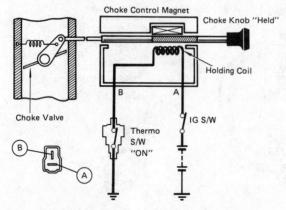

Choke return system—Cold position

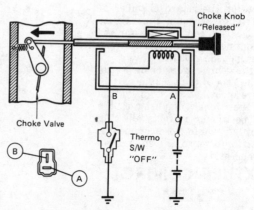

Choke return system—Hot position

for advance. Retarding the timing helps to reduce exhaust emissions, as well as making up for the lack of engine braking on models equipped with a throttle positioner.

SPARK CONTROL SYSTEM

Non-California Corolla models have a spark delay valve (SDV) in the distributor vacuum line. In 1978-79 all California 2T-C engines, all non-California 20R engines, and all 3K-C and 4M engines are equipped with the spark control system. The valve has a small orifice in it, which slows down the vacuum flow to the vacuum advance unit on the distributor. By delaying the vacuum to the distributor, a reduction in HC and CO emission is possible.

When the coolant temperature is between 95°F-140°F (1976-77) or 86°F (1978-79 2T-C) or 122°F (4M), 104°F (20R) or 95°F (1978-79 3K-C), a coolant temperature operated vacuum control valve is opened, allowing the distributor to receive undelayed, ported vacuum through a separate vacuum line. Above 95°F-140°F this line is blocked, and all ported vacuum must go through the spark delay valve.

Spark Control Valve Operating Temperatures

Closed	Degrees F
1976-77 non-Cal. Corolla	95
1978-79 3K-C	95
1978-79 4M	122
1978-79 20R	104
1978-79 2T-C	86
Open	
1976-77 non-Cal.	140
1978-79 3K-C	120
1978-79 4M	147
1978-79 20R	129
1978-79 2T-C	111

ENGINE MODIFICATIONS SYSTEM

Toyota also uses an assortment of engine modifications to regulate exhaust emissions. Most of these devices fall into the category of engine vacuum controls. There are three principal components used on the engine modifications system, as well as a number of smaller parts. The three major components are: a speed sensor; a computer (speed marker); and a vacuum switching valve.

The vacuum switching valve and computer circuit operates most of the emission control components. Depending upon year and engine usage, the vacuum switching valve and computer may operate the purge control for the evaporative emission control system; the transmission controlled spark (TCS) or speed controlled spark (SCS); the dual-diaphragm distributor; and the throttle positioner systems.

The major difference between the transmission controlled spark and speed controlled spark systems is in the manner in which system operation is determined.

Below a predetermined speed, or any gear other than fourth, the vacuum advance unit on the distributor is rendered inoperative or, on F engines, timing is retarded. By changing the distributor advance curve in this manner, it is possible to reduce emissions of oxides of nitrogen (NOx).

NOTE: *Some engines are equipped with a thermo-sensor so that the TCS on SCS system only operates when the coolant temperature is 140°-212° F.*

Aside from determining the conditions outlined above, the vacuum switching valve computer circuit operates other devices in the emission control system.

The computer acts as a speed marker; at certain speeds it sends a signal to the vacuum switching valve which acts as a gate, opening and closing the emission control system vacuum circuits.

Power Valve Control System 2F

In order to minimize CO while ensuring good driveability, the Land Cruiser 2F engines for California are equipped with a power valve control system on the carburetor.

Dependent on coolant temperature, the power system opens or closes turning the VSV "On." It is also influenced by the amount of pressure on the accelerator pedal. Various combinations of pedal pressure and temperature will result in more or less fuel available for use. Any breakdown in the system requires replacement of the part involved.

High Altitude Compensation System

For all engines to be sold in areas over 4,000 ft. in altitude, a system has been installed to automatically lean out the fuel mixture by supplying additional air. This also results in lower emissions.

Low atmospheric pressure allows the bellows in the system to expand

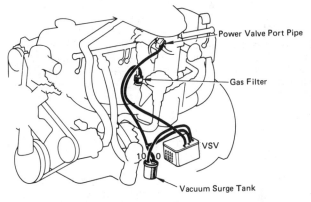

Power valve control system Calif. 2F engine—Typical

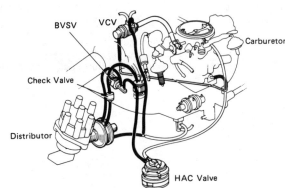

Spark control system (SC)—Typical

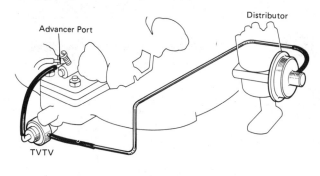

Spark control system (SC) Calif.—Typical

Toyota

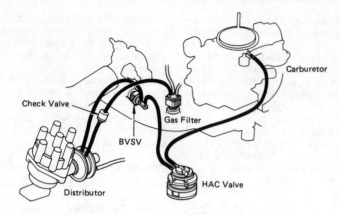

High altitude compensation system—Typical

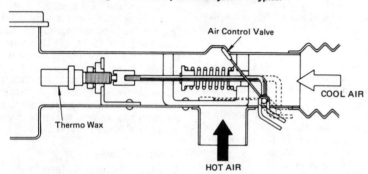

Automatic hot air intake system (HAI)—Typical

and close a port, allowing more air to enter from different sources.

In the 2T-C and 20R engines, this also results in a timing advance to improve driveability.

All parts in this system must be replaced. The only adjustment available is in the timing.

Hot Air Intake—All Engines

In order to keep the temperature of the air drawn into the carburetor as constant as possible, all engines (1977) are equipped with a Hot Air Intake System (HAI).

In all engines but the 2F, the system depends on a thermo valve to control the temperature. In the 2F (Land Cruiser), the temperature and valve are controlled by thermo wax.

At normal temperatures the air is drawn through the inlet in the air filter. When the temperature drops, the valve switches position, opening the way for air to be drawn from around the exhaust manifold.

When inspecting, check all hoses for poor connections or damage and visually check the air control valve in the air duct.

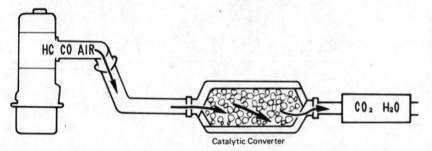

Catalaytic converter operation—Typical

NOTE: *When checking valve movement, do not push too strongly on the control face.*

Should there be a malfunction, replace the part involved.

CATALYTIC CONVERTERS

All Toyota vehicles sold in this country are equipped with a catalytic converter, except the 1976 Hi-Lux and Land Cruiser sold in California and the 1976 Mark IIs sold in the United States.

The catalysts are made of noble metals (platinum and palladium) which are bonded to individual pellets. These catalysts cause the HC and CO to break down into water and carbon dioxide (CO_2) without taking part in the reaction; hence, a catalyst life of 50,000 miles may be expected under normal conditions.

An air pump is used to supply air to the exhaust system to aid in the reaction. A thermosensor, inserted into the converter, shuts off the air supply if the catalyst temperature becomes excessive.

The same sensor circuit also causes a dash warning light labeled "EXH TEMP" to come on when the catalyst temperature gets too high.

NOTE: *It is normal for the light to come on temporarily if the car is being driven downhill for long periods of time (such as descending a mountain).*

The light will come on and stay on if the air injection system is malfunctioning or if the engine is misfiring.

Catalyst Precautions

1. Use only unleaded fuel.
2. Avoid prolonged idling; the engine should run no longer than 20 minutes at curb idle, nor longer than 10 minutes at fast idle.
3. Reduce the fast idle speed, by quickly depressing and releasing the accelerator pedal, as soon as the coolant temperature reaches 120°F.
4. Do not disconnect any spark plug leads while the engine is running.
5. Make engine compression checks as quickly as possible.
6. Do not dispose of the catalyst in a place where anything coated with grease, gas, or oil is present; spontaneous combustion could result.

Toyota

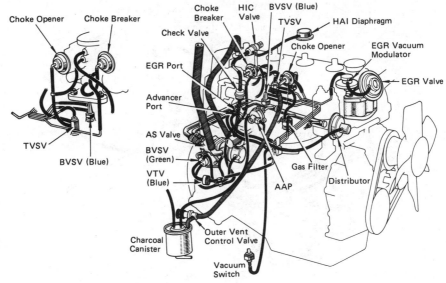

2T-C engine emission component layout—Typical

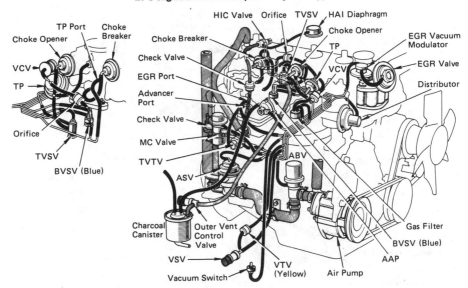

2T-C engine emission component layout Calif.—Typical

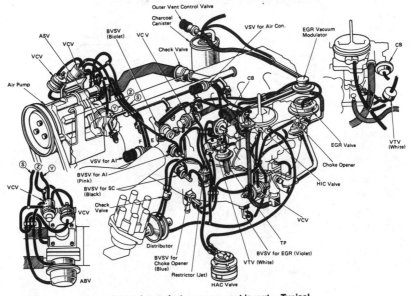

4M engine emission component layout—Typical

Toyota

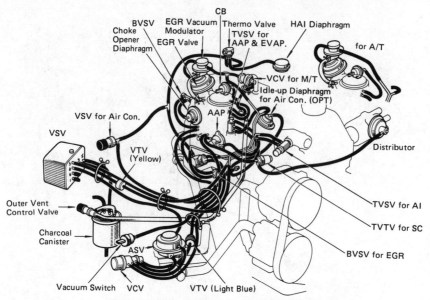

20R engine emission component layout—Typical

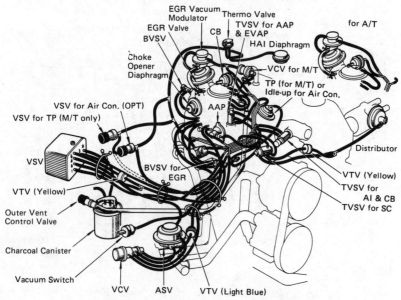

20R engine emission component layout Calif.—Typical

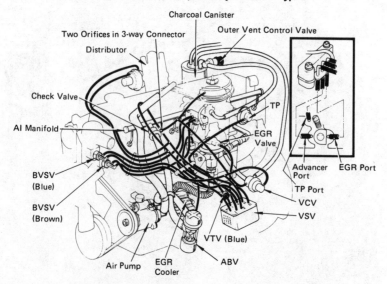

2F engine emission component layout—Typical

Toyota

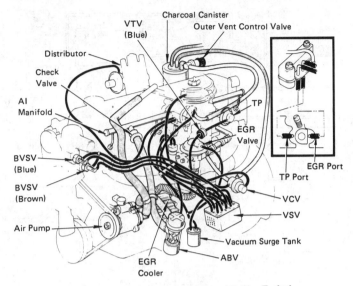

2F engine emission component layout Calif.—Typical

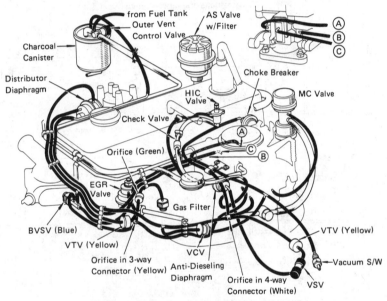

3K-C engine emission component layout—Typical

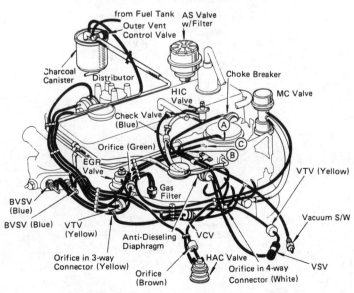

3K-C engine emission component layout Calif.—Typical

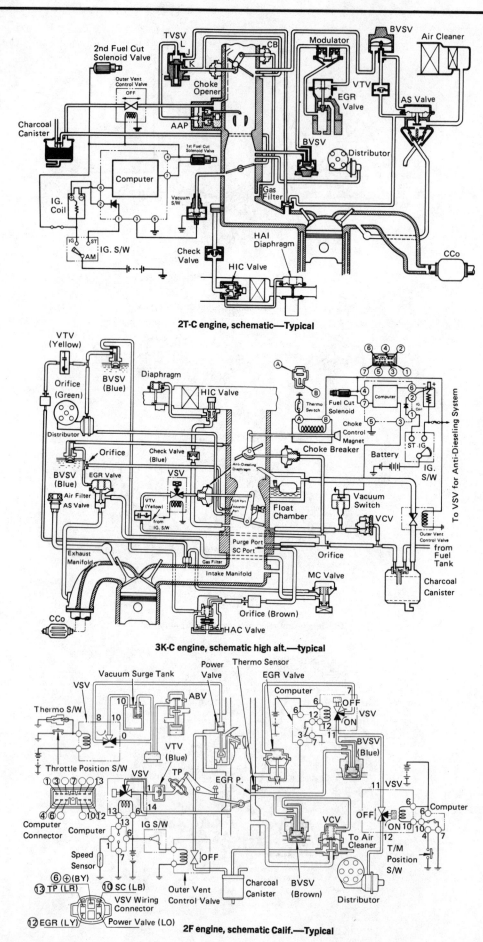

Toyota

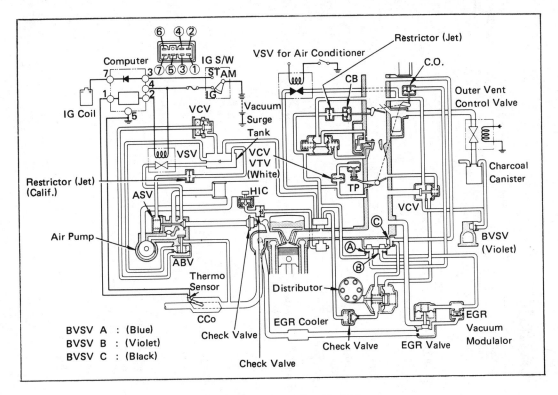

4M engine, schematic—Typical

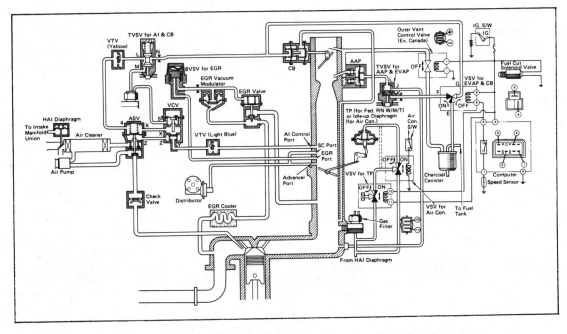

2OR engine, schematic U.S. series—Typical

TUNE-UP SPECIFICATIONS

Year	Code	Type	Common Designation	Spark Plugs Type	Spark Plugs Gap (in.)	Distributor Point Dwell (deg)	Distributor Point Gap (in.)	Ignition Timing (deg) MT	Ignition Timing (deg) AT	Fuel Pump Pressure (psi) @ 4000 rpm	Compression Pressure (psi)	Idle Speed (rpm) MT	Idle Speed (rpm) AT	Valve Clearance (in.) Cold In	Valve Clearance (in.) Cold Ex
1975	AJ	1	1600	W145M1 L288	.024	44-50	.016	5ATDC⑤	TDC⑤	28	85-135	875	875	.006	.006
	ED	2	1800	W145M2 N288	.024	44-50	.016	5ATDC⑤	5ATDC⑤	28	85-135	900	900	.006	.006
1976	AJ	1	1600	Bosch W145M1 Champ L288	.024	44-50	.016	5ATDC⑤	TDC⑤	28	85-135	875	925	.006	.006
	GD	2	2000	Bosch W145M2 Champ N288	.028	44-50	.016	7½BTDC ⑤	7½BTDC ⑤	28	85-135	900	950	.006	.006
1977	AJ	1	1600	Bosch M145M1 Champ N288	.028	44-50	.016	5ATDC⑤	5ATDC⑤	28	85-135	800-950	800-950	.006	.006
	GD	2	2000	Bosch M145M2 Champ N288	.028	44-50	.016	7½BTDC ⑤	7½BTDC ⑤	28	85-135	800-950	850-1000	.006	.006
1978	AJ	1	1600	Bosch W145M1 Champ L288	.028 .028	44-50	.016	5ATDC	5ATDC	28	85-135	800-950	800-950	.006	.006
	GE	2	2000	Bosch W145M2 Champ N288	.028 .028	44-50	.016	7½BTDC	7½BTDC	28	85-135	800-950	900-1000	Hydraulic	Hydraulic
1979	AJ	1	1600	Bosch W145M1 Champ L288	.028 .028	44-50	.016	5ATDC	5ATDC	28	85-135	800-950	800-950	.006	.006
	GE	2	2000	Bosch W145M2	.028	44-50	.016	7½BTDC ⑥	7½BTDC ⑥	28	85-135	800-950	900-1000	Hydraulic	Hydraulic

① At idle, throttle valve closed (Types 1 & 2), vacuum hose(s) on

② At idle, throttle valve closed (Types 1 & 2), vacuum hose(s) off

③ At 3,500 rpm, vacuum hose(s) off

④ From March 1973, vehicles with single diaphragm distributor (one vacuum hose); adjust timing to 7½° BTDC with hose disconnected and plugged. The starting serial numbers for those type 1 vehicles using the single diaphragm distributors are # 113 2674 897 (manual trans.) and 113 2690 032 (auto. stick shift)

⑤ Carbon canister hose at air cleaner disconnected; at idle; vacuum hose(s) on

MT Manual Transmission
AT Automatic Transmission
BTDC Before Top Dead Center
ATDC After Top Dead Center
⑥ 5ATDC—California

Volkswagen

ENGINE IDENTIFICATION CHART

Engine Code Letter	Type Vehicle	First Production Year	Last Production Year	Engine Type	Common Designation
AH (Calif.)	1	1972	1974	①	1600
AK	1	1973	1974	①	1600
AM (181)	1	1973	1974	①	1600
AJ	1	1975	In production	①	1600
CB	2	1972	1973	②	1700
CD	2	1973	1973	②	1700
AW	2	1974	1974	②	1800
ED	2	1975	1975	②	1800
GD	2	1976	1978	②	2000
GE	2	1979	In production	②	2000
U	3	1968	1973	②	1600
X	3	1972	1973	②	1600
EA	4	1972	1974	②	1700
EB (Calif.)	4	1973	1973	②	1700
EC	4	1974	1974	②	1800

① Fan driven by generator
② Fan driven by crankshaft

CHASSIS NUMBER CHART

Model Year	Vehicle	Model No.	Chassis Number From	Chassis Number To
1975	Beetle	11	115 2000 001	115 3200 000
	Beetle Convertible	15	155 2000 001	155 3200 000
	Thing	181	185 2000 001	185 3200 000
	Van	21	215 2000 001	215 2300 000
	Bus	22	225 2000 001	225 2300 000
	Camper, Kombi	23	235 2000 001	235 2300 000
1976	Beetle	11	116 2000 001	—
	Beetle Convertible	15	156 2000 001	—
	Bus	22	226 2000 001	—
	Camper, Kombi	23	236 2000 001	—
1977	Beetle	11	117 2000 001	—
	Beetle Convertible	15	157 2000 001	—
	Bus	22	227 2000 001	—
	Camper, Kombi	23	237 2000 001	—
1978	Beetle Convertible	15	158 2000 001	—
	Bus	22	228 2000 001	—
	Camper	23	238 2000 001	—
1979	Beetle Convertible	15	159 2000 001	—
	Bus	22	229 2000 001	—
	Camper	23	239 2000 001	—

Volkswagen

DASHER-RABBIT-SCIROCCO

TUNE-UP SPECIFICATIONS

Year, Model	Engine Displacement cm³	Spark Plugs		Distributor		Ignition Timing (deg)	Intake Valve Opens (deg)	Compression Pressure (psi)	Idle Speed (rpm)	Valve Clearance (in) ▲	
		Type	Gap (in.)	Point Dwell (deg)	Point Gap (in)					In	Ex
1974 Dasher	(1,471)	W175 T30 N8Y	0.024-0.028	44-50 ①	0.016	3 ATDC @ idle	4 BTDC	142-184	850-1000	0.008-0.012	0.016-0.020
1975 Dasher	(1,471)	W200 T30 N8Y	0.024-0.028	44-50	0.016	3 ATDC @ idle	4 BTDC	142-184	850-1000	0.008-0.012	0.016-0.020
1975 Scirocco, Rabbit	(1,471)	W200 T30 N8Y	0.024-0.028	44-50	0.016	3 ATDC @ idle	4 BTDC	142-184	900-1000	0.008-0.012	0.016-0.020
1976-79 Dasher	(1,588)	W215 T30 N7Y	0.024-0.028	44-50	0.016	3 ATDC @ idle	4 BTDC	142-184	850-1000	0.008-0.012	0.016-0.020
1976-77 Rabbit, Scirocco	(1,588)	W215 T30 N7Y	0.024-0.028	44-50	0.016	3 ATDC @ idle	4 BTDC	142-184	900-1000	0.008-0.012	0.016-0.020
1978-79 Rabbit	(1,457)	W175 T30 N8Y	0.024-0.028	44-50	0.016	3 ATDC @ idle	4 BTDC	142-184	850-1000	0.008-0.012	0.016-0.020
1978 Scirocco	(1,457)	W175 T30 N8Y	0.024-0.028	44-50	0.016	3 ATDC @ idle	4 BTDC	142-184	850-1000	0.008-0.012	0.016-0.020
1979 Scirocco	(1,588)	W175 T30 N8Y	0.024-0.028	44-50	0.016	3 ATDC @ idle	4 BTDC	142-184	850-1000	0.008-0.012	0.016-0.020

NOTE: The underhood specifications sticker often reflects tune-up specification changes made in production. Sticker figures must be used if they disagree with those in this chart.

① 47°-53°—California

▲ NOTE: Valve clearance need not be adjusted unless it varies more than 0.002 in. from specification.

DASHER-RABBIT-SCIROCCO

DIESEL TUNE-UP SPECIFICATIONS

Model	VALVE CLEARANCE (cold) ①		Intake valve opens (deg)	Injection pump setting (deg)	INJECTION NOZZLE PRESSURE (psi)		Idle speed (rpm) ③	Cranking compression pressure (psi)
	Intake (in.)	Exhaust (in.)			New	Used		
1977-79 Diesel Rabbit	0.008-0.012	0.016-0.020	N.A.	Align marks	1849	1706	770-870	398 minimum
1979 Dasher Diesel	0.008-0.012	0.016-0.020	N.A.	Align marks	1849	1706	770-870	398 minimum

① Warm clearance given—Cold clearance: Intake 0.006-0.010
Exhaust 0.014-0.018
Valve clearance need not be adjusted unless it varies more than 0.002 in. from specification.
N.A. Not Available

Volkswagen

EMISSION CONTROLS

VOLKSWAGEN—TYPES 1 AND 2
CRANKCASE VENTILATION SYSTEM

All models are equipped with a crankcase ventilation system.

Type 1 and 2 crankcase vapors are recirculated from the oil breather through a rubber hose to the air cleaner. The vapors then join the air/fuel mixture and are burned in the engine. Fuel injected cars mix crankcase vapors into the air/fuel mixture to be burned in the combustion chambers. Fresh air is forced through the engine to evacuate vapors and recirculate them into the oil breather, intake air distributor, and then to be burned.

The only maintenance required on the crankcase ventilation system is a periodic check. At every tune-up, examine the hoses for clogging or deterioration. Clean or replace the hoses as required.

EVAPORATIVE EMISSION CONTROL SYSTEM

Required by law, this system prevents raw fuel vapors from entering the atmosphere. The various systems for different models are similar. They consist of an expansion chamber, activated charcoal filter, and connecting lines. Fuel vapors are vented to the charcoal filter where hydrocarbons are deposited on the element. The engine fan forces fresh air into the filter when the engine is running. The air purges the filter and the hydrocarbons are forced into the air cleaner to become part of the air/fuel mixture and burned.

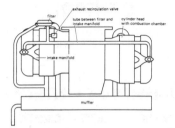

EGR system, 1975 and later Type 2

Maintenance of this system consists of checking the condition of the various connecting lines and the charcoal filter at 10,000 mile intervals. The charcoal filter, which is located under the engine compartment, should be replaced at 48,000 mile intervals.

AIR INJECTION SYSTEM

Type 2 vehicles, are equipped with the air injection system, or air pump as it is sometimes called. In this system, an engine driven air pump delivers fresh air to the engine exhaust ports. The additional air is used to promote afterburning of any unburned mixture as they leave the combustion chamber. In addition, the system supplies fresh air to the intake manifold during gear changes to provide more complete combustion of the air/fuel mixture.

Check the air pump belt tension and examine the hoses for deterioration as a regular part of your tune-up procedure. The filter element adjacent to the pump should be replaced every 18,000 miles or at least every two years.

EXHAUST GAS RECIRCULATION SYSTEM

Type 1

EGR is installed on all 1976 and later models. All applications use the element type filter and single stage EGR valve. Recirculation occurs during part throttle applications as before. The system is controlled by a throttle valve switch which measures throttle position, and an intake air sensor which reacts to engine vacuum. An odometer actuated EGR reminder light (on the dashboard) is used to inform the driver that it is time to service the EGR system. The reminder light measures elapsed mileage and lights at 15,000 mile intervals. A reset button is located behind the switch.

Type 2

All 1976 and later Type 2 models utilize an EGR system. A single stage EGR valve and element type filter are used on all applications. Recirculation occurs during part throttle openings, and is controlled by throttle position, engine vacuum, and engine compartment temperature. At 15,000

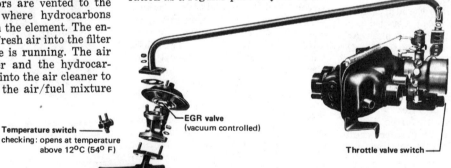

EGR system—1975 and later type 2

mile intervals, a dash mounted EGR service reminder light is activated to warn that EGR service is now due. A reset button is located behind the switch.

CATALYTIC CONVERTER SYSTEM

All 1976 and later Type 1 and 2 models sold in California are equipped with a catalytic converter. The converter is installed in the exhaust system, upstream and adjacent to the muffler.

Volkswagen

Catalytic converters change noxious emission of hydrocarbons (HC) and carbon monoxide (CO) into harmless carbon dioxide and water vapor. The reaction takes place inside the converter at great heat using platinum and palladium metals as the catalyst. If the engine is operated on lead-free fuel, they are designed to last 50,000 miles before replacement.

CRANKCASE VENTILATION

The crankcase ventilation system keeps harmful vapor byproducts of combustion from escaping into the atmosphere and prevents the building of crankcase pressure which can lead to oil leaking. Crankcase vapors are recirculated from the camshaft cover through a hose to the air cleaner. Here they are mixed with the air/fuel mixture and burned in the combustion chamber.

The only maintenance required on the crankcase ventilation system is a periodic check. At every tune up, examine the hoses for clogging or deterioration. Clean or replace the hoses as necessary.

EVAPORATIVE EMISSION CONTROL SYSTEM

This system prevents the escape of raw fuel vapors (unburned hydrocarbons or HC) into the atmosphere. The system consists of a sealed carburetor, unvented fuel tank filter cap, fuel tank expansion chamber, an activated charcoal filter canister and connector hoses. Fuel vapors which reach the filter deposit hydrocarbons on the surface of the charcoal filter element. Fresh air enters the filter when the engine is running and forces the hydrocarbons to the air cleaner where they join the air/fuel mixture and are burned.

Maintenance of the system requires checking the condition of the various connector hoses and the charcoal filter at 10,000 mile intervals. The charcoal filter should be replaced at 50,000 mile intervals.

DUAL DIAPHRAGM DISTRIBUTORS

The distributor is equipped with a vacuum retard diaphragm as well as vacuum advance.

Advance Diaphragm

1. Check the ignition timing.

DECELERATION CONTROL

All 1976 and later Type 2 models, as well as those 1976 and later Type 1 models equipped with manual transmission, are equipped with deceleration control to prevent an overly rich fuel mixture from reaching the exhaust. During deceleration, a vacuum valve (manual transmission) or electrical transmission switch (automatic transmission) opens, bypassing the closed throttle plate and allowing air to enter the combustion chambers.

EMISSION CONTROLS
DASHER - RABBIT - SCIROCCO

2. Remove the retard hose from the distributor and plug it. Increase the engine speed. The ignition timing should advance. If it doesn't, the vacuum unit is faulty and must be replaced.

Temperature Valve

1. Remove the temperature valve and place the threaded portion in hot water.
2. Create a vacuum by sucking on the angled connection.
3. The valve must be open above approximately 130°F.

EXHAUST GAS RECIRCULATION (EGR)

1975-76 Models

1975 models have an EGR filter and a 2-stage EGR valve. The first stage is controlled by the temperature valve. The second stage is controlled by the micro-switch on the carburetor throttle valve. The switch opens the valve when the throttle valve is open between 30°-67° (manual transmission) or 23°-63° (automatic transmission).

The EGR filter was discontinued on 1976 models but the 2-stage EGR valve was retained. On Federal vehicles, only the first stage is connected; California vehicles use both stages.

First stage EGR is controlled by engine vacuum and coolant temperature. The EGR valve is open above approximately 120°F. coolant temperature and below approximately 80° F. At idle and during full throttle acceleration (engine hot), there is no EGR since the engine vacuum is too low to open the valve.

The second stage is controlled by

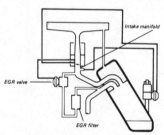

EGR system schematic

Resetting the EGR elapsed mileage odometer

Resetting the catalytic converter elapsed mileage odometer

temperature, engine vacuum and micro-switch on the carburetor throttle valve. Vacuum is always present at the second stage and the valve is opened at about 120°F. coolant temperature. When the throttle valve opens between 25° and 67°, the micro-switch activates the 2-way valve and allows engine vacuum to reach the second stage.

1977 and Later Models

The EGR valve on fuel injected models is controlled by a temperature valve and a vacuum amplifier. The valve is located at the front of the intake manifold.

Volkswagen

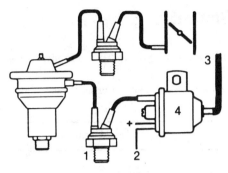

EGR operation-1975
1. Temperature valve
2. Two-way valve
3. EGR valve
4. To brake booster
5. To micro switch

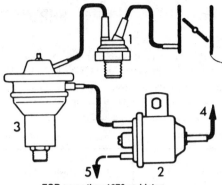

EGR operation-1976 and later
1. Temperature valve for EGR 2nd stage
2. To micro switch on throttle valve
3. Vacuum hose to brake booster
4. Two way valve

AIR INJECTION

This system includes pump, filter, check valve and anti-backfire (gulp) valve. The required maintenance on the air pump involves visually checking the pump, control valves, hoses and lines every 10,000 miles. Clean the pump filter silencer at this interval, and replace every 20,000 miles.

Air Pump

Clean or replace the air filter or air manifolds at the required intervals.

1. Blow compressed air into the anti-back fire valve in the direction of the air flow.
2. Start the engine. Exhaust gas should flow equally from each air inlet.
3. With the engine idling, block the relief valve air outlet—only a slight pressure should be felt if the system is operating properly.

Anti-Backfire Valve

1. Disconnect the air pump filter line from the anti-backfire valve.
2. Briefly disconnect the anti-backfire valve vacuum line with the engine running. There should be a noticeable vacuum.

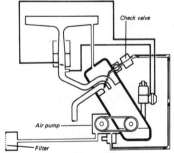

Air injection system schematic. The arrows indicate air flow

3. Replace the anti-backfire valve if the engine backfires.

Required maintenance involves replacing the catalyst when a malfunction occurs. Some early models may have a catalyst indicator light that will glow every 30,000 miles. The catalyst odometer can be reset by pushing the white button marked "CAT" on the mileage odometer in the engine compartment. Later models do not require the previous 30,000 mile check and can have the wire disconnected to disable the warning light.

CATALYTIC CONVERTER

All models are equipped with a catalytic converter located in the exhaust system.

Overheating of the catalytic converter is indicated by the CAT light in the speedometer flickering. This can temporarily be caused by a straining engine (trailer pulling, driving on steep grades, etc.) or high speed driving at high temperatures. Either easing the load or slowing down will stop the light from flickering.

More permanent causes include:
Misfiring (faulty or worn plugs),
Faulty ignition timing,
CO valve too high,
EGR not shutting off, or
Defective temperature sensor in converter.

Symptoms of a faulty converter include poor engine output, drop in the idle speed or continually stalling engine, rattle in the exhaust system (ceramic insert broken or loose) and high CO reading.

The converter is unbolted and removed from the car after disconnecting the temperature sensor.

Hold the converter up to a strong light and look through both ends, checking for blockages. If the converter is blocked, replace it.

Install the converter in the reverse order of removal. Reset the elapsed mileage odometer by pushing the white button marked "CAT".

CAUTION: *Do not drop or strike the converter assembly or damage to the ceramic insert will result.*

Damage and overheating of the catalytic converter, indicated by the flickering of the "CAT" warning light, can be caused by the following:
1. Engine misfire,
2. Improper ignition timing,
3. CO valve set too high,
4. Faulty air pump diverter valve,
5. Faulty temperature sensor, or
6. Engine strain caused by trailer hauling, high speed driving in hot weather.

A faulty converter is indicated by one of the following symptoms.
1. Poor engine performance.
2. The engine stalls.
3. Rattling in the exhaust system.
4. A CO reading greater than 0.4% at the tail pipe.

Resetting the CAT Warning Light

The CAT warning light in the speedometer should come on at 50,000 mile intervals to remind you to have the converter serviced.

The light can be reset by pushing the button marked CAT on the switch. The light on the speedometer should go out.

NOTE: *Vehicles equipped with the Solex 34 PICT-5 carburetor have no "CAT" warning light.*

Trouble Diagnosis — American Motors

SERVICE DIAGNOSIS

The following table lists causes of service problems in descending order of probability. It is more likely a problem results from the first listed "possible cause" than the tenth, for instance.

However, visual examination often leads directly to the correct solution and all service procedures should begin with a careful look at any suspected part or assembly.

Condition	Possible Cause	Correction
HARD STARTING (ENGINE CRANKS NORMALLY)	(1) Binding linkage, choke valve or choke piston.	(1) Repair as necessary.
	(2) Restricted choke vacuum and hot air passages, where applicable.	(2) Clean passages.
	(3) Improper fuel level.	(3) Adjust float level.
	(4) Dirty, worn or faulty needle valve and seat.	(4) Repair as necessary.
	(5) Float sticking.	(5) Repair as necessary.
	(6) Exhaust manifold heat valve stuck. (6- and 8-cylinder only).	(6) Lubricate or replace.
	(7) Faulty fuel pump.	(7) Replace fuel pump.
	(8) Incorrect choke cover adjustment.	(8) Adjust choke cover.
	(9) Inadequate unloader adjustment.	(9) Adjust unloader.
	(10) Faulty ignition coil.	(10) Test and replace as necessary.
	(11) Improper spark plug gap.	(11) Adjust gap.
	(12) Incorrect initial timing.	(12) Adjust timing.
	(13) Incorrect dwell (4-cylinder only).	(13) Adjust dwell.
	(14) Incorrect valve timing.	(14) Check valve timing; repair as necessary.
ROUGH IDLE OR STALLING	(1) Incorrect curb or fast idle speed.	(1) Adjust curb or fast idle speed.
	(2) Incorrect initial timing.	(2) Adjust timing to specifications.
	(3) Incorrect dwell (4-cylinder only).	(3) Adjust dwell.
	(4) Improper idle mixture adjustment.	(4) Adjust idle mixture.
	(5) Damaged tip on idle mixture screw.	(5) Replace mixture screw.
	(6) Improper fast idle cam adjustment.	(6) Adjust fast idle cam.
	(7) Faulty EGR valve operation.	(7) Test EGR system and replace as necessary.
	(8) Faulty PCV valve air flow.	(8) Test PCV valve and replace as necessary.
	(9) Exhaust manifold heat valve inoperative.	(9) Lubricate or replace heat valve as necessary.
	(10) Choke binding.	(10) Locate and eliminate binding condition.

American Motors — Trouble Diagnosis

SERVICE DIAGNOSIS—Continued

Condition	Possible Cause	Correction
ROUGH IDLE OR STALLING (Continued)	(11) Improper choke setting.	(11) Adjust choke.
	(12) Faulty TAC unit.	(12) Repair as necessary.
	(13) Vacuum leak.	(13) Check manifold vacuum and repair as necessary.
	(14) Improper fuel level.	(14) Adjust fuel level.
	(15) Faulty distributor rotor or cap.	(15) Replace rotor or cap.
	(16) Leaking engine valves.	(16) Check cylinder leakdown rate or compression, repair as necessary.
	(17) Incorrect ignition wiring.	(17) Check wiring and correct as necessary.
	(18) Faulty coil.	(18) Test coil and replace as necessary.
	(19) Clogged air bleed or idle passages.	(19) Clean passages.
	(20) Restricted air cleaner.	(20) Clean or replace air cleaner.
	(21) Faulty choke vacuum diaphragm.	(21) Repair as necessary.
FAULTY LOW-SPEED OPERATION	(1) Clogged idle transfer slots.	(1) Clean transfer slots.
	(2) Restricted idle air bleeds and passages.	(2) Clean air bleeds and passsages.
	(3) Restricted air cleaner.	(3) Clean or replace air cleaner.
	(4) Improper fuel level.	(4) Adjust fuel level.
	(5) Faulty spark plugs.	(5) Clean or replace spark plugs.
	(6) Dirty, corroded, or loose secondary circuit connections.	(6) Clean or tighten secondary circuit connections.
	(7) Faulty ignition cable.	(7) Replace ignition cable.
	(8) Faulty distributor cap.	(8) Replace cap.
	(9) Incorrect dwell (4-cylinder only).	(9) Adjust dwell.
FAULTY ACCELERATION	(1) Improper pump stroke.	(1) Adjust pump stroke.
	(2) Incorrect ignition timing.	(2) Adjust timing.
	(3) Inoperative pump discharge check ball or needle.	(3) Clean or replace as necessary.
	(4) Faulty elastomer valve.	(4) Replace valve.
	(5) Worn or damaged pump diaphragm or piston.	(5) Replace diaphragm or piston.
	(6) Leaking main body cover gasket.	(6) Replace gasket.
	(7) Engine cold and choke too lean.	(7) Adjust choke.
	(8) Improper metering rod adjustment (YF Model carburetor or BBD Model carburetor).	(8) Adjust metering rod.
	(9) Faulty spark plug(s).	(9) Clean or replace spark plug(s).

Trouble Diagnosis — American Motors

SERVICE DIAGNOSIS—Continued

Condition	Possible Cause	Correction
	(10) Leaking engine valves.	(10) Check cylinder leakdown rate or compression, repair as necessary.
	(11) Faulty coil.	(11) Test coil and replace as necessary.
FAULTY HIGH SPEED OPERATION	(1) Incorrect ignition timing.	(1) Adjust timing.
	(2) Excessive ignition point gap (4-cylinder only).	(2) Adjust dwell.
	(3) Defective TCS system.	(3) Test TCS system; repair as necessary.
	(4) Faulty distributor centrifugal advance.	(4) Check centrifugal advance and repair as necessary.
	(5) Faulty distributor vacuum advance.	(5) Check vacuum advance and repair as necessary.
	(6) Low fuel pump volume.	(6) Replace fuel pump.
	(7) Wrong spark plug gap; wrong plug.	(7) Adjust gap; install correct plug.
	(8) Faulty choke operation.	(8) Adjust choke.
	(9) Partially restricted exhaust manifold, exhaust pipe, muffler or tailpipe.	(9) Eliminate restriction.
	(10) Clogged vacuum passages.	(10) Clean passages.
	(11) Improper size or obstructed main jet.	(11) Clean or replace as necessary.
	(12) Restricted air cleaner.	(12) Clean or replace as necessary.
	(13) Faulty distributor rotor or cap.	(13) Replace rotor or cap.
	(14) Faulty coil.	(14) Test coil and replace as necessary.
	(15) Leaking engine valve(s).	(15) Check cylinder leakdown rate or compression, repair as necessary.
	(16) Faulty valve spring(s).	(16) Inspect and test valve spring tension and replace as necessary.
	(17) Incorrect valve timing.	(17) Check valve timing and repair as necessary.
	(18) Intake manifold restricted.	(18) Remove restriction or replace manifold.
	(19) Worn distributor shaft.	(19) Replace shaft.
MISFIRE AT ALL SPEEDS	(1) Faulty spark plug(s).	(1) Clean or replace spark plug(s).
	(2) Faulty spark plug cable(s).	(2) Replace as necessary.
	(3) Faulty distributor cap or rotor.	(3) Replace cap or rotor.
	(4) Faulty coil.	(4) Test coil and replace as necessary.
	(5) Trigger wheel too high (6- and 8-cylinder only).	(5) Set to specifications.

American Motors — Trouble Diagnosis

SERVICE DIAGNOSIS—Continued

Condition	Possible Cause	Correction
MISFIRE AT ALL SPEEDS (Continued)	(6) Incorrect dwell (4-cylinder only).	(6) Adjust dwell.
	(7) Faulty condenser (4-cylinder only).	(7) Replace condenser.
	(8) Primary circuit shorted or open intermittently.	(8) Trace primary circuit and repair as necessary.
	(9) Leaking engine valve(s).	(9) Check cylinder leakdown rate or compression, repair as necessary.
	(10) Faulty hydraulic tappet(s) (6- and 8-cylinder only).	(10) Clean or replace tappet(s).
	(11) Incorrect valve adjustment (4-cylinder only).	(11) Adjust valves.
	(12) Out-of-round or cracked tappets (4-cylinder only).	(12) Replace tappets.
	(13) Faulty valve spring(s).	(13) Inspect and test valve spring tension, repair as necessary.
	(14) Worn lobes on camshaft.	(14) Replace camshaft.
	(15) Vacuum leak.	(15) Check manifold vacuum and repair as necessary.
	(16) Improper carburetor settings.	(16) Adjust carburetor.
	(17) Fuel pump volume or pressure low.	(17) Replace fuel pump.
	(18) Blown cylinder head gasket.	(18) Replace gasket.
	(19) Intake or exhaust manifold passage(s) restricted.	(19) Pass chain through passages.
	(20) Wrong trigger wheel.	(20) Install correct wheel.
POWER NOT UP TO NORMAL	(1) Incorrect ignition timing.	(1) Adjust timing.
	(2) Faulty distributor rotor.	(2) Replace rotor.
	(3) Incorrect dwell (4-cylinder only).	(3) Adjust dwell.
	(4) Trigger wheel positioned too high or loose on shaft (6- and 8-cylinder only).	(4) Reposition or replace trigger wheel.
	(5) Incorrect spark plug gap.	(5) Adjust gap.
	(6) Faulty fuel pump.	(6) Replace fuel pump.
	(7) Incorrect valve timing.	(7) Check valve timing and repair as necessary.
	(8) Faulty coil.	(8) Test coil and replace as necessary.
	(9) Faulty ignition.	(9) Test cables and replace as necessary.
	(10) Leaking engine valves.	(10) Check cylinder leakdown rate or compression and repair as necessary.
	(11) Blown cylinder head gasket.	(11) Replace gasket.

Trouble Diagnosis — American Motors

SERVICE DIAGNOSIS—Continued

Condition	Possible Cause	Correction
	(12) Leaking piston rings.	(12) Check compression and repair as necessary.
	(13) Worn distributor shaft.	(13) Replace shaft.
INTAKE BACKFIRE	(1) Improper ignition timing.	(1) Adjust timing.
	(2) Incorrect dwell (4-cylinder only).	(2) Adjust dwell.
	(3) Faulty accelerator pump discharge	(3) Repair as necessary.
	(4) Improper choke operation.	(4) Repair as necessary.
	(5) Defective EGR CTO.	(5) Replace EGR CTO.
	(6) Defective TAC unit.	(6) Repair as necessary.
	(7) Lean fuel mixture.	(7) Check float level or manifold vacuum for vacuum leak. Remove sediment from bowl.
EXHAUST BACKFIRE	(1) Vacuum leak.	(1) Check manifold vacuum and repair as necessary.
	(2) Faulty diverter valve.	(2) Test diverter valve and replace as necessary.
	(3) Faulty choke operation.	(3) Repair as necessary.
	(4) Exhaust leak.	(4) Locate and eliminate leak.
PING OR SPARK KNOCK	(1) Incorrect ignition timing.	(1) Adjust timing.
	(2) Distributor centrifugal or vacuum advance malfunction.	(2) Check advance and repair as necessary.
	(3) Excessive combustion chamber deposits.	(3) Use combustion chamber cleaner.
	(4) Carburetor set too lean.	(4) Adjust carburetor.
	(5) Vacuum leak.	(5) Check manifold vacuum and repair as necessary.
	(6) Excessively high compression.	(6) Check compression and repair as necessary.
	(7) Fuel octane rating excessively low.	(7) Try alternate fuel source.
	(8) Heat riser stuck in heat ON position (6- and 8-cylinder only).	(8) Free-up or replace heat riser.
	(9) Sharp edges in combustion chamber.	(9) Grind smooth.
SURGING (CRUISING SPEEDS TO TOP SPEEDS)	(1) Low fuel level.	(1) Adjust fuel level.
	(2) Low fuel pump pressure or volume.	(2) Replace fuel pump.
	(3) Metering rod(s) not adjusted properly (YF Model Carburetor or BBD Model Carburetor).	(3) Adjust metering rod.
	(4) Improper PCV valve air flow.	(4) Test PCV valve and replace as necessary.

Buick — Trouble Diagnosis

SERVICE DIAGNOSIS

PCV SYSTEM

Condition	Possible Cause	Correction
Slow, unstable idle, frequent stalling.	1. Valve completely plugged or stuck.	1. Replace valve.
	2. Restricted filter	1. Replace filter, clean system.
Oil in air cleaner.	1. PCV system plugged.	1. Replace valve.
	2. Leak in closed ventilation system.	1. Clean system as required.
		2. Inspect for leaks to atmosphere and correct as necessary.

THERMOSTATIC AIR CLEANER SYSTEM

Condition	Possible Cause	Correction
Hesitation, sag and stalling during cold start.	1. Thermac damper door in full open snorkel position.	1. Perform tests on thermac system as described.
Poor fuel economy.	1. Thermac damper door in closed snorkel position.	1. Perform tests on thermac system as described.

CATALYTIC CONVERTER

Condition	Possible Cause	Correction
Exhaust system noisy.	1. Exhaust pipe joints Loose at catalytic converter.	1. Tighten clamps at joint.
	2. Catalytic converter ruptured.	1. Replace catalytic converter.
	3. Loose or missing catalyst replacement plug.	1. Tighten or replace (Recharge catalyst as necessary).
Poor car performance.	1. Failed catalytic converter.	1. Replace catalytic converter. Ignition system should also be diagnosed and repairs made if necessary.
B-B size particles coming out of tailpipe.	1. Failed catalytic converter.	1. Replace catalytic converter. Ignition system should also be diagnosed and repairs made if necessary.

Trouble Diagnosis — Buick

SERVICE DIAGNOSIS—Continued

CHOKE AIR MODULATOR SYSTEM

Condition	Possible Cause	Correction
Poor driveability after warm-up (choke not releasing).	1. Choke air modulator system plugged.	1. Locate restriction and correct as necessary.
Poor driveability during warm-up (choke releasing too soon).	1. Modulator valve stuck in open position.	1. Replace valve.

EGR SYSTEM

Condition	Possible Cause	Correction
Engine idles abnormally rough and/or stalls.	1. EGR valve vacuum hoses misrouted.	1. Check EGR valve vacuum hose routing. Correct as required.
	2. Leaking EGR valve.	1. Check EGR valve for corcect operation.
	3. Idle speed misadjusted.	1. Set idle RPM per engine label specification. Remove EGR vacuum hose from valve and observe effect on engine. Replace valve if speed is affected, reset RPM to specification and reconnect hose.
	4. Improper carburetor signal to EGR valve at idle.	1. Check vacuum signal from carburetor EGR port with engine at stabilized operating temperature and at curb idle speed. If signal is more than 2.0 in hg. vacuum, proceed to carburetor idle diagnosis.
	5. Failed EFE-EGR thermal vacuum switch.	1. Check vacuum signal into switch from carburetor EGR port with engine at normal operating temperature and at curb idle speed. Then check vacuum signal out of switch to EGR valve. If the two vacuum signals are not equal within $\pm 1/2$ in. hg. vacuum across them, proceed to EFE-EGR thermal vacuum switch diagnosis. Replace switch as required.
	6. EGR valve gasket failed or loose EGR attaching bolts.	1. Check EGR attaching bolts for tightness. Tighten as required. If not loose, remove EGR valve and inspect gasket. Replace as required.

Buick — Trouble Diagnosis

SERVICE DIAGNOSIS—Continued

Condition	Possible Cause	Correction
	7. Improper carburetor vacuum signal at idle.	Check vacuum from carburetor to back pressure EGR valve with engine at stabilized operating temperature and at curb idle speed. If vacuum is more than 1.0 in Hg., refer to engine performance diagnosis (idle diagnosis).
Engine runs rough on light throttle acceleration, poor part load performance and poor fuel economy.	1. EGR valve vacuum hose misrouted.	1. Check EGR valve vacuum hose routing. Correct as required.
	2. Failed EFE-EGR thermal vacuum switch.	1. Same as Step 5 above under "Engine Idles Abnormally Rough" and/or stalls.
	3. EGR flow unbalanced due to deposit accumulation in EGR passages or under carburetor.	1. Clean EGR passages of all deposits.
	4. Sticky or binding EGR valve.	1. Remove EGR valve and inspect for proper operation. Clean, repair or replace as required.
	5. Wrong or no EGR gaskets.	1. Check and correct as necessary.
	6. Exhaust system restricted causing excessive back pressure.	1. Inspect exhaust system for restriction and replace or repair as required.
	7. Control valve blocked or air flow restricted	1. Check internal control valve function per service procedure.
Engine stalls on decelerations.	1. Restriction in EGR vacuum line.	1. Check EGR vacuum lines for kinks, bends, etc. Remove or replace hoses as required. Check EGR vacuum control valve function. Check EGR valve for excessive deposits causing sticky or binding operation. Clean or repair as required.
	2. Sticking or binding EGR valve.	1. Remove EGR valve and inspect, then clean or repair as required.
	3. Control valve blocked or air flow restricted.	1. Check internal control valve function per service procedure.
Part throttle engine detonation.	1. Insufficient exhaust gas recirculation flow during part throttle accelerations.	1. Check EGR valve hose routing. Check EGR valve operation. Repair or replace as required. Check EFE/EGR thermal vacuum switch as listed under "Engine Idles Ab-

Trouble Diagnosis
Buick

SERVICE DIAGNOSIS—Continued

Condition	Possible Cause	Correction
		normally Rough and/or Stalls". Replace switch as required. Check EGR passages and valve for excessive deposits. Clean as required.
	2. Control valve blocked or air flow restricted.	1. Check internal control valve function per service procedure.
	(NOTE: Detonation can be caused by several other engine variables. Perform ignition and carburetor related diagnosis.)	
Engine starts but immediately stalls when cold.	1. EGR valve hoses misrouted.	1. Check EGR valve hose routings.
	2. EGR system malfunctioning when engine is cold.	1. Perform check to determine if EFE-EGR thermal vacuum switch is operational. Replace as required.
	3. Control valve blocked or air flow restricted.	1. Check internal control valve function per service procedure.
	(NOTE: Stalls after start can also be caused by carburetor problems. Refer to performance diagnosis).	

EFE SYSTEM

Condition	Possible Cause	Correction
Poor Operation during warm-up such as-rough idle, stumble, etc.	1. No vacuum to vacuum actuator during warm-up period for cold start.	1. Check vacuum source for vacuum of 8" hg. or above. Correct improper vacuum hose routing, leak in connecting system, diaphragm, or EFE/EGR-TVS. Replace failed EFE/EGR-TVS.
	2. EFE valve linkage bent or binding.	1. Repair EFE valve linkage.
	3. EFE valve linkage disconnected.	1. Reconnect linkage.
	4. EFE valve shaft frozen in bearing.	1. Replace EFE valve.
	5. EFE valve loose on shaft.	1. Replace EFE valve.

CLOSED EFE VALVE

Condition	Possible Cause	Correction
Poor Operation after warm-up, rough idle, lack of high speed performance, surge, misses at all speeds	1. Failed EFE/EGR-TVS -vacuum present at vacuum actuator.	1. Replace EFE/EGR-TVS.

Buick — Trouble Diagnosis

SERVICE DIAGNOSIS—Continued

Condition	Possible Cause	Correction
	2. EFE valve asm. shaft frozen in bearing.	1. Replace EFE valve.
	3. EFE valve to housing interference.	1. Repair EFE valve.
	4. Vacuum actuator linkage bent or binding.	1. Repair EFE valve linkage.
	5. EFE valve separated from shaft.	1. Repair EFE valve linkage.
Noisy EFE valve asm.	1. Linkage stop failed.	1. Repair linkage stop tab.
	2. No vacuum actuator linkage over travel.	1. Replace vacuum actuator.
	3. Valve loose on shaft.	1. Replace EFE valve.
	4. Shaft loose in bushing, or bushing loose in housing.	1. Replace EFE valve.

EFE-EGR THERMO VACUUM SWITCH

Condition	Possible Causes	Correction
Rough idle or stall during warm-up.	1. No vacuum to EFE vacuum actuator with engine coolant temperature below 120°F. ±3°F. (49°C.±2°) for 231 California and 350 49 State. 90°F. (32°C.) 231 and 196 49 State.	1. Check vacuum source for vacuum of 8" hg. or above.
		2. Correct improper vacuum hose routing, leak in connecting system, or EFE vacuum actuator diaphragm. Replace if required.
		3. Failed EFE-EGR thermo vacuum switch. Replace.
	2. Vacuum to EGR valve below 120° ±3°F. (49°.±2°) for 231 California and 350 49 State. 90°F. (32°C.) 231 and 196 49 State.	1. Correct improper vacuum hose routing if necessary.
		2. Failed EFE-EGR thermo vacuum switch. Replace.

Trouble Diagnosis — Buick

SERVICE DIAGNOSIS—Continued

Condition	Possible Causes	Correction
Rough idle, lack of performance, surge after warm-up period.	1. Vacuum to EFE vacuum actuator with engine coolant temperature above 120°F. ±3°F. (49°C.±2°) for 231 California and 350 49 State. 90°F (32°C.) 231 and 196 49 State.	1. Correct improper vacuum hose routing.
	2. Failed EFE-EGR thermo vacuum switch. Replace.	
Improper EGR operation.	1. Vacuum to EGR valve with engine coolant temperature below 120°F. ±3°F. (49°C.±2°) for 231 California and 350 49 State. 90°F(32°C.) 231 and 196 49 state.	1. Correct improper vacuum hose routing if necessary.
	2. Failed EFE-EGR thermo vacuum switch. Replace.	

AIR System - V8 Engines

NOTE: The AIR system is not completely noiseless. Under normal conditions, noise rises in pitch as engine speed increases. To determine if excessive noise is the fault of the air injection system, disconnect the drive belt and operate the engine. If noise now does not exist, proceed with diagnosis.

Condition	Possible Cause	Correction
Excessive belt noise	1. Loose belt 2. Seized pump	Tighten to spec. 2. Replace pump.
Excessive pump noise. Chirping	1. Insufficient break-in	1. Run car 10-15 miles at turnpike speeds - recheck.
Excessive pump noise, chirping, rumbling, or knocking	1. Leak in hose	1. Locate source of leak using soap solution and correct.
	2. Loose hose	2. Reassemble and replace or tighten hose clamp.
	3. Hose touching other engine parts.	3. Adjust hose position.
	4. Diverter valve inoperative	4. Replace diverter valve.
	5. Check valve inoperative	5. Replace check valve.
	6. Pump mounting fasteners loose	6. Tighten mounting screws as specified.
	7. Pump failure	7. Replace pump.

Buick — Trouble Diagnosis

SERVICE DIAGNOSIS—Continued

Condition	Possible Cause	Correction
No air supply (accelerate engine to 1500 rpm and observe air flow from hoses. If the flow increases as the rpm's increase, the pump is functioning normally. If not, check possible cause)	1. Loose drive belt 2. Leaks in supply hose 3. Leak at fitting(s) 4. Diverter valve leaking 5. Diverter valve inoperative 6. Check valve inoperative 7. Pump pressure relief plug leaking or damaged.	1. Tighten to specs. 2. Locate leak and repair or replace as required. 3. Tighten or replace clamps. 4. If air is expelled through diverter muffler with engine at idle, replace diverter valve. 5. Usually accompanied by backfire during deceleration. Replace diverter valve. 6. Blow through hose toward air manifold. If air passes, function is normal. If air can be sucked from manifold, replace check valve. 7. Replace pressure relief plug.
Centrifugal filter fan damaged or broken.	1. Mechanical damage	1. Replace centrifugal filter fan.
Poor idle or driveability.	1. A defective AIR pump cannot cause poor idle or driveability.	1. Do NOT replace AIR pump.

AIR SYSTEM - V6 Engines

Condition	Possible Cause	Correction
No air supply (accelerate engine to 1500 rpm and observe air flow from hoses. If the flow increases as the rpm's increase, the pump is functioning normally. If not, check possible cause.)	1. Loose drive belt. 2. Leaks in supply hose. 3. Leak at fittings. 4. Air expelled through by-pass valve. 4a. Connect a vacuum line directly from engine manifold vacuum to by-pass valve. 4b. Connect vacuum line from engine manifold vacuum source to by-pass valve through vacuum differential valve directly, by passing the differential vacuum delay and separator valve. 5. Check valve inoperative. 6. Pump failure.	1. Tighten to specifications. 2. Locate leak and repair. 3. Tighten or replace clamps. 4a. If this corrects the problem go to step b. If not, replace air by-pass valve. 4b. If this corrects the problem, check differential vacuum delay and separator valve and vacuum source line for plugging. Replace as required. If it doesn't, replace vacuum differential valve. 5. Disconnect hose and blow through hose toward check valve. If air passes, function is normal. If air can be sucked from check valve, replace check valve. 6. Replace pump.

Trouble Diagnosis — Buick

SERVICE DIAGNOSIS—Continued

AIR SYSTEM - V6 Engines

Condition	Possible Cause	Correction
Excessive pump noise, chirping, rumbling, knocking, loss of engine performance.	1. Leak in hose. 2. Loose hose. 3. Hose touching other engine parts. 4. Vacuum differential valve inoperative. 5. By-pass valve inoperative. 6. Pump mounting fasteners loose. 7. Pump failure. 8. Check valve inoperative. 9. Loose pump outlet elbow on rear cover of pump. Turbo engines.	1. Locate source of leak using soap solution and correct. 2. Reassemble and replace or tighten hose clamp. 3. Adjust hose position. 4. Replace vacuum differential valve. 5. Replace by-pass valve. 6. Tighten mounting screws as specified. 7. Replace pump. 8. Replace check valve. 9. Torque bolts to 16 N·m (12 lb. ft.)
Excessive belt noise.	1. Loose belt. 2. Seized pump.	1. Tighten to spec. 2. Replace pump.
Excessive pump noise. Chirping.	1. Insufficient break-in 2. Loose pump outlet elbow on rear cover of pump. Turbo engines.	1. Run vehicle 10-15 miles at interstate speeds and recheck. 2. Torque bolts to 16 N·m (12 lb. ft.)
Centrifugal filter fan damaged or broken.	1. Mechanical damage	1. Replace centrifugal filter fan.
Exhaust tube bent or damaged.	1. Mechanical damage.	1. Replace exhaust tube.
Poor idle or driveability.	1. A defective AIR system cannot cause poor idle or driveability.	1. Do not replace AIR system.

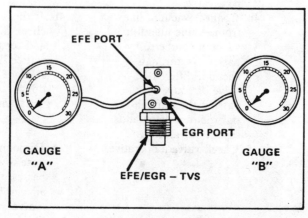

Vacuum Gage Hook-Up

Chevrolet
Trouble Diagnosis

SERVICE DIAGNOSIS
EXHAUST GAS RECIRCULATION SYSTEM DIAGNOSIS CHART

Condition	Possible Cause	Correction
Engine idles abnormally rough and/or stalls.	EGR valve vacuum hoses misrouted.	Check EGR valve vacuum hose routing. Correct as required.
	Leaking EGR valve.	Check EGR valve for correct operation.
	EGR valve gasket failed or loose EGR attaching bolts.	Check EGR attaching bolts for tightness. Tighten as required. If not loose, remove EGR valve and inspect gasket. Replace as required.
	EGR thermal control valve and/or EGR-TVS.	Check vacuum into valve from carburetor EGR port with engine at normal operating temperature and at curb idle speed. Then check the vacuum out of the EGR thermal control valve to EGR valve. If the two vacuum readings are not equal within ± 1/2 in Hg. (1.7 kPa), then proceed to EGR vacuum control diagnoses.
	Improper vacuum to EGR valve at idle.	Check vacuum from carburetor EGR port with engine at stabilized operating temperature and at curb idle speed. If vacuum is more than 1.0 in. Hg., refer to carburetor idle diagnosis.
Engine runs rough on light throttle acceleration, poor part load performance and poor fuel economy.	EGR valve vacuum hose misrouted.	Check EGR valve vacuum hose routing. Correct as required.
	Failed EGR vacuum control valve.	Same as listing in "Engine Idles Rough" condition.
	EGR flow unbalanced due to deposit accumulation in EGR passages or under carburetor.	Clean EGR passages of all deposits.
	Sticky or binding EGR valve.	Remove EGR valve and inspect. Clean or replace as required.
	Wrong or no EGR gaskets.	Check and correct as required.
(Vehicle with back pressure EGR valve.) Engine stalls on decelerations.	Control valve blocked or air flow restricted.	Check internal control valve function per service procedure.
	Restriction in EGR vacuum line.	Check EGR vacuum lines for kinks bends, etc. Remove or replace hoses as required. Check EGR vacuum control valve function.
		Check EGR valve for excessive deposits causing sticky or binding operation. Clean or repair as required.
	Sticking or binding EGR valve.	Remove EGR valve and inspect clean or repair as required.
(Vehicle with a back pressure EGR valve.) Part throttle engine detonation.	Control valve blocked or air flow restricted.	Check internal control valve function per service procedure.
	Insufficient exhaust gas recirculation flow during part throttle accelerations.	Check EGR valve hose routing. Check EGR valve operation. Repair or replace as required. Check EGR thermal control valve and/or EGR-TVS as listed in "Engine Idles Rough" section. Replace valve as required. Check EGR passages and valve for excessive deposit. Clean as required.
(Vehicle with a back pressure EGR valve.)	Control valve blocked or air flow restricted.	Check internal control valve function per service procedure.
(NOTE: Detonation can be caused by several other engine variables. Perform ignition and carburetor related diagnosis.)		
Engine starts but immediately stalls when cold.	EGR valve hoses misrouted.	Check EGR valve hose routings.
	EGR system malfunctioning when engine is cold.	Perform check to determine if the EGR thermal control valve and/or EGR-TVS are operational. Replace as required.
(Vehicle with a back pressure EGR valve.)	Control valve blocked or air flow restricted.	Check internal control valve function per service procedure.
(NOTE: Stalls after start can also be caused by carburetor problems.)		

Trouble Diagnosis

Chevrolet • Oldsmobile • Pontiac

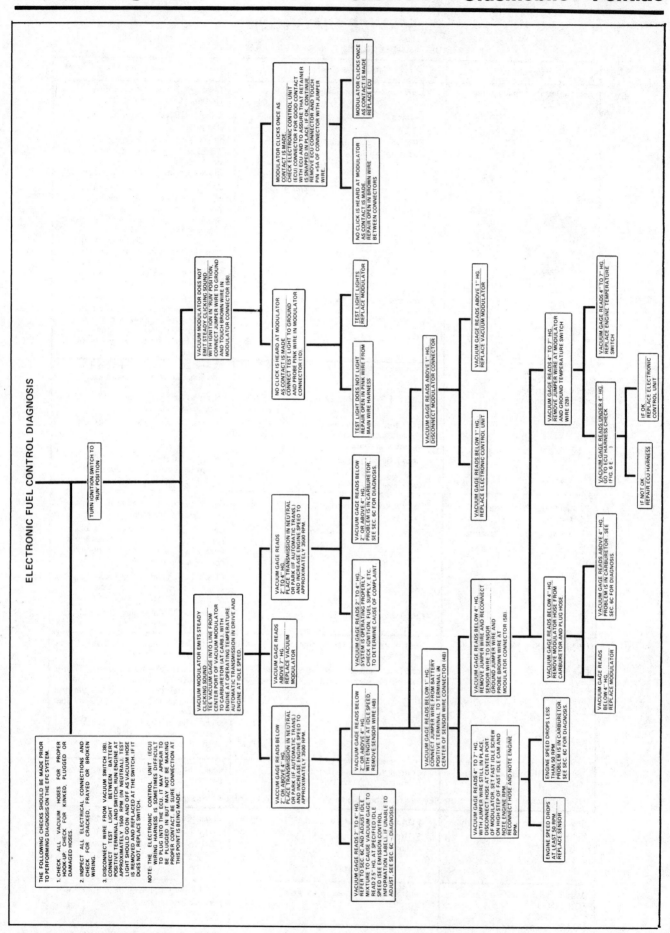

Chrysler — Trouble Diagnosis

SERVICE DIAGNOSIS

Condition	Possible Cause	Correction
EGR VALVE STEM DOES NOT MOVE ON SYSTEM TEST. (ALL TESTS MUST BE MADE WITH ENGINE RUNNING CONTINUOUSLY FOR AT LEAST 70 SECONDS.)	(a) Cracked, leaking, disconnected or plugged hoses.	(a) Verify correct hose connections and leak check and confirm that all hoses are open. If defective hoses are found, replace hose harness.
	(b) Defective EGR valve.	(b) Disconnect hose harness from EGR valve. Connect external vacuum source, 10" Hg or greater, to valve diaphragm while checking valve movement. If no valve movement occurs, replace valve. If valve opens, approx. 1/8" travel, clamp off supply hose to check for diaphragm leakage. Valve should remain open 30 seconds or longer. If leakage occurs, replace valve. If valve is satisfactory, evaluate control system.
EGR VALVE STEM DOES NOT MOVE ON SYSTEM TEST, OPERATES NORMALLY ON EXTERNAL VACUUM SOURCE.	(a) Defective CCEGR valve.	(a) Verify that temperature in radiator top tank is above 68°F. Then disconnect hoses from CCEGR valve on radiator and bypass the valve with a short length of 3/16" tubing. If normal movement of the EGR valve stem is restored, replace the CCEGR valve.
	(b) Defective time delay solenoid valve or time delay module.	(b) For Venturi Vacuum Control Systems, disconnect orange and unstriped hoses at time delay solenoid valve and by-pass the valve with a short length of 3/16" tubing. For Ported Vacuum Control System, disconnect blue, and yellow striped hoses at time delay solenoid valve and by-pass the valve with a short length of 1/8" tubing. If normal movement of the EGR valve is restored, reconnect hoses to valve and disconnect the electrical plug from solenoid. If EGR valve stem does not move on system test, the EGR delay solenoid valve is defective and should be replaced. If the EGR valve stem does move, test the time delay module. See "EGR Time Delay System Test".
	(c) Defective control system—Plugged passages.	(c) **Ported Vacuum Control System:** Remove carburetor and inspect port (slot type) in throttle bore and associated vacuum passages in carburetor throttle body including limiting orifice at hose end of passages. Use suitable solvent to remove deposits and check for flow with light air pressure. Normal operation should be restored to ported vacuum control EGR system.
	(d) Defective vacuum control unit (amplifier).	(d) **Venturi Vacuum Control System:** Remove venturi signal hose from nipple on carburetor. With engine operating at idle, apply a vacuum of 1 to 2" Hg to venturi signal hose. Engine speed should drop a minimum of 150 rpm and EGR valve stem should **visibly** move 1/8" or more. If this does not occur, replace vacuum control unit.

Trouble Diagnosis — Chrysler

SERVICE DIAGNOSIS

	Possible Cause	Correction
EXCESSIVE BELT NOISE	(a) Loose belt.	(a) Tighten belt (see Group 7 "Cooling System").
	(b) Seized pump.	(b) Replace pump.
EXCESSIVE PUMP NOISE, CHIRPING	(a) Insufficient break-in.	(a) Recheck for noise after 1000 miles of operation.
EXCESSIVE PUMP NOISE CHIRPING, RUMBLING, OR KNOCKING	(a) Leak in hose.	(a) Locate source of leak using soap solution and correct.
	(b) Loose hose.	(b) Reassemble and replace or tighten hose clamp.
	(c) Hose touching other engine parts.	(c) Adjust hose position.
	(d) Diverter valve inoperative.	(d) Replace diverter valve.
	(e) Check valve inoperative.	(e) Replace check valve.
	(f) Pump mounting fasteners loose.	(f) Tighten mounting screws as specified.
	(g) Pump failure.	(g) Replace pump.
NO AIR SUPPLY (ACCELERATE ENGINE TO 1500 RPM AND OBSERVE AIR FLOW FROM HOSES. IF THE FLOW INCREASES AS THE RPM'S INCREASE, THE PUMP IS FUNCTIONING NORMALLY. IF NOT, CHECK POSSIBLE CAUSE.	(a) Loose drive belt.	(a) Tighten to specifications.
	(b) Leaks in supply hose.	(b) Locate leak and repair or replace as required.
	(c) Leak at fitting(s).	(c) Tighten or replace clamps.
	(d) Diverter valve leaking.	(d) If air is expelled through diverter exhaust with vehicle at idle, replace diverter valve.
	(e) Diverter valve inoperative.	(e) Usually accompanied by backfire during deceleration. Replace diverter valve.
	(f) Check valve inoperative.	(f) Replace check valve.

EGR DIAGNOSIS

NOTE: ALL TESTS MUST BE MADE WITH FULLY WARM ENGINE RUNNING CONTINUOUSLY FOR AT LEAST TWO MINUTES

Condition	Possible Cause	Correction
EGR VALVE STEM DOES NOT MOVE ON SYSTEM TEST.	(a) Cracked, leaking, disconnected or plugged hoses.	(a) Verify correct hose connections and leak check and confirm that all hoses are open. If defective hoses are found, replace hose harness.
	(b) Defective EGR valve.	(b) Disconnect hose harness from EGR valve. Connect external vacuum source, 10" Hg or greater, to valve diaphragm while checking valve movement. If no valve movement occurs, replace valve. If valve opens, approx. 1/8" travel, clamp off supply hose to check for diaphragm leakage. Valve should remain open 30 seconds or longer. If leakage occurs, replace valve. If valve is satisfactory, evaluate control system.
EGR VALVE STEM DOES NOT MOVE ON SYSTEM TEST, OPERATES NORMALLY ON EXTERNAL VACUUM SOURCE.	(a) Defective thermal control valve.	(a) Disconnect CCEGR valve and bypass the valve with a short length of 3/16" tubing. If normal movement of the EGR valve is restored, replace the thermal valve. On systems having two CCEGR valves, each should be tested separately.
	(b) Defective time delay solenoid valve or time delay module.	(b) For Venturi Vacuum Control System, disconnect orange and blue hoses at time delay solenoid valve and by-pass the valve with a short length of 1/8" tubing. On models with timed idle enrichment disconnect the orange and unstripped hoses from the solenoid and bypass with short length of 3/16" tubing. If normal movement of the EGR valve is restored, reconnect hoses to valve and disconnect the electrical plug from solenoid. If EGR valve stem does not move on system test, the EGR delay solenoid valve is defective and should be replaced. If the EGR valve stem does move, the time delay module should be tested as described in paragraph D "Timer (EGR Time Delay System)".

SERVICE DIAGNOSIS—Continued

Condition	Possible Cause	Correction
	(c) Defective control system—Plugged passages.	(c) **Ported Vacuum Control System:** Remove carburetor and inspect port (slot type) in throttle bore and associated vacuum passages in carburetor throttle body including limiting orifice at hose end of passages. Use suitable solvent to remove deposits and check for flow with light air pressure. Normal operation should be restored to ported vacuum control EGR system.
	(d) Defective vacuum control unit (amplifier)	(c) **Venturi Vacuum Control System:** Remove venturi signal hose from nipple on carburetor. With engine operating at idle, apply a vacuum of approx. 2" Hg. to venturi signal hose. Engine speed should drop a minimum of 150 rpm and EGR valve stem should visibly move 1/8" or more. If this does not occur, **replace** vacuum amplifier.
	(e) Plugged carburetor venturi signal passage.	(d) If vacuum amplifier operates normally in previous test, a plugged vacuum tap to carburetor venturi is indicated. Use suitable carburetor solvent to remove deposits from passage and use light air pressure to verify that passage is clean.

NOTE: Do not use drills or wires to clean carburetor control passages as calibration of precision control orifices may be altered resulting in unsatisfactory vehicle operation.

Condition	Possible Cause	Correction
ENGINE WILL NOT IDLE, DIES OUT ON RETURN TO IDLE OR IDLE IS VERY ROUGH OR SLOW. EGR VALVE OPEN AT IDLE.	(a) Control system defective.	(a) Disconnect hose from EGR valve and plug hose. If idle is unsatisfactory, replace EGR valve. If idle is satisfactory, reconnect hose to EGR valve, disconnect venturi signal hose from carburetor. If idle is satisfactory, clean venturi tap per para. 2-5. If unsatisfactory, replace vacuum control unit. (amplifier).
ENGINE WILL NOT IDLE, DIES OUT ON RETURN TO IDLE OR IDLE IS ROUGH OR SLOW. EGR VALVE CLOSED AT IDLE.	(a) High EGR valve leakage in closed position.	(a) If removal of vacuum hose from EGR valve does not correct rough idle, remove EGR valve and inspect to insure that poppet is seated. Clean deposits if necessary or replace EGR valve if found defective.
DEEP SAGS AND/OR PASSOUTS AND/OR DIEOUTS DURING FIRST 30 SECONDS.	(a) Defective EGR time delay system.	(a) Stop engine, disconnect and reconnect the hose to the EGR valve, then restart. (1) Immediately open the throttle and observe EGR valve stem for motion. If it moves during the first 35 seconds (black color code) or 90 seconds (red color code) after starting, the EGR time delay system is defective. (2) Check hose connections to the time delay solenoid valve and timer. (3) If OK, disconnect the electrical plug from the solenoid valve. (4) Energize the solenoid valve by grounding either terminal and connecting the other terminal to the positive battery post. (5) Disconnect and reconnect hose to EGR valve. If the EGR valve stem moves on the system test, the solenoid valve is defective and should be replaced. (6) If the EGR valve does not move, the time delay module should be tested.

Trouble Diagnosis — Oldsmobile

ENGINE PERFORMANCE DIAGNOSIS CHARTS

Introduction

Engine Performance Diagnosis procedures are guides that will lead to the most probable causes of engine performance complaints. They consider all of the parts of the fuel, ignition, and mechanical systems that could cause a particular complaint, and then outline repairs in a logical sequence.

Each Sympton is defined, and it is vital that the correct one be selected based on the complaints reported or found.

Review the Symptoms and their definition to be sure that only the correct terms are used.

The words used may not be what you are used to in all cases, but because these terms have been used interchangeably for so long, it was necessary to decide on the most common usage and then define them. If the definition is not understood, and the exact Symptom is not used, the Diagnostic procedure will not work.

It is important to keep two facts in mind:

1. The procedures are written to diagnose problems on cars that have "run well at one time" and that time and wear have created the condition.

2. All possible causes cannot be covered, particularly with regard to emission controls that affect vacuum advance. If doing the work prescribed does not correct the complaint, then either the wrong Symptom was used, or a more detailed analysis will have to be made.

All of the Symptoms can be caused by worn out or defective parts such as Spark Plugs, Ignition Wiring, etc. If time and/or mileage indicate that parts should be replaced, it is recommended that it be done.

Symptom	Definition
Dieselling	Engine continues to run after the switch is turned off. It runs unevenly and may make knocking noises. The exhaust stinks.
Detonation	A mild to severe ping, usually worse under acceleration. The engine makes sharp metallic knocks that change with throttle opening. Sounds like pop corn popping.
Stalls	The engine quits running. It may be at idle or while driving.
Rough Idle	The engine runs unevenly at idle. If bad enough, it may make the car shake.
Miss	Steady pulsation or jerking that follows engine speed, usually more pronounced as engine load increases. Not normally felt above 1500 RPM or 30 mph. The exhaust has a steady spitting sound at idle or low speed.
Hesitates	Momentary lack of response as the accelerator is depressed. Can occur at all car speeds. Usually most severe when first trying to make the car move, as from a stop sign. May cause the engine to stall if severe enough.
Surges	Engine Power variation under steady throttle or cruise. Feels like the car speeds up and slows down with no change in the accelerator pedal. Can occur at any speed.
Sluggish	Engine delivers limited power under load or at high speed. Won't accelerate as fast as normal; loses too much speed going up hills; or has less top speed than normal.
Spongy	Less than the anticipated response to increased throttle opening. Little or no increase in speed when the accelerator pedal is pushed down a little to increase cruising speed. Continuing to push the pedal down will finally give an increase in speed.
Poor Gas Mileage	Self describing.
Cuts Out	Temporary complete loss of power. The engine quits at sharp, irregular intervals. May occur repeatedly, or intermittently. Usually worse under heavy acceleration.

Oldsmobile   **Trouble Diagnosis**

ENGINE PERFORMANCE DIAGNOSIS CHARTS (CONT'D)

CONDITION	POSSIBLE CAUSE	CORRECTION
Engine starts and runs but misses rough or stalls at idle speed only.	Engine vacuum leak.	Check engine vacuum connections and hoses for leaks. Repair or replace as necessary.
	Spark plug malfunction. (Fouled, cracked or incorrect gap.)	Inspect, clean, and adjust or install new plugs as necessary.
	Ignition wire malfunction	Clean, check continuity and inspect wires, replace if brittle, cracked or worn. Refer to HEI diagnosis.
	HEI malfunction	Refer to HEI diagnosis and/or engine electrical in Section 12.
	EGR malfunction	Refer to EGR diagnosis.
	PCV malfunction.	Refer to PCV diagnosis.
	Engine valve leakage.	Refer to Section 6C (compression check).
	Intake manifold or cylinder head gasket leaks.	Check intake manifold gaskets and seals for leaks. Make necessary repairs or replacements.
	Carburetor malfunction.	Refer to carburetor diagnosis.
Engine starts and runs, misses at all speeds.	Spark plug malfunction.	Inspect, clean, and adjust or replace as required.
	HEI malfunction.	Refer to HEI diagnosis and/or engine electrical in Section 12.
	Engine valve leakage.	Refer to Section 6C (compression check).
	EGR malfunction.	Refer to EGR diagnosis.
	EFE valve malfunction.	Make necessary repairs or replacements.
	Malfunctioning fuel pump or fuel system.	Refer to fuel pump diagnosis and/or fuel delivery system function.
	Faulty carburetion (including fuel filter).	Refer to carburetor diagnosis.
	Contaminated fuel.	Remove and install fresh fuel. Clean fuel system and carburetor as required. Refer to fuel system diagnosis.
Engine starts and runs but misses at high speeds.	Spark plug malfunction.	Inspect, clean and adjust or install new plugs.
	Engine valve leakage.	Refer to Section 6C (compression check).
	HEI malfunction.	Refer to HEI diagnosis and/or engine electrical in Section 12.

Trouble Diagnosis — Oldsmobile

ENGINE PERFORMANCE DIAGNOSIS CHART CONT'D

CONDITION	POSSIBLE CAUSE	CORRECTION
	Ignition wire malfunction.	Clean, check resistance, and inspect wires, replace if brittle, cracked or worn.
	Malfunctioning fuel pump or fuel system.	Refer to fuel pump diagnosis and/or fuel delivery system.
	Faulty carburetion (including fuel filters).	Refer to carburetor and/or fuel system diagnosis.
	EFE valve malfuction.	Refer to EFE diagnosis.
	PCV malfunction.	Refer to PCV diagnosis.
	Faulty carburetor inlet air temperature regulation.	Refer to air cleaner diagnosis.
	Exhaust system restricted.	Make necessary repairs or replacements.

HIGH ENERGY IGNITION (H.E.I.) SYSTEM DIAGNOSIS CHART

CONDITION	POSSIBLE CAUSE	CORRECTION
Engine cranks but will not start	Low battery.	Charge battery and/or check generator. See Section 12.
	No spark at spark plugs.	Disconnect one spark plug lead from spark plug, hold 1/4" from dry area of engine block while cranking engine. If spark occurs while cranking, check condition of spark plugs (if plugs are okay, problem is in fuel system). If no spark occurs: a. Refer to HEI diagnosis.
	a. Open circuit between "Bat" terminal and battery.	b. Repair open circuit between "Bat" terminal and battery.
	Engine timing out of adjustments.	Reset timing to specifications. For procedure to rough-set timing on an engine that won't run, see setting ignition timing.
	Internal engine problems.	Refer to engine diagnosis.
	Out of gas.	Put a supply of gas in tank and start engine.
	Malfunctioning carburetor.	Refer to carburetor diagnosis.
	Malfunctioning fuel pump.	Refer to fuel system diagnosis.
	Battery cables loose or corroded.	Tighten and/or clean cables.
Engine starts but stops when ignition switch is released to "RUN POSITION"	An "OPEN" in the ignition circuit or a defective ignition switch.	Use a 12 volt test lamp and check IGN-1 terminal of ignition switch in "RUN POSITION". a. If lamp lights, locate and repair "OPEN" in circuit to HEI "Bat" terminal. This includes pink wire from switch connector to cowl connector. Black wire with double white stripe from cowl connector to distributor connector. b. If lamp does not light, replace ignition switch.

Oldsmobile Trouble Diagnosis

HIGH ENERGY IGNITION SYSTEM CHART CONT'D

Engine will not crank.	Loose or corroded wire or cable connections.	Inspect for and clean and/or tighten connections as necessary.
	Discharged battery.	Refer to diagnosis in battery section.
	Starter assembly.	Test and/or repair starter as necessary. See Section 12.
	Neutral start switch malfunctioning out of adjustment or poor connection.	Make certain connector terminals are clean and connector is properly installed.
		Connect 12 volt test lamp to purple wire with white stripe and ground. Lamp should light in start position with shift in neutral or park. If lamp lights, ignition and neutral start switch are okay.
		If lamp did not light on purple wire with white stripe, connect it to light green wire with double black stripe. If lamp now lights in start position in neutral or park, adjust or replace neutral start switch. If lamp did not light, see following ignition switch test.
	Loose connection at or defective ignition switch.	Inspect connector to assure clean terminals and proper connection.
		Using 12 volt test lamp check both purple wires at neutral start switch. If lamp does not light at either wire, repair "OPEN" circuit in purple wire with white stripe to the ignition switch and/or replace ignition switch as required.
	"OPEN" circuit in wiring to solenoid or defective solenoid.	Using 12 volt test lamp between purple wire at solenoid and ground, lamps should light in start position in neutral or park.
		a. If lamp does not light, locate and correct "OPEN" in circuit. b. If lamp lights replace solenoid.
	Burned out fusible link.	Using 12 volt test lamp between No. 10 red wire at cowl connector and ground, lamp should light.
		a. If lamp does not light, replace fusible link.

Trouble Diagnosis — Oldsmobile

HIGH ENERGY IGNITION SYSTEM CHART CONT'D

Engine runs rough, poor power and fuel economy.	Faulty spark plugs.	Inspect, clean and adjust or adjust and install new plugs.
	Incorrect timing.	Adjust timing to specifications.
	Inoperative vacuum advance. (Engines equipped with vaccum advance)	With engine running at part throttle, hold finger on vacuum advance rod and remove vacuum hose. Rod should move toward distributor. If rod does not move, replace vacuum advance unit.
	Faulty centrifugal advance on engine.	Check centrifugal advance on engine.
	HEI distributor.	Refer to HEI diagnosis. See Section 12.
	Faulty plug wires.	Clean, check resistance, and inspect wires for brittle, cracked or loose insulation condition. Also inspect for burned, corroded terminals. Clean or replace wire assemblies as necessary including deteriorated nipples or boots.
	Faulty carburetion.	Refer to carburetion diagnosis.
	Faulty engine parts such as valves, rings, etc.	Refer to engine diagnosis.

CARBURETOR DIAGNOSIS PROCEDURE CHART

The following diagnostic procedures are directed toward carburetor related problems and their effects on car performance. It is understood that other systems of the car can also cause similar problems and should be investigated in conjunction with the carburetor. In all instances, the complaint item should be verified by competent service personnel. The problem areas described are:

1. Engine cranks normally. Will not start.
2. Engine starts and stalls.
3. Engine starts hard.
4. Engine idles abnormally and/or stalls.
5. Inconsistent engine idle speeds.
6. Engine diesels (after-run) upon shut off.
7. Engine hesitates on acceleration.
8. Engine has less than normal power at low speeds.
9. Engine has less than normal power on heavy acceleration or at high speed.
10. Engine surges.
11. Fuel economy complaints.

Oldsmobile **Trouble Diagnosis**

CARBURETOR DIAGNOSIS PROCEDURE CHART CONT'D

CONDITION	POSSIBLE CAUSE	CORRECTION
Engine Cranks Normally - Will Not Start.	Improper starting procedure used.	Check with the customer to determine if proper starting procedure is used, as outlined in the Owner's Manual.
	Choke valve not operating properly.	Adjust the choke thermostatic coil to specification. Check the choke valve and/or linkage as necessary. Replace parts if defective. If caused by foreign material and gum, clean with suitable non-oil base solvent. NOTE: After any choke system work, check choke vacuum break settings and correct as necessary.
	No fuel in carburetor.	Remove fuel line at carburetor. Connect hose to fuel line and run into metal container. Remove the wire from the "bat" terminal of the distributor. Crank over engine - if there is no fuel discharge from the fuel line, test fuel pump as outlined in Section 6B. If fuel supply is okay, check the following: a. Inspect fuel inlet filter. If plugged, replace. b. If fuel filter is okay, remove air horn and check for a bind in the float mechanism or a sticking inlet needle. If okay, adjust float as specified.
	Engine flooded. To check for flooding, remove the air cleaner with the engine immediately shut off and look into the carburetor bores. Fuel will be dripping off nozzles.	Remove the air horn. Check fuel inlet needle and seat for proper seal. If a needle and seat tester is not available, apply vacuum to the needle seat with needle installed. If the needle is leaking, replace. Check float for being loaded with fuel, bent float hanger or binds in the float arm. A solid float can be checked for fuel absorption by lightly squeezing between fingers. If wetness appears on surface or float feels heavy (check with known good float), replace the float assembly. If foreign material is in fuel system, clean the system and replace fuel filters as necessary. If excessive foreign material is found, completely disassemble and clean.

Trouble Diagnosis — Oldsmobile

CARBURETOR DIAGNOSIS PROCEDURE CHART CONT'D

Engine Starts - Will Not Keep Running	Fuel pump.	Check fuel pump pressure and volumn, replace if necessary.
	Idle speed.	Adjust idle to specifications.
	Choke heater system malfunctioning (may cause loading).	Chec vacuum supply at hot air inlet to choke housing. Should be not less than manifold vacuum minus 3" Hg. with engine running at idle. (Exc. IMV)
		Check for plugged, restricted, or broken heat tubes.
		Check routing of all hot air parts.
	Loose, broken or incorrect vacuum hose routing.	Check condition and routing of all vacuum hoses - correct as necessary.
	Engine does not have enough fast idle speed when cold.	Check for free movement of fast idle cam. Clean and/or realign as necessary.
	Choke vacuum break units are not adjusted to specification or are defective.	Adjust both vacuum break assemblies to specification. If adjusted okay, check the vacuum break units for proper operation as follows:
		To check the vacuum break units, apply a constant vacuum source of at least 10" Hg., plungers should slowly move inward and hold vacuum. If not, replace the unit.
		Always check the fast idle cam adjustment when adjusting vacuum break units.
	Choke valve sticking and/or binding.	Clean and align linkage or replace if necessary. Readjust all choke settings, see Section 6M, if part replacement or realignment is necessary.
	Insufficient fuel in carburetor.	Check fuel pump pressure and volume.
		Check for partially plugged fuel inlet filter. Replace if contaminated.
		Check the float level adjustment and for binding condition. Adjust as specified.
(NOTE: The EGR system diagnosis should also be performed.)		
Engine Starts Hard (Cranks Normally)	Loose, broken or incorrect vacuum hose routing.	Check condition and routing of all vacuum hoses - correct as necessary.
	Incorrect starting procedure.	Check to be sure customer is using the starting procedure outlined in Owner's Manual.
	Malfunction in accelerator pump system.	Check accelerator pump adjustment and operation.
		Check pump discharge ball for sticking or leakage.

CARBURETOR DIAGNOSIS PROCEDURE CHART CONT'D

	Choke valve not closing.	Adjust choke thermostatic coil. Check choke valve and linkage for binds and alignment. Clean and repair or replace as necessary.
	Vacuum breaks misadjusted or malfunctioning.	Check for adjustment and function of vacuum breaks. Correct as necessary.
	Insufficient fuel in bowl.	Check fuel pump pressure and volume. Check for partially plugged fuel inlet filter. Replace if dirty. Check float mechanism. Adjust as specified.
	Flooding.	Check float and needle and seat for proper operation.
	Where used, check to see if vent valve is inoperative or misadjusted.	Check for operation and adjustment of vent valve (if used).
	Slow engine cranking speed.	Refer to starting system diagnosis.

(NOTE: The EGR system diagnosis should also be performed.)

Engine Idles Abnormally	Incorrect idle speed.	Reset idle speed per instructions on underhood label.
	Air leaks into carburetor bores beneath throttle valves, manifold leaks, or vacuum hoses disconnected or installed improperly.	Check all vacuum hoses and restrictors leading into the manifold or carburetor base for leaks or being disconnected. Install or replace as necessary. Torque carburetor to manifold bolts to 15 ft. lbs. (L-6), 10 ft. lbs. (exc. L-6). Using a pressure oil can, spray light oil or kerosene around manifold to head surfaces and carburetor throttle body. NOTE: Do not spray at throttle shaft ends. If engine RPM changes, tighten or replace the carburetor or manifold gaskets as necessary.
	Clogged or malfunctioning PCV system.	Check PCV system. Clean and/or replace as necessary.
	Carburetor flooding.	Remove air horn and check float adjustments.
	Check by using procedure outlined under "Engine Flooded".	Check float needle and seat for proper seal. If a needle and seat tester is not available, apply vacuum to the needle seat with needle installed. If the needle is leaking or damaged, replace. Check float for being loaded with fuel. Check for bent float hanger or binds in the float arm. A solid float can be checked for fuel. Check for bent float hanger or binds in the float arm.

Trouble Diagnosis — Oldsmobile

CARBURETOR DIAGNOSIS PROCEDURE CHART CONT'D

		A solid float can be checked for fuel absorption by lightly squeezing between fingers. If wetness appears on surface or float feels heavy (check with known good float), replace the float assembly.
		If foreign material is found in the carburetor, clean the fuel system and carburetor. Replace fuel filters as necessary.
	Restricted air cleaner element.	Replace as necessary.
	Idle system plugged or restricted.	Clean per section 6M.
	Incorrect idle mixture adjustment.	Readjust per specified procedure (Section 6M).
	Defective idle stop solenoid, idle speed-up solenoid or wiring.	Check solenoid and wiring.
	Throttle blades or linkage sticking and/or binding.	Check throttle linkage and throttle blades (primary and secondary) for smooth and free operation. Correct problem areas.

(NOTE: The EGR system diagnosis should also be performed.)

Engine Diesels (After Run – Upon Shut Off)	Loose, broken or improperly routed vacuum hoses.	Check condition and routing of all vacuum hoses. Correct as necessary.
	Incorrect idle speed.	Reset idle speed per instructions on label in engine compartment.
	Malfunction of idle stop solenoid, idle speed-up solenoid or dashpot.	Check for correct operation of idle solenoid. Check for sticky or binding solenoid.
	Excessively lean idle mixture caused by air leaks into carburetor beneath throttle valves, manifold vacuum leaks, or failed PCV system.	See corrections listed causes 2 and 3 under "Engine Idles Abnormally and/or Stalls".
	Fast idle cam not fully off.	Check fast idle cam for freedom of operation. Clean, repair, or adjust as required. Check choke heated air tubes for routing, fittings being tight or tubes plugged. Check choke linkage for bending. Clean and correct as necessary.
	Excessively lean condition caused by maladjusted carburetor idle mixture.	Adjust carburetor idle mixture as described in Section 6M.
	Ignition timing retarded.	Set to specifications.

Oldsmobile
Trouble Diagnosis

CARBURETOR DIAGNOSIS PROCEDURE CHART CONT'D

Engine Hesitates On Acceleration	Loose, broken or incorrect vacuum hose routing	Check condition and routing of all vacuum hoses - correct or replace.
	Accelerator pump not adjusted to specification or inoperative.	Adjust accelerator pump, or replace.
	Inoperative accelerator pump system.	Remove air horn and check pump cup. If cracked, scored or distorted, replace the pump plunger.
	NOTE: A quick check of the pump system can be made as follows: With the engine off, look into the carburetor bores and observe pump shooters while briskly opening throttle lever. A full stream of fuel should emit from each pump shooter.)	Check the pump discharge ball for proper seating and location.
	Foreign matter in pump passages.	Clean and blow out with compressed air.
	Float level too low.	Check and reset float level to specification.
	Front vacuum break diaphragm not functioning properly.	Check adjustment and operation of vacuum break diaphragm.
	Air valve malfunction.	Check operation of secondary air valve. Check spring tension adjustment.
	Power enrichment system not operating correctly.	Check for binding or stuck power piston(s) - correct as necessary.
	Inoperative air cleaner heated air control.	Check operation of thermostatic air cleaner system.
	Fuel filter dirty or plugged.	Replace filter and clean fuel system as necessary.
	Distributor vacuum or mechanical advance malfunctioning.	Check for proper operation.
	Timing not to specifications.	Adjust to specifications.
	Choke coil misadjusted (cold operation.)	Adjust to specifications
	EGR valve stuck open.	Inspect and clean EGR valve.
Engine Has Less Than Normal Power At Normal Accelerations.	Loose, broken or incorrect vacuum hose routing.	Check condition and routing of all vacuum hoses.
	Clogged or defective PCV system.	Clean or replace as necessary.
	Choke sticking.	Check complete choke system for sticking or binding.
		Clean and realign as necessary.
		Check adjustment of choke thermostatic coil.
		Check connections and operation of choke hot air system.

447

Trouble Diagnosis   Oldsmobile

CARBURETOR DIAGNOSIS PROCEDURE CHART CONT'D

		Check jets and channels for plugging; clean and blow out passages.
	Clogged or inoperative power system.	Remove air horn and check for free operation of power pistons.
	Air cleaner temperature regulation improper.	Check regulation and operation of air cleaner system.
	Transmission malfunction.	Refer to transmission diagnosis.
	Ignition system malfunction.	Check ignition timing. Reset to specification.
		Refer to H.E.I. diagnosis.
	Excessive brake drag.	Refer to Section 5.
	Exhaust system.	Check for restrictions. Correct as required.

(NOTE: An engine tune-up should be conducted in conjunction with the carburetor diagnosis. The EGR system diagnosis should also be performed.)

Less Than Normal Power On Heavy Acceleration Or At High Speed	Carburetor throttle valves not going wide open. Check by pushing accelerator pedal to floor.	Correct throttle linkage to obtain wide open throttle in carburetor.
	Secondary throttle lockout not allowing secondaries to open.	Check for binding or sticking lockout lever.
		Check for free movement of fast idle cam.
		Check choke heated air system for proper and tight connections plus flow through system.
		Check adjustment of choke thermostatic coil.
	Spark plugs fouled, incorrect gap.	Clean, regap, or replace plugs.
	Plugged air cleaner element.	Replace element.
	Air valve malfunction. (Where applicable)	Check for free operation of air valve.
		Check spring tension adjustment. Make necessary adjustments and corrections.
	Contaminated fuel inlet filter.	Replace with a new filter element.
	Insufficient fuel to carburetor.	Check fuel pump and system, run pressure and volume test.
	Vapor lock.	Eliminate cause.
	Power enrichment system not operating correctly.	Remove the air horn and check for free operation of both power piston(s), clean and correct as necessary.
	Choke closed or partially closed.	Free choke valve or linkage.

CARBURETOR DIAGNOSIS PROCEDURE CHART CONT'D

		Check for loose jets.
	Float level too low.	Check and reset float level to specification.
	Transmission malfunction.	Refer to transmission diagnosis.
	Ignition system malfunction.	Check ignition timing. Reset to specification. Refer to H.E.I. diagnosis.
	Fuel metering jets restricted.	If the fuel metering jets are restricted and excessive amount of foreign material is found in the fuel bowl, the carburetor should be completely disassembled and cleaned.
	Fuel pump.	Check fuel pump pressure and volumn, inspect lines for leaks and restrictions.
	Exhaust system.	Check for restrictions. Correct as required.

(NOTE: Complete engine tune-up should be performed in conjunction with carburetor diagnosis.)

Engine Surges	Loose, broken or incorrect vacuum hose routing.	Check condition and routing of all vacuum hoses. Correct as necessary.
	PCV system clogged or malfunctioning.	Check PCV system. Clean or replace as necessary.
	Loose carburetor, EGR or intake manifold bolts and/or leaking gaskets.	Torque carburetor to manifold bolts to 15 ft. lbs. (L-6), 10 ft. lbs. (All exc. L-6). Using a pressure oil can, spray light oil or kerosene around manifold to head mounting surface and carburetor base. If engine RPM changes, tighten or replace the carburetor or manifold gaskets as necessary. Check EGR mounting bolt torque.
	Low or erratic fuel pump pressure.	Check fuel delivery and pressure.
	Contaminated fuel.	Check for contaminants in fuel. Clean system if necessary.
	Fuel filter plugged.	Check and replace as necessary.
	Float level too low.	Check and reset float level to specification.
	Malfunctioning float and/or needle and seat.	Check operation of system. Repair or replace as necessary.
	Power piston stuck or binding.	Check for free movement of power piston(s). Clean and correct as necessary.
	Fuel jets or passages plugged or restricted.	Clean and blow out with compressed air.
	Ignition system malfunction.	Check ignition timing. Correct as necessary.
	Exhaust system.	Check for restrictions. Correct as necessary.

(NOTE: EGR system diagnosis should also be performed.)

Trouble Diagnosis — Oldsmobile

CARBURETOR DIAGNOSIS PROCEDURE CHART CONT'D

Fuel Economy Complaints	Customer driving habits.	Run mileage test with customer driving if possible. Make sure car has 2000-3000 miles for the "break-in" period.
	Loose, broken or improperly routed vacuum hoses.	Check condition of all vacuum hose routings. Correct as necessary.
	Engine needs complete tune-up.	Check engine compression, examine spark plugs; if fouled or improperly gapped, clean and regap or replace. Check ignition wire condition and check and reset ignition timing. Replace air cleaner element if dirty. Check for restricted exhaust system and intake manifold for leakage. Check carburetor mounting bolt torque. Check vacuum and mechanical advance.
	Fuel leaks.	Check fuel tank, fuel lines and fuel pump for any fuel leakage.
	High fuel level in carburetor.	Check fuel inlet needle and seat for proper seal. Test, using suction from a vacuum source. If needle is leaking, replace.
		Check for loaded float. Reset float level to specification.
		If excessive foreign material is present in the carburetor bowl, the carburetor should be cleaned.
	Power system in carburetor not functioning properly. Power piston(s) sticking or metering rods out of jets.	Remove air horn and check for free movement of power piston(s). Clean and correct as necessary.
	Choke system.	Check choke heated air tubes for routing and/or plugging which would restrict hot air flow to choke housing. Check choke linkage for binding. Clean or repair as required. Check adjustment of thermostatic coil. Readjust to specification as required.
	Plugged air cleaner element.	Replace element.
	Exhaust system.	Check for restrictions. Correct as required.
	Low tire pressure or incorrect tire size.	Inflate tires to specifications and use correct size tires.
	Transmission malfunction.	Refer to transmission diagnosis.

THERMOSTATIC AIR CLEANER SYSTEM DIAGNOSIS CHART

CONDITION	POSSIBLE CAUSE	CORRECTION
Hesitation, sag, and stalling during cold start.	Failure of air cleaner in cold air mode.	Inspect and correct as required all parts connected with the air cleaner system.
Fuel Economy Complaints	Failure of air cleaner in hot air mode.	Inspect and correct as required all parts connected with the air cleaner.

Oldsmobile — Trouble Diagnosis

EXHAUST GAS RECIRCULATION SYSTEM CHART

CONDITION	POSSIBLE CAUSE	CORRECTION
Engine idles abnormally rough and/or stalls	EGR valve vacuum hoses misrouted.	Check EGR valve vacuum hose routing. Correct as required.
	Leaking EGR valve.	Check EGR valve for correct operation.
	Incorrect idle speed.	Set idle RPM per engine label specification. Remove EGR vacuum hose from valve and observe effect on engine RPM. If speed is affected, reset RPM to specification and reconnect hose.
	EGR valve gasket failed or loose EGR attaching bolts.	Check EGR attaching bolts for tightness. Tighten as required. If not loose, remove EGR valve and inspect gasket. Replace as required.
	EGR thermal control valve and/or EGR-TVS.	Check vacuum into valve from carburetor EGR port with engine at normal operating temperature and at curb idle speed. Then check the vacuum out of the EGR thermal control valve to EGR valve. If the two vacuum readings are not equal within ± 1/2 in. Hg., then proceed to EGR vacuum control diagnosis. See Section 6C.
	Improper vacuum to EGR valve at idle.	Check vacuum from carburetor EGR port with engine at stabilized operating temperature and at curb idle speed. If vacuum is more than 1.0 in. Hg., refer to carburetor idle diagnosis.
Cars with a back pressure transducer valve.	Improper carburetor vacuum to exhaust back pressure transducer at idle.	Check vacuum from carburetor to exhaust back pressure transducer with engine at stabilized operating temperature and at curb idle speed. If vacuum is more than 1.0 in. Hg., refer to carburetor idle diagnosis.
Engine runs rough on light throttle acceleration, poor part load performance and poor fuel economy	EGR valve vacuum hose misrouted.	Check EGR valve vacuum hose routing. Correct as required.
	Failed EGR vacuum control valve.	Same as listing in "Engine Idles Rough" condition.
	EGR flow unbalanced due to deposit accumulation in EGR passages or under carburetor.	Clean EGR passages of all deposits.
	Sticky or binding EGR valve.	Remove EGR valve and inspect. Clean or replace as required.
Cars with a back pressure transducer valve (No.'s 5, 6 and 7)	Failed exhaust back pressure transducer valve.	Functional check per procedure.
	Wrong or no EGR gaskets.	Check and correct as required.
	Exhaust system restricted causing excessive back pressure.	Inspect exhaust system for restriction and replace or repair as required.

Trouble Diagnosis — Oldsmobile

EXHAUST GAS RECIRCULATION SYSTEM CHART CONT'D

CONDITION	POSSIBLE CAUSE	CORRECTION
Engine stalls on decelerations.	Restriction in EGR vacuum line.	Check EGR vacuum lines for kinks, bends, etc. Remove or replace hoses as required. Check EGR vacuum control valve function.
		Check EGR valve for excessive deposits causing sticky or binding operation. Clean or repair as required.
	Sticking or binding EGR valve.	Remove EGR valve and inspect clean or repair as required.
Part throttle engine detonation	Insufficient exhaust gas recirculation flow during part throttle accelerations.	Check EGR valve hose routing. Check EGR valve operation. Repair or replace as required. Check EGR thermal control valve and/or EGR-TVS as listed in "Engine Idles Rough" section. Replace valve as required. Check EGR passages and valve for excessive deposit Clean as required.
	Exhaust back pressure transducer failed.	Check function per service procedure. (Section 6C)

(NOTE: Detonation can be caused by several other engine variables. Perform ignition and carburetor related diagnosis.)

CONDITION	POSSIBLE CAUSE	CORRECTION
Engine starts but immediately stalls when cold	EGR valve hoses misrouted.	Check EGR valve hose routings.
	EGR system malfunctioning when engine is cold.	Perform check to determine if the EGR thermal control valve and/or EGR-TVS are operational. Replace as required.
Cars with a back pressure transducer valve.	Exhaust back pressure transducer failed.	Check function per service manual procedure.

(NOTE: Stalls after start can also be caused by carburetor problems. Refer to carburetor diagnosis section.)

CATALYTIC CONVERTER DIAGNOSIS CHART

CONDITION	POSSIBLE CAUSE	CORRECTION
Exhaust system noisy	Exhaust pipe joints loose at catalytic converter.	Tighten clamps at joints.
	Catalytic converter ruptured.	Replace catalytic converter.
		Ignition system and AIR system (if used) should also be diagnosed and repairs made if necessary.

Oldsmobile — Trouble Diagnosis

EARLY FUEL EVAPORATION (EFE) SYSTEM DIAGNOSIS CHART

CONDITION	POSSIBLE CAUSE	CORRECTION
Operation during warm-up such as high idle, stumble.	No vacuum to vacuum actuator during warm-up period for cold start.	Check vacuum source for vacuum of 8" Hg or above. Repair vacuum hose routing, leak in connecting system, diaphragm, or EFE-EGR TVS.
	EFE valve linkage bent or binding.	Repair EFE valve linkage.
	EFE valve linkage disconnected.	Reinstall linkage.
	EFE valve shaft frozen in bearing.	Replace EFE valve.
	EFE valve loose on shaft.	Replace EFE valve.
Poor operation after warm-up, rough idle, lack of high speed performance or surge.	Malfunctioning EFE switch.	Replace EFE switch.
	EFE valve asm shaft frozen in bearing (Closed valve).	Replace EFE valve.
	Valve to housing interference.	Repair EFE valve.
	Vacuum actuator linkage bent or binding.	Repair EFE valve linkage.
	Valve to casting interference.	Adjust valve position.
Noisy EFE valve assembly.	Linkage stop failed.	Replace EFE valve.
	Incorrect vacuum line routing.	See vacuum line routing chart, Section 6C.
	No vacuum actuator linkage over travel.	Replace vacuum actuator.
	EFE-CV reversed or inoperative.	Reverse or replace EFE-Check Valve.
	Valve loose on shaft.	Replace EFE valve.
	Shaft loose in bushing, or bushing loose in housing.	Replace EFE valve.
Engine overheating or EFE Valve won't open.	EFE-TVS not switching or EFE vent in switch plugged.	Unplug EFE vent or replace EFE-TVS switch.

FUEL SYSTEM DIAGNOSIS CHART

CONDITION	POSSIBLE CAUSE	CORRECTION
Car feels like it is running out of gas-surging occurs in mid-speed range	Plugged fuel filters.	Remove and replace filters.
	Faulty fuel pump.	Perform diagnostic tests on the fuel pump as described in Section 6B. Remove and replace fuel pump as required.
	Foreign material in fuel system or kinked fuel pipes or hoses.	Inspect pipes and hoses for kinks and bends, blow out to check for plugging. Remove and replace as required.
Engine starts but will not continue to run or will run but surges and back fires.	Faulty fuel pump.	Perform diagnostic tests on the fuel pump as described in Section 6B. Remove and replace fuel pump as required.

Trouble Diagnosis — Oldsmobile

FUEL SYSTEM DIAGNOSIS CHART CONT'D

Condition	Possible Cause	Correction
Engine will not start	Faulty fuel pump.	Perform diagnostic tests on the fuel pump as described in Section 6B. Remove and replace fuel pump as required.

EVAPORATION EMISSION CONTROL SYSTEM DIAGNOSIS CHART

CONDITION	POSSIBLE CAUSE	CORRECTION
Fuel odor	Vapor leak from evap. system.	Inspect and correct as necessary fuel and evap. hoses and pipes, fuel sender sealing gasket, fuel cap.

POSITIVE CRANKCASE VENTILATION SYSTEM DIAGNOSIS CHART

CONDITION	POSSIBLE CAUSE	CORRECTION
Rough idle	PCV valve stuck open.	Test - remove and replace PCV system as required.
Oil in air cleaner	PCV system plugged.	Test - remove and replace PCV system as required.
	Leak in closed ventilation system.	Check and correct as necessary, for leaks to atmosphere of the closed crankcase ventilation system.
	Oil return holes in cylinder head restricted.	Remove valve covers, inspect and clean as required.
	Valve cover oil baffle restricted.	Remove, inspect and repair baffle as required.

(NOTE: See P.C.V. Diagnosis Chart using CT-3 Tester, Section 6C.)

EXCESSIVE ENGINE OIL CONSUMPTION DIAGNOSIS CHART

POSSIBLE CAUSE	CORRECTION
External oil leaks at: Rocker Arm Covers, Timing Chain Cover Oil Pan, Gasket Between Oil Pan and Flywheel Housing Intake Manifold Gasket.	Tighten attaching bolts. If leaks persist, remove cover (or pan), check sealing surfaces for burrs, scoring or distorted cover flanges. Make sure oil level is not overfull.
Improper reading of dipstick	Car may not be level when taking reading. Insufficient oil "drain-back" time allowed after stopping engine (three minutes must be allowed). Dipstick may not be completely pushed down against stop. Dipstick may be bent.
Oil viscosity too light	Use recommended SAE viscosity for prevailing temperatures.

EXCESSIVE ENGINE OIL CONSUMPTION
DIAGNOSIS CHART CONT'D

High-speed driving following normal slow-speed city driving	When principal use of automobile is city driving, crankcase dilution from condensation occurs. High-speed and temperatures will remove water, resulting in what appears to be rapid lowering of oil level. Inform customer of this fact.
Valve guides and/or valve stem seals worn, loose or broken	Ream out guides and install service valves with oversize stems and new valve stem seals.
Piston rings not "broken in"	Allow engine to accumulate at least 4,000 miles before attempting any engine disassembly to correct for oil consumption.
Trailer hauling	Because engine oil temperatures are higher during trailer hauling and oil is, therefore, thinner, higher than normal oil consumption can be expected. Inform customer.
Bore scoring and wear	Check air cleaner and element for leakage.
Ring scuffing	Repair or replace parts as necessary. Check type of oil being used by customer. Recommended oil is G.M. 6136 oil (high grade).

AIR INJECTOR REACTOR SYSTEM (A.I.R.) DIAGNOSIS CHART

(NOTE: The AIR system is not completely noiseless. Under normal conditions, noise rises in pitch as engine speed increases. To determine if excessive noise is the fault of the air injection system, disconnect the drive belt and operate the engine. If noise now does not exist, proceed with diagnosis.

CONDITION	POSSIBLE CAUSE	CORRECTION
Excessive belt noise	Loose belt	Tighten to spec.
	Seized pump	Replace pump.
Excessive pump noise. Chirping	Insufficient break-in	Run car 10-15 miles at turnpike speeds — recheck.
Excessive pump noise, chirping, rumbling, or knocking	Leak in hose	Locate source of leak using soap solution and correct.
	Loose hose	Reassemble and replace or tighten hose clamp.
	Hose touching other engine parts.	Adjust hose position.
	Diverter valve inoperative	Replace diverter valve.
	Check valve inoperative	Replace check valve.
	Pump mounting fasteners loose	Tighten mounting screws as specified.
	Pump failure	Replace pump.

Trouble Diagnosis — Oldsmobile

AIR INJECTOR REACTOR SYSTEM DIAGNOSIS CHART CONT'D

No air supply (accelerate engine to 1500 rpm and observe air flow from hoses. If the flow increases as the rpm's increase, the pump is functioning normally. If not, check possible cause.	Loose drive belt	Tighten to specs.
	Leaks in supply hose	Locate leak and repair or replace as required.
	Leak at fitting(s)	Tighten or replace clamps.
	Diverter valve leaking	If air is expelled through diverter muffler with engine at idle, replace diverter valve.
	Diverter valve inoperative	Usually accompanied by backfire during deceleration. Replace diverter valve.
	Check valve inoperative	Blow through hose toward air manifold. If air passes, function is normal. If air can be sucked from manifold, replace check valve.
	Pump pressure relief plug leaking or damaged.	Replace pressure relief plug.
Centrifugal filter fan damaged or broken.	Mechanical damage	Replace centrifugal filter fan.
Poor idle or driveability.	A defective AIR pump cannot cause poor idle or driveability.	Do NOT replace AIR pump.

POSITIVE CRANKCASE VENTILATION SYSTEM (PCV) DIAGNOSIS CHART

USING CT-3 TESTER

WINDOW READING	PROBABLE TROUBLE	CORRECTION
GREEN	System Satisfactory Vent valve partially plugged. Blow-by close to capacity of valve	Check valve
YELLOW	Tester hose kinked or blocked Crankcase not sealed properly Tester "selector knob" set incorrectly Vent-valve partially plugged Slight kink in CT-2 tester hose	Reposition or clean hose Check tester plugs and other seal-off points Check setting Check vent valve Reposition tester hose
YELLOW-GREEN	Slight engine blow-by Crankcase not sealed properly Tester "selector knob" set incorrectly Vent valve partially or fully plugged	Check vent valve Check tester plugs and other seal-off points Check setting Check vent valve
RED-YELLOW	Engine blow-by exceeds valve capacity Rubber vent hose collapsed or plugged	Engine overhaul indicated Clean or replace hose
RED	Vent valve plugged Vent valve stuck at engine off position Rubber vent hose collapsed or plugged Extreme engine blow-by	Check vent valve Check vent valve Replace hose Engine requires major overhaul

Brakes and Wheel Alignment Section Index

Disc Brakes 461
  Performance Diagnosis 463
  Service 461
  Specifications 466
    Domestic Cars 466
    Import Cars 490
    Import Trucks 490

Drum Brakes 457
  Performance Diagnosis 462
  Service 457
  Specifications 466
    Domestic Cars 466
    Import Cars 490
    Import Trucks 490

Wheel Alignment 464
  Front Suspension Geometry 464
  Specifications 478
    Domestic Cars 478
    Import Cars 496
    Import Trucks 496

Specifications 466
  Brakes—Domestic Cars 466
  Brakes—Import Cars 490
  Brakes—Import Trucks 490
  Wheel Alignment—Domestic Cars 478
  Wheel Alignment—Import Cars 496
  Wheel Alignment—Import Trucks 496

BRAKE SYSTEM TUNEUP PROCEDURE

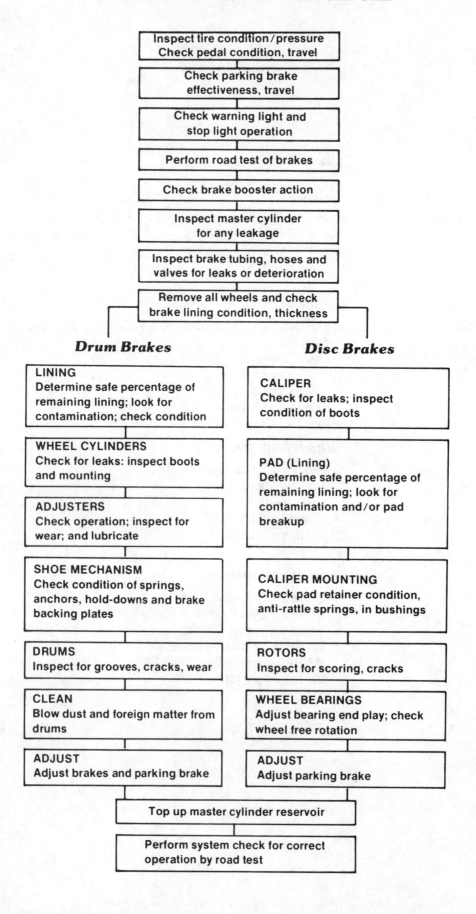

Brake Service

DRUM BRAKE SERVICE

Most cars, except Chevette, use self energizing drum brakes with automatic adjusters. Utilization of the frictional force to increase the pressure of shoes against the drum is called *self-energizing* action. Utilization of force in one shoe to apply the opposite shoe is called *servo* action.

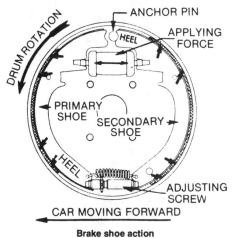

Brake shoe action

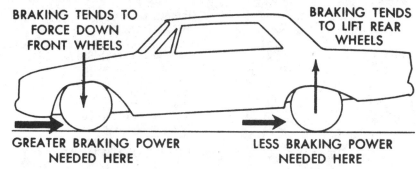

When a car's brakes are applied, the front of the car dips down and the back of the car rises up. The major braking effort is needed at the front of the vehicle.

Brake Lining

Brake lining is made of asbestos impregnated with special compounds to bind the asbestos fibers together. Some linings are woven of asbestos threads and fine copper wire. With a few exceptions the brake lining material is molded asbestos fibers ground up, pressed into shape and either riveted or bonded onto the brake shoe.

The primary shoe, sometimes called leading or forward brake shoe, is the shoe that faces toward the front of the car.

The secondary shoe, sometimes called the trailing or reverse brake shoe, faces the rear of the car.

Backing Plate

Thorough brake work starts at the brake backing plate. Check the brake area for any indication of lubricant leakage. If the leakage is brake fluid replace or rebuild wheel cylinder. If it is wheel bearing grease replace the inner bearing seal. It may be necessary to replace the axle bearing or seal. To check the backing plate mounting, tap the plate clockwise and counter-clockwise. If movement occurs in either direction remove the backing plate and check for worn bolts or elongated bolt holes. Replace worn parts. A loose backing plate can usually be detected listening for a "clicking" sound when applying the brakes while the car is moved forward and backward.

Wheel Cylinder

Wheel cylinders should be inspected for leakage. Carefully inspect the boots. If they are torn, cut, heat cracked or show evidence of leakage the wheel cylinder should be replaced or overhauled. Don't gamble. If the cylinder doesn't look healthy, replace or rebuild.

Wheel cylinder inspection

INSPECTION

1. Wash all parts in clean denatured alcohol. If alcohol is not available, use specified brake fluid. Dry with compressed air.
2. Replace scored pistons. Always replace the rubber cups and dust boots.
3. Inspect the cylinder bore for score marks or rust. If either condition is present, the cylinder bore must be honed. However, the cylinder should not be honed more than 0.003 inch beyond its original.
4. Check the bleeder hole to be sure that it is open.

ASSEMBLY

1. Apply a coating of heavy-duty brake fluid to all internal parts.
2. Thread the bleeder screw into the cylinder and tighten securely.
3. Insert the return spring, cups, and pistons into their respective positions in the cylinder bore. Place a boot over each end of the cylinder.

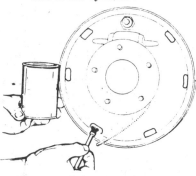

Lubricate the backing plate pads sparingly. The paste should be paper thin—no globs.

Adjusting Screw Assembly

Disassemble the adjusting screw assembly. Using an electric wire brush clean up the threads. Lubricate with brake fluid and reassem-

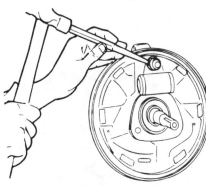

Testing backing plate mounting

Adjusting screw assembly—typical

457

Brake Service

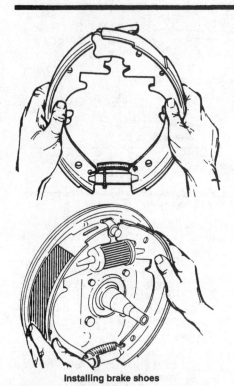

Installing brake shoes

ble the unit. Turn the threads all the way in by hand. If the threads bind at any point replace the unit.

Installing Brake Shoes

1. Preassemble the brake shoes, adjusting screw assembly and spring, plus the packing lever assembly (rear brakes only).
2. Spread the assembly; place it on the backing plate. Make sure wheel cylinder sockets are in the proper position.
3. Install the retainer pin and spring on both shoes.
4. Install the shoe guide.
5. Install the adjusting cable.
6. Install parking link and spring (rear only).
7. Install primary retracting spring.
8. Install secondary retracting spring.

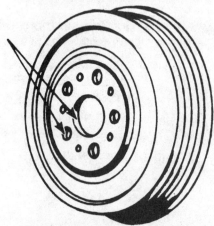

Web thickness: Full cast—3/16"–1/4"; composite—approximately 1/8"

Brake Drums

BRAKE DRUM TYPES

The FULL-CAST drum has a cast iron web (back) of 3/16 to 1/4 inch thickness (passenger car sizes) whereas the COMPOSITE drum has a steel web approximately 1/8 inch thick. These two types of drums, with few exceptions are not interchangeable.

BRAKE DRUM DEPTH

Place a straightedge across the drum diameter on the open side. The actual drum depth is the measurement at a right angle from the straightedge to that part of the web which mates against the hub mounting flange.

Brake drum depth measurement

ALUMINUM DRUMS

When replaced by other types, aluminum drums must be replaced in pairs.

METALLIC BRAKES

Drums designed for use with standard brake linings should not be used with metallic brakes.

BOLT CIRCLE

The circumference on which the centers of the wheel bolt holes are located around the drum-hub center is the bolt circle. It is shown as a double number (example: 6-5½). The first digit indicates the number of holes. The second number indicates the bolt circle diameter.

REMOVING TIGHT DRUMS

Difficulty removing a brake drum can be caused by shoes which are expanded beyond the drum's inner ridge, or shoes which have cut into, and ridged the drum. In either case back off the adjuster to obtain sufficient clearance for removal.

BRAKE DRUM INSPECTION

The condition of the brake drum surface is just as important as the surface to the brake lining. All drum surfaces should be clean, smooth, free from hard spots, heat checks, score marks and foreign matter imbedded in the drum surface. They should not be out of round, bellmouthed or barrel shaped. It is recommended that all drums be first checked with a drum micrometer to see if they are within oversize limits. If drum is within safe limits, even though the surface appears smooth, it should be turned not only to assure a true drum surface but also to remove any possible contamination in the surface from previous brake linings, road dusts, etc.

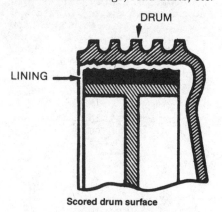

Scored drum surface

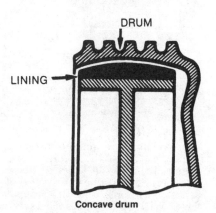

Concave drum

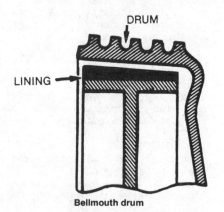

Bellmouth drum

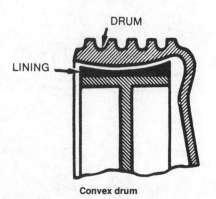

Convex drum

Brake Service

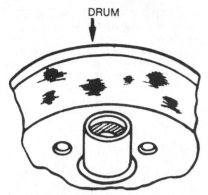
Hard or chill spots

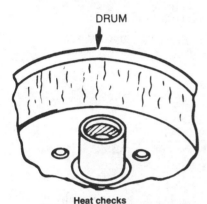
Heat checks

Oversize drum

Measuring inside drum diameter

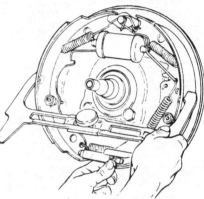
Measuring outside shoe diameter

Too much metal removed from a drum is unsafe and may result in:
1. Brake to fade due to the thin drum being unable to absorb the heat generated.
2. Poor and erratic brake due to distortion of drums.
3. Noise due to possible vibration caused by thin drums.
4. A cracked or broken drum on a severe or very hard brake application.

Brake drum run-out should not exceed .005". Drums turned to more than .060" oversize are unsafe and should be replaced with new drums, except for some heavy ribbed drums which have an .080" limit. It is recommended that the diameters of the left and right drums on any one axle be within .010" of each other. In order to avoid erratic brake action when replacing drums, it is always good to replace the drums on both wheels at the same time.

If the drums are true, smooth up any slight scores by polishing with fine emery cloth. If deep scores or grooves are present, which cannot be removed by this method, then the drum must be turned.

Sanding brake drums

Adjusting Drum Brakes
Preliminary Adjustment

1. Set a brake shoe adjustment gauge at .030 inch less than the brake drum diameter.
2. Center the gauge over the shoes at the greatest lining thickness and run out the adjuster until the new lining touches the gauge.
3. Install the drum.

NOTE: Tight clearance can aggravate normal seating problems. Most service technicians prefer to set the initial brake adjustment with a gauge and then brake the vehicle backward and forward allowing the shoes and drums to seek the correct running clearance. Complete seating normally occurs with 1000 driving miles.

Adjusting Drum Brakes
Routine Adjustment

1. Use a brake adjusting tool to expand the brake shoes against the drum. Raising the tool

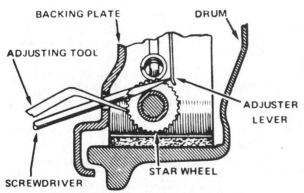
Shoe adjustment—all domestic cars except Chrysler and Chevette

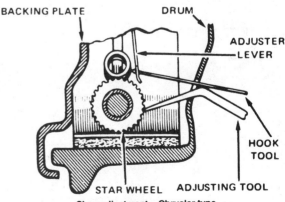
Shoe adjustment—Chrysler type

Brake Service

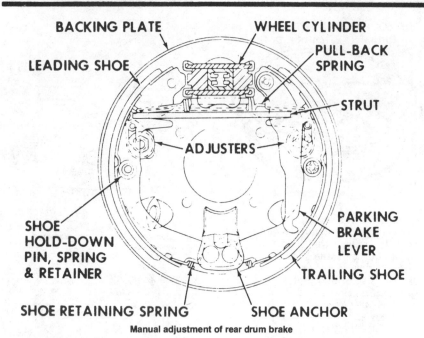

Manual adjustment of rear drum brake

handle turns the star wheel adjuster in the proper direction to expand the shoes. Turn the adjuster until a heavy drag is felt while turning the wheel.

2. Depress the brake pedal hard several times and recheck wheel drag. Continue to depress brake pedal and recheck drag until a true heavy drag is obtained.
3. Turn the star wheel adjuster in the opposite direction until the wheel turns freely.
4. Drive the car, braking forward and backward allowing the self-adjusters to obtain the best running clearance.

NOTE: Exceptions to the preceding are General Motor's "H" body cars (Astre, Monza, Skyhawk, Starfire, Vega). These brakes are automatically adjusted when the parking brake is applied. After brake service apply and release the parking brake until the brakes are correctly adjusted.

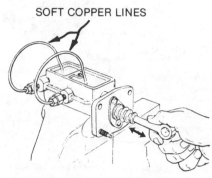

Bleeding the master cylinder (off-car)

BRAKE SYSTEM BLEEDING

If the master cylinder has been replaced it is more practical and safe to bleed most of the air out at the master cylinder. This can be done either on or off the car and prevents great masses of air from being passed through the system.

Manual Bleeding

1. Fill the master cylinder with new fluid of the correct type.
2. On cars with power brakes pump the brake pedal several times to remove all vacuum from the power unit.
3. Pump the brake pedal to pressurize the system and, while holding the pedal down, release the hydraulic pressure at the wheel cylinder bleeder valve. The pedal must be held depressed until the bleeder valve is closed to prevent air from entering the system.
4. Repeat until a steady, clear (no air bubbles) flow of fluid is seen at the wheel cylinder.

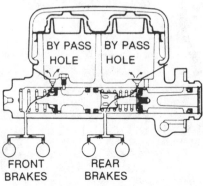

Full pedal return clears the bypass holes

CHILTON CAUTION: The bleeder valve at the wheel cylinder must be closed at the end of each stroke, and before the brake pedal is released, to insure that no air can enter the system. It is also important that the brake pedal be returned to the full up position so the piston in the master cylinder moves back enough to clear the bypass outlets.

Pressure Bleeding

Pressure bleeding equipment should be diaphragm type; placing a diaphragm between the pressurized air supply and the brake fluid. This prevents moisture and other contaminants from entering the hydraulic system.

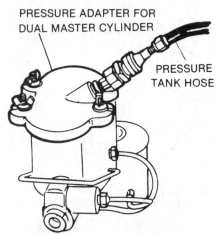

Pressure bleed connection at the master cylinder

NOTE: Front disc/rear drum equipped vehicles use a metering valve which closes off pressure to the front brakes under certain conditions. These systems contain manual release actuators which must be engaged to pressure bleed the front brakes.

1. Connect the tank hydraulic hose and adapter to the master cylinder.
2. Close hydraulic valve on the bleeder equipment.
3. Apply air pressure to the bleeder equipment.

CHILTON CAUTION: Follow equipment manufacturer's recommendations for correct air pressure.

4. Open the valve to bleed air out of the pressure hose to the master cylinder.

NOTE: Never bleed this system using the secondary piston stop-screw on the bottom of many master cylinders.

5. Open the hydraulic valve and bleed each wheel cylinder. Bleed rear brake system first when bleeding both front and rear systems.

460

Brake Service

FLUSHING HYDRAULIC BRAKE SYSTEMS

Hydraulic brake systems must be totally flushed if the fluid becomes contaminated with water, dirt or other corrosive chemicals. To flush, simply bleed the entire system until *all* fluid has been replaced with the correct type of new fluid.

DISC BRAKE SERVICE

Caliper disc brakes can be divided into three types: the four-piston, fixed-caliper type; the single-piston, floating-caliper type, and the single-piston sliding-caliper type.

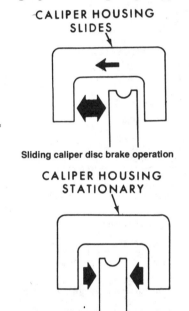

Sliding caliper disc brake operation

Floating caliper disc brake operation

In the four piston type (two in each side of the caliper) braking effect is achieved by hydraulically pushing both shoes against the disc sides.

With the single piston floating-caliper type the inboard shoe is pushed hydraulically into contact with the disc, while the reaction force thus generated is used to pull the outboard shoe into frictional contact (made possible by letting the caliper move slightly along the axle centerline).

In the sliding caliper (single piston) type, the caliper assembly slides along the machined surfaces of the anchor plate. A steel key located between the machined surfaces of the caliper and the machined surfaces of the anchor plate is held in place with either a retaining screw or two cotter pins. The caliper is held in place against the anchor plate with one or two support springs.

Inspection

Disc pads (lining and shoe assemblies) should be replaced in axle sets (both wheels) when the lining on any pad is worn to $1/16$ in. at any point. *If lining is allowed to wear past $1/16$ in. minimum thickness severe damage to disc may result.*

NOTE: State inspection specifications take precedence over these general recommendations.

Note that disc pads in floating caliper type brakes may wear at any angle, and measurement should be made at the narrow end of the taper. Tapered linings should be replaced if the taper exceeds $1/8$ in. from end to end (the difference between the thickest and thinnest points).

CAUTION: To prevent costly paint damage, remove some brake fluid (don't re-use) from the reservoir and install the reservoir cover before replacing the disc pads. When replacing the pads, the piston is depressed and fluid is forced back through the lines to squirt out of the fluid reservoir.

When the caliper is unbolted from the hub do not let it dangle by the brake hose; it can be rested on a suspension member or wired onto the frame. All disc brake systems are inherently self-adjusting and have no provision for manual adjustment.

Servicing the Caliper Assembly

1. Raise the vehicle on a hoist and remove the front wheels.
2. Working on one side at a time only, disconnect the hydraulic inlet line from the caliper and plug the end. Remove the caliper mounting bolts or pins, and shims, (if used) and slide the caliper off the disc.
3. Remove the disc pads from the caliper. If the old ones are to be reused, mark them so that they can be reinstalled in their original positions.
4. Open the caliper bleed screw and drain the fluid. Clean the outside of the caliper and mount it in a vise with padded jaws.

CAUTION: When cleaning any brake components, use only brake fluid or denatured (Isopropyl) alcohol. Never use a mineral-based solvent, such as gasoline or paint thinner, since it will swell and quickly deteriorate rubber parts.

5. Remove the bridge bolts, separate the caliper halves, and remove the two O-ring seals from the transfer holes.
6. Pry the lip on each piston dust boot from its groove and remove the piston assemblies and springs from the bores. If necessary, air pressure may be used to force the pistons out of the bores, using care to prevent them from popping out of control.
7. Remove the boots and seals from the pistons and clean the pistons in brake fluid. Blow out the caliper passages with an air hose.
8. Inspect the cylinder bores for scoring, pitting, or corrosion. Corrosion is a pitted or rough condition not to be confused with staining. Light rough spots may be removed by rotating crocus cloth, using finger pressure, in the bores. Do not polish with an in and out motion or use any other abrasive.
9. If the pistons are pitted, scored, or worn, they must be replaced. A corroded or deeply scored caliper should also be replaced.
10. Check the clearance of the pistons in the bores using a feeler gauge. Clearance should be 0.002-0.006 in. If there is excessive clearance the caliper must be replaced.
11. Replace all rubber parts and lubricate with brake fluid. Install the seals and boots in the grooves in each piston. The seal should be installed in the groove closest to the closed end of the piston with the seal lips facing the closed end. The lip on the boot should be facing the seal.
12. Lubricate the piston and bore with brake fluid. Position the piston return spring, large coil first, in the piston bore.
13. Install the piston in the bore, taking great care to avoid damaging the seal lip as it passes the edge of the cylinder bore.
14. Compress the lip on the dust boot into the groove in the caliper. Be sure the boot is fully seated in the groove, as poor sealing will allow contaminants to ruin the bore.
15. Position the O-rings in the cavities around the caliper transfer holes, and fit the caliper halves together. Install the bridge bolts (lubricated with brake fluid) and be sure to torque to specification.
16. Install the disc pads in the caliper and remount the caliper on the hub. Connect the brake line to the caliper and bleed the brakes. Replace the wheels. Recheck the brake fluid level, check the brake pedal travel, and road test the vehicle.

461

Brake Service

DRUM BRAKE PERFORMANCE DIAGNOSIS

The Condition	The Possible Cause	The Corrective Action
PEDAL GOES TO FLOOR	(a) Fluid low in reservoir. (b) Air in hydraulic brake system. (c) Improperly adjusted brake. (d) Leaking wheel cylinders. (e) Loose or broken brake lines. (f) Leaking or worn master cylinder. (g) Excessively worn brake lining.	(a) Fill and bleed master cylinder. (b) Fill and bleed hydraulic brake system. (c) Repair or replace self-adjuster as required. (d) Recondition or replace wheel cylinder and replace both brake shoes. (e) Tighten all brake fittings or replace brake line. (f) Recondition or replace master cylinder and bleed hydraulic system. (g) Reline and adjust brakes.
SPONGY BRAKE PEDAL	(a) Air in hydraulic system. (b) Improper brake fluid (low boiling point). (c) Excessively worn or cracked brake drums. (d) Broken pedal pivot bushing.	(a) Fill master cylinder and bleed hydraulic system. (b) Drain, flush and refill with brake fluid. (c) Replace all faulty brake drums. (d) Replace nylon pivot bushing.
BRAKES PULLING	(a) Contaminated lining. (b) Front end out of alignment. (c) Incorrect brake adjustment. (d) Unmatched brake lining. (e) Brake drums out of round. (f) Brake shoes distorted. (g) Restricted brake hose or line. (h) Broken rear spring.	(a) Replace contaminated brake lining. (b) Align front end. (c) Adjust brakes and check fluid. (d) Match primary, secondary with same type of lining on all wheels. (e) Grind or replace brake drums. (f) Replace faulty brake shoes. (g) Replace plugged hose or brake line. (h) Replace broken spring.
SQUEALING BRAKES	(a) Glazed brake lining. (b) Saturated brake lining. (c) Weak or broken brake shoe retaining spring. (d) Broken or weak brake shoe return spring. (e) Incorrect brake lining. (f) Distorted brake shoes. (g) Bent support plate. (h) Dust in brakes or scored brake drums.	(a) Cam grind or replace brake lining. (b) Replace saturated lining. (c) Replace retaining spring. (d) Replace return spring. (e) Install matched brake lining. (f) Replace brake shoes. (g) Replace support plate. (h) Blow out brake assembly with compressed air and grind brake drums.
CHIRPING BRAKES	(a) Out of round drum or eccentric axle flange pilot.	(a) Repair as necessary, and lubricate support plate contact areas (6 places).
DRAGGING BRAKES	(a) Incorrect wheel or parking brake adjustment. (b) Parking brakes engaged. (c) Weak or broken brake shoe return spring. (d) Brake pedal binding. (e) Master cylinder cup sticking. (f) Obstructed master cylinder relief port. (g) Saturated brake lining. (h) Bent or out of round brake drum.	(a) Adjust brake and check fluid. (b) Release parking brakes. (c) Replace brake shoe return spring. (d) Free up and lubricate brake pedal and linkage. (e) Recondition master cylinder. (f) Use compressed air and blow out relief port. (g) Replace brake lining. (h) Grind or replace faulty brake drum.
HARD PEDAL	(a) Brake booster inoperative. (b) Incorrect brake lining. (c) Restricted brake line or hose. (d) Frozen brake pedal linkage.	(a) Replace brake booster. (b) Install matched brake lining. (c) Clean out or replace brake line or hose. (d) Free up and lubricate brake linkage.
WHEEL LOCKS	(a) Contaminated brake lining. (b) Loose or torn brake lining. (c) Wheel cylinder cups sticking. (d) Incorrect wheel bearing adjustment.	(a) Reline both front or rear of all four brakes. (b) Replace brake lining. (c) Recondition or replace wheel cylinder. (d) Clean, pack and adjust wheel bearings.
BRAKES FADE (HIGH SPEED)	(a) Incorrect lining. (b) Overheated brake drums. (c) Incorrect brake fluid (low boiling temperature). (d) Saturated brake lining.	(a) Replace lining. (b) Inspect for dragging brakes. (c) Drain, flush, refill and bleed hydraulic brake system. (d) Reline both front or rear or all four brakes.
PEDAL PULSATES	(a) Bent or out of round brake drum.	(a) Grind or replace brake drums.
BRAKE CHATTER AND SHOE KNOCK	(a) Out of round brake drum. (b) Loose support plate. (c) Bent support plate. (d) Distorted brake shoes. (e) Machine grooves in contact face of brake drum. (Shoe Knock). (f) Contaminated brake lining.	(a) Grind or replace brake drums. (b) Tighten support plate bolts to proper specifications. (c) Replace support plate. (d) Replace brake shoes. (e) Grind or replace brake drum. (f) Replace either front or rear or all four linings
BRAKES DO NOT SELF ADJUST	(a) Adjuster screw frozen in thread. (b) Adjuster screw corroded at thrust washer. (c) Adjuster lever does not engage star wheel. (d) Adjuster installed on wrong wheel.	(a) Clean and free-up all thread areas. (b) Clean threads and replace thrust washer if necessary. (c) Repair, free up or replace adjusters as required. (d) Install correct adjuster parts.

Brake Service

DRUM BRAKE PERFORMANCE DIAGNOSIS

The Condition	The Possible Cause	The Corrective Action
NOISE—Groan—Brake noise emanating when slowly releasing brakes (creep-groan).	(a) Not detrimental to function of disc brakes—no corrective action required. (Indicate to operator this noise may be eliminated by slightly increasing or decreasing brake pedal efforts.)	
RATTLE—Brake noise or rattle emanating at low speeds on rough roads, (front wheels only).	(a) Shoe anti-rattle spring missing or not properly positioned. (b) Excessive clearance between shoe and caliper.	(a) Install new anti-rattle spring or position properly. (b) Install new shoe and lining assemblies.
SCRAPING	(a) Mounting bolts too long. (b) Loose wheel bearings.	(a) Install mounting bolts of correct length. (b) Readjust wheel bearings to correct specifications.
FRONT BRAKES HEAT UP DURING DRIVING AND FAIL TO RELEASE	(a) Operator riding brake pedal. (b) Stop light switch improperly adjusted. (c) Sticking pedal linkage. (d) Frozen or seized piston. (e) Residual pressure valve in master cylinder. (f) Power brake malfunction.	(a) Instruct owner how to drive with disc brakes. (b) Adjust stop light to allow full return of pedal. (c) Free up sticking pedal linkage. (d) Disassemble caliper and free up piston. (e) Remove valve. (f) Replace.
LEAKY WHEEL CYLINDER	(a) Damaged or worn caliper piston seal. (b) Scores or corrosion on surface of cylinder bore.	(a) Disassembly caliper and install new seat. (b) Disassemble caliper and hone cylinder bore. Install new seal.
GRABBING OR UNEVEN BRAKE ACTION	(a) Causes listed under "Pull". (b) Power brake malfunction.	(a) Corrections listed under 'Pull'. (b) Replace.
BRAKE PEDAL CAN BE DEPRESSED WITHOUT BRAKING EFFECT	(a) Air in hydraulic system or improper bleeding procedure. (b) Leak past primary cup in master cylinder. (c) Leak in system. (d) Rear brakes out of adjustment. (e) Bleeder screw open.	(a) Bleed system. (b) Recondition master cylinder. (c) Check for leak and repair as required. (d) Adjust rear brakes. (e) Close bleeder screw and bleed entire system.
EXCESSIVE PEDAL TRAVEL	(a) Air, leak, or insufficient fluid in system or caliper. (b) Warped or excessively tapered shoe and lining assembly. (c) Excessive disc runout. (d) Rear brake adjustment required. (e) Loose wheel bearing adjustment. (f) Damaged caliper piston seal. (g) Improper brake fluid (boil). (h) Power brake malfunction.	(a) Check system for leaks and bleed. (b) Install new shoe and linings. (c) Check disc for runout with dial indicator. Install new or refinished disc. (d) Check and adjust rear brakes. (e) Readjust wheel bearing to specified torque. (f) Install new piston seal. (g) Drain and install correct fluid. (h) Replace.
BRAKE ROUGHNESS OR CHATTER (PEDAL PUMPING)	(a) Excessive thickness variation of braking disc. (b) Excessive lateral runout of braking disc. (c) Rear brake drums out-of-round. (d) Excessive front bearing clearance.	(a) Check disc for thickness variation using a micrometer. (b) Check disc for lateral runout with dial indicator. Install new or refinished disc. (c) Reface rear drums and check for out-of-round. (d) Readjust wheel bearings to specified torque.
EXCESSIVE PEDAL EFFORT	(a) Brake fluid, oil or grease on linings. (b) Incorrect lining. (c) Frozen or seized pistons. (d) Power brake malfunction.	(a) Install new shoe linings as required. (b) Remove lining and install correct lining. (c) Disassemble caliper and free up pistons. (d) Replace.
PULL	(a) Brake fluid, oil or grease on linings. (b) Unmatched linings. (c) Distorted brake shoes. (d) Frozen or seized pistons. (e) Incorrect tire pressure. (f) Front end out of alignment. (g) Broken rear spring. (h) Rear brake pistons sticking. (i) Restricted hose or line. (j) Caliper not in proper alignment to braking disc.	(a) Install new shoe and linings. (b) Install correct lining. (c) Install new brake shoes. (d) Disassemble caliper and free up pistons. (e) Inflate tires to recommended pressures. (f) Align front end and check. (g) Install new rear spring. (h) Free up rear brake pistons. (i) Check hoses and lines and correct as necessary. (j) Remove caliper and reinstall. Check alignment.

Brake Service

WHEEL ALIGNMENT

Front wheel alignment is the position of the front wheels relative to each other and to the vehicle. It is determined, and must be maintained to provide safe, accurate steering, directional stability, and minimum tire wear. Many factors are involved in wheel alignment, and adjustments are provided to return those that might change due to normal wear to their original value. The factors which determine wheel alignment are dependent on one another; therefore, when one of the factors is adjusted, the others must be adjusted to compensate.

Descriptions of these factors and their effects on the car are provided below. Adjustment specifications for each model year are given in the charts.

Camber

Camber angle is the number of degrees that the centerline of the wheel is inclined from the vertical when viewed from the front. A small degree of positive camber reduces loading of the outer wheel bearing, and allows for easier steering.

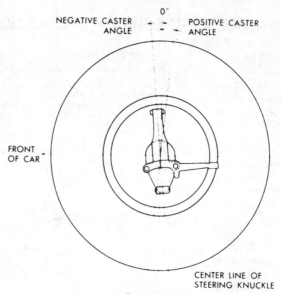

Caster angle

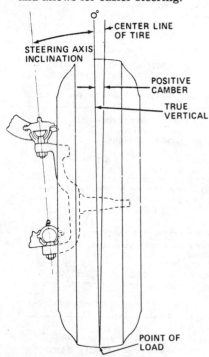

Camber and steering axis inclination angles

Steering Axis Inclination

Steering axis inclination is the number of degrees that a line drawn through the steering knuckle pivots is inclined to the vertical, when viewed from the front of the car. This, in combination with caster, is responsible for directional stability and self-centering of the steering. As the steering knuckle swings from lock to lock, the spindle generates an arc (see illustration), the high point being the straight ahead position of the wheel. Due to this arc, as the wheel turns, the front of the car is raised. The weight of the car acts against this lift, and attempts to return the spindle to the high point of the arc, resulting in self-centering when the steering wheel is released, and straight line stability.

Caster

Caster angle is the number of degrees that a line drawn through the steering knuckle pivots is inclined from the vertical, toward the front or rear of the car. A small degree of positive caster improves directional stability and decreases susceptibility to cross winds or road surface deviations.

Included Angle

Included angle is the sum of the camber angle and the steering axis inclination. This angle is determined by the design of the steering knuckle forging and must remain constant. Therefore, if a different camber angle is necessary to make the included angle on both sides identical, a bent spindle or steering knuckle is indicated. When indicated, the damaged suspension member must be replaced, to permit accurate front wheel alignment. Since steering knuckle damage is most commonly due to impact on the lower portion of the wheel (i.e., hitting curb), the side with the greater included angle (camber angle same on each side) will often be found to have a bent spindle.

Toe

Toe is the difference of the distance between the centers of the front and rear of the front wheels, measured at spindle height. It is most commonly measured in inches, but is occasionally referred to as an

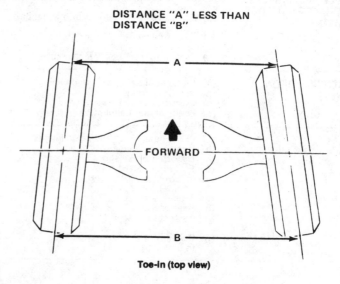

Toe-in (top view)

Brake Service

angle between the wheels. Toe-in indicates that the front of the tires are closer together than the rear; toe-out is the opposite condition. Toe-in compensates for the tendency of the wheels to deflect out while in motion. Due to this tendency, the wheels of a car with properly adjusted toe-in are traveling straight forward when the car itself is moving straight forward, resulting in directional stability and minimum tire wear. Front wheel drive and four wheel drive cars are normally set with toe-out, to compensate for the drive axles' tendency to pull the front wheels together.

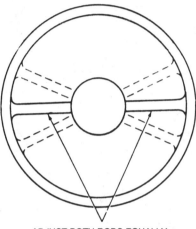

ADJUST BOTH RODS EQUALLY
TO MAINTAIN NORMAL SPOKE POSITION
Steering wheel spoke alignment

Steering wheel spoke misalignment is often an indication of incorrect front end alignment. Care should be exercised when aligning the front end to maintain steering wheel spoke position. When adjusting the tie rod ends, adjust each an equal amount (in the opposite direction) to increase or decrease toe. If, following toe adjustment, further adjustments are necessary to center the steering wheel spokes, adjust the tie rod ends an equal amount in the same direction.

Steering Radius

When a car is negotiating a turn, the outer wheel follows the path of a circle of a larger radius than the inner wheel. For this reason, the inner wheel must be steered to a somewhat larger angle than the outer wheel. This value (known as the Ackerman effect) is designed into the steering linkage; therefore, if alignment is adjusted properly, and the steering radius (or toe-out on turns) appears to be incorrect, it is indicated that the steering arms or the linkage is bent.

Tracking

Tracking is the relationship between the paths traveled by the front and rear wheels when the vehicle is traveling in a straight line. When a car is tracking correctly, the path of the rear wheels will duplicate, or evenly straddle the path of the front wheels. Observing the car from the rear as it is driven away in a straight line will often make incorrect tracking evident.

If incorrect tracking is indicated, check as follows: Drop a plumb line from each lower ball joint, and from a point at each end of the rear axle, and mark the points on the ground with chalk. Measure these points from front to rear and diagonally. If the diagonal measurements are different (a tolerance of +¼″ is acceptable), but the longitudinal measurements are the same, the frame is swayed (diamond shaped). If the diagonal and longitudinal measurements are both different, the rear axle is misaligned. If both diagonal and longitudinal measurements are different, but the car does not appear to be tracking incorrectly, a kneeback condition is indicated. Kneeback implies that one side of the front suspension is bent or pushed back. It is possible to align the front end to specifications, and, if kneeback exists, have very poor handling characteristics.

Ride Height Adjustment

This adjustment is required before adjusting front end alignment on cars with torsion bar front suspension.

NOTE: The car must be on a level floor with the gas tank full and the tires properly inflated. There should be no unusual loads in the car.

TIRE WEAR PATTERNS

CONDITION	RAPID WEAR AT SHOULDERS	RAPID WEAR AT CENTER	CRACKED TREADS	WEAR ON ONE SIDE	FEATHERED EDGE	BALD SPOTS	SCALLOPED WEAR
EFFECT							
CAUSE	UNDER-INFLATION OR LACK OF ROTATION	OVER-INFLATION OR LACK OF ROTATION	UNDER-INFLATION OR EXCESSIVE SPEED*	EXCESSIVE CAMBER	INCORRECT TOE	UNBALANCED WHEEL OR TIRE DEFECT*	LACK OF ROTATION OF TIRES OR WORN OR OUT-OF-ALIGNMENT SUSPENSION
CORRECTION	ADJUST PRESSURE TO SPECIFICATIONS WHEN TIRES ARE COOL; ROTATE TIRES			ADJUST CAMBER TO SPECIFICATIONS	ADJUST TOE-IN TO SPECIFICATIONS	DYNAMIC OR STATIC BALANCE WHEELS	ROTATE TIRES AND INSPECT SUSPENSION

*HAVE TIRE INSPECTED FOR FURTHER USE.

Domestic Cars

DISC BRAKE SPECIFICATIONS — AMC PASSENGER CARS AND JEEP

All measurements given in inches

YEAR	MODEL	CALIPER TYPE	CALIPER MOUNT BOLT TORQUE (ft/lbs)	MAXIMUM ALLOWABLE ROTOR SCORING	DISC PAD MINIMUM THICKNESS	ROTOR LATERAL RUNOUT	ROTOR MINIMUM ALLOWABLE THICKNESS	ROTOR THICKNESS VARIATION
'74	All Passenger Cars	Floating①	35	0.009	②	0.005③	0.940	0.0005
	All Jeep	Floating①	35	0.015	②	0.005③	1.230④	0.0005
'75	All Passenger Cars	Sliding⑤	15⑥	0.009	②	0.003③	1.130⑦	0.0005
	All Jeep	Floating①	35	0.015	②	0.005③	1.230④	0.0005
'76	All Passenger Cars	Sliding⑤	15⑥	0.009	②	0.003③	1.120	0.0005
	All Jeep	Floating①	35	0.015	②	0.005③	1.230④	0.0005
'77	All Passenger Cars	Sliding⑤	15⑥	0.009	②	0.003③	0.81⑧	0.0005
	CJ	Sliding⑤	15⑥	0.009	②	0.005③	1.120	0.001
	Cherokee, Wagoneer	Floating①	35	0.015	②	0.005③	1.215	0.001
'78	All Passenger Cars	Sliding⑤	15⑥	0.009	②	0.003③	0.81⑧	0.0005
	CJ	Sliding⑤	15⑥	0.009	②	0.005③	1.120	0.001
	Cherokee, Wagoneer	Floating①	35	0.015	②	0.005③	1.215	0.001
'79	All Passenger Cars	Sliding⑤	15⑥	0.009	②	0.003③	0.81	0.0005
	CJ	Sliding⑤	15⑥	0.009	②	0.005③	0.815	0.001
	Cherokee, Wagoneer	Floating①	35	0.015	②	0.005③	1.215	0.001
'80	All Passenger Cars	Sliding⑤	15⑥	0.009	②	0.003③	0.81	0.0005
	CJ	Sliding⑤	15⑥	0.009	②	0.005③	0.815	0.001
	Cherokee, Wagoneer	Floating①	35	0.015	②	0.005③	1.215	0.001

① Kelsey Hayes single piston
② Replace when lining is same thickness as metal shoe
③ Maximum rate of change should not exceed 0.001 in.
④ Discard at 1.215 in.
⑤ Bendix single piston
⑥ Caliper support key retaining screw
⑦ Discard at 1.120 in.
⑧ All models shown except: Matador—1.12 in.

BRAKE SPECIFICATIONS — AMC PASSENGER CARS AND JEEP

All measurements given in inches

YEAR	MODEL	MASTER CYLINDER BORE DIAMETER Disc	MASTER CYLINDER BORE DIAMETER Drum	WHEEL CYLINDER BORE DIAMETER Disc	WHEEL CYLINDER BORE DIAMETER Front Drum	WHEEL CYLINDER BORE DIAMETER Rear	BRAKE DISC OR DRUM DIAMETER Front Disc	BRAKE DISC OR DRUM DIAMETER Front Drum	BRAKE DISC OR DRUM DIAMETER Rear
74	Gremlin, Hornet—6 cyl.	①	1.00	2.275	1.125	0.875③	11.0	9.0	9.0
	Gremlin, Hornet—8 cyl.	①	1.00	2.275	1.187	0.875	11.0	10.0	10.0
	Matador	①	1.00	2.275	1.094②	0.9375	11.0	10.0	10.0
	Ambassador	①	1.00	2.750	—	0.9375	11.0	—	10.0
	AMX, Javelin	①	1.00	2.750	1.125②	0.875	11.0	9.0④	9.0④
	CJ	—	1.00	—	1.125	0.9375	—	11.0	11.0
	Cherokee, Wagoneer	1.125	1.00	2.9375	1.125	0.9375	12.0	11.0	11.0
75	Gremlin, Hornet—6 cyl.	1.00⑤	1.00	3.100	1.125	0.875③	10.75	9.0	9.0
	Gremlin, Hornet—8 cyl.	1.00⑤	1.00	3.100	1.187	0.875	10.75	10.0	10.0
	Matador	1.125⑤	—	3.100	—	0.875⑥	10.75	—	10.0
	Pacer	1.00⑤	1.00	3.100	1.0937	0.875	10.75	10.0	9.0
	CJ	—	1.00	—	1.125	0.9375	—	11.0	11.0
	Cherokee Wagoneer	1.125	1.00	2.9375	1.125	0.9375	12.0	11.0	11.0

Brake Specifications

BRAKE SPECIFICATIONS — AMC PASSENGER CARS AND JEEP

All measurements given in inches

YEAR	MODEL	MASTER CYLINDER BORE DIAMETER Disc	MASTER CYLINDER BORE DIAMETER Drum	WHEEL CYLINDER BORE DIAMETER Front Disc	WHEEL CYLINDER BORE DIAMETER Front Drum	WHEEL CYLINDER BORE DIAMETER Rear	BRAKE DISC OR DRUM DIAMETER Front Disc	BRAKE DISC OR DRUM DIAMETER Front Drum	BRAKE DISC OR DRUM DIAMETER Rear
76	Gremlin, Hornet—6 cyl.	1.00 ⑤	1.00	3.100	1.125	0.875 ③	10.75	9.0	9.0
	Gremlin, Hornet—8 cyl.	1.00 ⑤	1.00	3.100	1.187	0.875	10.75	10.0	10.0
	Matador	1.125 ⑤	—	3.100	—	0.875 ⑥	10.75	—	10.0
	Pacer	1.00 ⑤	1.00	3.100	1.0937	0.875	10.75	10.0	9.0
	CJ	—	1.00	—	1.125	0.9375	—	11.0	11.0
	Cherokee, Wagoneer	1.125	1.00	2.9375	1.125	0.9375	12.0	11.0	11.0
77	Gremlin, Hornet—6 cyl.	1.00	—	2.60	—	0.812	10.82	—	10.0
	Gremlin, Hornet—8 cyl.	1.00	—	2.60	—	0.812	10.82	—	10.0
	Matador	1.125	—	3.10	—	0.875 ⑥	10.82	—	10.0
	Pacer	1.00	—	2.60	—	0.812	10.82	—	10.0
	CJ	1.00	1.00	2.9375	1.125	0.9375	12.0	11.0	11.0
	Cherokee, Wagoneer	1.125	—	2.9375	—	0.9375	12.0	—	11.0
78	All except Matador	0.960	—	2.60	—	0.812 ⑦	10.80 ⑧	—	10.0
	Matador	1.125	—	3.10	—	0.875 ⑥	10.82	—	10.0
	CJ	1.00	—	3.10	—	0.875	11.70	—	11.0
	Cherokee, Wagoneer	1.125	—	2.9375	—	0.9375	12.0	—	11.0
79	Spirit, Concord—4 cyl.	0.94	—	2.640	—	0.94	10.27	—	9.0 ⑩
	Concord, Pacer, Spirit—6 & 8 cyl.	0.94	—	2.640	—	0.81	10.80	—	10.0
	AMX	0.94	—	2.640	—	0.94	10.80	—	10.0
	CJ	1.00	—	2.6	—	0.875	11.70	—	10.0
	Cherokee, Wagoneer	1.125	—	2.937	—	0.937	12.0	—	11.0
80	Spirit, Concord—4 cyl.	0.94	—	2.640	—	0.94	10.27	—	9.0 ⑩
	Concord, Pacer, Spirit—6 & 8 cyl.	0.94	—	2.640	—	0.81	10.80	—	10.0
	AMX	0.94	—	2.640	—	0.94	10.80	—	10.0
	CJ	1.00	—	2.6	—	0.875	11.70	—	10.0
	Cherokee, Wagoneer	1.125	—	2.937	—	0.937	12.0	—	11.0

— Not Applicable
① Power brakes—1.125; non-power brakes—1.062
② Six cylinder shown; eight cylinder—1.187
③ Hornet shown; Gremlin—0.812
④ Six cylinder shown; eight cylinder—10.0
⑤ Power assisted shown; non-power—1.0625
⑥ Sedan and Coupe shown; station wagon—0.9375
⑦ All models shown except: 4cyl. Gremlin—0.94
⑧ All models shown except: 8 cyl. AMX—10.30, 4 cyl. Gremlin—10.27
⑨ All models shown except: 4 cyl. Gremlin—9.0
⑩ Spirit shown; Concord—10.0

BRAKE SPECIFICATIONS — BUICK EXCEPT SKYHAWK AND 1980 SKYLARK

All readings in inches

YEAR	MODEL	MASTER CYLINDER BORE DIAMETER	CALIPER OR WHEEL CYLINDER Front	CALIPER OR WHEEL CYLINDER Rear	BRAKE DRUM/ROTOR DIAMETER Front	BRAKE DRUM/ROTOR DIAMETER Rear	ROTOR RUNOUT	ROTOR ALLOWABLE MINIMUM MACHINED THICKNESS	ROTOR THICKNESS VARIATION MAXIMUM
'74	Apollo w/drum	1	1⅛	⅞	9.5	9.5	—	—	—
	w/disc	1⅛	2.9375	⅞	—	9.5	.004	.980	.0005
	Century, Regal	1⅛ ①	2.9375	⅞	—	9.5	.004	.980	.0005

467

Domestic Cars

BRAKE SPECIFICATIONS — BUICK EXCEPT SKYHAWK AND 1980 SKYLARK

All readings in inches

YEAR	MODEL	MASTER CYLINDER BORE DIAMETER	CALIPER OR WHEEL CYLINDER Front	CALIPER OR WHEEL CYLINDER Rear	BRAKE DRUM/ROTOR DIAMETER Front	BRAKE DRUM/ROTOR DIAMETER Rear	ROTOR RUNOUT	ROTOR ALLOWABLE MINIMUM MACHINED THICKNESS	ROTOR THICKNESS VARIATION MAXIMUM
'74	LeSabre, Riviera, Electra	1⅛	2.9375	15/16	—	11.0	.005	1.230	.0005
	Estate Wagon	1⅛	2.9375	1	—	12.0	.005	1.230	.0005
'75	Apollo	1⅛	2.9375	⅞	—	9.5	.004	.980	.0005
	Century, Regal	1⅛①	2.9375	⅞	—	9.5	.004	.980	.0005
	Century Sta. Wgn.	1⅛①	2.9375	15/16	—	11.0	.004	.980	.0005
	LeSabre, Riviera, Electra	1⅛	2.9375	15/16	—	11.0	.005	1.230	.0005
	Estate Wagon	1⅛	2.9375	1	—	12.0	.005	1.230	.0005
'76	Skylark	1⅛	2.9375	⅞	—	9.5	.004	.980	.0005
	Century, Regal	1⅛②	2.9375	15/16	—	11.0	.004	.980	.0005
	LeSabre Riviera Electra	1⅛	2.9375	1	—	11.0	.005	1.230	.0005
	Estate Wagon	1⅛	2.9375	1	—	12.0	.005	1.230	.0005
'77	Skylark	1⅛	2.9375	15/16	—	9.5	.004	.980	.0005
	Century, Regal	1⅛②	2.9375	15/16	—	11.0	.004	.980	.0005
	LeSabre,	1⅛	2.9375	⅞	—	9.5	.004	.980	.0005
	Electra, Riviera Estate Wagon	1⅛	2.9375	15/16	—	11.0	.004	.980	.0005
'78	Skylark	1⅛	2.9375	15/16	—	9.5	.004	.980	.0005
	Century, Regal	.94③	2.43	¾	—	9.45	.004	.980	.0005
	LeSabre	1⅛	2.9375	⅞	—	9.5	.004	.980	.0005
	Electra, Riviera, Estate Wagon	1⅛	2.9375	15/16	—	11.0	.004	.980	.0005
'79	Skylark	1⅛①	2.9375	15/16	—	9.5	.005	.965	.0005
	Century, Regal	.94③	2.43	¾	—	9.45	.004	.965	.0005
	LeSabre, Riviera	1⅛	2.9375	⅞	—	9.5	.005	.965	.0005
	Electra, Estate Wagon	1⅛	2.9375	15/16	—	11.0	.005	.965	.0005
	Riviera	.945	2.50	¾	—	9.45	.005	.965	.0005
'80	Century, Regal	.94③	2.43	¾	—	9.45	.004	.965	.0005
	LeSabre, Riviera	1⅛	2.9375	⅞	—	9.5	.005	.965	.0005
	Electra, Estate Wagon	1⅛	2.9375	15/16	—	11.0	.005	.965	.0005
	Riviera	.945	2.50	¾	—	9.45	.005	.965	.0005

— Not applicable
① Manual brake—1.0
② Manual brake—15/16
③ Manual brake—0.87

Brake Specifications

BRAKE SPECIFICATIONS — CADILLAC

YEAR	MODEL	MASTER CYLINDER BORE DIAMETER	CALIPER OR WHEEL CYLINDER Front	CALIPER OR WHEEL CYLINDER Rear	BRAKE DRUM/ ROTOR* DIAMETER Front	BRAKE DRUM/ ROTOR* DIAMETER Rear	DRUM MAXIMUM MACHINED DIAMETER	ROTOR RUNOUT	ROTOR ALLOWABLE MINIMUM MACHINED THICKNESS	ROTOR THICKNESS VARIATION MAXIMUM
74	All Exc. Eldorado	1.125	2¹⁵/₁₆	¹⁵/₁₆ ①	11.74	12.0	12.060	.005	1.220	.0005
	Eldorado	1.125	2¹⁵/₁₆	¹⁵/₁₆	11.0	11.0	11.060	.008	1.190	.0005
75	All Exc. Eldorado	1.125	2¹⁵/₁₆	¹⁵/₁₆ ①	11.74	11.0	12.060	.005	1.220	.0005
	Eldorado	1.125	2¹⁵/₁₆	¹⁵/₁₆	11.0	11.0	11.060	.008	1.190	.0005
76	Seville	1.125	2¹⁵/₁₆	¹⁵/₁₆	11.0	11.0	11.060	.005	.980	.0005
	All Exc. Eldorado and Seville	1.125	2¹⁵/₁₆	¹⁵/₁₆ ①	11.74	12.0	12.060	.005	1.220	.0005
	Eldorado	1.125	2¹⁵/₁₆	2½	11.0	11.0	—	.008	1.190	.0005
77	Seville, Brougham	1.125	2¹⁵/₁₆	2½	11.74	11.14	—	F-.005 R-.0029	F-.980 R-.911	F-.0005 R-.0005
	DeVille	1.125	2¹⁵/₁₆	1	11.74	11.0	11.060	.005	.980	.0005
	Eldorado	1.125	2¹⁵/₁₆	2½	11.0	11.0	—	.008	1.190	.0005
	Limousine and Comm. Chassis	1.125	2¹⁵/₁₆	¹⁵/₁₆	11.74	12.0	12.060	.005	1.220	.0005
78	Seville, Brougham	1.125	2¹⁵/₁₆	2½	11.74	11.14	—	F-.005 R-.0029	F-.980 R-.911	F-.0005 R-.0005
	DeVille	1.125	2¹⁵/₁₆	1	11.74	11.0	11.060	.005	.980	.0005
	Eldorado	1.125	2¹⁵/₁₆	2½	11.0	11.0	—	.008	1.190	.0005
	Limousine and Comm. Chassis	1.125	2¹⁵/₁₆	¹⁵/₁₆	11.74	12.0	12.060	.005	1.220	.0005
79	Seville	1.125	2¹⁵/₁₆	2½	11.7	11.0	—	.005	F-.980 R-.911	.0005
	DeVille	1.125	2¹⁵/₁₆	2½ ①	11.7	11.0	11.060	.005	.980	.0005
	Eldorado	1.0	2½	2⅛	10.4	9.5	—	.005	1.036	.0005
	Comm. Chassis	1.125	2¹⁵/₁₆	2½	11.7	12.0	1.285	.005	1.220	.0005
80	Seville	1.0	2½	2⅛	10.4	9.5	—	.005	1.036	.0005
	DeVille	1.125	2¹⁵/₁₆	2½	11.7	11.0	11.060	.005	.980	.0005
	Eldorado	1.0	2½	2⅛	10.4	9.5	—	.005	1.036	.0005
	Comm. Chassis	1.125	2¹⁵/₁₆	2½	11.7	12.0	12.060	.005	1.220	.0005

* Rotor Working Outer Diameter
F = Front
R = Rear
— Not Applicable
① Comm. chassis—1"

DISC BRAKE SPECIFICATIONS — CHEVROLET EXCEPT CHEVETTE, CITATION, MONZA, VEGA, CORVETTE

YEAR	CALIPER TYPE	MOUNTING BOLT TORQUE	DISC PAD ORIGINAL THICKNESS	MFR.'S RECOMMENDED DISC PAD MINIMUM THICKNESS	ROTOR RUNOUT	ROTOR ALLOWABLE MINIMUM MACHINED THICKNESS	ROTOR THICKNESS VARIATION MAXIMUM
74-76	Sliding	35	—	¹/₃₂ ①	.005	.980	.0005
77	Sliding	35	—	.020 ①	.005	.980	.0005
78-80	Sliding	35	—	.020 ①	.004	.980	.0005

— Not applicable
① Above rivet head or thickness of remaining lining on bonded pads

Domestic Cars

DISC BRAKE SPECIFICATIONS — CORVETTE

YEAR	CALIPER TYPE	MOUNTING BOLT TORQUE	HOUSING BOLT TORQUE	MFR. RECOMMENDED DISC PAD MINIMUM THICKNESS	ROTOR RUNOUT	ROTOR ALLOWABLE MINIMUM MACHINED THICKNESS	ROTOR THICKNESS VARIATION MAXIMUM
74-80	F Fixed	70	130	.020①	.005	1.230	.0005
	R Fixed	70	60	.020①	.005	1.230	.0005

① .020 over rivet thickness.
F = Front brakes
R = Rear brakes

BRAKE SPECIFICATIONS — CHEVROLET, CAMARO, CHEVELLE, MONTE CARLO

MODEL	YEAR	MASTER CYLINDER BORE DIAMETER Manual	MASTER CYLINDER BORE DIAMETER Power	CALIPER OR WHEEL CYLINDER Front	CALIPER OR WHEEL CYLINDER Rear	BRAKE DRUM/ROTOR DIAMETER Front	BRAKE DRUM/ROTOR DIAMETER Rear
Camaro	74-77	1.0	1.125	2.9375	⅞	—	9.5
	78	1.0	1.125	2.9375	15/16	—	9.5
	79-80	1.0	1.125	2.9375	15/16	—	9.5
Chevelle exc Station Wagon	74-75	1.0	1.125	2.9375	⅞	—	9.5
Chevelle Station Wagon	74-75	1.0	1.125	2.9375	15/16	—	11.0
Chevelle	76-77	.9375	1.125	2.9375	15/16①	—	11.0
	78	.87	.94	2.500	¾	—	9.5
	79-80	.87②	.94③	2.500	¾	—	9.5
Chevrolet Full Size exc. S.W.	74-76	1.125	1.125	2.9375	15/16	—	11.0
Caprice, Impala	77		1.125	2.9375	⅞	—	9.5
	78		1.125	2.9375	⅞	—	9.5
	79-80		1.125	2.9375	⅞	—	9.5
Chevrolet Full Size Sta. Wgn.	74-76	1.125	1.125	2.9375	1.0	—	12.0
Caprice, Impala Sta. Wgn.	77		1.125	2.9375	15/16	—	11.0
	78		1.125	2.9375	15/16	—	11.0
	79-80		1.125	2.9375	15/16	—	11.0
Monte Carlo	74-75		1.125	2.9375	⅞	—	9.5
	76-77		1.125	2.9375	15/16	—	11
	78	.87	.94	2.500	¾	—	9.5
	79-80	.87②	.94③	2.500	¾	—	9.5

— Not applicable
① Manual brakes use 1 inch rear wheel cylinder
② V6 engine without air conditioning; manual or power brakes
③ V8 and V6 engines with air conditioning

BRAKE SPECIFICATIONS — CHEVROLET NOVA AND CORVETTE

MODEL	YEAR	MASTER CYLINDER BORE DIAMETER Manual	MASTER CYLINDER BORE DIAMETER Power	CALIPER OR WHEEL CYLINDER Front	CALIPER OR WHEEL CYLINDER Rear	BRAKE DRUM/ROTOR DIAMETER Front	BRAKE DRUM/ROTOR DIAMETER Rear
Nova w/drum brks.	74	1.0	1.0	1⅛	⅞	9.5	9.5

Brake Specifications

BRAKE SPECIFICATIONS — CHEVROLET NOVA AND CORVETTE

MODEL	YEAR	MASTER CYLINDER BORE DIAMETER		CALIPER OR WHEEL CYLINDER		BRAKE DRUM/ROTOR DIAMETER	
		Manual	Power	Front	Rear	Front	Rear
Nova w/disc brks.	74		1.125	2.9375	7/8	—	9.5
	75	1.00	1.125	2.9375	7/8	—	9.5
	76-77	1.00	1.125	2.9375	15/16	—	9.5
	78	1.00	1.125	2.9375	15/16	—	9.5
	79-80	1.00	1.125	2.9375	15/16	—	9.5
Corvette	74-77	1.00	1.125	1.875	1.375	11.75	11.75
	78-80	1.125	1.125	1.875	1.375	11.75	11.75

BRAKE SPECIFICATIONS — CHEVETTE

YEAR	MASTER CYLINDER BORE DIAMETER	CALIPER OR WHEEL CYLINDER		BRAKE DRUM/ROTOR DIAMETER		ROTOR RUNOUT	ROTOR THICKNESS VARIATION MAXIMUM
		Front	Rear	Front	Rear		
76	.750	1.875	.750	9.68	7.88	.005	.0005
77	.750	1.875	.750	9.68	7.88	.005	.0005
78	.750	1.875	.750	9.68	7.88	.005	.0005
79	.750	1.875	.750	9.68	7.88	.005	.0005
80	.750	1.875	.750	9.68	7.88	.005	.0005

BRAKE SPECIFICATIONS — CHRYSLER CORP. EXCEPT HORIZON AND OMNI

YEAR	MODEL	MASTER CYLINDER BORE DIAMETER	WHEEL CYLINDER Front	WHEEL CYLINDER Rear	BRAKE DRUM/ROTOR DIAMETER Front	BRAKE DRUM/ROTOR DIAMETER Rear	DISC SPECIFICATIONS Resurfacing Min. Thickness	DISC SPECIFICATIONS Parallel Variation	DISC SPECIFICATIONS Runout Maximum
74-75	VALIANT AND DART W/DRUM BRKS.	1 1/32"	1 1/8"	13/16"	10"	9"	—	—	—
	VALIANT AND DART W/DISC BRKS.	15/16"①	2.602"-2.604"	15/16"	10.98"	10"	.940"	.0005"	.004"
	FURY, CORONET, CORDOBA AND CHARGER W/STD. BRKS.	1 1/32"②	2.755"④	15/16"	10.98"	10"	.940"	.0005"	.004"
	FURY, CORONET, CORDOBA, AND CHARGER W/H.D. BRKS.	1 1/32"②	2.755"④	15/16"	10.98"	11"	.940"	.0005"	.004"
	MONACO, GRAN FURY, CHRYSLER	1 1/32"	3.102"-3.104"	15/16"	11.75"	11"	1.180"	.0005"	.004"
	IMPERIAL	1 1/16"	3.102"-3.104"	2.602"-2.604"	11.75"	11.63"	1.180" F .940" R	.0005"	.004"
76	DART & VALIANT W/DRUM BRKS.	1.031"	15/16"	15/16"③	10"	10"⑤	—	—	—
	DART, VALIANT, ASPEN, VOLARE W/MANUAL DISC BRKS.	1.031"	2.754"-2.756"	15/16"	10.98"	10"⑧	.940"	.0005"	.004"
	DART, VALIANT, ASPEN, VOLARE, W/POWER DISC BRKS.	.937"⑥	2.754"-2.756"	15/16"	10.98"	10"⑧	.940"	.0005"	.004"
	CORDOBA, CHARGER, CORONET, FURY UP TO 1-1-76	1.031"⑦	2.754"-2.756"	15/16"	10.98"	11"	.940"	.0005"	.004"
	CORDOBA, CHARGER, CORONET, FURY AFTER 1-1-76	1.031"	2.754"-2.756"	15/16"	11.75"	11"	.940"	.0005"	.004"
	GRAN FURY, MONACO AND CHRYSLER	1.031"	3.102"-3.104"	15/16"	11.75"	11"	1.180"	.0005"	.004"

Domestic Cars

BRAKE SPECIFICATIONS — CHRYSLER CORP. EXCEPT HORIZON AND OMNI

YEAR	MODEL	MASTER CYLINDER BORE DIAMETER	WHEEL CYLINDER Front	WHEEL CYLINDER Rear	BRAKE DRUM/ROTOR DIAMETER Front	BRAKE DRUM/ROTOR DIAMETER Rear	DISC SPECIFICATIONS Resurfacing Min. Thickness	DISC SPECIFICATIONS Parallel Variation	Runout Maximum
77	ASPEN, VOLARE DIPLOMAT, LEBARON	1.031"	2.754"-2.756"	15/16"	10.98"	10"⑧	.970"	.0005"	.004"
	FURY, MONACO, CORDOBA, CHARGER SE	1.031"	2.754"-2.756"	15/16"	11.75"	11"	.970"	.0005"	.004"
	ROYAL MONACO, GRAN FURY CHRYSLER	1.031"	3.102"-3.104"	15/16"	11.75"	11"	1.210"	.0005"	.004"
78	ASPEN, VOLARE, DIPLOMAT, LEBARON	1.031"	2.754"-2.755	15/16"	10.98"	10"⑨	.970"	.0005"	.004"
	FURY, MONACO, CHARGER, CORDOBA, MAGNUM	1.031"	2.754"-2.755	15/16"	11.75"	10"⑨	.970"	.0005"	.004"
	CHRYSLER	1.031"	3.102"-3.104"	15/16"	11.75"	11"	1.210"	.0005"	.004"
79-80	ASPEN, VOLARE, DIPLOMAT, LEBARON	1.031"	2.75"	.9375"	10.82"	10"	.970"	.0005"	.004"
	MAXNUM, CORDOBA ST. REGIS, CHRYSLER, LA-SCALA	1.031"	2.75"	.9375"	11.58"	10"⑩	.970"	.0005"	.004"

— Not applicable
① W/manual disc brakes 1 1/32"
② W/manual disc brakes 1"
③ 13/16" for 6 cyl. Dart and Valiant vehicles built before Jan. 1, 1976
④ Maximun after honing
⑤ 9" dia. rear drum on 6 cyl. Dart and Valiant before Jan. 1, 1976
⑥ Aspen and Volare 1.031"
⑦ W/manual disc brakes on Coronet and Fury 1.00"
⑧ Aspen and Volare wagons have 11" rear brakes
⑨ 11" drum on police and/or taxi Aspen, Volare, Monaco and Fury. Also on Cordoba, Magnum, Charger, Monaco and Fury equipped w/9¼" rear axle (12 bolt rear cover)

BRAKE SPECIFICATIONS — CHRYSLER CORP. HORIZON AND OMNI

YEAR	MODEL	MASTER CYLINDER BORE DIAMETER	CALIPER OR WHEEL CYLINDER BORE DIAMETER Front	CALIPER OR WHEEL CYLINDER BORE DIAMETER Rear	BRAKE DRUM/ROTOR DIAMETER Front	BRAKE DRUM/ROTOR DIAMETER Rear	DISC SPECIFICATIONS Resurfacing Min. Thickness	DISC SPECIFICATIONS Parallel Variation	Runout Maximum
ALL	ALL	.877"	1.89"	.628"	9"	7.87"	.461①	.0005"	.004"

NOTE: Do not hone master cylinder bore
① Discard thickness is .431"

FRONT DISC BRAKE SPECIFICATIONS — FORD MOTOR CO. SUB-COMPACT

YEAR	MODEL	CALIPER TYPE	MOUNTING BOLT TORQUE (ft/lbs)	DISC PAD ORIGINAL THICKNESS	MFR.'S RECOMMENDED DISC PAD MINIMUM THICKNESS	ROTOR RUNOUT	ROTOR ALLOWABLE MINIMUM MACHINED THICKNESS	ROTOR THICKNESS VARIATION MAXIMUM
74-76	Mustang II, Pinto, Bobcat	Ford Sliding	②	—	1/32①③	.003	.810	.0005
77-80	Mustang II, Pinto, Bobcat	Ford Sliding	②	.410⑤	1/8④	.003	.810	.0005
79-80	Mustang/Capri	Ford Sliding	35⑥	.410	1/8④	.003	.810	.0005

① From rivets or shoe to lining surface
② Anchor plate mounting bolts:
  Upper—90-120
  Lower—55-75
  Caliper key retainer screw—14
③ 1976—1/8
④ From shoe surface
⑤ Outer pad shown, inner .403
⑥ Caliper locating pins

Brake Specifications

FRONT DISC BRAKE SPECIFICATIONS — FORD MOTOR CO. FULL-SIZE

YEAR	MODEL	CALIPER TYPE	MOUNTING BOLT TORQUE	MFR.'S RECOMMENDED DISC PAD MINIMUM THICKNESS	ROTOR RUNOUT	ROTOR ALLOWABLE MINIMUM MACHINED THICKNESS	ROTOR THICKNESS VARIATION MAXIMUM
'74–'77	Ford & Mercury Full-Size Cars	Ford Sliding	90-120	1/8 ①④	.003	1.120	.0005 ②
'74–'77	Lincoln Continental, Mark IV, V & Ford Thunderbird	Ford Sliding	90-120	1/8 ①④	.003	1.120	.00025
'78	Ford & Mercury Full Size Cars	Ford Sliding	90-120	1/8 ①	.003	1.120	.0005 ⑤
'78–'79	Lincoln Continental and Mark V	Ford Sliding	90-120	1/8 ①	.003	1.120	.0005
'79 ⑥	Ford and Mercury Full Size Cars	Ford Pin-slider	40-60 ③	1/8 ①	.003	1.120	.0005 ⑤
'80	Lincoln Continental and Mark VI	Ford Sliding	90-120	1/8 ①	.003	1.120	.0005

① From rivets or shoe to lining surface
② Mercury 1975-1977, .0004"
③ Combination locating and mounting pin
④ 1974-1975: 1/3"
⑤ Mercury .0004"
⑥ Note: This brake system is new for 1979 and features an aluminum caliper assembly. It is extremely important to check the torque required to rotate the hub and rotor assembly after mounting the caliper assembly to the spindle. The torque required to rotate the hub and disc must not exceed 65 inch/lbs.

DISC BRAKE SPECIFICATIONS — FORD MOTOR CO. COMPACT AND INTERMEDIATE

YEAR	MODEL	CALIPER TYPE	MOUNTING BOLT TORQUE	MFR.'S RECOMMENDED DISC PAD MINIMUM THICKNESS	ROTOR RUNOUT	ROTOR ALLOWABLE MINIMUM MACHINED THICKNESS	ROTOR THICKNESS VARIATION MAXIMUM
'74–'78	Comet, Maverick Granada, Monarch, Versailles	Ford Sliding	Upper: 90-120 Lower: 55-75	1/8 ①②	.003	F.810 R.895	.0005
'74–'78	Cougar, LTD II, Montego Thunderbird, Torino, Elite	Ford Sliding	90-120	1/8 ①②	.003	1.120	.0005
'78	Fairmont, Zephyr	Ford Sliding	30-40 ③	1/8 ①	.003	.810	.0005
'79	Granada, Monarch Versailles	Ford Sliding	Upper: 90-120 Lower: 55-75	1/8 ①	.003	F.810 R.895	.0005
	Cougar, LTD II Thunderbird	Ford Sliding	90-120	1/8 ①	.003	.810	.0005
	Fairmont, Zephyr	Ford Sliding	30-40 ③	1/8 ①	.003	.810	.0005
'80	Granada, Monarch Versailles	Ford Sliding	Upper: 90-120 Lower: 55-75	1/8 ①	.003	F.810 R.895	.0005
	Cougar, LTD II Thunderbird						
	Fairmont, Zephyr	Ford Sliding	30-40 ③	1/8 ①	.003	.810	.0005

① From rivets or shoe to lining surface
② 1974 & 1975 1/32"
③ Caliper locating pin torque
F = front
R = rear

Domestic Cars

REAR DISC BRAKE SPECIFICATIONS — FORD MOTOR CO. FULL-SIZE

YEAR	MODEL	CALIPER TYPE	MOUNTING BOLT TORQUE	MFR.'S RECOMMENDED DISC PAD MINIMUM THICKNESS	ROTOR RUNOUT	ROTOR ALLOWABLE MINIMUM MACHINED THICKNESS	ROTOR THICKNESS VARIATION MAXIMUM
'75-77	Mercury Full-Size Cars	Ford Sliding	90-120	1/32 ①②	.004	.895	.0005
'75-77	Lincoln Continental, Mark IV, V & Ford Thunderbird	Ford Sliding	90-120	1/8 ①②	.004	.895	.0004
'78-80	Ford & Mercury Full Size Cars	Ford Sliding	75-95 ③	1/8	.004	.895	.0005
'78-79	Lincoln & Mark V	Ford Sliding	75-95 ③	1/8	.004	.895	.0004
'80	Lincoln & Mark VI						

① From top of rivets
② 1975: 1/32
③ Caliper end retainer bolt torque

BRAKE SPECIFICATIONS — GM ASTRE, MONZA, SKYHAWK, STARFIRE, SUNBIRD, VEGA

All readings in inches

YEAR	MODEL	MASTER CYLINDER BORE DIAMETER	CALIPER OR WHEEL CYLINDER Front	CALIPER OR WHEEL CYLINDER Rear	BRAKE DRUM/ROTOR DIAMETER Front	BRAKE DRUM/ROTOR DIAMETER Rear
74	Vega	.750	1.875	3/4	10	9.0
75	Astre, Monza, Vega, Starfire, Skyhawk	.750	1.875	3/4	10	9.0
76	Sunbird, Monza, Starfire	.875	2.50	11/16	9.8	9.5
	Skyhawk	.875	1.875	13/16	9.8	9.5
	Astre	.750	1.875	11/16	9.8	9.0
	Vega	.750	1.875	11/16	9.8	9.5
77	Vega, Astre	.750	1.875	11/16	9.8	9.5
	Sunbird, Starfire, Monza, Skyhawk	.875	2.50	11/16	9.8	9.5
78	Sunbird, Starfire, Monza, Skyhawk	.875	2.50	11/16	9.8	9.5
79	Sunbird, Starfire, Monza, Skyhawk	.875	2.50	11/16	9.8	9.5
80	Sunbird, Starfire, Monza, Skyhawk	.875	2.50	11/16	9.8	9.5

Brake Specifications

DISC BRAKE SPECIFICATIONS — GM ASTRE, MONZA, SKYHAWK, STARFIRE, SUNBIRD, VEGA

All readings in inches

YEAR	MODEL	CALIPER TYPE	MOUNTING BOLT TORQUE	DISC PAD ORIGINAL THICKNESS	MFR.'S RECOMMENDED DISC PAD MINIMUM THICKNESS	ROTOR RUNOUT	ROTOR ALLOWABLE MINIMUM MACHINED THICKNESS	ROTOR THICKNESS VARIATION MAXIMUM
74	Vega	Sliding	①	.370	②	.005	.455	.0005
75	Vega, Monza, Astre, Starfire, Sunbird, Skyhawk	Sliding	①	.370	②	.005	.455	.0005
76	Vega, Astre	Sliding	①	.370	②	.005	.455	.0005
	Monza, Sunbird, Starfire, Skyhawk	Sliding	①	.430	②	.005	.830	.0005
77	Vega, Astre	Sliding	①	.370	③	.004	.455	.0005
	Monza, Sunbird, Starfire, Skyhawk	Sliding	①	.430	③	.004	.830	.0005
78	Monza, Sunbird, Starfire, Skyhawk	Sliding	①	.430	③	.004	.830	.0005
79	Monza, Sunbird, Starfire, Skyhawk	Sliding	①	.430	③	.004	.830	.0005
80	Monza, Sunbird, Starfire, Skyhawk	Sliding	①	.430	③	.004	.830	.0005

① Pin type mounting
② 1/32" over the rivet head or shoe
③ .030 over the rivet head or shoe

BRAKE SPECIFICATIONS — GM X-BODY—CITATION, OMEGA, PHOENIX, SKYLARK

All readings in inches

YEAR	MODEL	MASTER CYLINDER BORE DIAMETER	CALIPER OR WHEEL CYLINDER		BRAKE DRUM/ROTOR DIAMETER	
			Front	Rear	Front	Rear
'80	All	.874	.689	.689	11	7.8

DISC BRAKE SPECIFICATIONS — GM X-BODY—CITATION, OMEGA, PHOENIX, SKYLARK

YEAR	MODEL	CALIPER TYPE	MOUNTING BOLT TORQUE (ft/lbs)	MFR.'S RECOMMENDED DISC PAD MINIMUM THICKNESS	ROTOR RUNOUT	ROTOR ALLOWABLE MINIMUM MACHINED THICKNESS	ROTOR THICKNESS VARIATION MAXIMUM
'80	All	Sliding	28	①	.005	.965	.0005

① Replace when lining is worn to within .030" of any rivet

Domestic Cars

BRAKE SPECIFICATIONS — OLDSMOBILE EXCEPT STARFIRE AND 1980 OMEGA

YEAR	MODEL		MASTER CYLINDER BORE DIAMETER	CALIPER OR WHEEL CYLINDER Front	Rear	BRAKE DRUM/ROTOR DIAMETER Front	Rear	ROTOR RUNOUT	ROTOR MINIMUM MACHINED THICKNESS	ROTOR THICKNESS VARIATION MAXIMUM
74	Omega	Drum	1.0	1⅛	⅞	9.5	9.5	—	—	—
		Disc	1.125	2.9375	⅞	—	9.5	.004	.980	.0005
	Cutlass		1.125	2.9375	⅞	—	9.5	.004	.980	.0005
	Vista Cruiser		1.125	2.9375	⅞	—	11.0	.004	.980	.0005
	88 & 98 Pass.		1.125	2.9375	15/16	—	11.0	.005	1.230	.0005
	88 & 98 Sta. Wgn.		1.125	2.9375	1	—	12.0	.005	1.230	.0005
	Toronado		1.125	2.9375	15/16	—	11.0	.002	1.185	.0005
75	Omega	Drum	1.0	1⅛	⅞	9.5	9.5	—	—	—
		Disc	1.125	2.9375	⅞	—	9.5	.004	.980	.0005
	Cutlass		1.125	2.9375	⅞	—	9.5	.004	.980	.0005
	Vista Cruiser		1.125	2.9375	⅞	—	11.0	.004	.980	.0005
	88 & 98 Pass.		1.125	2.9375	15/16	—	11.0	.005	1.230	.0005
	88 & 98 Sta. Wgn.		1.125	2.9375	1	—	12.0	.005	1.230	.0005
	Toronado		1.125	2.9375	15/16	—	11.0	.002	1.185	.0005
76	Omega		1.125	2.9375	⅞	—	9.5	.004	.980	.0005
	Cutlass	Man	.9375	2.9375	15/16	—	11.0	.004	.980	.0005
		Pwr.	1.125	2.9375	1	—	11.0	.004	.980	.0005
	Vista Cruiser		1.125	2.9375	1	—	11.0	.004	.980	.0005
	88 & 98 Pass.		1.125	2.9375	15/16	—	11.0	.005	1.230	.0005
	88 & 98 Sta. Wgn.		1.125	2.9375	1	—	12.0	.005	1.230	.0005
	Toronado		1.125	2.9375	15/16	—	11.0	.002	1.185	.0005
77	Omega		1.125	2.9375	⅞	—	9.5	.004	.980	.0005
	Cutlass, Vista Cruiser	Man.	.9375	2.9375	15/16	—	11.0	.005	.980	.0005
		Pwr.	1.125	2.9375	1	—	11.0	.005	.980	.0005
	88 & 98, Custom Cruiser		1.125	2.9375	15/16	—	11.0	.005	.980	.0005
	Toronado		1.125	2.9375	15/16	—	11.0	.002	1.185	.0005
78	Omega		1.125	2.9375	⅞	—	9.5	.004	.980	.0005
	Cutlass, Cutlas Wagon	Man.	.87	2.50	¾	—	9.5	.004	.980	.0005
		Pwr.	.94	2.50	¾	—	9.5	.004	.980	.0005
	88		1.125	2.9375	15/16	—	9.5	.004	.980	.0005
	98 Sta. Wgn.		1.125	2.9375	15/16	—	11.0	.004	.980	.0005
	Toronado		1.125	2.9375	15/16	—	11.0	.002	1.185	.0005
79	Omega		1.125	2.9375	⅞	—	9.5	.004	.980	.0005
	Cutlass, Cutlass Wagon	Man.	.87	2.43	¾	—	9.5	.004	.980	.0005
		Pwr.	.94	2.43	¾	—	9.5	.004	.980	.0005
	88		1.125	2.9375	15/16	—	9.5	.004	.980	.0005
	98 & Sta. Wgn.		1.125	2.9375	15/16	—	11.0	.004	.980	.0005
	Toronado		.945	2.50	¾	—	9.5	.002	1.185	.0005
80	Omega	See GM X-Body Compact Front Drive Car.								
	Cutlass, Cutlass Wagon	Man.	.87	2.43	¾	—	9.5	.004	.980	.0005
		Pwr.	.94	2.43	¾	—	9.5	.004	.980	.0005
	88		1.125	2.9375	15/16	—	9.5	.004	.980	.0005
	98 & Sta. Wgn.		1.125	2.9375	15/16	—	11.0	.004	.980	.0005
	Toronado		.945	2.50	¾	—	9.5	.002	1.185	.0005

Brake Specifications

BRAKE SPECIFICATIONS — PONTIAC EXCEPT ASTRE, SUNBIRD AND 1980 PHOENIX

YEAR	MODEL	MASTER CYLINDER BORE DIAMETER	CALIPER OR WHEEL CYLINDER Front	CALIPER OR WHEEL CYLINDER Rear	BRAKE DRUM/ROTOR DIAMETER Front	BRAKE DRUM/ROTOR DIAMETER Rear
74	Ventura	1.0⑤	1.125①	.875	9.5③	9.5
75-77	Ventura, Phoenix	1.125④	2.9375	.875②	11.0	9.5
78-79	Phoenix	1.125④	2.9375	.938	11.0	9.5
73	LeMans, Grand AM Firebird, Grand Prix	1.125	2.9375	.875	11.0	9.5
74-75	LeMans, Grand AM Firebird, Grand Prix	1.125④	2.9375	.875⑥⑧	11.0	9.5⑦⑨
76-77	LeMans, Firebird Grand Prix	1.125⑩	2.9375	.937⑭	11.0	11.0⑪
78	Firebird	1.125④	2.9375	.938	11.0	9.5
	Lemans,	.945⑫	2.50	.75	10.5	11.0
	Grand Prix	.945⑫	2.50	.75	10.5	11.0
79	Firebird	1.125④	2.9375	.938	11.0	9.5
	Lemans	.945⑫	2.50	.75	10.5	9.5
	Grand Prix	.945⑫	2.50	.75	10.5	9.5
73-76	Full Size Pontiac	1.125	2.9375	.9375⑬	11.86	11.0
	Full Size Wagon	1.125	2.9375	1.00	11.86	12.0
77	Full Size Pontiac	1.125	2.9375	.875	11.0	11.0
	Full Size Wagon	1.125	2.9375	.9375	11.0	11.0
78	Full Size Pontiac	1.125	2.9375	.9375⑬	11.86	11.0
	Full Size Wagon	1.125	2.9375	.9375	11.0	11.0
79-80	Full Size Pontiac	1.125	2.9375	.875⑮	11.0	9.5⑮
	Full Size Wagon	1.125	2.9375	.975⑯	11.0	11.0⑯

① —2.937 in. caliper
② 1976-77 w/disc brakes .938
③ 11" rotor
④ 1" w/o power brakes
⑤ W/disc brakes 1.125" bore
⑥ 74—Lemans Wagon and some 4 dr Sedans
⑦ 74 Lemans Wagon and some 4 dr Sedans—11.0"
⑧ 75 Grand Prix—.9375"
⑨ 75 Grand Prix—11.0"
⑩ Lemans w/o pwr. brakes—.9375" Firebird w/o pwr. brakes—1"
⑪ Firebird—11"
⑫ Lemans w/o power brks—.866"
⑬ 1976—1"
⑭ Firebird—.938
⑮ With 4.75" wheel bolt circle
⑯ With 5.0" wheel bolt circle

Domestic Cars

WHEEL ALIGNMENT SPECIFICATIONS — AMC PASSENGER CARS

YEAR	STEERING TYPE	MAKE/MODEL	CASTER Range (deg)	CASTER Pref Setting (deg)	CAMBER Range (deg)	CAMBER Pref Setting (deg)	TOE-IN (in)	STEERING AXIS INCLIN. (deg)	WHEEL RATIO PIVOT (deg) Inner Wheel	WHEEL RATIO PIVOT (deg) Outer Wheel
74–77	MAN/PWR	All Series	½°P to 1½°P	1°P	①	②	1/16" to 3/16"	7¾°	25°	22°
78–80	MAN/PWR	All except Pacer	0°P to 2°P	1°P	①	②	1/16" to 3/16"	7¾°	25°	22°
78–80	MAN/PWR	Pacer	1°P to 3°P	2°P	①	②	1/16" to 3/16"	7¾"	25°	22°

① Left: 1/8°P to 5/8°P; Right: 0° to ½°P
② Left: 3/8°P; Right: 1/8°P
N=negative—P=Positive

WHEEL ALIGNMENT SPECIFICATIONS — AMC JEEP

YEAR/MODEL	CASTER Pref. Setting (deg)	CAMBER Pref. Setting (deg)	TOE-IN (in)	STEERING Axis Inclination (deg)	TURNING ANGLE (deg)
ALL CJ	3	1.5	3/64–3/32"	8½	31°–32°
All Wagoneer, Cherokee	4	1.5	3/64–3/32"	8½	37°–38°

NOTE: Equalize air pressure in tires before checking alignment

WHEEL ALIGNMENT SPECIFICATIONS — BUICK EXCEPT SKYHAWK AND 1980 SKYLARK

YEAR	MODEL	STEERING TYPE	CASTER Range (deg)	CAMBER Range (deg) R	CAMBER Range (deg) L	TOE-IN (in.)	STEERING AXIS INCLIN. (deg)	WHEEL RATIO PIVOT (deg) Inner Wheel	WHEEL RATIO PIVOT (deg) Outer Wheel R	WHEEL RATIO PIVOT (deg) Outer Wheel L
'74	Apollo	Man./Pwr.	0 To +1	–¼ To +¾	–¼ To +¾	3/16 ± 1/16	9②			
	Century, Century Luxus, Regal	Man.	–½ To +½	0 To +1	+½ To +1½	1/16 ± 1/16	9⅝	20	19³/₁₆①	18¹³/₁₆①
		Pwr.	–½ To +½	0 To +1	+½ To +1½	1/16 ± 1/16	9⅝	20	19	18¹¹/₁₆
	LeSabre, Estate Wgn., Electra, Centurion, Riviera	Pwr.	+½ To +1½	0 To +1	+½ To +1½	1/16 ± 1/16	10½	20	18½	18½
'75	Apollo, Skylark	Man.	–1½ To –½	+¼ To +1¼	+¼ To +1¼	0 To ⅛	10③			
		Pwr.	+½ To +1½	+¼ To +1¼	+¼ To +1¼	0 To ⅛	10③			
	Century, Century Luxus, Regal	Man.	+1½ To +2½	0 To +1	+½ To +1½	0 To ⅛	9⅝	20	19³/₁₆①	18¹³/₁₆①
		Pwr.	+1½ To +2½	0 To +1	+½ To +1½	0 To ⅛	9⅝	20	19	18¹¹/₁₆
	LeSabre, Estate Wgn., Electra, Centurion, Riviera	Pwr.	+½ To +2½	0 To +1	+½ To +1½	0 To ⅛	10½	20	18½	18½
'76	Skylark	Man.	–1½ To –½	+¼ To +1¼	+¼ To +1¼	0 To ⅛	10③			
		Pwr.	+½ To +1½	+¼ To +1¼	+¼ To +1¼	0 To ⅛	10③			
	Century, Regal	Man./Pwr.	+1½ To +2½	0 To +1	+½ To +1½	0 To ⅛	8④			
	LeSabre, Estate Wgn., Electra, Riviera	Pwr.	+1° To +2	0 To +1	+½ To +1½	0 To ⅛	9.585④⑤			
'77	Skylark	Man.	–1½ To –½	+¼ To +1¼	+¼ To +1¼	0 To ⅛	10③			
		Pwr.	+½ To +1½	+¼ To +1¼	+¼ To +1¼	0 To ⅛	10③			
	Century, Regal	Pwr.	+1½ To +2½⑥	0 To +1	+½ To +1½	0 To ⅛	8④			
	LeSabre, Estate Wgn., Electra, Riviera	Pwr.	+2½ To +3½	+¼ To +1¼	+¼ To +1¼	1/16 To 3/16	9.585④⑤			

WHEEL ALIGNMENT SPECIFICATIONS — BUICK EXCEPT SKYHAWK AND 1980 SKYLARK

YEAR	MODEL	STEERING TYPE	CASTER Range (deg)	CAMBER Range (deg) R	CAMBER Range (deg) L	TOE-IN (in.)	STEERING AXIS INCLIN. (deg)	WHEEL RATIO PIVOT (deg) Inner Wheel	WHEEL RATIO PIVOT (deg) Outer Wheel R	WHEEL RATIO PIVOT (deg) Outer Wheel L
'78	Skylark	Man.	−1±0.5	+0.8±0.5	+0.8±0.5	+⅛±1/16	10③			
		Pwr.	+1±0.5	+0.8±0.5	+0.8±0.5	+⅛±1/16	10③			
	Century, Regal	Man.	+1±0.5	+0.5±0.5	+0.5±0.5	+⅛±1/16	8④			
		Pwr.	+3±0.5	+0.5±0.5	+0.5±0.5	+⅛±1/16	8④			
	Le Sabre, Estate Wgn. Electra, Riviera	Pwr.	+3±0.5	+0.8±0.5	+0.8±0.5	+⅛±1/16	9.585④⑤			
'79	Skylark	Man.	−1±0.5	+0.8±0.5	+0.8±0.5	+⅛±1/16	10③			
		Pwr.	+1±0.5	+0.8±0.5	+0.8±0.5	+⅛±1/16	10③			
	Century, Regal	Man./Pwr.	+3±0.5	+0.5±0.5	+0.5±0.5	+⅛±1/16	8④			
	LeSabre, Estate Wgn. Electra	Pwr.	+3±0.5	+0.8±0.5	+0.8±0.5	+⅛±1/16⑦	9.585④⑤			
	Riviera	Pwr.	+2.5±2.0	0±1.5	0±1.5	0±5/16	—			
'80	Century, Regal	Man./Pwr.	+3±0.5	+0.5±0.5	+0.5±0.5	+⅛±1/16	8④			
	Electra, Estate Wgn., LeSabre	Pwr.	+3±0.5	+0.8±0.5	+0.8±0.5	+⅛±1/16⑦	9.585④⑤			
	Riviera	Pwr.	+2.5±2.0	0±1.5	0±1.5	0±5/16	—			

① Applies to station wagon with or without power steering
② At 0.5° camber
③ At 0.75° camber
④ At 0.1° camber
⑤ 10° 43′ at 0° 53′ camber for station wagon
⑥ Figure shown is for radial tires. For bias tires set at +½ to +1½
⑦ 1/16 for Estate Wagon

WHEEL ALIGNMENT SPECIFICATIONS — CADILLAC

YEAR	STEERING TYPE	CASTER Range (deg)	CASTER Pref Setting (deg)	CAMBER Range (deg)	CAMBER Pref Setting (deg)	TOE-IN (in)	STEERING AXIS INCLIN. (deg)	WHEEL RATIO PIVOT (deg) Inner Wheel	WHEEL RATIO PIVOT (deg) Outer Wheel
SEVILLE									
76	PWR	1½P-2½P	2P	LH-⅛P-⅞P RH-⅛N-⅝P	½P ¼P	1/16	6	NA	NA
77	PWR	1½P-2½P	2P	⅜N-⅜P	0	1/16	10°35′	NA	NA
78	PWR	1½P-2½P	2P	⅜N-⅜P	0	1/16	10°35′	NA	NA
79	PWR	1½P-2½P	2P	⅜N-⅜P	0	1/16	10°35′	NA	NA
80	PWR	½N-½P	0	⅜N-⅜P	0	1/16N-1/16P	11	NA	NA
ELDORADO									
74	PWR	½N-½P	0	LH-⅜N-⅜P RH-⅛N-⅝P	0 ¼P	1/16N-1/16P	11	20	18⅛
75	PWR	½N-½P	0	LH-⅜N-⅜P RH-⅛N-⅝P	0 ¼P	1/16N-1/16P	11	20	18⅛
76	PWR	½N-½P	0	LH-⅜N-⅜P RH-⅝N-⅛P	0 ¼N	1/16N-1/16P	11	20	18⅛
77-78-79	PWR	½N-½P	0	⅜N-⅜P	0	1/16N-1/16P	11	NA	NA
80	PWR	½N-½P	0	⅜N-⅜P	0	1/16N-1/16P	11	NA	NA

Domestic Cars

WHEEL ALIGNMENT SPECIFICATIONS — CADILLAC

YEAR	STEERING TYPE	CASTER Range (deg)	CASTER Pref Setting (deg)	CAMBER Range (deg)	CAMBER Pref Setting (deg)	TOE-IN (in)	STEERING AXIS INCLIN. (deg)	WHEEL RATIO PIVOT (deg) Inner Wheel	WHEEL RATIO PIVOT (deg) Outer Wheel
CADILLAC EXCEPT SEVILLE AND ELDORADO									
74	PWR	½N-½P	0	LH-⅜N-⅜P RH-⅝N-⅛P	0 ¼N	⅛P-¼P	6	20	18
75	PWR	½N-½P	0	LH-⅜N-⅜P RH-⅝N-⅛P	0 ¼N	⅛P-¼P	6	20	18
76	PWR	½N-½P ①	0	LH-⅜N-⅜P RH-⅝N-⅛P	0 ¼N	⅛P-¼P	6	20	18
77	PWR	2½P-3½P	3P	⅛P-⅞P	½P	1/16N-1/16P	10.59	N.A.	N.A.
78	PWR	2½P-3½P	3P	⅛P-⅞P	½P	1/16N-1/16P	10.59	N.A.	N.A.
79-80	PWR	2½P-3½P	3P	⅛P-⅞P	½P	1/16N-1/16P	10.59	N.A.	N.A.

① Fleetwood 75,—1½N-½N
N.A. = Not Available
P = Positive
N = Negative
LH = Left Hand Wheel
RH = Right Hand Wheel

WHEEL ALIGNMENT SPECIFICATIONS — CHEVROLET, CAMARO, CHEVELLE, NOVA, MONTE CARLO

YEAR	MODEL	STEERING TYPE	CASTER Range (deg)	CAMBER Range (deg) L	CAMBER Range (deg) R	TOE-IN (in)	STEERING AXIS INCLIN. (deg)
74	Chevrolet	Pwr.	+1±½	+1±½	+½±½	1/16±1/16	9.59
	Monte Carlo	Pwr.	+5±½	+1±½	+½±½	1/16±1/16	9.6
	Chevelle	Man.	−½±½	+1±½	+½±½	1/16±1/16	9.6
	Chevelle	Pwr.	0±½	+1±½	+½±½	1/16±1/16	9.6
	Nova	Man./Pwr.	+½±½	+¼±½		3/16±1/16	9
	Camaro Std	Man./Pwr.	0±½	+1±½		3/16±1/16	10.35
	Camaro Z-28	Pwr.	−1±½	+¾±½		3/16±1/16	10.35
75	Chevrolet	Pwr.	+1½±½	+1±½	+½±½	1/16±1/16	9.11
	Monte Carlo	Pwr.	+5±½	+1±½	+½±½	1/16±1/16	9.6
	Chevelle	Man.	+1±½	+1±½	+½±½	1/16±1/16	9.6
	Chevelle	Pwr.	+2±½	+1±½	+½±½	1/16±1/16	9.6
	Nova	Man.	−1±½	+¾±½		1/16±1/16	10
	Nova	Pwr.	+1±½	+¾±½		1/16±1/16	10
	Camaro	Pwr.	0±½	+1±½		1/16±1/16	10.35
76	Chevrolet	Pwr.	+1±½ ①	+1±½	+½±½	1/16±1/16	9.11
	Monte Carlo	Pwr.	+5±½	+1±½	+½±½	1/16±1/16	9.6
	Chevelle	Man.	+1±½	+1±½	+½±½	1/16±1/16	9.6
	Chevelle	Pwr.	+2±½	+1±½	+½±½	1/16±1/16	9.6
	Nova	Man.	−1±½	+¾±½		1/16±1/16	10
	Nova	Pwr.	+1±½	+¾±½		1/16±1/16	10
	Camaro	Pwr.	+1±½	+1±½		1/16±1/16	10.35
77	Caprice, Impala	Pwr.	+3±0.5	+0.8±0.5		3/16±1/16	9.785
	Monte Carlo	Pwr.	+5±0.5	+1±0.5	+0.5±0.5	1/16±1/16	9.6
	Chevelle	Man.	+1±0.5	+1±0.5	+0.5±0.5	1/16±1/16	9.6
	Chevelle	Pwr.	+1±0.5 ②	+1±0.5	+0.5±0.5	1/16±1/16	9.6
	Nova	Man.	−1±0.5	+0.8±0.5		1/16±1/16	10
	Nova	Pwr.	+1±0.5	+0.8±0.5		1/16±1/16	10
	Camaro	Pwr.	+1±0.5	+1±0.5		1/16±1/16	10.35

Wheel Alignment Specifications

WHEEL ALIGNMENT SPECIFICATIONS

YEAR	MODEL	STEERING TYPE	CASTER Range (deg)	CAMBER Range (deg) L	R	TOE-IN (in)	STEERING AXIS INCLIN. (deg)
78-80	Caprice, Impala	Pwr.	+3±0.5	+0.8±0.5		1/8±1/16	9.785
	Malibu, Monte Carlo	Man.	+1±0.5	+0.5±0.5		1/8±1/16	7.86
	Malibu, Monte Carlo	Pwr.	+3±0.5	+0.5±0.5		1/8±1/16	7.86
	Nova	Man.	−1±0.5	+0.8±0.5		1/8±1/16	10
	Nova	Pwr.	+1±0.5	+0.8±0.5		1/8±1/16	10
	Camaro	Pwr.	+1±0.5	+1±0.5		1/8±1/16	10.35

① With radial tire—+1 1/2 ± 1/2 ② With radial tire—+2 ± 1/2

WHEEL ALIGNMENT SPECIFICATIONS — CORVETTE

YEAR	STEERING TYPE	MODEL	CASTER Range (deg)	CASTER Pref Setting (deg)	CAMBER Range (deg)	CAMBER Pref Setting (deg)	TOE-IN (in.)	STEERING AXIS Inclin. (deg)	WHEEL PIVOT RATIO (deg) Inner Wheel	Outer Wheel
1974	Man/Pwr	Corvette	1/2°P-1 1/2°P①	1°P	1/4°P-1 1/4°P②	3/4°P	3/32" to 5/32"②	7 3/4°	N.A.	N.A.
1975	Man/Pwr	Corvette	1/2°P-1 1/2°P①	1°P	1/4°P-1 1/4°P③	3/4°P	1/32" to 3/32"③	7 3/4°	N.A.	N.A.
1976	Man	Corvette	1/2°P-1 1/2°P	1°P	1/4°P-1 1/4°P	3/4°P	3/16"-5/16"	7.68°	N.A.	N.A.
	Pwr	Corvette	1 3/4°P-2 3/4°P	2 1/4°P	1/4°P-1 1/4°P	3/4°P	3/16"-5/16"	7.68°	N.A.	N.A.
1977	Man/Pwr	Corvette	2°P-3°P	2 1/2°P	1/4°P-1 1/4°P	3/4°P	0"-1/32"	7.68	N.A.	N.A.
1978-80	Man/Pwr	Corvette	+2 1/4° ± 1/4°	2.4°P	+3/4° ± 1/2°	.709P	+1/4" ± 1/16"④	7.68°	N.A.	N.A.

① With power steering—1 3/4°P to 2 3/4°P
② Rear wheel alignment: camber—7/8 N ± 1/4; toe-in—2/32 ± 1/32
③ Rear wheel alignment: camber—11/16 N ± 1/4; toe-in—0 ± 1/32
④ Rear wheel alignment: camber 0.874 ± 1/4; toe-in 0 ± 1/32

WHEEL ALIGNMENT SPECIFICATIONS — CHEVETTE

YEAR	CASTER (deg)	CAMBER (deg)	TOE-IN
76	P 4 1/2 ± 1/2	P 1/4 ± 1/2	1/16" ± 3/64"
77	P 4 1/2 ± 1	P 1/5 ± 2/5	.12° ± .08°
78	P 4 1/2 ± 1	P 1/5 ± 2/5	.1° ± .05°①
79	P 4 1/2 ± 1	P 1/5 ± 2/5	.1° ± .05°①
80	P 4 1/2 ± 1	P 1/5 ± 2/3	.1° ± .05°①

① Each wheel

WHEEL ALIGNMENT SPECIFICATIONS — CHRYSLER CORP. EXCEPT HORIZON AND OMNI

YEAR	STEERING TYPE	MAKE/MODEL	CASTER Range (deg)	CASTER Pref Setting (deg)	CAMBER Left Wheel	CAMBER Right Wheel (deg)	TOE-IN (in)	STEERING AXIS INCLIN. (deg)	WHEEL RATIO PIVOT (deg) Inner Wheel	Outer Wheel
74	MANUAL	ALL MODELS	1 3/4 N-1/2 P	1/2 N	0°-1P	1/4 N to 3/4 P	1/16-1/4	7 1/2③	20	18.3④
	POWER	ALL MODELS	1/2 N-1 3/4 P	3/4 P	0°-1P	1/4 N to 3/4 P	1/16-1/4	7 1/2③	20	18.3④

Domestic Cars

WHEEL ALIGNMENT SPECIFICATIONS — CHRYSLER CORP. EXCEPT HORIZON AND OMNI

YEAR	STEERING TYPE	MAKE/MODEL	CASTER Range (deg)	CASTER Pref Setting (deg)	CAMBER Left Wheel (deg)	CAMBER Right Wheel (deg)	TOE-IN (in)	STEERING AXIS INCLIN. (deg)	WHEEL RATIO PIVOT (deg) Inner Wheel	WHEEL RATIO PIVOT (deg) Outer Wheel
75	MANUAL	ALL MODELS	1¾N-½P	½N	0°-1P	¼N to ¾P	1/16-¼	7½ ⑤	20	18.3 ⑥
	POWER	ALL MODELS	½N-1¾P	¾P	0°-1P	¼N to ¾P	1/16-¼	7½ ⑤	20	18.3 ⑥
76	MANUAL	ALL EXCEPT VOLARE & ASPEN	1¾N-½P	½N	0°-1P	¼N to ¾P	1/16-¼	7½ ⑤	20	18.3 ⑥
	POWER	ALL EXCEPT VOLARE ASPEN	½N-1¾P	¾P	0°-1P	¼N to ¾P	1/16-¼	7½ ⑤	20	18.3 ⑥
	MANUAL & POWER	VOLARE & ASPEN	1½P-3¾P	2½P	0°-1P	¼N to ¾P	1/16-¼	8	20	18
77	MANUAL	ALL EXCEPT VOLARE, ASPEN DIPLOMAT, LEBARON	1¾N-¾P	½N	0°-1P	¼N to ¾P	1/16-¼	8	20	18.3 ⑧
	POWER	ALL EXCEPT VOLARE, ASPEN, DIPLOMAT, LEBARON	½N-2P	¾P	0°-1P	¼N to ¾P	1/16-¼	8 ⑦	20	18.3 ⑧
	MANUAL & POWER	VOLARE, ASPEN DIPLOMAT, LEBARON	1½P-3¾P	2½P	0°-1P	¼N to ¾P	1/16-¼	8	20	18
78	MANUAL	ALL EXCEPT VOLARE, ASPEN DIPLOMAT, LEBARON	1¾N-¾P	½N	0°-1P	¼N to ¾P	1/16-¼	8	20	18
	POWER	ALL EXCEPT VOLARE, ASPEN DIPLOMAT, LEBARON	½N-2P	¾P	0°-1P	¼N to ¾P	1/16-¼	8 ⑨	20	18 ⑩
	MANUAL & POWER	VOLARE, ASPEN DIPLOMAT, LEBARON	1½P-3¾P	2½P	0°-1P	¼N to ¾P	1/16-¼	8	20	18
79	POWER	CHRYSLER, ALL; DODGE ST. REGIS, MAGNUM XE	½N-2P	¾P	0°-1P	¼N to ¾P	1/16-¼	8	20	18
	MANUAL & POWER	VOLARE, ASPEN, DIPLOMAT, LEBARON	½P-3¾P	2½P	0°-1P	¼N to ¾P	1/16-¼	8	20	18
80	POWER	CHRYSLER, ALL; DODGE ST. REGIS, MAGNUM XE LA-SCALA	½N-2P	¾P	0°-1P	¼N to ¾P	1/16-¼	8	20	18
	MANUAL & POWER	VOLARE, ASPEN, DIPLOMAT, LEBARON	½P-3¾P	2½P	0°-1P	¼N to ¾P	1/16-¼	8	20	18

NOTE: Set front suspension height before aligning any Chrysler Corporation car.

P = Positive N = Negative

① Satellite, Coronet, Charger 8°; Imperial 9°;
② Valiant, Dart, Barracuda, Challenger 17.5°; Satellite, Coronet, Charger 18.5°; Imperial 17.9°;
③ Satellite, Coronet, Charger 8°; Fury, Polara, Monaco, Chrysler and Imperial, 9°;
④ Valiant, Dart, 18.5°; Barracuda, Challenger, 18.4°; Coronet, Charger, Satellite, 18°;
⑤ Fury, Coronet, Charger se, Corodoba 8°; Gran Fury, Monaco, Chrysler and Imperial, 9°;
⑥ Dart and Valiant, 18.5°; Fury, Coronet, Charger se and Cordoba, 18°;
⑦ Royal Monaco, Gran fury and Chrysler, 9°;
⑧ Fury, Monaco, Charger se, Cordoba, 18°;
⑨ Chrysler 9°;
⑩ Chrysler 18.3°;

Wheel Alignment Specifications

FRONT END HEIGHT SPECIFICATIONS

YEAR	MAKE & MODEL	HEIGHT (in)
74①	VALIANT AND DART	1⅞
	BARRACUDA AND CHALLENGER	1⅛
	SATELLITE, CORONET, AND CHARGER	1⅞
	FURY, POLARA, MONACO, CHRYSLER	1
	IMPERIAL	1
75②	VALIANT AND DART	10¹⁵⁄₁₆
	FURY, CORONET, CHARGER SE, AND CORDOBA, ALL EXC. WAG.	10¾
	FURY AND CORONET WAGONS	11¼
	GRAN FURY, MONACO, CHRYSLER AND IMPERIAL	10⅛
76②	VALIANT AND DART	10¹⁵⁄₁₆
	FURY, CORONET, CHARGER SE AND CORDOBA, ALL EXC. WAG.	10¾
	FURY AND CORONET WAGON	11¼
	VOLARE AND ASPEN	10¼
	GRAN FURY, MONACO, CHRYSLER	10⅛
77②	VOLARE, ASPEN, DIPLOMAT, LEBARON	10¼
	MONACO, CHARGER SE, FURY CORDOBA, ALL EXC. WAG.	10¾
	MONACO AND FURY WAGON	11¼
	ROYAL MONACO, GRAN FURY AND CHRYSLER	10⅛
78②	VOLARE, ASPEN, DIPLOMAT LEBARON	10¼
	MONACO, FURY, CHARGER SE, MAGNUM XE, CORDOBA ALL EXC. WAGON	10¾
	FURY & MONACO WAGONS	11¼
	CHRYSLER	10⅛
79-80②	VOLARE, ASPEN, DIPLOMAT LEBARON	10¼
	MAGNUM XE, ST. REGIS, CHRYSLER	10¾

NOTE: Before front suspension height is set, vehicle should be devoid of cargo and passengers. The tire pressure should be equalized and set to specifications, and the gasoline tank should be full. If the gasoline tank cannot be filled, weight should be added to compensate for the lack of gasoline.

①1974 MODELS:
Valiant, Dart, Barracuda, and Challenger; measurement shown is the difference between "a"—the distance from the lowest point of the control arm adjusting blade to the floor and "b"—the lowest point of the steering knuckle arm at its center line, to the floor. Measure one side at a time. 74 All models except Valiant, Dart, Barracuda & Challenger; measurement shown is the difference between "a"—the distance from the lowest point of the front torsion bar anchor at the rear of the lower control arm flange to the floor and "b"—the lowest point of the ball joint housing to the floor. Measure one side at a time. ("a" minus "b"= measurement shown)

②1975-1980 ALL MODELS EXCEPT VOLARE, ASPEN DIPLOMAT AND LEBARON:
Measurement shown is the distance from the lowest point of lower control arm torsion anchor at a point 1 inch from the rear face of the anchor, to the ground.
1976-1980 VOLARE, ASPEN, DIPLOMAT AND LEBARON;
Measurement shown is the distance from the lowest point of the lower control arm inner pivot pushing to the floor

Domestic Cars

WHEEL ALIGNMENT SPECIFICATIONS — FORD MOTOR CO. SUB-COMPACT

YEAR	STEERING TYPE		CAMBER Range (deg)	CAMBER Prefer (deg)	TOE Range (in)	TOE Prefer (in)
'78	MAN/PWR	FRONT	¼N to ¾P	5/16P	5/32" OUT to ⅛" IN	1/16" OUT
		REAR	1½N to ½N	1N	5/32" OUT to 11/32" IN	3/32" IN
'79	MAN/PWR	FRONT	¼N to ¾P	5/16P	5/32" OUT to ⅛" IN	1/16" OUT
		REAR	1½N to ½N	1N	5/32" OUT to 11/32" IN	3/32" IN
'80	MAN/PWR	FRONT	¼N to ¾P	5/16P	5/32" OUT to ⅛" IN	1/16" OUT
		REAR	1½N to ½N	IN	5/32" OUT to 11/32" IN	3/32" IN

MAN—Manual Steering
PWR—Power Steering
N—Negative
P—Positive

WHEEL ALIGNMENT SPECIFICATIONS — FORD MOTOR CO. COMPACT

YEAR	STEERING TYPE	MODEL	CASTER Range (deg)	CASTER Pref Setting (deg)	CAMBER Range (deg)	CAMBER Pref Setting (deg)	TOE-IN (in.)	STEERING AXIS Inclin. (deg)	WHEEL PIVOT RATIO (deg) Inner Wheel	WHEEL PIVOT RATIO (deg) Outer Wheel
'74–'75	Man/Pwr	Comet, Monarch Maverick, Granada	2½°N to 1½°P	½°N	¾°N to 1¼°P	¼°P	1/16" to ⅜"	6¾°	20°	18.39°
'76	Man/Pwr	Comet, Maverick, Granada, Monarch	1¼°N to ¼°P	½°N	½°N to 1°P	¼°P	0" to ⅜"	6¾°	20°	18.39°
'77	Man/Pwr	Comet, Maverick, Granada, Monarch	2°N-1°P	½°N	¾°N-1¼°P	¼°P	0"-⅜"	6¾°	20°	18.39°
'78	Man/Pwr	Fairmont, Zephyr	⅛°P-1⅝°P	⅞°P	⅜°N-1⅛°P	⅜°P	3/16"-7/16"	—	20°	19.47°
	Man/Pwr	Monarch, Granada Versailles	1¼°N-¼°P	½°N	½°N-1°P	¼°P	0"-¼"	6¾°	20°	18.20°①
'79	Man/Pwr	Fairmont, Zephyr	⅛°P-1⅝°P	⅞°P	⅜°N-1⅛°P	⅜°P	3/16"-7/16"	—	20°	19.47
	Man/Pwr	Monarch, Granada Versailles	1¼°N-¼°P	½°N	½°N-1°P	¼°P	0"-¼"	6¾°	20°	18.20°①
'80	Man/Pwr	Fairmont, Zephyr	⅛°P-1⅝°P	⅞°P	⅜°N-1⅛°P	⅜°P	3/16"-7/16"	—	20°	19.47
	Man/Pwr	Monarch, Granada Versailles	1¼°N-¼°P	½°N	½°N-1°P	¼°P	0"-¼"	6¾°	20°	18.20°①

① With power steering shown; w/o p/s 18.43°

WHEEL ALIGNMENT SPECIFICATIONS — FORD MOTOR CO. FULL-SIZE

YEAR	STEERING TYPE	MAKE/MODEL	CASTER Range (deg)	CASTER Pref Setting (deg)	CAMBER Range (deg)	CAMBER Pref Setting (deg)	TOE-IN (in)	STEERING AXIS INCLIN. (deg)	WHEEL RATIO PIVOT (deg) Inner Wheel	WHEEL RATIO PIVOT (deg) Outer Wheel
'74	Pwr	Ford, Mercury	0 to 4P	2P	L ½N to 1½P / R ¾N to 1¼P	L ½P / R ¼P	3/16	9½	20	18.75
	Pwr	Thunderbird	½P to 3½P	2P	L 0 to 2P / R ½N to 1½P	L 1P / R ½P	3/16	9	20	18.07
	Pwr	Mark IV	¼P to 3¼P	1¾P	L ¼N-1¾P / R ¾N-1¼P	L ¾P / R ¼P	3/16	7¾	20	18.07
	Pwr	Continental	½N to 3½P	1½P	½N to 1½P	½P	⅛	9½	20	18.16

Wheel Alignment Specifications

WHEEL ALIGNMENT SPECIFICATIONS — FORD MOTOR CO. FULL-SIZE

YEAR	STEERING TYPE	MAKE/MODEL	CASTER Range (deg)	CASTER Pref Setting (deg)	CAMBER Range (deg)	CAMBER Pref Setting (deg)	TOE-IN (in)	STEERING AXIS INCLIN. (deg)	WHEEL RATIO PIVOT Inner Wheel (deg)	WHEEL RATIO PIVOT Outer Wheel (deg)
'75–'76	Pwr	Ford, Mercury	1¼P to 2¾P	2P	L ¼N to 1¼P / R ½N to 1P	L ½P / R ¼P	3/16	9.44	20	18.72
	Pwr	Thunderbird	3¼P to 4¾P	4P	L ¼N to 1¼P / R ½N to 1P	L ½P / R ¼P	3/16	9.008	20	18.09
	Pwr	Mark IV	1¼P to 2¾P	2P	L ¼N to 1¼P / R ½N to 1P	L ½P / R ¼P	3/16	9.49	20	18.09
	Pwr	Continental	1¼P to 2¾P	2P	L ¼N to 1¼P / R ½N to 1P	L ½P / R ¼P	1/8	9.50	20	18.16
77	Pwr	Ford, Mercury	1¼P to 2¾P	2P	L ¼N to 1¼P / R ½N to 1P	L ½P / R ¼P	3/16	9.44	20	18.72
	Pwr	Mark V	1¼P to 2¾P	2P	L ¼N to 1¼P / R ½N to 1P	L ½P / R ¼P	3/16	9.49	20	18.09
	Pwr	Continental	1¼P to 2¾P	2P	L ¼N to 1¼P / R ½N to 1P	L ½P / R ¼P	1/8	9.50	20	18.16
'78	Pwr	Ford, Mercury	1¼P to 2¾P	2P	L ¼N to 1¼P / R ½N to 1P	L ½P / R ¼P	3/16	9.44	20	18.72
	Pwr	Mark V	3¾P to 4¾P	4P	L ¼N to 1¼P / R ½N to 1P	L ½P / R ¼P	3/16	9.49	20	18.09
	Pwr	Continental	1¼P to 2¾P	2P	L ¼N to 1¼P / R ½N to 1P	L ½P / R ¼P	1/8	9.50	20	18.09
'79	Pwr	Ford, Mercury	3¾P to 2¼P	3P	1¼P to ¼N	½P	1/16	9.44	20	18.72
	Pwr	Mark V	3¾P to 4¾P	4P	L ¼N to 1¼P / R ½N to 1P	L ½P / R ¼P	3/16	9.49	20	18.09
	Pwr	Continental	1¼P to 2¾P	2P	L ¼N to 1¼P / R ½N to 1P	L ½P / R ¼P	1/8	9.50	20	18.09
'80	Pwr	Ford, Mercury	3¾P to 2¼P	3P	1¼P to ¼N	½P	1/16	9.44	20	18.72
	Pwr	Mark VI	①	①	①	①	①	①	①	①
	Pwr	Continental	①	①	①	①	①	①	①	①

① Not available at press date.

WHEEL ALIGNMENT SPECIFICATIONS — FORD MOTOR CO. SMALL CARS

YEAR	STEERING TYPE	MODEL	CASTER Range (deg)	CASTER Pref Setting (deg)	CAMBER Range (deg)	CAMBER Pref Setting (deg)	TOE-IN (in.)	STEERING AXIS Inclin. (deg)	WHEEL PIVOT RATIO Inner Wheel (deg)	WHEEL PIVOT RATIO Outer Wheel (deg)
'74	Man/Pwr	Mustang II	⅜°N to 1⅞°P	⅞°P	½°N to 1½°P	½°P	0 to ¼	9.763	20	18.84
	Man/Pwr	Pinto	¾°N to 3¼°P	1¼°P	¼°N to 1¾°P	¾°P	1/8 to 3/8	10.018	20	18.84
	Man/Pwr	Sta. Wgn.	½°N to 3½°P	1½°P	¼°N to 1¾°P	¾°P	1/8 to 3/8	10.018	20	18.84
'75	Man/Pwr	Bobcat	½°P to 2°P	1¼°P	0 to 1½°P	¾°P	1/8 to 3/8	9.763	20	18.84
	Man/Pwr	Mustang II	⅛°P to 1⅝°P	⅞°P	¼°N to 1¼°P	½°P	0-¼	9.763	20	18.84
	Man/Pwr	Pinto	½°P to 2°P	1¼°P	0-1½°P	¾°P	1/8 to 3/8	9.763	20	18.84
	Man/Pwr	Sta. Wgn.	¾°P to 2¼°P	1½°P	0 to 1½°P	¾°P	1/8 to 3/8	9.763	20	18.84
'76	Man/Pwr	Bobcat, Pinto	½°P to 2°P	1¼°P	0° to 1½°P	¾°P	1/8 to 3/8	10.018	20	18.84
	Man/Pwr	Bobcat, Pinto Sta Wgn	¾°P to 2¼°P	1½°P	0° to 1½°P	¾°P	1/8 to 3/8	10.018	20	18.84
	Man/Pwr	Mustang II	⅛°P to 1⅝°P	⅞°P	¼°N to 1⅛°P	½°P	0 to ⅛	9.763	20	18.84

WHEEL ALIGNMENT SPECIFICATIONS

FORD MOTOR CO. SMALL CARS

YEAR	STEERING TYPE	MODEL	CASTER Range (deg)	CASTER Pref Setting (deg)	CAMBER Range (deg)	CAMBER Pref Setting (deg)	TOE-IN (in.)	STEERING AXIS Inclin. (deg)	WHEEL PIVOT RATIO (deg) Inner Wheel	WHEEL PIVOT RATIO (deg) Outer Wheel
'77	Man/Pwr	Bobcat, Pinto	¼°P to 1¾°P	1°P	¼°N to 1¼°P	½°P	0 to ¼	10.018	20	18.84
	Man/Pwr	Bobcat, Pinto Sta Wgn	½°N to 1°P	¼°P	¼°N to 1¼°P	½°P	0 to ¼	10.018	20	18.84
	Man/Pwr	Mustang II	⅛°P to 1⅝°P	⅞°P	¼°N to 1¼°P	½°P	0 to ¼	9.763	20	18.84
'78	Man/Pwr	Bobcat, Pinto	¼°P to 1¾°P	1°P	¼°N to 1¼°P	½°P	0 to ¼	10.018	20	18.84
	Man/Pwr	Bobcat, Pinto Sta Wgn	½°N to 1°P	¼°P	¼°N to 1¼°P	½°P	0 to ¼	10.018	20	18.84
	Man/Pwr	Mustang II	⅛°P to 1⅝°P	⅞°P	¼°N to 1¼°P	½°P	0 to ¼	9.763	20	18.84
'79	Man/Pwr	Bobcat, Pinto	¼°P to 1¼°P	1°P	¼°N to 1¼°P	½°P	0 to ¼	10.018	20	18.84
	Man/Pwr	Bobcat, Pinto Sta Wgn	½°N to 1°P	¼°P	¼°N to 1¼°P	½°P	0 to ¼	10.018	20	18.84
	Man/Pwr	Mustang, Capri	1¾°P to ¼°P	1°P	1°P to ½°N	¼°P	⅕ to ⅖	9.763	—	—
'80	Man/Pwr	Bobcat, Pinto	¼°P to 1¼°P	1°P	¼°N to 1¼°P	½°P	0 to ¼	10.018	20	18.84
	Man/Pwr	Bobcat, Pinto Sta Wgn	½°N to 1°P	¼°	¼°N to 1¼°P	½°P	0 to ¼	10.018	20	18.84
	Man/Pwr	Mustang, Capri	1¾°P to ¼°P	1°P	1°P to ½°N	¼°P	⅕ to ⅖	9.763	—	—

WHEEL ALIGNMENT SPECIFICATIONS

FORD MOTOR CO. MID-SIZE

YEAR	STEERING TYPE	MODEL	CASTER Range (deg)	CASTER Pref Setting (deg)	CAMBER Range (deg)	CAMBER Pref Setting (deg)	TOE-IN (in)	STEERING AXIS Inclin. (deg)	WHEEL PIVOT RATIO (deg) Inner Wheel	WHEEL PIVOT RATIO (deg) Outer Wheel
'74–'75	Man/Pwr	Cougar, Montego, Torino	½°P to 3½°P	2°P	②	③	0″ to ⅜″	9°	20°	18°
'76–'77	Pwr	Cougar	3¼°P to 4¾°P	4°P	④	⑤	0″ to ⅜″	9°	20°	18°
	Pwr	Montego, Torino	3¼°P to 4¾°P	4°P	④	⑤	0″ to ⅜″	9°	20°	18°
	Pwr	LTD II	3¼°P to 4¾°P	4°P	④	⑤	0″ to ⅜″	9°	20°	17¾°
	Pwr	Thunderbird	3¼°P to 4¾°P	4°P	④	⑤	1/16″ to 7/16″	9½°	—	—
'78–'79	Pwr	Cougar, LTD II Thunderbird	3¼°P to 4¾°P	4°P	④	⑤	0″ to ¼″	9°⑥	20°	18.06°
'80	Pwr	Cougar, LTD II Thunderbird	3¼°P to 4¾°P	4°P	④	⑤	0″ to ¼″	9″⑥	20°	18.06°

— Not applicable
① Manual steering—17½°; Power steering—17½°
② Left—⅜°N to 1⅝°P
Right—⅞°N to 1⅛°P
③ Left—⅝°P
Right—⅛°P
④ Left—¼°N to 1¼°P
Right—½°N TO 1°P
⑤ Left—½°P
Right—¼°P
⑥ Thunderbird—9.49°

WHEEL ALIGNMENT SPECIFICATIONS

GM ASTRE, MONZA, SKYHAWK, STARFIRE, SUNBIRD, VEGA

YEAR	MAKE/MODEL	CASTER Range (deg)	CAMBER Range (deg)	TOE-IN (in)	STEERING① AXIS INCLIN. (deg)
74–80	All	.75 ± .50	.25 ± .50	1/16 ± 1/16	8.55

① At 25°

Wheel Alignment Specifications

WHEEL ALIGNMENT SPECIFICATIONS
GM X-BODY—CITATION, OMEGA, PHOENIX, SKYLARK

YEAR	MAKE/MODEL	CAMBER (deg)	TOE-IN (deg)
'80	All	+.50 ± .50	+.10 ± .10

WHEEL ALIGNMENT SPECIFICATIONS — OLDSMOBILE EXCEPT STARFIRE AND 1980 OMEGA

YEAR	MAKE/MODEL	STEERING TYPE	CASTER Range (deg)	CAMBER Range (deg) L	CAMBER Range (deg) R	TOE-IN (in)	STEERING AXIS INCLIN. (deg)	WHEEL RATIO PIVOT (deg) Inner Wheel	WHEEL RATIO PIVOT (deg) Outer Wheel
'74	Omega		½P ± ½		¼P ± ½	3/16 ± 1/16	9	—	—
	Cutlass EXC Y-78 Salon		0 ± ½	1P ± ½	½P ± ½	1/16 ± 1/16	10°34'	—	—
	Y-78 Salon		2P ± ½	1P ± ½	½P ± ½	1/16 ± 1/16	10°34'	—	—
	88 & 98		1P ± ½	1P ± ½	½P ± ½	1/16 ± 1/8	9.6	—	—
	Toronado		2N ± ½	¼P ± ½	¼N ± ½	0 ± 1/16	11	—	—
'75	Omega	Man.	1N ± ½		¾P ± ½	1/16 ± 1/16	10°35'	—	—
		Pwr.	1P ± ½		¾P ± ½	1/16 ± 1/16	10°35'	—	—
	Cutlass		2P ± ½	1P ± ½	½P ± ½	1/16 ± 1/16	10°35'	—	—
	88 & 98		1½P ± ½	1P ± ½	½P ± ½	1/16 ± 1/16	10°35'	—	—
	Toronado		0	¼P ± ½	¼N ± ½	0 ± 1/16	11	—	—
'76	Omega	Man.	1N ± ½		¾P ± ½	1/16 ± 1/16	10°35'	—	—
		Pwr.	1P ± ½		¾P ± ½	1/16 ± 1/16	10°35'	—	—
	Cutlass		2P ± ½	1P ± ½	½P ± ½	1/16 ± 1/16	10°35'	—	—
	88 & 98		1½P ± ½	1P ± ½	½P ± ½	1/16 ± 1/16	10°35'	—	—
	Toronado		0	¼P ± ½	¼N ± ½	0 ± 1/16	11	—	—
'77	Omega	Man.	1N ± ½		¾P ± ½	1/16 ± 1/16	10°35'	—	—
		Pwr.	1P ± ½		¾P ± ½	1/16 ± 1/16	10°35'	—	—
	Cutlass		2P ± ½	1P ± ½	½P ± ½	1/16 ± 1/16	10°35'	—	—
	88 & 98		3P ± ½	¾P ± ½	¾P ± ½	1/8 ± 1/16	10°35'	—	—
	Toronado		0	¼P ± ½	¼N ± ½	0 ± 1/16	11	—	—
'78	Omega	Man.	1N ± ½		0.8P ± .5	+1/8 ± 1/16	10°35'	—	—
		Pwr.	1P ± ½		0.8P ± .5	+1/8 ± 1/16	10°35'	—	—
	Cutlass	Man.	1P ± 0.5		0.5P ± 0.5	+1/8 ± 1/16	6.98	—	—
		Pwr.	3P ± 0.5		0.5P ± 0.5	+1/8 ± 1/16	6.98	—	—
	88 & 98		3P ± 0.5		0.8P ± 0.5	+1/8 ± 1/16	10°35'	—	—
	Toronado		0	0.3P ± 0.5	−0.3 ± 0.5	0 ± 1/16	11	—	—
'79	Omega	Man.	1N ± ½		¾P ± ¾	1/16 ± 1/8	10°35'	—	—
		Pwr.	1P ± ½		¾P ± ¾	1/16 ± 1/8	10°35'	—	—
	Cutlass		1.0P ± 0.5		0.5P ± 0.5	0.1 ± 0.05	6.98	—	—
	88 & 98		3.0P ± 0.5		0.8P ± 0.5	0.12 ± 0.06	10°35'	—	—
	Toronado		2.5P ± 0.5		0 ± 0.5	0 ± 0.05	11	—	—
'80	Omega		See GM X-Body Front Drive Compact Car						
	Cutlass		1.0P ± 0.5		0.5P ± 0.5	0.1 ± 0.05	6.98	—	—
	88 & 98		3.0P ± 0.5		0.8P ± 0.5	0.12 ± 0.06	10°35'	—	—
	Toronado		2.5P ± 0.5		0 ± 0.5	0 ± 0.05	11	—	—

Domestic Cars

WHEEL ALIGNMENT SPECIFICATIONS — PONTIAC EXCEPT ASTRE, SUNBIRD AND 1980 PHOENIX

YEAR	STEERING TYPE	MODEL	CASTER Range (deg)	CASTER Pref Setting (deg)	CAMBER Range (deg)	CAMBER Pref Setting (deg)	TOE-IN (in.)	STEERING AXIS INCLIN. (deg)	WHEEL RATIO PIVOT (deg) Inner Wheel	WHEEL RATIO PIVOT (deg) Outer Wheel	WHEEL LUG TORQUE Ft/Lbs
74	ALL	Ventura	0 to 1P	½P	¼N to ¾ P	¼P	⅛ to ¼	9	20		70
	ALL	Firebird	½N to ½P	0	½P to 1½P	1P	⅛ to ¼	10.35	20		80
	PWR	Grand Prix	2½P to 3½P	3P	L-½P to 1½P / R-0 to 1P	L-1P / R-½P	0 to ⅛	10.35	20	(RT) 19 3/16 / (LT) 18 13/16	80
	MAN	Lemans Ser.	1½N to ½N	1N	L-½P to 1½P / R-0 to 1P	L-1P / R-½P	0 to ⅛	10.35	20	(RT) 19 3/16 / (LT) 18 13/16	80
	PWR	Lemans, Grand AM	½N to ½P	0	L-½P to 1½P / R-0 to 1P	L-1P / R-½P	0 to ⅛	10.35	20	(RT) 19 3/16 / (LT) 18 13/16	80
	PWR	Catalina, Bonneville, Grand Ville	½P to 1½P	1P	L-½P to 1½P / R-0 to 1P	L-1P / R-½P	0 to ⅛	10.35	20	18½	75
75	MAN	Ventura	1½N to ½N	1N	¼P to 1¼P	¾P	0 to ⅛	10.85	20	18½	70
	PWR	Ventura	½P to 1½P	1P	¼P to 1¼P	¾P	0 to ⅛	10.85	20	18½	70
	ALL	Firebird	½N to ½P	0	½P to 1½P	1P	0 to ⅛	10.35	20	18¾	70
	PWR	Grand Prix	2½P to 3½P	3P	L-½P to 1½P / R-0 to 1P	L-1P / R-½P	0 to ⅛	10.35	20	(LT) 19 3/16 / (RT) 18 13/16	80
	MAN	Lemans Ser.	½P to 1½P	1P	L-½P to 1½P / R-0 to 1P	L-1P / R-½P	0 to ⅛	10.35	20	(LT) 19 3/16 / (RT) 18 13/16	80
	PWR	Lemans, Grand AM	1½P to 2½P	2P	L-½P to 1½P / R-0 to 1P	L-1P / R-½P	0 to ⅛	10.35	20	(LT) 19 3/16 / (RT) 18 13/16	80
	PWR	Catalina Bonneville Grand Ville	½P to 2½P	1½P	L-½P to 1½P / R-0 to 1P	L-1P / R-½P	0 to ⅛	10.35	20	18½	75
76	MAN	Ventura	1½N to ½N	1N	¼P to 1¼P	¾P	0 to ⅛	10	20	18½	70
	PWR	Ventura	½P to 1½P	1P	¼P to 1¼P	¾P	0 to ⅛	10	20	18½	70
	ALL	Firebird	½N to ½P	0	½P to 1½P	1P	0 to ⅛	10.35	20	18¾	80
	PWR	Grand Prix	2½P to 3½P	3P	L-½P to 1½P / R-0 to 1P	L-1P / R-½P	0 to ⅛	10.35	20	(LT) 19 3/16 / (RT) 18 13/16	80
	MAN	Lemans Ser.	½P to 1½P	1P	L-½P to 1½P / R-0 to 1P	L-1P / R-½P	0 to ⅛	10.35	20	(LT) 19 3/16 / (RT) 18 13/16	80
	PWR	Lemans, Grand AM	1½P to 2½P	2P	L-½P to 1½P / R-0 to 1P	L-1P / R-½P	0 to ⅛	10.35	20	(LT) 19 3/16 / (RT) 18 13/16	80
	PWR	Catalina Bonneville	1P to 2P	1½P	L-½P to 1½P / R-0 to 1P	L-1P / R-½P	0 to ⅛	10.35	20	18½	75
77	MAN	Ventura	1½N to ½N	1N	3/10P to 1 3/10P	⅘P	0 to ⅛	10	—	—	70
	PWR	Ventura	½P to 1½P	1P	3/10P to 1 3/10P	⅘P	0 to ⅛	10	—	—	70
	ALL	Firebird	½P to 1½P	1P	½P to 1½P	1P	0 to ⅛	10.35	—	—	80
	PWR	Grand Prix	4½P to 5½P	5P	L-½P to 1½P / R-0 to 1P	L-1P / R-½P	0 to ⅛	10.35	—	—	80
	MAN	Lemans Ser.	½P to 1½P	1P	L-½P to 1½P / R-0 to 1P	L-1P / R-½P	0 to ⅛	10.35	—	—	80
	PWR	Lemans Ser w/radial tires / Lemans Ser. w/belted tires	1½P to 2½P / ½P to 1½P	RA 2P / BT 1P	L-½P to 1½P / R-0 to 1P	L-1P / R-½P	0 to ⅛	10.35	—	—	80
	PWR	Catalina, Bonneville	2½P to 3½P	3P	3/10P to 1 3/10P	⅘P	⅛ to ¼	10.35	—	—	75
78	MAN	Phoenix	1½N to ½N	1N	3/10P	⅘P	1/16 to 3/16	10	—	—	70
	PWR	Phoenix	½P to 1½P	1P	3/10P to 1 3/10P	⅘P	1/16 to 3/16	10	—	—	70
	ALL	Firebird	½P to 1½P	1P	½P to 1½P	1P	1/16 to 3/16	10.35	—	—	80
	MAN	Lemans Ser.	½P to 1½P	1P	0 to 1P	½P	1/16 to 3/16	8	—	—	80
	PWR	Lemans, Grand Prix	2½P to 3½P	3P	0 to 1P	½P	1/16 to 3/16	8	—	—	80

Wheel Alignment Specifications

WHEEL ALIGNMENT SPECIFICATIONS — PONTIAC EXCEPT ASTRE, SUNBIRD AND 1980 PHOENIX

YEAR	STEERING TYPE	MODEL	CASTER Range (deg)	CASTER Pref Setting (deg)	CAMBER Range (deg)	CAMBER Pref Setting (deg)	TOE-IN (in.)	STEERING AXIS INCLIN. (deg)	WHEEL RATIO PIVOT (deg) Inner Wheel	WHEEL RATIO PIVOT (deg) Outer Wheel	WHEEL LUG TORQUE Ft/Lbs
	PWR	Catalina, Bonneville	2½P to 3½P	3P	3/10P to 1 3/10P	4/5P	1/16 to 3/16	10.35	—	—	75
79	MAN	Phoenix	1½N to ½P	1N	½P to 1½P	1P	1/16 to 3/16	10	—	—	70
	PWR	Phoenix	½P to 1½P	1P	½P to 1½P	1P	1/16 to 3/16	10	—	—	70
	ALL	Firebird	½P to 1½P	1P	½P to 1½P	1P	1/16 to 3/16	10.35	—	—	80
	MAN	Lemans Ser.	½P to 1½P	1P	0 to 1P	½P	1/16 to 3/16	8	—	—	80
	PWR	Lemans Ser. Grand Prix	2½P to 3½P	3P	0 to 1P	½P	1/16 to 3/16	8	—	—	80
	PWR	Catalina, Bonneville	2½P to 3½P	3P	3/10P to 1 3/10P	4/5P	1/16 to 3/16	10.35	—	—	75
80	ALL	Firebird	½P to 1½P	1P	½P to 1½P	1P	1/16 to 3/16	10.35	—	—	70
	MAN	Lemans Ser.	½P to 1½P	1P	0 to 1P	½P	1/16 to 3/16	8	—	—	70
	PWR	Lemans Ser. Grand Prix	2½P to 3½P	3P	0 to 1P	½P	1/16 to 3/16	8	—	—	
	PWR	Catalina, Bonneville	2½P to 3½P	3P	3/10P to 1 3/10P	4/5P	1/16 to 3/16	10.35	—	—	75

N= Negative
P= Positive
LT= Left turn
RT= Right turn
L= Left wheel
R= Right wheel
RA= Radial tires
BT= Belted tires
— Not Available

Import Cars

Note: If hubs must be removed, refer to manufacturers service manual for current installation procedure

IDENTIFICATION MAKE AND MODEL	YEAR	DISC ROTOR Runout (in.)	Minimum Thickness (in.)	Parellelism (in.)	BRAKE DRUM New Dia. (in.)	Max. Mach. Oversize (in.)	Max. Oversize Due to Wear (in.)	Lug Nut Torque (ft/lbs)
AUDI								
Fox	79-78	.002	.413	—	7.87	7.90	7.913	65
Fox	77-73	.004	.413	.0008	7.87	7.90	7.913	58
100 Series	77-70	.0047	.196	.0012	7.87	7.90	7.913	58
5000	79-78	.004	.807	.0008	9.005	9.094	9.134	80
AUSTIN								
Marina	75-73	.006	—	.001	8.0	8.03	8.05	60-65
America	71-70	.006	—	.001	8.0	8.03	8.05	50
BMW								
320i	79-77	.008[2]	.827	.0008	9.842	9.882	—	59-65
528i	79							60-66
633I	79							60-66
733i Front	79-78	.006[2]	.827	.0008	—	—	—	59-65
733i Rear	79-78	.006[2]	.354	.0008	—	—	—	59-65
530i	78-77	.008 Frt.	.827 Frt.	.0008 Frt.	—	—	—	59-65
530i	78-77	.008 Rr.	.334 Rr.	.0008 Rr.	—	—	—	59-65
530i	76-75	.008 Frt.	.460 Frt.	.0008 Frt.	—	—	—	59-65
530i	76-75	.008 Rr.	.334 Rr.	.0008 Rr.	—	—	—	59-65
630 CSi	78-77	.008 Frt.	.827 Frt.	.0008 Frt.	—	—	—	59-65
630 CSi	78-77	.008 Rr.	.709 Rr.	.0008 Rr.	—	—	—	59-65
2002, 2002 tii	76-70	.008	.354[1]	.0008	9.06	9.10	9.130	59-65
3.0 Si	76	.008 Frt.	.827 Frt.	.0008 Frt.	—	—	—	60-67
3.0 Si	76	.008 Rr.	.827 Rr.	.0008 Rr.	—	—	—	60-67
3.0 Series	75-72	.008 Frt.	.827 Frt.	.0008 Frt.	—	—	—	60-67
3.0 Series	75-72	.008 Rr.	.709 Rr.	.0008 Rr.	—	—	—	60-67
2800 (w/4 Wheel Discs)	71-70	.008 Frt.	.461 Frt.	.0008 Frt.	—	—	—	60-67
2800 (w/4 Wheel Discs)	71-70	.008 Rr.	.335 Rr.	.0008 Rr.	—	—	—	60-67
2800 (w/front discs)	71-70	.008	.461	.0008	9.84	9.88	—	60-67
2500	71-70	.008 Frt.	.461 Frt.	.0008 Frt.	—	—	—	60-67
2500	71-70	.008 Rr.	.335 Rr.	.0008 Rr.	—	—	—	60-67
1600	71-70	.008	.354	.0008	7.87	7.91	—	59-65

[1]2002 tu - thickness .460" [2]on-car

BUICK OPEL								
All	79-76	.006	.338	—	9.00	9.040	9.060	50-68
All	75-73	.006	.339	—	9.00	9.040	9.060	50
All	72	.004	.465	.0004	9.060	9.090	—	65
All Except 1100 cc	74-70	.004	.394	.0006	9.060	9.090	—	65[1]
Opel 1100 cc	71-70	.004	.394	.0006	7.870	7.900	—	65

[1]1972 - 1900 torque is 72 FT/LBS

CHALLENGER								
W/rear drums	79-78	.006	.45[1]	—	9.00	9.060	9.080	51-58[2]
W/4 wheel disc-front	79-78	.006	.45[1]	—	—	—	—	51-58[2]
W/4 wheel disc-rear	79-78	.0065	.33	—	—	—	—	51-58[2]

[1]Pin type caliper; sliding caliper min.thickness .430 in. [2]Steel wheels; aluminum wheel 58-72 ft/lbs.

DATSUN								
310	79	.0024	.339	.0012	8.00	—	8.050	58-72
200 SX	79-78	.0047	.331	.0012	9.00	—	9.060	58-65[1]
210	79	.0047	.331	.0012	8.00	—	8.050	58-72
280ZX Front	79	.0039	.709	.0012	—	—	—	—
280ZX Rear	79	.0059	.339	.0012	—	—	—	—
810	79	.0059	.413	.0012	9.00	9.060	—	58-65
510	79-78	.0047	.331	.0028	9.00	9.060	—	58-65[1]
510	73-72	.0048	.331	—	9.00	9.039	—	58-65
510	71-70	.0024	.331	—	9.00	9.039	—	58-65
B210	78	.0047	.331	—	8.00	8.051	—	58-65
B210	77-73	.0047	.331	—	8.00	8.051	—	58-65

Brake Specifications

Note: If hubs must be removed, refer to manufacturers service manual for current installation procedure

IDENTIFICATION MAKE AND MODEL	YEAR	Runout (in.)	DISC ROTOR Minimum Thickness (in.)	Parellelism (in.)	BRAKE DRUM New Dia. (in.)	Max. Mach. Oversize (in.)	Max. Oversize Due to Wear (in.)	Lug Nut Torque (ft/lbs)
DATSUN								
F10	78-77	.0059	.339	—	8.00	8.051	—	58-65
280Z	78-75	.0039	.413	—	9.00	9.06	—	58-65
260Z	74-73	.0059	.414	—	9.00	9.055	—	58-65
240Z	73-70	.0059	.414	.0028	9.00	9.055	—	58-65
610	77-73	.0048	.331	—	9.00	9.055	—	58-65
710	77-76	.0047	.331	—	9.00	9.055	—	58-65
710	75-74	.0024	.331	—	9.00	9.055	—	58-65
810	78	.0047	.331	—	9.00	—	9.060	58-65
810	77	.0059	.413	—	9.00	—	9.060	58-65
B110 (1200)	73-71	.0012	.331	—	8.00	8.051	—	58-65[1]

[1]1972-1970 - torque is 61.5 to 72.3 FT/LBS   [2]Steel wheels; aluminun wheels 58-72 ft/lbs.

DODGE COLT								
Colt (exc. Hatchback)	79	.006	.45[2]	—	9.00	9.080	—	51-58[3]
Dodge Colt Hatchback	79	.006	.45[2]	—	7.1	—	7.2	51-58[2]
All	78-74	.006	.450[1]	—	9.00	9.050	9.079	51-58
	73-71	.006	.330	—	9.00	9.050	9.079	51-58

[1]1978 Station wagon .43"   [2]Pin type caliper; sliding caliper .430 in.   [3]Steel wheels, aluminum wheels 58-72 ft/lbs.

FIAT								
X1/9	79-74	.0059[1]	.354[1]	.0019[1]	—	—	—	51
128 Series	79-72	.006	.354	.0019	7.30	7.3315	7.3554	51
Spider 2000	79	.0059[1]	.354	.0019[1]	—	—	—	51
Brava	79	.006	.350	.0019	9.000	9.030	9.0551	65
Strada	79	—	.368	—	7.293-7.304	7.336	7.355	
131	78-75	.006	.350	.0019	9.00	9,030	9.0551	65
124 Series All	78-70	.0059[1]	.354[1]	.0019[1]	—	—	—	51[2]
850 Sedan	73-70	—	—	—	7.30	7.332	7.355	51
850 Coupe	73-70	.0047	.374	—	7.30[3]	7.332[3]	7.355[3]	51
850 Spec. Coupe/Spi.	73-70	.006	.354	—	7.30	7.332	7.355	51

[1]Front & rear   [2]1977 - torque is 65 FT/LBS   [3]Drums front & rear

FIESTA								
	79	.006	.34	—	7.00	—	—	63-85
	78	.006	.34	—	7.00	—	—	52-74

HONDA								
Civic (CVCC) Exc. Wag.	79-76	.0059	.343	.0028	7.08	7.13	7.15	51-65
Civic (CVCC) Wag.	79-76	.0059	.343	.0028	7.87	7.91	7.93	51-65
Civic	79-73	.0064	.354	.0028	7.08	7.13	7.15	—
Accord	79-76	.0059	.433	.0028	7.08	7.13	7.15	51-65
Prelude	79	NA	NA	NA	NA	NA	NA	80
Civic CVCC	77-75	.0059	.343	.0028	7.08	7.13	7.15	51-65
Civic CVCC Wagon	77-75	.0059	.343	.0028	7.87	7.91	7.93	51-65
600 Coupe	72-70	.0039	.354	.0028	7.08	7.13	7.15	51-65

MAZDA								
GLC	79-77	.0024	.4724	—	7.874	—	7.914	65-80
RX7	79	.004	.6693	—	7.874	—	7.914	65-80
626	79	NA	NA	NA	9.00	NA	NA	NA
Cosmo	77-76	.0024	.6693 Frt.	—	—	—	—	65-72
Cosmo	77-76	.0024	.3543 Rr.	—	—	—	—	65-72
Mizer	77	.0039	.3937	—	7.874	7.9135	—	65-72
808	76-74	.0039	.3937	—	7.874	7.9135	—	65-72
808	73-72	.0030	.394	—	7.874	7.9135	—	65
RX3SP	77	.0039	.3937	—	7,874	7.9135	—	65-72
RX3	76-74	.0039	.3937	—	7.874	7.9135	—	65-72
RX3	73-72	.0030	.394	—	7.8741	7.9135	—	65
RX4	77-74	.0039	.4331	—	8.999	9.0395	—	65-72

Import Cars

Note: If hubs must be removed, refer to manufacturers service manual for current installation procedure

IDENTIFICATION MAKE AND MODEL	YEAR	Runout (in.)	DISC ROTOR Minimum Thickness (in.)	Parellelism (in.)	BRAKE DRUM New Dia. (in.)	Max. Mach. Oversize (in.)	Max. Oversize Due to Wear (in.)	Lug Nut Torque (ft/lbs)
MERCEDES-BENZ								
450 6.9	79-78	.0047	.789	.0007	—	—	—	72-75
450 Rear	79-78	.0047	.326	.0007	—	—	—	72-75
450 Series Front	79-77	.0047	.811	.0007	—	—	—	72-75
450 Series Rear	79-77	.0059	.3351	.0007	—	—	—	72-75
280E Front	79-77	.0047	.435	.0007	—	—	—	72-75
280E Rear	79-77	.0047	.3351	.0007	—	—	—	72-75
280SE Front	79-77	.0047	.811	.0007	—	—	—	74-75
280SE Rear	79-77	.0047	.3351	.0007	—	—	—	74-75
280CE Front	79-78	.0047	.417	.0007	—	—	—	74-75
280CE Rear	79-78	.0059	.3267	.001	—	—	—	74-75
450 All Frt.	76-73	.0047	.789	—	—	—	—	74
All Rr.	76-73	.0047	.355	—	—	—	—	74
300 SEL 4.5 Frt.	73-72	—	.709	.001	—	—	—	74
Rr.	73-72	—	.370	.001	—	—	—	74
350 SL Frt.	72	.0047	.789	—	—	—	—	74
Rr.	72	.0047	.355	—	—	—	—	74
300 SEL/8 Frt.	71-70	—	.653	.001	—	—	—	74
Rr.	71-70	—	.370	.001	—	—	—	74
300 SEL/8-6.3 Frt.	71-70	—	.709	.001	—	—	—	74
& 300 SEL/9 Rr.	71-70	—	.709	.001	—	—	—	74
280, 280 C Frt.	76-73	.0047	.435	—	—	—	—	74
Rr.	76-73	.0047	.355	—	—	—	—	74
280S, 280SE/8, SL/8, Frt.	76-75, 71-70	—	.653	.001	—	—	—	74
Rr.	76-75, 71-70	—	.370	.001	—	—	—	74
280 SE/SEL Frt.	73-70	—	.709	.001	—	—	—	74
Rr.	73-70	—	.370	.001	—	—	—	74
250 All Frt.	72-70	.0047	.435	—	—	—	—	74
All Rr.	72-70	.0047	.355	—	—	—	—	74
230 Frt.	78-77	.0047	.435	.0007	—	—	—	72
Rr.	78-77	.0059	.3351	.0007	—	—	—	72
230 Frt.	76-74	.0047	.435	—	—	—	—	72
230 Rr.	76-74	.0047	.355	—	—	—	—	72
220 Frt.	72	.0047	.435	—	—	—	—	74
Rr.	72	.0047	.355	—	—	—	—	74
220/8 Frt.	71-70	.0047	.460	—	—	—	—	74
Rr.	71-70	.0047	.370	—	—	—	—	74
MERCEDES DIESELS								
300 CD	79-78	.0047	.417	—	—	—	—	72-75
300 CD Rear	79-78	.0059	.3267	—	—	—	—	72-75
300 SD	79-78	.0047	.811	—	—	—	—	72-75
300 SD Rear	79-78	.0059	.3351	—	—	—	—	72-75
300 D, 240 D Front	79-77	.0047	.435	.0007	—	—	—	72-75
300 D, 240 D Rear	79-77	.0059	.335	.0007	—	—	—	72-75
300 D, 240 D	76-74	.0047	.435	—	—	—	—	74
		.0047	.355	—	—	—	—	74
220 D	73-70	.0047	.435	—	—	—	—	74
220 D	73-70	.0047	.355	—	—	—	—	74
MERCURY CAPRI								
All	77-70	.0035	—	—	9.00	9.050	—	50-55
MG								
MGB	79-77	.006	—	—	10.000	—	—	60-65
MGB	76-70	.003	.300	.001	10.00	—	—	60-65
Midget	79-70	.006	.390	—	7.00	—	—	44-46

Brake Specifications

Note: If hubs must be removed, refer to manufacturers service manual for current installation procedure

IDENTIFICATION MAKE AND MODEL	YEAR	Runout (in.)	DISC ROTOR Minimum Thickness (in.)	Parellelism (in.)	BRAKE DRUM New Dia. (in.)	Max. Mach. Oversize (in.)	Max. Oversize Due to Wear (in.)	Lug Nut Torque (ft/lbs)
PLYMOUTH ARROW								
Arrow Rear Drum	79	.006	.450[2]	—	9.00	9.060	9.080	51-58[3]
Arrow Rear Disc	79	.006	.450[2]	—	9.00	—	—	51-58[3]
W/Rear Disc	79	.0065	.330	—	—	—	—	51-58[3]
Champ	79	.006	.450	—	7.1	—	7.2	51-58[3]
	78-76	.006	.45[1]	—	9.00	9.050	9.079	51-59

[1]1978 Station Wagon 0.43"   [2]Pin type caliper; sliding caliper .430 in.   [3]Steel wheels; aluminum wheels 58-72 ft/lbs.

PORSCHE								
911 Turbo Front	79-78	.008	1.205	.001	—	—	—	94
911 Turbo Rear	79-78	.008	1.050	.001	—	—	—	94
911 CS Front	79-78	.008	.807	.0012	—	—	—	94
911 CS Rear	79-78	.008	.787	.0012	—	—	—	94
924	79	.008	.453	.0008	9.055	—	9.114	94
928 Front	79-78	—	.708	—	—	—	—	94
928 Rear	79-78	—	.708	—	—	—	—	94
911	77-70	.008	.708	.0012	—	—	—	94
912 E Frt.	76	.008	.433	.001	—	—	—	94
Rr.	76	.008	.355	.001	—	—	—	94
924	78-77	.008	.450	.0008	9.055	—	—	94
914	76-70	.008	.394	.008	—	—	—	94
914 Exc. 914/6	71-70	.008 Frt.	.394 Frt.	.008 Frt.	—	—	—	94
	71-70	.008 Rr.	.335 Rr.	.0008 Rr.	—	—	—	94
914/6	71-70	.008 Frt.	.708 Frt.	.0008 Frt.	—	—	—	94
	71-70	.008 Rr.	.355 Rr.	.0008 Rr.	—	—	—	94

[1]New disc thickness

RENAULT								
LeCar	79-77	.008	.354	.0004	7.096	Don't mach.	9.035	40-45
17 Gordini	79-77	.008	.354	.0004	9.000	Don't mach.	9.035	45-60
17 GTL	77	.008	.354	.0004	7.087	Don't mach.	7.136	45-60
12 Except Wagon	77	.008	.354	.0004	7.087	Don't mach.	7.136	45-60
12 Wagon	77	.008	.354	.0004	9.000	Don't mach.	9.035	45-60
17 Coupe/Conv.	77	.008	.354	.0004	9.000	Don't mach.	9.035	45-60

SAAB								
900 Front	79	.004	.461	.0006	—	—	—	65-80
900 Disc Rear		.004	.374	.0006	—	—	—	65-80
99 Front	78-75	.004	.461	.0006	—	—	—	65-80
99 Disc Rear	78-75	.004	.374	.0006	—	—	—	65-80
99	74-70	.008	.374	—	—	—	—	65-80

SAPPORO								
W/Rear Drums	79-78	.006	.45[1]	—	9.000	9.060	9.080	51-58[2]
W/4 Wheel Disc-Front	79-78	.006	.45[1]	—	—	—	—	51-58[2]
W/4 Wheel Disc-Rear	79-78	.0065	.43	—	—	—	—	51-58[2]

[1]Pin type caliper; sliding caliper .430 in.   [2]Steel wheels; aluminum wheels 58-72 ft/lbs.

SUBARU								
1600	79-76	.006	.33	—	7.09	7.12	7.50	58-72
1400	76-73	.006	.33	—	7.09	—	7.17	40-54
1300	72	.006	.33	—	7.09	—	7.17	40-54
1100 Frt.	71-70	—	—	—	8.00	—	8.04	40-54
1100 Rr.	71-70	—	—	—	7.09	—	7.17	40-54
360	70	—	—	—	6.69	—	6.71	—

TOYOTA								
Corolla (1200)	79	.006	.354	—	7.953	—	8.000	65-86
Corolla (1600)	79-71	.006	.354	—	9.000	—	9.080	66-86

Import Cars

Note: If hubs must be removed, refer to manufacturers service manual for current installation procedure

IDENTIFICATION MAKE AND MODEL	YEAR	DISC ROTOR Runout (in.)	Minimum Thickness (in.)	Parellelism (in.)	BRAKE DRUM New Dia. (in.)	Max. Mach. Oversize (in.)	Max. Oversize Due to Wear (in.)	Lug Nut Torque (ft/lbs)
TOYOTA								
Celica	79-77	.006	.450	—	9.000	—	9.080	66-86
Supra Front	79	.006	.450	—	—	—	—	66-86
Supra Rear	—	.006	.350	—	9.000	—	9.080	66-86
Corona	79-74	.006	.450	—	9.000	—	9.080	66-86
Cressida	79-78	.006	.450	—	9.000	—	9.080	66-86
Corolla (3KC)	78-71	.0059	.354	—	7.874	—	7.953	65-87
Corolla (2TC)	77-71	.0059	.354	—	9.00	—	9.079	65-87[1]
Corolla w/Drums (3KC)	70	—	—	—	7,874 Frt.	—	7.953 Frt.	58-75
w/Drums (3KC)	70	—	—	—	7.874 Rr.	—	7.953 Rr.	58-75
Celica, Carina	76-71	.006	.354	—	9.00	—	9.079	65-87
Corona	73-70	.006	.355	—	9.01	—	9.085	65-87
Mark II	76-72	.006	.453	—	9.00	—	9.079	65-87
Mark II	71-70	.006	.374	—	9.00	—	9.079	65-87
Crown	71	.006	.453	—	10.00	—	10.080	65-87
	70	.006	.420	—	9.055	—	9.134	65-87

[1] '73-71 lug nut torque 58-75

TRIUMPH								
TR-7	79-75	—	.375	—	8.00	—	8.050	60-75
TR-7 5 speed	79-77	—	.375	—	9.00	—	9.050	60-75
Spitfire	79-70	—	—	—	7.00	—	—	48
TR-6	76-70	—	—	—	9.00	—	—	80
GT-6	73-70	—	—	—	8.00	—	—	48
Stag	73-72	.006	—	—	9.00	—	—	—

[1] New rotor thickness

VOLVO								
240, 260 Front	79-75	.004	.557	.0012	—	—	—	70-100
240, 260 Rear	79-75	.006	.331	.0012	—	—	—	70-100
164	75-72	.004 Frt.	.900 Frt.	.0012 Frt.	—	—	—	70-100
164	75-72	.006 Rr.	.331	.0012 Rr.	—	—	—	70-100
164	71	.004 Frt.	.517 Frt.	.0012 Frt.	—	—	—	70-100
164	71	.006 Rr.	.331 Rr.	.0012 Rr.	—	—	—	70-100
164	70	.004 Frt.	.457 Frt.	.0012 Frt.	—	—	—	70-100
164	70	.006 Rr.	.331 Rr.	.0012 Rr.	—	—	—	70-100
140 (B20E)	74	.004 Frt.	.557 Frt.	.0012 Frt.	—	—	—	70-100
140 (B20E)	74	.006 Rr.	.331 Rr.	.0012 Rr.	—	—	—	70-100
140 (B20A, B20B, B20F)	74-71	.004 Frt.	.457 Frt.	.0012 Frt.	—	—	—	70-100
140 (B20A, B20B, B20F)	74-71	.006 Rr.	.331 Rr.	.0012 Rr.	—	—	—	70-100
140 (B20A, B20B)	70	.004 Frt.	.457 Frt.	.0012 Frt.	—	—	—	70-100
140 (B20A, B20B)	70	.006 Rr.	.331 Rr.	.0012 Rr.	—	—	—	70-100
1800 ES	73-70	.004 Frt.	.52 Frt.	.0012 Frt.	—	—	—	70-100
1800 ES	73-70	.006 Rr.	.331 Rr.	.0012 Rr.	—	—	—	70-100
VOLKSWAGEN								
Rabbit	79-75	.004	.413	.0008	7.086	7.105	7.125	65
Scirocco	79-75	.004	.413	.0008	7.086	7.105	7.125	65
Rabbit Front Drum	79-77	—	—	—	9.059	9.079.	9.098	65
Rabbit Rear Drum	79-77	—	—	—	7.086	7.105	7.125	65
Convertible (S. Beetle)	79-71	—	—	—	9.768	9.803	9.822 Front	94
Convertible (S. Beetle)	79-71	—	—	—	9.055	9.094	9.114 Rear	94
Dasher	79-78	.002	.393	—	7.850	7.900	7.913	65
Beetle Frt.	77-70	—	—	—	9.059 Frt.	9.098 Frt.	9.114 Frt.	87-94
Beetle Rr.	77-70	—	—	—	9.055 Rr.	9.094 Rr.	9.114 Rr.	87-94
Super Beetle Frt.	74-71	—	—	—	9.768 Frt.	9.803 Frt.	9.823 Frt.	87-94
Super Beetle Rr.	74-71	—	—	—	9.055 Rr.	9.094 Rr.	9.114 Rr.	87-94

Brake Specifications

Note: If hubs must be removed, refer to manufacturers service manual for current installation procedure

IDENTIFICATION MAKE AND MODEL	YEAR	DISC ROTOR Runout (in.)	Minimum Thickness (in.)	Parellelism (in.)	BRAKE DRUM New Dia. (in.)	Max. Mach. Oversize (in.)	Max. Oversize Due to Wear (in.)	Lug Nut Torque (ft/lbs)
VOLKSWAGEN								
LaGrande Bug Frt.	75	—	—	—	9.768 Frt.	9.803 Frt.	9.823 Frt.	87-94
LaGrande Bug Rr.	75	—	—	—	9.055 Rr.	9.094 Rr.	9.114 Rr.	87-94
Karmann Ghia	74-70	.008	.315	.001	9.055	9.094	9.114	87-94
Dasher	77	.004	.41	.0008	7.85	7.894	7.913	65
Fast or Squareback								
Type 3	73-71	.008	.374	.0008	9.768	9.803	9.823	87-94
Type 3	70	.008	.315	.0008	9.768	9.803	9.823	89-94
Type 4	74-71	.008	.374	.0008	9.768	9.803	9.823	87-94
TRUCKS								
CHEVROLET LUV								
LUV	79-78	.005	.653	.003	10.000	10.059	10.079	65
LUV	77-76	.005	.668	.003	10.00	10.059	10.079	65
LUV	75-72	—	—	—	10.00	10.059	10.079	65
DODGE D50								
D50	79	.006	.720	—	9.500	—	9.579	51-58
FORD COURIER								
All	79-77	.0039	.4331	—	10.2362	10.2445	10.2756	58-65
All	76-72	—	—	—	10.236[1]	10.2445[1]	10.2756[1]	58-65

[1]Front & rear

DATSUN								
620	79-78	.0059	.413	—	10.00	—	10.060	58-72
620	77-72	—	—	—	10.00	—	10.059	58-65
521	72-70	—	—	—	10.00	—	10.040	62-73
MAZDA								
B2000	79	.0039	—	—	10.2364	—	10.2758	58-65
B1600	77	.0039	—	—	10.2364	—	10.2758	58-65
B1600	76-72	—	—	—	10.2364	—	10.2758	58-65
Rotary Pickup	77-74	.0039	.4331	—	10.2364	—	10.2758	58-65
PLYMOUTH ARROW								
Arrow Pickup	79	.006	.720	—	9.500	—	9.579	51-58
TOYOTA								
Pickup	79	.006	.450	—	10.000	10.060	10.080	66-86
Pickup	78-75	.0059	.453	—	10.00	10.060	10.080	65-87
Pickup	74	—	—	—	10.00[1]	10.060[1]	10.080[1]	65-87
Pickup RN22 & 27	73-72	—	—	—	10.00[1]	10.060[1]	10.080[1]	65-87
Pickup RN12 & 14	72-70	—	—	—	9.06[1]	—	9.13[1]	65-87
Land Cruiser	77-76	.0047	.74	—	11.40	—	11.50	65-87
Land Cruiser	75	.0059	.79	—	11.40	—	11.50	65-87
Land Cruiser	74	—	—	—	11.40[1]	—	11.50[1]	65-87
Land Cruiser	73-70	—	—	—	11.40[1]	—	11.50[1]	66-68

[1]Front & rear [2]Rear drum wear limit 11.50 inches

VOLKSWAGEN								
Type II	79-78	.004	.512	—	9.921	9.960	9.980	94
Type II Van	70	—	—	—	9.842[1]	9.882[1]	9.900[1]	94

[1]Front & rear

Import Cars

+ mark (°) indicates degrees   mark (') indicates minutes, 60' = 1°   * (-) indicates TOE-OUT

YEAR	MAKE AND MODEL	CASTER+ (DEGREES°) 60' = 1° RANGE	PREF.	CAMBER+ (DEGREES°) 60' = 1° RANGE	PREF.	TOE-IN*
	AUDI					
79-76	Fox	+30' ± 30'	+30'	+30' ± 30'	+30'	+10' ± 15'
75-73	Fox	+30' ± 25'	+30'	+30' ± 30'	+30'	+10' ± 10'
79	5000	-10' ± 40'	-10'	-30' ± 30'	-30'	+5' to -10'
78	5000	-15' ± 15'	-15'	+30' ± 30'	+30'	7-1/2' ± 7-1/2'
77-75	100 series	-30' ± 15' (S)	-30'	0° ± 30'	0°	.04" ± .16"
77-75	100 series	0° ± 15' (P)	0°	0° ± 30'	0°	.04" ± .16"
74-73	100 series	-10' ± 15' (S)	-10'	+15' ± 30'	+15'	0 ± .06"
74-73	100 series	+20' ± 15' (P)	+20'	+15' ± 30'	+15'	0 ± .06"
72-70	100 series	+6' ± 20'	+6'	+11' ± 20'	+11'	0 to (-) .08" [1]
72-70	Super 90	+10' ± 20'	+10'	+15' ± 20'	+15'	0 to (-) .08" [1]

(S) - Standard steering   (P) - Power steering   [1] TOE-OUT

VEHICLE LOADED

YEAR	MAKE AND MODEL	CASTER RANGE	PREF.	CAMBER RANGE	PREF.	TOE-IN
	BMW					
79-78	733i	+9° ± 30'	+9°	0° ± 30'	0°	+.020" +.040 -.020
79-78	320i	+8°20' ± 30'	+8°20'	0° ± 30°	0'	0.059" +.040 -.020
79	528i		+7°40'		1/2°	+1/16"
79	633i		+7°40'		0	+1/16"
78-75	530i	+7° 40' ± 30'	+7° 40'	0° ± 30' [1]	0° [1]	+.0591"
78-77	630 CSi	+7°40' ± 30'	+7°40'	0° ± 30'	0°	+.0591"
76-70	2002, 2002 tii	+4° ± 30'	+4°	0° to +1°	+30'	+.039 to +.099"
76	3.0 Si	+7° 40' ± 30'	+7° 40'	0° to +1°	+30'	+.0591"
75	3.0 series	+9° 40' ± 30'	+9° 40'	0° to +1°	+30'	+.0591"
74-72	3.0 series	+9° 30' ± 30'	+9° 30'	0° ± 30'	0°	+.0591"
71-70	2800	+9° 30' ± 30'	+9° 30'	0° ± 30'	0°	+.039"
71-70	2500	+9° 30' ± 30'	+9° 30'	0 ± 30'	0°	+.039"
71-70	1600	+4° ± 30'	+4°	0° ± 30'	0°	0° to +.080"
70	2000	+4° ± 30'	+4°	0° ± 30'	0°	+.040

[1] 1976 and 1975 30' ± 30' - prefer 30'

YEAR	MAKE AND MODEL	CASTER RANGE	PREF.	CAMBER RANGE	PREF.	TOE-IN
	BUICK OPEL					
79	Buick Opel	+4-1/2° ± 1-1/2°	+4-1/2°	+1/4° ± 1/2°	+1/4°	+1/16" ± 3/64
78	All	—	5°	—	0°	+1/8"
77-76	All	+5° ± 1°30'	—	0° ± 1°	—	+1/8" ± 3/32"
75	All	+3° to +6°	—	+15' to -1° 15'	—	+1/8" ± 1/32"
74	All	+3° to +6°	—	-1° ± 30'	-1°	+1/8" to +3/16"
73	1900 & Manta	+3° 130' to x6°30'	—	-1° ± 30'	-1°	+1/8" to +3/16"
73	GT	+3° ± 1°	3°	1° ± 30'	+1°	+1/32" to +1/8"
72	Opel	+2° ± 1°	—	—	+1° 30'	+1/32" to +1/8"
72	1900	+3° 30' to +6° 30'	—		-1° 30'	+1/8" to +3/16"
72	GT	+3° ± 1°	—		+1°30'	+1/32" to 1/8"
71	Opel	+2° ± 1°	—	+1° ± 30'	—	+1/32" to +1/8"
71	1900	+3°30' ± 1°	—	1° 30'	—	+1/8" to +3/16"
71	GT	+2° ± 1°	—	+1° ± 30'	—	+1/32" to +1/8"
70	Except GT	+2° ± 1°	—	+1° ± 30'	—	+1/32" to +1/8"
70	GT	+3° ± 1°	—	+1° ± 30'	—	+1/32" to +1/8"
	DATSUN					
79	310	+25' to 1°55'	+25'	+15' to 1° 45'	+15'	+0 to .08"
79-78	200SX	+1°05' to 2°35'	+1°05'	+20' to 1°50'	+20'	+0.08" to 0.16"
77	200 SX	+1°2' to +2°18'	—	+30' to +1°30'	—	+.079" to +.157"
79	210	+1°40' to 3°10' [1]	+1°40'	+0' to 1°30'	+0'	+0.04 to 0.12"
79	280ZX	+4°5' to 5°35'	+4°5'	-30' to 1°	-30'	0.04" to 0.12"
79-78	810	+1°10' to 2°40'	+1°10'	+0' to 1°30'	+0'	+0 to 0.08"
79-78	510 Sedan, Hatchback	+1°05' to 2°35'	+1°05'	-15' to +1°15'	-15'	+0.04" to 0.12"
79-78	510 Station Wagon	+55' to 2°35'	—	+5' to 1°35'	—	+0.04 to 0.12"
78	F-10	+15' to 1°45'	—	50' to 2°15'	—	+.12" to +.22" [1]

Wheel Alignment Specifications

+ mark (°) indicates degrees   mark (') indicates minutes, 60' = 1°   * (-) indicates TOE-OUT

YEAR	MAKE AND MODEL	CASTER+ (DEGREES°) 60' = 1° RANGE	PREF.	CAMBER+ (DEGREES°) 60' = 1° RANGE	PREF.	TOE-IN*
77	F-10	+20' to +1°50'	—	+50' to +2°20'	—	+.20" to +.28" [1]
78	B210	+1° to +2°30'	—	+25' to +1°55'	—	+.08" to 0.16"
77-75	B210	+1° to +2°30'	—	+25' to +1°55'	—	+.079" to +.157"
74-73	B210	+1° to +2°15'	—	+40' to +1°4'	—	+.079" to +.157"
77-75	610	+1°05' to +2°35'	—	+1°15' to +2°45'	—	+.24" to +.31" [3]
75-74	610	+1°15' to +2°45'	—	+1°15' to +2°45' [4]	—	+.43" to +.55"
73	610 Sedan & Hdtop	+45' to +2°15'	—	+1° to +2°30'	—	+.236" to +.354"
73	610 Wagon	+55' to +2°25'	—	+1°10' to +2°40'	—	+.315" to +.433"
77-76	710	+1°5' to +2°35'	—	+1°15' to +2°45'	—	+.24 to +.31 [3]
75	710	+1°10' to +2°40'	—	+1°25' to +2°55'	—	+.315 to +.433 [5]
78-75	280Z	+2°3' to +3°33'	—	+18' to +1°48' [2]	—	+0" to +.118"
74	260Z	+2°54' ± 45'	—	+46' ± 45'	—	+.079" to +.197"
	DATSUN					
73	240Z	+2°55' ± 45'	—	+50' ± 45'	—	+1/16" to +7/32"
72-70	240Z	+2°55' ± 45'	—	+50' ± 30'	—	+.079" to +.197"
73	B110 (1200)	+1°40' ± 30'	—	+5' to +2°5'	—	+.16" to +.24"
73	B110 (1200)	+1°40' ± 30'	—	+1°5' ± 30'	—	+.20" ± .04"
73-72	510 Except Wagon	—	+1°35'	—	+25'	+.118" to +.236"
73	510 Wagon	—	+1°40'	—	+35'	+.158" to +.276"
73	510 Wagon W/HD Susp.	—	+1°5'	—	+20'	+.118" to +.236"
72	510 Wagon	—	+1°5'	—	+10'	+.079" to +.197"
71-70	510 Except Wagon	—	+1°40'	—	+1°15'	+.079" to +.118"
71-70	510 Wagon	—	+2°	—	+1°10'	+.118" to +.236"

[1] w/bias ply tires; w/radial tires 0" to .079"
[2] 2 seat; 4 seat, + 21' to 1°51'
[3] w/bias ply tires; w/radial tires +.16" to +.24"
[4] Hardtop; wagon +1°30' to +3° camber
[5] 1974 models toe is +.559 to +.669
[6] Except station wagon 1°55' to 3°25'

	DODGE					
79-78	Challenger	+2°38' ± 1/2°	+2°38'	+1°28' ± 1/2°	+1°28'	+0.08" to .35"
	DODGE COLT					
79-78	Coupe & Sedan	+2°05' ± 1/2°	+2°05'	+1° ± 1/2°	+1°	+0.08" to 0.24"
79-78	Station Wagon	+2°38' ± 1/2°	+2°38'	+1°28' ± 1/2°	+1°28'	+0.08" to 0.35"
79	Hatchback	+50' ± 20'	+50'	+1/2° ± 1/2°	+1/2°	-.08" to 0.16"
77	Coupe, Sedan	+2°5' ± 30'	—	+1° ± 45'	—	+.08 to +.23"
77	Hardtop, Wagon	+1°9' ± 30'	—	+51' ± 30'	—	+.08" to +.23"
76-74	All	+1°15' ± 30'	—	+50' ± 30'	—	+.08" to +.23"
73-71	All	+1°15' ± 30'	—	+1° ± 30'	—	+.08" to +.23"
FIAT		**VEHICLE LOADED**				
79	X1/9	+6°30' to +7°30'		-1°-0		+.079" to +.236"
79	128 Sedan	+1°10' to +2°10'		+1°10' to +2°10'		-.118" to +.039"
79	128 Hatchback	+1°10' to +2°10'		+50' to +1°50'		-.059" to +.100"
79	Spider 2000	+2°40' to +3°40'		-20' to +40'		+.157" to +.314"
79	Brava All	+3°15' to 4°15'		+25' to +1°25'		+.157" to +.314"
79	Strada	+1°30' to +2°30'	+2°	+1°10' to +2°10'	+1°40'	-0.177" to -0.098"
78	X1/9	+6°30' to 7°30'	—	-1° to 0	—	+.06" to +.25"
77	X1/9	+6°30' to +7°30'	+7°	-1° to 0	0°	+.079" to +.236"
76-75	A1/9	+6°30' to +7°30'	+7°	-1° ± 20'	-1°	+.394" to +.197"
74	X1/9	+6°30' to +7°30'	+7°	-1° ± 30'	-1°	+.394" to +.197"
78	131	+3°15' to +4°15'	—	+15' to 1°15'	—	+.156" to +.310"
77	131	+3°15' to +4°15'	—	+25' to 1°25'	—	+.157" to +.314"
76-75	131	+4° to +5°	—	+0° to +1°	—	+.08" to +.16
78	124 Spider	+2°45' to +3°45'	—	-15' to +45'	—	+.157" to +.314"
77	124 Spider/Coupe	+2°40' to +3°40'	—	-20' to +40'	—	+.157" to +.314"
76-73	124 Spider/Coupe	+3°30' ± 30'	—	+30' ±30'	—	+.12" ± .039"
72-70	124 Spider/Coupe	+3°20' to +4°	—	+30' ± 20'	—	+.118" to +.039"
74-73	124 Special Wagon	+3°30' ± 30'	—	+30' ± 30'	—	+.04 to +.20"

Import Cars

YEAR	MAKE AND MODEL	CASTER+ (DEGREES°) 60' = 1° RANGE	PREF.	CAMBER+ (DEGREES°) 60' = 1° RANGE	PREF.	TOE-IN*
72-70	124 Sedan & Wagon	+3° 20' to 4°	—	+30' ± 20'	—	+.118" ± .039"
78	128 3 Dr. H/B	+1° 15' to +2° 15'	—	+30' to +1° 30'	—	-.125 to +.03125
78	128 Sedan	+1° 5' to +2° 15'	—	+1° 15' to +2° 15'	—	-.125 to +.03125
77	128 Sedan & Wagon	+1° 10' to +2° 10'	—	+1° 10' to +2° 10'	—	-.118 to +.039"
76-75	128 Sedan & Wagon	+2° 15' ± 30'	—	+1° 30' ± 30'	—	-.08" to +.08"
74	128 Sedan	+2° 15' ± 30'	—	+2° ± 30'	—	-.08" to +.08"
73-72	128 Sedan	+2° to 2° 30'	—	+40' to 1° 20'	—	0" ± .039"
77	128 Coupe	+1° 10' to +2° 10'	—	+50' to +1° 50'	—	-.059" to +.100"
76-75	128 Sport Coupe	+2° 15' ± 30'	—	+45' ± 30'	—	0" ± .079"
74	128 Sport L	+2° 15' ± 30'	—	1° ±30'	—	0" ± .079"
73-70	850	+9° ±1°	—	+2° 10' ± 15'	—	+.079" to .157"
	FIESTA					
79		+1° 20' to -0° 25'	0° 20'	+3° 15' to 1° 15'	+2° 15'	-0.14" ± 0.04"
78		0° 20' ± 45'	—	2° 45' ± 30'	2° 15'	-.010" to .040"
	HONDA					
79-78	Civic Sedan	+3/4° ± 1/2°	+3/4°	+1/2° ± 1/2°	+1/2°	+0.040" ± .080"
79-78	Civic Wagon	+1/2° ± 1/2°	+1/2°	+1/2° ± 1/2°	+1/2°	+0.040" ± .080"
79	Accord	+1-1/4° ± 1/2°	+1-1/4°	+1/2° ± 1/2°	+1/2°	+0.040" ± 0.120"
79	Prelude	+1° 30' ± 1°	+1° 30'	0 ± 1°	0	0"
78	Accord		+1° 45'	—	+45'	0
77-76	Accord CVCC	—	+1° 50'	—	+40'	-.04"
78	Civic CVCC-Sedan	—	+45'	—	+30'	-.04"
78	Civic CVCC-Wagon	—	+30'	—	+30'	0"
77	Civic CVCC	—	+40'	—	+30'	-.04"
76-75	Civic CVCC	2° ± 30'	—	+30' ± 30'	—	+.039" ± .039"
78	Civic	+45'	—	+30'	—	-.031
77-73	Civic	—	+1° 45'	—	+30'	-.04"
72-70	600	1°	—	1°	—	-.078
	MAZDA					
79	GLC		+1° 40'	—	+45'	+.24"
79	RX7	+4° 3' ± 45'	+4° 3'	+1° 10' ± 30'	+1° 10'	+0 to 0.24"
79	626	+3°	—	+1° 15'	—	+0.12" ± .012"
78	GLC	+1° 35' ± 45'	—	+45' ± 1°	—	+.25"
77	GLC	+1° 35' ± 45'	—	+40' ± 1°	—	0 to +.24"
78	Cosmo (P)	+2° 25' ± 45'	—	1° ± 1°	—	+.24"
78	Cosmo (M)	+1° 45' ± 45'	—	1° ± 1°	—	+.24"
77-76	Cosmo	+1° 50' ± 45' [1]	—	+1° ± 1°	—	0 to +.24"
77-76	Mizer, 808-(1300)	+1° 25' ± 45' [2]	—	+50' ± 1° [3]	—	0 to +.24"
74-73	808 (1300)	+1° 31' ± 45'	—	+1° 5' ± 1°	—	0 to +.24"
77	808 (1600)	+1° 50' ± 45' [4]	—	+1° 5' ± 1°	—	0 to +.24"
76	808 (1600)	+1° 35' ± 45' [5]	—	+1° ± 1°	—	0 to +.24"
75	808 (1600)	+1° 25' ± 45'	—	+45' ± 1°	—	0 to +.24"
74-72	808 (1600)	+1° 15' ± 45'	—	+40' ± 1°	—	0 to +.24"
78	RX 3SP	+2° 15' ± 45'	—	+1° ± 1°	—	+.25°
77	RX3SP	+2° 10' ± 45'	—	+1° 5' ± 1°	—	0 to +.24"
76-75	RX3	+1° 55' ± 45'	—	+55' ± 1°	—	0 to +.24"
74-72	RX3	+1° 15' ± 45'	+1° 15	+40' ± 1°	+40'	0 to +.24"
78	RX4 Wagon	+1° 45' ± 45'	—	+1° 1/4 ± 1°	—	+.24"
78	RX4 Sedan	+1° 45' ± 45'	—	+1° ± 1°	—	+.24"
77-76	RX4 Sedan	+1° 49' ± 45'	—	+1° 2' ± 1°	—	0 to +.24"
77-76	RX4 Hardtop	+1° 49' ± 45'	—	+59' ± 1°	—	0 to +.24"
77	RX4 Wagon	+1° 49' ± 45'	—	+1° 17' ± 1°	—	0 to ±.24"
75-74	RX4	+2° ± 45'	—	+1° ± 1°	—	0 to +.24"
74-71	RX2	—	+1° 3'	—	-15'	+.16" to +.08"

(P) Power steering
(S) Standard steering

[1] w/power steering caster is 2° 15' ± 45'
[2] Coupe + 1° 45' ± 45'
[3] Wagon + 1° ± 1°
[4] Coupe; caster is 2° 45' ± 45'
[5] Coupe; caster is 1° 50' ± 45' Wagon; caster is + 1° 45' ± 45'

YEAR	MAKE AND MODEL	CASTER RANGE	PREF.	CAMBER RANGE	PREF.	TOE-IN
	MERCEDES-BENZ					
79	450 SEL	+10° ± 30'	+10°	-10' ± 10'	-10'	+0.12" ± .04"

Wheel Alignment Specifications

YEAR	MAKE AND MODEL	CASTER+ (DEGREES°) 60' = 1° RANGE	PREF.	CAMBER+ (DEGREES°) 60' = 1° RANGE	PREF.	TOE-IN*
73-71	Corolla 1200 (3KC)	+1° 40' to +2° 20'	—	+20' to 1° 20'	—	+1/16" to +3/16"
73-71	Corolla 1600 (2TC)	+1° 35' to +2° 15' [4]	—	+30' to 1° 30'	—	+1/16" to +3/16"
70	Corolla 1200 (3KC)	—	+1°	—	+1° 30'	+1/16" to +3/16"
	TOYOTA					
79	Celica	+1° 45' ± 30'	+1° 45'	+1° 05' ± 30'	+1° 05'	+.04" ± .04"
78-77	Celica (2OR)	+1° 15' to 2° 15'	—	+30' to 1° 30'	—	0" to +.08"
76	Celica Except GT (2OR)	+30' to +1° 30' [5]	—	+30' to 1° 30' [5]	—	+.04" to +.12"
76	Celica GT	+1° 20' to +2° 20'	—	+20' to 1° 20'	—	+.04" to +.12"
75	Celica (2OR)	+30' to +1° 30'	—	+30' to 1° 30'	—	+.20 to +.28
74	Celica (18R-C)	+30' to +1° 30'	—	+30' to 1° 30'	—	+13/64" to +1/4"
73-71	Celica	+30' to +1° 30'	—	+30' to 1° 30'	—	+.20" to +.28"
79-78	Cressida	+50' ± 30'	—	+1° 14' ± 30'	—	.12" ± .04"
79	Corona	+1° 45' ± 30' [8]	—	+1° ± 30'	—	.04" ± 04"
78-77	Corona	+20' to +1° 20'	—	+5' to +1° 5'	—	+.04" to +.12" [5]
76-74	Corona	+20' to +1° 20'	—	+5' to +1° 5'	—	+.04" to +.12"
73-70	Corona	-10' to +50'	—	—	+11° 20'	+3/16" to +1/4"
79	Supra	+1° 45' ± 30'	+1° 45'	+50' ± 30'	+50'	+.04"
76	Mark II	+1° 45' to +2° 45'	—	+35' to 1° 35'	—	+0" to +.12" [6]
75-72	Mark II Except Wagon	-5' to +55'	—	+35' to 1° 35'	—	+3/16" to +1/4"
75-72	Mark II Wagon	10' to +1° 10'	—	+35' to 1° 35'	—	+3/16" to +1/4" [7]
71-70	Mark II	-5' to +45'	—	+45' to 1° 45'	—	+.08" to +.16"
71	Crown Except Wagon	-15' to +1° 30'	—	-5' to +55'	—	+1/8" to +3/16"
71	Crown Wagon	-2° 20' to -1° 20'	—	-5' to +55'	—	+1/8" to +3/16"
70	Crown	-1° to 0°	—	-5' to +55'	—	+1/8" to +3/16"
73-71	Carina	+30' to +1° 30'	—	+30' to 1° 30'	—	+.20" to +.28"

[1] w/bias tires; w/radial tires 0" to +.08"
[2] Wagon caster +45' to 1°25'
[3] Hardtop camber +15' to 1°15'; wagon camber +10' to 1°10'
[4] Wagon caster +1°20' to 2°
[5] w/radial tires; w/bias tires +.12" to +.20"
[6] w/radial tires; w/bias tires +.16" to +.24"
[7] 1974 models; toe is +5/32" to 1/4"
[8] Sedan; wagon +1°30' ± 30'

YEAR	MAKE AND MODEL	CASTER RANGE	PREF.	CAMBER RANGE	PREF.	TOE-IN
	TRIUMPH					
79-78	TR-8	NA	—	NA	—	+0 to +.0625"
79-75	TR-7	+3° 30' ± 1°	+3° 30'	-15 ± 1°	—	0 to +1/16"
79-77	Spitfire	—	+4°	—	+3°	+1/16" to 1/8"
76-70	Spitfire	+4° ± 1°	—	+3° ± 1°	—	+1/16" to +1/8"
76-70	TR-6	+2° 45' ± 1°	—	+15' ± 1°	—	+1/16" to +1/8"
73-70	GT-6	—	+4°	—	+3°	+1/16" to +1/8"
73-72	Stag (M.T., No A.C.)	—	+2°	—	+1° 15'	+1/16" to +1/8"

(NA) Not adjustable

YEAR	MAKE AND MODEL	CASTER RANGE	PREF.	CAMBER RANGE	PREF.	TOE-IN
	VOLVO					
79	240 Manual Steering	+2 to +3°	—	0° to -1°	—	+.18" ± .06"
79	240 Power Steering, 260 Series	+3° to +4°	—	0° to 1°	—	+.12" ± .06"
78-76	240 & 260 Series	+2° 30' ± 30'	—	+1° to +1° 30'	—	+.18" ± .06" [1]
75	242, 244, 245	+2° to +3°	—	+1° to +1° 30'	—	+.177" ± .06" [1]
75-74	164	+1° 30' to +2° 30'	—	0° to +30'	—	+.08" to +.20"
73-70	164	+0° to 1°	—	0° to +30'	—	+.08" to +.20" [2]
74	140	+1° 30' to +2° 30'	—	0° to +30' [3]	—	+.08" to +.20"
73	140	+1° to +2°	—	0° to +30'	—	+.08" to +.20"
72-70	140	0° to +1°	—	0° to +30'	—	+.08" to +.20"
73	1800 ES	+2° to +2° 30'	—	0° to +30'	—	0 to +.12"
72	1800 E, ES	0° to +1°	—	0° to +30'	—	0 to +.12"
71-70	1800 E	0° to +1°	—	0° to +30'	—	0 to +.16"

[1] w/power steering; .12" ± .06"
[2] 1970 & 71 toe-in; +0 to +1.6"
[3] w/power steering; +2° to +3°

YEAR	MAKE AND MODEL	CASTER RANGE	PREF.	CAMBER RANGE	PREF.	TOE-IN
	VOLKSWAGEN					
79-75	Rabbit	+1° 50' ± 30'	+1° 50'	+20' ± 30'	+20'	-15' +10' -15'
79-75	Scirocco	+1° 50' ± 30'	+1° 50'	+20' ± 30'	+20'	-15' +10' -15'
79-71	Convertible (S. Beetle)	+2° ± 35'	+2°	+1° +20' -40'	—	+30' ± 15'

Import Cars

YEAR	MAKE AND MODEL	CASTER+ (DEGREES°) 60' = 1° RANGE	PREF.	CAMBER+ (DEGREES°) 60' = 1° RANGE	PREF.	TOE-IN*
	VOLKSWAGEN					
79	Dasher	+30' ± 30'	+30'	+30' ± 30'	+30'	+10' ± 15'
78	Dasher Type 32/33	+30' ± 30'	—	-40' ± 40'	—	0 ± 50'
77-76	Dasher Type 32/33	0° to +10	—	0° to +1°	—	+10' ± 15'
75-74	Dasher Type 32/33	0° to +1°	—	-5 to +55'	—	+10' ± 15'
77-70	Beetle Type 1	+3°20' ± 1°	—	+30' ± 20'	—	+1/16" to +7/32"
74-70	Karmann Ghia	+3°20' ± 1°	—	+30' ± 20'	—	+1/16" to +7/32"
74-71	Super Beetle	+2° ± 35'	—	+20' to +1°20'	—	+30' ± 15'
75	LaGrande	+2° ± 35'	—	+20' to +1°20'	—	+30' ± 15'
78-71	Convertible (S. Beetle)	+2° ± 35'	—	+20' to +1°20'	—	+30' ± 15'
73-70	Fast Squareback Type 3	+4° ± 40'	—	+1°20' ± 20'	—	+.118" to +.259"
74-71	Type 4	+1°45' ± 35'	—	+40' to +1°35'	—	+20' ± 15'
	TRUCKS					
	CHEVROLET LUV					
79-76	Luv 4x2	-10' ± 1° [1]	-10'	+30' ± 3/4° [2]	+30'	+0" ± 1/8" [3]
79	Luv 4x4	+20' ± 1 [4]	+20'	+35' ± 3/4° [5]	+35'	+0" ± 1/8" [6]
75-72	Luv	—	20'	—	+1°	1/8"

[1] Service check, resetting -10 ± 1/2°
[2] Service check, resetting -30 ± 1/2°
[3] Service check, resetting 0 ± 1/10"
[4] Service check, resetting +20 ± 1/2°
[5] Service check, resetting +35' ± 1/2°
[6] Service check, resetting 0" ± 1/16"

YEAR	MAKE AND MODEL	CASTER RANGE	PREF.	CAMBER RANGE	PREF.	TOE-IN
	COURIER FORD					
79-78	Courier	+3/4° to +1-1/4°	—	+1/2°-1-1/4°	—	+0" to +1/4"
77-73		+45' to +1°15'	—	+1° to +1°45'	—	0 to +1/4"
72		+45' to +1°15'	—	+1° to +1°45'	—	+5/64 to +1/8"
	DATSUN					
79-78	620	+35' to +2°05'	—	-15'-1°15'	—	+.20" to +.28"
77	620	+1°50' ± 45'	—	+1°15' ± 1°	—	+.079" to +.118"
76-73	620	+1°50' ± 45'	—	+1°15' ± 1°	—	+.039" to +.197"
72-70	521	—	+1°50'	—	+1°	+.236" to +.354"
	DODGE					
79	D-50	+3° ± 1°	—	+1° ± 30'	—	+.08" to +.35"
	MAZDA					
79	B2000	—	+10	—	+45'	+.24"
77-76	Pickup, Rotary	+1°57' ± 20'	—	+15' ± 20'	—	0" to +.24"
75	Pickup, Rotary	+1°12' ± 20'	—	+15' ± 20'	—	0" to +.24"
74	Pickup, Rotary	+1°57' ± 20'	—	+15' ± 20'	—	0" to +.24"
78	B1800, Pickup	+1° ± 15'	—	+45 ± 15'	—	0 to +.24"
77-72	B1600, Pickup	+1° ± 20'	—	+1°23' ± 20'	—	0" to .24"
	PLYMOUTH					
79	Arrow Pickup	+3° ± 1°	+1° ± 30'	—		+.08" to +.35"
	TOYOTA					
79	Pick-up	+30' ± 30'	+30'	+1°05' ± 30'	—	+.2/4 ± .04"
78-76	Pickup	0° to +1°	—	+30' to +1°30'	—	+.20" to +.27"
75	Pickup	-1°15' to +15'	—	+15' to +1°45'	—	+.08" to +.275"
74	Pickup	-1°15' to +15'	—	+15' to +1°45'	—	+15/64" to +1/4"
73-72	Pickup (RN 22 or 27)	-1°15' to +15'	—	+15' to +1°45'	—	+3/32" to +9/32"
72-70	Pickup (RN 12 or 14)	—	-20'	—	+1°	+1/4"
78-70	Land Cruiser	—	+1°	—	+1°	+.12" to +.20"
	SUBARU					
70	360 Van	13°1°	+13°10'	+1°30' to +2°30'	—	+.47" to +.67"
	VOLKSWAGEN					
79-70	Type 2	+3° ± 40'	—	+40' ± 25'	—	+15' ± 15'

Wheel Alignment Specifications

YEAR	MAKE AND MODEL	CASTER+ (DEGREES°) 60' = 1° RANGE	PREF.	CAMBER+ (DEGREES°) 60' = 1° RANGE	PREF.	TOE-IN*
78-77	450 SEL	+9° 30' ± 30'	—	0° ± 10'	—	+.12" ± .04"
76-73	450 SEL/SE	+10° ± 30'	—	-10° ± 10'	—	+.039" to +.157"
79	450 SL/SLC	+3° 40' ± 20'	+3° 40'	+0°+10'-20'	—	+.08" ± .04"
78-73	450 SL/SLC	+3° 15' ± 20'	—	-20' to 10'	—	+.039" to +.118"
79	450 (6.9 Liter)	+10° 15' ± 30'	+10° 15'	-20' ± 10'	-20'	+.12" ± .04"
79	280 CE/E	+8° 45' ± 30'	+8° 45'	0° ± 10'	0°	+.12" ± .04"
79-78	280SE	+10° ± 30'	+10'	-10° ± 10'	-10'	+.12" ± .04"
79-78	300 CD/TD	+8° 45' ± 30'	+8° 45'	0 ± 10'	0	+.12" ± .04"
	MERCEDES-BENZ					
79	300SD	+10° ± 30'	+10'	-10 ± 10'	-10'	+.12" ± .04"
79-78	240D	+8° 45' ± 30'	+8° 45'	0° ± 10'	0°	+.12" ± .04"
79	300D	+8° 45' ± 30'	+8° 45'	0° ± 10'	0°	+.12" ± .04"
78	450 (6.9 Liter)	+10° 15' ± 30'	—	+15' ± 15'	—	+.125" ± .031"
77	450 (6.9 Liter)	+9° 45' ± 30'	—	-20' ± 10'	—	+.12" ± .04"
78	280 CE	+8° 45' ± 30'	—	0° ± 15'	—	+.125' ± .031"
77	280 E	+8° 15' ± 30'	—	0° ± 10'	—	+.12" ± .04"
77	280 SE	+9° 30' ± 30'	—	0° ± 10'	—	+.12" ± .04"
76-73	280	+2° 30' ± 20' [1]	—	0° ± 10'	—	+.039" to +.157"
76-73	280 C	+2° 30' ± 20' [1]	—	0° ± 10'	—	+.039" to +.157"
76-75	280 S	+10° ± 30' [2]	—	-10' ± 10'	—	+.039" to +.157"
73-72	280 SEL	+3° 30' ± 15' [2]	—	+10' to +30'	—	+.039" to +.118"
71-70	280 S	+3° 30' ± 15' [2]	—	+10' to +30'	—	+.039" to +.118"
71-70	280 SL/8	+3° 30' ± 15' [2]	—	+10' to +30'	—	+.039" to +.118"
71-70	280 SL/9	+3° 30' ± 15' [2]	—	+10' to +30'	—	+.039" to +.118"
72	250	+2° 30' ± 20' [1]	—	0° ± 10'	—	+.039" to +.157"
72-70	250/8	+2° 30' ± 20' [1]	—	0° ± 10'	—	+.039" to +.157"
73-72	300 SEL	+3° 30' ± 15' [2]	—	0° to +20'	—	+.039" to +.118"
71-70	300 SEL-8	+3° 30' ± 15'	—	0° to +20'	—	+.039" to +.118"
72	350 SL	+3° 15' ± 20'	—	-20' to +10'	—	+.039" to +.118"
77	230	+8° 15' ± 30'	—	0° ± 10'	—	+.12" ± .04"
76-74	230	+2° 30' ± 20'	—	0° ± 10'	—	+.039" to +.157"
73-70	220/8	+2° 30' ± 20' [1]	—	0° ± 10'	—	+.039" to +.118"
73-72	220SE	+3° 30' ± 15' [2]	—	+10' to 30'	—	+.039" to +.118"
77	240D	+8° 15' ± 30'	—	0° ± 10'	—	+.12" ± .04"
76-74	240D	+2° 30' ± 20' [1]	—	0° ± 10'	—	+.039 to +.157"
78-77	300D	+8° 15' ± 30'	—	0° ± 10'	—	+.12" ± .04"
76-75	300D	+2° 30' ± 20' [1]	—	0° ± 10'	—	+.039 to +.157"
73-72	220D	+2° 30' ± 20' [1]	—	0° ± 10'	—	+.039" to +.118"
71-70	220D	+2° 30' ± 20' [1]	—	0° ± 10'	—	+.078 to +.157"

[1] w/power steering 3°30' ± 20'   [2] w/power steering 4° ± 15'

YEAR	MAKE AND MODEL	CASTER RANGE	PREF.	CAMBER RANGE	PREF.	TOE-IN*
	MG					
79-70	MGB	+5° to +7° 15'	—	-15' to +1° 15'	—	+.063" to +.0938"
79-70	Midget	—	+3°	—	+45'	+.125"

YEAR	MAKE AND MODEL	CASTER RANGE	PREF.	CAMBER RANGE	PREF.	TOE-IN*
	MERCURY CAPRI					
77-76	Capri II	+1° to +2° 15'	—	+45' to 2° 15'	—	+.25" to +.50" [1]
74	Capri	+30' to +1° 30'	—	+30' to +30'	—	0 to +.25"
73	Capri	+45' to +1° 45'	—	—	0°	0 to +.25"
72	Capri	+30' to +1° 30'	—	-15' to -45'	—	+.09 to +.15"
71-70	Capri	+30' to +1° 30'	—	-30' to +30'	-30'	+.09 to +.15"

[1] 1976 toe is 0" to .281"

YEAR	MAKE AND MODEL	CASTER RANGE	PREF.	CAMBER RANGE	PREF.	TOE-IN*
	PLYMOUTH ARROW					
79-78	Coupe & Sedan	+2° 05' ± 1/2°	+2° 05'	+1° ± 1/2°	+1°	+.08" to +.24"
78	Wagon	+2° 38' ± 30'	—	+1° 28' ± 30'	—	+.08 to +.35"
77	Coupe & Sedan	+2° 5' ± 30'	—	+1° ± 45'	—	+.08 to +.23"

Import Cars

YEAR	MAKE AND MODEL	CASTER+ (DEGREES°) 60' = 1° RANGE	PREF.	CAMBER+ (DEGREES°) 60' = 1° RANGE	PREF.	TOE-IN*
77	Hdtop & Wagon	+1°9' ± 30'	—	+51' ± 30'	—	+.08" to +.23"
76	All	+1°45' ± 30'	—	+1° ± 30'	—	+.08" to +.23"
	PLYMOUTH CHAMP					
79	Champ	+50' ± 20'	+50'	+1/2° ± 1/2°	+1/2	-.08" to .16"
	PORSCHE					
79-78	911 Carrera (Turbo)	+6°5' ± 15'	+6°5'	+30' ± 10'	+30'	0
79-77	911 CS	+6° ± 15'	+6°5'	+30' ± 10'	+30'	0
79-76	924	+2°45' ± 30'	+2°45' ± 30	-20' ± 15	-20'	0° +5' -15'
79-78	928	+3°30' ± 30'	+3°30'	-30' ± 10'	-30'	0° ± 5'
77	911	+6°5' ± 15'	—	+30' ± 10'	—	0
76-72	911	+6°5' ± 15'	—	0° ± 10'	—	0
71	911T	+6°45' ± 45'	—	0° ± 10'	—	0
71	911 E, S	+6°45' ± 45'	—	0° ± 20'	—	+40'
70	911	+6°45' ± 45'	—	0° ± 20'	—	+40'
76	912 E	+6°5' ± 15'	—	+30' ± 10'	—	0
75-70	914	+6° ± 30'	—	0° ± 20'	—	+20' ± 10'
	RENAULT					
79	LeCar	—	+6°	—	+1°30'	-3/64" to -3/16"
79	17 Gordini	—	+4°	—	+1°30'	-3/64" to -5/32"
78	LeCar, GTL	+12° to +13°	—	0° to +1°	—	+.093" to +.188"
77	LeCar	+13° ± 1°	—	—	+1°30'	+.047" to +.188"
77	17 GTL, 17	—	+1°30'	—	4	+.047" to +.156"
78	17 Gordini	—	+4°	—	+1°30'	+.093 to +.188"
77	12 Series	—	+1°30'	—	4°	+.047" to +.150
	SAAB					
79	900 Manual Steering	+1° ± 1/2°	—	+1/2° ± 1/2°	—	0" ± 0.04" [1]
79	900 Power Steering	+1° ± 1/2°	—	+2° ± 1/2°	—	0" ± 0.04" [1]
78-75	99 Manual Steering	+1° ± 1/2°	—	+1/2° ± 1/2°	—	0.04" ± .04" [1]
78-75	99 Power Steering	+1° ± 1/2°	—	+1/2° ± 1/2°	—	0" ± .04" [1]
74-72	99	+1° ± .5°	—	+.5° ± .5°	—	0" ± .04" [1]
71-70	99	+1.25° ± .25°	—	+.75° ± .25°	—	0" ± .04" [1]

[1] Measured at rims

YEAR	MAKE AND MODEL	CASTER RANGE	PREF.	CAMBER RANGE	PREF.	TOE-IN
	SAPPORO					
79-78		+2°38' ± 1/2°	+2°38'	+1°28' ± 1/2°	+1°28'	+0.08" to +0.35"
	SUBARU					
79	4WD Brat, 4WD Wagon	-50' ± 45'	-50'	+2°10' ± 45'	+2°10'	+.24" to .47"
79	DL Sedan, Coupe, FE Coupe, GF Hardtop	-50 ± 45'	-50	+1°30' ± 45'	+1°30'	+.08" to .32"
79	DL Wagon	-10 ± 45'	-50'	+1°45' ± 45'	+1°30'	+.08" to .32"
78	4WD, Brat	-1°35' to -5'	—	+1°15' ± 45'	—	+.25 to .468"
78	DL & GF Sedan, Coupe, HT	-1°35' to -5'	—	+1°45' ± 45'	—	+.08" to +.312"
78	DL Sta. Wagon	-55' to +35'	—	+1°45' ± 45'	—	+.08" to +.312"
77	DL & GF Coupe	+50' ± 45'	—	+1°30' ± 45'	—	+.08" to .32"
77	DL Sta. Wagon	+10' ± 45'	—	+1°45' ± 45'	—	+.08" to .32"
77	DL 4 Wheel Drive	+50' ± 45'	—	+2°10' ± 45'	—	+.24" to .47"
76-72	DL & GF Coupe	+45' ± 45'	—	+1°30' ± 30'	—	+.08" to .32"
76-72	4 WD	+45' ± 45'	—	+2°30' ± 30'	—	+.24" to .47"
71-70	360 Sedan	—	+13°17'	—	+2°	+.47" to .63"
71-70	1100 Sedan	—	+2°	—	1°50'	.2
	1100 Wagon	+1°30' to +2°	—	+1°20' to +1°50'	—	+.079" to +.314"
	TOYOTA					
79	Corolla (1200 & 1600)	+1°50' ± 30'	+1°50'	+1° ± 30	+1°	+.08" to +.16" [1]
78-76	Corolla	+1°20' to +2°20'	—	+30' to +1°30'	—	+.08 to +.16 [2]
75	Corolla	+1°31' to +2°10' [2]	—	+20' to +10°20' [3]	—	+.04"
74	Corolla 1200 (3KC)	+1°40' to +2°20'	—	+20' to +1°20'	—	+1/16" to +3/16"
74	Corolla 1600 (2TC)	+1°20' to +2°	—	+30' to +1°30'	—	+1/16" to +3/16"